Calculus
INTRODUCTORY EDITION
Volume 2

AL SHENK
University of California- San Diego

with
Lee van de Wetering and Carol Shenk

Boston San Francisco New York
London Toronto Sydney Tokyo Singapore Madrid
Mexico City Munich Paris Cape Town Hong Kong Montreal

Senior Acquisitions Editor: Laurie Rosatone
Project Manager: Jennifer Albanese
Sr. Production Supervisor: Peggy McMahon
Print Buyer: Evelyn Beaton
Sr. Prepress Supervisor: Caroline Fell
Technical Art Supervisor: Joe Vetere
Senior Designer: Barbara T. Atkinson
Cover Design: Gretchen Lally
Cover Photo: © Dewitt Jones/Corbis
This book was typeset by the author.

Photo/Illustration Credits
p.2, fig. 10.1.1, From Invitation to Sailing by A. Brown, Copyright 1968: Simon and Schuster; p.8, fig. 10.1.21, From Orthopaedic Biomechanics by Frankel and Burstein, Copyright 1971: Lea and Febiger; p.10, fig. 10.1.27, From Orthopaedic Biomechanics by Frankel and Burstein, Copyright 1971: Lea and Febiger; p.11, fig. 10.1.30, From Engineering Mechanics, by Pytel and Kuisalaas, Copyright 1994: HarperCollins; p.20, fig. 10.1.47, From Engineering Mechanics, by Pytel and Kuisalaas, Copyright 1994: HarperCollins; p.20, fig.10.1.48, From Engineering Mechanics, by Pytel and Kuisalaas, Copyright 1994: HarperCollins; p.20, fig. 10.1.49, From Orthopaedic Biomechanics by Frankel and Burstein, Copyright 1971: Lea and Febiger; p.20, fig. 10.1.50, From Total Sailing, by Hall, Copyright 1980: Barmes and Company; p.77, fig. 10.4.56, From Flammarion Book of Astronomy by Allen and Unwin LTD, Copyright 1964: Simon and Schuster; p.80, fig. 10.4.64, Culver Pictures, Inc.; p.94, fig. 10.5.15, From Flammarion Book of Astronomy by Allen and Unwin LTD, Copyright 1964: Simon and Schuster; p.160, fig. 11.1.68, From Environmental Geoscience by Strahler and Strahler, Copyright 1973: Hamilton Publishing Company; p.181, fig. 11.2.17, From Physical Geography by Strahler and Strahler, Copyright 1997: Wiley; p.192, fig. 11.3.10 and 11, From Zoogeography of the Sea by Elkman, Copyright 1953: Sidwich and Jackson; p.205, fig. 11.3.33, From Submarine Canyons and Other Sea Valleys by Shepard and Dill, Copyright 1966: Rand McNally; p.208, fig. 11.3.40, Physical Geography by Strahler and Strahler, Copyright 1997: Wiley; p.208, fig. 11.3.41, Physical Geography by Strahler and Strahler, Copyright 1997: Wiley; p.209, fig. 11.3.44, National Academy of Sciences (NAS)

ISBN 0-201-70345-9

Copyright © 2001 Addison Wesley Longman, Inc. All rights reserved. No part of this publication may be reproduced, stored in a retrieval system, or transmitted, in any form or by any means electronic, mechanical, photocopying, recording, or otherwise, without the prior written permission of the publisher. Printed in the United States of America.

1 2 3 4 5 6 7 8 9 10 — CRS — 02020100

CALCULUS, VOLUME 2
Preliminary Edition
Al Shenk

Introduction v
To the instructor vii

Chapter 10. Vectors and Curves.

10.1. Vectors in the plane .... p. 1
  Vectors in the xy-plane. Displacement and position vectors. The unit vectors **i** and **j**. Force vectors. Orthogonal projections and components of vectors. The dot product. Angles between vectors. Perpendicular vectors.

10.2. Vectors in space .... p. 23
  Rectangular coordinates in space. The Pythagorean Theorem in xyz-space. Vectors in xyz-space. Dot products, components, and projections. The unit vectors **i**, **j**, and **k**. Direction cosines and direction angles.

10.3. Lines and planes in space .... p. 35
  Parametric equations of lines in xyz-space. Equations of planes in xyz-space. The cross product of two vectors. The distance from a point to a plane. The scalar triple product.

10.4. Parametric equations of curves and velocity vectors .... p. 56
  Parametric equations of curves in the xy-plane and in xyz-space. Vector representations of curves. Limits of vector-valued functions. Derivatives of vector-valued functions. Velocity vectors and speed. Lengths of parametric curves. Relative velocity vectors.

10.5. Acceleration, force, and curvature .... p. 83
  Integrating vector-valued functions. Acceleration vectors. Newton's Law of Motion. Unit tangent and normal vectors to plane curves. Curvature of plane curves. Tangential and normal components of acceleration in the plane. Kepler's Lawsd of planetary motion and Newton's Law of gravity. Velocity and acceleration vectors in polar coordinates.

Responses to Questions .... p. 109
Answers and outlines of solutions to Tune-up Exercises and selected Problems ... p. 117

Chapter 11. Derivatives with Two or More Variables.

11.1. Functions of two variables .... p. 137
  Functions of two variables and their domains, ranges, graphs and level curves. Limits and continuity of functions with two variables.

11.2. Partial derivatives with two variables .... p. 165
  First-order partial derivatives. Estimating first-order partial derivatives from tables and level curves. The Chain Rules with two variables. Higher-order partial derivatives.

11.3. Directional derivatives and tangent planes .... p. 187
  Directional derivatives. Linear functions of two variables. Tangent planes to graphs of functions of two variables. Differentials and error estimates with two variables.

11.4. Derivatives with three or more variables .... p. 210
  Functions with three variables and their level surfaces. Quadric surfaces. Partial derivatives, directional derivatives, gradients, linear approximations, and differentials with three variables. Tangent planes to level surfaces in xyz-space. n-dimensional Euclidean space. Partial derivatives of functions with four or more variables.

11.5. Maxima and minima with two or three variables. .... p. 224
  Local maxima and minima and the First-Derivative Test with two and three variables. The Second-Derivative Test with two variables. Lagrange multipliers with two and three variables.

Typeset by $\mathcal{A}_{\mathcal{M}}\mathcal{S}$-TEX

Responses to questions .... p. 247
Answers and outlines of solutions to Tune-up Exercises and selected Problems ... p. 255

Chapter 12. Multiple Integrals.

12.1. Double integrals. .... p. 273
    Double integrals in rectangular coordinates. Volumes and two-dimensional density. Centers of gravity of plates. Average value of $f(x, y)$.

12.2. Triple integrals. .... p. 298
    Triple integrals in rectangular coordinates. Three-dimensional density. Centers of gravity of solids. Average value of $f(x, y, z)$.

12.3. Integrals in polar, cylindrical, and spherical coordinates. .... p. 311
    Double integrals in polar coordinates. Triple integrals in cylindrical and spherical coordinates. Spherical coordinates and geography.

12.4. Other changes of variables: Jacobians .... p. 332
    Affine changes of variables in double and triple integrals. Jacobians. General changes of variables in double and triple integrals.

Responses to questions .... p. 349
Answers and outlines of solutions to Tune-up Exercises and selected Problems .. p. 355

Chapter 13. Vector Analysis.

13.1. Vector fields and line integrals. .... p. 369
    Vector fields. Streamlines of velocity fields. Flux and work for constant velocity and force fields. Line integrals in the plane and in space. Flux and work for general fields. Average value on a curve.

13.2. The Fundamental Theorems in the plane .... p. 394
    Green's Theorem. Circulation and Stokes' Theorem in the plane. The Divergence Theorem in the plane. Gradient vector fields, potential functions, and path-independent line integrals in the plane.

13.3. Surface integrals and the Fundamental Theorems in space .... p. 414
    Surface integrals. Surface area. Average values on surfaces. Weights and centers of gravity of surfaces. Gauss' Theorem. Divergence and the Divergence Theorem in space. The curl and Stokes' Theorem in space. Gradient vector fields, potential functions, and path-independent line integrals in space.

Responses to questions .... p. 439
Answers and outlines of solutions to Tune-up Exercises and selected Problems ... p. 445

Chapter 14. Further Topics in Differential Equations.

14.1. First-order linear and exact differential equations. .... p. 455
    First-order linear and exact differential equations and applications.

14.2. Second-order linear equations with constant coefficients .... p. 465
    Second-order linear differential equations with constant coefficients. Applications to vibrating springs and electric circuits..

14.3. Power series solutions of linear differential equations .... p. 483
    Power series solutions of first- and second-order linear differential equations.

Responses to questions .... p. 493
Answers and outlines of solutions to Tune-up Exercises and selected Problems ... p. 497
Index .... p. 505

INTRODUCTION

"Most of calculus is easy, once you figure it out."

The above statement, which is the basic premise of this text, might come as a surprise to you and requires some explanation.

First, the statement does not say that calculus is easy. It is not. Calculus is a complex and abstract subject that has been developed over the past 2500 years and is used to study a wide variety of topics, including the effects of gravity and air resistance on the velocity and acceleration of vehicles and projectiles, the motion of planets and satellites, areas and volumes, probability, chemical reaction rates, air currents and ocean waves, electricity and magnetism, radioactive decay, the effects of supply and demand on market prices, and the growth of organisms and populations.

Second, the statement speaks of "figuring out calculus" and not of "learning calculus" to make an important point. When someone tells you how to get to a restaurant during your first visit to a city, you learn the directions so you can follow them, but you have to figure out how to interpret those directions as you find your way to the restaurant. Moreover, after you are familiar with the city and have been to the restaurant several times, your understanding of how to find it is based more on your experience than on what you were told initially.

The process of mastering mathematics is similar. You come to understand a new mathematical idea by working with it and by figuring out what it means to you based on your experience and on the ways you formulate and remember abstract ideas. You need to do much more than listen to your instructor and study the explanations given in your textbook.

To do your best, you need to approach every new mathematical topic by asking yourself, "What is this really about?" and "How can I explain this so it makes more sense to me?" If you approach the topic as a mystery to be unraveled, rather than just as information and procedures to be learned, you will bring out the best of your intellectual abilities. Then, as you work with the topic and the new concepts and techniques it entails, it will become clearer and more meaningful to you until it is, in fact, easy to understand. At that point your sense of the topic may well differ considerably from the explanations in your textbook or lectures.

This book is designed to help you to figure out calculus—with the assistance of your instructor—so it can be a useful tool in your later work and studies.

You will see that, in addition to solved examples as are found in other mathematics textbooks, most discussions of ideas, procedures, and problems in this book are interspersed with questions for you to answer to supply key steps of reasoning and calculations. The goal is to keep you actively involved in the logical development and to allow you to formulate your own ways of looking at the topics as you study.

To take full advantage of the questions, pause in your reading to spend time on each one. Think carefully about your response. Then, when you are satisfied with your conclusion, compare it with the response given at the end of the chapter. If you have trouble with a question, do not read the response in the book right away. Continue reading the text and return to the question later, when you can approach it with fresh ideas.

The tune-up exercises at the end of each section cover the basic techniques that are needed to apply the ideas and results covered in the section. Be sure you can work each of these types of exercises before you begin the regular problems that follow.

You will also notice that many of the regular problems do not look exactly like the examples in the body of the text, and that some require precalculus procedures, techniques from earlier in the textbook, or other lines of reasoning not covered in the section. This has not been done to make the problems difficult—most of them are not—but to help you develop your overall ability to read, analyze, and solve problems.

Arrange to study with other students in your class. You can solve harder problems and write better solutions by working together, and discussing ideas with others is a natural way to improve your own understanding.

In many cases you will figure out how to solve a problem after you have formulated and tested several strategies for working it. Morover, in addition to thinking about the "big picture," you need to concentrate on the details and pay close attention to the logic of going from each step to the next. As is suggested in the diagram on the right, you need to write down carefully the results of each step and think about what you have achieved and where you hope you are heading before you proceed.[†]

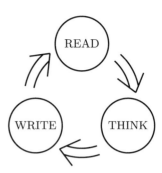

You probably follow this procedure to a certain extent whevever you solve a problem successfully. Suppose, for example, you are asked to find the radius of a circle of area 5π square feet. You might say to yourself, "Because the area of a circle of radius r is πr^2, I need to solve the equation, $5\pi = \pi r^2$." After writing down and reading this equation, you might say, "I need to divide both sides of the equation by π," and write "$5 = r^2$." Reading this might lead you to think, "If I take the square root of both sides to find r, I would have $r = \pm\sqrt{5}$, but a a radius cannot be negative," and you give the answer, "$r = \sqrt{5}$."

It is especially important that you also concentrate on each step when you are working a type of problem that is not familiar to you. Paying close attention to the details and to the logical flow of your reasoning will not only help you avoid mistakes. It can also help you develop a strategy for solving the entire problem as you work on it.

[†]Diagram suggested by Rick LeBorne

TO THE INSTRUCTOR

My goal in writing this textbook has been to provide a more effective tool for achieving traditional objectives in introductory calculus classes by incorporating teaching principles and techniques that have been developed in the calculus reform movement and through research in mathematics education. This book is intended for math, science, and engineering students but can be used in lower-level classes by skipping some of the harder topics and avoiding the hardest problems.

This text is very different from my earlier calculus book, which, like most calculus texts of the 1970's and 1980's, was based on the premise that students could learn effectively by studying material that was presented in a formal deductive manner. The discussions in most textbooks of that period proceeded from definitions to theorems to descriptions of problem-solving procedures. Students were expected to master the subject by studying representative types of exercises that were solved in the textbook and lectures, practiced on homework, and tested on examinations. The assumption was that learning to solve these, what we now call "template" problems, would be an efficient way for them to learn concepts, theory, and techniques.

Most of us who ran our courses in this manner probably felt that our teaching was reasonably successful. The average students could produce solutions of the template problems on our exams to earn—with generous partial credit—their B's and C's. Moreover, the best students seemed to understand the material and had some success with the occasional non-template exercises that we posed for them.

Now it appears that we were perhaps not doing as good a job as we thought; we certainly were not always getting all we could out of our students. Our students were generally overwhelmed by and ignored our presentations of theory, and the template-problem approach allowed them to rely unduly on pattern recognition in working exercises. We did not give them very much incentive to question, analyze, and organize the subject for themselves or much opportunity to develop robust understanding of concepts and procedures by applying basic ideas in new contexts.

Another common mistake we made, as authors and lecturers, was to concentrate on presenting our analysis of the material without taking into account students' learning processes. We tried to do as much of the thinking for them as possible, instead of training them to think for themselves—with the negative effect of limiting how much intellectual effort they gave to our courses and, thereby, limiting how much they could learn.

What can we do to make our classes more effective—without increasing our work loads or lowering our standards? What can we do in our lectures, what sorts of homework can we assign, and what types of questions can we put on our examinations to help students maximize their potential for understanding calculus? How can we get students to work harder and think more? How can we structure our courses so that those students who try to do their best have positive and genuinely creative learning experiences?

First, there is a lot we do not have to change. A great deal of calculus involves learning basic skills, such as finding derivatives and integrals, and many of these are suitably taught and tested with routine exercises. But, what more can we do to get our students to move beyond the ability to work familiar problems to the the point where they have a command of calculus as a flexible tool for working with a broad range of applications in new contexts, and how can we make such changes and still give homework and examinations that are reasonable and fair?

The approach taken in this text—and that I recommend instructors use in their classes—is to maintain a balance between "reaction" and "reflection" problems, where a reaction problem is one that a well-prepared student can solve immediately, and a reflection problem requires that the student reflect on what is being asked, on how the problem might be solved, or on an interpretation or application of the results. The primary goal of our template-problem courses was to train students to master specific types of reaction problems. The inclusion of reflection problems in lectures, homework, and examinations gives students more opportunity to develop their mathematical abilities and can generate more interest and vitality in our classes.

Most reflection problems in this book are not technically difficult, but because they ask questions for which there are no matching solved examples in the text, they take students out of a template-application mode and encourage them to think more abstractly and creatively. Many of these problems require additional reasoning or deal with formulas, data, or graphs from other disciplines. Some use precalculus techniques or techniques from previous sections, and others preview concepts to be presented later.

This book has a number of features, in addition to the emphasis on reflection problems, that set it apart from most other calculus texts:

- *Students need access to graphing calculators or computers.*
  Many of the questions, examples, and problems require a calculator or a computer with calculus software. These remarkable tools make it much easier for students to grasp ideas and make them better problem solvers. Seeing the geometric and numerical aspects of their work helps them avoid errors and leaves them with a much richer understanding of what they are doing.

- *Questions in the text lead students to be more active learners.*
  Discussions of theory and techniques, sample calculations, and applications are interspersed with questions which students answer to supply details and interpretations. Responses are given at the end of the chapters for them to check the accuracy of their calculations and reasoning. This process focuses their attention on important concepts and techniques and gives them opportunities to see the mathematics from their own points of view, while providing them enough feedback so they can study effectively on their own. Solved examples are used as appropriate to consolidate students' understanding, to introduce new techniques, and to illustrate technical notation and terminology.

- *Tune-up exercises develop basic skills*
  While the basic goal of the text is to develop students' conceptual understanding and problem-solving abilities, they also need practice with basic operations. Tune-up exercises, which students are expected to master before they attempt the regular problems, cover the easier skills. Other problems provide practice with more difficult procedures, such as differentiation, integration, and the sketching of graphs.

- *The text gives students extensive resources for studying on their own.*
  By the time students have worked through all of the questions in a section and solved all of that section's tune-up exercises, they have spent a substantial amount of time in structured study. Then they can consolidate and expand their skills and understanding by doing their homework and working other problems whose answers or outlines of solutions are given in the text.

- *Lecture time can be saved by assigning questions as homework.*
  Instructors can have students study each topic before it is covered in class by requiring that they turn in, as part of their homework, detailed responses to some or all of the questions on that topic in the text. This leads the students to come to class better prepared to discuss the material with comprehension and frees instructors from having to spend very much time introducing new material.

- *Instructors can use problems from the text in their lectures.*
  The preparation of nonroutine problems—especially those that are based on graphs or data from specific applications or that involve calculators or computers—can require a great deal of time and energy. The text contains a wide variety of problems suitable for use in lectures, freeing instructors to spend their class-preparation time on other aspects of teaching, such as planning ways to stimulate student interest and develop their overall understanding. Moreover, if the students can read the problems from their books, instructors do not have to waste class time writing them on the board for the students to copy.

- ***Ample problem sets are included.***
  The wide variety of problems in some sections are included to meet the requirements of different classes, to give instructors more choice of problems to use in their lectures, and so instructors can vary their assignments from year to year. The problems are in some cases organized by topic but are generally ordered by level of difficulty, with the most challenging exercises at the end.

- ***Marginal notes help instructors plan their classes***
  Footnotes at the beginning of each section indicate, when appropriate, how the section could be divided into two or more teaching units. Lists of the corresponding tune-up rxercises and problems are also given with suggestions of questions that might be assigned to prepare students for lectures and of problems that serve well for in-class discussions. In addition, a classification by level of difficulty and topic is given after each tune-up exercise and problem.[†]

- ***The** Instructor's Handbook **will be a solution manual and instructor's guide.***
  The *Instructor's Handbook*, which will be supplied to users of the book, will contain answers and outlines of solutions of all tune-up exercises and problems and more detailed suggestions for using the book.

This text is the result of eight years of experimenting with a variety of teaching materials and techniques in my large calculus classes at the University of California at San Diego. I found that by using drafts of this book, I got much more positive feedback from my students than I had with more traditional textbooks. The students said they liked the question-response format and the variety of applications and that they learned more in my course than they had in their previous math classes. I also had much more fun in my lectures, since I spent most of my time discussing problems with the students rather than just presenting material on the board.

In developing this text I have been guided by my students' performance and by their comments and suggestions. In my writing, I have also kept in mind my understanding of other instructors' needs and interests, in the hope that the book will be as flexible and useful as possible.

Al Shenk, Mathematics Department, UCSD

La Jolla, CA 92093 (ashenk@ucsd.edu)

[†]These marginal notes will appear only in the instructors' desk copies and not in students' copies of the forthcoming First Edition.

CHAPTER 10
VECTORS AND CURVES

Often a number and a direction appear together in the description of a geometric or physical situation, as when we say that one town is five miles to the northwest of another town or that gravity exerts a downward force of thirty pounds on an object. In such cases we can treat the number and direction together as a VECTOR. We study vectors in two and three dimensions and some of their applications in Sections 10.1 and 10.2. In Section 10.3, we use vectors to find equations for lines and planes in xyz-space. We study curves defined by PARAMETRIC EQUATIONS and VELOCITY VECTORS in Section 10.4. The effects of forces on objects with curved trajectories are analyzed in Section 10.5, where we discuss Newton's Law of motion, curvature of plane curves, and the tangential and normal components of acceleration vectors in a plane. Section 10.5 ends with a brief discussion of Newton's Law of gravity and Kepler's laws of planetary motion.

Section 10.1

Vectors in the plane

OVERVIEW: *In this section we will use vectors in an xy-plane to study positions and relative positions of points and constant forces. Then we will see how the* DOT PRODUCT *of vectors is used to find* ORTHOGONAL PROJECTIONS *of vectors and the* COMPONENT *of one vector in the direction of another nonzero vector, to determine whether vectors are perpendicular, and to find angles between vectors.*

Topics:

- *Vectors in a coordinate plane*
- *Displacement and position vectors*
- *The unit vectors* **i** *and* **j**
- *Adding vectors and multiplying vectors by real numbers*
- *Force vectors*
- *The orthogonal (perpendicular) projection of a vector on a line*
- *The component of one vector in the direction of another*
- *The dot product and angles between vectors*

Vectors in a coordinate plane

We begin with two Questions dealing with positions and relative positions that will serve as background for the definition of vectors.

Sailboats cannot sail directly into the wind, but they can sail in any direction that makes an angle of at least 45° with the direction from which the wind is blowing. To move upwind, they have to tack back and forth. The sailboat in Figure 1, for example, moves to the east, into the wind, by first tacking toward the left from the origin O in the xy-plane to point P, then tacking toward the right, from point P to point Q, and then tacking toward the left, from point Q to point R.[1]

Question 1   Point P in Figure 1 is 100 yards east and 100 yards north of the origin O; point Q is 75 yards east and 75 yards south of P; and point R is 50 yards east and 50 yards north of Q. How far is R east and north of O?

---

[1]Drawing adapted from *Introduction to Sailing* by A. Brown, New York, NY: Simon and Schuster, 1968, p. 44. The first and third tacks in Figure 1 are called "starboard tacks" because in those cases the wind is blowing from the starboard (right) side of the boat. The second tack is a "port" tack because then the wind is blowing from the port (left) side.

♦ SUGGESTIONS TO INSTRUCTORS: The main topics in this section can be divided into two parts, which can be covered in two or more class meetings.

Part 1 (Vectors, sums and real multiples of vectors, position and displacement vectors, **i** and **j**, force vectors; Tune-up Exercises T1–T6, T15, Problems 1–28): Assign Questions 1 through 6 to be started before the class meeting. Question 3, Example 7, Question 6, and Problems 11, 12, 16, and 18 are good for in-class discussion.

Part 2 (Projections, components, dot products, angles between vectors, perpendicular vectors; Tune-up Exercise T7–T14, Problems 29–62): Assign Questions 7 through 10 to be started before the class meeting. Example 8 and Problems 27, 30, 34, 37, and 44 are good for in-class discussion.

FIGURE 1

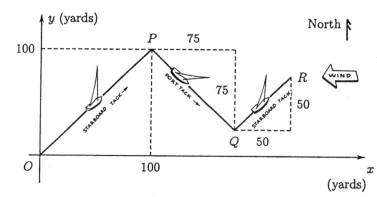

Question 2 One boat sails from point S to point T in Figure 2. The point S is 25 yards east and 50 yards north of the origin O, and T is 50 yards east and 30 yards north of S. Later another boat sails from S to U in Figure 3, where the direction from S to U is the same as the direction from S to T in Figure 1 and the distance from S to U is twice the distance from S to T. How far is U east and north of the origin?

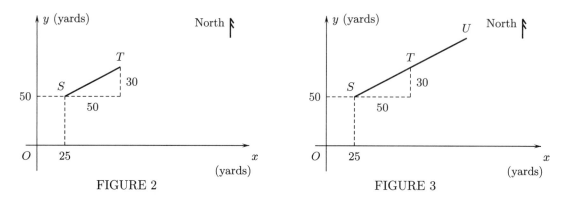

FIGURE 2 FIGURE 3

As we will see below, the calculations in the responses to Questions 1 and 2 can be carried out more systematically by using vectors.

Definition 1 *A* VECTOR **A** *represents a nonnegative number and, if the number is not zero, a direction. The number associated with the vector* **A** *is called its* LENGTH *or* MAGNITUDE *and is denoted* $|\mathbf{A}|$.

Vectors are generally denoted in printed materials by bold-faced letters, like the **A** in Definition 1, or by symbols with arrows over them, as in the symbol $\overrightarrow{PQ}$ that we will use below for displacement vectors. In hand-written materials, vectors are usually represented by letters with arrows over them, as in the symbol $\overrightarrow{A}$.

Vectors are represented in drawings by arrows, where in each case the length of the arrow is the magnitude of the vector, measured with a scale that might or might not be the scale used on the coordinate axes. The direction of the arrow is the direction associated with the vector (Figure 4). The same vector can be represented by different arrows in different locations (Figure 5), provided that the different arrows are parallel, have the same lengths, and point in the same direction.

10.1 Vectors in the plane

FIGURE 4   FIGURE 5

If a vector is represented by an arrow in an xy-plane as in Figure 6, then the x- and y-COMPONENTS of the vector are the changes in the x- and y-coordinates from the base to the tip of the arrow, measured with the same scale as used for the arrow. If the x-component of $\mathbf{A}$ is A_1 and the y-component is A_2, as shown in Figure 6 with positive A_1 and A_2, we write $\mathbf{A} = \langle A_1, A_2 \rangle$.[†] The length of a vector can be calculated from its x- and y-components by using the Pythagorean Theorem:

$$|\mathbf{A}| = |\langle A_1, A_2 \rangle| = \sqrt{(A_1)^2 + (A_2)^2}. \tag{1}$$

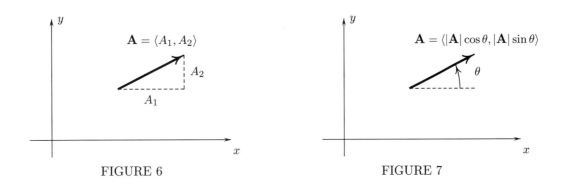

FIGURE 6   FIGURE 7

A nonzero vector $\mathbf{A}$ in an xy-plane can also be described by giving its length and its ANGLE OF INCLINATION, which is an angle θ from the positive x-direction (Figure 7). Then the formula

$$\mathbf{A} = \langle |\mathbf{A}| \cos\theta, |\mathbf{A}| \sin\theta \rangle \tag{2}$$

for the vector in terms of its length and an angle of inclination follows from the definitions of the sine and cosine (see Appendix 2 of Volume 1).

Example 1 Find the x- and y-components of the vector $\mathbf{B}$ of length 10 with angle of inclination $\frac{5}{6}\pi$.

SOLUTION By formula (2), $\mathbf{B} = \langle 10\cos\left(\frac{5}{6}\pi\right), 10\sin\left(\frac{5}{6}\pi\right) \rangle = \langle 10\left(-\frac{1}{2}\sqrt{3}\right), 10\left(\frac{1}{2}\right) \rangle = \langle -5\sqrt{3}, 5 \rangle$ (Figure 8). □

Question 3 Give an angle of inclination of the vector $\mathbf{C} = \langle 3, 4 \rangle$.

[†]Angular brackets are used in this notation to distinguish vectors from points.

FIGURE 8

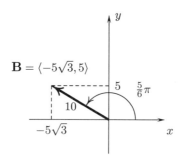

The ZERO VECTOR is denoted **O**. It has zero length and no direction, and in a coordinate plane has zero components: $\mathbf{O} = \langle 0, 0 \rangle$.

Adding vectors and multiplying vectors by numbers

Two vectors in a coordinate plane are added by adding their components, and multiplying a vector in a coordinate plane by a real number yields the vector whose components are the products of the number with the components of the original vector:

Definition 2 For any vectors $\mathbf{A} = \langle A_1, A_2 \rangle$ and $\mathbf{B} = \langle B_1, B_2 \rangle$ and any number k,

$$\mathbf{A} + \mathbf{B} = \langle A_1, A_2 \rangle + \langle B_1, B_2 \rangle = \langle A_1 + B_1, A_2 + B_2 \rangle \tag{3}$$

$$k\mathbf{A} = k\langle A_1, A_2 \rangle = \langle kA_1, kA_2 \rangle. \tag{4}$$

Notice that with Definition 2, $\mathbf{A} + \mathbf{B} = \mathbf{B} + \mathbf{A}$ and $a(b\mathbf{A} + c\mathbf{B}) = ab\mathbf{A} + ac\mathbf{B}$ for any vectors $\mathbf{A}$ and $\mathbf{B}$ in a plane and any numbers a, b, and c.

Equation (3) has two geometric interpretations that are illustrated in Figures 9 and 10 for vectors with positive components: If we place the base of $\mathbf{B}$ at the tip of $\mathbf{A}$, then the vector $\mathbf{A} + \mathbf{B}$ is given by the arrow from the base of $\mathbf{A}$ to the tip of $\mathbf{B}$, as shown in Figure 9. We can also find $\mathbf{A} + \mathbf{B}$ geometrically by placing the bases of $\mathbf{A}$ and $\mathbf{B}$ together, as in Figure 10, and completing the parallelogram with those vectors as sides. Then $\mathbf{A} + \mathbf{B}$ is represented by the arrow with base at the bases of $\mathbf{A}$ and $\mathbf{B}$ that goes along the diagonal of the parallelogram to the opposite vertex.

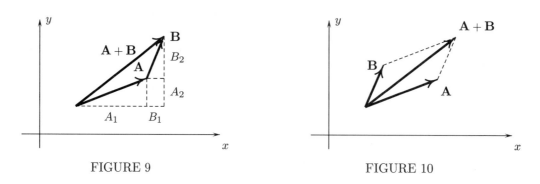

FIGURE 9   FIGURE 10

Equation (4) is illustrated in Figure 11. Multiplying the vector $\mathbf{A}$ by a positive number k yields a vector with the same direction as $\mathbf{A}$ whose length is k multiplied by the length of $\mathbf{A}$, and multiplying $\mathbf{A}$ by a negative number m yields a vector with the opposite direction as $\mathbf{A}$ whose length is $|m|$ multiplied by the length of $\mathbf{A}$.

10.1 Vectors in the plane

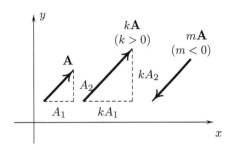

FIGURE 11

Example 2 Write $3\langle 4, -1\rangle - 2\langle 10, -5\rangle$ in the form $\langle A_1, A_2\rangle$.

SOLUTION We use Definition 2: $3\langle 4, -1\rangle - 2\langle 10, -5\rangle = \langle 3(4), 3(-1)\rangle + \langle -2(10), -2(-5)\rangle$
$= \langle 12, -3\rangle + \langle -20, 10\rangle = \langle -8, 7\rangle.$ □

Displacement and position vectors

Relative positions of points in an xy-plane can be described with DISPLACEMENT VECTORS. The displacement vector from one point $P(x_0, y_0)$ to a second point $Q(x_1, y_1)$ is represented by an arrow with base P and tip Q. Its components are obtained by subtracting the coordinates of Q from the coordinates of P

$$\overrightarrow{PQ} = \langle x_1 - x_0, y_1 - y_0\rangle. \tag{5}$$

This is illustrated in Figure 12 in a case where $x_1 - x_0$ and $y_1 - y_0$ are positive.

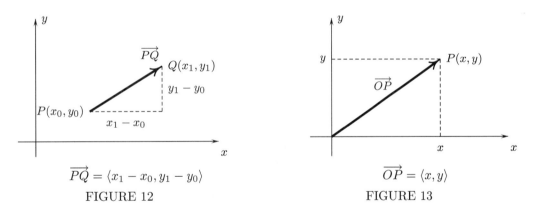

FIGURE 12   FIGURE 13

Points can be located in an xy-plane by using POSITION VECTORS. The position vector of the point $P(x, y)$ is the displacement vector $\overrightarrow{OP}$

$$\overrightarrow{OP} = \langle x, y\rangle \tag{6}$$

from the origin to the point (Figure 13). The components of the position vector are the coordinates of the point.

Now we can show how the calculations in Questions 1 and 2 can be made using vectors.

Example 3 Use vectors to find how far the point R in Figure 1 is to the east and north of the origin O.

SOLUTION Figure 14 shows the position vector $\overrightarrow{OP}$ and the displacement vectors $\overrightarrow{PQ}$ and $\overrightarrow{PR}$ with the points $P, Q,$ and R from Figure 1.

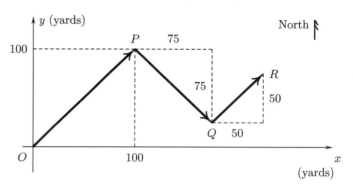

FIGURE 14

As can be seen in Figure 14, the position vector $\overrightarrow{OR}$ is the sum of $\overrightarrow{OP}, \overrightarrow{PQ}$ and $\overrightarrow{QR}$. Also, $\overrightarrow{OP} = \langle 100, 100 \rangle$, $\overrightarrow{PQ} = \langle 75, -75 \rangle$, and $\overrightarrow{QR} = \langle 50, 50 \rangle$. Therefore,

$$\overrightarrow{OR} = \overrightarrow{OP} + \overrightarrow{PQ} + \overrightarrow{QR}$$
$$= \langle 100, 100 \rangle + \langle 75, -75 \rangle + \langle 50, 50 \rangle$$
$$= \langle 100 + 75 + 50, 100 - 75 + 50 \rangle = \langle 225, 75 \rangle.$$

Because the position vector $\overrightarrow{OR}$ of R is $\langle 225, 75 \rangle$, the point R has xy-coordinates $(225, 75)$ and is 225 yards east and 75 yards north of O, as you saw in answering Question 1. □

Example 4 Use vectors to find the coordinates of the point U in Figure 3.

SOLUTION Figure 15 shows the position vector $\overrightarrow{OS}$ and the displacement vector $\overrightarrow{ST}$ with S and T from Figure 3. Figure 16 shows $\overrightarrow{OS}$ and $\overrightarrow{SU}$. Because the direction from S to U is the same as the direction from S to T and the distance from U to S is twice the distance from T to S, $\overrightarrow{SU} = 2\overrightarrow{ST}$.

From Figure 15 we can see that $\overrightarrow{OS} = \langle 25, 50 \rangle$ and $\overrightarrow{ST} = \langle 50, 30 \rangle$. Hence,

$$\overrightarrow{OU} = \overrightarrow{OS} + \overrightarrow{SU} = \overrightarrow{OS} + 2\overrightarrow{ST}$$
$$= \langle 25, 50 \rangle + 2\langle 50, 30 \rangle = \langle 25, 50 \rangle + \langle 100, 60 \rangle$$
$$= \langle 25 + 100, 50 + 60 \rangle = \langle 125, 110 \rangle.$$

The coordinates of U are $(125, 110)$, as you found in answering Question 2. □

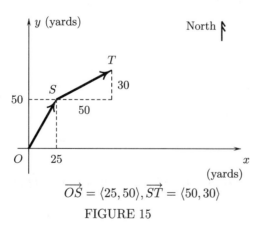

$\overrightarrow{OS} = \langle 25, 50 \rangle, \overrightarrow{ST} = \langle 50, 30 \rangle$

FIGURE 15

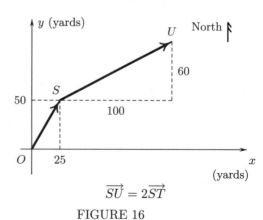

$\overrightarrow{SU} = 2\overrightarrow{ST}$

FIGURE 16

10.1 Vectors in the plane

In the next Question we will use position and displacement vectors to find one vertex of a parallelogram from the coordinates of its other three vertices.

Question 4 Three vertices of the parallelogram $PRSQ$ in Figure 17 are $P = (3,4), Q = (7,8)$, and $R = (12,2)$. **(a)** Give the x- and y-components of the vectors $\overrightarrow{OP}, \overrightarrow{PQ}$, and $\overrightarrow{PR}$ in Figure 18. **(b)** Why is $\overrightarrow{RS}$ equal to $\overrightarrow{PQ}$? **(c)** Use the vectors from parts (a) and (b) to find the x- and y-components of $\overrightarrow{OS}$. **(d)** What are the coordinates of S?

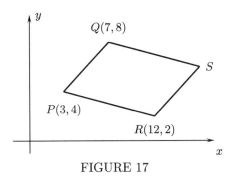

FIGURE 17

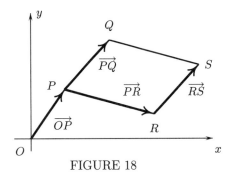

FIGURE 18

The unit vectors i and j

The vectors $\langle 1, 0 \rangle$ and $\langle 0, 1 \rangle$ of length 1 in the directions of the positive x- and y-axes in Figure 19 are denoted **i** and **j**, respectively. Because their length is 1, they are called UNIT VECTORS. We can use these vectors in place of angular brackets to express $\mathbf{A} = \langle A_1, A_2 \rangle$ in terms of its x-and y-components by writing

$$\mathbf{A} = \langle A_1, A_2 \rangle = A_1 \langle 1, 0 \rangle + A_2 \langle 0, 1 \rangle = A_1 \mathbf{i} + A_2 \mathbf{j}.$$

A geometric interpretation of this formula is shown in Figure 20 for a case with positive A_1 and A_2.

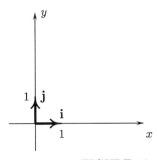

FIGURE 19

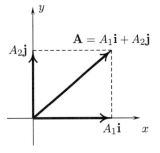

FIGURE 20

Example 5 Express $5(3\mathbf{i} - 2\mathbf{j}) + 6(-\mathbf{i} + 3\mathbf{j})$ in the form $A_1 \mathbf{i} + A_2 \mathbf{j}$.

SOLUTION $5(3\mathbf{i} - 2\mathbf{j}) + 6(-\mathbf{i} + 3\mathbf{j}) = [5(3) + 6(-1)]\mathbf{i} + [5(-2) + 6(3)]\mathbf{j} = 9\mathbf{i} + 8\mathbf{j}.$ □

Force vectors

A force can be represented by a vector whose length is the strength of the force and whose direction is the direction in which the force is exerted. If the force is applied to an object, then the base of the force vector is positioned at the point on the object where the force is applied. Figure 21 shows, for example, the force **F** exerted by the ground by a person pushing on the stationary tip of a crutch.[2]

---

[2] Drawing adapted from *Orthhopædic Biomechanics* by V. Frankel and A. Burstein, London: Lea & Febiger, 1971, pp. 6-7.

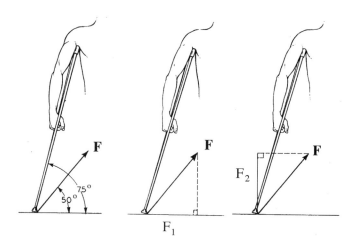

FIGURE 21

Imagine that each of the three drawings in Figure 21 is in an xy-plane that has the usual orientation of axes and that contains the vector **F** and the vertical line through its base. Then $\mathbf{F} = \langle F_1, F_2 \rangle$ with F_1 as in the center drawing and F_2 as in the drawing on the right.

Question 5   The magnitude of the force vector **F** in Figure 21 is 80 pounds and its angle of inclination is 50°. Give exact and approximate decimal values of its components F_1 and F_2.

Example 6   What is the force of the person on the tip of the crutch Figure 21?

SOLUTION   In Question 5 you saw that the force of the ground on the crutch is
$\mathbf{F} = \langle 80\cos\left(\frac{5}{18}\pi\right), 80\sin\left(\frac{5}{18}\pi\right)\rangle \doteq \langle 51.4, 61.3 \rangle$ pounds.[†] Because the tip of the crutch is stationary, the force of the person on the tip of the crutch balances the force of the ground on the tip of the crutch and is the negative $-\mathbf{F} = \langle -80\cos\left(\frac{5}{18}\pi\right), -80\sin\left(\frac{5}{18}\pi\right)\rangle \doteq \langle -51.42, -61.28 \rangle$ of the force by the ground (Figure 22). □

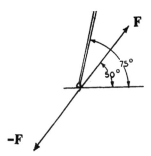

FIGURE 22

Force vectors generally have different effects if they are placed in different positions. Imagine, for instance, that Figures 23 through 25 show a square metal plate that is to be dragged along a floor by a constant force **F** parallel to the floor. Because of friction with the floor, the plate would twist clockwise if the force were applied at point P on the plate in Figure 23, and would twist counterclockwise if the force were applied at point Q in Figure 24.

[†]Force vectors are given the same units as used for their components and magnitudes.

10.1 Vectors in the plane

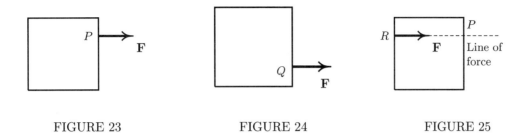

FIGURE 23   FIGURE 24   FIGURE 25

There is, however, one way that force vectors can be repositioned without changing their effects at a paricular moment. If the force is being applied to a RIGID BODY, i. e., an object like the metal plate in Figures 23 through 25 that can be moved but not deformed by the forces under consideration, then the base of the force vector can be placed at any point on the line through the original vector. This line is called the LINE OF FORCE of the original vector.

This principle is illustrated in Figure 25, which shows the line of force of the vector **F** from Figure 23. Applying the force **F** at point R on that line, as indicated in Figure 25, would have the same effect on the plate at that moment as applying the force at the original point P because R is on the line of force of **F** with its base at P.

Sums of force vectors

It is an empirical fact that two forces **F** and **G** applied at the same point P on an object have the same effect as their sum **F** + **G** (Figure 26). Because of this, the sum is called the RESULTANT of **F** and **G**.

FIGURE 26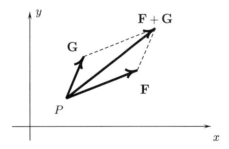

Example 7   Figure 27 shows a hospital patient's leg that is held in traction by a ten-pound weight at one end of a rope that passes through a fixed pulley and then through a free pulley fastened to the patient's leg. The other end of the rope is attached to a hook on the wall directly over the fixed pulley.[3] Suppose that the angle between the parts of the rope leading from the free pulley is 106°. What is the force by the rope on the free pulley?

---

[3] Drawing adapted from *Orthhopædic Biomechanics* by V. Frankel and A. Burstein, London: Lea & Febiger, 1971, p. 4.

10   Chapter 10: Vectors and Curves

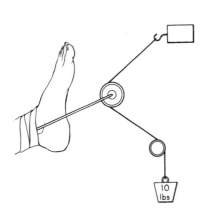
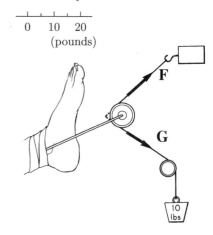

FIGURE 27   FIGURE 28

SOLUTION We let **F** be the force on the free pulley by the upper part of the rope and **G** the force by the lower part, as in Figure 28, where the lengths of the arrows are measured by the scale at the top of the drawing. We move these vectors to the left along their lines of force to the intersection of those lines and introduce xy-axes with their origin at the base of the vectors, as in Figure 29.

The ten pound weight at the end of the rope creates a tension of ten pounds all along the rope, so that the magnitudes of **F** and **G** are both ten pounds. Also, because the free pulley can move up and down, the y-component of **F** is the negative of the y-component of **G**. Consequently, the x-axis bisects the angle between the vectors in Figure 29. Hence, the angle of inclination of **F** is 53° or $\frac{53}{180}\pi$ radians, the angle of inclination of **G** is $-53°$ or $-\frac{53}{180}\pi$ radians, and

$$\mathbf{F} = 10\langle\cos\left(\tfrac{53}{180}\pi\right), \sin\left(\tfrac{53}{180}\pi\right)\rangle$$
$$\mathbf{G} = 10\langle\cos\left(-\tfrac{53}{180}\pi\right), \sin\left(-\tfrac{53}{180}\pi\right)\rangle = 10\langle\cos\left(\tfrac{53}{180}\pi\right), -\sin\left(\tfrac{53}{180}\pi\right)\rangle$$

and the resultant of **F** and **G** is

$$\mathbf{F}+\mathbf{G} = 10\langle\cos\left(\tfrac{53}{180}\pi\right), \sin\left(\tfrac{53}{180}\pi\right)\rangle + 10\langle\cos\left(\tfrac{53}{180}\pi\right), -\sin\left(\tfrac{53}{180}\pi\right)\rangle$$
$$=\langle 20\cos\left(\tfrac{53}{180}\pi\right), 0\rangle.$$

The force of the rope on the free pulley is horizontal, pointed to the right, and of magnitude $20\cos\left(\tfrac{53}{180}\pi\right) \doteq 12.04$ pounds. □

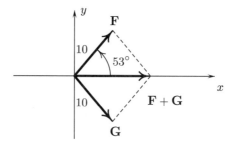

FIGURE 29

10.1 Vectors in the plane

Question 6   One man is lifting a boulder with a rod while another is pulling it with a rope as in Figure 30. **(a)** Find the x- and y-components of the two force vectors, with the usual orientation of axes. **(b)** Find the resultant of the two forces and the approximate decimal values of its magnitude and angle of inclination.[4]

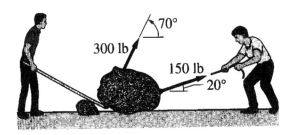

FIGURE 30

Orthogonal projections of vectors

The ORTHOGONAL PROJECTION or PERPENDICULAR PROJECTION of a vector $\mathbf{A}$ on a line L is the vector $\mathbf{A}_{\text{proj}}$ that is obtained by placing the base of $\mathbf{A}$ on the line L and taking $\mathbf{A}_{\text{proj}}$ to be the vector from the base of $\mathbf{A}$ to the foot of the perpendicular line from the tip of $\mathbf{A}$ to L (Figure 31).

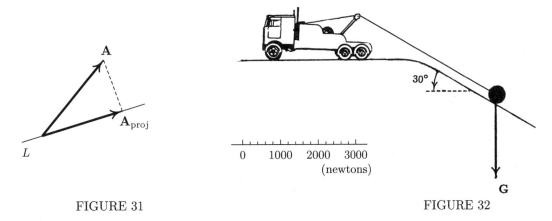

FIGURE 31   FIGURE 32

Example 8   Imagine that a chain from a tow truck, as in Figure 32, is holding a ball weighing 2000 newtons ($\approx$ 450 pounds) in a fixed position on a 30° incline. With the usual orientation of coordinate axes, the downward force of gravity $\mathbf{G}$ on the ball is $\langle 0, -2000 \rangle$ newtons and is represented by the arrow in Figure 32 with its length measured by the scale in that drawing. Find **(a)** the projection $\mathbf{G_T}$ of $\mathbf{G}$ on a line parallel to the incline and **(b)** the projection $\mathbf{G_N}$ on a line perpendicular to it.[†]

SOLUTION   **(a)** The vector $\mathbf{G}$ and its projections are shown in Figure 33. Because $\mathbf{G}$ is 2000 newtons long and the angle between $\mathbf{G}$ and $\mathbf{G_T}$ is 60° or $\frac{1}{3}\pi$ radians, the length of $\mathbf{G_T}$ is $2000 \cos\left(\frac{1}{3}\pi\right) = 2000\left(\frac{1}{2}\right) = 1000$ newtons. Then, because the angle of inclination of $\mathbf{G_T}$ is $-30°$ or $-\frac{1}{6}\pi$ radians, $\mathbf{G_T} = 1000\langle \cos\left(-\frac{1}{6}\pi\right), \sin\left(-\frac{1}{6}\pi\right)\rangle$
$= 1000\langle \frac{1}{2}\sqrt{3}, -\frac{1}{2}\rangle = \langle 500\sqrt{3}, -500\rangle$ newtons.

---

[4]Problem and drawing adapted from *Engineering Mechanics* by A. Pytel and J. Kuisalaas, New York, NY: HarperCollins, 1994, p. 39.

[†]The subscripts "T" and "N" are used here because $\mathbf{G_T}$ is a force tangent to the incline and $\mathbf{G_N}$ is a force normal (perpendicular) to it.

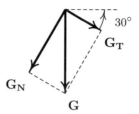

FIGURE 33

(b) The angle between **G** and $\mathbf{G_N}$ is $\frac{1}{6}\pi$ radians, so the length of $\mathbf{G_N}$ is $2000\cos\left(\frac{1}{6}\pi\right) = 2000\left(\frac{1}{2}\sqrt{3}\right) = 1000\sqrt{3}$ pounds. Then, since the angle of inclination of $\mathbf{G_N}$ is $-\frac{2}{3}\pi$ radians,

$$\mathbf{G_N} = 1000\sqrt{3}\langle\cos\left(-\tfrac{2}{3}\pi\right),\sin\left(-\tfrac{2}{3}\pi\right)\rangle$$
$$= 1000\sqrt{3}\langle-\tfrac{1}{2},-\tfrac{1}{2}\sqrt{3}\rangle = \langle-500\sqrt{3},-1500\rangle \text{ newtons.} \square$$

Because the projections $\mathbf{G_T}$ and $\mathbf{G_N}$ in Figure 33 are perpendicular, $\mathbf{G} = \mathbf{G_T} + \mathbf{G_N}$ and we can consider the force of gravity to be the sum of the tangential force $\mathbf{G_T}$ and the normal force $\mathbf{G_N}$. $\mathbf{G_T}$ is the force of gravity on the chain and $\mathbf{G_N}$ is the force of gravity perpendicular to the incline.

The component of one vector in the direction of another

Most results concerning orthogonal projections of vectors can also be expressed in terms of the (ORTHOGONAL) COMPONENT of one vector in the direction of another.

To find the component of **A** in the direction of a nonzero **B**, introduce a t-axis through **B** with its origin at the base of **B** and with distances measured using the scale for measuring the arrow representing **A**. Put the base of **A** at the base of **B**, and drop a perpendicular line from the tip of **A** to the t-axis, as in Figures 34 and 35. The component of **A** in the direction of **B** is the t-coordinate of the foot of the perpendicular line.

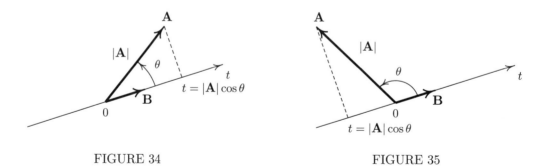

FIGURE 34   FIGURE 35

Figures 34 and 35 show that if **A** is not zero, then the component of **A** in the direction of **B** equals $|\mathbf{A}|\cos\theta$, where θ is an angle between the two vectors. Also, if **A** is the zero vector, then its component in the direction of **B** is the number zero, so we have the following definition.

Definition 3 *The* COMPONENT *of a vector* **A** *in the direction of a nonzero vector* **B** *is the number* 0 *if* $\mathbf{A} = \mathbf{O}$ *and is the number* $|\mathbf{A}|\cos\theta$, *where* θ *is an angle between* **A** *and* **B** *if* **A** *is not* **O**.

As can be seen in Figures 34 and 35, the component of **A** in the direction of **B** is positive if the smallest positive angle θ between **A** and **B** satisfies $0 \leq \theta < \frac{1}{2}\pi$ and is negative if $\frac{1}{2}\pi < \theta \leq \pi$. It is zero if θ is a right angle.

Question 7 (a) What is the component of **A** in the direction of **B** in Figure 34 if $|\mathbf{A}| = 6$ and $\theta = 0.6$ radians? (b) What is the component of **A** in the direction of **B** in Figure 35 if $|\mathbf{A}| = 6$ and $\theta = 2.15$ radians?

The dot product

In Example 8 and Question 6 we found projections and components of vectors by using their lengths and angles of inclination. If we are given instead the x- and y-components of the vectors, it is easier to use the DOT PRODUCT, as we will see below.

We can motivate the definition of the dot product with some calculations. Suppose that $\mathbf{A} = \langle A_1, A_2 \rangle$ and $\mathbf{B} = \langle B_1, B_2 \rangle$, as in Figure 36. The vector $\mathbf{A} - \mathbf{B} = \langle A_1 - B_1, A_2 - B_2 \rangle$ is given by the arrow from the tip of **B** to the tip of **A** (since adding $\mathbf{A} - \mathbf{B}$ to **B** gives **A**), and the square of the length of $\mathbf{A} - \mathbf{B}$ is

$$\begin{aligned}|\mathbf{A} - \mathbf{B}|^2 &= (A_1 - B_1)^2 + (A_2 - B_2)^2 \\ &= (A_1^2 - 2A_1B_1 + B_1^2) + (A_2^2 - 2A_2B_2 + B_2^2) \\ &= (A_1^2 + A_2^2) + (B_1^2 + B_2^2) - 2(A_1B_1 + A_2B_2) \\ &= |\mathbf{A}|^2 + |\mathbf{B}|^2 - 2(A_1B_1 + A_2B_2).\end{aligned} \quad (7)$$

We can also calculate $|\mathbf{A} - \mathbf{B}|^2$ another way. We introduce new uv-axes with the positive u-axis passing through **B** and we let θ be an angle from **A** to **B**, as in Figure 37. Relative to these coordinates $\mathbf{A} = \langle |\mathbf{A}|\cos\theta, |\mathbf{A}|\sin\theta \rangle$ and $\mathbf{B} = \langle |\mathbf{B}|, 0 \rangle$, so that[†]

$$\begin{aligned}|\mathbf{A} - \mathbf{B}|^2 &= (|\mathbf{A}|\cos\theta - |\mathbf{B}|)^2 + (|\mathbf{A}|\sin\theta - 0)^2 \\ &= (|\mathbf{A}|^2\cos^2\theta - 2|\mathbf{A}||\mathbf{B}|\cos\theta + |\mathbf{B}|^2) + |\mathbf{A}|^2\sin^2\theta \\ &= |\mathbf{A}|^2(\cos^2\theta + \sin^2\theta) + |\mathbf{B}|^2 - 2|\mathbf{A}||\mathbf{B}|\cos\theta \\ &= |\mathbf{A}|^2 + |\mathbf{B}|^2 - 2|\mathbf{A}||\mathbf{B}|\cos\theta.\end{aligned} \quad (8)$$

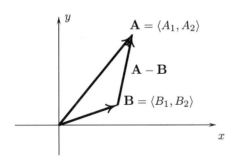

FIGURE 36

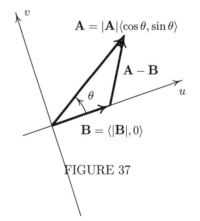

FIGURE 37

[†]Formula **(8)** is the Law of Cosines in vector notation.

Comparing equations (7) and (8) shows that $A_1B_1 + A_2B_2$ equals $|\mathbf{A}||\mathbf{B}|\cos\theta$, so we can use the first expression to find the second without finding θ. The first expression is called the DOT PRODUCT of $\mathbf{A}$ and $\mathbf{B}$ and is denoted $\mathbf{A}\cdot\mathbf{B}$:

Definition 4 The DOT PRODUCT of vectors $\mathbf{A} = \langle A_1, A_2\rangle$ and $\mathbf{B} = \langle B_1, B_2\rangle$ in a coordinate plane is the number

$$\mathbf{A}\cdot\mathbf{B} = A_1B_1 + A_2B_2. \tag{9}$$

Notice that with this definition,

$$\mathbf{A}\cdot\mathbf{B} = \mathbf{B}\cdot\mathbf{A} \tag{10}$$
$$\mathbf{A}\cdot(\mathbf{B}+\mathbf{C}) = \mathbf{A}\cdot\mathbf{B} + \mathbf{A}\cdot\mathbf{C} \tag{11}$$
$$(a\mathbf{A})(b\mathbf{B}) = (ab)(\mathbf{A}\cdot\mathbf{B}) \tag{12}$$

for any vectors $\mathbf{A}, \mathbf{B}$, and $\mathbf{C}$ in a plane and any numbers a and b.

The calculation above yields the following result.

Theorem 1 If neither $\mathbf{A}$ nor $\mathbf{B}$ is the zero vector, then

$$\mathbf{A}\cdot\mathbf{B} = |\mathbf{A}||\mathbf{B}|\cos\theta \tag{13}$$

where θ is an angle between $\mathbf{A}$ and $\mathbf{B}$.

Any angle θ between the vectors can be used in (13) because $\cos(-\theta + 2n\pi)$ and $\cos(\theta + 2n\pi)$ equal $\cos\theta$ for any integer n.

In Definition 3 we defined the component of $\mathbf{A}$ in the direction of a nonzero $\mathbf{B}$ to be the number $|\mathbf{A}|\cos\theta$. Combining this with Theorem 1 gives formula (14) in the next theorem.

Theorem 2 For $\mathbf{A}$ and nonzero $\mathbf{B}$ in a coordinate plane,

$$[\text{The component of } \mathbf{A} \text{ in the direction of } \mathbf{B}] = \frac{\mathbf{A}\cdot\mathbf{B}}{|\mathbf{B}|} \tag{14}$$

and

$$[\text{The projection of } \mathbf{A} \text{ on a line through } \mathbf{B}] = \frac{\mathbf{A}\cdot\mathbf{B}}{|\mathbf{B}|^2}\mathbf{B}. \tag{15}$$

Formula (15) follows from (14) because $\mathbf{B}/|\mathbf{B}|$ is the unit vector in the direction of $\mathbf{B}$.

Example 9 (a) Draw the vectors $\mathbf{A} = \langle -6, 3\rangle$ and $\mathbf{B} = \langle 2, 2\rangle$ in an xy-plane relative to equal scales on the axes. (b) Calculate $\mathbf{A}\cdot\mathbf{B}$. (c) Find the component of $\mathbf{A}$ in the direction of $\mathbf{B}$. (d) Find the smallest positive angle θ between $\mathbf{A}$ and $\mathbf{B}$. In parts (c) and (d) give exact and approximate decimal answers.

SOLUTION (a) Figure 38 shows the vectors with their bases at the origin, the value t on the t-axis that equals the component of $\mathbf{A}$ in the direction of $\mathbf{B}$, and the angle θ.

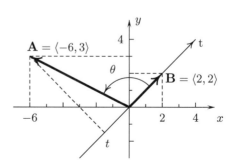

FIGURE 38

(b) $\mathbf{A} \cdot \mathbf{B} = \langle -6, 3 \rangle \cdot \langle 2, 2 \rangle = (-6)(2) + 3(2) = -6$.

(c) Since $|\mathbf{B}| = |\langle 2, 2 \rangle| = \sqrt{2^2 + 2^2} = \sqrt{8} = 2\sqrt{2}$, formula **(14)** shows that the component of $\mathbf{A}$ in the direction of $\mathbf{B}$ is $\dfrac{\mathbf{A} \cdot \mathbf{B}}{|\mathbf{B}|} = \dfrac{-6}{2\sqrt{2}} = \dfrac{-3}{\sqrt{2}} \doteq -2.1213$.

(d) Formula **(13)** implies that

$$\cos\theta = \frac{\mathbf{A} \cdot \mathbf{B}}{|\mathbf{A}||\mathbf{B}|}. \tag{16}$$

Here $|\mathbf{A}| = |\langle -6, 3 \rangle| = \sqrt{6^2 + 3^2} = \sqrt{45}$, and

$$\cos\theta = \frac{-6}{\sqrt{45}(2\sqrt{2})} = \frac{-3}{\sqrt{90}} = \frac{-1}{\sqrt{10}}.$$

The smallest positive angle with this cosine is

$$\theta = \cos^{-1}\left(\frac{-1}{\sqrt{10}}\right) \doteq 1.8925 \text{ radians. } \square$$

Question 8   Find the projection of $\mathbf{A}$ on the line through $\mathbf{B}$ with $\mathbf{A}$ and $\mathbf{B}$ from Example 9. Then draw the three vectors in the same xy-plane.

Perpendicular vectors

Because two nonzero vectors $\mathbf{A}$ and $\mathbf{B}$ in a plane are perpendicular if and only if the cosine of an angle θ between them is zero, they are perpendicular if and only if $\mathbf{A} \cdot \mathbf{B} = |\mathbf{A}||\mathbf{B}|\cos\theta$ is zero. For convenience in making general statements about perpendicular vectors, we will say that the zero vector $\mathbf{0}$ is perpendicular to all vectors. Then since $\mathbf{0} \cdot \mathbf{A} = 0$ for all $\mathbf{A}$, we have part (a) of the following result:

Rule 2   **(a)** *Two vectors $\mathbf{A}$ and $\mathbf{B}$ in a plane are perpendicular if and only if $\mathbf{A} \cdot \mathbf{B} = 0$.*

(b) *If $A_1 \neq 0$ and $A_2 \neq 0$, then the two vectors perpendicular to $\langle A_1, A_2 \rangle$ that have the same length as $\langle A_1, A_2 \rangle$ are the vectors*

$$\langle -A_2, A_1 \rangle \quad \text{and} \quad \langle A_2, -A_1 \rangle \tag{17}$$

obtained by interchanging the components of $\langle A_1, A_2 \rangle$ and multiplying one of them by -1.

Part (b) of this Rule follows from part (a) because $\langle A_1, A_2 \rangle \cdot \langle -A_2, A_1 \rangle = A_1(-A_2) + A_2(A_1) = 0$ and $\langle A_2, -A_1 \rangle$ is the negative of $\langle -A_2, A_1 \rangle$.

Question 9   Figure 39 shows the vectors $\langle -3, -1 \rangle$ and $\langle A_1, -2 \rangle$ with a positive number A_1. Find the value of A_1 such that the vectors are perpendicular. Then draw the two vectors.

Question 10   **(a)** Find the two vectors that have the same length as $\langle 1, 2 \rangle$ and are perpendicular to it. Draw the three vectors together.   **(b)** What are the two unit vectors that are perpendicular to $\langle 1, 2 \rangle$?

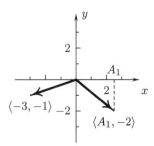

FIGURE 39

Principles and procedures

- Make drawings of the relevant vectors whenever you are dealing with angles of inclination or displacement vectors so you have visual checks on your answers.
- Displacement vectors are generally added geometrically by placing the base of one at the tip of another, as in Figure 9. Force vectors are generally added geometrically by placing their bases together, as in Figure 10.
- To obtain the unit vector with the same direction as a nonzero vector **A**, divide **A** by its length. Multiplying this unit vector by a positive number L then gives the vector of length L with the same direction as **A**.
- Use a drawing and the sine and cosine functions to find the projection of **A** on a line through a nonzero **B** or the component of **A** in the direction of **B** if **A** and **B** are given by their lengths and angles of inclination. Use the dot product if **A** and **B** are given by their x- and y-components.
- To remember the formula $\dfrac{\mathbf{A} \cdot \mathbf{B}}{|\mathbf{B}|}$ for the component of **A** in the direction of **B**, write it as the dot product $\mathbf{A} \cdot \left(\dfrac{\mathbf{B}}{|\mathbf{B}|}\right)$ of **A** with the unit vector $\dfrac{\mathbf{B}}{|\mathbf{B}|}$ in the direction of **B**. To remember the formula $\dfrac{\mathbf{A} \cdot \mathbf{B}}{|\mathbf{B}|^2}\mathbf{B}$ for the projection of **A** on a line through **B**, write it as the product $\left[\mathbf{A} \cdot \left(\dfrac{\mathbf{B}}{|\mathbf{B}|}\right)\right]\dfrac{\mathbf{B}}{|\mathbf{B}|}$ of the component of **A** in the direction of **B** and the unit vector in that direction.
- If **A** and **B** are not zero, then $\cos\theta = \dfrac{\mathbf{A} \cdot \mathbf{B}}{|\mathbf{A}||\mathbf{B}|}$ for any angle θ between the vectors. $\theta = \cos^{-1}\left(\dfrac{\mathbf{A} \cdot \mathbf{B}}{|\mathbf{A}||\mathbf{B}|}\right)$ is the smallest positive angle between the vectors because for any number t with $-1 \le t \le 1$, $\cos^{-1} t$ is the angle between 0 and π whose cosine is t.

Tune-Up Exercises 10.1♦

^AAnswer provided. ^OOutline of solution provided.

T1.^O (a) Express $4\langle 3, -2\rangle - 5\langle 10, 1\rangle$ in the form $\langle A_1, A_2\rangle$. (b) Write $-2(3\mathbf{i} + 4\mathbf{j}) + 10(\mathbf{i} - \mathbf{j})$ in the form $A_1\mathbf{i} + A_2\mathbf{j}$.

♦ Type 1, combining vectors

T2.^O Draw $\mathbf{A} = \langle 10, 20\rangle, \mathbf{B} = \langle 30, 10\rangle$, and $\mathbf{A} + \mathbf{B}$ together in an xy-plane.

♦ Type 1, combining vectors

T3.^O Figure 40 gives the x- and y-coordinates of a point P and the x- and y-components of the vectors $\overrightarrow{PQ}, \overrightarrow{QR}$, and $\overrightarrow{RS}$. What are the coordinates of S?

♦ Type 1, position and displacement vectors

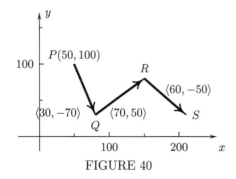

FIGURE 40

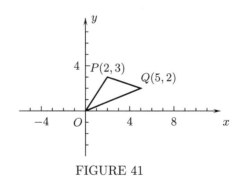

FIGURE 41

♦The Tune-up Exercises and Problems are classified by type and content. The types are (1) basic, reactive; (2) basic reflective; (3) intermediate, reactive; (4) intermediate, reflective; (5) advanced, reactive; (6) advanced, reflective; and (7) advanced, theoretical.

T4.⁰ Draw $\mathbf{A} = \langle 3, 5 \rangle$, $\mathbf{B} = \langle -5, 2 \rangle$, and $3\mathbf{A} - 2\mathbf{B}$ together in an xy-plane.

♦ Type 1, combining vectors

T5.⁰ Figure 41 shows a triangle OPQ with vertices $O(0,0), P(2,3)$, and $Q(5,2)$. **(a)** Draw the triangle ORS, where $\overrightarrow{OR} = 2\overrightarrow{OP}$ and $\overrightarrow{OS} = 2\overrightarrow{OQ}$. **(b)** Draw the triangle OTU, where $\overrightarrow{OT} = -\overrightarrow{OP}$ and $\overrightarrow{OU} = -\overrightarrow{OQ}$.

♦ Type 1, scalar multiples of vectors

T6.⁰ Figure 42 shows force vectors $\mathbf{F}, \mathbf{G}$, and $\mathbf{H}$ and their x- and y-components. Calculate their resultant $\mathbf{F} + \mathbf{G} + \mathbf{H}$ and add it to the drawing.

♦ Type 1, resultant vectors

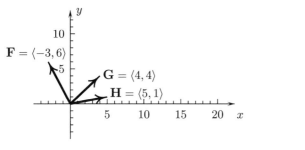

FIGURE 42

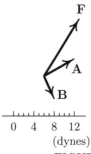

FIGURE 43

T7.⁰ **(a)** Calculate $\mathbf{A} \cdot \mathbf{B}$ for $\mathbf{A} = \mathbf{i} + 5\mathbf{j}$ and $\mathbf{B} = 6\mathbf{i} + 4\mathbf{j}$. **(b)** What is the component of $\mathbf{A}$ in the direction of $\mathbf{B}$? **(c)** Find the orthogonal projection $\mathbf{A}_{\text{proj}}$ of $\mathbf{A}$ on a line through $\mathbf{B}$ and draw the three vectors together. **(d)** What is the length of $\mathbf{A}_{\text{proj}}$, and why is it equal to the answer from part (b)?

♦ Type 1, dot products, components, and projections

T8.⁰ What is the smallest positive angle between the vectors $\mathbf{A}$ and $\mathbf{B}$ from Exercise T7?

♦ Type 1, angles between vectors

T9.⁰ Figure 43 shows a force vector $\mathbf{F}$ with its length measured by the scale in the drawing and two other vectors $\mathbf{A}$ and $\mathbf{B}$. What are the approximate components of $\mathbf{F}$ in the directions of $\mathbf{A}$ and $\mathbf{B}$?

♦ Type 1, components of a force vector

T10.⁰ Calculate $\mathbf{i} \cdot \mathbf{i}, \mathbf{i} \cdot \mathbf{j}$, and $\mathbf{j} \cdot \mathbf{j}$, and relate the results to the lengths of $\mathbf{i}$ and $\mathbf{j}$ and the angle between them.

♦ Type 1, dot products, i and j

T11.⁰ Give the exact x- and y-components **(a)** of the vector $\mathbf{A}$ of length 6 and angle of inclination $\frac{1}{4}\pi$ and **(b)** of the vector $\mathbf{B}$ of length 5 and angle of inclination 4. Then draw the two vectors.

♦ Type 1, angles of inclination

T12.⁰ Give the exact x- and y-components of the unit vector with the direction of $5\mathbf{i} - 12\mathbf{j}$.

♦ Type 1, unit vectors

T13.⁰ Give the two vectors of length 10 that are perpendicular to $\langle -3, 4 \rangle$. Then draw the three vectors.

♦ Type 1, perpendicular vectors

T14.⁰ Find an angle of inclination of $\langle -5, -3 \rangle$. Give an exact answer and its approximate decimal value.

♦ Type 2, angle of inclination

T15.⁰ Find numbers a and b such that $a\langle 3, -1 \rangle + b\langle 1, 2 \rangle = \langle 1, -12 \rangle$.

♦ Type 3, a vector equation

Problems 10.1

A Answer provided. **O** Outline of solution provided.

1.^{A} (a) Express $6\langle 3, -1\rangle - 2\langle 5, 4\rangle$ in the form $\langle A_1, A_2\rangle$. (b) Express $3(\mathbf{i} - 5\mathbf{j}) + 4(\mathbf{i} + \mathbf{j}) - 6(\mathbf{i} - \mathbf{j})$ in the form $A_1\mathbf{i} + A_2\mathbf{j}$.

♦ Type 1, combining vectors

2.^{A} Write $3\mathbf{A} - 4\mathbf{B} + 9\mathbf{C}$ in the form $A_1\mathbf{i} + A_2\mathbf{j}$ where $\mathbf{A} = 2\mathbf{i} - 4\mathbf{j}, \mathbf{B} = 10\mathbf{j}$, and $\mathbf{C} = -3\mathbf{i} + 4\mathbf{j}$.

♦ Type 1, combining vectors

3.^{A} Put $3(6\mathbf{A} - 5\mathbf{B}) - 4(2\mathbf{A} + 4\mathbf{B})$ in the form $\langle A_1, A_2\rangle$ where $\mathbf{A} = \langle 6, 5\rangle$ and $\mathbf{B} = \langle -2, -2\rangle$.

♦ Type 1, combining vectors

4.^{A} Write (a) $2\mathbf{C} - 3\mathbf{A} + \mathbf{B}$, (b) $4\mathbf{C} - \langle 3, 2\rangle$, and (c) $2(\mathbf{A} + \mathbf{B} + \langle 1, 3\rangle)$ in the form $\langle A_1, A_2\rangle$ where $\mathbf{A} = \langle 1, -1\rangle, \mathbf{B} = \langle -2, 4\rangle$ and $\mathbf{C} = \langle 2, 0\rangle$.

♦ Type 1, combining vectors

5.^{A} Find the exact x- and y-components of (a) the vector of length 5 with angle of inclination $\frac{2}{3}\pi$ and (b) the vector of length 7 with angle of inclination $\frac{5}{4}\pi$. Then draw the two vectors.

♦ Type 1, angles of inclination

6. Find the displacement vector $\overrightarrow{PQ}$ for (a^A) $P = (2, 5)$ and $Q = (4, -3)$, (b^A) $P = (0, -1)$ and $Q = (-7, -6)$, and (c) $P = (-1, 1)$ and $Q = (1, 1)$.

♦ Type 1, displacement vectors

7. The displacement vector $\overrightarrow{PQ}$ has length 10 and angle of inclination -0.5 radians, and the point P is $(-5, 5)$. Give the exact x- and y-coordinates of Q.

♦ Type 2, displacement vectors

8. Find the vertex S opposite Q in the parallelogram $PQRS$ for (a^A) $P = (3, 4), Q = (1, 1)$, and $R = (5, 2)$ and for (b) $P = (3, 1), Q = (5, 3)$, and $R = (7, -4)$.

♦ Type 2, displacement vectors

9. Find the fourth vertex R in each of the three parallelograms with vertices $P = (1, 2), Q = (2, 0)$, and $R = (4, 4)$.

♦ Type 2, displacement vectors

10.^{A} Draw the vectors (a) $-\mathbf{i} + \mathbf{j}$, (b) $\mathbf{i} - \mathbf{j}$, and (c) $4\mathbf{i} - 3\mathbf{j}$ and find their angles of inclination.

♦ Type 1, angle of inclination

11.^{A} What are the length and angle of inclination of $2\langle -3, 1\rangle + 4\langle 1, -1\rangle$?

♦ Type 1, angle of inclination

12.^{A} What vector of length 7 has the same direction as $\overrightarrow{PQ}$ where $P = (-5, -3)$ and $Q = (4, -8)$?

♦ Type 1, changing the length of a vectors

13. Find numbers s and t (a^A) such that $s\langle 2, 5\rangle + t\langle 3, 2\rangle = \langle 6, -7\rangle$ and (b) such that $s\langle 5, 10\rangle + t\langle 1, -3\rangle = \langle 7, 4\rangle$.

♦ Type 3, a vector equation

14.^{A} Find $\mathbf{A}$ and $\mathbf{B}$ such that $3\mathbf{A} + 4\mathbf{B} = \mathbf{i} + \mathbf{j}$ and $\mathbf{A} - 2\mathbf{B} = 2\mathbf{i} - 3\mathbf{j}$.

♦ Type 3, a vector equation

15. Solve the following equations for $\mathbf{A}, \mathbf{B}$, and $\mathbf{C}$: (a) $2\mathbf{A} + 3\langle 4, 6\rangle = \langle 2, -1\rangle$, (b^A) $\mathbf{B} - \langle 1, 3\rangle = 3\mathbf{B} + \langle 4, 6\rangle$, and (c) $\mathbf{C} - \mathbf{i} + \mathbf{j} = 4(3\mathbf{i} - 2\mathbf{j} - 3\mathbf{C})$.

♦ Type 3, a vector equation

16.^{A} A sailor tacks toward the northeast from point P to point R and then tacks toward the northwest to Q (Figure 44). The point Q is 300 meters north and 100 meters east of P. How far does the sailor travel on each of the tacks?

♦ Type 4, a vector equation

17. A skater skates in the direction of the vector $\langle 4, 3\rangle$ from point P to point R and then in the direction of $\langle -4, 2\rangle$ to Q, where $\overrightarrow{PQ} = \langle 0, 50\rangle$ meters (Figure 45). How far is R from P and how far is R from Q?

♦ Type 4, a vector equation

10.1 Vectors in the plane

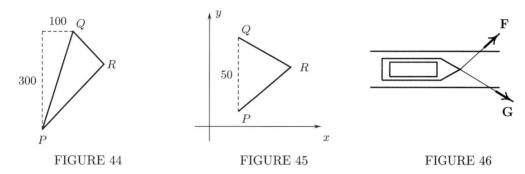

FIGURE 44 FIGURE 45 FIGURE 46

18.[A] The boat in Figure 46 is being pulled into a dock by two ropes. The force **F** by the upper rope has the direction of the vector $\langle 1, 1\rangle$ with the usual orientation of coordinate axes, and the force **G** by the lower rope is in the direction $\langle 3, -2\rangle$. The total force on the boat by the two ropes is $\langle 50, 0\rangle$ pounds. What are the magnitudes of the forces by the two ropes?

♦ Type 4, a vector equation

19.[A] An object in an xy-plane is being pulled by two ropes. One exerts a force in the direction of $\langle -2.5\rangle$ and the other a force in the direction of $\langle 3, 2\rangle$. The sum of the forces is $\langle 0, 190\rangle$ pounds. What are the magnitudes of the forces by the ropes?

♦ Type 4, a vector equation

20.[A] The force of the wind on a rubber raft is twenty-five pounds toward the southwest. What force must be exerted by the raft's motor for the combined force of the wind and motor to be twenty pounds toward the south?

♦ Type 4, a vector equation

21. The forces $\mathbf{F} = \langle 2, 1\rangle, \mathbf{G} = \langle -1, 3\rangle$ and $\mathbf{C} = \langle 1, 4\rangle$ (dynes) are applied at the same point. What is their combined force and what are its magnitude and angle of inclination?

♦ Type 2, force vectors

22. The resultant of two forces in an xy-plane is $3\mathbf{i} - 7\mathbf{j}$ newtons. One of the forces is $2\mathbf{i} - \mathbf{j}$ newtons. What are the magnitude and angle of inclination of the other force?

♦ Type 2, force vectors

23. The forces $\mathbf{F} = 3\mathbf{i} - 2\mathbf{j}$ and $\mathbf{G} = 5\mathbf{i} + \mathbf{j}$ are applied at the same point on an object. What single force applied at that point would balance their their combined effect?

♦ Type 2, force vectors

24.[A] A twenty-five-pound weight is suspended by two wires. One wire makes an angle of 45° with the vertical and the tension in it (the magnitude of the force it exerts) is twenty pounds. What is the tension in the other rope and what angle does it make with the vertical? (Notice that the angle with the vertical is not the angle of inclination.)

♦ Type 4, force vectors

25. Figure 47 shows four forces, measured in newtons, that are applied to a ring. Find the magnitude and angle of inclination of their resultant.[5]

♦ Type 4, force vectors

---

[5]Problem and drawing adapted from *Engineering Mechanics* by A. Pytel and J. Kuisalaas, New York, NY: HarperCollins, 1994, p. 39.

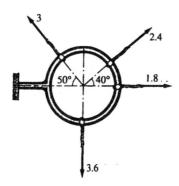

FIGURE 47   FIGURE 48

26. The resultant of the three forces on the eyebolt in Figure 48 is 1000**j** pounds, with the usual orientation of axes. **(a)** What is the angle of inclination of the nine-hundred-pound force? **(b)** What is the magnitude of the force **P**?[6]

♦ Type 4, force vectors

27. The force **F** exerted by the lower jaw in Figure 49 is directed upward and is the sum of the force $\mathbf{F_T}$ exerted by the temporalis muscles and the force $\mathbf{F_M}$ exerted by the masseter muscles[7]. **(a)** What is the angle between **F** and $\mathbf{F_T}$ if $|\mathbf{F}| = 5$ pounds, $|\mathbf{F_T}| = |\mathbf{F_M}|$, and the angle between **F** and $\mathbf{F_M}$ is 0.61 radians? **(b)** Find F_T and F_M.

♦ Type 4, force vectors

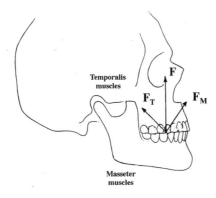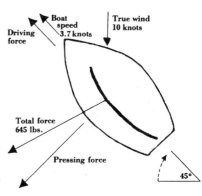

FIGURE 49   FIGURE 50

28. The boat in Figure 50 is tacking at an angle of 45° with the direction of the wind. The total force of the wind on the sail, which has a magnitude of 645 pounds and makes an angle of 119° degrees with the direction of the wind, is the sum of the DRIVING FORCE, in the direction of the boat's motion, and the perpendicular PRESSING FORCE.[8] What are the magnitudes of the driving and pressing forces?

♦ Type 4, force vectors

29.$^\mathbf{A}$ Calculate the dot products **(a)** $\langle 3, 1 \rangle \cdot \langle -6, 2 \rangle$, **(b)** $(2\mathbf{i} - 3\mathbf{j}) \cdot (-7\mathbf{j})$, **(c)** $(\mathbf{i} + \mathbf{j}) \cdot (5\mathbf{i} - 6\mathbf{j})$, and **(d)** $\langle -1, 6 \rangle \cdot \langle 4, -2 \rangle$.

♦ Type 1, dot products

30.$^\mathbf{O}$ What is the smallest positive angle between $\langle 2, 1 \rangle$ and $\langle -2, 3 \rangle$?

♦ Type 1, angle between vectors

---

[6] Problem and drawing adapted from *Engineering Mechanics* by A. Pytel and J. Kuisalaas, New York, NY: HarperCollins, 1994, p. 103.

[7] Drawing adapted from *Orthopædic Biomechanics* by V. Frankel and A. Burstein, London: Lea & Febiger, 1971, p. 419.

[8] Drawing adapted from *Total Sailing* by T. Hall, San Diego, CA: A. S. Barmes & Company, 1980, p. 87.

31. Give approximate decimal values of the smallest positive angles (**a**[A]) between $5\mathbf{i}+3.6\mathbf{j}$ and $-4.33\mathbf{i}-0.034\mathbf{j}$, (**b**[A]) between $\langle 1078, 2931 \rangle$ and $\langle -2046, 7431 \rangle$, and (**c**) between $\mathbf{i}-e\mathbf{j}$ and $\ln(5)\mathbf{i}+3\mathbf{j}$.
   ♦ Type 2, angle between vectors

32. Give approximate decimal values of the two interior angles in the parallelogram $PQRS$ with vertices $P=(1,1), Q=(2,5), R=(4,3)$ and $S=(3,-1)$.
   ♦ Type 1, angle between vectors

33.[A] Give the two unit vectors in an xy-plane that are parallel to the line $y=2x+1$.
   ♦ Type 2, unit vectors

34.[A] Give the two unit vectors in an xy-plane that are parallel to the tangent line to $y=x^3$ at $x=1$.
   ♦ Type 2, unit vectors

35. Give the two unit vectors in an xy-plane that tangent to $y=\sin x$ at $x=\frac{1}{3}\pi$.
   ♦ Type 2, unit vectors

36. Find the unit vector that points in the direction $\frac{1}{4}\pi$ counterclockwise from the direction of $\langle 1,1 \rangle$. (Use a drawing.)
   ♦ Type 4, unit vectors

37.[O] Find the two vectors of length 15 that are perpendicular to $\langle 5,6 \rangle$ in an xy-plane.
   ♦ Type 2, perpendicular vectors

38. Find numbers A such that $\langle 3,4 \rangle$ and $\langle A,2 \rangle$ (**a**) are parallel and (**b**) are perpendicular.
   ♦ Type 2, parallel and perpendicular vectors

39. Find all vectors perpendicular (**a**[A]) to $\langle 2,-5 \rangle$, (**b**) to $\langle 0,1 \rangle$, and (**c**) to $\langle 4,0 \rangle$.
   ♦ Type 2, perpendicular vectors

40. Find the constant k such that $\langle k,3 \rangle$ and $\langle 5,-2 \rangle$ are perpendicular.
   ♦ Type 2, perpendicular vectors

41. Find a vector $\mathbf{A}$ perpendicular to $\mathbf{B}=\langle 3,2 \rangle$ such that $\mathbf{A}+\mathbf{B}$ has a zero y-component.
   ♦ Type 2, perpendicular vectors

42. (**a**) For what values of the parameter $b \geq -3$ does the smallest positive angle θ between $\langle 3,4 \rangle$ and $\langle -4,b \rangle$ satisfy $0 < \theta < \frac{1}{2}\pi$? (**b**) For what values of $b \geq -3$ does θ satisfy $\frac{1}{2}\pi < \theta < \pi$?
   ♦ Type 4, angles between vectors

43. Find the vector $\mathbf{A}$ parallel to $\mathbf{B}=\langle 1,5 \rangle$ such that $\mathbf{A}+\mathbf{B}$ has component $12\sqrt{2}$ in the direction of $\langle 1,1 \rangle$.
   ♦ Type 4, parallel vectors

44. Vectors $\mathbf{A}, \mathbf{B}, \mathbf{C}$, and $\mathbf{D}$ and a scale for measuring their lengths are shown in Figure 51. Give approximate values of the components of (**a**[A]) $\mathbf{A}$, (**b**[A]) $\mathbf{B}$, and (**c**) $\mathbf{C}$ in the direction of $\mathbf{D}$.
   ♦ Type 1, components of vectors

FIGURE 51
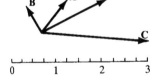

45. Find the orthogonal projection of $\mathbf{A}$ on lines parallel to $\mathbf{B}$ (**a**[A]) for $\mathbf{A}=\mathbf{i}-3\mathbf{j}, \mathbf{B}=3\mathbf{i}+4\mathbf{j}$, (**b**[A]) for $\mathbf{A}=\langle 2,4 \rangle, \mathbf{B}=\langle -2,1 \rangle$, and (**c**) for $\mathbf{A}=\mathbf{i}, \mathbf{B}=-4\mathbf{i}+2\mathbf{j}$,
   ♦ Type 3, projections of vectors

46. For each of the following pairs of vectors, find the component of $\mathbf{A}$ in the direction of $\mathbf{B}$ and the orthogonal projection of $\mathbf{A}$ on lines parallel to $\mathbf{B}$. Then draw the three vectors with a common base: (**a**[A]) $\mathbf{A}=\langle 1,4 \rangle, \mathbf{B}=\langle 2,1 \rangle$, (**b**) $\mathbf{A}=6\mathbf{i}+8\mathbf{j}, \mathbf{B}=4\mathbf{i}+3\mathbf{j}$, (**c**[A]) $\mathbf{A}=\langle -6,-3 \rangle, \mathbf{B}=\langle 1,1 \rangle$, and (**d**) $\mathbf{A}=-\mathbf{i}+6\mathbf{j}, \mathbf{B}=\mathbf{i}-6\mathbf{j}$.
   ♦ Type 3, projections of vectors

47. (**a**) Explain why three forces applied at the same point are in equilibrium if and only if the arrows representing them can be translated to form the sides of a triangle with the tip of each vector at the base of another. (**b**) What is the corresponding result for four force vectors in a plane?
   ♦ Type 4, resultant forces

48. Show that two nonzero vectors $\mathbf{A}$ and $\mathbf{B}$ are perpendicular if and only if $|\mathbf{A}+t\mathbf{B}| \geq |\mathbf{A}|$ for all numbers t.
   ♦ Type 4, lengths of vectors

49.[A] (a) Find the coordinates of the point one-third of the way from $(2, 4)$ along the line segment to $(6, 3)$.
(b) Find the coordinates of the point one-fifth of the way from $(1, -3)$ along the line segment to $(7, 0)$.

◆ Type 4, weighted averages of vectors

50. Two ropes at angles of $\frac{1}{8}\pi$ and $\frac{5}{16}\pi$ with the vertical are supporting a 500 newton weight. What is the tension (magnitude of the force) in each rope? Use the identity $\sin\alpha\cos\beta - \cos\alpha\sin\beta = \sin(\alpha - \beta)$ to simplify the answer.

◆ Type 4, resultant forces

51.[A] Use vectors to show that the line segment joining the midpoints of two sides of any triangle is parallel to the third side and half as long.

◆ Type 6, a vector proof

52. Use vectors to show that a parallelogram is a rhombus if and only if its diagonals are perpendicular.

◆ Type 6, a vector proof

53. Use vectors to show that the diagonals of any parallelogram bisect each other.

◆ Type 6, a vector proof

54. Use vectors to show that the sum of the squares of the lengths of the sides of any parallelogram equals the sum of the squares of the lengths of its diagonals.

◆ Type 6, a vector proof

55.[A] Use vectors to show that the line segment joining the midpoints of the nonparallel sides of a trapezoid is parallel to the other two sides and that its length is the average of lengths of those sides.

◆ Type 6, a vector proof

56. The point P in an xy-plane lies one-fourth of the distance along the line segment from Q to R. Use vectors to express the coordinates of P in terms of the coordinates of Q and R.

◆ Type 6, a vector proof

57. Use vectors to show that the midpoints of the sides of any quadrilateral are the vertices of a parallelogram.

◆ Type 6, a vector proof

58. Use vectors to prove that the altitudes of any triangle (the lines through the vertices and perpendicular to the opposite sides) intersect at a point.

◆ Type 6, a vector proof

59. Use vectors to prove that the medians of any triangle (the line segments from the midpoints of the sides to the opposite vertices) intersect at a point that is two-thirds of the distance from each vertex to the midpoint of the opposite side.

◆ Type 6, a vector proof

60. Use vectors to prove that the sum of the distances from any point inside an equilateral triangle to the three sides is constant.

◆ Type 6, a vector proof

61. Use vectors to show that if M is the midpoint of side AC in the triangle of Figure 52 and T is one-third of the distance from the vertex B to the vertex C, then P is three-fourths of the distance from A to T.

◆ Type 6, a vector proof

FIGURE 52

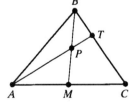

62. Use trigonometric identities to find the unit vector that is a counterclockwise angle of $\frac{1}{4}\pi$ from $\langle 3, 4 \rangle$.

◆ Type 4, using a trigonometric identity

Section 10.2

Vectors in space

OVERVIEW: *In this section we first discuss how objects in three-dimensional xyz-space can be represented in drawings by* PARALLEL PROJECTION. *Then we begin our study of vectors in xyz-space by modifying definitions and results from Section 10.1 concerning vectors in xy-planes. Working with vectors in space does not involve many new ideas but is somewhat more difficult because vectors have three instead of two components and it is more difficult to visualize vectors in space.*

Topics:

- **Rectangular coordinates in space**
- **The Pythagorean Theorem in space**
- **Vectors in space**
- **The dot product in space**
- **Components and projections of vectors**
- **The unit vectors i, j, and k**
- **Direction cosines and direction angles**

Rectangular coordinates in space

To use rectangular xyz-coordinates in three-dimensional space, we introduce mutually perpendicular x-, y-, and z-axes intersecting at their origins, as in Figure 1. These axes form xy-, xz-, and yz-coordinate planes, which divide the space into eight OCTANTS.

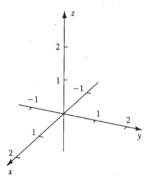

FIGURE 1

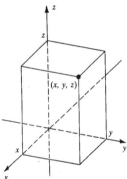

FIGURE 2

The coordinates (x, y, z) of a point in space are determined by planes that pass through the point and are perpendicular to the coordinate axes. The plane through the point and perpendicular to the x-axis intersects it at the x-coordinate of the point; the plane through the point and perpendicular to the y-axis intersects it at the y-coordinate; and the plane through the point and pependicular to the z-axis intersects it at the z-coordinate. If none of the coordinates are zero, then these three planes through the point and the coordinate planes form a rectangular box with the origin $(0,0,0)$ at one corner of the box and the point (x, y, z) at the opposite corner, as shown in Figure 2 for a case with positive x, y, and z.

All our drawings of xyz-space, like those in Figures 1 and 2, will be obtained by PARALLEL PROJECTION, where points in space are projected onto the plane of the drawing along parallel lines, called the LINES OF PROJECTION. All of the points on one line of projecton appear as one point in the drawing, and because the lines of projection are parallel, our drawings will not show any perspective: the size of the image of an object will not change when the object is moved to the front or back. Also, parallel lines in space will appear as parallel lines in the drawings.

♦ SUGGESTIONS TO INSTRUCTORS: This section deals with extensions to three dimensions of definitions and results concerning vectors in the xy-plane from Section 10.1. It can be covered in one or more lectures. Have students start work on Questions 1 through 5 before class. Example 3 and Problems 2, 4, 5, 6, 9, 13, 17, 23, and 29 are good for in-class discussions.

We will always have the positive z-axis point up and orient the positive x- and y-axes so that, viewed from above, the positive x-axis can be moved into the position of the positive y-axis by a counterclockwise rotation of $90°$. We will usually have the positive y-axis point to the right and the positive x-axis down and to the left, as in Figures 1 and 2, but occasionally we will find it more convenient to use different orientations of the x- and y-axes, as in Figure 3 and 4.

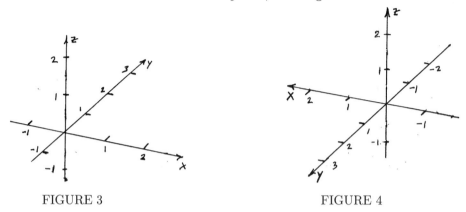

FIGURE 3   FIGURE 4

The projections that led to Figures 1 through 4 are called ORTHOGONAL (PERPENDICULAR) PROJECTIONS because in each case the lines of projection are perpendicular to the drawings. You might prefer to use an OBLIQUE PROJECTION as in Figure 5, where the yz-plane in space is parallel to the plane of the drawing and the lines of projection go from the upper right in the front to the lower left in the back, rather than being perpendicular to the drawing.

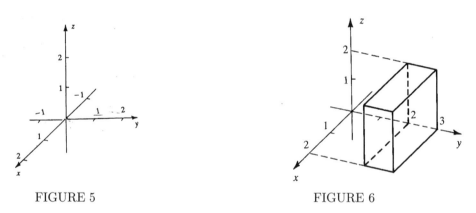

FIGURE 5   FIGURE 6

Example 1   Use orthogonal projection with the axes drawn as in Figure 1 to sketch the box consisting of all points (x, y, z) with $0 \leq x \leq 2, 2 \leq y \leq 3$, and $0 \leq z \leq 2$. What are the coordinates of the eight corners of the box?

SOLUTION   The box is drawn in Figure 6. The corners of its base are $(2,2,0), (2,3,0), (0,3,0)$, and $(0,2,0)$. The corners of its top are $(2,2,2), (2,3,2), (0,3,2)$, and $(0,2,2)$. □

The Pythagorean Theorem and the distance between two points

If a rectangular box has length a, width b, and height c, as in Figure 7, then, by the Pythagorean Theorem for a right triangle, the length of a diagonal of its base is $\sqrt{a^2 + b^2}$. Then, because the diagonal of the box is the hypotenuse of a right triangle with base of length $\sqrt{a^2 + b^2}$ and height c (Figure 8), its length is the square root of $[\sqrt{a^2 + b^2}]^2 + c^2 = a^2 + b^2 + c^2$. This gives the Pythagorean Theorem in space:

$$\begin{bmatrix} \text{The length of a diagonal of a} \\ \text{rectangular box with sides } a, b, \text{ and } c \text{ is} \end{bmatrix} = \sqrt{a^2 + b^2 + c^2}. \tag{1}$$

10.2 Vectors in space,

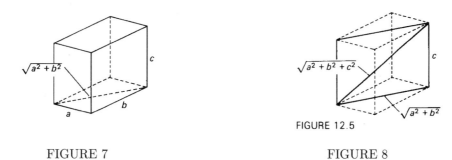

FIGURE 7   FIGURE 8

Example 2 What is the length of the diagonals of the box in Figure 6?

SOLUTION   Because the sides of the box are of length 1, 2, and 2, the length of each of its four diagonals is $\sqrt{1^2 + 2^2 + 2^2} = 3$. □

Because points $P = (x_1, y_1, z_1)$ and $Q = (x_2, y_2, z_2)$ in xyz-space are at diagonally opposite corners of a rectangular box with sides of lengths $|x_2 - x_1|, |y_2 - y_1|$, and $|z_2 - z_1|$ (Figure 9), the distance $\overline{PQ}$ between the points is

$$\overline{PQ} = \sqrt{(x_2 - x_1)^2 + (y_2 - y_1)^2 + (z_2 - z_1)^2}. \tag{2}$$

Here, as in any application of this distance formula, we assume that the scales on the coordinate axes in xyz-space are equal, regardless of what scales are used in associated drawings.

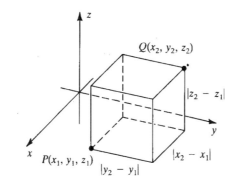

FIGURE 9

Question 1   What is the distance between the points $(4, 5, 0)$ and $(-1, 3, 2)$?

Vectors in space

A nonzero vector in xyz- space, like a nonzero vector **A** in an xy-plane, represents a positive number and a direction. If we put the base of the vector at the origin, as in Figure 10, then the coordinates (A_1, A_2, A_3) of its tip are the x-, y-, and z-components of the vector and we write $\mathbf{A} = \langle A_1, A_2, A_3 \rangle$.

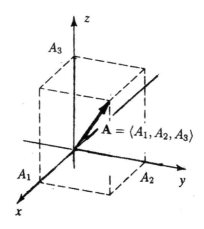

FIGURE 10

The Pythagorean Theorem in space shows that the length of $\langle A_1, A_2, A_3 \rangle$ is equal to the square root of the sum of the squares of its components:

$$|\mathbf{A}| = |\langle A_1, A_2, A_3 \rangle| = \sqrt{(A_1)^2 + (A_2)^2 + (A_3)^2}. \tag{3}$$

Notice that **(3)** looks like the formula $|\mathbf{A}| = |\langle A_1, A_2 \rangle| = \sqrt{(A_1)^2 + (A_2)^2}$ for the length of a vector in an xy-plane except that the term $(A_3)^2$ for the third component has been added. The rules for adding two vectors in space and multiplying a vector in space by a real number are also similar to those for vectors in a plane:

Definition 1 For any vectors $\mathbf{A} = \langle A_1, A_2, A_3 \rangle$ and $\mathbf{B} = \langle B_1, B_2, B_3 \rangle$ and any number k,

$$\mathbf{A} + \mathbf{B} = \langle A_1, A_2, A_3 \rangle + \langle B_1, B_2, B_3 \rangle = \langle A_1 + B_1, A_2 + B_2, A_3 + B_3 \rangle \tag{4}$$

$$k\mathbf{A} = k\langle A_1, A_2, A_3 \rangle = \langle kA_1, kA_2, kA_3 \rangle. \tag{5}$$

These operations have the same geometric interpretations as they do in an xy-plane: The sum $\mathbf{A} + \mathbf{B}$ can be obtained by placing the base of an arrow representing $\mathbf{B}$ at the tip of an arrow representing $\mathbf{A}$. Then $\mathbf{A} + \mathbf{B}$ is given by the arrow from the base of $\mathbf{A}$ to the tip of $\mathbf{B}$ (Figure 11). Or, we can place the bases of $\mathbf{A}$ and $\mathbf{B}$ together, as in Figure 12, and complete the parallelogram with those vectors as sides. Then $\mathbf{A} + \mathbf{B}$ is given by the arrow from the bases of the vectors to the opposite vertex.

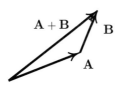

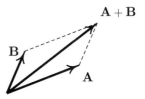

FIGURE 11   FIGURE 12

Also, as in a plane, multiplying the vector $\mathbf{A}$ by a positive number k yields a vector with the same direction as $\mathbf{A}$ whose length is k multiplied by the length of $\mathbf{A}$, and multiplying $\mathbf{A}$ by a negative number m yields a vector with the opposite direction as $\mathbf{A}$ whose length is $|m|$ multiplied by the length of $\mathbf{A}$ (Figure 13).

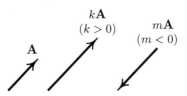

FIGURE 13

The position vector of a point (x, y, z) is $\langle x, y, z \rangle$ (Figure 14), and the displacement vector $\overrightarrow{PQ}$ from $P = (x_1, y_1, z_1)$ to $Q = (x_2, y_2, z_2)$ is

$$\overrightarrow{PQ} = \langle x_2 - x_1, y_2 - y_1, z_2 - z_1 \rangle \tag{6}$$

as shown in Figure 15 for a case where $x_2 - x_1, y_2 - y_1,$ and $z_2 - z_1$ are positive.

10.2 Vectors in space,

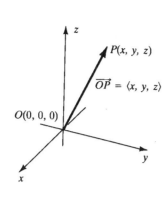

FIGURE 14

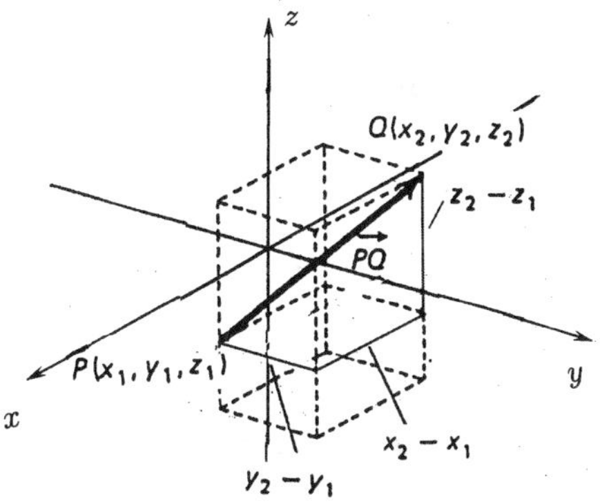

$$\overrightarrow{PQ} = \langle x_2 - x_1, y_2 - y_1, z_2 - z_1 \rangle$$
FIGURE 15

Example 3 Three adjacent vertices of a parallelogram $PQRS$ in space are $P = (1, 3, 2)$, $Q = (4, 5, 3)$, and $R = (2, -1, 0)$. What are the coordinates of the point S opposite Q?

SOLUTION We use the schematic sketch of the parallelogram in Figure 16 to guide our calculations. (An accurate drawing relative to xyz-axes is not needed.) Because $PQRS$ is a parallelogram, the displacement vectors $\overrightarrow{QP}$ and $\overrightarrow{RS}$ are equal and

$$\overrightarrow{RS} = \overrightarrow{QP} = \langle 1 - 4, 3 - 5, 2 - 3 \rangle = \langle -3, -2, -1 \rangle.$$

On the other hand, the position vector $\overrightarrow{OS}$ is equal to the sum of the position vector $\overrightarrow{OR}$ and the displacement vector $\overrightarrow{RS}$:

$$\overrightarrow{OS} = \overrightarrow{OR} + \overrightarrow{RS}$$
$$= \langle 2, -1, 0 \rangle + \langle -3, -2, -1 \rangle = \langle -1, -3, -1 \rangle$$

Therefore, $S = (-1, -3, -1)$. □

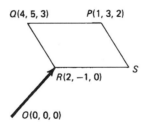

FIGURE 16

The dot product in space
The definition of the dot product in space is obtained from the definition $\langle A_1, A_2 \rangle \cdot \langle B_1, B_2 \rangle = A_1 B_1 + A_2 B_2$ for vectors in a plane by adding a term $A_3 B_3$ for the third components:

Deinition 2 The DOT PRODUCT of vectors $\mathbf{A} = \langle A_1, A_2, A_3\rangle$ and $\mathbf{B} = \langle B_1, B_2, B_3\rangle$ in xyz-space is the number

$$\mathbf{A}\cdot\mathbf{B} = A_1B_1 + A_2B_2. + A_3B_3. \tag{7}$$

The derivation of the following result, which was given in Section 10.1 for vectors in a plane, can be easily modified for vectors in space because it uses only the Pythagorean Theorem.

Theorem 1 If neither $\mathbf{A}$ nor $\mathbf{B}$ is the zero vector, then

$$\mathbf{A}\cdot\mathbf{B} = |\mathbf{A}||\mathbf{B}|\cos\theta \tag{8}$$

where θ is the angle between $\mathbf{A}$ and $\mathbf{B}$ (Figure 17).

We cannot give any meaning to negative angles between vectors in space because what is a clockwise rotation when viewed from one side is a counterclockwise rotation when viewed from the other side. Consequently, whenever we use Theorem 1 in space, we speak of θ as the angle between the vectors, not as the angle from one vector to the other, and we always choose θ to satisfy $0 \leq \theta \leq \pi$.

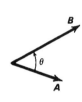

FIGURE 17   FIGURE 18

Example 4 Show that the triangle with vertices $(4, 1, -1), (3, 3, 5)$, and $(1, 0, 2)$ is a right triangle.

SOLUTION   We write $P = (4, 1, -1), Q = (3, 3, 5)$, and $R = (1, 0, 2)$, and use the schematic sketch of the triangle in Figure 18. The sides of the triangle are formed by

$$\overrightarrow{PQ} = \langle 3-4, 3-1, 5-(-1)\rangle = \langle -1, 2, 6\rangle$$
$$\overrightarrow{PR} = \langle 1-4, 0-1, 2-(-1)\rangle = \langle -3, -1, 3\rangle$$
$$\overrightarrow{QR} = \langle 1-3, 0-3, 2-5\rangle = \langle -2, -3, -3\rangle.$$

We calculate

$$\overrightarrow{PQ}\cdot\overrightarrow{PR} = \langle -1, 2, 6\rangle \cdot \langle -3, -1, 3\rangle$$
$$= (-1)(-3) + 2(-1) + 6(3) = 19 \tag{9}$$
$$\overrightarrow{PQ}\cdot\overrightarrow{QR} = \langle -1, 2, 6\rangle \cdot \langle -2, -3, -3\rangle$$
$$= (-1)(-2) + 2(-3) + 6(-3) = -22 \tag{10}$$
$$\overrightarrow{QR}\cdot\overrightarrow{PR} = \langle -2, -3, -3\rangle \cdot \langle -3, -1, 3\rangle$$
$$= (-2)(-3) + (-3)(-1) + (-3)(3) = 0. \tag{11}$$

Because **(11)** is zero, the vectors $\overrightarrow{QR}$ and $\overrightarrow{PR}$ are perpendicular and the triangle is a right triangle with the right angle at R. □

10.2 Vectors in space,

Example 5 Give the approximate decimal value of the angle at P in the triangle of Example 4.

SOLUTION We calculate $|\overrightarrow{PQ}| = |\langle -1, 2, 6\rangle| = \sqrt{1^2 + 2^2 + 6^2} = \sqrt{41}$ and $|\overrightarrow{PR}| = |\langle -3, -1, 3\rangle| = \sqrt{3^2 + 1^2 + 3^2} = \sqrt{19}$. In calculation **(9)** we found that $\overrightarrow{PQ} \cdot \overrightarrow{PR} = 19$. Therefore,

$$\cos([\text{The angle at } P]) = \frac{\overrightarrow{PQ} \cdot \overrightarrow{PR}}{|\overrightarrow{PQ}||\overrightarrow{PR}|} = \frac{19}{\sqrt{41}\sqrt{19}} = \frac{\sqrt{19}}{\sqrt{41}}.$$

The angle at P is $\cos^{-1}(\sqrt{19}/\sqrt{41}) \doteq 0.822$ radians. □

Question 2 Find the approximate decimal value of the angle at Q in the triangle of Example 4.

Components and projections of vectors

The formulas from Section 10.1 for the component of **A** in the direction of a nonzero **B** and for the projection of **A** on a line through **B** are based on Theorem 1 for the dot product and consequently are the same for vectors in space:

Rule 1 *For **A** and nonzero **B** in xyz-space,*

$$[\text{The component of } \mathbf{A} \text{ in the direction of } \mathbf{B}] = \frac{\mathbf{A} \cdot \mathbf{B}}{|\mathbf{B}|} \tag{12}$$

and

$$[\text{The projection of } \mathbf{A} \text{ on a line through } \mathbf{B}] = \frac{\mathbf{A} \cdot \mathbf{B}}{|\mathbf{B}|^2} \mathbf{B}. \tag{13}$$

Example 6 What is the component of the force vector $\mathbf{F} = \langle 2, -4, 0\rangle$ pounds in the direction of $\mathbf{A} = \langle 1, -2, -5\rangle$?

SOLUTION We calculate $\mathbf{F} \cdot \mathbf{A} = \langle 2, -4, 0\rangle \cdot \langle 1, -2, -5\rangle = 2(1) - 4(-2) + 0(-5) = 10$ and $|\mathbf{A}| = |\langle 1, -2, -5\rangle| = \sqrt{1^2 + 2^2 + 5^2} = \sqrt{30}$. By **(12)**, the component of **F** in the direction of **A** is

$$\frac{\mathbf{F} \cdot \mathbf{A}}{|\mathbf{A}|} = \frac{10}{\sqrt{30}} \text{ pounds.} □$$

Question 3 What is the orthogonal projection of **F** on a line through **A** for **F** and **A** of Example 6?

The unit vectors **i, j,** *and* **k**

In the last section we expressed vector $\langle A_1, A_2\rangle$ in the plane as $A_1\mathbf{i} + A_2\mathbf{j}$ where **i** and **j** are unit vectors in the directions of the positive x- and y-axes, respectively. In three dimensions, we also use a third unit vector **k** in the direction of the positive z-axis, as in Figure 19. Then $\langle A_1, A_2, A_3\rangle = A_1\mathbf{i} + A_2\mathbf{j} + A_3\mathbf{k}$ for any $A_1, A_2,$ and A_3 (Figure 20).

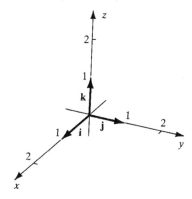

FIGURE 19

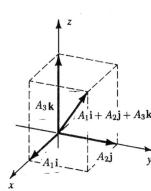

FIGURE 20

Direction cosines and direction angles

We can describe the direction of a nonzero vector **A** in an xy-plane by giving one angle, such as its angle θ from the positive x-direction (Figure 21). Then, as we saw in the last section,

$$\mathbf{A} = \langle |\mathbf{A}| \cos\theta, |\mathbf{A}| \sin\theta \rangle = |\mathbf{A}| \langle \cos\theta, \sin\theta \rangle. \tag{14}$$

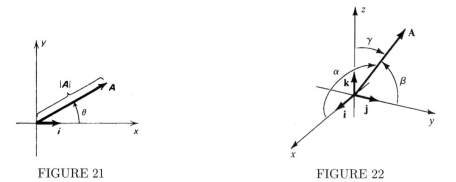

FIGURE 21   FIGURE 22

It takes, however, at least two angles to describe the direction of a nonzero vector **A** in space, and in many applications it is easiest to use three angles, namely, the angle α between **A** and the positive x-axis, the angle β between **A** and the positive y-axis, and the angle γ between **A** and the positive z-axis (Figure 22). These are called the DIRECTION ANGLES of the vector.

Question 4   Suppose that $\mathbf{A} = \langle A_1, A_2, A_3 \rangle$ in Figure 22. **(a)** Calculate $\mathbf{A} \cdot \mathbf{i}, \mathbf{A} \cdot \mathbf{j}$, and $\mathbf{A} \cdot \mathbf{k}$.
(b) Use Theorem 1 to express A_1, A_2, and A_3 in terms of $|\mathbf{A}|$ and the cosines of the angles α, β, and γ.

The results of Question 4 give the formula

$$\mathbf{A} = \langle |\mathbf{A}| \cos\alpha, |\mathbf{A}| \cos\beta, |\mathbf{A}| \cos\gamma \rangle = |\mathbf{A}| \langle \cos\alpha, \cos\beta, \cos\gamma \rangle \tag{15}$$

for the vector in terms of its length and its angles of inclination.

Notice that **(15)** uses cosines of the angles α, β, and γ with the positive x-, y-, and z-axes whereas **(14)** uses the cosine and sine of the angle θ with the positive x-axis. This is consistent because the sine of the angle θ in **(14)** is the cosine of the angle between **A** and the positive y-axis.

The numbers $\cos\alpha, \cos\beta$, and $\cos\gamma$ in **(15)** are called the DIRECTION COSINES of **A**. Because they are the components of the unit vector $\mathbf{u} = \langle \cos\alpha, \cos\beta, \cos\gamma \rangle$ they satisfy the Pythagorean Identity,

$$\cos^2\alpha + \cos^2\beta + \cos^2\gamma = 1. \tag{16}$$

Example 7   Find the direction cosines and the exact and approximate decimal values of the direction angles of $2\mathbf{i} - 2\mathbf{j} + \mathbf{k}$.

SOLUTION   Because $|\mathbf{A}| = |\langle 2, -2, 1 \rangle| = \sqrt{2^2 + 2^2 + 1} = \sqrt{9} = 3$, the unit vector in the direction of **A** is

$$\tfrac{1}{3}\mathbf{A} = \tfrac{1}{3}\langle 2, -2, 1 \rangle = \langle \tfrac{2}{3}, -\tfrac{2}{3}, \tfrac{1}{3} \rangle.$$

Therefore, the direction cosines are $\cos\alpha = \tfrac{2}{3}, \cos\beta = -\tfrac{2}{3}$, and $\cos\gamma = \tfrac{1}{3}$. The direction angles are

$$\alpha = \cos^{-1}\left(\tfrac{2}{3}\right) \doteq 0.84 \text{ radians}$$
$$\beta = \cos^{-1}\left(-\tfrac{2}{3}\right) \doteq 2.30 \text{ radians}$$
$$\gamma = \cos^{-1}\left(\tfrac{1}{3}\right) \doteq 1.23 \text{ radians. } \square$$

10.2 Vectors in space,

Example 8   Find the two unit vectors that make an angle $\frac{1}{3}\pi$ with the positive x-axis and an angle $\frac{3}{4}\pi$ with the positive y-axis.

SOLUTION   Here $\alpha = \frac{1}{3}\pi$ and $\beta = \frac{3}{4}\pi$, so the unit vectors are

$$\mathbf{u} = \langle \cos(\tfrac{1}{3}\pi), \cos(\tfrac{3}{4}\pi), \cos\gamma \rangle = \langle \tfrac{1}{2}, -\tfrac{1}{2}\sqrt{2}, \cos\gamma \rangle$$

with $\cos\gamma$ to be determined. Because $\mathbf{u}$ is a unit vector, $(\tfrac{1}{2})^2 + (\tfrac{1}{2}\sqrt{2})^2 + \cos^2\gamma = 1$ and

$$\cos^2\gamma = 1 - (\tfrac{1}{2})^2 - (\tfrac{1}{2}\sqrt{2})^2 = 1 - \tfrac{1}{4} - \tfrac{1}{2} = \tfrac{1}{4}.$$

Therefore, $\cos\gamma = \pm\sqrt{\tfrac{1}{4}} = \pm\tfrac{1}{2}$, and the unit vectors are $\langle \tfrac{1}{2}, -\tfrac{1}{2}\sqrt{2}, \tfrac{1}{2} \rangle$ and $\langle \tfrac{1}{2}, -\tfrac{1}{2}\sqrt{2}, -\tfrac{1}{2} \rangle$. □

Question 5   What is the third direction angle γ for each of the two unit vectors from Example 8?

Principles and procedures

- Use a lined plastic ruler to draw parallel lines in sketches of xyz-space.
- You can imagine that the origin of the axes in Figures 1 and 2 is in the planes of the drawings. Then, because orthogonal projection is used, the positive x-, y-, and z-axes are in front of the drawing.
- If oblique projection is used with the yz-plane parallel to the drawing as in Figure 5, then all objects in the yz-plane or in planes parallel to it appear unchanged in drawings. In particular, the images of all rectangles in such planes are rectangles in the drawing, whereas rectangles in other planes appear as nonrectangular parallelograms.
- You can imagine that there are equal scales on the x-, y-, and z-axes in space from which Figures 1 through 4 were obtained. The tickmarks at the integers in the drawings are then different distances apart because the angles between the lines of projection and the axes are different. We will, however, generally not make exact calculations of the scales on the axes, and you can pick the scales however you choose in hand-drawn sketches.
- You can position the x- and y-axes however you want when you draw a figure in space, except that the positive z-axis should always point up and the positive y-axis should always be a counterclockwise angle of 90° from the positive x-axis when viewed from above.
- The formulas for lengths of vectors and for the dot product of vectors in space are easy to remember because they are obtained from the formulas in an xy-plane by adding terms for the third components.
- The geometric interpretations of calculations involving ony two vectors—such as sums, dot products, components, and orthogonal projections—can be shown by drawing the plane that contains the two vectors.
- Geometric relationships involving three or more vectors that do not lie in a plane are often difficult to represent in drawings. In these cases, you often have to reason more abstractly to decide which steps are required in solving a problem.

Tune-Up Exercises 10.2♦

A Answer provided. **O** Outline of solution provided.

T1.^O Sketch the parallelogram with vertices $(2,0,0), (2,1,0), (0,1,2)$, and $(0,2,2)$ in xyz-space.

♦ Type 1, drawing

T2.^O Sketch the right circular cylinder of height 2 in xyz-space whose base is the circle of radius 1 with its center at the origin in the xy-plane.

♦ Type 1, drawing

T3.^O Sketch the right circular cone of height 3 in xyz-space whose base is the circle of radius 1 with its center at the origin in the xy-plane.

♦ Type 1, drawing

T4.^O Draw the sphere of radius 3 whose center is at the origin of xyz-space.

♦ Type 1, drawing

T5.^O (a) Explain why $\sqrt{x^2 + y^2}$ is the distance from (x,y,z) to the z-axis. Give formulas for the distances (b) from (x,y,z) to the y-axis, and (c) from (x,y,z) to the x-axis.

♦ Type 1, distances in space

T6.^A Find the distances from $(3,-4,-1)$ (a) to the z-axis, (b) to the x-axis, (c) to the y-axis, (d) to the xy-plane, (e) to the xz-plane, (f) to the yz-plane, and (g) to the origin.

♦ Type 1, distances in space

T7.^O Express $\overrightarrow{PQ}$ in the form $A_1\mathbf{i} + A_2\mathbf{j} + A_3\mathbf{k}$ (a) for $P = (0,3,2)$ and $Q = (1,4,6)$ and (b) for $P = (1,4,1)$ and $Q = (-1,-4,-1)$.

♦ Type 1, displacement vectors

T8.^O Find the x- and y-components of $\mathbf{A} + 2\mathbf{B}$, the magnitude of $\mathbf{A} + 2\mathbf{B}$, and the unit vector in the direction of $\mathbf{A} + 2\mathbf{B}$ (a) for $\mathbf{A} = \langle 2,4,0 \rangle$, $\mathbf{B} = \langle 1,1,1 \rangle$ and (b) for $\mathbf{A} = -4\mathbf{k}$, $\mathbf{B} = \mathbf{i} + \mathbf{j} - \mathbf{k}$.

♦ Type 1, combining vectors and unit vectos

T9.^O Calculate $\mathbf{A} \cdot \mathbf{B}$ (a) for $\mathbf{A} = \langle 2,5,-4 \rangle$, $\mathbf{B} = \langle 3,1,3 \rangle$ and (b) for $\mathbf{A} = \mathbf{i} - 3\mathbf{k}$, $\mathbf{B} = 2\mathbf{i} + \mathbf{j} - \mathbf{k}$.

♦ Type 1, dot products

T10.^O Find exact and approximate decimal values of the angles (a) between $\langle 1,3,4 \rangle$ and $\langle -2,0,1 \rangle$ and (b) between $2\mathbf{i} + 3\mathbf{j} + 4\mathbf{k}$ and $-3\mathbf{i} + 4\mathbf{j} - 5\mathbf{k}$.†

♦ Type 1, angles between vectors

T11.^O Find (a) the component of $\mathbf{A}$ in the direction of $\mathbf{B}$ and (b) the orthogonal projection of $\mathbf{A}$ on a line through $\mathbf{B}$ where $\mathbf{A} = \langle 3,5,2 \rangle$ and $\mathbf{B} = \langle 1,-1,-3 \rangle$.

♦ Type 1, components and projections

T12.^O Find the direction cosines and the exact and approximate decimal values of the direction angles of $\langle 2,-1,-2 \rangle$.

♦ Type 1, direction cosines and angles

♦The Tune-up Exercises and Problems are classified by type and content. The types are (1) basic, reactive; (2) basic reflective; (3) intermediate, reactive; (4) intermediate, reflective; (5) advanced, reactive; (6) advanced, reflective; and (7) advanced, theoretical.

†Recall that whenever we speak of an angle θ between vectors in three-dimensional space, we assume that $0 \leq \theta \leq \pi$.

10.2 Vectors in space,

Problems 10.2

[A]Answer provided. [O]Outline of solution provided.

In Problems 1 through 8 sketch the indicated geometric objects in xyz-space.

1.[O] The tetrahedron with vertices $(0,0,0), (2,0,0), (0,3,0),$ and $(0,0,4)$
♦ Type 1, drawing

2.[O] The triangle with vertices $(0,0,0), (0,4,0),$ and $(0,-2,2)$
♦ Type 1, drawing

3.[O] The box with corners $(1,0,0), (2,1,0), (1,2,0), (0,1,0), (1,0,3), (2,1,3), (1,2,3),$ and $(0,1,3)$
♦ Type 1, drawing

4.[A] The disk consisting of all points (x,y,z) with $x^2 + y^2 \leq 1$ and $z = 0$
♦ Type 1, drawing

5.[A] The disk consisting of all (x,y,z) with $y^2 + z^2 \leq 1$ and $x = 0$
♦ Type 1, drawing

6. The wedge-shaped region with straight edges joining the vertices $(0,-1,0), (0,0,0), (3,0,0), (3,-1,0), (3,0,-3),$ and $(0,0,-3)$
♦ Type 1, drawing

7. The rectangle consisting of all (x,y,z) such that $0 \leq x \leq 4, y = 0,$ and $0 \leq z \leq 3$
♦ Type 1, drawing

8. The hemisphere consisting of all (x,y,z) such that $x^2 + y^2 + z^2 = 1$ and $z \geq 0$
♦ Type 1, drawing

9.[A] **(a)** Draw the double cone consisting of all points whose distances to the x-axis are equal to half of their distances to the origin and whose distances to the yz-plane are ≤ 2. **(b)** What is the angle between the axis and elements of this cone?
♦ Type 2, drawing

10. **(a)** Draw the double cone consisting of all points whose distances to the z-axis are equal to their distances to the xy-plane and whose distances to the xy-plane are ≤ 5. **(b)** What is the angle between the axis and elements of this cone?
♦ Type 2, drawing

11.[A] Find the x-, y-, and z-components **(a)** of $\overrightarrow{PQ} + \overrightarrow{QR}$ and **(b)** of $\overrightarrow{PR}$ for $P = (1,2,3), Q = (3,1,0),$ and $R = (4,-2,1)$.
♦ Type 1, displacement vectors

12.[A] **(a)** What are the x-, y-, and z-components of the unit vector $\mathbf{u}$ with the same direction as $\langle 5, 1, -1 \rangle$? **(b)** Express the vector $\mathbf{B}$ of length 8 in the direction of $\mathbf{A} = 2\mathbf{i} - \mathbf{j} + 3\mathbf{k}$ in the form $B_1\mathbf{i} + B_2\mathbf{j} + B_3\mathbf{k}$.
♦ Type 2, vectors of given lengths

13.[O] Three vertices of the parallelogram $PQRS$ are $P = (6,8,-2), Q = (3,4,0),$ and $R = (4,-7,-10)$. What is the vertex S opposite P?
♦ Type 1, displacement vectors

14. Find the vertex S opposite Q in the parallelogram $PQRS$ with $P = (1,2,1), Q = (3,0,4),$ and $R = (4,-3,2)$.
♦ Type 1, displacement vectors

15.[A] Are the points $(3,0,1), (2,4,5),$ and $(-1,4,4)$ on the same line? Justify your answer.
♦ Type 2, parallel vectors

16.[O] A box is supported by three ropes attached to a hook on it. The ropes exert forces $\mathbf{F_1} = 20\mathbf{i} + 10\mathbf{j} + 15\mathbf{k}, \mathbf{F_2} = -30\mathbf{i} - 4\mathbf{j} + 6\mathbf{k},$ and $\mathbf{F_3} = 10\mathbf{i} - 6\mathbf{j} + 8\mathbf{k}$ pounds, relative to xyz-coordinates with the positive z-axis pointing up. How much does the box weigh?
♦ Type 2, resultant force vectors

17. Three ropes are supporting a ten-pound weight. Two of the ropes exert forces $\mathbf{F_1} = \langle 2,3,4 \rangle$ and $\mathbf{F_2} = \langle -1,-2,3 \rangle$ newtons in xyz-space with the positive z-axis pointing up. What is the tension in the third rope?
♦ Type 2, resultant force vectors

18. Give exact and approximate decimal values of the three angles in the triangle with vertices $P = (1,2,3), Q = (4,0,2),$ and $R = (2,1,3)$.
♦ Type 3, angles between vectors

19. Give exact and approximate decimal values of the three angles in the triangle with vertices $P = (2,3,-6), Q = (1,4,-4),$ and $R = (10,5,-3)$.
♦ Type 3, angles between vectors

20. Which of the angles in the triangle with vertices $P = (1, -2, 0), Q = (2, 1, -2)$, and $R = (6, -1, -3)$ is a right angle?
♦ Type 2, dot products

21.[A] Calculate the area of the right triangle with vertices $P = (3, 5, 1), Q = (2, 4, 3)$, and $R = (5, 7, 3)$.
♦ Type 2, dot products

22. (a) Find the number k such that the quadrilateral $PQRS$ with $P = (0, 1, 0), Q = (1, 3, k)$, $R = (3, k, k)$, and $S = (2, 0, 0)$ is a parallelogram. (b) Is it a rectangle? (c) Is it a square?
♦ Type 4, parallel and perpendicular vectors

23. Which pairs of the vectors $\mathbf{A} = \langle 4, -6, 2 \rangle, \mathbf{B} = \langle 1, 2, 4 \rangle$, and $\mathbf{C} = \langle -6, 9, -3 \rangle$ are parallel? Which pairs are perpendicular?
♦ Type 4, parallel and perpendicular vectors

24. Find the component of $\mathbf{A}$ in the direction of $\mathbf{B}$ ($\mathbf{a^A}$) for $\mathbf{A} = \langle 1, 2, 4 \rangle$ and $\mathbf{B} = \langle -1, 2, 5 \rangle$, ($\mathbf{b^A}$) for $\mathbf{A} = \mathbf{i} - 2\mathbf{j} + 10\mathbf{k}$ and $\mathbf{B} = -\mathbf{i} - \mathbf{k}$, (c) for $\mathbf{A} = \langle 4, 3, 2 \rangle$ and $\mathbf{B} = \langle -3, 1, -1 \rangle$.
♦ Type 3, components of vectors

25. Find the orthogonal projection of $\mathbf{A}$ on a line parallel to $\mathbf{B}$ ($\mathbf{a^A}$) for $\mathbf{A} = \langle 5, 4, 3 \rangle$ and $\mathbf{B} = \langle 1, 1, 2 \rangle$, ($\mathbf{b^A}$) for $\mathbf{A} = 5\mathbf{i} - 4\mathbf{k}$ and $\mathbf{B} = 6\mathbf{j} + \mathbf{k}$, and (c) for $\mathbf{A} = \langle 2, 3, 1 \rangle$ and $\mathbf{B} = \langle 2, -1, -1 \rangle$.
♦ Type 3, projections of vectors

26. Give the direction cosines and exact and approximate decimal values of the direction angles ($\mathbf{a^A}$) of $\langle 3, -2, -3 \rangle$ ($\mathbf{b^A}$) of $\langle 1, -2, 4 \rangle$, and (c) $\langle -4, 3, 2 \rangle$.
♦ Type 3, direction cosines and angles

27. What is the component of $\mathbf{i} + 2\mathbf{j} - 3\mathbf{k}$ in the direction from $(2, 5, 8)$ toward $(-2, 3, 9)$?
♦ Type 3, components of vectors

28. A grocer sells a gallons of milk, b pounds of potatoes, and c bags of flour. He makes p dollars profit per gallon of milk, q dollars profit per pound of potatoes, and r dollars profit per bag of flour. What does $\mathbf{A} \cdot \mathbf{B}$ represent if $\mathbf{A} = \langle a, b, c \rangle$ and $\mathbf{B} = \langle p, q, r \rangle$?
♦ Type 4, dot products

29. Which of the points $P = (3, -4, -4), Q = (1, 5, -3)$, and $R = (4, 4, 2)$ is closest to the z-axis and which is closest to the xy-plane?
♦ Type 2, distances in space

30. Find the constant k such that the projection of $\langle 6, 10, k \rangle$ on a line parallel to $\langle -3, 2, 1 \rangle$ is $\langle \frac{3}{7}, -\frac{2}{7}, -\frac{1}{7} \rangle$.
♦ Type 4, projections of vectors

31. Is the angle between $\langle a, b, c \rangle$ and $\langle 1, 1, 1 \rangle$ less than or greater than the angle between $\langle a, b, c \rangle$ and $\langle -1, -1, -1 \rangle$ if a, b, and c are positive numbers? Justify your answer.
♦ Type 4, angles between vectors

32. A rectangle has corners $(2, 1, 3), (3, 5, 9), (4, -1, -4)$, and $(5, 3, 10)$, not necessarily in that order. What is its area?
♦ Type 2, dot products

33. (a) Find the unit vector that makes equal angles with the x-, y-, and z-axes. (b) Give the approximate decimal value of the direction angles, measured in degrees, of the vector from part (a).
♦ Type 4, direction cosines and angles

34. Find the vector whose component in the direction of $\mathbf{i}$ is 1, whose component in the direction of $\mathbf{j} + \mathbf{k}$ is $2\sqrt{2}$, and whose component in the direction of $\mathbf{i} - 3\mathbf{j}$ is $\sqrt{10}$.
♦ Type 4, components of vectors

35. Find (a) the exact acute angle between the diagonal of a cube and an edge and (b) the exact acute angle between the diagonal of a cube and the diagonal of one of its sides.
♦ Type 4, angles between vectors

36. A rectangular room is 10 feet high, 12 feet wide, and 20 feet long. Two strings from the corners of the floor 12 feet apart are stretched to the diagonally opposite corners. What is the angle between the strings?
♦ Type 4, angles between vectors

37. Explain why the equation $\mathbf{A} \cdot \mathbf{B} = \mathbf{A} \cdot \mathbf{C}$ for vectors in space with $\mathbf{A} \neq \mathbf{0}$ does not imply that $\mathbf{B} = \mathbf{C}$. What geometric relationship does the equation imply?
♦ Type 4, dot products

38. Suppose you want to find a nonzero vector that makes acute angles α and β with the x- and y-axes, respectively. Explain, by considering intersections of cones, why there are two such vectors if $\alpha + \beta > \frac{1}{2}\pi$, one such vector if $\alpha + \beta = \frac{1}{2}\pi$, and no such vectors if $\alpha + \beta < \frac{1}{2}\pi$.
♦ Type 6, direction angles

39. Find the two unit vectors that make an angle of $60°$ with $\mathbf{A} = \langle -2, -2, 1 \rangle$ and an angle of $45°$ with $\mathbf{B} = \langle 0, -1, 1 \rangle$.
♦ Type 4, angles between vectors

Section 10.3

Lines and planes in space

OVERVIEW: *We begin this section by using vectors to find* PARAMETRIC EQUATIONS *of lines and equations of planes in xyz-space. Then we study the* CROSS PRODUCT *of two vectors in space and show how it is used to find a vector perpendicular to two given nonparallel vectors, to find areas of parallelograms, and, in the* SCALAR TRIPLE PRODUCT, *to find volumes of parallelepipeds and tetrahedra.*

Topics:

- *Parametric equations of lines in space*
- *Equations of planes in space*
- *The cross product of two vectors*
- *The distance from a point to a plane*
- *The scalar triple product*
- *Areas and volumes*

Parametric equations of lines in space

The most convenient way to describe a line in xyz-space is to give the coordinates of a point $P = (x_0, y_0, z_0)$ on it and a nonzero vector $\mathbf{v} = \langle a, b, c \rangle$ parallel to it, as in Figure 1.[†]

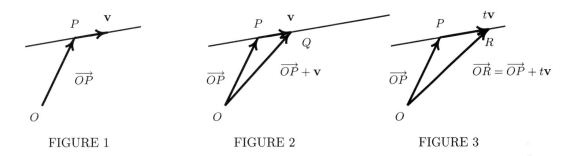

FIGURE 1   FIGURE 2   FIGURE 3

Question 1   Figure 2 shows the origin O, a point P, its position vector $\overrightarrow{OP}$, a vector $\mathbf{v}$, and the point Q with position vector $\overrightarrow{OP} + \mathbf{v}$ in xyz-space. Plot the point S with position vector $\overrightarrow{OP} + 2\mathbf{v}$ and the point T with position vector $\overrightarrow{OP} - \mathbf{v}$.

The points S, and T with position vectors $\overrightarrow{OP} + 2\mathbf{v}$ and $\overrightarrow{OP} - \mathbf{v}$. that you plotted in Question 1 are on the line passing through the point P that is parallel to the vector $\mathbf{v}$. You can imagine that the point R with position $\overrightarrow{OR} = \overrightarrow{OP} + t\mathbf{v}$ in Figure 3 is on the line for any number t and that as t ranges over all real numbers, the point R traverses the entire line. We let (x, y, z) be the coordinates of R, so that $\overrightarrow{OR} = \langle x, y, z \rangle$ and

---

[†]We use the symbol $\mathbf{v}$ for the vector parallel to the line because, as we will see in Section 10.4, if t represents time, then $\mathbf{v}$ is the velocity vector of a point moving on the line.

     ♦ SUGGESTIONS TO INSTRUCTORS: The main topics in this section can be divided into two parts, which can be covered in two or more class meetings.

Part 1 (Lines and planes in space, the cross product; Tune-up Exercises T1–T10, Problems 1–16, 20–26) Have students begin work on Questions 1 through 7 before class. Problems 1–5, 6, 9, 10 are good for in-class discussion.

Part 2 (The distance from a point to a plane in space, skew lines, intercept equations of planes, distance from a point to a line in an xy-plane, the scalar triple product; Tune-up Exercises T11-T13, Problems 17–19, 27–50) Have students begin work on Questions 8 through 12 before class. Problems 15, 16, 17 and 26 are good for in-class discussion.

$$\langle x, y, z \rangle = \overrightarrow{OR} = \overrightarrow{OP} + t\mathbf{v}$$
$$= \langle x_0, y_0, z_0 \rangle + t\langle a, bc \rangle \quad (1)$$
$$= \langle x_0 + at, y_0 + bt, z_0 + ct \rangle.$$

This vector equation gives a formula for the position vector of a general point $R = (x, y, z)$ on the line. We rewrite it as three equations, one for each component, and obtain the following result:

Rule 1 *The line through the point $P = (x_0, y_0, z_0)$ and parallel to the nonzero vector $\mathbf{v} = \langle a, b, c \rangle$ in xyz-space has the parametric equations*

$$\begin{cases} x = x_0 + at \\ y = y_0 + bt \\ z = z_0 + ct \end{cases} \quad (2)$$

where t ranges over all real numbers.

You might find it easier to remember equations **(2)** by recalling the vector equation **(1)** which is used in deriving them.

Equations **(2)** are called PARAMETRIC EQUATIONS of the line because they express the coordinates (x, y, z) of a variable point on the line as functions of the parameter t. It is frequently convenient to think of t as the time. Then equations **(2)** show how the line is generated as time varies over all real values.

Example 1 Give parametric equations for the line through the point $(6, 4, 3)$ and parallel to the vector $2\mathbf{i} + 5\mathbf{j} - 7\mathbf{k}$.

SOLUTION We set $P = (6, 4, 3)$ and $\mathbf{v} = 2\mathbf{i} + 5\mathbf{j} - 7\mathbf{k} = \langle 2, 5, -7 \rangle$, and let $R = (x, y, z)$ be a general point on the line. Then

$$\langle x, y, z \rangle = \overrightarrow{OR} = \overrightarrow{OP} + t\mathbf{v}$$
$$= \langle 6, 4, 3 \rangle + t\langle 2, 5, -7 \rangle$$
$$= \langle 6 + 2t, 4 + 5t, 3 - 7t \rangle.$$

The line has the parametric equations, $x = 6 + 2t, y = 4 + 5t, z = 3 - 7t$. □.

Rule 1 also enables us to find a point on a line and a vector parallel to it from its parametric equations: If we are given parametric equations of the form **(2)**, with at least one of the numbers a, b, and c not zero, then they are equations of the line through the point (x_0, y_0, z_0) and parallel to $\langle a, b, c \rangle$.

Question 2 Give a nonzero vector parallel to the line with parametric equations $x = 3 + 2t$, $y = 6, z = 5 - 8t$.

We can also describe a line by giving two points on it, as in the next Example, rather than one point and a parallel vector.

Example 2 Give parametric equations for the line through the points $P = (5, 3, 1)$ and $Q = (7, -2, 0)$.

SOLUTION Because P and Q are on the line, $\mathbf{v} = \overrightarrow{PQ} = \langle 7 - 5, -2 - 3, 0 - 1 \rangle = \langle 2, -5, -1 \rangle$ is parallel to it. The line consists of the points with position vectors

$$\langle x, y, z \rangle = \overrightarrow{OP} + t\mathbf{v} = \langle 5, 3, 1 \rangle + t\langle 2, -5, -1 \rangle = \langle 5 + 2t, 3 - 5t, 1 - t \rangle$$

as t ranges over all numbers. The line has the parametric equations,

10.3 Lines and planes in space,

$$x = 5 + 2t, y = 3 - 5t, z = 1 - t.$$

(As a check, we can verify that these equations give the point P for $t = 0$ and the point Q for $t = 1$.) □

Equations of planes in space

A vector $\mathbf{n} = \langle a, b, c \rangle$ is said to be perpendicular or NORMAL to a plane in xyz-space if it is perpendicular to all lines in the plane (Figure 4). If $\mathbf{n}$ is a nonzero normal vector to a plane and P is in the plane, then Q is in the plane if and only if $\mathbf{n}$ is perpendicular to $\overrightarrow{PQ}$ (Figure 5), and we can check whether the vectors are perpendicular by calculating their dot product.

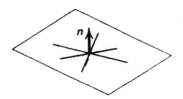
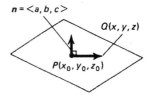

$\mathbf{n}$ is perpendicular to all lines in the plane.

FIGURE 4

Q is on the plane $\iff \mathbf{n} \cdot \overrightarrow{PQ} = 0.$

FIGURE 5

Example 3 The point $P = (2, -3, 1)$ is on a plane with normal vector $\mathbf{n} = \langle 3, 4, -1 \rangle$. Is the point $Q = (4, -4, 3)$ also on the plane?

SOLUTION We calculate $\overrightarrow{PQ} = \langle 4 - 2, -4 - (-3), 3 - 1 \rangle = \langle 2, -1, 2 \rangle$ and then

$$\mathbf{n} \cdot \overrightarrow{PQ} = \langle 3, 4, -1 \rangle \cdot \langle 2, -1, 2 \rangle = 3(2) + 4(-1) - 1(2) = 0.$$

The dot product is zero, so $\overrightarrow{PQ}$ is perpendicular to $\mathbf{n}$ and, therefore, parallel to the plane. Because P is on the plane, so is Q. □

Question 3 Is the point $S = (3, -2, 2)$ on the plane of Example 3?

The reasoning in Example 3 and Question 3 can be used to find an equation of the plane that contains the point $P = (x_0, y_0, z_0)$ and has the nonzero normal vector $\mathbf{n} = \langle a, b, c \rangle$. A variable point $R = (x, y, z)$ is on the plane if and only if the displacement vector $\overrightarrow{PR} = \langle x - x_0, y - y_0, z - z_0 \rangle$ is in the plane and consequently is perpendicular to the normal vector $\mathbf{n}$. Hence, $R = \langle x, y, z \rangle$ is on the plane if and only if

$$\mathbf{n} \cdot \overrightarrow{PR} = 0. \tag{3}$$

Since

$$\begin{aligned}\mathbf{n} \cdot \overrightarrow{PR} &= \langle x - x_0, y - y_0, z - z_0 \rangle \cdot \langle a, b, c \rangle \\ &= a(x - x_0) + b(y - y_0) + c(z - z_0)\end{aligned} \tag{4}$$

we have the following result:

Rule 2 The plane through the point $P = (x_0, y_0, z_0)$ and perpendicular to the nonzero normal vector $\mathbf{n} = \langle a, b, c \rangle$ has the equation,

$$a(x - x_0) + b(y - y_0) + c(z - z_0) = 0. \tag{5}$$

You can remember **(5)** directly or recall it when needed from formulas **(3)** and **(4)**.

Example 4 Give an equation of the plane through the point $(2, 3, 4)$ and perpendicular to the vector $\langle -6, 5, -4 \rangle$.

SOLUTION Set $P = (x_0, y_0, z_0) = (2, 3, 4)$ and $\mathbf{n} = \langle a, b, c \rangle = \langle -6, 5, -4 \rangle$. For $R = (x, y, z)$, $\overrightarrow{PR} = \langle x - 2, y - 3, x - 4 \rangle$, and R is on the plane if and only if $\mathbf{n} \cdot \overrightarrow{PR} = 0$. The plane has the equation

$$-6(x - 2) + 5(y - 3) - 4(z - 4) = 0. \ \square$$

Question 4 Give an equation for the plane through $(6, 10, -3)$ and perpendicular to the line $x = -3t, y = 6 + t, z = 4 - 7t$.

Rule 2 implies that any equation in the form,

$$ax + by + cz = k \tag{6}$$

where at least one of the numbers $a, b,$ or c is not zero is the graph of a plane with normal vector $\mathbf{n} = \langle a, b, c \rangle$.

To show that this is the case, we first note that we can pick a point $P = (x_0, y_0, z_0)$ that is on the plane, so that

$$ax_0 + by_0 + cz_0 = k. \tag{7}$$

(If a is not zero, we can set $y_0 = 0, z_0 = 0$, and solve for $x_0 = k/a$; if b is not zero, we can set $x_0 = 0$, $z_0 = 0$, and $y_0 = k/b$; and if c is not zero, we can set $x_0 = 0$, $y_0 = 0$, and $z_0 = k/c$.) Then subtracting equation **(7)** from equation **(6)** yields **(5)**, so the graph is the plane through P with normal vector $\langle a, b, c \rangle$.

Question 5 Give a nonzero vector normal to the plane $3x - 5y + 6z = 10$.

Example 5 Give an equation of the plane through the point $(1, -1, 2)$ that is parallel to the plane $3x - 5y + 6z = 10$.

SOLUTION We could use Rule 2 with the normal vector from Question 5, but there is an easier approach. The planes parallel to the plane $3x - 5y + 6z = 10$ have the equations $3x - 5y + 6z = k$ with constants k. Setting $x = 1, y = -1$, and $z = 2$ gives $3(1) - 5(-1) + 6(2) = k$, so that $k = 4$ and the plane is $3x - 5y + 6z = 4$. $\square$

The cross product of two vectors

There are two directions perpendicular to a nonzero vector $\mathbf{A} = \langle a, b \rangle$ in an xy-plane. They are the directions of the vectors $\langle -b, a \rangle$ and $\langle b, -a \rangle$ obtained by interchanging the components of $\langle a, b \rangle$ and multiplying one of them by -1 (Figure 6). The vector $\langle -b, a \rangle$ can be obtained geometrically by rotating $\mathbf{A}$ ninety degrees counterclockwise and $\langle b, -a \rangle$ can be obtained by rotating $\mathbf{A}$ ninety degrees clockwise.

Similarly, there are two directions perpendicular to two nonzero and nonparallel vectors $\mathbf{A}$ and $\mathbf{B}$ in xyz-space (Figure 7). We distinguish between these two directions by using the RIGHT-HAND RULE: A nonzero vector $\mathbf{C}$ perpendicular to $\mathbf{A}$ and $\mathbf{B}$ has the direction given by the right-hand rule from $\mathbf{A}$ toward $\mathbf{B}$ if, when the fingers of a right hand curl from $\mathbf{A}$ toward $\mathbf{B}$, as in Figure 8, the thumb points in the direction of $\mathbf{C}$.

10.3 Lines and planes in space,

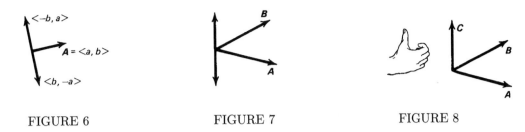

FIGURE 6   FIGURE 7   FIGURE 8

The CROSS PRODUCT $\mathbf{A} \times \mathbf{B}$ is a vector perpendicular to $\mathbf{A}$ and $\mathbf{B}$ whose direction is determined by the right-hand rule:

Definition 1 *The cross product $\mathbf{A} \times \mathbf{B}$ of nonzero and nonparallel vectors $\mathbf{A}$ and $\mathbf{B}$ in xyz-space is the vector perpendicular to $\mathbf{A}$ and $\mathbf{B}$ with direction determined by the right-hand rule from $\mathbf{A}$ toward $\mathbf{B}$ and whose length is*

$$|\mathbf{A} \times \mathbf{B}| = |\mathbf{A}||\mathbf{B}| \sin \theta \tag{8}$$

where θ is an angle with $0 < \theta < \pi$ between $\mathbf{A}$ and $\mathbf{B}$. (Figure 9). If $\mathbf{A}$ or $\mathbf{B}$ is the zero vector or they are parallel, then $\mathbf{A} \times \mathbf{B}$ is the zero vector.

FIGURE 9   FIGURE 10

In order to use the cross product, we need a formula for calculating $\mathbf{A} \times \mathbf{B}$ directly from the components of $\mathbf{A}$ and $\mathbf{B}$, and to derive such a formula, we need first to establish algebraic properties of the cross product from Definition 1.

We first observe that for any $\mathbf{A}$ and $\mathbf{B}$, $\mathbf{A} \times \mathbf{B}$ and $\mathbf{B} \times \mathbf{A}$ have the same length, since the same angle θ is used in (8) to define both lengths. Also, the direction determined by the right-hand rule from $\mathbf{B}$ to $\mathbf{A}$ (Figure 10) has the opposite direction from that determined by the right-hand rule from $\mathbf{A}$ toward $\mathbf{B}$ (Figure 9), so that $\mathbf{B} \times \mathbf{A} = -\mathbf{A} \times \mathbf{B}$. This is equation (9) in the following theorem. We will discuss the proofs of (10) and (11) at the end of the section.

Theorem 1 *For any vectors $\mathbf{A}, \mathbf{B}$, and $\mathbf{C}$ in xyz-space and any number k,*

$$\mathbf{B} \times \mathbf{A} = -\mathbf{A} \times \mathbf{B} \tag{9}$$

$$(k\mathbf{A}) \times \mathbf{B} = \mathbf{A} \times (k\mathbf{B}) = k(\mathbf{A} \times \mathbf{B}) \tag{10}$$

$$\mathbf{A} \times (\mathbf{B} + \mathbf{C}) = \mathbf{A} \times \mathbf{B} + \mathbf{A} \times \mathbf{C}. \tag{11}$$

Before we can use Theorem 1 to derive a formula for the cross product, we need to calculate the cross products of pairs of the unit vectors $\mathbf{i}, \mathbf{j}$, and $\mathbf{k}$.

Example 6 Express (a) $\mathbf{i} \times \mathbf{i}$, (b) $\mathbf{i} \times \mathbf{j}$, and (c) $\mathbf{i} \times \mathbf{k}$ in terms of $\mathbf{i}, \mathbf{j}, \mathbf{k}$, and $\mathbf{0}$.

SOLUTION (a) $\mathbf{i} \times \mathbf{i} = \mathbf{0}$ because $\mathbf{i}$ is parallel to itself.

(b) The angle θ between $\mathbf{i}$ and $\mathbf{j}$ is a right angle because the vectors are pependicular. Also $|\mathbf{i}| = 1$ and $|\mathbf{j}| = 1$, so

$$|\mathbf{i} \times \mathbf{j}| = |\mathbf{i}||\mathbf{j}| \sin \theta = 1$$

and $\mathbf{i} \times \mathbf{j}$ is a unit vector. The direction of $\mathbf{i} \times \mathbf{j}$, determined by the Right-Hand Rule from $\mathbf{i}$ toward $\mathbf{j}$, is the direction of the positive z-axis. (Figure 11). Therefore, $\mathbf{i} \times \mathbf{j} = \mathbf{k}$.

(c) The reasoning in part (b) shows that $\mathbf{i} \times \mathbf{k}$ is also a unit vector. The direction determined by the Right-Hand Rule from $\mathbf{i}$ toward $\mathbf{k}$ is the direction of the negative y-axis (Figure 11), so that $\mathbf{i} \times \mathbf{k} = -\mathbf{j}$. □

FIGURE 11

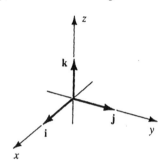

The solution of Example 6 gives the first three of the following basic formulas for cross products. The other six can be established similarly.

$$\begin{aligned} \mathbf{i} \times \mathbf{i} = \mathbf{0}, \quad & \mathbf{i} \times \mathbf{j} = \mathbf{k}, \quad & \mathbf{i} \times \mathbf{k} = -\mathbf{j}, \\ \mathbf{j} \times \mathbf{i} = -\mathbf{k}, \quad & \mathbf{j} \times \mathbf{j} = \mathbf{0}, \quad & \mathbf{j} \times \mathbf{k} = \mathbf{i}, \\ \mathbf{k} \times \mathbf{i} = \mathbf{j}, \quad & \mathbf{k} \times \mathbf{j} = -\mathbf{i}, \quad & \mathbf{k} \times \mathbf{k} = \mathbf{0}. \end{aligned} \quad (12)$$

Now consider any vectors $\mathbf{A} = A_1\mathbf{i} + A_2\mathbf{j} + A_3\mathbf{k}$ and $\mathbf{B} = B_1\mathbf{i} + B_2\mathbf{j} + B_3\mathbf{k}$. Theorem 1 implies that

$$\begin{aligned} \mathbf{A} \times \mathbf{B} &= (A_1\mathbf{i} + A_2\mathbf{j} + A_3\mathbf{k}) \times (B_1\mathbf{i} + B_2\mathbf{j} + B_3\mathbf{k}) \\ &= A_1B_1(\mathbf{i} \times \mathbf{i}) + A_1B_2(\mathbf{i} \times \mathbf{j}) + A_1B_3(\mathbf{i} \times \mathbf{k}) \\ &\quad + A_2B_1(\mathbf{j} \times \mathbf{i}) + A_2B_2(\mathbf{j} \times \mathbf{j}) + A_2B_3(\mathbf{j} \times \mathbf{k}) \\ &\quad + A_3B_1(\mathbf{k} \times \mathbf{i}) + A_3B_2(\mathbf{k} \times \mathbf{j}) + A_3B_3(\mathbf{k} \times \mathbf{k}). \end{aligned}$$

When we apply formulas (12) and simplify the result we obtain

$$\mathbf{A} \times \mathbf{B} = (A_2B_3 - A_3B_2)\mathbf{i} - (A_1B_3 - A_3B_1)\mathbf{j} + (A_1B_2 - A_2B_1)\mathbf{k}. \quad (13)$$

10.3 Lines and planes in space,

Calculating cross products with determinants

Because formula **(13)** is very difficult to remember, we will obtain the same result by using the notation of DETERMINANTS from linear algebra.

The 2×2 (two-by-two) determinant

$$\begin{vmatrix} x_1 & x_2 \\ y_1 & y_2 \end{vmatrix}$$

denotes the number $x_1 y_2 - x_2 y_1$ (Figure 12).

FIGURE 12

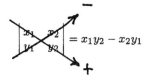

$= x_1 y_2 - x_2 y_1$

Then 3×3 determinants can be calculated with the formula,

$$\begin{vmatrix} x_1 & x_2 & x_3 \\ y_1 & y_2 & y_3 \\ z_1 & z_2 & z_3 \end{vmatrix} = x_1 \begin{vmatrix} y_2 & y_3 \\ z_2 & z_2 \end{vmatrix} - x_2 \begin{vmatrix} y_1 & y_3 \\ z_1 & z_3 \end{vmatrix} + x_3 \begin{vmatrix} y_1 & y_2 \\ z_1 & z_2 \end{vmatrix}. \tag{14}$$

Each of the determinants on the right of **(14)** is obtained by crossing out the row and column of one of the numbers in the first row of the 3×3 determinant. The expression on the right equals the first number in the first row of the original determinant, multiplied by the corresponding 2×2 determinant, minus the second number in the first row multiplied by the corresponding 2×2 determinant, plus the third number in the first row multiplied by the corresponding 2×2 determinant (Figure 13).

FIGURE 13

$$\begin{vmatrix} x_1 & x_2 & x_3 \\ y_1 & y_2 & y_3 \\ z_1 & z_2 & z_3 \end{vmatrix} = x_1 \begin{vmatrix} \cancel{x_1} & \cancel{x_2} & \cancel{x_3} \\ \cancel{y_1} & y_2 & y_3 \\ \cancel{z_1} & z_2 & z_3 \end{vmatrix} - x_2 \begin{vmatrix} \cancel{x_1} & \cancel{x_2} & \cancel{x_3} \\ y_1 & \cancel{y_2} & y_3 \\ z_1 & \cancel{z_2} & z_3 \end{vmatrix} + x_3 \begin{vmatrix} \cancel{x_1} & \cancel{x_2} & \cancel{x_3} \\ y_1 & y_2 & \cancel{y_3} \\ z_1 & z_2 & \cancel{z_3} \end{vmatrix}$$

Example 7 Evaluate $\begin{vmatrix} 3 & 2 & 4 \\ -1 & 0 & 6 \\ 5 & 1 & -2 \end{vmatrix}$.

SOLUTION By **(14)**, the determinant equals

$$3 \begin{vmatrix} 0 & 6 \\ 1 & -2 \end{vmatrix} - 2 \begin{vmatrix} -1 & 6 \\ 5 & -2 \end{vmatrix} + 4 \begin{vmatrix} -1 & 0 \\ 5 & 1 \end{vmatrix}$$
$$= 3[(0)(-2) - 6(1)] - 2[(-1)(-2) - 6(5)] + 4[(-1)(1) - 0(5)]$$
$$= 3(-6) - 2(-28) + 4(-1) = -18 + 56 - 4 = 34. \ \square$$

Question 6 Evaluate $\begin{vmatrix} 1 & 2 & -3 \\ 4 & 2 & 0 \\ -3 & 5 & 1 \end{vmatrix}$.

Now we can give a relatively easy procedure for calculating cross products:

Theorem 2 The cross product of vectors $\mathbf{A} = \langle A_1, A_2, A_3 \rangle$ and $\mathbf{B} = \langle B_1, B_2, B_3 \rangle$ is equal to the determinant

$$\mathbf{A} \times \mathbf{B} = \langle A_1, A_2, A_3 \rangle \times \langle B_1, B_2, B_3 \rangle = \begin{vmatrix} \mathbf{i} & \mathbf{j} & \mathbf{k} \\ A_1 & A_2 & A_3 \\ B_1 & B_2 & B_3 \end{vmatrix} \quad (15)$$

that is obtained by putting the unit vectors $\mathbf{i}, \mathbf{j}$, and $\mathbf{k}$ in the first row, the components of $\mathbf{A}$ in the second row, and the components of $\mathbf{B}$ in the third row.

To derive **(15)** we evaluate the determinant:

$$\begin{vmatrix} \mathbf{i} & \mathbf{j} & \mathbf{k} \\ A_1 & A_2 & A_3 \\ B_1 & B_2 & B_3 \end{vmatrix} = \begin{vmatrix} A_2 & A_3 \\ B_2 & B_3 \end{vmatrix} \mathbf{i} - \begin{vmatrix} A_1 & A_3 \\ B_1 & B_3 \end{vmatrix} \mathbf{j} + \begin{vmatrix} A_1 & A_2 \\ B_1 & B_2 \end{vmatrix} \mathbf{k}$$

$$= (A_2 B_3 - A_3 B_2) \mathbf{i} - (A_1 B_3 - A_3 B_1) \mathbf{j} + (A_1 B_2 - A_2 B_1) \mathbf{k}.$$

This is the expression for the cross product that we found in **(13)**, so the theorem is established.

Example 8 Find the components of the cross product $\langle 3, 1, -2 \rangle \times \langle 0, 4, 2 \rangle$.

SOLUTION Formula **(15)** yields

$$\langle 3, 1, -2 \rangle \times \langle 0, 4, 2 \rangle = \begin{vmatrix} \mathbf{i} & \mathbf{j} & \mathbf{k} \\ 3 & 1 & -2 \\ 0 & 4 & 2 \end{vmatrix} = \mathbf{i} \begin{vmatrix} 1 & -2 \\ 4 & 2 \end{vmatrix} - \mathbf{j} \begin{vmatrix} 3 & -2 \\ 0 & 2 \end{vmatrix} + \mathbf{k} \begin{vmatrix} 3 & 1 \\ 0 & 4 \end{vmatrix}$$

$$= [1(2) - (-2)(4)]\mathbf{i} - [3(2) - (-2)(0)]\mathbf{j} + [3(4) - 1(0)]\mathbf{k}$$

$$= 10\mathbf{i} - 6\mathbf{j} + 12\mathbf{k} = \langle 10, -6, 12 \rangle. \quad \square$$

As a partial check of the calculations in Example 8, we can find the dot products of each the given vectors with the calculated cross product:

$$\langle 3, 1, -2 \rangle \cdot \langle 10, -6, 12 \rangle = 3(10) + 1(-6) - 2(12) = 0$$

$$\langle 0, 4, 2 \rangle \cdot \langle 10, -6, 12 \rangle = 0(10) + 4(-6) + 2(12) = 0.$$

These dot products are both zero, as they should be, because the cross product is perpendicular to both of the original vectors.

Question 7 Give the components of $\mathbf{A} \times \mathbf{B}$ for $\mathbf{A} = 3\mathbf{i} - 2\mathbf{j} + \mathbf{k}$ and $\mathbf{B} = 6\mathbf{i} - 3\mathbf{k}$. As a partial check of your calculations, find the dot products of your result with $\mathbf{A}$ and with $\mathbf{B}$.

The distance from a point to a plane

Suppose that P is a point on a plane and $\mathbf{n}$ a nonzero vector normal to it. Then, as is shown in Figure 14, the distance from any point Q to the plane is the absolute value of the component of $\overrightarrow{PQ}$ in the direction of $\mathbf{n}$. With the formula $(\overrightarrow{PQ} \cdot \mathbf{n})/|\mathbf{n}|$ from Section 10.2 for that component, we obtain

$$\begin{bmatrix} \text{The distance from } Q \text{ to the plane} \\ \text{through } P \text{ with normal vector } \mathbf{n} \end{bmatrix} = \frac{|\overrightarrow{PQ} \cdot \mathbf{n}|}{|\mathbf{n}|}. \quad (16)$$

FIGURE 14

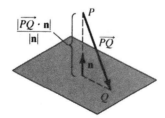

10.3 Lines and planes in space,

Example 9 How far is the point $(2, 1, 5)$ from the plane through $(1, -1, 4)$ with normal vector $\langle 2, 4, 1 \rangle$?

SOLUTION Set $P = (1, -1, 4), Q = (2, 1, 5)$, and $\mathbf{n} = \langle 2, 4, 1 \rangle$. Then $\overrightarrow{PQ} = \langle 2-1, -1-1, 5-4 \rangle = \langle 1, 2, 1 \rangle$ and, by **(16)**, the distance is

$$\frac{|\overrightarrow{PQ} \cdot \mathbf{n}|}{|\mathbf{n}|} = \frac{|\langle 1, 2, 1 \rangle \cdot \langle 2, 4, 1 \rangle|}{\sqrt{2^2 + 4^2 + 1^2}} = \frac{11}{\sqrt{21}}. \ \square$$

We will now use **(16)** to derive another formula for the distance for cases where we are given an equation of the form $ax + by + cz + d = 0$ for the plane rather than a point on it and a normal vector to it. Suppose that $P = (x_0, y_0, z_0)$, $Q = (x, y, z)$, and $\mathbf{n} = \langle a, b, c \rangle$ in **(16)**. Then

$$\begin{aligned} \overrightarrow{PQ} \cdot \mathbf{n} &= \langle x - x_0, y - y_0, z - z_0 \rangle \cdot \langle a, b, c \rangle \\ &= a(x - x_0) + b(y - y_0) + c(z - z_0) \\ &= ax + by + cz - (ax_0 + by_0 + cz_0). \end{aligned} \tag{17}$$

Question 8 Because $P = (x_0, y_0, z_0)$ is on the plane, its coordinates satisfy the equation $ax_0 + by_0 + cz_0 + d = 0$. Use this fact to simplify the formula on the right of **(17)**.

Since $|\mathbf{n}| = |\langle a, b, c \rangle| = \sqrt{a^2 + b^2 + c^2}$, formula **(16)** with the result of Question 8 establishes the folowing:

Rule 3 The distance from the point (x, y, z) to the plane $ax + by + cz + d = 0$ is

$$\frac{|ax + by + cz + d|}{\sqrt{a^2 + b^2 + c^2}}. \tag{18}$$

Notice that the distance **(18)** is zero if $ax + by + cz + d = 0$ and the point is on the plane.

Example 10 What is the distance from $(1, 2, -3)$ to the plane $2x - 3y + 4z - 5 = 0$?

SOLUTION By **(18)** with $(x, y, z) = (1, 2, -3)$ and $\langle a, b, c \rangle = \langle 2, -3, 4 \rangle$, the distance is

$$\frac{|2(1) - 3(2) + 4(-3) - 5|}{\sqrt{2^2 + 3^2 + 4^2}} = \frac{|-21|}{\sqrt{29}} = \frac{21}{\sqrt{29}}. \ \square$$

Skew lines

Two lines, such as the lines L_1 and L_2 in Figure 15, that are not parallel and do not intersect are said to be SKEW. Such lines lie in parallel planes. Figure 16 shows, for instance, the parallel planes containing the lines from Figure 15. As is shown in the drawing, the plane containing L_1 is determined by L_1 and the line L_3 parallel to L_2 that intersects L_1. The (shortest) distance between L_1 and L_2 is the (perpendicular) distance between the two planes.

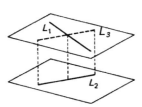
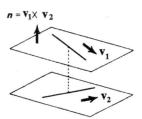

FIGURE 15   FIGURE 16   FIGURE 17

Example 11 (a) Show that the lines $L_1 : x = 7t, y = 2 + t, z = 4 - 3t$ and $L_2 : x = 3 - t$, $y = 5, z = 6 + 2t$ are not parallel. (b) Determine whether the lines intersect or are skew by calculating the distance between them.

SOLUTION (a) The lines are not parallel because $\mathbf{v}_1 = \langle 7, 1, -3 \rangle$ is parallel to L_1, $\mathbf{v}_2 = \langle -1, 0, 2 \rangle$ is parallel to L_2, and $\mathbf{v}_1$ is not equal to a constant multiplied by $\mathbf{v}_2$.

(b) The cross product $\mathbf{v}_1 \times \mathbf{v}_2$ is perpendicular to both lines and hence is a normal vector $\mathbf{n}$ to the parallel planes that contain them (Figure 17). We calculate its components:

$$\mathbf{n} = \mathbf{v}_1 \times \mathbf{v}_2 = \begin{vmatrix} \mathbf{i} & \mathbf{j} & \mathbf{k} \\ 7 & 1 & -3 \\ -1 & 0 & 2 \end{vmatrix}$$

$$= \mathbf{i} \begin{vmatrix} 1 & -3 \\ 0 & 2 \end{vmatrix} - \mathbf{j} \begin{vmatrix} 7 & -3 \\ -1 & 2 \end{vmatrix} + \mathbf{k} \begin{vmatrix} 7 & 1 \\ -1 & 0 \end{vmatrix}$$

$$= [1(2) - (-3)(0)]\mathbf{i} - [7(2) - (-3)(-1)]\mathbf{j} + [7(0) - 1(-1)]\mathbf{k}$$

$$= \langle 2, -11, 1 \rangle.$$

Setting $t = 0$ in the equations for L_1 shows that $P = (0, 2, 4)$ is on it and setting $t = 0$ in the equations for L_2 shows that $Q = (3, 5, 6)$ is on it. We use $\overrightarrow{PQ} = \langle 3 - 0, 5 - 2, 6 - 4 \rangle = \langle 3, 3, 2 \rangle$ and $\mathbf{n} = \langle 2, -11, 1 \rangle$ in **(16)** to see that the distance between the lines is

$$\frac{|\overrightarrow{PQ} \cdot \mathbf{n}|}{|\mathbf{n}|} = \frac{|\langle 3, 3, 2 \rangle \cdot \langle 2, -11, 1 \rangle|}{|\langle 2, -11, 1 \rangle|} = \frac{|-25|}{\sqrt{126}} = \frac{25}{\sqrt{126}}.$$

The distance is not zero, so the lines do not intersect and are skew. □

Intercept equations of planes

The x-, y-, and z-INTERCEPTS of a plane are the values where it intersects the x-, y-, and z-axes, respectively. If any of the intercepts is zero, then the plane passes through the origin and has an equation of the form $ax + by + cz = 0$. If the plane does not go through the origin and intersects all three axes, then it has the equation,

$$\frac{x}{A} + \frac{y}{B} + \frac{z}{C} = 1 \tag{19}$$

where A, B, and C are its intercepts. (Check this formula by setting $y = 0, z = 0$ and solving for the x-intercept A, by setting $x = 0, z = 0$ and solving for the y-intercept B, and by setting $x = 0, y = 0$ and solving for the z-intercept.) This plane can be shown by drawing the triangular portion of it with the three intercepts as vertices, as in Figure 18.

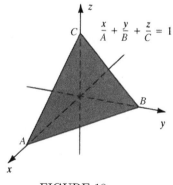

FIGURE 18

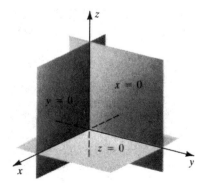

FIGURE 19

10.3 Lines and planes in space,

If the plane intersects two of the axes but is parallel to the third, then it has an equation in one of the forms,
$$\frac{x}{A} + \frac{y}{B} = 1, \quad \frac{x}{A} + \frac{z}{C} = 1, \quad \frac{y}{B} + \frac{z}{C} = 1 \tag{20}$$
where $A, B,$ and C are the x-, y-, and z-intercepts as appropriate.

If the plane is parallel to two of the axes, it is perpendicular to the third and has an equation of one of the forms,
$$x = A, y = B, z = C. \tag{21}$$
In particular, the yz-plane has the equation $x = 0$, the xz-plane has the equation $y = 0$, and the xy-plane has the equation $z = 0$ (Figure 19).

Question 9 Figure 20 shows the plane that is parallel to the y-axis and has x-intercept 4 and z-intercept 4. Give an equation for this plane.

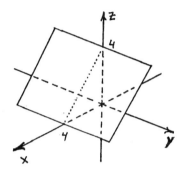

FIGURE 20

Normal vectors and distances to lines in an xy-plane

The equation $ax + by + cz + d = 0$ of a plane with normal vector $\langle a, b, c \rangle$ and formula (18) for distance from a point to a plane have counterparts for lines in planes that can be derived with the same reasoning.

Rule 4 (a) *The line through the point (x_0, y_0) and perpendicular to the nonzero normal vector $\mathbf{n} = \langle a, b \rangle$ (Figure 21) has the equation*
$$a(x - x_0) + b(y - y_0) = 0. \tag{22}$$

(b) *The distance from (x, y) to the line $ax + by + c = 0$ is*
$$\frac{|ax + by + c|}{\sqrt{a^2 + b^2}}. \tag{23}$$

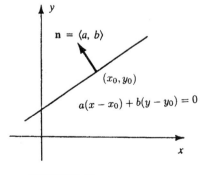

FIGURE 21

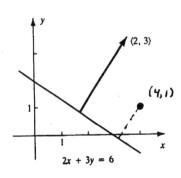

FIGURE 22

Example 12 Give a nonzero normal vector to the line $2x + 3y - 6 = 0$ in an xy-plane and the distance from $(4, 1)$ to the line.

SOLUTION By part (a) of Rule 4, the vector $\langle 2, 3 \rangle$ is normal to the line (Figure 22), and by **(23)**, the distance from $(4, 1)$ to it is $\dfrac{|2(4) + 3(1) - 6|}{\sqrt{2^2 + 3^2}} = \dfrac{5}{\sqrt{13}}.$ □

The scalar triple product

The number $\mathbf{A} \cdot (\mathbf{B} \times \mathbf{C})$ is called a SCALAR TRIPLE PRODUCT.† If $\mathbf{A} = \langle A_1, A_2, A_3 \rangle$ and $\mathbf{B} = \langle B_1, B_2, B_3 \rangle$, then

$$\mathbf{B} \times \mathbf{C} = \begin{vmatrix} \mathbf{i} & \mathbf{j} & \mathbf{k} \\ B_1 & B_2 & B_3 \\ C_1 & C_2 & C_3 \end{vmatrix}.$$

and because we can obtain $\mathbf{A} \cdot (\mathbf{B} \times \mathbf{C})$ by replacing the symbols $\mathbf{i}, \mathbf{j}$, and $\mathbf{k}$ in a formula for $\mathbf{B} \times \mathbf{C}$ in the formula that would be obtained by expanding this determinant, the scalar triple product is given by the determinant

$$\mathbf{A} \cdot (\mathbf{B} \times \mathbf{C}) = \begin{vmatrix} A_1 & A_2 & A_3 \\ B_1 & B_2 & B_3 \\ C_1 & C_2 & C_3 \end{vmatrix} \tag{24}$$

whose rows are the components of $\mathbf{A}, \mathbf{B}$, and $\mathbf{C}$ in that order.

Question 10 Calculate $\mathbf{A} \cdot (\mathbf{B} \times \mathbf{C})$ for $\mathbf{A} = \langle 3, 3, -1 \rangle, \mathbf{B} = \langle 4, 6, 5 \rangle$, and $\mathbf{C} = \langle 2, 2, -1 \rangle$.

Areas and volumes

If the nonzero vectors $\mathbf{A}$ and $\mathbf{B}$ are placed with their bases together, as in Figure 23, they form the sides of a parallelogram whose area equals the length $|\mathbf{A}|$ of its base multiplied by its height. Its height, on the other hand, equals $|\mathbf{B}| \sin \theta$, where θ is the angle with $0 \leq \theta \leq \pi$ between the vectors, so the area of the parallelogram equals the length $|\mathbf{A}||\mathbf{B}| \sin \theta$ of $\mathbf{A} \times \mathbf{B}$, as defined in **(8)**. This gives part (a) of the next rule. Part (b) follows because the area of the triangle is half the area of the parallelogram (Figure 24).

Rule 5 **(a)** *If the nonzero vectors $\mathbf{A}$ and $\mathbf{B}$ with their bases at the same point in xyz-space form two sides of a parallelogram, then*

$$[\text{The area of the parallelogram}] = |\mathbf{A} \times \mathbf{B}|. \tag{25}$$

(b) *If the vectors $\mathbf{A}$ and $\mathbf{B}$ with their bases at the same point in xyz-space form two sides of a triangle, then*

$$[\text{The area of the triangle}] = \tfrac{1}{2}|\mathbf{A} \times \mathbf{B}|. \tag{26}$$

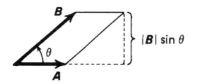

FIGURE 23 FIGURE 24

---

†$\mathbf{A} \cdot (\mathbf{B} \times \mathbf{C})$ is called a scalar triple product because it is a scalar (number) and to distinguish it from the vector triple product $\mathbf{A} \times (\mathbf{B} \times \mathbf{C})$ (see Problem 48).

10.3 Lines and planes in space,

If three vectors **A**, **B**, and **C** in space are not parallel to the same plane and we position them with their bases together, as in Figure 25, then they form three adjacent edges of the PARALLELEPIPED with four edges formed by vectors equal to **A**, four edges formed by vectors equal to **B**, and four edges formed by vectors equal to **C**.

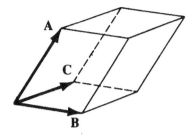
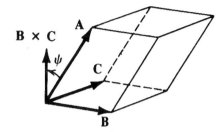

FIGURE 25 FIGURE 26

The volume of the parallelepiped in Figure 26 is equal to the area of its base multiplied by its height. The area of its base is $|\mathbf{B} \times \mathbf{C}|$ by **(25)**. We let ψ be the angle between **A** and $\mathbf{B} \times \mathbf{C}$ and suppose first that $0 \leq \psi \leq \frac{1}{2}\pi$, as in Figure 26. Because $\mathbf{B} \times \mathbf{C}$ is perpendicular to the plane the height of the parallelepiped equals $|\mathbf{A}|\cos\psi$, the triple product

$$\mathbf{A} \cdot (\mathbf{B} \times \mathbf{C}) = |\mathbf{A}||\mathbf{B} \times \mathbf{C}|\cos\psi$$

equals the volume of the parallelopiped. If $\frac{1}{2}\pi < \psi \leq \pi$, then $\cos\psi \leq 0$ and the triple product equals the negative of the volume of the parallelepiped. In either case, the volume of the parallelepiped equals the absolute value of the triple product. This gives part (a) of the next rule.

Rule 6 (a) *If the vectors* **A**, **B**, *and* **C** *with their bases at the same point in xyz-space form adjacent sides of a parallelepiped, then*

$$[\text{The volume of the parallelepiped}] = |\mathbf{A} \cdot (\mathbf{B} \times \mathbf{C})|. \tag{27}$$

(b) *If the vectors* **A**, **B**, *and* **C** *with their bases at the same point in xyz-space form adjacent sides of a tetrahedron (Figure 27), then*

$$[\text{The volume of the tetrahedron}] = \tfrac{1}{6}|\mathbf{A} \cdot (\mathbf{B} \times \mathbf{C})|. \tag{28}$$

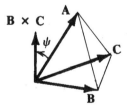

FIGURE 27

Part (b) of Rule 6 follows because the volume of the tetrahedron equals one-third of the product of the area of its base and height and the area of its base is half the area of the parallelogram with sides formed by **B** and **C**, so its volume is one-sixth of the volume of the parallelepiped

Example 13 What is the volume of the parallelepiped with vertex $P = (1,1,1)$ and adjacent vertices $Q = (4,4,0), R = (5,7,6)$, and $S = (3,3,0)$?

SOLUTION   Adjacent sides of the parallelepiped are formed by the vectors

$$\overrightarrow{PQ} = \langle 4-1, 4-1, 0-1 \rangle = \langle 3, 3, -1 \rangle$$
$$\overrightarrow{PR} = \langle 5-1, 7-1, 6-1 \rangle = \langle 4, 6, 5 \rangle$$
$$\overrightarrow{PS} = \langle 3-1, 3-1, 0-1 \rangle = \langle 2, 2, -1 \rangle.$$

You calculated the triple product $\overrightarrow{PQ} \cdot (\overrightarrow{PR} \times \overrightarrow{PS})$ in Question 10 to be -2. The volume of the parallelepiped is therefore $|-2| = 2$. □

Adding three vectors in space

Recall that the sum $\mathbf{A} + \mathbf{B}$ of two vectors in a plane or in space can be calculated geometrically by placing their bases together and drawing the parallelogram with the vectors as sides. Then $\mathbf{A} + \mathbf{B}$ is the vector with its base at the base of $\mathbf{A}$ and $\mathbf{B}$ and its tip at the diagonally opposite corner of the parallelogram.

Similarly, the sum $\mathbf{A} + \mathbf{B} + \mathbf{C}$ of three vectors in space can be obtained by placing the bases of the vectors together and drawing the parallelopiped with the vectors as sides, as in Figure 28. Then $\mathbf{A} + \mathbf{B} + \mathbf{C}$ is the vector with its base at the base of $\mathbf{A}, \mathbf{B}$, and $\mathbf{C}$ and its tip at the diagonally opposite corner of the parallelopiped, as shown in Figure 29.

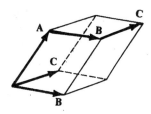

FIGURE 28

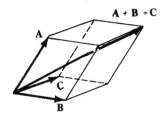

FIGURE 29

More on the proof of Theorem 1

Equation (**10**) in Theorem 1 is

$$(k\mathbf{A}) \times \mathbf{B} = \mathbf{A} \times (k\mathbf{B}) = k(\mathbf{A} \times \mathbf{B}). \tag{29}$$

This equation holds if $k = 0$, if $\mathbf{A}$ or $\mathbf{B}$ is the zero vector, or if $\mathbf{A}$ and $\mathbf{B}$ are parallel because in these cases all three vectors in the equations are zero. The three vectors are also equal if $k > 0$ and $\mathbf{A}$ and $\mathbf{B}$ are not zero and not parallel because then $\mathbf{kA}$ has the same direction as $\mathbf{A}$, $\mathbf{kB}$ has the same direction as $\mathbf{B}$, $|\mathbf{kA}| = k|\mathbf{A}|$, and $|\mathbf{kA}| = k|\mathbf{A}|$. We treat the case of negative k in the next Question.

Question 11   Show that for nonzero, nonparallel $\mathbf{A}$ and $\mathbf{B}$, the vectors $(-\mathbf{A}) \times \mathbf{B}$ and $-(\mathbf{A} \times \mathbf{B})$
(a) have the same length and (b) have the opposite direction of $\mathbf{A} \times \mathbf{B}$.

The Response to Question 11 shows that $(-\mathbf{A}) \times \mathbf{B} = -\mathbf{A} \times \mathbf{B}$, which combined with (**10**) for positive k shows that (**10**) is also valid for negative k.

Equation (**11**) in Theorem 1 is

$$\mathbf{A} \times (\mathbf{B} + \mathbf{C}) = \mathbf{A} \times \mathbf{B} + \mathbf{A} \times \mathbf{C}. \tag{30}$$

To derive this result, we suppose initially that $\mathbf{A}$ is a unit vector, and consider the plane pependicular to $\mathbf{A}$ through bases of $\mathbf{A}$ and $\mathbf{B}$ as in Figure 30. That plane contains the cross product $\mathbf{A} \times \mathbf{B}$ because it is perpendicular to $\mathbf{A}$ and $\mathbf{B}$.

10.3 Lines and planes in space,

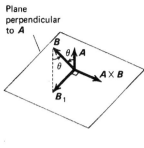
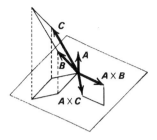

FIGURE 30   FIGURE 31

The ORTHOGONAL PROJECTION of **B** on the plane is the vector $\mathbf{B_1}$ in Figure 30, whose base is the base of **B** and whose tip is the foot of the perpendicular line from the tip of **B** to the plane.

Question 12 Express the length of the orthogonal projection $\mathbf{B_1}$ in Figure 30 in terms of the length of **B** and the angle θ between **A** and **B**.

Because $|\mathbf{A}| = 1$, the length $|\mathbf{B}|\sin\theta$ of $\mathbf{B_1}$ that you found in Question 12 equals the length $|\mathbf{A}||\mathbf{B}|\sin\theta$ of $\mathbf{A} \times \mathbf{B}$. We can therefore obtain $\mathbf{A} \times \mathbf{B}$ by rotating $\mathbf{B_1}$ in the plane of Figure 30 ninety degrees counterclockwise as viewed from above.

When we apply this projection and rotation procedure to the parallelogram formed by **B** and **C**, whose diagonal is formed by $\mathbf{B} + \mathbf{C}$ (Figure 31), we obtain the parallelogram in the plane with sides formed by $\mathbf{A} \times \mathbf{B}$ and $\mathbf{A} \times \mathbf{C}$, whose diagonal is formed by $\mathbf{A} \times \mathbf{B} + \mathbf{A} \times \mathbf{C}$. This diagonal also equals $\mathbf{A} \times (\mathbf{B} + \mathbf{C})$, as stated in **(11)**, because it can be obtained from the diagonal of the original parallelogram by the process of orthogonal projection and rotation. This completes the proof for the case of a unit vector **A**. The general case follows from **(29)**.

Principles and procedures

- Perhaps the best way to remember the parametric equations **(2)** for a line is to think of their vector formulation in **(1)**, and perhaps the best way to remember the equation **(5)** of a plane is to remember its vector derivation in **(3)** and **(4)**.
- You have to be very careful when you calculate determinants and cross products. Writing down every step in an orderly way without doing any arithmetic in your head—as in the solutions of Examples 7 and 8—can help you avoid mistakes. You can also check values of determinants with many calculators and computers.
- Check expressions $\langle C_1, C_2, C_3 \rangle$ for cross products $\mathbf{A} \times \mathbf{B}$ by calculating their dot products with **A** and **B**. Both dot products should be zero.
- The most frequent use of cross products is for finding the directions of the vectors perpendicular to two given vectors. The length of the cross product is required, however, in finding areas of parallelograms and triangles and volumes of parallelepipeds and tetrahedrons.
- To remember **(18)** for the distance from a point in space to the plane $ax + by + cz + d = 0$, note that the denominator is the length of the normal vector $\langle a, b, c \rangle$, and to remember **(23)** for the distance from a point in an xy-plane to the line $ax + by + c = 0$, note that the denominator is the length of the normal vector $\langle a, b \rangle$.
- You can draw a plane either by drawing a parallelogram that represents a rectangle in it with sides parallel to coordinate planes or, if a plane has nonzero intercepts, by drawing the triangle with those intercepts.

Tune-Up Exercises 10.3♦

A Answer provided. **O** Outline of solution provided.

T1.^O Give parametric equations of the line through the point $(3, 0, -2)$ and parallel to the vector $2\mathbf{i} - 5\mathbf{k}$.
♦ Type 1, equations of a line

T2.^O Give parametric equations of the line through the points $(-6, 4, 2)$ and $(3, 5, -1)$.
♦ Type 1, equations of a line

T3.^O Find an equation of the plane through the point $(1, 3, 4)$ and perpendicular to the vector $\langle 2, -4, 1 \rangle$.
♦ Type 1, equation of a plane

T4.^O Give parametric equations of the line through the point $(0, 4, 0)$ and perpendicular to the plane $4x - 5y + 6z = 2$.
♦ Type 1, equations of a line

T5.^O Find an equation of the plane through $(1, 2, 3)$ and perpendicular to the line $L: x = 3 - 5t$, $y = 4 + t, z = 6 + 2t$.
♦ Type 1, equation of a plane

T6.^O Find an equation for the plane through $(1, 1, 5)$ and parallel to the plane $6x = 5y$.
♦ Type 1, equation of a plane

T7.^O Evaluate the determinants **(a)** $\begin{vmatrix} 5 & 0 & 1 \\ 0 & 4 & -2 \\ 1 & 2 & 3 \end{vmatrix}$ and **(b)** $\begin{vmatrix} 1 & 1 & 2 \\ 2 & 1 & 0 \\ 1 & 3 & 1 \end{vmatrix}$.
♦ Type 1, determinants

T8.^O Find $\mathbf{A} \times \mathbf{B}$ for $\mathbf{A} = \langle 1, 3, 7 \rangle$ and $\mathbf{B} = \langle 6, -2, 0 \rangle$.
♦ Type 1, cross product

T9.^O Give parametric equations of the line that passes through $(1, 0, 3)$ and is perpendicular to $\mathbf{A} = \langle 2, -1, 1 \rangle$ and to $\mathbf{B} = \langle 3, 2, -2 \rangle$.
♦ Type 1, equations of a line

T10.^O Find parametric equations for the line through $(3, 2, 0)$ and parallel to the planes $x - 2y + 3z = 0$ and $4x - 5y - z = 0$.
♦ Type 1, equations of a line

T11.^O One corner of a parallelogram is $(1, 2, 3)$ and the adjacent corners are $(4, 3, 2)$ and $(0, 0, 1)$. Find its area.
♦ Type 1, area of a parallelogram

T12.^O Find the value of $\mathbf{A} \cdot (\mathbf{B} \times \mathbf{C})$ with $\mathbf{A} = \langle 1, 3, 2 \rangle, \mathbf{B} = \langle -1, 0, 4 \rangle$, and $\mathbf{C} = \langle -2, 1, 5 \rangle$.
♦ Type 1, triple product

T13.^O One vertex of a parallelepiped is $P = (1, 0, 2)$ and the three vertices adjacent to it are $Q = (4, 3, 1), R = (5, 6, 7)$, and $S = (3, 2, 1)$. Find the coordinates of the vertex T opposite P.
♦ Type 1, volume of a parallelepiped

♦The Tune-up Exercises and Problems are classified by type and content. The types are (1) basic, reactive; (2) basic reflective; (3) intermediate, reactive; (4) intermediate, reflective; (5) advanced, reactive; (6) advanced, reflective; and (7) advanced, theoretical.

10.3 Lines and planes in space,

Problems 10.3
^{A}Answer provided. ^{O}Outline of solution provided.

1. Evaluate the determinants $(\mathbf{a^O})$ $\begin{vmatrix} 1 & 3 & 0 \\ 2 & -1 & 1 \\ 0 & 4 & 5 \end{vmatrix}$, $(\mathbf{b^A})$ $\begin{vmatrix} 1 & 2 & 3 \\ 0 & 1 & 5 \\ 4 & 0 & 1 \end{vmatrix}$, $(\mathbf{c^A})$ $\begin{vmatrix} 5 & 0 & 0 \\ 0 & 6 & 4 \\ -3 & 2 & 8 \end{vmatrix}$,
   (d) $\begin{vmatrix} 10 & -10 & 20 \\ 10 & 0 & -10 \\ 20 & 20 & 10 \end{vmatrix}$, $(\mathbf{e^A})$ $\begin{vmatrix} -1 & 1 & 0 \\ 0 & 2 & 1 \\ 3 & 3 & 3 \end{vmatrix}$, and (f) $\begin{vmatrix} 1 & 2 & 3 \\ 4 & 5 & 6 \\ 7 & 8 & 9 \end{vmatrix}$.

   ◆ Type 1, determinants

2. Find the cross products $(\mathbf{a^O})$ $\langle 1, 4, -6 \rangle \times \langle 3, 2, -4 \rangle$, $(\mathbf{b^A})$ $\langle 3, 1, -2 \rangle \times \langle 1, 4, -0 \rangle$, $(\mathbf{c^A})$ $\langle 2, -1, 6 \rangle \times \langle 1, 3, -2 \rangle$, (d) $\langle -1, 0, 7 \rangle \times \langle 6, 4, -1 \rangle$, $(\mathbf{e^O})$ $(\mathbf{i} + \mathbf{j} - \mathbf{k}) \times (6\mathbf{i} - 7\mathbf{k})$, and (f) $(-\mathbf{i} + \mathbf{j} - 3\mathbf{k}) \times (2\mathbf{i} + 2\mathbf{j} - 5\mathbf{k})$.

   ◆ Type 1, cross products

3.^{A} Which of the following statements are true and which are false for all lines and planes in xyz-space: (a) Two lines parallel to a third line are parallel. (b) Two lines perpendicular to a third line are parallel. (c) Two planes parallel to a third plane are parallel. (d) Two planes perpendicular to a third plane are parallel. (e) Two lines parallel to a plane are parallel. (f) Two lines perpendicular to a plane are parallel. (g) Two planes parallel to a line are parallel. (h) Two planes perpendicular to a line are parallel. (i) Two lines either are parallel or intersect. (j) Two planes either are parallel or intersect. (k) A line and a plane either are parallel or intersect. (ℓ) Two nonparallel lines either intersect or there are parallel planes that contain them.

   ◆ Type 2, parallel and perpendicular lines and planes

4. Give parametric equations of $(\mathbf{a^O})$ the line through the point $(2, 4, 3)$ and parallel to the vector $\langle 4, 0, -7 \rangle$, $(\mathbf{b^O})$ the line through $(3, -4, 5)$ and perpendicular to the plane $6x - 4y = 10$, $(\mathbf{c^A})$ the line through $(5, 6, 7)$ and parallel to the line $x = 4, y = 6 - t, z = 9 + 2t$, (d) the line through the origin and perpendicular to the plane $3x - 4y + 5z - 18 = 0$, $(\mathbf{e^O})$ the line through $(6, 2, 3)$ and $(7, 0, -10)$, $(\mathbf{f^A})$ the line through $(0, 5, -4)$ and $(2, 0, 3)$, and (g) the line through $(2, 5, 6)$ and $(9, 4, -2)$.

   ◆ Type 1, equations of lines

5. Find equations $(\mathbf{a^O})$ the plane through the point $(3, 0, 8)$ and perpendicular to the vector $\langle 4, 0, 5 \rangle$, $(\mathbf{b^A})$ the plane through $(10, 5, 0)$ and perpendicular to the line $x = 3 + t, y = 4 - 2t, z = 3t$, and (c) the plane through $(0, 2, -5)$ and perpendicular to $\langle -1, 2, -3 \rangle$.

   ◆ Type 1, equations of planes

6. Give $(\mathbf{a^O})$ parametric equations of the line through $(6, 0, -3)$ and parallel to the planes $2x - 4y = 7$ and $3y + 5z = 0$, $(\mathbf{b^A})$ an equation for the plane through $(0, 1, 0)$ and parallel to $\mathbf{i} + \mathbf{j}$ and to $\mathbf{j} - \mathbf{k}$, $(\mathbf{c^O})$ an equation for the plane through $(5, -1, -2)$ and perpendicular to the planes $y - z = 4$ and $x + z = 3$, (d) an equation for the plane through $(2, 0, 0)$ and perpendicular to the planes $z = 4$ and $x + y + z = 0$, $(\mathbf{e^O})$ an equation for the plane through the points $(1, -1, 2), (4, -1, 5)$, and $(1, 2, -1)$, (f) an equation for the plane through $(2, 2, 4), (5, 6, 4)$, and $(1, 3, 5)$, $(\mathbf{g^O})$ an equation for the plane through $(1, 2, 3)$ and parallel to the plane $4x - y + 3z = 0$, and (h) an equation for the plane through the origin and parallel to the plane $3x - y + z = 1000$.

   ◆ Type 1, equations of lines and planes

7. Give $(\mathbf{a^O})$ the two unit vectors perpendicular to $\langle 4, 3, 1 \rangle$ and $\langle 5, 7, 2 \rangle$, $(\mathbf{b^A})$ the two unit vectors perpendicular to $\langle 5, 1, -2 \rangle$ and $\langle 0, 4, -1 \rangle$, and (c) the two vectors of length 10 that are perpendicular to $\mathbf{i} + 6\mathbf{j}$ and $2\mathbf{i} + \mathbf{j} + 3\mathbf{k}$.

   ◆ Type 1, perpendicular vectors

8.^{A} For what values of the constant α is the determinant $\begin{vmatrix} -3\alpha & -2\alpha & \alpha \\ 4 & 0 & 2 \\ -5 & 1 & 6 \end{vmatrix}$ equal to -78?

   ◆ v 2, determinants

9.[A] How are the constants a and b related if $\langle a, b, 1 \rangle$ is perpendicular to the line $x = 2t, y = t, z = -t$?
♦ Type 2, perpendicular vectors

10. How are the numbers a and b related if $\langle a, b, 5 \rangle$ is parallel to the plane $4x - 2y + z = 5$?
♦ Type 2, vectors parallel to a plane

11. What are the numbers a and b if $\langle a, b, 5 \rangle$ is perpendicular to the plane $4x - 2y + z = 5$?
♦ Type 2, vectors perpendicular to a plane

12.[O] What are b and c if $\langle 2, b, c \rangle$ is parallel to $\langle 1, 3, 5 \rangle$?
♦ Type 2, parallel vectors

13. (a) Show that $\mathbf{A} = \langle 1, 2, -1 \rangle$ is parallel to the plane $3x - y + z = 7$. (b) Find a nonzero vector $\mathbf{B}$ that is parallel to the plane and perpendicular to the vector $\mathbf{A}$ from part (a).
♦ Type 2, parallel and perpendicular vectors

14. Evaluate $\mathbf{A} \cdot (\mathbf{B} \times \mathbf{C})$, (a[O]) for $\mathbf{A} = \langle 3, 0, 1 \rangle, \mathbf{B} = \langle 1, 2, -1 \rangle, \mathbf{C} = \langle 0, 1, 6 \rangle$ (b[A]) for $\mathbf{A} = \langle 1, 2, 3 \rangle, \mathbf{B} = \langle -1, 0, 4 \rangle, \mathbf{C} = \langle 3, 2, 1 \rangle$, (c) for $\mathbf{A} = \langle 2, 0, 3 \rangle, \mathbf{B} = \langle 1, 6, 0 \rangle, \mathbf{C} = \langle -1, 4, 6 \rangle$ and (d) for $\mathbf{A} = \langle 1, 1, 2 \rangle, \mathbf{B} = \langle 3, 0, 4 \rangle, \mathbf{C} = \langle -1, 2, 2 \rangle$.
♦ Type 1, triple products

15. What are the areas of (a[O]) the parallelogram $PQRS$ with vertices $P = (1, 2, 4), Q = (6, 1, -1), R = (8, -3, -5)$, and $S = (3, -2, 0)$ and (b) the parallelogram $PQRS$ with vertices $P = (-1, 0, 10), Q = (3, 1, 9), R = (8, 7, 5)$, and $S = (4, 6, 6)$?
♦ Type 1, areas of parallelograms

16. Find the areas (a[O]) of the triangle with vertices $P = (2, 1, 0), Q = (3, 4, 5), R = (6, 1, 2)$ and (b) of the triangle with vertices $P = (5, 3, 4), Q = (1, 4, 6), R = (0, 2, 0)$.
♦ Type 1, areas of triangles

17.[O] One vertex of a parallelepiped is $P = (1, 1, 1)$ and the three adjacent vertices are $Q = (0, 0, -6), R = (8, 1, 2)$, and $S = (-4, 3, 1)$. (a) What is the vertex T opposite P? (b) What is the volume of the parallelepiped?
♦ Type 1, volumes of parallelepipeds

18. One vertex of a parallelepiped is $P = (-1, 2, -3)$ and the three adjacent vertices are $Q = (-1, -1, -3), R = (5, 3, -3)$, and $S = (4, 2, 0)$. (a) What is the vertex T opposite P? (b) What is the volume of the parallelepiped?
♦ Type 1, volumes of parallelepipeds

19. Find the volumes (a[O]) of the tetrahedron with vertices $(1, 2, 3), (2, 1, 4), (6, 0, 2), (4, 4, 4)$, (b[A]) of the tetrahedron with vertices $(0, 1, 0), (1, 0, 1), (0, 0, 1), (1, 1, 1)$, and (c) of the tetrahedron with vertices $(-2, 3, 2), (0, 1, 5), (1, 4, 4), (-1, 3, 8)$.
♦ Type 2, volumes of tetrahedrs

20.[A] Find an equation of each of the following planes and then show how it relates to the coordinate axes by drawing a triangular or rectangular portion of it that includes its intercepts: (a) the plane with x-intercept 2, y-intercept 3, and z-intercept 5, (b) the plane parallel to the x-axis with y-intercept 5 and z-intercept 2, (c) the plane parallel to the yz-plane with x-intercept 10.
♦ Type 1, equations and drawings of planes

21. Find the x-, y-, and z-intercepts of the plane given by each of the following equations. Then show how the plane relates to the coordinate axes by drawing a triangular or rectangular portion of it that includes its intercepts: (a) $-6x + 5y + 3z = 30$ (b) $2x + 5z = 10$, (c) $x + 5y = 15$, and (d) $y = 4$.
♦ Type 1, equations and drawings of planes

22. Give equations of the four planes that form the sides of the tetrahedron with vertices $(0, 0, 0), (3, 0, 0), (0, -4, 0)$, and $(0, 0, 5)$.
♦ Type 1, equations of planes

23. Show with sketches how three planes can divide xyz-space into four, six, seven, or eight regions.
♦ Type 4 or 6, allignments of planes

24. Give parametric equations of the line through the point $(a, b, 0)$ and parallel to the y-axis.
♦ Type 2, equations of a line

25. One vertex of a parallelopiped is $P = (0, 1, 1)$ and the three adjacent vertices are $Q = (0, 3, 2), R = (0, 2, -1)$, and $S = (\sqrt{5}, 1, 1)$. Show that the parallelepiped is a cube.
♦ Type 2, perpendicular vectors

10.3 Lines and planes in space,

26. The vertices of a quadrilateral are $P = (0,0,0), Q = (1,1,1), R = (2+a, 3+a, 4+a)$, and $S = (1+a, 2+a, 3+a)$, where a is a constant. **(a)** Show that the quadrilateral is a parallelogram. **(b)** Give a formula for its area in terms of a.
♦ Type 4, parallelogram and its area

27. **(a)** Use Rule 6 to show that the points P, Q, R, and S are in the same plane if and only if $\overrightarrow{PQ} \cdot (\overrightarrow{PR} \times \overrightarrow{PS})$ is zero. **(b^A)** Are $(1,2,-1), (3,3,-4), (2,2,1)$, and $(5,3,0)$ in the same plane? **(c)** Are $(3,2,1), (2,4,6), (4,0,1)$, and $(5,-2,7)$ in the same plane? **(d)** Are $(0,0,0), (1,2,3), (0,1,4)$, and $(2,0,1)$ in the same plane?
♦ Type 4, triple products

28. **(a)** Show that the lines $L_1 : x = 2-t, y = 3+t, z = 4-2t$ and $L_2 : x = -3+t, y = -1+2t, z = 9-3t$ intersect at $(0,5,0)$. **(b)** Give an equation for the plane that contains the lines from part (a).
♦ Type 4, intersecting lines

29. What are **(a^O)** the distance from the point $(0,1,0)$ to the plane $x - 3y + 4z = 100$, **(b)** the distance from $(1,4,-3)$ to the plane $x - y + 2z = 4$, and **(c)** the distance from $(2,4,-1)$ to the plane $z = 2x + y + 3$?
♦ Type 3, distances from points to planes

30. Which of the points $(3,6,2)$ and $(2,-1,12)$ is closer to the plane $2x - 3y + z - 4 = 0$?
♦ Type 3, distances from points to planes

31.^{A} Find the distance between the planes $2x + 2y + z = 5$ and $2x + 2y + z = 25$.
♦ Type 3, distances from points to planes

32. **(a^O)** Give a nonzero normal vector to the line $15x + 5y = 47$ in an xy-plane and the distance from the line to the point $(100, 200)$. **(b^A)** Give a nonzero normal vector to the line $4x - 3y = 10$ in an xy-plane and the distance from the line to the origin. **(c)** Give a nonzero normal vector to the line $y = 3x + 4$ in an xy-plane and the distance from the line the point $(-5, 10)$.
♦ Type 3, distances from points to lines in a plane

33.^{A} Find the distance between the lines $2x + 5y = 4$ and $2x + 5y = 5$ in a xy-plane.
♦ Type 4, distances between lines in a plane

34. **(a)** Find the distance between the lines $x - 3y = 10$ and $x - 3y = -6$ in an xy-plane. **(b)** Find the distance between the planes $x - 3y = 10$ and $x - 3y = -6$ in xyz-space.
♦ Type 4, distances between lines in xy-planes

35. Find the value(s) of the constant c such that the planes $2x + 3y + cz = 4$ and $2x + 3y + cz = 11$ are one unit apart.
♦ Type 4, distances between lines in a plane

36. Give parametric equations of **(a^O)** the line of intersection of the planes $x+y+z = 1$ and $x-2y-z = 1$, **(b^A)** the line of intersection of the planes $2x-y = 3$ and $3y-4z = 10$, and **(c)** the line of intersection of the planes $x + y - z = 0$ and $x - y + z = 0$.
♦ Type 4, intersections of planes

37. Show that $L_1 : x = 3 + 2t, y = 6 + t, z = 4 + 3t$ and $L_2 : x = -1 + 4t, y = 4 + 2t, z = -2 + 6t$ are the same line.
♦ Type 4, equivalent equations of a line

38.º Find the point where the lines $L_1 : x = 2t, y = 3 - t, z = -2 + 4t$ and $L_2 : x = 4 + t, y = 1 + 2t, z = 6$ intersect. (Replace t by s in the second set of equations and solve for t and s.)
♦ Type 4, intersections of lines

39. Give parametric equations of the line that passes through the origin and through the intersection of the lines $L_1 : x = 3 + 2t, y = -4t, z = -3 + t$ and $L_2 : x = 3 + 10t, y = -25 + 5t, z = 4 - 2t$.
♦ Type 4, intersections of lines

40. Give parametric equations of the line that is perpendicular to the lines $L_1 : x = 2 + t, y = 4 - t, z = 6 + 2t$ and $L_2 : x = 2 - t, y = 3 + 2t, z = 7 - 3t$ and passes through their point of intersection.
♦ Type 4, intersections of lines

41. What are the distances **(a^O)** between the lines $L_1 : x = 2 - t, y = 3 + 4t, z = 2t$ and $L_2 : x = -1 + t, y = 2, z = -1 + 2t$, and **(b)** between the lines $L_3 : x = 5, y = 3t, z = 4t$ and $L_4 : x = -2 + t, y = 0, z = 2 + 3t$?
♦ Type 4, distances between lines

42. If the line $L : x = t, y = 1 + 2t, z = -3 + 3t$ and the plane $3x + 2y - z = 4$ are parallel, find the distance between them. If not, find their point of intersection.
♦ Type 4, distances between lines

43. Use calculus to find the minimum distance from the point $(1, 2, 3)$ to the line $L : x = -1 + 2t, y = 2 + t, z = 2 + 3t$. (Minimize the square $f(t)$ of the distance between $(1, 2, 3)$ and $(-1 + 2t, 2 + t, 2 + 3t)$.)
♦ Type 4, distance from a point to a line in space

44. (a) Show that the line $L : x = 2t - 3, y = 4t - 2, z = 6$ is parallel to the plane $2x - y + z = 0$.
(b) Give an equation of the plane parallel to $2x - y + z = 0$ that contains the line L.
♦ Type 4, lines parallel to planes

45. Give an equation for the plane containing the line $L : x = 2t, y = 3t, z = 4t$ and the intersection of the planes $x + y + z = 0$ and $2y - z = 0$.
♦ Type 4, a plane containing given lines

46. Explain why the line with parametric equations $x = x_0 + at, y_0 + bt, z_0 + ct$ with nonzero a, b, and c can be described by the equations $\dfrac{x - x_0}{a} = \dfrac{y - y_0}{b} = \dfrac{z - z_0}{c}$. These are known as the SYMMETRIC EQUATIONS of the line.
♦ Type 4, symmetric equations of lines

47. Give equations for the locus of points $S = (x, y, z)$ equidistant from $P = (1, 0, 0), Q = (0, 1, 0)$, and $R = (0, 0, 1)$.
♦ Type 4, a line as a locus of points

48. (The vector triple product). Derive the identity $\mathbf{A} \times (\mathbf{B} \times \mathbf{C}) = (\mathbf{C} \cdot \mathbf{A})\mathbf{B} - (\mathbf{B} \cdot \mathbf{A})\mathbf{C}$ in the special case of $\mathbf{A} = \mathbf{i}$.
♦ Type 4, the vector triple product

49. Prove that the equations $\mathbf{A} \times \mathbf{B} = \mathbf{A} \times \mathbf{C}$ and $\mathbf{A} \cdot \mathbf{B} = \mathbf{A} \cdot \mathbf{C}$ with $\mathbf{A} \neq \mathbf{0}$ imply that $\mathbf{B} = \mathbf{C}$.
♦ Type 4, a proof with dot and cross products

50. Show that if adjacent sides of a parallelogram are formed by nonzero vectors $\mathbf{A}$ and $\mathbf{B}$, then the area of the parallelogram equals $\sqrt{|\mathbf{A}|^2|\mathbf{B}|^2 - (\mathbf{A} \cdot \mathbf{B})^2}$.
♦ Type 6, a proof with dot and cross products

51. (Cramer's rule) The system of three linear equations in three unknowns,

$$\begin{cases} a_1 x + b_1 y + c_1 z = d_1 \\ a_2 x + b_2 y + c_2 z = d_2 \\ a_3 x + b_3 y + c_3 z = d_3 \end{cases} \tag{31}$$

can be written

$$x\mathbf{A} + y\mathbf{B} + z\mathbf{C} = \mathbf{D} \tag{32}$$

with the vectors $\mathbf{A}, \mathbf{B}, \mathbf{C}$, and $\mathbf{D}$ written as columns:

$$\mathbf{A} = \begin{bmatrix} a_1 \\ a_2 \\ a_3 \end{bmatrix}, \mathbf{B} = \begin{bmatrix} b_1 \\ b_2 \\ b_3 \end{bmatrix}, \mathbf{C} = \begin{bmatrix} c_1 \\ c_2 \\ c_3 \end{bmatrix}, \text{ and } \mathbf{D} = \begin{bmatrix} d_1 \\ d_2 \\ d_3 \end{bmatrix}. \tag{33}$$

10.3 Lines and planes in space

(a) Suppose that $\mathbf{A}, \mathbf{B}$, and $\mathbf{C}$ are as in Figure 32, with $\mathbf{A}$ and $\mathbf{B}$ nonzero, nonparallel vectors in a horizontal plane, and $\mathbf{C}$ oriented so that $\mathbf{B} \times \mathbf{C}$ makes an acute angle with $\mathbf{A}$ and the triple product $\mathbf{A} \cdot (\mathbf{B} \times \mathbf{C})$ is positive. Also, suppose that **(32)** is satisfied with positive x, y, and z, so that $\mathbf{D}$ is formed by the diagonal of the parallelepiped in Figure 33. What is the volume of the parallelepiped in Figure 32? **(b)** Show that the height of the parallelepiped in Figure 33 equals z multiplied by the height of the parallelepiped in Figure 32. **(c)** Express the volume of the parallelepiped in Figure 34 first in terms of $\mathbf{A}, \mathbf{B}$, and $\mathbf{D}$ and then in terms of $\mathbf{A}, \mathbf{B}, \mathbf{C}$, and z.

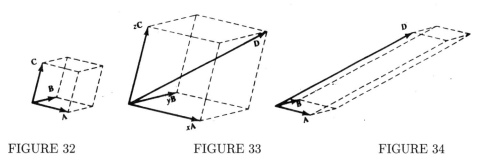

FIGURE 32 FIGURE 33 FIGURE 34

(c) Show that the relationship between the volumes of the parallelepipeds in Figures 32 and 34 yields the third of the following formulas. (The first two formulas can be derived similarly. These are known as CRAMER'S RULE and are used to solve equations **(31)**.)

$$x = \frac{\begin{vmatrix} d_1 & b_1 & c_1 \\ d_2 & b_2 & c_2 \\ d_3 & b_3 & c_3 \end{vmatrix}}{\begin{vmatrix} a_1 & b_1 & c_1 \\ a_2 & b_2 & c_2 \\ a_3 & b_3 & c_3 \end{vmatrix}}, \quad y = \frac{\begin{vmatrix} a_1 & d_1 & c_1 \\ a_2 & d_2 & c_2 \\ a_3 & d_3 & c_3 \end{vmatrix}}{\begin{vmatrix} a_1 & b_1 & c_1 \\ a_2 & b_2 & c_2 \\ a_3 & b_3 & c_3 \end{vmatrix}}, \quad z = \frac{\begin{vmatrix} a_1 & b_1 & d_1 \\ a_2 & b_2 & d_2 \\ a_3 & b_3 & d_3 \end{vmatrix}}{\begin{vmatrix} a_1 & b_1 & c_1 \\ a_2 & b_2 & c_2 \\ a_3 & b_3 & c_3 \end{vmatrix}}$$

♦ Type 7, Cramer's Rule from volumes

Section 10.4

Parametric equations of curves and velocity vectors

OVERVIEW: *In Section 10.3 we studied parametric equations of lines in xy-planes and in xyz-space. In this section we examine parametric equations of other types of curves. We define* LIMITS *and* DERIVATIVES *of vector-valued functions and use vector formulations of parametric equations to define* VELOCITY VECTORS *and* SPEED *of objects moving on curved paths. Then we give formulas for the lengths of parametric curves and discuss problems involving relative velocities of moving objects.*

Topics:

- *Parametric equations of curves*
- *Vector representations of curves*
- *Limits of vector-valued functions*
- *Derivatives of vector-valued functions*
- *Velocity vectors and speed*
- *Lengths of parametric curves*
- *Using relative velocity vectors*

Parametric equations of curves

A curve C in an xy-plane can be described by giving PARAMETRIC EQUATIONS

$$C : x = x(t), y = y(t), a \leq t \leq b$$

where $x(t)$ and $y(t)$ are functions defined in the closed interval $[a, b]$. The curve consists of all points $(x(t), y(t))$ for t in that interval. Similarly, a curve in xyz-space can be given by parametric equations

$$C : x = x(t), y = y(t), z = z(t), a \leq t \leq b.$$

In both cases the curve is considered to be ORIENTED in the direction of increasing t.

As a first example, we consider the curve

$$C : x = x(t), y = y(t), 0 \leq t \leq 3 \tag{1}$$

in an xy-plane, where $x(t)$ and $y(t)$ are the piecewise-linear functions whose graphs are shown in Figures 1 and 2.

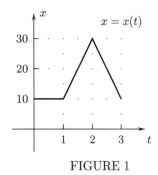

FIGURE 1

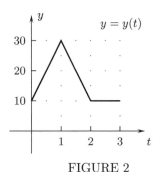

FIGURE 2

♦ SUGGESTIONS TO INSTRUCTORS: The main topics in this section can be divided into two parts. Less emphasis can be placed on the second part to save time if necessary.
Part 1 (Parametric equations, velocity vectors and speed, lengths of curves; Tune-up Exercises T1–T7, Problems 1–29) Have students begin work on Questions 1, 2, 3, 6, 7, and 8 before class. Problems 2, 8, 11, 12, 15, 18 are good for in-class discussion.
Part 2 (Reflective problems with parametric equations, deriving parametric equations of curves, and relative velocities; Tune-up Exercise T8, Problems 30–64) Have students begin work on Questions 4, 5, and 9 before class. Problems 19, 20, 23, 29, 30, 31, 32, 40, 53, and 56 are good for in-class discussion.

10.4 Parametric equations of curves and velocity vectors,

Question 1 Use the graphs in Figures 1 and 2 to complete Table 1 of values of the functions $x(t)$ and $y(t)$.

TABLE 1

t	0	1	2	3
$x = x(t)$	10			
$y = y(t)$	10			

Example 1 Draw the curve **(1)** with $x(t)$ and $y(t)$ from Figures 1 and 2.

SOLUTION The curve is the triangle in Figure 3, oriented clockwise. It consists of three line segments because $x(t)$ and $y(t)$ are linear for $0 \leq t \leq 1, 1 \leq t \leq 2$, and $2 \leq t \leq 3$. We plot the points $(x(t), y(t))$ for $t = 0, 1, 2, 3$ from the completed Table 1 and connect them with line segments. The arrows on the curve show its direction of orientation, which is from the point for $t = 0$ to the point for $t = 1$, to the point for $t = 2$, to the point for $t = 3$ (which happens to be the same point as for $t = 0$). □

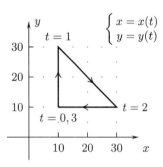

FIGURE 3

Question 2 Draw in an xy-plane the curve

$$C : x = x(t), y = y(t), 0 \leq t \leq 300 \tag{2}$$

where $x(t)$ and $y(t)$ are the piecewise-linear functions whose graphs are shown in Figures 4 and 5.

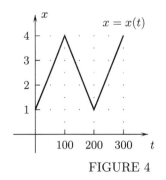

FIGURE 4

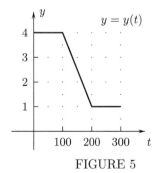

FIGURE 5

Question 3 Figures 6 and 7 show the graphs of $x(t) = t^3 - 9t$ and $y(t) = 2t^2$ for $-4 \leq t \leq 4$. **(a)** Generate the curve $C : x = t^3 - 9t$, $y = 2t^2$, $-4 \leq t \leq 4$ in the window $-36 \leq x \leq 36, -8 \leq y \leq 40$.[†] You should obtain the curve in Figure 8. What are the x- and y-coordinates of the beginning and the end of the curve? **(b)** At what two values of y does the curve intersect the y-axis?

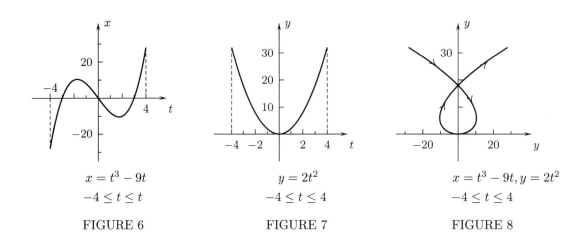

FIGURE 6   FIGURE 7   FIGURE 8

Circular motion at a constant speed

Recall from the definitions of $\sin\theta$ and $\cos\theta$ that the terminal side of an angle θ (radians) in an xy-plane intersects the circle $x^2 + y^2 = 1$ of radius 1 at the point $(\cos\theta, \sin\theta)$ (Figure 9).[‡] Consequently, the circle, oriented counterclockwise, has the parametric equations,

$$x = \cos\theta, \; y = \sin\theta, \; 0 \leq \theta \leq 2\pi \tag{3}$$

with θ as parameter. Similarly, the circle $x^2 + y^2 = r^2$ of radius r for any positive r (Figure 10) has the parametric equations

$$x = r\cos\theta, \; y = r\sin\theta, \; 0 \leq \theta \leq 2\pi. \tag{4}$$

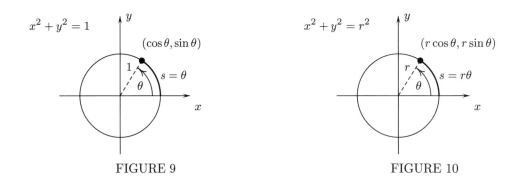

FIGURE 9   FIGURE 10

[†]To put a Texas Instruments Calculator in parametric-equation mode, press <MODE> and select <Par>, <Param>, or <Parametric> instead of <Func> or <Function>. Define $x(t)$ and $y(t)$ in the $Y=$ menu. In the window menu choose ranges of x and y and set $T\text{min} = a$ and $T\text{max} = b$ if the curve is to be generated for $a \leq t \leq b$. Tstep is the t-distance between plotted points. Making it smaller causes more points to be plotted and can improve the quality of a drawing, but can also cause the curve to be generated more slowly.

[‡]See Appendix 3 in Volume 1.

10.4 Parametric equations of curves and velocity vectors,

Because θ is measured in radians, the angle θ subtends an arc of length $s = \theta$ on the circle of radius 1 in Figure 9 and an arc of length $s = r\theta$ on the circle of radius r in Figure 10.

We will say that an object with a curved path has CONSTANT SPEED if the distance it travels is a linear function of the time, as in the next Example.

Example 2 A toy car travels at the constant speed of fifteen meters per minute counterclockwise around the circle $x^2 + y^2 = 100$ with a ten-meter radius in an xy-plane. At $t = 0$ (minutes) it is at $x = 10, y = 0$ on the positive x-axis. Give formulas for its coordinates $\bigl(x(t), y(t)\bigr)$ as functions of t.

SOLUTION Figure 11 shows the car's position when the line between it and the origin has rotated θ radians from the positive x-axis and $x = 10\cos\theta$ (meters), $y = 10\sin\theta$ (meters). Because the car has the constant speed of fifteen meters per minute and $s = 0$ at $t = 0$, the distance it travels from time 0 to time t is given by the linear function $s = 15t$ (meters). On the other hand, $s = 10\theta$, since the circle has radius 10, and therefore

$$\theta = \tfrac{1}{10}s = \tfrac{1}{10}(15t) = \tfrac{3}{2}t.$$

At time t, the car's x- and y-coordinates are $x = 10\cos(\tfrac{3}{2}t),\ y = 10\sin(\tfrac{3}{2}t)$ (Figure 12). □

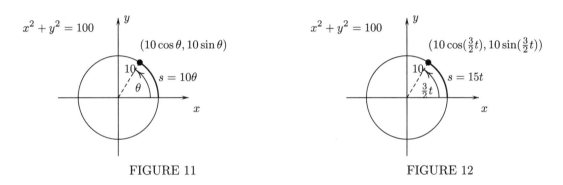

FIGURE 11 FIGURE 12

Question 4 Give formulas for the coordinates of the car of Example 2 in the case that its constant speed is 20π meters per minute.

Vector representations of curves

Parametric equations for curves can be given in vector form. For a curve in an xy-plane,

$$C : x = x(t),\ y = y(t),\ a \le t \le b$$

we let $\mathbf{R}(t)$ denote the position vector of the point $\bigl(x(t), y(t)\bigr)$ on the curve at the value t of the parameter (Figure 13) and write

$$C : \mathbf{R}(t) = x(t)\mathbf{i} + y(t)\mathbf{j},\ a \le t \le b. \tag{5}$$

Similarly, for a curve

$$C : x = x(t),\ y = y(t), z = z(t)\ a \le t \le b$$

in xyz-space, we let $\mathbf{R}(t)$ be the position vector of the point $(x(t), y(t), z(t))$ and write

$$C : \mathbf{R}(t) = x(t)\mathbf{i} + y(t)\mathbf{j} + z(t)\mathbf{k},\ a \le t \le b \tag{6}$$

(Figure 14). We refer to (5) and (6) as VECTOR-VALUED FUNCTIONS of the parameter t.

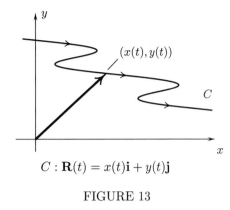

$C: \mathbf{R}(t) = x(t)\mathbf{i} + y(t)\mathbf{j}$

FIGURE 13

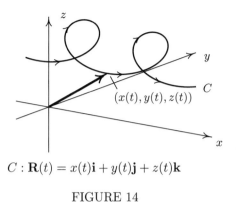

$C: \mathbf{R}(t) = x(t)\mathbf{i} + y(t)\mathbf{j} + z(t)\mathbf{k}$

FIGURE 14

We use position- and displacement-vector-valued functions in the next example.

Example 3 The point P in Figure 15 is on a circle of radius 3 whose center Q is on the circle of radius 4 with its center at the origin. Imagine that each circle is rotating at a constant rate and that, as t increases from 0 to 2π, the larger circle rotates one counterclockwise revolution and the smaller circle rotates three clockwise revolutions. Finally suppose that at $t = 0$, the point Q is at $(4,0)$ and P is at $(7,0)$ on the positive x-axis. Find parametric equations for the object's path with t as parameter and generate the path.

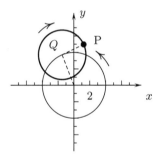

FIGURE 15

SOLUTION We let $\mathbf{R_1}(t)$ be the position vector of the center Q of the smaller circle and let $\mathbf{R_2}(t)$ be the displacement vector from the point Q to the point P, as in Figure 16. Because the larger circle has radius 4 and rotates one counterclockwise revolution as t increases from 0 to 2π and because Q is at $(4,0)$ at $t = 0$,

$$\mathbf{R_1}(t) = 4\cos t\, \mathbf{i} + 4\sin t\, \mathbf{j}. \tag{7}$$

Similarly, because the smaller circle has radius 3 and rotates clockwise three revolutions as t increases from 0 to 2π and since the point P is at $(7,0)$ at $t = 0$,

$$\mathbf{R_2}(t) = 3\cos(-3t)\, \mathbf{i} + 3\sin(-3t)\, \mathbf{j}. \tag{8}$$

The position vector of P at time t is $\mathbf{R}(t) = \mathbf{R_1}(t) + \mathbf{R_2}(t)$, so the parametric equations of its path are

$$\mathbf{R}(t) = [4\cos t + 3\cos(-3t)]\,\mathbf{i} + [4\sin t + 3\sin(-3t)]\,\mathbf{j}, 0 \leq t \leq 2\pi.$$

We generate the curve in a window with equal scales on the axes and including the square $-8 \leq x \leq 8, -8 \leq y \leq 8$.[†] The curve in Figure 17 is a HYPOTROCHOID.[1] □

[†]To obtain equal scales on a Texas Instruments TI-81, TI-82, or TI-83 calculator, choose the y-range to be approximately 1.5 multiplied by the x-range and execute the zoom-square command in the zoom menu. On a TI-85, TI-86, or TI-89 start with a window with the y-range equal to approximately twice the x-range, and on a TI-91 start with a window with the y-range equal to approximately 2.3 times the x-range and the execute the zoom-square command.

[1]*A Catalog of Special Plane Curves* by J. Lawrence, New York, NY: Dover Publications, Inc. 1972, p. 167.

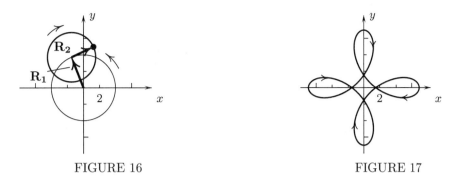

FIGURE 16   FIGURE 17

Question 5   If Example 3 is modified to have the smaller circle in Figure 15 rotate counterclockwise rather than clockwise, the curve that is generated is called an EPITROCHOID.[2] Give a vector equation for this curve and then generate it.

Limits of vector-valued functions

We say that the LIMIT of a vector-valued function $\mathbf{R}(t)$ as $t \to t_0$ is the vector $\mathbf{R_0}$ if when we place the bases of the vectors together, then the tip of $\mathbf{R}(t)$ approaches the tip of $\mathbf{R_0}$ as $t \to t_0$. This is illustrated in Figure 18 for vectors in the plane with their bases at the origin.

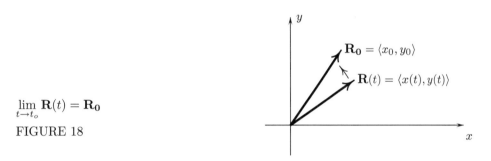

$$\lim_{t \to t_o} \mathbf{R}(t) = \mathbf{R_0}$$

FIGURE 18

Because the tip of $\mathbf{R}(t) = \langle x(t), y(t) \rangle$ is the point $(x(t), y(t))$ and the tip of $\mathbf{R_0} = \langle x_0, y_0 \rangle$ is the point (x_0, y_0), the limit of $\mathbf{R}(t)$ in Figure 18 is $\mathbf{R_0}$ if and only if $x(t) \to x_0$ and $y(t) \to y_0$ as $t \to t_0$. Similarly for vectors in space, the limit of $\mathbf{R}(t) = \langle x(t), y(t), z(t) \rangle$ is $\mathbf{R_0} = \langle x_0, y_0, z_0 \rangle$ as $t \to t_0$ if and only if $x(t) \to x_0, y(t) \to y_0$, and $z(t) \to z_0$ as $t \to t_0$. We state this principle as a Rule:

Rule 1   **(a)** For a vector-valued function $\mathbf{R}(t) = \langle x(t), y(t) \rangle$ in an xy-plane,

$$\lim_{t \to t_0} \mathbf{R}(t) = \lim_{t \to t_0} \langle x(t), y(t) \rangle = \left\langle \lim_{t \to t_0} x(t), \lim_{t \to t_0} y(t) \right\rangle$$

provided the limits of $x(t)$ and $y(t)$ exist and are finite.

(b) For a vector-valued function $\mathbf{R}(t) = \langle x(t), y(t), z(t) \rangle$ in xyz-space,

$$\lim_{t \to t_0} \mathbf{R}(t) = \lim_{t \to t_0} \langle x(t), y(t), z(t) \rangle = \left\langle \lim_{t \to t_0} x(t), \lim_{t \to t_0} y(t), \lim_{t \to t_0} z(t) \right\rangle$$

if the limits of $x(t), y(t)$, and $z(t)$ exist and are finite.

The two-sided limit in Rule 1 can be replaced by one-sided limits $t \to t_0^+$ or $t \to t_0^-$ or by limits as $t \to \infty$ or $t \to -\infty$.

---

[2] Ibid, p. 163

Example 4 Find $\lim_{t \to 1} \langle t^2 - 3, e^{3t}, \ln t \rangle$.

SOLUTION We consider the three components of the vector-valued function separately. As $t \to 1$, the first component $t^2 - 3$ tends to $1^2 - 3 = -2$, the second component e^{3t} tends to $e^{3(1)} = e^3$, and the third component $\ln t$ tends to $\ln(1) = 0$. Therefore,

$$\lim_{t \to 1} \langle t^2 - 3, e^{3t}, \ln t \rangle = \langle -2, e^3, 0 \rangle. \ \square$$

Question 6 Find $\mathbf{R_0} = \lim_{t \to 3} \mathbf{R}(t)$ with $\mathbf{R}(t) = \langle -t, t^2 - 5 \rangle$. Then draw $\mathbf{R_0}$ and $\mathbf{R}(t)$ for a value of t close to 3 with their bases at the origin in an xy-plane.

Derivatives of vector-valued functions

The derivative of a vector-valued function $\mathbf{R}(t)$ is defined by a formula similar to that used in defining the derivative

$$\frac{df}{dx}(x) = \lim_{\Delta x \to 0} \frac{f(x + \Delta x) - f(x)}{\Delta x}$$

of a real-valued function $f(x)$:

Definition 1 *For a function $\mathbf{R}(t)$ whose values are vectors in an xy-plane or in xyz-space,*

$$\frac{d\mathbf{R}}{dt}(t) = \lim_{\Delta t \to 0} \frac{\mathbf{R}(t + \Delta t) - \mathbf{R}(t)}{\Delta t}. \tag{9}$$

Suppose first that $\mathbf{R}(t) = \langle x(t), y(t) \rangle$ and that the derivatives $\frac{dx}{dt}(t)$ and $\frac{dy}{dt}(t)$ are defined. Then Definition 1 and Rule 1 give

$$\frac{d}{dt}\langle x(t), y(t) \rangle = \lim_{\Delta t \to 0} \frac{\langle x(t + \Delta t), y(t + \Delta t) \rangle - \langle x(t), y(t) \rangle}{\Delta t}$$

$$= \lim_{\Delta t \to 0} \frac{\langle x(t + \Delta t) - x(t), y(t + \Delta t) - y(t) \rangle}{\Delta t}$$

$$= \lim_{\Delta t \to 0} \left\langle \frac{x(t + \Delta t) - x(t)}{\Delta t}, \frac{y(t + \Delta t) - y(t)}{\Delta t} \right\rangle$$

$$= \left\langle \lim_{\Delta t \to 0} \frac{x(t + \Delta t) - x(t)}{\Delta t}, \lim_{\Delta t \to 0} \frac{y(t + \Delta t) - y(t)}{\Delta t} \right\rangle$$

$$= \left\langle \frac{dx}{dt}(t), \frac{dy}{dt}(t) \right\rangle.$$

Thus, the derivative of the vector-valued function is obtained by differentiating its components. The same principle applies to $\mathbf{R}(t) = \langle x(t), y(t), z(t) \rangle$, so we have the following:

Theorem 1 **(a)** *At any t where the derivatives $\frac{dx}{dt}(t)$ and $\frac{dy}{dt}(t)$ are defined,*

$$\frac{d}{dt}\langle x(t), y(t) \rangle = \left\langle \frac{dx}{dt}(t), \frac{dy}{dt}(t) \right\rangle. \tag{10}$$

(b) *At any t where the derivatives $\frac{dx}{dt}(t), \frac{dy}{dt}(t),$ and $\frac{dz}{dt}(t)$ are defined,*

$$\frac{d}{dt}\langle x(t), y(t), z(t) \rangle = \left\langle \frac{dx}{dt}(t), \frac{dy}{dt}(t), \frac{dz}{dt}(t) \right\rangle. \tag{11}$$

10.4 Parametric equations of curves and velocity vectors,

Example 5   Find the derivative, $\dfrac{d}{dt}\langle t^2 - 3,\ e^{3t},\ \ln t\rangle$.

SOLUTION   Again, we can treat the three components separately. The t-derivative of $t^2 - 3$ is $2t$; the t-derivative of e^{3t} is $e^{3t}\dfrac{d}{dt}(3t) = 3e^{3t}$ by the Chain Rule; and the t-derivative of $\ln t$ is $1/t$. Consequently, $\dfrac{d}{dt}\langle t^2 - 3,\ e^{3t},\ \ln t\rangle = \langle 2t,\ 3e^{3t},\ 1/t\rangle$. □

Question 7   What is the derivative $\mathbf{R}'(t)$ of $\mathbf{R}(t) = 2\cos t\,\mathbf{i} + 3\sin t\,\mathbf{j}$?

We say that a vector-valued function $\langle x(t), y(t)\rangle$ or $\langle x(t), y(t), z(t)\rangle$ is CONTINUOUS at a value of t or in an interval if each of its components is continuous at that value of t or in that interval. In particular, because each component is continuous at any point where it has a derivative (Theorem 1 of Section 2.4), a vector-valued function is continuous at any point where it has a derivative.

Velocity vectors and speed

Suppose that an object whose path is the curve C in Figure 19 in an xy-plane is at the point with position vector $\mathbf{R}(t)$ at time t. Then at time $t + \Delta t$ for a small positive Δt, the object is at the nearby point with position vector $\mathbf{R}(t + \Delta t)$, and the displacement vector $\mathbf{R}(t + \Delta t) - \mathbf{R}(t)$ points in the direction of the object's motion from one point to the other on the curve (Figure 20).

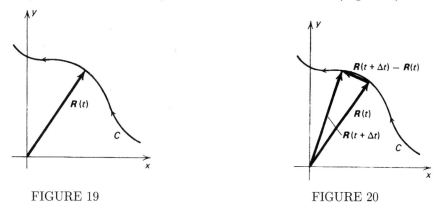

FIGURE 19   FIGURE 20

Dividing $\mathbf{R}(t + \Delta t) - \mathbf{R}(t)$ by the positive number Δt to form the difference quotient,

$$\frac{\mathbf{R}(t + \Delta t) - \mathbf{R}(t)}{\Delta t} \tag{12}$$

changes the length of the vector but not its direction. Consequently, the difference quotient (12) is a vector along the secant line through the two points that points in the direction of the object's motion (Figure 21). The vector (12) also lies along a secant line and points in the direction of the object's motion for $\Delta t < 0$ because then $\mathbf{R}(t + \Delta t) - \mathbf{R}(t)$ points in the opposite direction and dividing it by a negative number reverses its direction.

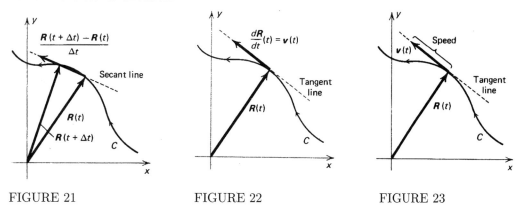

FIGURE 21   FIGURE 22   FIGURE 23

Now, suppose that the derivative $\dfrac{d\mathbf{R}}{dt}(t)$ exists. Then $\mathbf{R}(t)$ is continuous at t and the tip of $\mathbf{R}(t + \Delta t)$ approaches the tip of $\mathbf{R}(t)$ as $\Delta t \to 0$. Moreover, if the derivative is not the zero vector, then the secant line approaches a line through the tip of $\mathbf{R}(t)$ parallel to $\dfrac{d\mathbf{R}}{dt}(t)$. We define the line to be the tangent line to the curve and refer to $\dfrac{d\mathbf{R}}{dt}(t)$ as the VELOCITY VECTOR $\mathbf{v}(t)$ of the moving object at that point (Figure 22) and we define the length of the velocity vector to be the object's SPEED (Figure 23). The same reasoning can be applied to vector-valued functions $\mathbf{R}(t)$ in xyz-space and leads to the next Definition.

Definition 2 (a) *If an object's position vector at time t is $\mathbf{R}(t)$, then its* VELOCITY VECTOR *at time t is the derivative,*

$$\mathbf{v}(t) = \frac{d\mathbf{R}}{dt}(t) \tag{13}$$

and the length of this vector is the object's SPEED *at time t:*

$$[\text{Speed}] = |\mathbf{v}(t)| = \left|\frac{d\mathbf{R}}{dt}(t)\right|. \tag{14}$$

(b) *If the velocity vector* (13) *is not* $\mathbf{0}$, *then the* TANGENT LINE *to the curve $C : \mathbf{R} = \mathbf{R}(t)$ at the point with position vector $\mathbf{R}(t)$ is the line through the point and parallel to the velocity vector.*

We also use the terms "velocity vector" and "speed," as defined in (13) and (14), when t is a parameter that is not the time.

Notice that even though the terms "velocity" and "speed" are generally interchangeable in everyday English, they are given different meanings in calculus. In Chapter 3 we defined the velocity of an object moving on an s-axis to be the derivative $v(t) = \dfrac{ds}{dt}(t)$ of its position $s(t)$ on the axis at time t. The velocity could be a positive number, a negative number, or zero, depending on whether the object was moving in the positive or negative direction on the s-axis. And, in (13) we define velocity as a vector, whose direction is the direction of an object's motion. In contrast, an object's speed, as defined in (14), is always a nonnegative number. It indicates how fast an object is moving but not its direction.

To explain definition (14) of speed, we consider the length

$$\left|\frac{\mathbf{R}(t + \Delta t) - \mathbf{R}(t)}{\Delta t}\right| = \frac{|\mathbf{R}(t + \Delta t) - \mathbf{R}(t)|}{|\Delta t|} \tag{15}$$

of the difference quotient in (12). As is shown in Figure 20, the vector $\mathbf{R}(t+\Delta t) - \mathbf{R}(t)$ runs from the object's position at time t to its position at time $t + \Delta t$, so that the numerator $|\mathbf{R}(t+\Delta t) - \mathbf{R}(t)|$ on the right of (15) is the net distance the object travels between times t and $t+\Delta t$, and the denominator is the time $|\Delta t|$ taken. The length $|\mathbf{v}(t)|$ of the velocity vector is the limit, as $\Delta t \to 0$, of (15). We define this to be the object's speed because it is the limit, as $\Delta t \to 0$, of the net distance the object travels divided by the time taken.

Example 6 An ant has the position vector $\mathbf{R}(t) = t^3 \mathbf{i} + t^2 \mathbf{j}$ (meters) in an xy-plane at time t (minutes) for $-1 \leq t \leq 1$. **(a)** Give an equation in x and y for its path. **(b)** Find its velocity vector and speed at $t = \tfrac{1}{2}$ and draw the ant's path with that velocity vector.

SOLUTION **(a)** The ant's coordinates at time t are $x = t^3$ and $y = t^2$. Solving the first equation for $t = x^{1/3}$ and substituting the result in the second gives the equation $y = x^{2/3}$ for its path in terms of x and y.

10.4 Parametric equations of curves and velocity vectors,

(b) The ant's velocity vector is

$$\mathbf{v}(t) = \frac{d}{dt}(t^3)\mathbf{i} + \frac{d}{dt}(t^2)\mathbf{j} = 3t^2\mathbf{i} + 2t\mathbf{j} \text{ meters per minute}$$

which equals $3(\frac{1}{2})^2\mathbf{i} + 2(\frac{1}{2})\mathbf{j} = \frac{3}{4}\mathbf{i} + \mathbf{j}$ at $t = \frac{1}{2}$. Its speed at that time is $|\frac{3}{4}\mathbf{i} + \mathbf{j}| = \sqrt{(\frac{3}{4})^2 + 1^2} = \frac{5}{4}$ meters per minute. The curve and velocity vector are shown in Figure 24. □

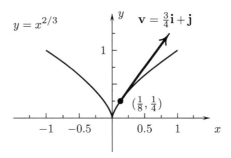

FIGURE 24

Motion on an ellipse

As we saw above, the circle of radius a in Figure 25 has the parametric equations $x = a\cos t, y = a\sin t$, $0 \leq t \leq 2\pi$. If we multiply all of the y-coordinates of points on the circle by the number b/a with $0 < b < a$, the circle is compressed in the y-direction and we obtain the ellipse in Figure 26.[†] This process with $b > a$ would stretch the circle the y-direction, resulting in an ellipse that is taller than wide. In either case, the ellipse has the parametric equations,

$$x = a\cos t, y = b\sin t, 0 \leq t \leq 2\pi \tag{16}$$

and the equation $\dfrac{x^2}{a^2} + \dfrac{y^2}{b^2} = 1$ in terms of x and y. The last equation follows from **(16)** because $\dfrac{(a\cos t)^2}{a^2} + \dfrac{(b\sin t)^2}{b^2} = \cos^2 t + \sin^2 t = 1.$

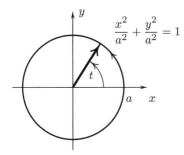

$x = a\cos t, y = a\sin t$

FIGURE 25

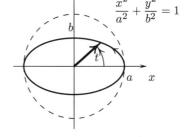

$x = a\cos t, y = b\sin t$

FIGURE 26

[†]Notice that because of the compression in the y-direction that is used to transform the circle in Figure 25 to the ellipse in Figure 26, t is the radian measure of the angle in the circle Figure 25 but is greater than the radian measure of the angle corresponding to the value t of the parameter for the ellipse in Figure 26.

Example 7 (a) Find the velocity vector to the ellipse $C: x = 5\cos t, y = 3\sin t, 0 \le t \le 2\pi$ at $t = \frac{1}{4}\pi$. Then draw the ellipse and the velocity vector. (b) Give parametric equations for the tangent line to the ellipse at $t = \frac{1}{4}\pi$. (c) Give an equation in terms of x and y for the tangent line from part (b).

SOLUTION (a) Set $\mathbf{R}(t) = \langle 5\cos t, 3\sin t\rangle$. The velocity vector at t is

$$\mathbf{v}(t) = \frac{d}{dt}\langle 5\cos t, 3\sin t\rangle = \langle -5\sin t, 3\cos t\rangle \tag{17}$$

and, since $\sin(\frac{1}{4}\pi)$ and $\cos(\frac{1}{4}\pi)$ are both equal to $1/\sqrt{2}$,

$$\mathbf{v}(\tfrac{1}{4}\pi) = \langle -5\sin(\tfrac{1}{4}\pi), 3\cos(\tfrac{1}{4}\pi)\rangle = \frac{\langle -5, 3\rangle}{\sqrt{2}}. \tag{18}$$

The ellipse and velocity vector $\mathbf{v}(\frac{1}{4}\pi)$ are shown in Figure 27. The base of the velocity vector is at the tip of $\mathbf{R}(\frac{1}{4}\pi) = \dfrac{\langle 5,3\rangle}{\sqrt{2}} \doteq \langle 3.54, 2.12\rangle$ and its tip is at the tip of

$$\mathbf{R}(\tfrac{1}{4}\pi) + \mathbf{v}(\tfrac{1}{4}\pi) = \frac{\langle 5,3\rangle}{\sqrt{2}} + \frac{\langle -5,3\rangle}{\sqrt{2}} = \frac{\langle 0,6\rangle}{\sqrt{2}} \doteq \langle 0, 4.24\rangle$$

on the y-axis.

(b) The tangent line passes through the point with coordinates $x = 5/\sqrt{2}, y = 3/\sqrt{2}$ at the tip of $\mathbf{R}(\frac{1}{4}\pi)$ and is parallel to $\mathbf{v}(\frac{1}{4}\pi)$, which is given in (18). The tangent line therefore has the parametric equations,

$$x = \frac{5}{\sqrt{2}} - \left(\frac{5}{\sqrt{2}}\right)t,\ y = \frac{3}{\sqrt{2}} + \left(\frac{3}{\sqrt{2}}\right)t. \ \square$$

(c) The slope of the line segment from the base of $\mathbf{v}(\frac{1}{4}\pi) = \dfrac{\langle -5,3\rangle}{\sqrt{2}}$ to its tip on the tangent line is $\dfrac{3/\sqrt{2}}{-5/\sqrt{2}} = -\frac{3}{5}$ and the tangent line passes through the point $(5/\sqrt{2}, 3/\sqrt{2})$ at the tip of $\mathbf{R}(\frac{1}{4}\pi)$ so it has the equation

$$y = \frac{3}{\sqrt{2}} - \frac{3}{5}\left(x - \frac{5}{\sqrt{2}}\right). \ \square$$

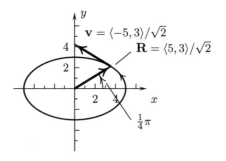

FIGURE 27

Question 8 Use geometric reasoning (a) to find the value of t where the velocity vector $\mathbf{v}(t)$ to the ellipse in Example 7 equals $-\mathbf{v}(\frac{1}{4}\pi)$ and (b) to find the two values of t where $\mathbf{v}(t)$ is vertical. (c) Draw the ellipse and the three velocity vectors from parts (a) and (b).

10.4 Parametric equations of curves and velocity vectors,

Lengths of curves

The formula,

$$[\text{Length}] = \int_a^b \sqrt{1 + \left[\frac{df}{dx}(x)\right]^2}\, dx \qquad (19)$$

from Section 4.6 for the length of the graph $y = f(x), a \leq x \leq b$ of a function $f(x)$ with a continuous derivative in the interval $[a, b]$, has the following generalization for curves given by parametric equations.

Definition 3 *Suppose that a curve is given by $C : \mathbf{R} = \mathbf{R}(t), a \leq t \leq b$, where the vector-valued function $\mathbf{R}(t)$ is continuous in the finite closed interval $[a, b]$, its derivative $\mathbf{v}(t) = \dfrac{d\mathbf{R}}{dt}(t)$ exists at all except possibly a finite number of points in the interval, and the integral of $\left|\dfrac{d\mathbf{R}}{dt}(t)\right|$ from a to b is defined. Then the length of the curve is equal to that integral:*

$$[\text{Length}] = \int_a^b |\mathbf{v}(t)|\, dt = \int_a^b \left|\frac{d\mathbf{R}}{dt}(t)\right| dt. \qquad (20)$$

In the case of a curve $C : \mathbf{R}(t) = \langle x(t), y(t) \rangle$ in an xy-plane, formula **(20)** gives

$$[\text{Length}] = \int_a^b \sqrt{\left[\frac{dx}{dt}(t)\right]^2 + \left[\frac{dy}{dt}(t)\right]^2}\, dt \qquad (21)$$

and for a curve $C : \mathbf{R}(t) = \langle x(t), y(t), z(t) \rangle$ in space,

$$[\text{Length}] = \int_a^b \sqrt{\left[\frac{dx}{dt}(t)\right]^2 + \left[\frac{dy}{dt}(t)\right]^2 + \left[\frac{dz}{dt}(t)\right]^2}\, dt. \qquad (22)$$

Formula **(19)** is the special case of **(21)** with $x(t) = t, y(t) = f(t)$.

We are led to **(21)** by using arbitrary partition $a = t_0 < t_1 < t_2 < \cdots < t_N = b$ of $[a, b]$ that includes any points where $x(t)$ and $y(t)$ do not have derivatives. We approximate the curve by the polygonal line through the successive points $(x(t_j), y(t_j))$ on it. The length of the approximation is

$$\sum_{j=1}^{N} \sqrt{[x(t_j) - x(t_{j-1})]^2 + [y(t_j) - y(t_{j-1})]^2}$$

which, by the Mean-Value Theorem of Section 3.1 applied to each of the functions in each subinterval equals

$$\sum_{j=1}^{N} \sqrt{\left[\frac{dx}{dt}(c_j)\right]^2 + \left[\frac{dy}{dt}(d_j)\right]^2}\, \Delta t_j \qquad (23)$$

where c_j and d_j are points in the jth subinterval and Δt_j is its length. We take the integral **(21)** to be the length of the original curve because it is the limit of the sums **(23)** as the number of subintervals in the partition tends to ∞ and their widths tend to zero. Formula **(22)** can be derived by including the third component.

Example 8   Give a definite integral that equals the exact length of the curve $C: x = t^3 - 9t$, $y = 2t^2, -4 \leq t \leq 4$. Then use a calculator or computer algorithm to find its approximate decimal value.

SOLUTION   Here $x(t) = t^3 - 9t$ and $y(t) = 2t^2$, so that $\dfrac{dx}{dt} = 3t^2 - 9$, $\dfrac{dy}{dt} = 4t$ and, by **(21)** the length of the curve is

$$\int_{-4}^{4} \sqrt{(3t^2 - 9)^2 + (4t)^2}\, dt \doteq 121.92. \ \square$$

Formula **(20)** for the length of a curve provides another reason for defining the length of an object's velocity vector to be its speed. Suppose that the object has position vector $\mathbf{R}(t)$ at time t for $a \leq t \leq b$ and that its velocity vector $\mathbf{v}(t)$ is continuous in $[a,b]$. Then the distance the object travels from time a to time T is the length,

$$s(T) = \int_{a}^{T} |\mathbf{v}(t)|\, dt$$

of its path for $a \leq t \leq T$. The Fundamental Theorem of Calculus for derivatives of integrals (Theorem 1 of Section 6.3) gives, for $a < T < b$,

$$\frac{ds}{dT}(T) = \frac{d}{dT}\int_{a}^{T} |\mathbf{v}(t)|\, dt = |\mathbf{v}(T)|.$$

Thus, the object's speed $|\mathbf{v}(T)|$ is the rate of change with respect to time of the distance it has traveled.

Using relative velocities

Just as we can use displacement vectors to show relative positions of points, we can use relative velocity vectors to study relative motion.

The drawing in Figure 28 represents a cruiseship that is sailing straight ahead with the constant velocity of five feet per second and a passenger at point P who is walking across the ship with constant velocity of two feet per second, relative to the ship. We suppose there is no current, so the water is stationary.

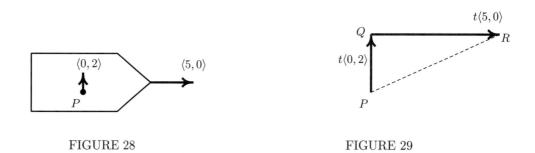

FIGURE 28   FIGURE 29

Question 9   The passenger's displacement vector relative to the ship after t seconds is $t\langle 0, 2\rangle$ feet from P to Q in Figure 29, while, relative to the water, the displacement vector of the point Q on the ship is $t\langle 5, 0\rangle$ feet to point R, so that, relative to the water, the passenger is at point R at time t. **(a)** What is the displacement vector from the passenger's position at time 0, relative to the water, to his position at time t, relative to the water? **(b)** What is the passenger's velocity relative to the water? **(c)** What is his speed relative to the water?

10.4 Parametric equations of curves and velocity vectors

Question 9 illustrates, in a case with constant velocities, a general principle concerning relative velocities: Any equation relating position vectors and relative position vectors of fixed or moving points as the time t varies can be differentiated with respect to t to obtain a relationship between the corresponding velocities and relative velocities. For instance, if points P and Q are moving in an xy-plane or xyz-space and O is the origin, then at any time

$$\overrightarrow{OQ} = \overrightarrow{OP} + \overrightarrow{PQ}$$

and differentiating this equation with respect to the time t yields

$$\frac{d}{dt}(\overrightarrow{OQ}) = \frac{d}{dt}(\overrightarrow{OP}) + \frac{d}{dt}(\overrightarrow{PQ}).$$

This equation states that the vector velocity of Q equals the vector velocity of P plus the vector velocity of Q relative to P. We use similar reasoning in the next example.

Example 9   A car at point P in Figure 30 is traveling toward the east on one road, while another car is traveling north-east on another road. Imagine that the roads are in an xy-plane with the usual orientation, so the positive y-axis is pointing north. At one moment the first car's vector velocity is $\mathbf{v}_1 = \langle 60, 0 \rangle$ miles per hour and the second car's vector velocity is $\mathbf{v}_2 = \langle 20\sqrt{2}, 20\sqrt{2} \rangle$ miles per hour. **(a)** What are the cars' speeds at that moment? **(b)** Based on Figure 30, does the vector velocity of the car at Q relative to the car at P at that moment have a positive or a negative x-component? Does it have a positive or a negative y-component? **(c)** What is the vector velocity of the car at Q relative to the car at P?

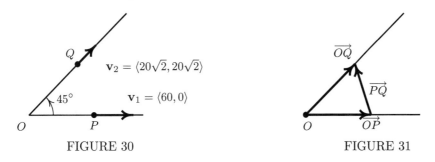

FIGURE 30   FIGURE 31

SOLUTION   **(a)** The speed of the car at P is $|\mathbf{v}_1| = |\langle 60, 0 \rangle| = 60$ miles per hour, and the speed of the car at Q is $|\mathbf{v}_2| = |\langle 20\sqrt{2}, 20\sqrt{2} \rangle| = \sqrt{(20\sqrt{2})^2 + (20\sqrt{2})^2} = 40$ miles per hour.
(b) The vector velocity of the car at Q relative to the car at P has a negative x-component because the car at Q is moving more slowly to the right than is the car at P. It has a positive y-component because the car at Q is moving up and the car at P is moving to the right.
(c) The velocity of the car at Q relative to the car at P is the t-derivative of the displacement vector $\overrightarrow{PQ} = \overrightarrow{OQ} - \overrightarrow{OP}$ in Figure 31. Since the t-derivative of the position vector $\overrightarrow{OP}$ of the car at P is $\mathbf{v}_1$ and the t-derivative of the position vector $\overrightarrow{OQ}$ of the car at Q is $\mathbf{v}_2$, the velocity of the car at Q relative to the car at P is

$$\frac{d}{dt}(\overrightarrow{PQ}) = \frac{d}{dt}(\overrightarrow{OQ}) - \frac{d}{dt}(\overrightarrow{OP}) = \mathbf{v}_2 - \mathbf{v}_1$$
$$= \langle 20\sqrt{2}, 20\sqrt{2} \rangle - \langle 60, 0 \rangle$$
$$= \langle 20\sqrt{2} - 60, 20\sqrt{2} \rangle \doteq \langle -31.7, 28.3 \rangle \text{ miles per hour. } \square$$

Principles and procedures

- Whenever you are comparing graphs of $x(t)$ and $y(t)$ in tx- and ty-planes with the parametric curve $x = x(t), y = y(t)$ in an xy-plane, notice that the point at t on the parametric curve moves to the right when $x(t)$ is increasing, moves to the left when $x(t)$ is decreasing, moves up when $y(t)$ is increasing, and moves down when $y(t)$ is decreasing.

- The simplest curved path is a circle. Accordingly, examples of motion on circles are useful for illustrating motion in curved paths, as in Example 2.

- It is easy to understand limits, continuity, and derivatives of vector-valued functions because you just deal with each component separately. The associated geometric concepts, on the other hand, involve genuinely new ideas. Be sure you understand them also. In particular, be sure you understand why a nonzero vector velocity, defined as the t-derivative of a position vector, is parallel to the tangent line to the object's path and points in the direction of the object's motion and why the length of the vector velocity is defined to be the object's speed.

- Be sure you understand the distinction between (1) the scalar velocity of an object moving on an s-axis as studied in earlier chapters, (2) the vector velocity of an object moving in an xy-plane or in xyz-space, and (3) the speed of an object moving in an xy-plane or in xyz-space.

- If a curve in an xy-plane has the nonzero velocity vector $\mathbf{v} = \langle v_1, v_2 \rangle$ at a point P and v_1 is not zero, then the tangent line at P has slope v_2/v_1 and $\phi = \tan^{-1}(v_2/v_1)$ is the angle of inclination of the tangent line with $-\frac{1}{2}\pi < \phi < \frac{1}{2}\pi$. If v_1 is zero, the tangent line is vertical.

- You can think of displacement vectors as relative position vectors. They are used in deriving parametric equations of certain curves, as in Example 3, and to study relative velocities, as in Question 9 and Example 9.

Tune-Up Exercises 10.4♦

A Answer provided. **O** Outline of solution provided.

T1.^O (a) An object is at $x = x(t), y = y(t)$ at time t for $0 \leq t \leq 3$, where $x(t)$ and $y(t)$ are the piecewise-linear functions whose graphs are given in Figures 32 and 33. What are the object's velocity vector and speed for $0 < t < 1$, for $1 < t < 2$, and for $2 < t < 3$? (b) Draw its path in an xy-plane with equal scales on the axes with the velocity vectors on the curve at $t = 0.25, t = 1.25,$ and $t = 2.25$, using the scale in Figure 34 to measure the components.

♦ Type 1, drawing a parametric curve

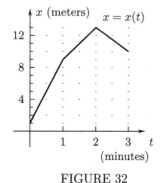

FIGURE 32

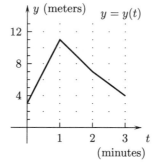

FIGURE 33

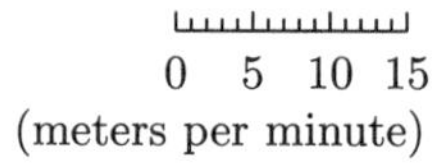

FIGURE 34

♦The Tune-up Exercises and Problems are classified by type and content. The types are (1) basic, reactive; (2) basic reflective; (3) intermediate, reactive; (4) intermediate, reflective; (5) advanced, reactive; (6) advanced, reflective; and (7) advanced, theoretical.

10.4 Parametric equations of curves and velocity vectors,

T2.⁰ What is $\lim_{t \to 1} \langle 3t^2, \ln t, \cos(\pi t) \rangle$?

♦ Type 1, a limit of a vector

T3.⁰ What is $\dfrac{d}{dt}(te^t \mathbf{i} - \tan^{-1} t \, \mathbf{j})$?

♦ Type 1, a derivative of a vector

T4.⁰ (a) Find the velocity vectors at $t = -5$ and at $t = 5$ for the curve $C: x = \frac{1}{6}t^2, y = \sin t, -7 \le t \le 7$.
(b) Generate the curve from part (a) in the window $-1 \le t \le 8, -1.5 \le y \le 1.5$, copy it on your paper, and draw the velocity vectors from part (a), using the scale on the x-axis to measure their x-components and the scale on the y-axis to measure their y-components.

♦ Type 1, velocity vectors

T5.⁰ An ice skater has position vector $\mathbf{R}(t) = \langle x(t), y(t) \rangle$ (meters) in an xy-plane at time t (minutes) for $0 \le t \le 6$, where $x(t)$ and $y(t)$ are the differentiable functions whose graphs are shown in Figures 35 and 36. (a) Draw the skater's approximate path by first plotting the points at $t = 0, 1, 2, 3, \ldots, 6$. (b) Use approximate tangent lines to the curves in Figures 32 and 33 to estimate the skater's vector velocity at $t = 2$ and draw it with the path, using the scales on the axes to measure the components of the vector.

♦ Type 1, an approximate velocity vector

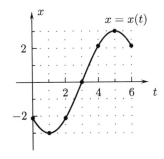
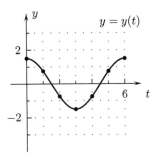

FIGURE 35   FIGURE 36

T6.⁰ An airplane is at $x = 300t, y = -400t, z = 3$ (miles) at time t (hours) in xyz-space with the positive z-axis pointing up and the origin on the ground. What are its vector velocity and speed in terms of t?

♦ Type 1, velocity and speed for a curve in space

T7.⁰ Give a definite integral that equals the length of the ellipse $x = 3\cos t, y = 7\sin t, 0 \le t \le 2\pi$. Then use a calculator- or computer-algorithm to find the approximate value of the integral.

♦ Type 1, length of a curve

T8.⁰ If P, Q, and R are points moving in xyz-space such that at a particular moment the vector velocity of P is $\langle 7, 3, 1 \rangle$ yards per minute, the vector velocity of Q relative to P is $\langle 2, 1, 3 \rangle$ yards per minute, and the vector velocity of R relative to Q is $\langle 1, 1, 1 \rangle$ yards per minute, what is the vector velocity of R at that moment?

♦ Type 3, relative velocity

Problems 10.4

A Answer provided. **O** Outline of solution provided.

1.**A** Draw the curve $C : x = x(t), y = y(t), 1 \leq t \leq 8$, where $x(t)$ and $y(t)$ are the piecewise-linear functions whose graphs are given in Figures 37 and 38.

♦ Type 1, drawing a parametrized curve

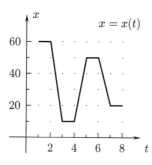 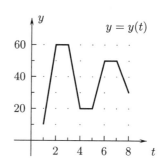

FIGURE 37   FIGURE 38

2. Draw the curve $C : x = x(t), y = y(t), -2 \leq t \leq 2$, where $x(t)$ and $y(t)$ are the piecewise-linear functions whose graphs are given in Figures 39 and 40.

♦ Type 1, drawing a parametrized curve

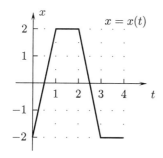 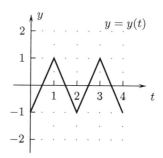

FIGURE 39   FIGURE 40

3.**A** Draw the curve $C : x = x(t), y = y(t), 1 \leq t \leq 8$, where $x(t)$ and $y(t)$ are the continuous functions whose graphs are given in Figures 41 and 42. (Plot the points at $t = 0, \pm 1, \pm 2, \pm 3, \pm 4$, and ± 5.)

♦ Type 1, drawing a parametrized curve

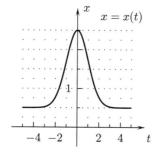 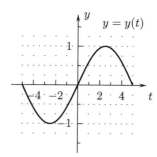

FIGURE 41   FIGURE 42

10.4 Parametric equations of curves and velocity vectors,

4. Draw the curve $C : x = x(t), y = y(t), 1 \leq t \leq 8$, where $x(t)$ and $y(t)$ are the differentiable functions whose graphs are given in Figures 43 and 44. (Plot the points at $t = 0, 1, 2, 3, 4, 5, 6$.) Then draw on your sketch of C the approximate velocity vectors at $t = 2$ and $t = 4$.

♦ Type 1, drawing a parametrized curve

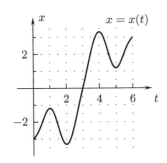

FIGURE 43   FIGURE 44

5. Find (a^A) $\lim\limits_{t \to \pi} \langle 6\cos(\tfrac{1}{3}t), 4\sin(\tfrac{1}{4}t) \rangle$, (b^A) $\lim\limits_{t \to 0^+} \langle \sqrt{t}, \sin(5t^3), 3t + 5 \rangle$, and (c) $\lim\limits_{t \to e} [(\ln t)\mathbf{i} + (t^3)\mathbf{j}]$.

♦ Type 1, limits of vectors

6. Find the derivatives, (a^A) $\dfrac{d}{dt}\langle t\sin t, e^{t^2} \rangle$, (b^A) $\dfrac{d}{dt}[(\sqrt{t^2 + 4})\mathbf{i} + (e^{2t})\mathbf{j} - \mathbf{k}]$, and

(c) $\left[\dfrac{d}{dt}\langle t, t^2, t^3 \rangle\right]_{t=4}$.

♦ Type 1, derivatives of vectors

7. What are the velocity vectors for the position vectors (a^A) $\mathbf{R}(t) = \langle \ln(2t + 1), \tan^{-1} t, \sin^{-1} t \rangle$, (b^A) $\mathbf{R}(t) = \langle \sec t, \csc t \rangle$, and (c) $\mathbf{R}(t) = \sin(3t)\mathbf{i} - \cos(4t)\mathbf{j}$?

♦ Type 1, velocity vectors

8.^{A} Generate $C : x = 6t - \tfrac{1}{2}t^3, y = \tfrac{1}{4}t^4 - \tfrac{13}{4}t^2 + 9, -3 \leq t \leq 3$ in the window $-12 \leq x \leq 12, -4 \leq y \leq 12$ and copy it on your paper. Then find the corresponding velocity vectors at $t = -2, t = 0$, and $t = 1$ and put them in your sketch, using the scales on the axes to measure their components.

♦ Type 1, velocity vectors

9. Generate $C : x = 2\cos t, y = \sin(3t), 0 \leq t \leq 2\pi$ in the window $-2.5 \leq x \leq 2.5, -3.2 \leq y \leq 1.8$ and copy it on your paper. Then find the corresponding velocity vectors at $t = \tfrac{1}{3}\pi, t = \pi$, and $t = \tfrac{7}{4}\pi$ and put them in your sketch, using the scales on the axes to measure their components.

♦ Type 1, velocity vectors

10. Generate $C : x = t + \cos(\pi t), y = \sin(\pi t), 0 \leq t \leq 2\pi$ in the window $-0.5 \leq x \leq 5.5, -2 \leq y \leq 2$ and copy it on your paper. Then find the values of t where the tangent line to the curve is horizontal.

♦ Type 2, velocity vectors

11.^{A} (a) At time t (minutes) for $0 \leq t \leq 16$ a snail is at $x = \sqrt{t}, y = \sin(\pi\sqrt{t})$ in an xy-plane with distances measured in feet. Generate its path in the window $-0.4 \leq x \leq 4.4, -2.1 \leq y \leq 1.5$ and copy it on your paper. (b) What are the snail's vector velocities at $t = 1$ and $t = 9$? Draw them in your sketch, using the scales on the axes to measure their components. (c) How much faster is the snail traveling at $t = 1$ than at $t = 9$?

♦ Type 2, velocity vectors and speed

12. An object's position vector in an xy-plane is $\mathbf{R}(t) = \langle e^t, t \rangle$ (feet) at time t (seconds) for $-2 \leq t \leq 2$. (a) Generate its path in the window $-1 \leq x \leq 8, -3.5 \leq y \leq 2.5$ and copy it on your paper. (b) The graph is a portion of the graph $y = f(x)$ of a function of x. What is the function? (c) Find the object's velocity vector at $t = 1$ and draw it in your sketch, using the scales on the axes to measure its components. (d) When is the object's speed $\sqrt{e^6 + 1}$ feet per second?

♦ Type 2, velocity vectors and speed

13.^A (a) Generate the curve $\mathbf{R}(t) = \langle 4 + 4\cos(-t), -3 + 3\sin(-t)\rangle$ for $0 \leq t \leq 2\pi$ in a window with equal scales on the axes and copy it on your paper. (b) What type of curve is it and in which direction is it oriented? (c) What are its x- and y-intercepts and at what values of t do they occur?

♦ Type 2, ellipse and its intercepts

14.º Give an equation in x and y for the tangent line to the ellipse $C: x = 3\cos t, y = 2\sin t$ at $t = \frac{1}{3}\pi$. Then draw the ellipse and the tangent line.

♦ Type 2, tangent line

15. An object is at $(x(t), y(t), z(t))$ in xyz-space at time t, where $x(t), y(t),$ and $z(t)$ are the differentiable functions whose graphs are in Figures 45 through 47. What is the object's approximate speed at $t = 1$?

♦ Type 2, estimating speed on a curve in space

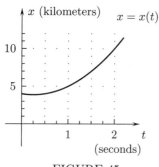
FIGURE 45

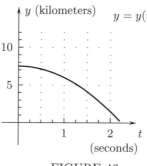
FIGURE 46

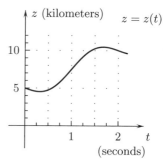
FIGURE 47

16.^A Find the exact length of the curve $x = \frac{1}{3}t^3, y = \frac{1}{2}t^2, 0 \leq t \leq 2$.

♦ Type 3, length of a curve

17. Use your computer or calculator to find the approximate lengths (a^O) of $x = t^4 - t^2, y = t^5 - t^3$, $-2 \leq t \leq 2$, (b^A) of $x = \sin t, y = te^t, z = t^3 - 3t, 0 \leq t \leq 10$, (c) of $x = t^2, y = t^3, z = t - 2\cos t$, $-5\pi \leq t \leq 5\pi$, and (d) $x = t^{-1}, y = t^{-2}, z = t^{-3}, 1 \leq t \leq 10$.

♦ Type 3, approximate lengths of curves

18. Figure 48 shows the BOWDITCH CURVE $x = \cos(\frac{11}{12}t), y = \sin t, 0 \leq t \leq 24\pi$. Use a computer or calculator algorithm to find its approximate length.

♦ Type 3, approximate length of a curve

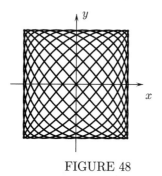
FIGURE 48

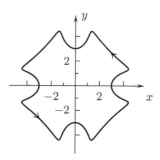
FIGURE 49

19. Use your computer or calculator to find the approximate coordinates of the points on the curve $x = \cos t[4 - \cos(6t)], y = \sin t[4 + \cos(6t)]$ in Figure 49 that are the farthest to the right.

♦ Type 2, finding approximate extreme points

10.4 Parametric equations of curves and velocity vectors,

20. (a) Show that an object that is at $x = t^n, y = -t^{-n}$ at time $t > 0$ with a positive constant n moves from left to right on the curve $y = -x^{-1}$ in Figure 50. (b) Find the value of n such that the object's vector velocity at $t = 1$ is $\langle 2, 2 \rangle$, as shown in the drawing.

♦ Type 2, finding a parameter from the velocity

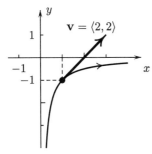

FIGURE 50

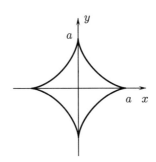

FIGURE 51

21. (a) Show that the ASTROID $x = a\cos^3 t, y = a\sin^3 t, 0 \leq t \leq 2\pi$ in Figure 51 has the equation $x^{2/3} + y^{2/3} = a^{2/3}$. (b) Find a formula for its exact length in terms of the positive parameter a.

♦ Type 2, using the Pythagorean Theorem and finding a length with a parameter

22.[A] An object is at $(\cos t, \sin t, \frac{1}{3}t)$ in xyz-space at time t seconds. Distances are measured in meters. (a) Explain why its path is a HELIX (spiral). (b) Give formulas for its vector velocity and speed in terms of t. (c) What is the exact total distance it travels for $0 \leq t \leq 6\pi$?

♦ Type 4, tangent lines and speed

23. (a) Generate TSCHIRNHAUSEN'S CUBIC $x = t^3 - 3, y = \frac{1}{3}t^3 - t$ in a window with equal scales on the axes that includes the square $-4 \leq x \leq 4, -4 \leq y \leq 4$ and copy it on your paper. (b) Find the exact coordinates of the lowest point on the curve with $x < 0$.

♦ Type 4, finding the lowest point on a curve

24.[A] An object's coordinates in an xy-plane with distances measured in meters are $x = 2\sec t, y = 2\csc t$ for $0 < t < \frac{1}{2}\pi$. (a) Show that it moves on the CROSS CURVE $\dfrac{4}{x^2} + \dfrac{4}{y^2} = 1$ in Figure 52. (b) Where is it at $t = -\frac{1}{4}\pi$? (c) What is its speed then?

♦ Type 2, using the Pythagorean Theorem and speed

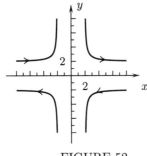

FIGURE 52

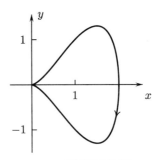

FIGURE 53

25.[A] At time t an object is at $x = 1 + \sin t, y = \cos t(1 + \sin t)$ on a PIRIFORM[4] in Figure 53. (a) Give equations in x and y for the tangent lines to the path at $(1, 1)$ and at $(1, -1)$. (b) What is the object's speed when it is at the origin?

♦ Type 2, tangent lines and speed in the plane

---

[4] First studied by De Lonchamps in 1886

26. Find equations in x and y (**a^O**) for the tangent line to $x = t^3 - t, y = t^2 - 4t^4$ at $t = 1$, (**b^A**) for the tangent line to $x = e^{2t}, y = t^2$ at $t = 0$, (**c^A**) $x = \sin(3t), y = -2\cos t$ at $t = \frac{1}{4}\pi$, and (**d**) for the tangent line to the ellipse $x = 6\cos t, y = 4\sin t$ at $t = \frac{1}{4}\pi$.

♦ Type 2, tangent line in the plane

27. Give parametric equations (**a^A**) for the tangent line to $x = \sin t, y = \sin(2t), z = \sin(3t)$ at $t = 0$ and (**b**) for the tangent line to $x = e^{5t}, y = e^{6t}, z = e^{7t}$ at $t = 0$.

♦ Type 4, parametric equations of tangent line in space

28.^{A} An mouse is at $x = \ln t, y = \frac{1}{4}e^t, 0.1 \le t \le 3$ at time t (seconds) in an xy-plane with distances measured in yards. (**a**) Generate its path in a window with equal scales on the axes and including the square $-3 \le x \le 3, -1 \le y \le 5$ and copy it on your paper. (**b**) Use your calculator or computer to find the mouse's approximate minimum speed and the approximate coordinates of the point where it occurs.

♦ Type 4, finding an approximate minimum speed

29. Figure 54 shows the Cissoid of Dioclese, $x = \sin^2 t, y = \tan t \sin^2 t, -\frac{1}{2}\pi < t < \frac{1}{2}\pi$. (**a**) What is its vertical asymptote and why is it an asymptote? (**b**) Show that the curve has the equation $y^2 = \dfrac{x^3}{1-x}$ in terms of x and y.

♦ Type 4, asymptote and xy-equation of a curve

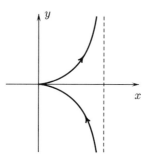

FIGURE 54

30.º A car is traveling east at 60 miles per hour when a boy throws a ball out a window with speed 10 miles per hour toward the north, relative to the car. What are the ball's velocity and speed relative to the ground when he lets it go? Use an xy-plane with **i** pointing east.

♦ Type 3, relative velocity

31.^{A} A passenger train is traveling 50 miles per hour toward the northeast when a ball is rolled from right to left across the aisle with the speed of one mile per hour relative to the train. What is the ball's velocity relative to the ground at that time? Use an xy-plane with **j** pointing north.

♦ Type 3, relative velocity

32. As an aircraft carrier is sailing northwest with a speed of 25 knots, an officer is walking from right to left across the ship with a speed of three knots relative to the ship. What is her speed relative to the water?

♦ Type 3, relative velocity

33. As you are sailing your boat directly south with a speed of ten knots, the wind gauge on your boat shows that the wind's velocity relative to the boat is two knots toward the west. What is the wind's velocity relative to the stationary water? (Use an xy-plane with **j** pointing north.)

♦ Type 3, relative velocity

34.^{A} A red ant is at $x = 2 + \cos t, y = 2 + \sin t$ on the circle $(x-2)^2 + (y-2)^2 = 1$ and a black ant is at $x = 2 + \cos(3t - \frac{3}{2}\pi), y = 2 + \sin(3t - \frac{3}{2}\pi)$ on the same circle at time t. Draw both ants' velocity vectors at $t = \frac{3}{4}\pi$ on the circle in separate drawings, and explain the result.

♦ Type 4, velocity vectors

10.4 Parametric equations of curves and velocity vectors,

35. (a) Give formulas for the x- and y-coordinates of the earth and the planet Mercury under the simplifying assumptions that the orbits are circles in an xy-plane with the sun at the origin and distances measured in millions of miles; that the radius of the earth's orbit is 93 million miles; that the radius of Mercury's orbit is 36 million miles; that the earth's year is 365 days long; and that Mercury's year is 88 days long. Have both planets move counterclockwise with Mercury on the positive x-axis and the earth on the negative x-axis at $t = 0$ (days). (b) Give formulas for the path of Mercury as viewed from the earth, i. e., with the earth at the origin. Generate this curve for $0 \leq t \leq 365$ in a window with equal scales on the axes that includes the square $-150 \leq x \leq 150, -150 \leq y \leq 150$. You should obtain the curve in Figure 55. (The actual path of Mercury, as viewed from the earth duing one year is shown in Figure 56.[3])

♦ Type 4, relative velocity

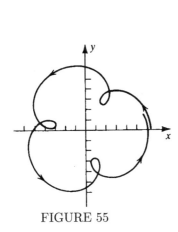

FIGURE 55

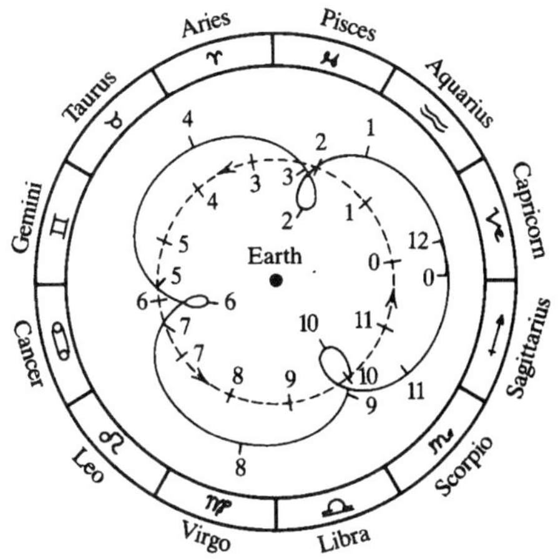

FIGURE 56

36. A ship is sailing with a speed of 30 knots (nautical miles per hour) at an angle $30°$ north of east in an xy-plane with **i** pointing east and distances measured in nautical miles. A sailor on the boat is at $x = x_0 + \frac{1}{300}\cos(300t), y = y_0 + \frac{1}{300}\sin(300t)$ at time t (hours), where (x_0, y_0) is a point on the boat. (a) What is his path and how fast is he walking? (b) What is his vector velocity at $t = 1$, relative to the stationary water?

♦ Type 3, relative velocity

37. Figure 57 shows the velocity and direction of the apparent wind, as measured on a sailboat, and of the true wind relative to the ground.[4] What is the boat's speed? (Use the Law of Cosines.)

♦ Type 4, relative velocity

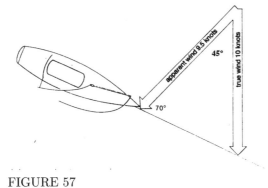

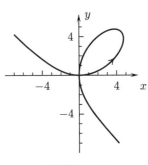

FIGURE 57

FIGURE 58

---

[3] Adapted from *The Flammarion Book of Astronomy*, London: George Allen and Unwin, Ltd., 1964, p. 255.
[4] Drawing adapted from *Racing*, G. Constable, editor, New York, NY: Time, inc. 1976, p. 26.

38. At time t (hours) for $0 \leq t \leq \pi$ a sleepwalker is at $x = 2\cos t, y = \cos t$ in an xy-plane with distances measured in miles. **(a)** Generate his path and describe it. **(b)** What are his maximum and minimum speeds?

♦ Type 3, ellipse and maximum/minimum speed

39. The LOGARITHMIC SPIRAL with polar equation $r = e^{\theta/6}, 0 \leq \theta \leq 6\pi$ has the parametric equations $x = e^{\theta/6}\cos\theta, y = e^{\theta/6}\sin\theta, 0 \leq \theta \leq 6\pi$. **(a)** Generate it in a window with equal scales on the axes and containing the square $-24 \leq x \leq 24, -24 \leq y \leq 24$ and draw it on your paper. **(b)** Calculate the exact length of the curve.

♦ Type 4, exact length of a curve

40. **(a)** What happens to a point on the FOLIUM OF DESCARTES, $x = \dfrac{9t}{1+t^3}, y = \dfrac{9t^2}{1+t^3}$ in Figure 56 as $t \to \pm\infty$? **(b)** Find the two values of t where the curve intersects the line $y = x$ and the velocity vectors at those points. **(c)** Show that the curve has the equation $x^3 + y^3 = 9xy$ in x and y.

♦ Type 4, limits on a curve and rectangular coordinates

41. Find the maximum speed on the ellipse $x = a\cos t, y = b\sin t$ with $0 < b < a$. (Use the Pythagorean Theorem $\cos^2 t = 1 - \sin^2 t$ and analyze the formula for the speed, not its derivative.)

♦ Type 3, maximum speed

42. **(a)** Generate the curve $x = \sin(2t), y = \sin t$ in a window with equal scales on the axes and copy it on your paper. **(b)** Find the maximum speed by analyzing a formula for it.

♦ Type 4, maximum speed

43. Figure 59 shows the curve $x = 2\cos t[4 - \cos(6t)], y = \sin t[4 + 4\cos(6t)]$. Use your calculator or computer to find the approximate coordinates of the highest points on the small loops at the left and right sides.

♦ Type 3, finding approximate extreme points

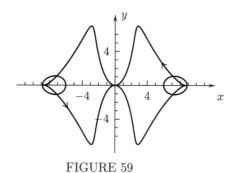

FIGURE 59

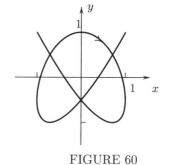

FIGURE 60

44. The curve $x = \cos(3t), y = \cos(4t)$ in Figure 60 is generated for $0 \leq t \leq 2\pi$ because the smallest common period of $\cos(3t)$ and $\cos(4t)$ is 2π. It is also generated, however, for $0 \leq t \leq \pi$. Explain.

♦ Type 6, picking the domain of the parameter

45.^A The BOWDITCH CURVE $x = 8\sin(\tfrac{3}{4}t), y = 7\sin t$ in shown in Figure 61. **(a)** What is the smallest number b such that the entire curve is generated for $0 \leq t \leq b$? **(b)** A point generating the curve passes through the origin once from the right and once from the left in the time interval from part (a). What are the velocity vectors at those two times?

♦ Type 6, picking the domain of the parameter

10.4 Parametric equations of curves and velocity vectors,

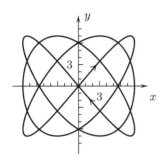

FIGURE 61

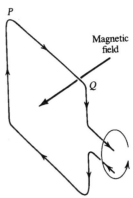

FIGURE 62

46. Alternating current is generated by rotating an armature wrapped with wires in a magnetic field. If the loop of wire in Figure 62, for example, is rotated with constant velocity in a uniform magnetic field, the current in the loop is proportional to the vertical velocity of the segment PQ and flows from P to Q when PQ is moving up and from Q to P when PQ is moving down. Show that the current can be given in terms of the time t with a sine or cosine.
♦ Type 6, why alternating current is sinusoidal

47. Suppose that the position vector $\mathbf{R}(t)$ of an object satisfies the differential equation, $\dfrac{d\mathbf{R}}{dt}(t) = \mathbf{A} \times \mathbf{R}(t)$, where $\mathbf{A}$ is a nonzero constant vector. Show that the object's acceleration vector is always perpendicular to $\mathbf{A}$ and that its speed is constant.
♦ Type 6, a vector differential equation

48. (a) Generate the HYPOTROCHOID $x = 6\cos t + 5\cos(3t), y = 6\sin t - 5\sin(3t)$ in a window with equal scales on the axes and containing the square $-12 \leq x \leq 12, -12 \leq y \leq 12$ and draw it on your paper.
(b) Use the trigonometric identities $\cos(3t) = 4\cos^3 t - 3\cos t$, and $\sin(3t) = 3\sin t - 4\sin^3 t$ to find the exact x-intercepts of the curve.
♦ Type 4, using a trigonometric identity to find intercepts

49.º Find the intersections of the curves $C_1 : x = t^3 + 6, y = 3t + 4$ and $C_2 : x = t^2 - 3, y = 3t - 5$. (Change the parameter in C_2 to u.)
♦ Type 4, intersection of curves

50.^A At time t (hours) one airplane is at $x = 300t, y = 1670t + 10t^2, z = 500 + 60t$ in xyz-space with distances measured in kilometers, and another airplane is at $x = 100 + 100t, y = -80t + 1840, z = 280t$.
(a) Do the planes collide? (b) Do their paths intersect? If so, at what point?
♦ Type 4, intersection of curves

51. At time t (seconds) one mosquito is at $x = 3t - 3, y = 9(t-1)^2$ in an xy-plane with distances measured in feet, and another mosquito is at $x = t^6, y = 2t^3$. Show that the mosquitos do not collide but their paths intersect twice.
♦ Type 4, intersection of curves

52. Each of the following curves is the graph of a function $y = y(x)$. In each case give formulas for $\dfrac{dy}{dx}$ and $\dfrac{d^2y}{dx^2}$ in terms of t: (a^O) $x = e^{-t} - t^3, y = \sin t$, (b^A) $x = t\ln t, y = t^5$, (c) $x = \tan^{-1} t + t, y = e^t - t$.
♦ Type 6, finding x-derivatives of y from parametric equations

53.[A] The line $y = kx$ in Figure 63 intersects the circle $x^2 + (y - \frac{1}{2})^2 = \frac{1}{4}$ at the point R and intersects the line $y = 1$ at the point S. As k varies, the corner P in the right triangle PRS generates a curve known as the VERSIERA OF AGNESI or the WITCH OF AGNESI.[5] Give parametric equations for the curve with k as parameter, and then give an equation for the curve in x and y.

♦ Type 4, deriving parametric equations

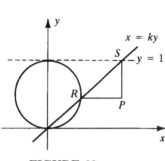

FIGURE 63

FIGURE 64

54. When the circle of radius 1 in Figure 65 is rolled to the right along the x-axis, the point P that is initially at the origin generates the CYCLOID in Figure 66. **(a)** After the circle has rotated t radians, it is tangent to the x-axis at the point T in Figure 66. Explain why $T = (t, 0)$. **(b)** Express the coordinates of the center C of the circle and the displacement vector $\overrightarrow{CP}$ in Figure 66 in terms of t and use the results to give formulas for the coordinates (x, y) of P. Then generate the curve in a window with equal scales on the axes that includes the rectangle $0 \leq x \leq 16, 0 \leq y \leq 2$ and copy it on your paper.

♦ Type 4, deriving parametric equations

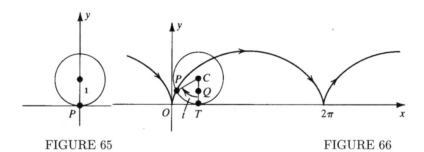

FIGURE 65   FIGURE 66

55. Figure 67 shows a circle of radius $\frac{1}{4}$ that rolls inside the unit circle. At $t = 0$ the point P at the tip of the vector **B** is at $(1, 0)$ on the x-axis. The drawing shows its position when the angle of inclination of vector **A** is t radians for a positive value of t. **(a)** Explain why the smaller circle rotates clockwise $4t$ radians when **A** rotates t radians counterclockwise, and consequently, the angle of inclination of **B** is $-3t$. **(b)** Express **A** and **B** in terms of t and use the results to give parametric equations of the HYPOCYCLOID that the point P generates as t goes from 0 to 2π. Then generate the curve in a window with equal scales on the axes and copy it on your paper. **(c)** Use the trigonometric identities $\cos^3 \alpha = \frac{3}{4} \cos \alpha + \frac{1}{4} \cos(3\alpha), \sin^3 \alpha = \frac{3}{4} \sin \alpha - \frac{1}{4} \sin(3\alpha)$ to show that the curve from part (b) is the ASTROID $x^{2/3} + y^{2/3} = 1$.

♦ Type 4, deriving parametric equations

---

[5] This curve was first given this derivation by Maria Gaetana Agnesi (1718–1799) (Figure 64) who wrote one of the first popular calculus texts. She named the curve a "versiera" after the Latin "versiera" for "to turn." Cobson, a British mathematician, who published a translation of her book from Italian into English in 1802, gave the name "witch of Agnesi" to the curve because "versiera" is also an Italian word for "she devil."

10.4 Parametric equations of curves and velocity vectors,

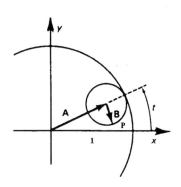

FIGURE 67   FIGURE 68

56. A point P is fastened to a circle of radius 1 at the distance 2 from its center, and that circle is rolling around a circle of radius 3 with its center at the origin, as in Figure 68. **(a)** Explain why the smaller circle rotates $4t$ radians counterclockwise with the position vector of its center rotates t radians counterclockwise. **(b)** Find parametric equations of the EPITROCHOID that is generated by the point P as t goes from 0 to 2π. Then generate the curve in a window with equals scales on the axes and copy it on your paper.

♦ Type 4, deriving parametric equations

57.^A (CAS†) When a circle of radius 1 rolls inside a circle of radius 3, a point on the circumference of the smaller circle generates the hypocycloid in Figure 69, which is also called a DELTOID. The curve has the parametric equations $x = 2\cos^2 t + 2\cos t - 1, y = 2\sin t - 2\sin t \cos t$. Find the exact maxima and minima of x and y on the curve.

♦ Type 4 with CAS, Level 6 without CAS, finding extreme points

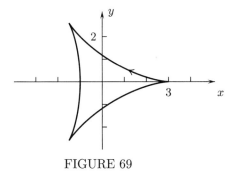 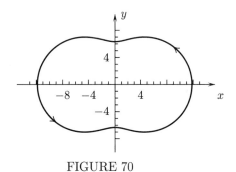

FIGURE 69   FIGURE 70

58.^A (CAS) Figure 70 shows the HIPPOPEDE with parametric equations $x = 2\cos t\sqrt{35 - 25\sin^2 t}$, $y = 2\sin t\sqrt{35 - 25\sin^2 t}$. Find the exact slope of its tangent lines at the two points where it intersects the line $y = x$.

♦ Type 4 with CAS, Level 6 without CAS, tangent lines

59. Give parametric equations of the CURTATE CYCLOID that is generated by a point P on a wheel that is rotating on level ground if the radius of the wheel is b and the distance from P to the center of the wheel is a with $0 < a < b$. Let the level ground be the x-axis and have P be under the center of the circle when $x = 0$.

♦ Type 4, deriving parametric equations

60. (CAS) What is the exact length of the spiral $x = t\cos t, y = t\sin t, 0 \le t \le 6\pi$?

♦ Type 4 with CAS, Level 6 without CAS, length of a curve

---

†Problems with this label can be worked best by utilizing a computer- or calculator-algebra system.

61. (CAS) What is the slope of the tangent line to $x = \cos t[4+\cos(6t)], y = -\sin t[4+\cos(6t)], 0 \leq t \leq 2\pi$ in Figure 71 at the two points where it intersects the line $y = x$?

◆ Type 4 with CAS, Level 6 without CAS, tangent lines

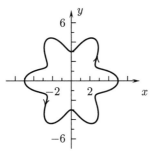

FIGURE 71

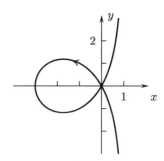

FIGURE 72

62. (CAS) MACLAURIN'S TRISECTRIX in Figure 72 has the parametric equations $x = 1 - 4\cos^2 t$, $y = \tan t(1 - 4\cos^2 t), -\frac{1}{2}\pi < t < \frac{1}{2}\pi$. **(a)** Show that it has a vertical asymptote. **(b)** Find its velocity vectors at the origin.

◆ Type 4 with CAS, Level 6 without CAS, asymptote and velocity vectors

63. (CAS) **(a)** Generate the curve $x = \dfrac{6t - 5t^2}{1+t^2}, y = \dfrac{2t^3 - 3t + 6}{1+t^2}, -4.6 \leq t \leq 7.8$ and copy it on your paper. **(a)** Show that it intersects itself at $x = -3$. **(c)** Find the exact maximum value of x on the curve.

◆ Type 4 with CAS, Level 6 without CAS, intersection and extreme point

64. The hanging cable in Figure 73 has constant density ρ pounds per foot and has the shape of the graph $y = f(x)$. Consequently, the portion of the cable for $0 \leq x \leq X$ for $X > 0$ weighs
$$w(X) = \rho \int_0^X \sqrt{1 + [f'(x)]^2}\, dx$$ pounds. **(a)** Use the Fundamental Theorem of Calculus for derivatives of integrals in Section 6.5 to give a formula for $w'(X)$ in terms of $f'(X)$. **(b)** Suppose that the horizontal tension in the cable is the positive constant k (pounds) and that the force on the end of the portion of the cable for $0 \leq x \leq X$ is tangent to the cable at the end. Show that $f'(X) = w(X)/k$. **(c)** Use the results of parts (a) and (b) to show that $f(x)$ satisfies the differential equation $f''(x) = \dfrac{\rho}{k}\sqrt{1 + [f'(x)]^2}$. **(d)** Let $m(x) = f'(x)$ be the slope of the cable at x, so that $\dfrac{dm}{dx} = \dfrac{\rho}{k}\sqrt{1+m^2}$. Solve this differential equation by separating variables. Use the initial condition $m(0) = 0$ so the lowest point on the cable is at $x = 0$, and make the substitution $m = \sinh \theta$ in the integral. **(e)** Integrate the result of part (d) to show that $f(x) = \dfrac{k}{\rho}\cosh\left(\dfrac{\rho x}{k}\right) + C$ with a constant C. This curve is called a CATENARY.

◆ Type 6, a separable differential equation using hyperbolic functions

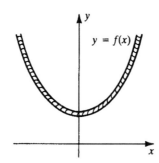

FIGURE 73

Section 10.5

Acceleration, force, and curvature

OVERVIEW: *In this section we first define integrals of vector-valued functions and illustrate how position vectors can be obtained from velocity vectors and velocity vectors from* ACCELERATION VECTORS *by integration. Then we discuss* NEWTON'S LAW OF MOTION *in vector form. We next define unit* TANGENT *and* NORMAL VECTORS *and* CURVATURE *of plane curves and use them to study how forces on objects moving in a plane affect their speeds and the shapes of their paths. At the end of the section we show that* NEWTON'S LAW OF GRAVITY *implies* KEPLER'S LAWS *of planetary motion for circular orbits and describe the representation of acceleration vectors in polar coordinates that are used to deal with general orbits of planets.*

Topics:

- *Integrating vector-valued functions*
- *Acceleration vectors*
- *Newton's Law of motion*
- *Unit tangent and normal vectors to plane cures*
- *Curvature of plane curves*
- *Tangential and normal components of acceleration in the plane*
- *Kepler's Laws of planetary motion and Newton's Law of gravity*
- *Velocity and acceleration vectors in polar coordinates*

Integrating vector-valued functions

We begin with an example of finding an object's position from its velocity vector.

Example 1 A robot moving in an xy-plane with distances measured in meters is at $(50,0)$ at $t = 0$ (minutes) and its velocity vector is $\mathbf{v}(t) = \langle -100\sin(2t),\ 60\cos(2t)\rangle$ (meters per minute) at time t for $0 \leq t \leq \pi$. Find a formula for its position vector $\mathbf{R}(t)$ in terms of t for $0 \leq t \leq \pi$. Then generate and describe the object's path.

SOLUTION We write $\mathbf{R}(t) = \langle x(t), y(t)\rangle$ with $x(t)$ and $y(t)$ the robot's coordinates at time t. Then its velocity vector $\mathbf{v}(t) = \langle -100\sin(2t),\ 60\cos(2t)\rangle$ equals $\dfrac{d\mathbf{R}}{dt}(t) = \left\langle \dfrac{dx}{dt}(t),\ \dfrac{dy}{dt}(t)\right\rangle$, so that

$$\frac{dx}{dt}(t) = -100\sin(2t),\ \frac{dy}{dt}(t) = 60\cos(2t). \tag{1}$$

We are also given that

$$x(0) = 50,\ y(0) = 0. \tag{2}$$

The first of equations (1) yields

$$\begin{aligned} x(t) &= -\int 100\sin(2t)\, dt = -100\left(\tfrac{1}{2}\right)\int \sin u\, du \\ &= 50\cos u + C_1 = 50\cos(2t) + C_1 \end{aligned} \tag{3}$$

♦ SUGGESTIONS TO INSTRUCTORS: The main topics in this section can be divided in three parts, which can be covered in three or more class meetings. Part 3 can be skipped, if necessary, to save time.
Part 1 (Integrating vector-valued functions, acceleration vectors, and Newton's law in vector form in the plane and in space; Tune-up Exercises T1–T3, Problems 1–13, 37–48, 61–65) Have students begin Questions 1 and 2 and Tune-up Exercises T1, T2, and T3 before class. Problems 1d, 3c, 5a, 6, 9, 12 are good for in-class discussion.
Part 2 (Unit tangent and normal vectors, curvature, circles of curvature, and tangential and normal components of acceleration in the plane; Tune-up Exercises T4–T7, Problems 14–36, 49–60, 66–68) Have students begin Questions 3 through 7 and Tune-up Exercises T4, T5, and T6 before class. Problems 16, 19, 22, 26, and 29 are good for in-class discussion.
Part 3 (Kepler's Laws and Newton's Law of gravity; Problems 69–70) Have students begin Question 8 and 9 before class. Problems 69 and 70 could be discussed in class.

where we have used the substitution $u = 2t, du = 2\,dt$ in the integral. Then the first of equations **(2)** shows that $x(0) = 50\cos(0) + C_1 = 50 + C_1$ equals 50, so that $C_1 = 0$ and $x(t) = 50\cos(2t)$.

Similarly, the second of equations **(1)** gives, with the same substitution,

$$y(t) = \int 60\cos(2t)\,dt = 60\left(\tfrac{1}{2}\right)\int \cos u\,du \qquad (4)$$
$$= 30\sin u + C_2 = 30\sin(2t) + C_2.$$

By the second of equations **(2)**, $y(0) = 30\sin(0) + C_2 = C_2$ is 0, so that $C_2 = 0$ and $y(t) = 30\sin(2t)$. The position vector of the robot is $\mathbf{R}(t) = \langle 50\cos(2t),\ 30\sin(2t)\rangle$ at time t, and its path is the ellipse in Figure 1. □

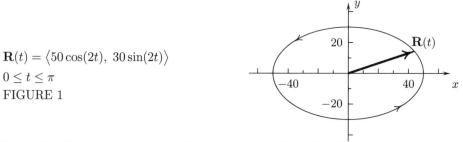

$\mathbf{R}(t) = \langle 50\cos(2t),\ 30\sin(2t)\rangle$
$0 \le t \le \pi$
FIGURE 1

In working Example 1, we found the components $x(t)$ and $y(t)$ of the robot's position vector by integrating the components of its velocity vector. We can interpret these calculations as integration of the velocity vector itself by employing the following definition, which states that integrals of vector-valued functions are found by integrating their components.

Definition 1 **(a)** If $x(t)$ and $y(t)$ have antiderivatives (indefinite integrals) at t, then

$$\int \langle x(t), y(t)\rangle\,dt = \left\langle \int x(t)\,dt,\ \int y(t)\,dt \right\rangle \qquad (5)$$

and if the definite integrals of $x(t)$ and $y(t)$ from $t = a$ to $t = b$ are defined, then

$$\int_a^b \langle x(t), y(t)\rangle\,dt = \left\langle \int_a^b x(t)\,dt,\ \int_a^b y(t)\,dt \right\rangle. \qquad (6)$$

(a) If $x(t), y(t)$, and $z(t)$ have antiderivatives at t, then

$$\int \langle x(t), y(t), z(t)\rangle = \left\langle \int x(t)\,dt,\ \int y(t)\,dt,\ \int z(t)\,dt \right\rangle \qquad (7)$$

and if the definite integrals of $x(t), y(z)$ and $z(t)$ from $t = a$ to $t = b$ are defined, then

$$\int_a^b \langle x(t), y(t), z(t)\rangle\,dt = \left\langle \int_a^b x(t)\,dt,\ \int_a^b y(t)\,dt,\ \int_a^b z(t)\,dt \right\rangle. \qquad (8)$$

This vector-based definition enables us to express calculations more concisely. For instance, equations **(3)** and **(4)** in the solution of Example 1 can be combined into the one vector equation,

$$\langle x(t), y(t)\rangle = \int \langle -100\sin(2t), 60\cos(2t)\rangle\,dt = \langle 50\cos(2t) + C_1, 30\sin(2t) + C_2\rangle$$
$$= \langle 50\cos(2t), 30\sin(2t)\rangle + \langle C_1, C_2\rangle.$$

10.5 Acceleration, force, and curvature

Because differentiation and integration of vector-valued functions $\mathbf{R}(t)$ are carried out component by component, the basic principles relating derivatives and integrals of real-valued functions apply to vector-valued functions. In particular, the indefinite integral $\int \mathbf{R}(t)\, dt$ is an antiderivative of $\mathbf{R}(t)$:

$$\frac{d}{dt}\int \mathbf{R}(t)\, dt = \mathbf{R}(t). \tag{9}$$

Also, if $\mathbf{R}(t)$ is continuous in an interval containing a and b and the integral $\int_a^b \frac{d\mathbf{R}}{dt}(t)\, dt$ is defined, then, by the Fundamental Theorem from Section 4.2 applied to each component,

$$\int_a^b \frac{d\mathbf{R}}{dt}(t)\, dt = \mathbf{R}(b) - \mathbf{R}(a). \tag{10}$$

Example 2 Find a formula for the antiderivative, $\int \langle t^2 + 3, 5t^4 \rangle\, dt$.

SOLUTION (a) To avoid confusion, we integrate each component by writing

$$\int (t^2 + 3)\, dt = \tfrac{1}{3}t^3 + 3t + C_1$$

$$\int 5t^4\, dt = t^5 + C_2$$

with C_1 and C_2 arbitrary constants. Then we combine the results to obtain

$$\int \langle t^2 + 3, 5t^4 \rangle\, dt = \langle \tfrac{1}{3}t^3 + 3t + C_1, t^5 + C_2 \rangle = \langle \tfrac{1}{3}t^3 + 3t, t^5 \rangle + \mathbf{C}$$

where $\mathbf{C} = \langle C_1, C_2 \rangle$ is an arbitrary constant vector. □

Example 3 Evaluate the definite integral, $\int_0^1 \langle t^2 + 3, 5t^4 \rangle\, dt$.

SOLUTION The first component of the integral is

$$\int_0^1 (t^2 + 3)\, dt = \left[\tfrac{1}{3}t^3 + 3t\right]_0^1 = [\tfrac{1}{3}(1^3) + 3(1)] - [\tfrac{1}{3}(0^3) + 3(0)] = \tfrac{10}{3}$$

and the second component is

$$\int_0^1 5t^4\, dt = \left[t^5\right]_0^1 = 1^5 - 0^5 = 1.$$

Consequently, $\int_0^1 \langle t^2 + 3, 5t^4 \rangle\, dt = \langle \tfrac{10}{3}, 1 \rangle.$ □

Question 1 What are the components of (a) $\int \left(e^t\, \mathbf{i} + \cos t\, \mathbf{j} + \frac{1}{t}\, \mathbf{k}\right) dt$ and (b) $\int_1^2 \left(e^t\, \mathbf{i} + \cos t\, \mathbf{j} + \frac{1}{t}\, \mathbf{k}\right) dt$?

Acceleration vectors

In earlier chapters we defined the acceleration $a(t)$ of an object moving on an s-axis to be the derivative $\dfrac{dv}{dt}(t)$ of its velocity, which is the second derivative $\dfrac{d^2s}{dt^2}(t)$ of its position $s(t)$ at time t. Acceleration vectors are defined similarly. If an object has position vector $\mathbf{R}(t)$ and velocity vector $\mathbf{v}(t) = \dfrac{d\mathbf{R}}{dt}(t)$ at time t, then its ACCELERATION VECTOR $\mathbf{a}(t)$ at time t is the first derivative of the velocity vector and the second derivative of the position vector:[†]

$$\mathbf{a}(t) = \frac{d\mathbf{v}}{dt}(t) = \frac{d^2\mathbf{R}}{dt^2}(t). \tag{11}$$

Example 4 Find $\mathbf{a}(t)$ for $\mathbf{R}(t) = \langle t^3, 5t^4, 6t - t^2 \rangle$.

SOLUTION Because each calculation can be done in one step, we work directly with the vectors. The velocity vector is $\mathbf{v}(t) = \dfrac{d}{dt}\langle t^3, 5t^4, 6t - t^2 \rangle = \langle 3t^2, 20t^3, 6 - 2t \rangle$, and the acceleration vector is $\mathbf{a}(t) = \dfrac{d}{dt}\langle 3t^2, 20t^3, 6 - 2t \rangle = \langle 6t, 60t^2, -2 \rangle$. □

The vector form of Newton's Law of motion

In earlier chapters we studied the effect of forces on the motion of objects that moved on s-axes by using Newton's law, $F(t) = ma(t)$, where $F(t)$ is the total force on the object in the positive s-direction at time t, m is the object's mass, and $a(t)$ is its (scalar) acceleration, with these quantities measured in compatible units.[‡]

To study motion on curves, we use Newton's law in the vector form,

$$\mathbf{F}(t) = m\mathbf{a}(t) \tag{12}$$

where the vector $\mathbf{F}(t)$ represents the total force on an object and $\mathbf{a}(t)$ is its acceleration vector at time t.

Example 5 The only forces for $0 \leq t \leq 5$ (seconds) on a two-kilogram balloon moving in xyz-space with the positive z-axis pointing up are the horizontal force of the wind $8\mathbf{i} + \tfrac{1}{10}t\mathbf{j}$ (newtons) and the downward force of gravity. At $t = 0$ the balloon is at $x = 0, y = 10, z = 200$ (meters) and its velocity vector is $3\mathbf{i} - 2\mathbf{j} + 10\mathbf{k}$ (meters per second). Give a formula for its position vector in terms of t.

SOLUTION The mass m of the balloon is 2 kilograms and the acceleration due to gravity g, measured in meters per second^2, is 9.8, so the balloon's weight is $2(9.8) = 19.6$ newtons and the vector force of gravity on it is $-19.6\,\mathbf{k}$ newtons. Adding this to the force $8\mathbf{i} + \tfrac{1}{10}t\mathbf{j}$ of the wind shows that the total force on the balloon at time t is $\mathbf{F}(t) = 8\mathbf{i} + \tfrac{1}{10}t\mathbf{j} - 19.6\,\mathbf{k}$ newtons. Its acceleration vector at time t is therefore

$$\mathbf{a}(t) = \frac{1}{m}\mathbf{F}(t) = \tfrac{1}{2}\langle 8, \tfrac{1}{10}t, -19.6 \rangle = \langle 4, \tfrac{1}{20}t, -9.8 \rangle \text{ meters per second}^2. \tag{13}$$

We can find the position vector $\mathbf{R}(t)$ of the balloon either by studying the components in **(13)** separately and combining the results or by using vector notation. Because the calculations that are required here are fairly simple, we will use vectors.

[†] We will occasionally refer to the acceleration $a(t)$ of an object moving on an s-axis as the SCALAR ACCELERATION to distinguish it from acceleration vectors. The term "scalar" is used here because $a(t)$ is a number, not a vector.

[‡] Recall that to use Newton's law, the time should be measured in seconds and distances, forces, and masses should be measured in feet, pounds, and slugs; in meters, newtons, and kilograms; or in centimeters, dynes, and grams. Also weight w can be converted to mass m or vice-versa by using the equation $w = mg$ with $g = 32$ if the units are feet, pounds, and slugs, $g = 9.8$ if the units are meters, newtons, and kilograms, and $g = 980$ if the units are centimeters, dynes, and grams.

10.5 Acceleration, force, and curvature

When we take the antiderivative of **(13)**, we obtain

$$\mathbf{v}(t) = \int \mathbf{a}(t)\, dt = \int \langle 4, \tfrac{1}{20}t, -9.8 \rangle\, dt$$
$$= \left\langle \int 4\, dt, \int \tfrac{1}{20}t\, dt, \int -9.8\, dt \right\rangle \qquad (14)$$
$$= \langle 4t + C_1, \tfrac{1}{40}t^2 + C_2, -9.8t + C_3 \rangle.$$

with constants C_1, C_2, C_3. Then $\mathbf{v}(0) = \langle C_1, C_2, C_3 \rangle$. We are given that $\mathbf{v}(0) = \langle 3, -2, 10 \rangle$, so that $C_1 = 3, C_2 = -2, C_3 = 10$, and by **(14)**

$$\mathbf{v}(t) = \langle 4t + 3, \tfrac{1}{40}t^2 - 2, -9.8t + 10 \rangle \text{ meters per second.}$$

Taking another antiderivative yields

$$\mathbf{R}(t) = \int \mathbf{v}(t)\, dt = \int \langle 4t + 3, \tfrac{1}{40}t^2 - 2, -9.8t + 10 \rangle$$
$$= \left\langle \int (4t + 3)\, dt, \int (\tfrac{1}{40}t^2 - 2)\, dt, \int (-9.8t + 10)\, dt \right\rangle \qquad (15)$$
$$= \langle 2t^2 + 3t + C_4, \tfrac{1}{120}t^3 - 2t + C_5, -4.9t^2 + 10t + C_6 \rangle.$$

with other constants C_4, C_5, and C_6. This gives $\mathbf{R}(0) = \langle C_4, C_5, C_6 \rangle$, and because we are told that $x(0) = 0, y(0) = 10$, and $z(0) = 200$, $C_4 = 0, C_5 = 10, C_6 = 200$ and by **(15)**

$$\mathbf{R}(t) = \langle 2t^2 + 3t, \tfrac{1}{120}t^3 - 2t + 10, -4.9t^2 + 10t + 200 \rangle \text{ meters.} \quad \square$$

Question 2 The total force on a five-gram ball moving in an xy-plane with distances measured in centimeters is $\mathbf{F} = \langle -15\cos t, -15\sin t \rangle$ (dynes) at time t (seconds) for $0 \leq t \leq 11$. At $t = 0$ the ball is at $x = 6, y = 3$ and its velocity vector is $\langle 1, 4 \rangle$ centimeters per second. **(a)** Give a formula for the ball's position vector in terms of t and generate its path. **(b)** Why, based on the formulas for the force or position vectors, is it plausible that the path has loops?

Unit tangent and normal vectors to a plane curve

If a plane curve has the parametric equations $\mathbf{R}(t) = \langle x(t), y(t) \rangle$, then at any point P where the velocity vector $\mathbf{v}(t) = \langle x'(t), y'(t) \rangle$ is not zero, the vector

$$\mathbf{T}(t) = \frac{\mathbf{v}(t)}{|\mathbf{v}(t)|} \qquad (16)$$

is parallel to the tangent line at P and points in the direction of the curve's orientation. We refer to $\mathbf{T}(t)$ as the UNIT TANGENT VECTOR to the curve at that point. The unit NORMAL VECTOR $\mathbf{N}(t)$ is the unit vector perpendicular to $\mathbf{T}(t)$ that points to its left. To find its components, we interchange the components of $\mathbf{T}(t) = \langle a, b \rangle$ and then multiply the first by -1 to obtain $\mathbf{N}(t) = \langle -b, a \rangle$ (Figure 2).

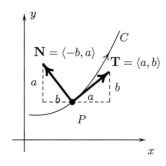
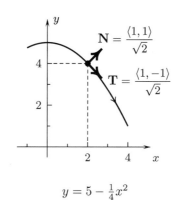

FIGURE 2   FIGURE 3

Example 6 Find the x- and y-components of **T** and **N** at $x = 2$ on the parabola $y = 5 - \frac{1}{4}x^2$, $0 \leq x \leq 4$, oriented from left to right. Then draw the vectors with the curve, using the scales on the axes to measure the components.

SOLUTION The curve is oriented from left to right if we use x as parameter. The position vector at x is then $\mathbf{R}(x) = \langle x, 5 - \frac{1}{4}x^2 \rangle$, for which

$$\mathbf{v}(x) = \frac{d}{dx}\langle x, 5 - \tfrac{1}{4}x^2\rangle = \langle 1, -\tfrac{1}{2}x\rangle.$$

Setting $x = 2$ gives $\mathbf{v}(2) = \langle 1, -1\rangle$. The length of this velocity vector is $|\langle 1, -1\rangle| = \sqrt{2}$, so the unit tangent vector is $\mathbf{T} = \dfrac{\langle 1, -1\rangle}{\sqrt{2}}$. Interchanging the components and then multiplying the first component by -1 gives the unit normal vector $\mathbf{N} = \dfrac{\langle 1, 1\rangle}{\sqrt{2}}$.
The curve and the two vectors are shown in Figure 3. □

Curvature of plane curves

In order to study how forces on objects moving in an xy-plane relate to the shapes of their paths, we need a way to measure how rapidly a curve is bending and whether it is bending to the right or left. We use the CURVATURE of a curve. We suppose that the curve has parametric equations,

$$C : x = x(s), y = y(s)$$

with arclength s along the curve as parameter, and we suppose that $x(s)$ and $y(s)$ have continuous derivatives. We let $\mathbf{T}(s)$ denote the unit tangent vector at $\big((x(s), y(s))\big)$ on the curve and let $\phi(s)$ (phi) be the angle of orientation of $\mathbf{T}(s)$ (Figure 4). Then we make the following definition.

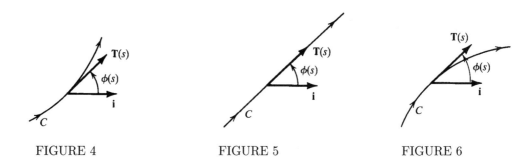

FIGURE 4   FIGURE 5   FIGURE 6

10.5 Acceleration, force, and curvature

Definition 2 *The* CURVATURE *κ (kappa) of $C : x = x(s), y = y(s)$ at $\big((x(s), y(s))\big)$ is the derivative of the angle ϕ of inclination of the unit tangent vector with respect to the arclength s:*

$$\kappa(s) = \frac{d\phi}{ds}(s). \tag{17}$$

The curvature is zero if the curve is a straight line and ϕ is constant, as in Figure 5. If the curvature is positive then $\phi(s)$ is increasing and the curve is bending to the left, as in Figure 4. If the curvature is negative, then $\phi(s)$ is decreasing and the curve is bending to the right, as in Figure 6. Moreover, the curve bends sharply where $|\kappa(s)|$ is large and bends gradually where $|\kappa(s)|$ is small.

In the case of the graph $C : y = y(x)$ of a function, the slope $y'(x)$ of the tangent line at the point with x-coordinate x is the tangent of the angle ϕ, so that we can use $\phi = \tan^{-1}[y'(x)]$. This gives the equation,

$$\frac{d\phi}{dx} = \frac{d}{dx}\{\tan^{-1}[y'(x)]\} = \frac{1}{1 + [y'(x)]^2} \frac{d}{dx}[y'(x)] = \frac{y''(x)}{1 + [y'(x)]^2}. \tag{18}$$

We assume that $y(x)$ has continuous first and second derivatives. Then by Rule 5 of Section 4.6, the length of the graph for $a \leq x \leq X$ with a fixed number a is

$$s(X) = \int_a^X \sqrt{1 + [y'(x)]^2}\, dx.$$

We assume that $y'(x)$ is continuous. Then by the Fundamental Theorem for derivatives of integrals (Theorem 1 of Section 6.5)

$$\frac{ds}{dX}(X) = \frac{d}{dX}\int_a^X \sqrt{1 + [y'(x)]^2}\, dx = \sqrt{1 + [y'(X)]^2}.$$

Then, because $\dfrac{ds}{dx}(x)$ is the positive number $\sqrt{1 + [y'(x)]^2}$, $s(x)$ has an inverse function $x(s)$ whose derivative $\dfrac{dx}{ds}$ equals $\dfrac{1}{\sqrt{1 + [y'(x)]^2}}$, and with the Chain Rule, **(18)** yields

$$\frac{d\phi}{ds} = \frac{d\phi}{dx}\frac{dx}{ds} = \frac{y''(x)}{\{1 + [y'(x)]^2\}^{3/2}}.$$

We have established the following result.

Theorem 1 *In any interval where $y(x)$ has a continuous second derivative, the curvature of the graph $C : y = y(x)$, oriented from left to right, at $\big(x, y(x)\big)$ is*

$$\kappa(x) = \frac{y''(x)}{\{1 + [y'(x)]^2\}^{3/2}}. \tag{19}$$

Example 7 What is the curvature at the origin of the parabola $y = -x^2$, oriented from left to right?

SOLUTION For $y(x) = -x^2$, we have $y'(x) = -2x$ and $y''(x) = -2$. Since $y'(0) = 0$, and $y''(0) = -2$, the curvature **(19)** in this case is $\kappa(0) = \dfrac{-2}{(1 + 0^2)^{3/2}} = -2$. □

The derivation of Theorem 1 can be modified to give the following result for curves given by parametric equations:

Theorem 2 *The curvature at $\bigl(x(t), y(t)\bigr)$ on the curve $C : x = x(t), y = y(t)$ is*

$$\kappa(t) = \frac{x'(t)y''(t) - y'(t)x''(t)}{\{[x'(t)]^2 + [y'(t)]^2\}^{3/2}}. \tag{20}$$

Example 8   What is the curvature at $t = \tfrac{1}{2}\pi$ of the ellipse $x = 3\cos t, y = 5\sin t$?

SOLUTION   For $x(t) = 3\cos t$ and $y(t) = 5\sin t$, we have $x'(t) = -3\sin t, x''(t) = -3\cos t, y'(t) = 5\cos t$ and $y''(t) = -5\sin t$. At $t = \tfrac{1}{2}\pi$, these have the values $x' = -3, x'' = 0, y' = 0$, and $y'' = -5$, so the curvature **(20)** is

$$\kappa(\tfrac{1}{2}\pi) = \frac{(-3)(-5) - 0(0)}{(3^2 + 0^2)^{3/2}} = \tfrac{15}{27} = \tfrac{5}{9}. \ \square$$

Question 3   What is the curvatature **(a)** of the circle $x = \rho\cos t, y = \rho\sin t$ of radius $\rho > 0$ oriented counterclockwise and **(b)** of the circle $x = \rho\cos(-t), y = \rho\sin(-t)$ of radius ρ oriented clockwise?

Circles of curvature

Recall that the tangent line at a point P on a curve is the line that best approximates the curve near that point and that the tangent line is the limiting position of the secant line through P and a nearby point Q as Q approaches P. If the curvature of the curve is not zero at P, then we can obtain a better approximation of the curve near P by using its CIRCLE OF CURVATURE at that point, which is defined as follows.

Definition 3   *If the curvature of a curve C at the point P is not zero, then the circle of curvature of the curve at P (Figure 7) is the limiting position of the circle through points $P, Q,$ and R on the curve (Figure 8) as Q and R approach P.*

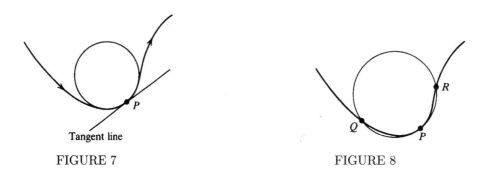

FIGURE 7   FIGURE 8

You saw in Question 3 that the curvature of a circle of radius ρ equals $1/\rho$ if the circle is oriented counterclockwise and equals $-1/\rho$ if the curve is oriented clockwise. In Problems 66 through 68 below we describe a proof of the following Theorem, which shows that curvatures and circles of curvatures for general plane curves are related similarly. The proof is based on l'Hopital's Rule and also shows that the circle through $P, Q,$ and R in Definition 3 approaches the circle of curvature.

10.5 Acceleration, force, and curvature

Theorem 3 *The* CIRCLE OF CURVATURE *at a point P on a curve in the xy-plane at a point where the curvature κ is not zero is the circle of radius $1/|\kappa|$ that has the same tangent line at P as the curve and is to the left of curve, relative to the tangent vector T if κ is positive and to the right of the curve if κ is negative. The radius $\rho = 1/|\kappa|$ is the* RADIUS OF CURVATURE *and the center of the circle is the* CENTER OF CURVATURE *at P.*

Question 4 The centers of curvature at the points P and Q on the curve in Figure 9 are indicated by the nearby dots. What is the approximate curvature of the curve **(a)** at P and **(b)** at Q?

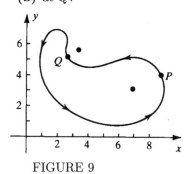
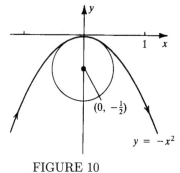

FIGURE 9   FIGURE 10

Example 9 Draw the parabola $y = -x^2$ of Example 7 and its circle of curvature at the origin.

SOLUTION In Example 7 we found that the curvature of the parabola at the origin is -2, so the radius of curvature there is $1/|-2| = \frac{1}{2}$. The circle of curvature is tangent to the x-axis at the origin and is below the x-axis because κ is negative and the curve is oriented toward the right. Its center is at the point $(0, -\frac{1}{2})$ (Figure 10). □

It was relatively easy to find the center of curvature in Example 9 because the corresponding tangent line was horizontal. Centers of curvature can be found in cases where the tangent line is not horizontal or vertical with the formula

$$\overrightarrow{OC} = \mathbf{R} + \frac{1}{\kappa}\mathbf{N} \tag{21}$$

for the position vector of the center C. Here $\mathbf{R}$ is the position vector of the point on the curve, and κ and $\mathbf{N}$ are the curvature and unit normal vector at that point (Figure 11).

FIGURE 11

Question 5 Find the center of curvature of the parabola $y = \frac{1}{2}x^2$ at $x = 1$.

Tangential and normal components of acceleration in the plane

If no force is applied to an object moving in an xy-plane, then its acceleration vector is zero, its velocity vector is constant, and it moves in a straight line at a constant speed. In order for it to speed up or slow down or for its path to bend, force must be applied to it. We will see below how it is the component of the force tangent to the object's path that makes the object speed up or slow down and the component of the force normal to the path that causes the path to bend.

To derive the formula that displays these facts, we suppose that $s = s(t)$ is the distance along the path that the object has traveled. Then ds/dt is the object's speed and its velocity vector is the product of its speed and the unit tangent vector to its path:

$$\mathbf{v} = \frac{ds}{dt}\mathbf{T}. \qquad (22)$$

Differentiating this equation with respect to t, using a vector form of the Product Rule, gives the acceleration vector,

$$\mathbf{a} = \frac{d\mathbf{v}}{dt} = \frac{d^2s}{dt^2}\mathbf{T} + \frac{ds}{dt}\frac{d\mathbf{T}}{dt}. \qquad (23)$$

To find the derivative of the tangent vector, we write it in terms of its angle of inclination ϕ as $\mathbf{T} = \langle \cos\phi, \sin\phi \rangle$. Then by **(15)**, the corresponding unit normal vector is $\mathbf{N} = \langle -\sin\phi, \cos\phi \rangle$ so that

$$\frac{d\mathbf{T}}{d\phi} = \mathbf{N}. \qquad (24)$$

Question 6 Express in symbols the Chain-Rule statement that the rate of change of $\mathbf{T}$ with respect to t equals its rate of change with respect to ϕ, multiplied by the rate of change of ϕ with respect to s, multiplied by the rate of change of s with respect to t.

With the formula from Question 6, equation **(23)** gives

$$\mathbf{a} = \frac{d^2s}{dt^2}\mathbf{T} + \frac{ds}{dt}\left(\frac{d\mathbf{T}}{d\phi}\frac{d\phi}{ds}\frac{ds}{dt}\right).$$

Then, since $\dfrac{d\mathbf{T}}{d\phi} = \mathbf{N}$ by **(24)** and $\dfrac{d\phi}{ds} = \kappa$ by Definition 2, we obtain the following result:

Theorem 4 Suppose that an object moving in an xy-plane is at $x = x(t), y = y(t)$ for $a \leq t \leq b$, that $x(t)$ and $y(t)$ have continuous second derivatives, and that the object's path can also be given by parametric equations with arclength as parameter and with continuous second derivatives. Then the object's acceleration vector at time t can be expressed in the form,

$$\mathbf{a} = \frac{d^2s}{dt^2}\mathbf{T} + \kappa\left(\frac{ds}{dt}\right)^2\mathbf{N}. \qquad (25)$$

The quantity $\dfrac{ds}{dt}$ in **(25)** is the object's nonnegative speed and $\dfrac{d^2s}{dt^2}$ is the rate of change of the speed with respect to time. This is the rate at which the object is speeding up if it is positive, and is the negative of the rate at which the object is slowing down if it is negative.

Because $\mathbf{T}$ and $\mathbf{N}$ are perpendicular unit vectors, the quantity $\dfrac{d^2s}{dt^2}$ in **(25)** is the tangential component of the object's acceleration vector and $\kappa\left(\dfrac{ds}{dt}\right)^2$ is the normal component.

One way to interpret formula **(25)** is to compare it to the special cases of motion on a line and motion at a constant speed around a circle. In the case of motion on a line, $\kappa = 0$ and the acceleration vector equals the object's scalar acceleration $\dfrac{d^2s}{dt^2}$, as defined in earlier chapters for motion on an s-axis, multiplied by the unit tangent vector. In the case of motion at a constant speed around a circle, $\dfrac{d^2s}{dt^2} = 0$ and the acceleration vector equals the constant $\kappa\left(\dfrac{ds}{dt}\right)^2$, multiplied by the unit normal vector. Thus, in any case, the acceleration vector **(25)** at each point is equal to the

10.5 Acceleration, force, and curvature

acceleration vector of an object moving on the tangent line with scalar acceleration $\dfrac{d^2s}{dt^2}$, plus the acceleration vector of an object moving around the circle of curvature with constant speed $\dfrac{ds}{dt}$.

Example 10   A ball swung on a string is moving counterclockwise around the circle $x^2 + y^2 = 9$ in an xy-plane with distances measured in meters. The ball is going 2 meters per second and is speeding up 3 meters per second^2 when it is at the point $(3\cos(1), 3\sin(1))$. Draw its path and its acceleration vector at that point, using the scales on the axes to measure the components.

SOLUTION   Since the path is a circle of radius 3 oriented counterclockwise, its curvature κ is the constant $\frac{1}{3}$. By formula **(25)**, the ball's acceleration vector at $(3\cos(1), 3\sin(1))$ is

$$\mathbf{a} = \dfrac{d^2s}{dt^2}\mathbf{T} + \kappa\left(\dfrac{ds}{dt}\right)^2 \mathbf{N} = 3\mathbf{T} + \tfrac{1}{3}(2^2)\mathbf{N} = 3\mathbf{T} + \tfrac{4}{3}\mathbf{N}$$

with $\mathbf{T}$ and $\mathbf{N}$ the unit tangent and normal vectors at the point. We use the scales on the axes to draw the projection $\mathbf{a_T} = 3\mathbf{T}$ of the acceleration vector in the direction of $\mathbf{T}$ and its projection $\mathbf{a_N} = \tfrac{4}{3}\mathbf{N}$ in the direction of $\mathbf{N}$. Then we draw their sum $\mathbf{a} = \mathbf{a_T} + \mathbf{a_N}$ across the diagonal of the rectangle they form, as in Figure 12. □

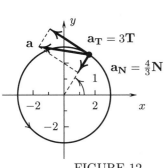

FIGURE 12

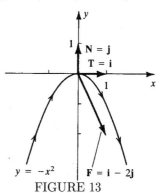

FIGURE 13

Question 7   (a) What are the components of $\mathbf{T}$ and $\mathbf{N}$ at $t = 1$ in Figure 12? (b) Find the x- and y-components of the acceleration vector in Figure 12.

Example 11   An eight-pound cat is moving from left to right along the curve $y = -x^2$ in an xy-plane with distances measured in feet. The total force on the cat is $\langle 1, -2 \rangle$ pounds when it is at the origin. What is the cat's speed and at what rate is it speeding up or slowing down when it is at the origin?

SOLUTION   Because the acceleration of gravity is 32 feet per second^2, the cat's mass m is $\tfrac{8}{32} = \tfrac{1}{4}$ slug and, by Newton's Law, the cat's acceleration vector at the origin is

$$\mathbf{a} = \dfrac{1}{m}\mathbf{F} = \dfrac{1}{1/4}(\mathbf{i} - 2\mathbf{j}) = 4\mathbf{i} - 8\mathbf{j}. \qquad (26)$$

On the other hand, the unit tangent vector $\mathbf{T}$ at the origin is $\mathbf{i}$ and the unit normal vector $\mathbf{N}$ is $\mathbf{j}$, as shown in Figure 13, and we saw in Example 7 that the curvature κ of the parabola at the origin is -2. Therefore, equation **(25)** gives

$$\mathbf{a} = \dfrac{d^2s}{dt^2}\mathbf{i} - 2\left(\dfrac{ds}{dt}\right)^2 \mathbf{j}. \qquad (27)$$

Comparing **(26)** and **(27)** shows that $\dfrac{d^2s}{dt^2} = 4$ and $-2\left(\dfrac{ds}{dt}\right)^2 = -8$. The second equation gives $\dfrac{ds}{dt} = 2$ since $\dfrac{ds}{dt}$ is nonnegative. The cat is traveling 2 feet per second and is speeding up 4 feet per second² at the origin. □

Kepler's Laws and Newton's Law of gravity

In 1609 and 1619 the German astronomer Johann Kepler (1571–1630) published his three laws of planetary motion:

1. The planets have elliptical orbits with the sun at one focus.

2. The line segment joining a planet to the sun sweeps out equal areas in equal times (see Figure 14).

3. The ratio T^2/d^3 is the same for all planets, where d is the average of the planet's closest and farthest distances from the sun and T is its period (the time it takes for it to make one revolution).

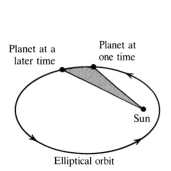

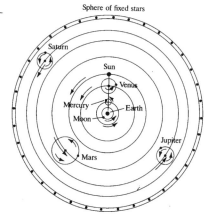

Ptolemy's theory of the solar system

FIGURE 14   FIGURE 15

Kepler's laws were the beginning of modern astronomy. Most earlier astronomers thought that the orbits of the planets were circles or curves formed by circles turning on circles. The Egyptian astronomer Ptolemy (ca. 150 AD), for example believed that the sun went around the earth in an elliptical path and the planets in epicycles—circles turning on circles (Figure 15[1]). Kepler formulated his laws by studying voluminous data collected by the Danish astronomer Tycho Brahe (1546–1601). Brahe obtained his data without the aid of a telescope, which was not invented until the beginning of the seventeenth century.

In 1687, Isaac Newton published his *Principia Mathematica*, in which he showed that Kepler's laws could be derived from the inverse-square law of gravitation and that Kepler's laws imply the inverse-square law. The inverse-square law (which is now commonly referred to as "Newton's law of gravity") states that two forces with masses M and m attract each other with a force directed along the line segment between them and of magnitude

$$\frac{GmM}{r^2} \tag{28}$$

where G is a universal constant.[†] Versions of inverse square laws had been proposed earlier to explain the motion of the planets, but Newton was the first to state the law in its definitive form and to carry out the mathematics to analyze it—a process that required that he develop much of what is know known as calculus.

---

[1]Adapted from *The Flammarion Book of Astronomy*, London: George Allen and Unwin Ltd., 1964, p. 253.

[†]If, for instance, M and m are measured in kilograms, r in meters, and the force in newtons, then $G \doteq 6.670 \times 10^{-11}$ newton-meters²/kilogram².

Circular orbits

We can use formula (**25**) for the tangential and normal components of acceleration to show that Newton's inverse-square law implies Kepler's second and third laws in the special case of a planet with a circular orbit. We suppose that the orbit of a planet is a circle of radius r with its center at the sun. We put the sun at the origin of an xy-plane containing the orbit and assume that the planet goes around it counterclockwise as in Figure 16.

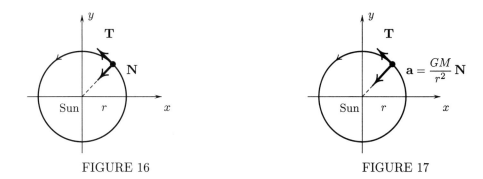

FIGURE 16   FIGURE 17

Figure 16 shows the unit tangent and normal vectors at a point on the circle (drawn with a different scale than is used for distances in the drawing). Because **N** points toward the center of the circle, the force of gravity on the planet is $\mathbf{F} = \dfrac{GmM}{r^2}\mathbf{N}$ with G the universal constant of gravitation, M the mass of the sun, and m the mass of the planet. Dividing **F** by the mass of the planet gives its acceleration vector (drawn with another scale in Figure 17),

$$\mathbf{a} = \frac{GM}{r^2}\mathbf{N}. \tag{29}$$

Comparing this equation with (**25**) shows that for all values of the time t,

$$\frac{d^2s}{dt^2} = 0 \quad \text{and} \quad \kappa\left(\frac{ds}{dt}\right)^2 = \frac{GM}{r^2}. \tag{30}$$

Example 12 (a) Give a formula in terms of r for the speed a planet would have if it had a circular orbit of radius r with the sun at the center. (b) Give a formula in terms of r for the period T of the planet from part (a).

SOLUTION (a) With coordinates as in Figure 16, the curvature of the orbit is $\kappa = 1/r$ and the second of equations (**30**) gives $\dfrac{1}{r}\left(\dfrac{ds}{dt}\right)^2 = \dfrac{GM}{r^2}$, so that $\left(\dfrac{ds}{dt}\right)^2 = \dfrac{GM}{r}$ and the planet's speed is $\dfrac{ds}{dt} = \sqrt{\dfrac{GM}{r}}$.

(b) The planet's period T is the circumference $2\pi r$ of its path divided by its speed, so the formula from part (a) gives

$$T = \frac{2\pi r}{\sqrt{GM/r}} = \frac{2\pi r^{3/2}}{\sqrt{GM}} \text{ seconds.} \ \square$$

Question 8 (a) Why is Kepler's First Law satisfied by the planet of Example 12? (b) Use the results of Example 12 to show that Kepler's Second Law is satisfied in this case. (c) Use the results of Example 12 to show that the quantity T^2/d^3 in Kepler's Third Law would be the same for all planets with circular orbits.

General orbits

The first step in showing that Newton's inverse-square law implies Kepler's laws is to show that each orbit lies in a plane through the sun. Suppose that the total force vector on a planet is given by Newton's law. We introduce coordinates in xyz-space with the origin at the sun and assume that the position vector $\mathbf{R}(t)$ of the planet has continuous first and second derivatives and is never zero. We also assume that at time $t = 0$ the velocity vector $\mathbf{v} = \dfrac{d\mathbf{R}}{dt}$ is not pointed to or away from the origin, since if it were the planet would go directly into the sun.

We use a Product Rule for cross products to calculate

$$\frac{d}{dt}(\mathbf{v} \times \mathbf{R}) = \frac{d\mathbf{v}}{dt} \times \mathbf{v} + \mathbf{v} \times \frac{d\mathbf{R}}{dt} = \mathbf{a} \times \mathbf{R} + \mathbf{v} \times \mathbf{v}. \tag{31}$$

Question 9 Explain why both vectors on the right of **(31)** are equal to $\mathbf{0}$ for all values of t.

Because $\dfrac{d}{dt}(\mathbf{v} \times \mathbf{R}) = \mathbf{0}$ for all t, as shown in the Response to Question 9, $\mathbf{v} \times \mathbf{R} = \mathbf{C}$ with a constant vector $\mathbf{C}$. The constant vector is not zero because $\mathbf{v}$ and $\mathbf{R}$ are nonzero and nonparallel at $t = 0$. Consequently, the position vector $\mathbf{R}$, which goes from the sun to the planet, is always perpendicular to $\mathbf{C}$, so that the planet is always in the plane through the origin perpendicular to $\mathbf{C}$.

Polar forms of position, velocity, and acceleration vectors

Now that we know that the orbit of the planet is in a plane, we put xy-coordinates in the plane with the sun at the origin and use polar coordinates $[r, \theta]$.

We define perpendicular unit vectors $\mathbf{u_r}$ and $\mathbf{u}_\theta$ at the point $[r, \theta]$ by having $\mathbf{u_r}$ point in the direction away from the origin and having $\mathbf{u}_\theta$ be perpendicular to $\mathbf{u_r}$ and point to its left, as in Figure 18. These vectors depend only on the angle θ and are given by

$$\begin{aligned}\mathbf{u_r}(\theta) &= \cos\theta\,\mathbf{i} + \sin\theta\,\mathbf{j} \\ \mathbf{u}_\theta(\theta) &= -\sin\theta\,\mathbf{i} + \cos\theta\,\mathbf{j}.\end{aligned} \tag{32}$$

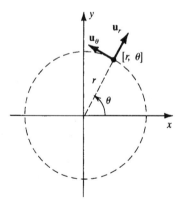

FIGURE 18

The next theorem gives formulas position, velocity, and acceleration vectors in terms of polar coordinates and the vectors $\mathbf{u_r}$ and $\mathbf{u}_\theta$.

10.5 Acceleration, force, and curvature

Theorem 5 *At any time t where an object's polar coordinates $r(t)$ and $\theta(t)$ have continuous second derivatives and $r(t)$ is not zero, the object's position, velocity, and acceleration vectors are*

$$\mathbf{R} = r\mathbf{u_r} \tag{33}$$

$$\mathbf{v} = \frac{dr}{dt}\mathbf{u_r} + r\frac{d\theta}{dt}\mathbf{u_\theta} \tag{34}$$

$$\mathbf{a} = \left[\frac{d^2r}{dt^2} - r\left(\frac{d\theta}{dt}\right)^2\right]\mathbf{u_r} + \left[r\frac{d^2\theta}{dt^2} + 2\frac{dr}{dt}\frac{d\theta}{dt}\right]\mathbf{u_\theta}. \tag{35}$$

Equation **(33)** follows from the definition of polar coordinates. Equations **(34)** and **(35)** can then be derived by using the Product and Chain Rules with the formulas,

$$\frac{d}{d\theta}\mathbf{u_r} = \mathbf{u_\theta}, \quad \text{and} \quad \frac{d}{d\theta}\mathbf{u_\theta} = -\mathbf{u_r} \tag{36}$$

that come from definitions **(32)**.

Newton's inverse-square law states that a planet's acceleration vector is

$$\mathbf{a} = -\frac{GM}{r^2}\mathbf{u_r}$$

and comparing this with **(35)** yields the scalar equations

$$r\frac{d^2\theta}{dt^2} + 2\frac{dr}{dt}\frac{d\theta}{dt} = 0 \tag{37}$$

$$\frac{d^2r}{dt^2} - r\left(\frac{d\theta}{dt}\right)^2 = -\frac{GM}{r^2}. \tag{38}$$

These equations for a general orbit take the place of the much simpler equations

$$\frac{d^2s}{dt^2} = 0 \quad \text{and} \quad \frac{1}{r}\left(\frac{ds}{dt}\right)^2 = \frac{GM}{r^2}$$

that we used above to study the case of a circular orbit. Equation **(37)** can be used to derive Kepler's Second Law and then **(38)** can be used to derive his First Law (see Problems 69 and 70.)

Principles and Procedures

- Integrals of vector-valued functions, like their derivatives, can be calculated with vector notation or by dealing with each component separately and then combining the results. You might find that it is easier to deal with the components separately if the calculations required are very complex. Vector calculations, on the other hand, are often easier to understand.

- Integrating a velocity vector gives the corresponding position vector with an unknown vector constant of integration in any time interval where the position vector is continuous, and integrating an acceleration vector gives the corresponding velocity vector with an unknown vector constant of integration in any time interval where the velocity vector is continuous. The vector constants of integration can be determined from values of the position or velocity vectors at one value of the time.

- Use Newton's law $\mathbf{F} = m\mathbf{a}$, with m the object's mass, to relate the total force vector $\mathbf{F}$ on an object to its acceleration vector $\mathbf{a}$ at a particular time.

- Any units can be used for time and distance in calculations involving only position, velocity and acceleration vectors. Compatible units for time, distance, mass, and force (seconds, feet, slugs, and pounds; seconds, centimeters, grams, and dynes; or seconds, meters, kilograms, and newtons) must be used, however, with Newton's law $\mathbf{F} = m\mathbf{a}$.

- Think of curvature of plane curves in terms of circles of curvature. If the curvature κ is positive at a point, then it equals $1/\rho$ with ρ the radius of the circle of curvature at that point and the center of the circle of curvature is to the left of the unit tangent vector at the point. If κ is negative, then it equals $-1/\rho$ and the center of the circle of curvature is to the right of the unit tangent vector.
- The radius of curvature can be considered to be infinite at any point where the curvature is zero.
- Notice that the formula $\kappa(x) = \dfrac{y''(x)}{\{1+[y'(x)]^2\}^{3/2}}$ for the curvature of the graph $y = y(x)$ of a function oriented from left to right becomes $\kappa(x) = y''(x)$ at any point where the tangent line is horizontal and $y'(t) = 0$, and that $|\kappa(x)| < |y''(x)|$ where the tangent line is not horizontal.
- To remember the sign in the formula $\kappa(t) = \dfrac{x'(t)y''(t) - y'(t)x''(t)}{\{[x'(t)]^2 + [y'(t)]^2\}^{3/2}}$ for the curvature of a parametrized curve, compare it with the formula for the graph of the function by setting $x(t) = t$ for which $x'(t) = 1$ and $x''(t) = 0$, so that $\kappa(t) = \dfrac{(1)y''(t) - y'(t)(0)}{\{1^2 + [y'(t)]^2\}^{3/2}} = \dfrac{y''(t)}{\{1 + [y'(t)]^2\}^{3/2}}$.
- Use the formula $\overrightarrow{OC} = \mathbf{R} + \dfrac{1}{\kappa}\mathbf{N}$ for the position vector of a center of curvature to find center's of curvature in cases where the tangent line is not vertical or horizontal (see Figure 11).
- To generate the circle of curvature with center $C = (C_1, C_2)$ and radius $\rho = 1/|\kappa|$ use a window with equal scales on the axes and the parametric equations $x = C_1 + \rho\cos t, y = C_2 + \rho\sin t$.
- To remember the formula $\mathbf{a} = \dfrac{d^2s}{dt^2}\mathbf{T} + \kappa\left(\dfrac{ds}{dt}\right)^2\mathbf{N}$, recall that $\dfrac{d^2s}{dt^2}\mathbf{T}$ would be the acceleration vector of an object moving on a line with tangent vector $\mathbf{T}$ and scalar acceleration $\dfrac{d^2s}{dt^2}$ and that $\kappa\left(\dfrac{ds}{dt}\right)^2\mathbf{N}$ would be the acceration vector of an object going at the constant speed $\dfrac{ds}{dt}$ around a circle with curvature κ.
- To draw an aceleration vector determined by the formula $\mathbf{a} = \dfrac{d^2s}{dt^2}\mathbf{T} + \kappa\left(\dfrac{ds}{dt}\right)^2\mathbf{N}$, pick a convenient scale for measuring lengths of acceleration vectors. Next, draw $\dfrac{d^2s}{dt^2}\mathbf{T}$ tangent to the curve and $\kappa\left(\dfrac{ds}{dt}\right)^2\mathbf{N}$ normal to it. Then draw $\mathbf{a}$ as the diagonal of the rectangle formed by the first two vectors.
- To find approximate values of $\dfrac{d^2s}{dt^2}$ and $\kappa\left(\dfrac{ds}{dt}\right)^2$ from an arrow representing $\mathbf{a}$, measure its projections in the directions of $\mathbf{T}$ and $\mathbf{N}$ by drawing a rectangle with $\mathbf{a}$ as diagonal and one side tangent to the curve at the point being considered.
- For an object's path to turn, its acceleration vector has to point toward the inside of the curve. The acceleration vector points in front the normal line if the object is speeding up and behind the normal line if the object is slowing down.

Tune-Up Exercises 10.5

^A*Answer provided.* ^O*Outline of solution provided.*

T1.^O Find the acceleration vector $\mathbf{a}(t)$ for an object with position vector $\mathbf{R}(t) = \langle t^5, -t^7 \rangle$ at time t.
♦ Type 1, finding an acceleration vector

T2.^O Find the antiderivative $\int \langle t^3, -t^4, t^5 \rangle \, dt$.
♦ Type 1, integrating a vector-valued function

T3.^O The total force vector on a five-gram object moving in an xy-plane with distances measured in centimeters is $\langle 10t, 5t^2 \rangle$ dynes at t seconds. The object is at $x = 1, y = 0$ and its velocity vector is $\langle 2, 1 \rangle$ centimeters per second at $t = 0$. Give a formula for its position vector in terms of t.
♦ Type 1, finding position from an acceleration vector

T4.^O (a) What is the curvature of the circle $x = 5 + 3\cos t, y = 6 + 3\sin t$? (b) What is the curvature of the circle $x = 7 + \frac{1}{2}\cos(-2t), y = -7 + \frac{1}{2}\sin(-2t)$?
♦ Type 1, curvature

T5^O Consider the curve $y = \ln x$, oriented from left to right. (a) What are its unit tangent and normal vectors at $x = 1$? Draw them with the curve. (b) What is its curvature at $x = 1$? (c) What is its radius of curvature at $x = 1$? (d) What is its center of curvature at $x = 1$? (e) Generate the curve and the circle of curvature at $x = 1$ in a window with equal scales on the axes that includes the square $-3 \le x \le 7, -7 \le y \le 7$.
♦ Type 1, tangent and normal vectors, curvature, circle of curvature

T6.^O The curvature of a toy car's path at point P is 2 meters^{-1} and when the car is at that point its acceleration vector is $\mathbf{a} = -7\mathbf{T} + 18\mathbf{N}$ meters per second^2, where $\mathbf{T}$ and $\mathbf{N}$ are the unit tangent vectors. (a) How fast is the car moving when it is at P? (b) At what rate is the car speeding up or slowing down at P?
♦ Type 3, tangential and normal components of acceleration

T7.^O Figure 19 shows an insect's path and the tangent and normal lines to the path at a point P, where the radius of curvature is four centimeters. When the insect is at P, it is moving six centimeters per second and is speeding up five centimeters per second^2. Draw its approximate acceleration vector at that point, using the scales on the axes to measure the components.
♦ Type 3, approximate tangential and normal components of acceleration

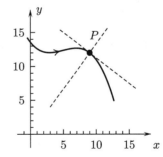

FIGURE 19

♦The Tune-up Exercises and Problems are classified by type and content. The types are (1) basic, reactive; (2) basic reflective; (3) intermediate, reactive; (4) intermediate, reflective; (5) advanced, reactive; (6) advanced, reflective; and (7) advanced, theoretical.

Problems 10.5
AAnswer provided. **O**Outline of solution provided.

1. Find $\dfrac{d\mathbf{R}}{dt}(t)$ and $\dfrac{d^2\mathbf{R}}{dt^2}(t)$ where (**a^O**) $\mathbf{R}(t) = te^t\,\mathbf{i} - \sin t\,\mathbf{j}$, (**b^A**) $\mathbf{R}(t) = \langle \ln(3t), \tan^{-1}(3t)\rangle$, (**c^A**) $\mathbf{R}(t) = \cos(3t)\,\mathbf{i} + \sin(3t)\,\mathbf{j} + t\,\mathbf{k}$, and (**d**) $\mathbf{R}(t) = \langle t^3, -\ln t, \sin^{-1} t\rangle$
   ♦ Type 1, derivatives of vectors

2. Find the velocity vector, speed, and acceleration vector (**a^O**) of an object that is at $x = \sin(2t)$, $y = 3\cos(4t)$ (millimeters) in an xy-plane at time t (minutes), (**b^A**) of an object that is at $x = 3t^2 + 1, y = \sin t, z = \cos^2 t$ (feet) in xyz-space at time t (seconds), (**c**) of an object that is at $x = \sin(2t), y = \sin(3t), z = \sin(4t)$ (meters) in xyz-space at time t (seconds), and (**d**) of an object that is at $x = e^t, y = t^3 - t^2, z = \ln(t^2 + 1)$ (yards) in xyz-space at time t (seconds).
   ♦ Type 1, velocity and acceleration vectors

3. Find formulas for the antiderivatives, (**a^O**) $\displaystyle\int \langle \sin t, \cos t, e^t\rangle\,dt$, (**b^A**) $\displaystyle\int \langle t\cos(t^2), t^2\cos(t^3)\rangle\,dt$, and (**c**) $\displaystyle\int \left(\dfrac{1}{t^2+1}\mathbf{i} - \dfrac{t}{t^2+1}\mathbf{j}\right) dt$.
   ♦ Type 1, integrals of vector-valued functions

4. Find formulas for the position vectors from the velocity vectors and particular values of the position vectors, (**a^O**) where $\mathbf{v}(t) = \langle t, -t^2\rangle$ and $\mathbf{R}(1) = \langle 1, 0\rangle$, (**b^A**) where $\mathbf{v}(t) = \langle \sin(3t), \cos(4t)\rangle$ and $\mathbf{R}(0) = \langle 4, 5\rangle$, (**c**) where $\mathbf{v}(t) = \langle e^{3t}, -e^{-4t}\rangle$ and $\mathbf{R}(0) = \langle 5, 5\rangle$, and (**d**) where $\mathbf{v}(t) = \langle 3t^{-1}, 5t^{-1}\rangle$ and $\mathbf{R}(1) = \langle 1, 1\rangle$.
   ♦ Type 1, finding position from velocity vectors

5. Find formulas for the position vectors from the acceleration vectors and particular values of the position and velocity vectors, (**a^O**) where $\mathbf{a}(t) = \langle t^4, t^3, t^2\rangle$, $\mathbf{v}(0) = \langle 1, 2, 3\rangle$, $\mathbf{R}(0) = \langle 7, 6, 5\rangle$, (**b^A**) where $\mathbf{a}(t) = \langle \sin(2t), 0, -\sin(2t)\rangle$, $\mathbf{v}(0) = \langle 0, 5, 0\rangle$, $\mathbf{R}(0) = \langle 5, 0, 0\rangle$, (**c**) where $\mathbf{a}(t) = \langle 0, 0, 1\rangle$, $\mathbf{v}(10) = \langle 0, 1, 0\rangle$, and $\mathbf{R}(10) = \langle 1, 0, 0\rangle$, and (**d**) where $\mathbf{a}(t) = \langle e^t, e^{2t}, e^{3t}\rangle$, $\mathbf{v}(1) = \langle e, \tfrac{1}{2}e^2, \tfrac{1}{3}e^3\rangle$, $\mathbf{R}(1) = \langle e, \tfrac{1}{4}e^2, \tfrac{1}{9}e^3\rangle$.
   ♦ Type 1, finding position from acceleration vectors

6.º An arrow is moving in xyz-space with the positive z-axis pointing up and with distances measured in meters. The arrow is at the origin at $t = 0$ (seconds) and its highest point is $(250, 0, 490)$. Give a formula for its position vector under the assumption that there is no air resistance.
   ♦ Type 1, motion under the force of gravity with no air resistance

7. A ball is moving in xyz-space with the positive z-axis pointing up and with distances measured in feet. The ball is at the point $(3, 0, 2)$ at $t = 0$ (seconds) and its velocity vector is $\langle 5, 0, 5\rangle$ feet per second at $t = 2$. Give a formula for its position vector under the assumption that there is no air resistance.
   ♦ Type 1, motion under the force of gravity with no air resistance

8.^{A} At $t = 1$ hours a boat is at $(10, -10)$ in a horizontal xy-plane with distances measured in nautical miles and its velocity vector is $\langle 5, -1\rangle$ knots (nautical miles per hour). Its acceleration vector for $t \geq 1$ is $\langle 3\sqrt{t}, t^{-2}\rangle$ knots per hour. Where is it for $t \geq 1$?
   ♦ Type 1, finding position from an acceleration vector

9.^{A} The total vector force acting on a three-gram object in an xy-plane with distances measured in centimeters is $\mathbf{F} = (6 + 9\sin t)\,\mathbf{i} + (t^2 - \cos t)\,\mathbf{j}$ dynes at time t (seconds). At $t = 0$ the object has the velocity vector $\langle -3, 0\rangle$ centimeters per second and is at $(5, -6)$. Where is it for $t > 0$?
   ♦ Type 3, finding position from a force vector

10. A ninety-six-pound object is moving in xyz-space with distances measured in feet. The total vector force acting on it at time t (seconds) is $\mathbf{F} = \langle 3t, -2, 0\rangle$ pounds, and it is at $(3, 2, 1)$ and has velocity vector $\langle 1, -2, 1\rangle$ feet per second at $t = 1$. Give formulas for its velocity and position vectors in terms of t.
   ♦ Type 3, finding position from a force vector

10.5 Acceleration, force, and curvature

11. An object moving in an xy-plane with distances measured in meters is at $(5, -6)$ and its velocity vector is $3\mathbf{i}$ meters per minute at time $t = 1$ (minutes). Its acceleration vector is $2t\mathbf{i} - t^3\mathbf{j}$ meters per minute^2 at time t. Give formulas for its xy-coordinates in terms of t.

◆ Type 1, finding position from an acceleration vector

12. At $t = 1$ (seconds), a five-kilogram ball is at the origin in xyz-space with distances measured in meters and its velocity vector is $\langle 0, 0, -1 \rangle$ meters per second. The total force acting on it at time t is $\mathbf{F} = \langle 30, 10t, 10t^{-3} \rangle$ newtons. Give formulas for its velocity and position vectors in terms of t.

◆ Type 3, finding position from a force vector

13. At $t = 0$ (seconds), a 16 pound ball is at at rest at the origin in an xy-plane with distances measured in feet. The vector force $\mathbf{F}(t)$ is applied to the ball for $t \geq 0$ is such that its coordinates at time t are $x = t - \sin t, y = 1 - \cos t$. **(a)** Give a formula for $\mathbf{F}(t)$. **(b)** Generate the ball's path (a CYCLOID) in a window with equal scales on the axes, copy the drawing on your paper, and add the force vectors at $t = \frac{1}{2}\pi, \pi, \frac{3}{2}\pi$, and π, using the scales on the axes to measure the components.

◆ Type 3, finding a force vector from a position vector

14. What are the unit tangent and normal vectors, $\mathbf{T}$ and $\mathbf{N}$ (**a**$^\mathbf{O}$) at $x = 2$ on the cubic $y = \frac{1}{4}x^3$, oriented from left to right, (**b**$^\mathbf{A}$) at $t = \pi$ on the curve $x = \sin t, y = t + 3\cos t$, and (**c**) at $t = -2$ on $y = 6/x$, oriented from left to right?

◆ Type 1, unit tangent and normal vectors

15. Draw, without making any calculations, the unit tangent and normal vectors at $t = \frac{1}{4}\pi, t = \pi$ and $t = \frac{7}{4}\pi$ **(a)** on the circle $x = 2\cos t, y = 2\sin t$ and **(b)** on the circle $x = 2\cos(-t), y = 2\sin(-t)$.

◆ Type 1, unit tangent and normal vectors

16.$^\mathbf{A}$ Figure 20 shows a curve C, points P, Q, and R on it, and the centers of curvature at the three points. Find the approximate curvature of the curve at P, Q, and R. Then trace the curve and draw the unit tangent vectors $\mathbf{T}$ and $\mathbf{N}$ at the three points, using the scales on the axes to measure their lengths.

◆ Type 1, approximate curvature

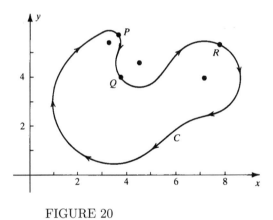

FIGURE 20   FIGURE 21

17. Find the approximate curvature at P, Q, and R on the curve in Figure 21. Then trace the curve and draw the unit tangent vectors $\mathbf{T}$ and $\mathbf{N}$ at the three points, using the scales on the axes to measure their lengths.

◆ Type 1, approximate curvature

18.$^\mathbf{A}$ Find the curvature and radius of curvature of the ellipse $x = 5\cos t, y = 3\sin t$ at $t = 0$. Then generate the ellipse and its circle of curvature at that point, using a window with equal scales on the axes and draw them on your paper.

◆ Type 3, circle of curvature

19.$^\mathbf{A}$ Find the curvature and radius of curvature of the curve $x = \frac{1}{3}t^3, y = \frac{1}{2}t^2$ at $t = 1$. Then generate the curve for $-2.4 \leq t \leq 2.4$ and its circle of curvature at that point, using a window with equal scales on the axes that includes the square $-3 \leq x \leq 6, -6 \leq y \leq 3$, and draw them on your paper.

◆ Type 3, circle of curvature

20.^A Find the center of curvature of the parabola $y = \frac{1}{2}x^2 - 4x$ at $x = 2$. Then generate the parabola and its circle of curvature at that point, using a window with equal scales on the axes that includes the rectangle $-12 \leq x \leq 28, -12 \leq y \leq 12$, and draw them on your paper.

♦ Type 3, circle of curvature

21. Find the center of curvature of $y = 2x - x^3$ at $x = 1$. Then generate the curve and its circle of curvature at that point, using a window with equal scales on the axes that includes the square $-1.8 \leq x \leq 1.8, -1.8 \leq y \leq 1.8$, and copy them on your paper.

♦ Type 3, circle of curvature

22. Show that the hyperbola $y = 1/x$ has the same radius of curvature at $x = -1$ as at $x = 1$. Then generate the hyperbola and its circles of curvature at $x = \pm 1$, using a window with equal scales on the axes that includes the square $-1.8 \leq x \leq 1.8, -1.8 \leq y \leq 1.8$, and copy them on your paper.

♦ Type 4, circles of curvature

23. Show that the centers of curvature of $y = \frac{1}{2}x^4 - x^2$ at $x = -1, x = 0$, and $= 1$ are all tangent to the x-axis. Then generate the curve and the circles of curvature, using a window with equal scales on the axes that includes the rectangle $-1.8 \leq x \leq 1.8, -1.2 \leq y \leq 1.2$, and copy them on your paper.

♦ Type 4, circles of curvature

24. Explain why the curvature of $y = f(x)$ is positive where the graph is concave up and negative where the graph is concave down.

♦ Type 2, circles of curvature

25. For what values of x does $y = \tan^{-1} x$ have positive curvature?

♦ Type 2, curvature

26.º A 96-pound object is moving from left to right on the curve $y = e^{-x}$ in an xy-plane with distances measured in feet. When it is at $x = 0$, its speed is 6 feet per second and it is speeding up $2\sqrt{2}$ feet per second^2. What is the total vector force on it at that point?

♦ Type 3, tangential and normal components of acceleration

27.^A An ant is traveling counterclockwise around the ellipse $\frac{1}{36}x^2 + \frac{1}{64}y^2 = 1$ in an xy-plane with distances measured in centimeters. When it is at $(-6, 0)$, its vector acceleration is $\langle 6, 4 \rangle$ centimeters per second^2. What is its speed and at what rate is it speeding up or slowing down at that point?

♦ Type 4, tangential and normal components of acceleration

28. At time t (seconds) a ten-kilogram ball is at $x = t \cos t, y = t \sin t$ on a spiral in an xy-plane with distances measured in meters. When it is at the origin it is traveling three meters per second and is speeding up seven meters per second^2. What is the total vector force on it at that point?

♦ Type 4, tangential and normal components of acceleration

29. At a point P on an object's path in an xy-plane with distances measured in feet, the unit tangent vector is $\langle -1, 5 \rangle / \sqrt{26}$, the curvature of the path is $-\frac{1}{6}$, and the object's vector acceleration is $\sqrt{26}(\mathbf{i} + \mathbf{j})$. What is the object's speed and at what rate is it speeding up or slowing down when it is at P?

♦ Type 4, tangential and normal components of acceleration

30.^A An object is traveling counterclockwise around the ellipse $\frac{1}{9}x^2 + \frac{1}{4}y^2 = 1$ in an xy-plane with distances measured in feet. When it reaches the point $(0, -2)$, its vector acceleration is $\langle 3, 5 \rangle$ feet per second^2. What is the object's speed and at what rate is it speeding up or slowing down at that time?

♦ Type 4, tangential and normal components of acceleration

31. A sixteen-pound dog has vector velocity $4\mathbf{j}$ feet per second and the total vector force on it is $\mathbf{i} - 3\mathbf{j}$ pounds when it reaches the origin in an xy-plane with distances measured in feet. Find the curvature of its path at that moment.

♦ Type 4, tangential and normal components of acceleration

32. A man weighing 160 pounds is running counterclockwise around the circle $x^2 + y^2 = 25$ in an xy-plane with distances measured in feet. When he gets to the point $(3, 4)$, he is running five feet per second and is speeding up at the rate of ten feet per second^2. What is his acceleration vector at that point?

♦ Type 3, tangential and normal components of acceleration

10.5 Acceleration, force, and curvature

33.º A toy car is traveling around the curve in Figure 22. **(a)** What is the approximate curvature of its path at the point P? (The corresponding center of curvature is shown with a dot.) **(b)** The angle of inclination of the unit tangent vector at P is $\frac{1}{4}\pi$ and the car's vector acceleration at P is $\langle 5, -3 \rangle / \sqrt{2}$ feet per second. Approximately how fast is it going and at what rate is it speeding up or slowing down at P?

♦ Type 4, tangential and normal components of acceleration

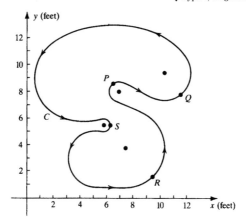

FIGURE 22

34.^A When the toy car traveling around the curve in Figure 22 is at the point Q, its acceleration vector is $-\mathbf{T} + 2\mathbf{N}$, with $\mathbf{T}$ and $\mathbf{N}$ the unit tangent vectors at that point. Approximately how fast is the car going and at what rate is it speeding up or slowing down at Q? (The center of curvature at Q is shown by a dot.)

♦ Type 4, tangential and normal components of acceleration

35. When the toy car traveling around the curve in Figure 22 is at the point R, it is neither speeding up nor slowing down and its speed is three feet per second. Give the x- and y-components of its vector acceleration at that point. (The center of curvature at R is shown by another dot, and the angle of inclination of the unit tangent vector at R is $\frac{1}{4}\pi$.)

♦ Type 4, tangential and normal components of acceleration

36. When the toy car from Problems 33–35 is at point S its speed is zero and it is speeding up at the rate of two feet per second^2. The car's mass is six slugs, the unit tangent vector at S is vertical, and the center of curvature at S is shown by another dot. Give the approximate x- and y-components of the total vector force acting on the car at that point.

♦ Type 4, tangential and normal components of acceleration

37.^A A ball is thrown at time $t = 0$ (seconds) from the point $(0, 98)$ in an xy-plane with the positive y-axis pointing up and with distances measured in meters. Its velocity vector at $t = 0$ is $\langle 3, 4.9 \rangle$ meters per second. Where does its path cross the x-axis if there is no air resistance and the only force on it is gravity?

♦ Type 4, motion under gravity with no air resistance

38.^A A ten-kilogram box is dropped from an airplane at time $t = 0$ (seconds) when it is directly over a farmhouse. The plane is flying toward the north at the constant speed of 100 meters per second and at the constant elevation of 490 meters. How far from the farmhouse does the box hit the ground if the only forces on the box are the force of gravity and the force of the wind, which is $10\mathbf{i} + 50\mathbf{j}$ newtons at time t, where $\mathbf{i}$ points toward the east and $\mathbf{j}$ points north?

♦ Type 4, motion under gravity with no air resistance

39. At time $t = 0$ (seconds), a rock is thrown into the air from the origin in xyz-space with the positive z-axis pointing up and distances measured in meters. Suppose there is no air resistance and that the rock reaches its highest point at $t = 1$ and hits the ground at the point $(31, 16, 0)$. Give formulas for its coordinates in terms of t from $t = 0$ until it hits the ground.

♦ Type 4, motion under gravity with no air resistance

40. An arrow is shot horizontally at $t = 0$ seconds from the point $(0, 0, 100)$ in xyz-space with the positive z-axis pointing up and distances measured in meters, passes through the point $(60, 40, 21.6)$. How long after it is shot does it hit the ground if there is no air resistance?

♦ Type 4, motion under gravity with no air resistance

41. An ball is thrown into the air in xyz-space with the positive z-axis pointing up and distances measured in feet. It passes through the point $(40, 30, 50)$ at $t = 0$ and $t = 1$ seconds. Describe and give formulas for its path under the assumption that there is no air resistance.

♦ Type 4, motion under gravity with no air resistance

42. An arrow in xyz-space with the positive z-axis pointing up and distances measured in meters passes through the origin at $t = 0$ and intersects the x-axis at $x = 100$ five seconds later. Give its coordinates for $0 \leq t \leq 5$ under the assumption that there is no air resistance.

♦ Type 4, motion under gravity with no air resistance

43. An arrow in xyz-space with the positive z-axis pointing up and distances measured in feet has the velocity vector $10\mathbf{i}$ feet per second at $t = 3$, when it is at $(50, 40, 200)$, which is its highest point. Give formulas in terms of t its coordinates while it is in the air under the assumption that there is no air resistance.

♦ Type 4, motion under gravity with no air resistance

44. A ten-pound ball in xyz-space with the positive z-axis pointing up and distances measured in feet is at $(50, 40, 300)$ and has the velocity vector $\langle 25, -40, 32 \rangle$ feet per second at $t = 0$. What are its coordinates at its highest point if there is no air resistance?

♦ Type 4, motion under gravity with no air resistance

45.^A A two-thousand-kilogram car is moving in an xy-plane with distances measured in meters. Its velocity at time t (seconds) is $\langle 40 - 6t, 50 - t \rangle$ meters per second and at $t = 2$ it is at $x = 100, y = 100$. Give formulas **(a)** for its position vector and **(b)** for the total force vector on it in terms of t.

♦ Type 4, finding position and force from acceleration

46. By using the inertial navigation system on a space station, it is determined that the station's acceleration at time t (seconds) for $0 \leq t \leq 10$ is $\langle 6t, \frac{1}{25}t^3, -20 \rangle$ kilometers per hour^2 with respect to a fixed xyz-coordinate system. It is known that the the station's velocity is $\langle 50, -40, 30 \rangle$ kilometers per hour at $t = 0$. **(a)** What is the net distance the station travels for $0 \leq t \leq 10$ (the distance between its positions at $t = 0$ and $t = 10$)? **(b)** Express as an integral the total distance it travels for $0 \leq t \leq 10$ and find its approximate decimal value.

♦ Type 4, finding total distance from acceleration

47. At $t = 0$ (hours) a car is at $(2,3)$ in an xy-plane with distances measured in miles, and at $t = 1$ it is at $(5,0)$. Its vector acceleration at time t is $\langle 20t^3, 2 \rangle$ miles per hour^2. Give a formula for its vector velocity in terms of t.

♦ Type 4, finding velocity from accleration and two positions

48. At $t = 0$ (minutes) a ball is at $(1,0)$ in an xy-plane with distances measured in yards and its vector velocity is $\langle -1, 1 \rangle$ yards per minute. Its vector acceleration at time t for $t \geq 0$ is $\langle 2(t+1)^{-3}, -2t(t^2+1)^{-2} \rangle$ yards per minute. What happens to the ball as $t \to \infty$?

♦ Type 4, finding a limit of a velocity

49. Show that the curvature of $y = e^x$ tends to zero as $x \to \infty$ and as $x \to -\infty$.

♦ Type 4, finding a limit of a curvature

50. What is the curvature of $x = e^t, y = 2e^t$? Give a geometric explanation of the result.

♦ Type 4, interpreting curvature

51. Find the maximum and minumum curvature of $y = \sin x$. (Study the formula for κ, not its derivative.) Then draw the curve and circles of curvature at points where the curvature is a maximum and a minimum.

♦ Type 4, maximum and minimum curvature

52. Show that the least radius of curvature of $y = \ln x$ is $\frac{3}{2}\sqrt{3}$.

♦ Type 4, minimum radius of curvature

10.5 Acceleration, force, and curvature

53. What is the total vector force on a three-kilogram object at a point where its vector velocity is $\langle 2, -1 \rangle$ meters per second if it is speeding up six meters per second^2 and the curvature of its path is $-\frac{1}{5}$ meters^{-1}?
♦ Type 4, tangential and normal components of acceleration

54. What is the maximum magnitude of the force of friction that is required to keep a car on a circular racetrack of radius 300 feet if the car weighs 3200 pounds and travels at the constant speed of 200 feet per second?
♦ Type 4, tangential and normal components of acceleration

55. A satellite is travelling at a constant speed in a circular orbit 400 miles above the surface of the earth where the acceleration due to gravity is 30 feet per second^2. What is its speed? Use 4000 miles for the radius of the earth. (One mile is 5280 feet.)
♦ Type 4, tangential and normal components of acceleration

56. Suppose that a dime weighing k pounds will not slide off a turning phonograph record provided the force on it is no greater than $\frac{1}{9}k$ pounds. How far from the center of a record turning at 45 revolutions per minute can a dime be without sliding off?
♦ Type 4, tangential and normal components of acceleration

57. Explain why a car is just as likely to skid traveling at a constant speed of $2v_0$ around a circular track of radius ρ as at the constant speed of v_0 around a circular track of radius $\frac{1}{4}\rho$.
♦ Type 6, tangential and normal components of acceleration

58. What is the maximum magnitude of the force required to cause an object weighting two pounds to move at the constant speed of three feet per second on the parabola $y = x^2$ in an xy-plane with distances measured in feet?
♦ Type 6, tangential and normal components of acceleration

59. Word-class jogger Rainier Schein is running in a horizontal xy-plane with distances measured in meters and with the vector $\mathbf{j}$ pointing north. At one point he is running three meters per second toward the north and the total vector force on him is $36\mathbf{i} + 160\mathbf{j}$ newtons. He weighs 80 kilograms. At what rate is he speeding up or slowing down at that point? What is the curvature of his path at that point? Is he turning right or left then?
♦ Type 4, tangential and normal components of acceleration

60. Show that the curvature of $x = t\cos t, y = 2\sin t$ in Figure 23 is zero only at the origin.
♦ Type 4, tangential and normal components of acceleration

FIGURE 23
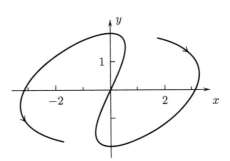

61. Show that if the vector acceleration of an object is $\mathbf{0}$ for all t, then the object moves on a straight line at a constant speed.
♦ Type 4, integrating acceleration

62.$^\mathbf{A}$ A boy throws a ball at an angle of 45° with the horizontal and with an initial speed of 30 feet per second toward a building 35 feet away. The ball leaves the boy's hand from three feet above the ground. **(a)** Will the ball reach the building if there is no air resistance? If so, when? **(b)** Will the ball reach the building if there is air resistance? If so, when? If not, why not?
♦ Type 6, motion under gravity with no air resistance

63. A bead slides down a ramp that makes an angle of 5° with the horizontal. There is no friction or air resistance, so the only force exerted by the ramp is the force that cancels the component of the force of gravity perpendicular to the ramp. How far does the bead slide in six seconds if its initial speed is zero?

♦ Type 4, motion under gravity with no air resistance

64. A cannon with muzzle velocity M feet per second is fired at an angle ψ with the horizontal ($0 < \psi < \frac{1}{2}\pi$). Suppose there is no air resistance. **(a)** What is the range of the cannon? **(b)** What angle ψ gives the maximum range? **(c)** Find two angles of fire such that if the muzzle velocity is 2000 feet per second a shell from the cannon hits the ground 62,500 feet away. (Use the trigonometric identity $2 \sin \psi \cos \psi = \sin(2\psi)$.)

♦ Type 4, motion under gravity with no air resistance

65. A baseball player throws a ball with an initial speed of 60 feet per second from a height of four feet above the ground and at an angle of 30° with the horizontal. How high does it go and at what distance does it hit the ground if there is no air resistance?

♦ Type 4, motion under gravity with no air resistance

(Circles of curvature) In the next three Problems we outline a proof, based on l'Hopital's Rule, that if a curve $x = x(t), y = y(t)$ has nonzero curvature at the point $P = (x(t_0), y(t_0))$ then the circle through P and points $Q = (x(t_1), y(t_1))$ and $R = (x(t_2), y(t_2))$ on the curve with $t_2 < t_0 < t_1$ approaches the circle of curvature at P as t_1 and t_2 tend to t_0. To simplify notation, we choose the parameter t so that $t_0 = 0$.

66. Suppose that $x(0) = 0$ and $y(0) = 0$, so that P is the origin, that $y'(0) = 0$, so the curve is tangent to the x-axis at P and that $y''(0) \neq 0$, so the curvature is not zero at P. Suppose also that $x''(t)$ and $y''(t)$ are continuous, that the curve is symmetric about the y-axis, and that the points Q and R are symmetric about the y-axis, as in Figure 24. Let $Q = (x(t), y(t))$ be the point on the curve for a t close to 0. **(a)** Explain why the center of the circle through P, Q, and R is the point U where the perpendicular bisectors of the chords PQ and PR intersect. **(b)** Find the slope and midpoint of the chord PQ and use these numbers to give an equation for the perpendicular bisector SU. **(c)** Express the y-coordinate y_I of U in terms of $x(t)$ and $y(t)$. **(c)** Use l'Hopital's Rule twice to show that $y_I \to \dfrac{[x'(0)]^2}{y''(0)}$ as $t \to 0$ and explain why the limit is the reciprocal of the curvature in this case.

♦ Type 7, a curvature proof

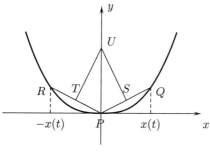

FIGURE 24

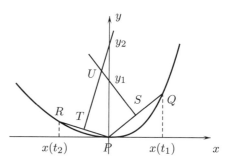

FIGURE 25

67. We again suppose that $x(0) = 0, y(0) = 0, y'(0) = 0$, and $y''(0) \neq 0$, as in Problem 66, but drop the requirement that the curve be symmetric about the y-axis and consider points $Q = (x(t_1), y(t_1))$ and $R = (x(t_2), y(t_2))$ with $t_2 < 0 < t_1$ that are not necessarily symmetric about the y-axis. Then the center of the circle through P, Q, and R is the point U in Figure 25, where the perpendicular bisectors of the chords PQ and PS intersect. The derivation in Problem 66 shows that the y-intercepts, y_1 and y_2 of the perpendicular bisectors tend to $\dfrac{[x'(0)]^2}{y''(0)}$ as t_1 and t_2 tend to 0. Why does this imply that the y-coordinate of U has the same limit, so that $\dfrac{[x'(0)]^2}{y''(0)}$ is the reciprocal of the curvature at P in this case as well.

♦ Type 7, a curvature proof

10.5 Acceleration, force, and curvature

68. If $P = (x_0, y_0)$ is not the origin or the curve is not tangent to the x-axis at P, we can introduce new XY-coordinates, $X = x\cos\alpha - y\sin\alpha - x_0, Y = x\sin\alpha + y\cos\alpha - y_0$, with the angle α from the positive x-axis to the positive X-axis chosen so that P is at $X = 0, Y = 0$ and the curve is tangent to the X-axis at $t = 0$. By the results of Problems 66 and 67, the curvature of the curve at P is $\dfrac{Y''(0)}{[X'(0)]^2}$. Show that this gives formula **(20)** in Theorem 2.

♦ Type 7, a curvature proof

We saw above that Newton's inverse-square law of gravity implies that each planet moves in a plane containing the sun. The proofs outlined in the next two Problems complete the demonstration that Newton's Law implies Kepler's first two laws.[6]

69. Suppose that the orbit of a planet moving in an xy-plane with the sun at the origin has the polar equation $r = r(\theta)$, where $r(\theta)$ is not zero and has a continuous second derivative. By Rule 1 of Section 9.3, the area swept out by a line segment from the sun to the planet when the line segment rotates counterclockwise from an angle of inclination θ_0 to the angle of inclination Θ is $A(\Theta) = \displaystyle\int_{\theta_0}^{\Theta} \tfrac{1}{2}[r(\theta)]^2 \, d\theta$, so that by the Fundamental Theorem of Calculus for derivatives of integrals in Section 6.5, $\dfrac{dA}{d\Theta}(\Theta) = \tfrac{1}{2}[r(\Theta)]^2$, or, with Θ replaced by θ,

$$\frac{dA}{d\theta}(\theta) = \tfrac{1}{2}[r(\theta)]^2. \tag{39}$$

(a) Suppose that the planet's polar coordinates at time t are $[r(t), \theta(t)]$, and that $A(t)$ is the area swept out by the line segment between it and the sum from time t_0 to time t. Use **(39)** to express dA/dt in terms of r and $d\theta/dt$. **(b)** Use the Product and Chain Rules with the result of part (a) to express d^2A/dt^2 in terms of $r, dr/dt, d\theta/dt$, and $d^2\theta/dt^2$. Use this with formula **(35)** to show that if the total force on the planet is directed toward the sun, then dA/dt is constant.

♦ Type 7, Kepler's and Newton's Laws

70. Suppose that the conditions of Problem 69 are satisfied and choose coordinates so the planet is closest to the sun and is on the positive x axis at $t = 0$ as in Figure 26. Let $\mathbf{v}(t)$ be the planet's velocity vector at time t Then $\mathbf{v}(0) = c\mathbf{j}$ with a constant c. We write the result of Problem 69 as

$$r^2 \frac{d\theta}{dt} = k \tag{40}$$

with a positive constant k. Newton's law can be written in the form

$$\frac{d\mathbf{v}}{dt} = -\frac{\mu}{r^2}\mathbf{u_r} \tag{41}$$

with a positive constant μ that depends on the masses of the planet and the sun. We let $\mathbf{u_r}$ and $\mathbf{u_\theta}$ be the unit polar vectors **(32)** at the planet's position. Then the second of formulas **(36)** and the Chain Rule show that $\dfrac{d\mathbf{u_\theta}}{dt} = -\dfrac{d\theta}{dt}\mathbf{u_r}$. **(a)** Use this with equations **(40)** and **(41)** to show that

$$\frac{d\mathbf{v}}{dt} = \frac{\mu}{k}\frac{d\mathbf{u_\theta}}{dt}. \tag{42}$$

---

[6] The reasoning in Problem 70 is adapted from "Inverse-square orbits: a simple treatment" by D. Richmond, *Selected Papers on Calculus*, Washington, DC: Mathematical Association of America, 1969, pp. 164–167.

(b) Integrating **(42)** with respect to t yields $\mathbf{v} = \dfrac{\mu}{k}\mathbf{u}_\theta + \mathbf{C}$ with a constant vector C. Since $\mathbf{v}(0) = c\mathbf{j}$, $\mathbf{C}$ equals $\dfrac{\mu e}{k}\mathbf{j}$ with a positive constant e, and

$$\mathbf{v} = \frac{\mu}{k}\mathbf{u}_\theta + \frac{\mu e}{k}\mathbf{j}. \tag{43}$$

Calculate $\mathbf{u}_\theta \cdot \mathbf{v}$ using formulas **(34)** and **(43)** for $\mathbf{v}$ and equate the results. Then use **(40)** to show that $r = \dfrac{p}{1 + e\cos\theta}$ with $p = k^2/\mu$, so that by formula **(4)** of Section 9.3, the orbit of the planet is a conic section with eccentricity e and the sun at a focus. The orbit is an ellipse because it is closed.

◆ Type 7, Kepler's and Newton's Laws

FIGURE 26

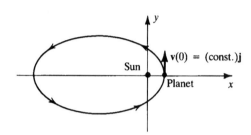

RESPONSES TO QUESTIONS IN CHAPTER 10
Responses 10.1

Response 1   The x-coordinate of O is 0, the change in the x-coordinate from O to P is 100, the change in the x-coordinate from P to Q is 75, and the change in the x-coordinate from Q to R is 50. • The x-coordinate of R is $0 + 100 + 75 + 50 = 225$. • The y-coordinate of O is 0, the change in the y-coordinate from O to P is 100, the change in the y-coordinate from P to Q is -75, and the change in the y-coordinate from Q to R is 50. • The y-coordinate of R is $0 + 100 - 75 + 50 = 75$.

Response 2   Because U and T are in the same direction from S and T is twice as far from S as T, the change in the x-coordinate from S to U is twice the change in the x-coordinate from S to T, and the change in the y-coordinate from S to U is twice the change in the y-coordinate from S to T. • The x-coordinate of U is $25 + 2(50) = 125$, and the y-coordinate of U is $50 + 2(30) = 110$. •
U has coordinates $(125, 110)$ and is 125 yards east and 110 yards north of O.

Response 3   The tangent of the angle of inclination of $\mathbf{C} = \langle 3, 4 \rangle$ is $\frac{4}{3}$ (Figure R 3). • We can use $\theta = \tan^{-1}(\frac{4}{3})$ as the angle of inclination.

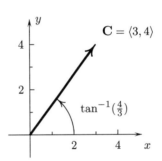

Figure R3

Response 4   (a) $\overrightarrow{OP} = \langle 3, 4 \rangle$ • $\overrightarrow{PQ} = \langle 7-3, 8-4 \rangle = \langle 4, 4 \rangle$ • $\overrightarrow{PR} = \langle 12-3, 2-4 \rangle = \langle 9, -2 \rangle$
(b) $\overrightarrow{RS} = \overrightarrow{PQ} = \langle 4, 4 \rangle$ because the line segment RS has the same length as PR and the direction from P to Q is the same as the direction from R to S because $PRSQ$ is a parallelogram..
(c) $\overrightarrow{OS} = \overrightarrow{OP} + \overrightarrow{PR} + \overrightarrow{RS} = \langle 3, 4 \rangle + \langle 9, -2 \rangle + \langle 4, 4 \rangle = \langle 3+9+4, 4-2+4 \rangle = \langle 16, 6 \rangle$
Or: $\overrightarrow{OS} = \overrightarrow{OR} + \overrightarrow{RS} = \langle 12, 2 \rangle + \langle 4, 4 \rangle = \langle 12+4, 2+4 \rangle = \langle 16, 6 \rangle$
(d) $S = (16, 6)$

Response 5   Since $50°$ is $\frac{50}{180}\pi = \frac{5}{18}\pi$ radians, $\mathbf{F} = 80 \langle \cos\left(\frac{5}{18}\pi\right), \sin\left(\frac{5}{18}\pi\right) \rangle \doteq \langle 51.42, 61.28 \rangle$ pounds

Response 6   (a) Let $\mathbf{F}$ be the force exerted by the man with the rod and $\mathbf{G}$ the force exerted by the man with the rope. • $\mathbf{F} = 300 \langle \cos\left(\frac{7}{18}\pi\right), \sin\left(\frac{7}{18}\pi\right) \rangle$ pounds •
$\mathbf{G} = 150 \langle \cos\left(\frac{1}{9}\pi\right), \sin\left(\frac{1}{9}\pi\right) \rangle$ pounds
(b) $\mathbf{F} + \mathbf{G} = \langle 300\cos\left(\frac{7}{18}\pi\right) + 150\cos\left(\frac{1}{9}\pi\right), 300\sin\left(\frac{7}{18}\pi\right) + 150\sin\left(\frac{1}{9}\pi\right) \rangle \doteq$
$\langle 243.56, 333.21 \rangle$ pounds • $|\mathbf{F} + \mathbf{G}| \doteq \sqrt{(243.56)^2 + (333.21)^2} \doteq 412.74$ pounds •
[Angle of inclination] $\doteq \tan^{-1}(333.21/243.56) \doteq 0.94$ radians

Response 7   (a) The component of $\mathbf{A}$ in the direction of $\mathbf{B}$ in Figure 34 is $6\cos(0.6) \doteq 4.95$.
(b) The component of $\mathbf{A}$ in the direction of $\mathbf{B}$ in Figure 35 is $6\cos(2.15) \doteq -3.28$.

Response 8 $|\mathbf{B}|^2 = 8$ • $\mathbf{A}_{\text{proj}} = \dfrac{\mathbf{A} \cdot \mathbf{B}}{|\mathbf{B}|^2} \mathbf{B} = \dfrac{-6}{8} \langle 2, 2 \rangle = \langle -\frac{3}{2}, -\frac{3}{2} \rangle$ • Figure R8

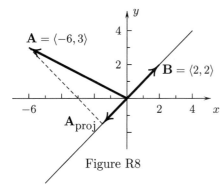

Figure R8

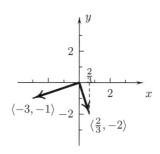

Figure R9

Response 9 $\langle A_1, -2 \rangle$ is perpendicular to $\langle -3, -1 \rangle$ if $\langle A_1, -2 \rangle \cdot \langle -3, -1 \rangle = -3A_1 + (-2)(-1)$ is zero. • This is the case if $A_1 = \frac{2}{3}$ • Figure R9

Response 10 (a) $\langle -2, 1 \rangle$ and $\langle 2, -1 \rangle$ • Figure R10
(b) $|\langle -2, 1 \rangle| = \sqrt{(-2)^2 + 1^2} = \sqrt{5}$ • The two unit vectors perpendicular to $\langle 1, 2 \rangle$ are $\dfrac{\langle -2, 1 \rangle}{\sqrt{5}}$ and $\dfrac{\langle 2, -1 \rangle}{\sqrt{5}}$.

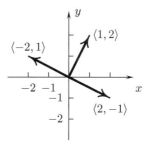

Figure R10

Responses 10.2

Response 1 The distance between $(4, 5, 0)$ and $(-1, 3, 2)$ is
$$\sqrt{(-1-4)^2 + (3-5)^2 + (2-0)^2} = \sqrt{5^2 + 2^2 + 2^2} = \sqrt{33}.$$

Response 2 $\overrightarrow{PQ} \cdot \overrightarrow{QR} = -22$ from calculation **(10)** in Example 4 • $\overrightarrow{QP} = -\overrightarrow{PQ}$ •
$\overrightarrow{QP} \cdot \overrightarrow{QR} = -\overrightarrow{PQ} \cdot \overrightarrow{QR} = 22$ • $|\overrightarrow{QP}| = |\overrightarrow{PQ}| = \sqrt{41}$ from Example 5 •
$|\overrightarrow{QR}| = |\langle -2, -3, -3 \rangle| = \sqrt{2^2 + 3^2 + 3^2} = \sqrt{22}$ •
$$\cos([\text{The angle at } Q]) = \frac{\overrightarrow{QP} \cdot \overrightarrow{QR}}{|\overrightarrow{QP}||\overrightarrow{QR}|} = \frac{22}{\sqrt{41}\sqrt{22}} = \frac{\sqrt{22}}{\sqrt{41}} \bullet$$
$[\text{The angle at } Q] = \cos^{-1}\left(\dfrac{\sqrt{22}}{\sqrt{41}}\right) \doteq 0.749$ radians

Response 3 The orthogonal projection of $\mathbf{F}$ on a line through $\mathbf{A}$ is
$\dfrac{\mathbf{F} \cdot \mathbf{A}}{|\mathbf{A}|^2} \mathbf{A} = \dfrac{10}{30} \mathbf{A} = \dfrac{1}{3} \langle 1, -2, -5 \rangle = \langle \dfrac{1}{3}, -\dfrac{2}{3}, -\dfrac{5}{3} \rangle$ pounds.

Response 4 (a) $\mathbf{A} \cdot \mathbf{i} = \langle A_1, A_2, A_3 \rangle \cdot \langle 1, 0, 0 \rangle = A_1$ •
$\mathbf{A} \cdot \mathbf{j} = \langle A_1, A_2, A_3 \rangle \cdot \langle 0, 1, 0 \rangle = A_2$ •
$\mathbf{A} \cdot \mathbf{k} = \langle A_1, A_2, A_3 \rangle \cdot \langle 0, 0, 1 \rangle = A_3$

(b) α is the angle between **A** and **i**, β is the angle between **A** and **j**, and γ is the angle between **A** and **k**. •
$A_1 = |\mathbf{A}||\mathbf{i}|\cos\alpha = |\mathbf{A}|\cos\alpha$ •
$A_2 = |\mathbf{A}||\mathbf{j}|\cos\beta = |\mathbf{A}|\cos\beta$ •
$A_3 = |\mathbf{A}||\mathbf{k}|\cos\gamma = |\mathbf{A}|\cos\gamma$

Response 5 For $\langle \frac{1}{2}, -\frac{1}{2}\sqrt{2}, \frac{1}{2}\rangle$: $\cos\gamma = \frac{1}{2}$ and $\gamma = \cos^{-1}(\frac{1}{2}) = \frac{1}{3}\pi$ •
For $\langle \frac{1}{2}, -\frac{1}{2}\sqrt{2}, -\frac{1}{2}\rangle$: $\cos\gamma = -\frac{1}{2}$ and $\gamma = \cos^{-1}(-\frac{1}{2}) = \frac{2}{3}\pi$

Responses 10.3

Response 1 The point S with position vector $\overrightarrow{OP} + 2\mathbf{v}$ is in Figure R1a and the point T with position vector $\overrightarrow{OP} - \mathbf{v}$ is in Figure R1b.

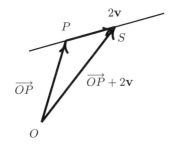

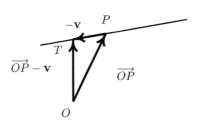

Figure R1a Figure R1b

Response 2 Write the equations $x = 3 + 2t, y = 6, z = 5 - 8t$ of the line in vector form $\langle x, y, z\rangle = \langle 3 + 2t, 6, 5 - 8t\rangle = \langle 3, 6, 5\rangle + t\langle 2, 0, -8\rangle$ to see that $\langle 2, 0, -8\rangle$ is parallel to it.

Response 3 For $P = (2, -3, 1)$ and $S = (3, -2, 2)$: $\overrightarrow{PS} = \langle 3 - 2, -2 - (-3), 2 - 1\rangle = \langle 1, 1, 1\rangle$ and

$$\mathbf{n} \cdot \overrightarrow{PS} = \langle 3, 4, -1\rangle \cdot \langle 1, 1, 1\rangle = 3(1) + 4(1) - 1(1) = 6.$$

Because the dot product is not zero, $\overrightarrow{PS}$ is not perpendicular to **n** and, therefore, is not parallel to the plane. • S is not on the plane because P is on it.

Response 4 Set $P = (6, 10, -3)$. • The vector $\mathbf{n} = \langle -3, 1, -7\rangle$ is parallel to the line and hence perpendicular to the plane. • The point $R = (x, y, z)$ is on the plane if and only if $\mathbf{n} \cdot \overrightarrow{PR} = 0$. • The plane has the equation $-3(x - 6) + (y - 10) - 7(z + 3) = 0$.

Response 5 $\mathbf{n} = \langle 3, -5, 6\rangle$

Response 6
$$\begin{vmatrix} 1 & 2 & -3 \\ 4 & 2 & 0 \\ -3 & 5 & 1 \end{vmatrix} = 1\begin{vmatrix} 2 & 0 \\ 5 & 1 \end{vmatrix} - 2\begin{vmatrix} 4 & 0 \\ -3 & 1 \end{vmatrix} + (-3)\begin{vmatrix} 4 & 2 \\ -3 & 5 \end{vmatrix}$$
$$= 1[2(1) - 0(5)] - 2[4(1) - 0(-3)] - 3[4(5) - 2(-3)]$$
$$= 1(2) - 2(4) - 3(26) = -84$$

Response 7 $\langle 3, -2, 1\rangle \times \langle 6, 0, -3\rangle = \begin{vmatrix} \mathbf{i} & \mathbf{j} & \mathbf{k} \\ 3 & -2 & 1 \\ 6 & 0 & -3 \end{vmatrix} = \mathbf{i}\begin{vmatrix} -2 & 1 \\ 0 & -3 \end{vmatrix} - \mathbf{j}\begin{vmatrix} 3 & 1 \\ 6 & -3 \end{vmatrix} + \mathbf{k}\begin{vmatrix} 3 & -2 \\ 6 & 0 \end{vmatrix}$
$= [(-2)(-3) - 1(0)]\mathbf{i} - [3(-3) - 1(6)]\mathbf{j} + [3(0) - (-2)(6)]\mathbf{k} = 6\mathbf{i} + 15\mathbf{j} + 12\mathbf{k}$
Partial check: $\langle 6, 15, 12\rangle \cdot \langle 3, -2, 1\rangle = 6(3) + 15(-2) + 12(1) = 18 - 30 + 12 = 0$
$\langle 6, 5, 12\rangle \cdot \langle 6, 0, -3\rangle = 6(6) + 5(0) + 12(-3) = 36 + 0 - 36 = 0$

112   Chapter 10: Vectors and Curves

Response 8 $ax_0 + by_0 + cz_0 = -d$ • $\overrightarrow{PQ} \cdot \mathbf{n} = ax + by + cz + d$

Response 9 $\dfrac{y}{4} + \dfrac{z}{4} = 1$ or $y + z = 4$

Response 10 $\langle 3, 3, -1 \rangle \cdot (\langle 4, 6, 5 \rangle \times \langle 2, 2, -1 \rangle)$

$$= \begin{vmatrix} 3 & 3 & -1 \\ 4 & 6 & 5 \\ 2 & 2 & -1 \end{vmatrix} = 3 \begin{vmatrix} 6 & 5 \\ 2 & -1 \end{vmatrix} - 3 \begin{vmatrix} 4 & 5 \\ 2 & -1 \end{vmatrix} + (-1) \begin{vmatrix} 4 & 6 \\ 2 & 2 \end{vmatrix}$$

$= 3[6(-1) - 5(2)] - 3[4(-1) - 5(2)] - [4(2) - 6(2)]$

$= 3(-16) - 3(-14) - (-4) = -2$

Response 11   (a) If θ is the angle between $\mathbf{A}$ and $\mathbf{B}$, then the angle between $-\mathbf{A}$ and $\mathbf{B}$ is $\pi - \theta$. (Figure R11) • $|(-\mathbf{A}) \times \mathbf{B}| = |-\mathbf{A}||\mathbf{B}|\sin(\pi - \theta) = |\mathbf{A}||\mathbf{B}|\sin\theta = |\mathbf{A} \times \mathbf{B}|$
(b) $(-\mathbf{A}) \times \mathbf{B} = -\mathbf{A} \times \mathbf{B}$ because the right-hand rule from $-\mathbf{A}$ to $\mathbf{B}$ gives the opposite direction from that given by the right-hand rule from $\mathbf{A}$ to $\mathbf{B}$.

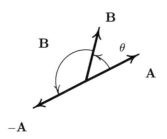

Figure R11

Response 12 $|\mathbf{B}_1| = |\mathbf{B}| \sin \theta$

Responses 10.4

Response 1   The completed table is below.

t	0	1	2	3
$x = x(t)$	10	10	30	10
$y = y(t)$	10	30	10	10

Response 2   Make the table below of values of $x(t)$ and $y(t)$, plot the corresponding points in an xy-plane, connect them by line segments, and put arrowheads on the curve to show its orientation (Figure R2).

t	0	100	200	300
$x = x(t)$	1	4	1	4
$y = y(t)$	4	4	1	1

Response 3   (a) The beginning of the curve is at $t = -4$ where $x = (-4)^3 - 9(-4) = -28$ and $y = 2(-4)^2 = 32$. • The end of the curve is at $t = 4$ where $x = 4^3 - 9(4) = 28$ and $y = 2(4)^2 = 32$.

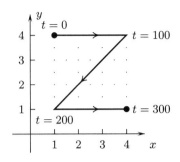
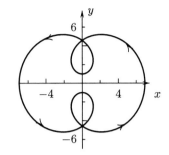

$x = x(t),\ y = y(t)$
$0 \le t \le 300$
Figure R2

$\mathbf{R}(t) = [4\cos t + 3\cos(3t)]\,\mathbf{i}$
$\quad + [4\sin t + 3\sin(3t)]\,\mathbf{j}$
$0 \le t \le 2\pi$
Figure R5

(b) The curve intersects the y axis where $x(t) = t^3 - 9t$ is zero. • $t^3 - 9t = t(t-3)(t+3)$ is zero at $t = 0$, where $y = 2(0^2) = 0$, and at $t = \pm 3$, where $y = 2(\pm 3)^2 = 18$.

Response 4 Use Figure 11 in Section 10.4 again: $x = 10\cos\theta,\ y = 10\sin\theta$. • The car travels $s = 20\pi t$ meters, counterclockwise, in t minutes. • Since the radius is 10, $\theta = \tfrac{1}{10}s = \tfrac{1}{10}(20\pi t) = 2\pi t$. • $x = 10\cos(2\pi t),\ y = 10\sin(2\pi t)$

Response 5 Use $\mathbf{R_1}(t) = 4\cos t\,\mathbf{i} + 4\sin t\,\mathbf{j}$ from Example 3 and $\mathbf{R_2}(t) = 3\cos(3t)\,\mathbf{i} + 3\sin(3t)\,\mathbf{j}$ to have the smaller circle rotate counterclockwise. • $\mathbf{R}(t) = \mathbf{R_1}(t) + \mathbf{R_2}(t)$ • $C: \mathbf{R}(t) = [4\cos t + 3\cos(3t)]\,\mathbf{i} + [4\sin t + 3\sin(3t)]\,\mathbf{j},\ 0 \le t \le 2\pi$ • Figure R5

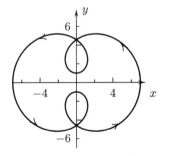

Figure R5

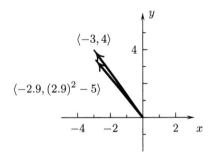

Figure R6

Response 6 $\lim_{t\to 3}\langle -t,\ t^2 - 5\rangle = \langle -3,\ (-3)^2 - 5\rangle = \langle -3,\ 4\rangle$ • Figure R6 shows $\mathbf{R_0} = \langle -3, 4\rangle$ and $\langle -t,\ t^2 - 5\rangle$ at $t = 2.9$.

Response 7 $\mathbf{R}'(t) = \dfrac{d}{dt}(2\cos t\,\mathbf{i} + 3\sin t\,\mathbf{j}) = \dfrac{d}{dt}(2\cos t)\,\mathbf{i} + \dfrac{d}{dt}(3\sin t)\,\mathbf{j} = -2\sin t\,\mathbf{i} + 3\cos t\,\mathbf{j}$

Response 8 (a) The velocity vector at $t = \tfrac{1}{4}\pi + \pi = \tfrac{5}{4}\pi$ is the negative of $\mathbf{v}(\tfrac{1}{4}\pi)$ because the ellipse is symmetric about the origin.
(b) The velocity vector is vertical at $t = 0$ and $t = \pi$ where the ellipse crosses the x-axis and has vertical tangent lines.
(c) Use formula (17) for $\mathbf{v}$: $\mathbf{v}(\tfrac{5}{4}\pi) = \langle -5\sin(\tfrac{5}{4}\pi),\ 3\cos(\tfrac{5}{4}\pi)\rangle = \dfrac{\langle 5, -3\rangle}{\sqrt{2}}$, $\mathbf{v}(0) = \langle 0, 3\rangle$, and $\mathbf{v}(\pi) = \langle 0, -3\rangle$ • Figure R8

Response 9 (a) The passenger's position at time t relative to the stationary water is the point R in Figure 29, so his displacement vector from time 0 to time t is

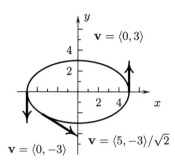

Figure R8

$\overrightarrow{PR} = t\langle 0,2\rangle + t\langle 5,0\rangle$.

(b) His velocity vector relative to the water is $\dfrac{d}{dt}(t\langle 0,2\rangle + t\langle 5,0\rangle) = \langle 0,2\rangle + \langle 5,0\rangle = \langle 5,2\rangle$ feet per second.

(c) [The passenger's speed relative to the water] $= |\langle 5,2\rangle| = \sqrt{5^2 + 2^2} = \sqrt{29} \doteq 5.39$ feet per second.

Responses 10.5

Response 1   **(a)** $\int e^t\, dt = e^t + C_1$ • $\int \cos t\, dt = \sin t + C_2$ • $\int \dfrac{1}{t}\, dt = \ln|t| + C_3$ •
$\int (e^t\mathbf{i} + \cos t\,\mathbf{j} + \dfrac{1}{t}\mathbf{k})\, dt = (e^t + C_1)\mathbf{i} + (\sin t + C_2)\mathbf{j} + (\ln|t| + C_3)\mathbf{k}$

(b) $\int_1^2 e^t\, dt = \left[e^t\right]_1^2 = e^2 - e$ • $\int_1^2 \cos t\, dt = \left[\sin t\right]_1^2 = \sin(2) - \sin(1)$ •
$\int_1^2 \dfrac{1}{t}\, dt = \left[\ln|t|\right]_1^2 = \ln(2)$ •
$\int_1^2 (e^t\mathbf{i} + \cos t\,\mathbf{j} + \dfrac{1}{t}\mathbf{k})\, dt = (e^2 - e)\mathbf{i} + [\sin(2) - \sin(1)]\mathbf{j} + \ln(2)\mathbf{k}$

Response 2   **(a)** $m = 5$ • $\mathbf{a} = \dfrac{1}{m}\mathbf{F} = \tfrac{1}{5}\langle -15\cos t, -15\sin t\rangle = \langle -3\cos t, -3\sin t\rangle$ •
$x''(t) = -3\cos t$ • $x'(t) = -3\sin t + C_1$ • $x'(0) = 1 \Longrightarrow C_1 = 1$ •
$x'(t) = -3\sin t + 1$ • $x(t) = 3\cos t + t + C_2$ • $x(0) = 6 \Longrightarrow C_2 = 3$ •
$x(t) = 3\cos t + t + 3$ •
$y''(t) = -3\sin t$ • $y'(t) = 3\cos t + C_3$ • $y'(0) = 4 \Longrightarrow C_3 = 1$ •
$y'(t) = 3\cos t + 1$ • $y(t) = 3\sin t + t + C_4$ • $y(0) = 3 \Longrightarrow C_4 = 3$ •
$y(t) = 3\sin t + t + 3$ • Figure R2
(b) One explanation: The loops in the path are plausible because the components of the force oscillate. • Another explanation: The loops in the path are plausible because the terms $3\cos t$ in $x(t)$ and $3\sin t$ in $y(t)$ oscillate between 3 and -3.

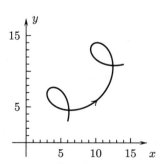

Figure R2

Response 3 (a) $x' = -\rho \sin t, x'' = -\rho \cos t$ • $y' = \rho \cos t, y'' = -\rho \sin t$ •
$$\kappa = \frac{x'y'' - y'x''}{[(x')^2 + (y')^2]^{3/2}} = \frac{\rho^2(\cos^2 t + \sin^2 t)}{[\rho^2(\sin^2 t + \cos^2 t)]^{3/2}} = \frac{1}{\rho}$$
(b) $x' = -\rho \sin(-t)\frac{d}{dt}(-t) = \rho\sin(-t), x'' = \rho\cos(-t)\frac{d}{dt}(-t) = -\rho\cos(-t)$ •
$y' = \rho\cos(-t)\frac{d}{dt}(-t) = -\rho\cos(-t), y'' = \rho\sin(-t)\frac{d}{dt}(-t) = -\rho\sin(-t)$
$$\kappa = \frac{x'y'' - y'x''}{[(x')^2 + (y')^2]^{3/2}} = \frac{-\rho^2[\sin^2(-t) + \cos^2(-t)]}{[\rho^2(\cos^2(-t) + \sin^2(-t))]^{3/2}} = -\frac{1}{\rho}$$

Response 4 (a) The radius of curvature ρ at P (the distance between the dots) is approximately 2 and the curve is bending toward the left there. • $\kappa = 1/\rho \approx \frac{1}{2}$ at P
(b) The radius of curvature ρ at Q is approximately 1 and the curve is bending toward the right there. • $\kappa = -1/\rho \approx -1$ at Q

Response 5 Set $y(x) = \frac{1}{2}x^2$. • $y'(x) = x, y''(x) = 1$ • $y(1) = \frac{1}{2}, y'(1) = 1, y''(1) = 1$ •
$\kappa(1) = \frac{y''(1)}{\{1 + [y'(1)]^2\}^{3/2}} = \frac{1}{2^{3/2}}$ • $\mathbf{R}(1) = \langle 1, \frac{1}{2}\rangle$ • $\mathbf{v}(1) = \langle 1, 1\rangle$ •
$\mathbf{T} = \langle 1, 1\rangle/\sqrt{2}$ • $\mathbf{N} = \langle -1, 1\rangle/\sqrt{2}$ • With C the center of curvature:
$\overrightarrow{OC} = \mathbf{R}(1) + \frac{1}{\kappa(1)}\mathbf{N} = \langle 1, \frac{1}{2}\rangle + 2\sqrt{2}\left[\frac{\langle -1, 1\rangle}{\sqrt{2}}\right] = \langle 1, \frac{1}{2}\rangle + \langle -2, 2\rangle = \langle -1, \frac{5}{2}\rangle$ •
$C = (-1, \frac{5}{2})$ • (The circle of curvature is shown in Figure R5.)

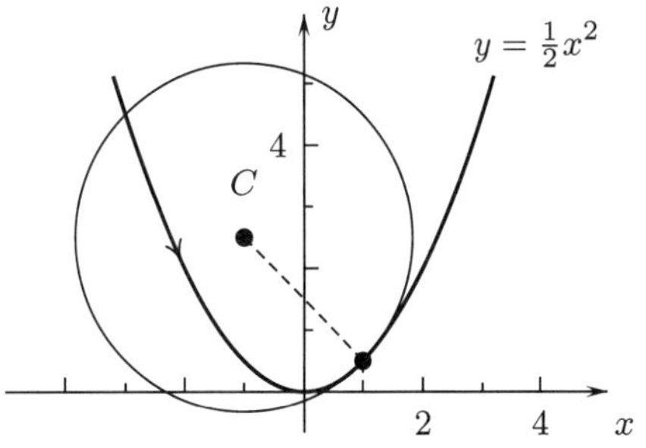

Figure R5

Response 6 $\dfrac{d\mathbf{T}}{dt} = \dfrac{d\mathbf{T}}{d\phi}\dfrac{d\phi}{ds}\dfrac{ds}{dt}$

Response 7 (a) $\mathbf{T}$ is the unit vector $\langle -\sin(1), \cos(1)\rangle$ perpendicular to and pointing to the left from the position vector $\langle 3\cos(1), 3\sin(1)\rangle$ • $\mathbf{N}$ is the unit vector $\langle -\cos(1), -\sin(1)\rangle$ perpendicular to and pointing to the left from $\mathbf{T}$.
(b) $\mathbf{a} = 3\mathbf{T} + \frac{4}{3}\mathbf{N} = 3\langle -\sin(1), \cos(1)\rangle + \frac{4}{3}\langle -\cos(1), -\sin(1)\rangle$
$= \langle -3\sin(1) - \frac{4}{3}\cos(1), 3\cos(1) - \frac{4}{3}\sin(1)\rangle$

Response 8 (a) Kepler's First Law holds because we assume the orbit is a circle with the sun at the center, and a circle is an ellipse.
(b) The line segment from the planet to the sun sweeps out equal areas in equal times and Kepler's Second Law holds because by part (a) of Example 12 the planet travels around the circle at a constant speed.
(c) The planet is always a distance r from the sun, so $d = r$ in Kepler's Third law. Also $T = \dfrac{2\pi r^{3/2}}{\sqrt{GM}}$ by part (b) of Example 12, so $\dfrac{T^2}{d^3} = \dfrac{1}{r^3}\left[\dfrac{(2\pi)^2 r^3}{GM}\right] = \dfrac{(2\pi)^2}{GM}$
is independent of r.

Response 9 $\mathbf{a}$ has the same direction as the total force vector $\mathbf{F}$, which by Newton's law is toward the origin. • $\mathbf{a} \times \mathbf{R} = \mathbf{0}$ because $\mathbf{a}$ and $\mathbf{R}$ are parallel. •
$\mathbf{v} \times \mathbf{v} = \mathbf{0}$ because any cross product of a vector with itself is $\mathbf{0}$.

Answers 10.1

CHAPTER 10 ANSWERS

Tune-up exercises 10.1

T1. (a) $\langle 4(3) - 5(10), 4(-2) - 5(1)\rangle = \langle -38, -13\rangle$
(b) $[-2(3) + 10(1)]\mathbf{i} + [-2(4) + 10(-1)]\mathbf{j} = 4\mathbf{i} - 18\mathbf{j}$

T2. $\mathbf{A} + \mathbf{B} = \langle 40, 30\rangle$ • Figure T2

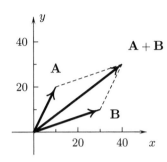
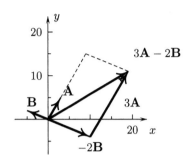

Figure T2   Figure T4

T3. $\overrightarrow{OS} = \overrightarrow{OP} + \overrightarrow{PQ} + \overrightarrow{QR} + \overrightarrow{RS} = \langle 210, 30\rangle$ • $S = (210, 30)$

T4. $3\mathbf{A} - 2\mathbf{B} = \langle 3(3) - 2(-5), 3(5) - 2(2)\rangle = \langle 19, 11\rangle$ • Figure T4

T5. (a) $R = (4, 6), S = (10, 4)$
(b) $T = (-2, -3), U = (-5, -2)$ • Figure T5

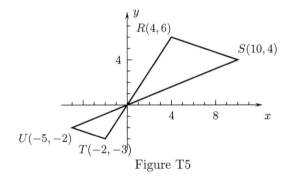
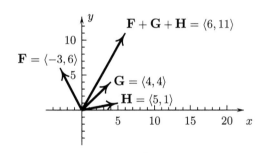

Figure T5   Figure T6

T6. [Resultant force] $= \mathbf{F} + \mathbf{G} + \mathbf{H} = \langle -3 + 4 + 5, 6 + 4 + 1\rangle = \langle 6, 11\rangle$ • Figure T6

T7. (a) $\mathbf{A} \cdot \mathbf{B} = \langle 1, 5\rangle \cdot \langle 6, 4\rangle = 26$
(b) $|\mathbf{B}| = \sqrt{6^2 + 4^2} = 2\sqrt{3^2 + 2^2} = 2\sqrt{13}$ • [Component of $\mathbf{A}$ in the direction of $\mathbf{B}$] $= \dfrac{\mathbf{A} \cdot \mathbf{B}}{|\mathbf{B}|} = \sqrt{13}$
(c) $\mathbf{A}_{\text{proj}} = \dfrac{\mathbf{A} \cdot \mathbf{B}}{|\mathbf{B}|^2}\mathbf{B} = \dfrac{26}{4(13)}\langle 6, 4\rangle = \langle 3, 2\rangle$ • Figure T7

T8. $|\mathbf{A}| = \sqrt{1^2 + 5^2} = \sqrt{26}$ • $\cos\theta = \dfrac{\mathbf{A} \cdot \mathbf{B}}{|\mathbf{A}||\mathbf{B}|} = \dfrac{1}{\sqrt{2}}$ • $\theta = \cos^{-1}(1/\sqrt{2}) = \tfrac{1}{4}\pi$

T9. Draw perpendicular lines from the tip of $\mathbf{F}$ to the lines through $\mathbf{A}$ and $\mathbf{B}$, as in Figure T9, and let a and b be the distances from the feet of these perpendicular lines to the bases of the vectors. • $a \approx 11, b \approx 6$ • The component of $\mathbf{F}$ in the direction of $\mathbf{A}$ is approximately 11 dynes and its component in the direction of $\mathbf{B}$ is approximately -6 dynes.

T10. $\mathbf{i} \cdot \mathbf{i} = \langle 1, 0\rangle \cdot \langle 1, 0\rangle = 1(1) + 0(0) = 1$ •
$\mathbf{i} \cdot \mathbf{j} = \langle 1, 0\rangle \cdot \langle 0, 1\rangle = 1(0) + 0(1) = 0$ •
$\mathbf{j} \cdot \mathbf{j} = \langle 0, 1\rangle \cdot \langle 0, 1\rangle = 0(0) + 1(1) = 1$ •

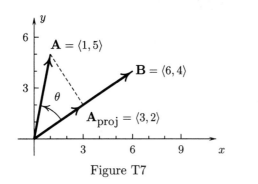

Figure T7

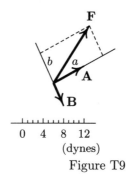

Figure T9

$\mathbf{i} \cdot \mathbf{i}$ and $\mathbf{j} \cdot \mathbf{j}$ are 1 because $\mathbf{i}$ and $\mathbf{j}$ are unit vectors. •
$\mathbf{i} \cdot \mathbf{j}$ is zero because these vectors are perpendicular.

T11. (a) $\mathbf{A} = 6\langle\cos(\tfrac{1}{4}\pi), \sin(\tfrac{1}{4}\pi)\rangle = 6\left\langle \dfrac{1}{\sqrt{2}}, \dfrac{1}{\sqrt{2}} \right\rangle = \langle 3\sqrt{2}, 3\sqrt{2}\rangle$
(b) $\mathbf{B} = 5\langle\cos(4), \sin(4)\rangle = \langle 5\cos(4), 5\sin(4)\rangle$ •
$\mathbf{A} \doteq \langle 4.24, 4.24\rangle$ and $\mathbf{B} \doteq \langle -3.27, -3.78\rangle$ • Figure T11

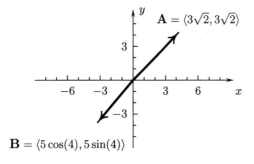

Figure T11

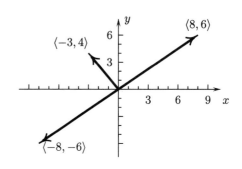

Figure T13

T12. $|5\mathbf{i} - 12\mathbf{j}| = \sqrt{5^2 + 12^2} = 13$ • $\dfrac{5\mathbf{i} - 12\mathbf{j}}{|5\mathbf{i} - 12\mathbf{j}|} = \tfrac{5}{13}\mathbf{i} - \tfrac{12}{13}\mathbf{j}$

T13. Interchange the components and multiply one by -1 to get the perpendicular vectors $\langle 4, 3\rangle$ and $\langle -4, -3\rangle$ • $|\langle 4, 3\rangle| = 5$ • Unit vectors: $\langle \tfrac{4}{5}, \tfrac{3}{5}\rangle$ and $\langle -\tfrac{4}{5}, -\tfrac{3}{5}\rangle$ •
Vectors of length 10: $\langle 8, 6\rangle$ and $\langle -8, -6\rangle$ • Figure T13

T14. The positive acute angle α in Figure T14 equals $\tan^{-1}(\tfrac{3}{5})$ • $\theta = \tan^{-1}(\tfrac{3}{5}) + \pi$

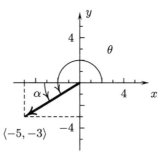

Figure T14

15. $a\langle 3, -1\rangle + b\langle 1, 2\rangle = \langle 3a + b, -a + 2b\rangle$ • Solve $\begin{cases} 3a + b = 1 \\ -a + 2b = -12 \end{cases}$ •
Add the first equation to -2 times the second: $-7a = -14$ • $a = 2, b = -5$

Problems 10.1

1. (a) $\langle 8,-14\rangle$ (b) $\mathbf{i}-5\mathbf{j}$
2. $-21\mathbf{i}-16\mathbf{j}$
3. $\langle 122,112\rangle$
4. (a) $\langle -1,7\rangle$ (b) $\langle 5,-2\rangle$ (c) $\langle 0,12\rangle$
5. (a) $\langle -\frac{5}{2},\frac{5}{2}\sqrt{3}\rangle$ (b) $\langle -\frac{7}{2}\sqrt{2},-\frac{7}{2}\sqrt{2}\rangle$ • Figure A5

Figure A5

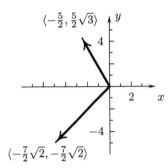

6. (a) $\overrightarrow{PQ}=\langle 2,-8\rangle$ (b) $\overrightarrow{RS}=\langle 7,-5\rangle$
8. (a) $S=(7,5)$
10. (a) Figure A10a • $\theta=\frac{3}{4}\pi$
    (b) Figure A10b • $\theta=\frac{7}{4}\pi$
    (c) Figure A10c • $\alpha=\tan^{-1}(\frac{3}{4})$ • $\theta=-\tan^{-1}(\frac{3}{4})$

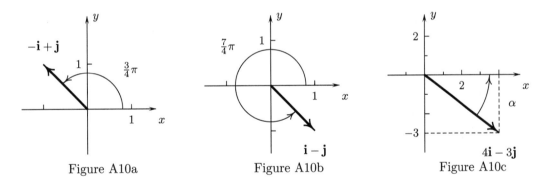

Figure A10a   Figure A10b   Figure A10c

11. [Length] $=2\sqrt{2}$ • [Angle of inclination] $=\frac{5}{4}\pi$
12. $\frac{1}{\sqrt{106}}\langle 63,-35\rangle$
13. (a) $s=-3, t=4$
14. $\mathbf{A}=\mathbf{i}-\mathbf{j}, \mathbf{B}=-\frac{1}{2}\mathbf{i}+\mathbf{j}$
15. (b) $\mathbf{B}=\langle -\frac{5}{2},-\frac{9}{2}\rangle$
16. He sails $200\sqrt{2}\doteq 282.8$ meters to the northeast and then $100\sqrt{2}\doteq 141.1$ meters to the northeast.
18. The magnitude of the force on the upper rope is $20\sqrt{2}\doteq 28.28$ pounds and the magnitude of the force on the lower rope is $10\sqrt{13}\doteq 36.06$ pounds.
19. The magnitude of the force on the left rope is $30\sqrt{29}\doteq 161.6$ pounds and the magnitude of the force on the right rope is $20\sqrt{13}\doteq 72.1$ pounds.
20. $\langle \frac{25}{2}\sqrt{2},\frac{25}{2}\sqrt{2}-20\rangle$ pounds.

24. The magnitude of the force (tension) on the second rope is $\sqrt{(10\sqrt{2})^2 + (25 - 10\sqrt{2})^2} \doteq 17.83$ pounds. • The angle between the second rope and the vertical is $\tan^{-1}\left(\dfrac{10\sqrt{2}}{25 - 10\sqrt{2}}\right) \doteq 0.916$ radians

29. (a) -16 (b) 21 (c) -1 (d) -16

30. $\cos\theta = \dfrac{\langle 2,1\rangle \cdot \langle -2,3\rangle}{|\langle 2,1\rangle||\langle -2,3\rangle|} = \dfrac{-1}{\sqrt{65}}$ • $\theta = \cos^{-1}\left(\dfrac{-1}{\sqrt{65}}\right) \doteq 1.695$ radians

33. $\pm\dfrac{\langle 1,2\rangle}{\sqrt{5}}$

34. $\pm\dfrac{\langle 1,3\rangle}{\sqrt{10}}$

37. $\pm\dfrac{\langle -90,75\rangle}{\sqrt{61}}$

39. (a) $\langle 5t, 2t\rangle$ with arbitrary numbers t

44. (a) ≈ 0.8 (b) ≈ 0

45. (a) $-\dfrac{9}{25}\langle 3,4\rangle$ (b) $\mathbf{0}$

46. (a) [Component of $\mathbf{A}$ in the direction of $\mathbf{B}$] $= \dfrac{6}{\sqrt{5}}$ •
[Projection of $\mathbf{A}$ on lines parallel to $\mathbf{B}$] $= \dfrac{6}{5}\langle 2,1\rangle$ • Figure A46a
(c) [Component of $\mathbf{A}$ in the direction of $\mathbf{B}$] $= -\dfrac{9}{\sqrt{2}}$ •
[Projection of $\mathbf{A}$ on lines parallel to $\mathbf{B}$] $= -\dfrac{9}{2}\langle 1,1\rangle$ • Figure A46c

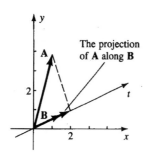

Figure A46a

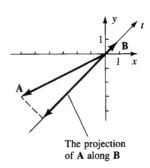

Figure A46c

49. (a) $\left(\dfrac{10}{3}, \dfrac{11}{3}\right)$ (b) $\left(\dfrac{11}{5}, -\dfrac{12}{5}\right)$

51. Let the triangle be PQR, let M be the midpoint of PR, and let N be the midpoint of PQ. • $\overrightarrow{PN} = \frac{1}{2}\overrightarrow{PQ}$ and $\overrightarrow{PM} = \frac{1}{2}\overrightarrow{PR}$ • $\overrightarrow{MN} = \overrightarrow{PN} - \overrightarrow{PM} = \frac{1}{2}(\overrightarrow{PQ} - \overrightarrow{PR}) = \frac{1}{2}\overrightarrow{RQ}$

55. Let the parallel sides of the trapezoid be formed by $\mathbf{B}$ and $t\mathbf{B}$ with a positive constant t, and let one of the other sides be formed by $\mathbf{A}$, where $\mathbf{A}$ and $\mathbf{B}$ have their bases at the origin in an xy-plane. • The midpoints M and N of the nonparallel sides have position vectors $\overrightarrow{ON} = \frac{1}{2}\mathbf{A}$ and $\overrightarrow{OM} = \frac{1}{2}(\mathbf{A} + t\mathbf{B} + \mathbf{B}) = \frac{1}{2}[\mathbf{A} + (t+1)\mathbf{B}]$. • $\overrightarrow{MN} = \frac{1}{2}(1+t)\mathbf{B}$ is parallel to $\mathbf{B}$ and its length is the average of the lengths of $\mathbf{B}$ and $t\mathbf{B}$.

Answers 10.2

Tune-up exercises 10.2

T1. Draw the axes, plot the vertices, and draw the sides. • Figure T1

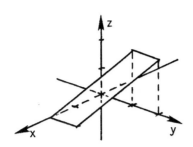

Figure T1

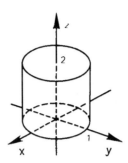

Figure T2

T2. Draw the axes, draw an ellipse to represent the circle $x^2 + y^2 = 1$ in the xy-plane, draw the sides, and draw an ellipse for the top. • Figure T2

T3. Draw the axes, draw an ellipse to represent the circle $x^2 + y^2 = 1$ in the xy-plane, plot the top at $z = 3$ on the z-axis, and draw lines connecting the top to the sides of the base. • Figure T3

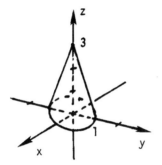

Figure T3

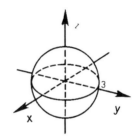

Figure T4

T4. Draw the axes, draw a circle with its center at the origin for the profile of the sphere, and add an ellipse for its circular intersection with the xy-plane. • Figure T4

T5. (a) $\sqrt{x^2 + y^2}$ is the distance from (x, y, z) to the z-axis because it is the distance from $(x, y, 0)$ to the origin $(0, 0, 0)$. (b) $\sqrt{x^2 + z^2}$ is the distance from (x, y, z) to the y-axis because it is the distance from $(x, 0, z)$ to the origin. (c) $\sqrt{y^2 + z^2}$ is the distance from (x, y, z) to the x-axis because it is the distance from $(0, y, z)$ to the origin.

T6. (a) 5. (b) $\sqrt{17}$. (c) $\sqrt{10}$. (d) $|z| = 1$. (e) $|y| = 4$. (f) $|x| = 3$. (g) $\sqrt{26}$.

T7. (a) $\overrightarrow{PQ} = \langle 1 - 0, 4 - 3, 6 - 2 \rangle = \langle 1, 1, 4 \rangle = \mathbf{i} + \mathbf{j} + 4\mathbf{k}$ (b) $\overrightarrow{PQ} = \langle -1 - 1, -4 - 4, -1 - 1 \rangle = \langle -2, -8, -2 \rangle = -2\mathbf{i} - 8\mathbf{j} - 2\mathbf{k}$

T8. (a) $\mathbf{A} + 2\mathbf{B} = \langle 2, 4, 0 \rangle + 2\langle 1, 1, 1 \rangle = \langle 2 + 2(1), 4 + 2(1), 0 + 2(1) \rangle = \langle 4, 6, 2 \rangle$ •
$|\mathbf{A} + 2\mathbf{B}| = |\langle 4, 6, 2 \rangle| = 2|\langle 2, 3, 1 \rangle| = 2\sqrt{2^2 + 3^2 + 1^2} = 2\sqrt{14}$ •
$\mathbf{u} = \dfrac{\mathbf{A} + 2\mathbf{B}}{|\mathbf{A} + 2\mathbf{B}|} = \dfrac{\langle 4, 6, 2 \rangle}{2\sqrt{14}} = \left\langle \dfrac{2}{\sqrt{14}}, \dfrac{3}{\sqrt{14}}, \dfrac{1}{\sqrt{14}} \right\rangle$
(b) $\mathbf{A} + 2\mathbf{B} = \langle 0, 0, -4 \rangle + 2\langle 1, 1, -1 \rangle = \langle 0 + 2(1), 0 + 2(1), -4 + 2(-1) \rangle = \langle 2, 2, -6 \rangle$ •
$|\mathbf{A} + 2\mathbf{B}| = |\langle 2, 2, -6 \rangle| = 2|\langle 1, 1, -3 \rangle| = 2\sqrt{1^2 + 1^2 + 3^2} = 2\sqrt{11}$ •
$\mathbf{u} = \dfrac{\mathbf{A} + 2\mathbf{B}}{|\mathbf{A} + 2\mathbf{B}|} = \dfrac{\langle 2, 2, -6 \rangle}{2\sqrt{11}} = \dfrac{1}{\sqrt{11}}\mathbf{i} + \dfrac{1}{\sqrt{11}}\mathbf{j} + \dfrac{-3}{\sqrt{11}}\mathbf{k}$

T9. (a) $\mathbf{A}\cdot\mathbf{B} = \langle 2,5,-4\rangle\cdot\langle 3,1,3\rangle = 2(3)+5(1)+(-4)(3) = -1$
(b) $\mathbf{A}\cdot\mathbf{B} = \langle 1,0,-3\rangle\cdot\langle 2,1,-1\rangle$
$= 1(2)+0(1)+(-3)(-1) = 5$

T10. (a) Set $\mathbf{A} = \langle 1,3,4\rangle$ and $\mathbf{B} = \langle -2,0,1\rangle$ • $\mathbf{A}\cdot\mathbf{B} = 1(-2)+3(0)+(4)(1) = 2$ •
$|\mathbf{A}| = \sqrt{1^2+3^2+4^2} = \sqrt{26}$ • $|\mathbf{B}| = \sqrt{(-2)^2+0^2+1^2} = \sqrt{5}$ • $\cos\theta = \dfrac{\mathbf{A}\cdot\mathbf{B}}{|\mathbf{A}||\mathbf{B}|} = \dfrac{2}{\sqrt{5}\sqrt{26}}$ •
$\theta = \cos^{-1}\left(\dfrac{2}{\sqrt{130}}\right) \doteq 1.39$ radians
(b) (a) Set $\mathbf{A} = \langle 2,3,4\rangle$ and $\mathbf{B} = \langle -3,4,-5\rangle$ • $\mathbf{A}\cdot\mathbf{B} = 2(-3)+3(4)+(4)(-5) = -14$ •
$|\mathbf{A}| = \sqrt{2^2+3^2+4^2} = \sqrt{29}$ • $|\mathbf{B}| = \sqrt{(-3)^2+4^2+(-5)^2} = \sqrt{50}$ • $\cos\theta = \dfrac{\mathbf{A}\cdot\mathbf{B}}{|\mathbf{A}||\mathbf{B}|} = \dfrac{-14}{\sqrt{50}\sqrt{29}}$
• $\theta = \cos^{-1}\left(\dfrac{-14}{\sqrt{1450}}\right) \doteq 1.947$ radians

T11. (a) $\mathbf{A}\cdot\mathbf{B} = \langle 3,5,2\rangle\cdot\langle 1,-1,-3\rangle = 3(1)+5(-1)+(2)(-3) = -8$ •
$|\mathbf{B}| = |\langle 1,-1,-3\rangle| = \sqrt{1^2+1^2+3^2} = \sqrt{11}$ •
[Component of $\mathbf{A}$ in the direction of $\mathbf{B}$] $= \dfrac{\mathbf{A}\cdot\mathbf{B}}{|\mathbf{B}|} = \dfrac{-8}{\sqrt{11}}$
(b) [Orthogonal projection of $\mathbf{A}$ on a line through $\mathbf{B}$] $= \dfrac{\mathbf{A}\cdot\mathbf{B}}{|\mathbf{B}|^2}\mathbf{B} = -\dfrac{8}{11}\langle 1,-1,-3\rangle$

T12. $|\langle 2,-1,-2\rangle| = \sqrt{2^2+1^2+2^2} = 3$ • The unit vector in the direction of $\langle 2,-1,-2\rangle$ is $\langle \tfrac{2}{3},-\tfrac{1}{3},-\tfrac{2}{3}\rangle$
• Direction cosines: $\cos\alpha = u_1 = \tfrac{2}{3}$, $\cos\beta = u_2 = -\tfrac{1}{3}$, $\cos\gamma = u_3 = -\tfrac{2}{3}$ •
Direction angles: $\alpha = \cos^{-1}(\tfrac{2}{3}) \doteq 0.841$, $\beta = \cos^{-1}(\tfrac{1}{3}) \doteq 1.911$, $\gamma = \cos^{-1}(-\tfrac{2}{3}) \doteq 2.301$ radians

Problems 10.2

1. Draw the axes, plot the vertices, and draw the triangular front of the (four-sided) tetrahedron. • Figure A1

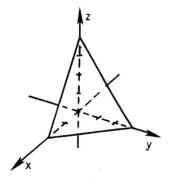
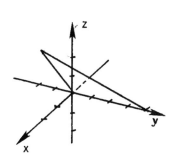

Figure A1

Figure A2

2. Draw the axes, plot the vertices, and draw the sides. • Figure A2
3. Draw the axes, plot the four corners of the base and join them, draw the vertical edges the same lengths, and draw the top. • Figure A3

Answers 10.2

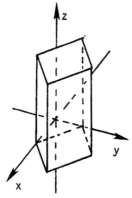

Figure A3

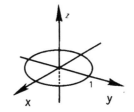

Figure A4

4. Figure A4
5. Figure A5

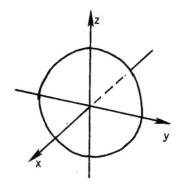

Figure A5

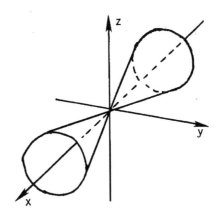

Figure A9

9. (a) Figure A9 (b) $\sin^{-1}(\frac{1}{2}) = \frac{1}{6}\pi$

11. (a) $\overrightarrow{PQ} + \overrightarrow{QR} = \langle 3, -4, -2 \rangle$ (b) $\overrightarrow{PR} = \langle 3, -4, -2 \rangle$

12. (a) $\mathbf{u} = \dfrac{\langle 5, 1, -1 \rangle}{\sqrt{27}} = \langle \dfrac{5}{\sqrt{27}}, \dfrac{1}{\sqrt{27}}, \dfrac{-1}{\sqrt{27}} \rangle$ (b) $\mathbf{B} = 8\dfrac{\mathbf{A}}{|\mathbf{A}|} = \dfrac{16}{\sqrt{14}}\mathbf{i} - \dfrac{8}{\sqrt{14}}\mathbf{j} + \dfrac{24}{\sqrt{14}}\mathbf{k}$

13. Make a schematic sketch as in Figure A13 with O the origin $(0,0,0)$. • The sides RS and QP are equal and parallel. • $\overrightarrow{RS} = \overrightarrow{QP} = \langle 6-3, 8-4, -2-0 \rangle = \langle 3, 4, -2 \rangle$ • $\overrightarrow{OS} = \overrightarrow{OR} + \overrightarrow{RS} = \langle 4, -7, -10 \rangle + \langle 3, 4, -2 \rangle = \langle 7, -3, -12 \rangle$ • $S = (7, -3, -12)$

Figure A13
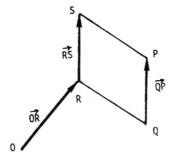

15. Set $P = (3,0,1)$, $Q = (2,4,5)$, and $R = (-1,4,4)$. • $\overrightarrow{PQ} = \langle 2-3, 4-0, 5-1\rangle = \langle -1,-4,4\rangle$ is not a constant times $\overrightarrow{QR} = \langle -1-2, 4-4, 4-5\rangle = \langle -3,0,-1\rangle$ so $\overrightarrow{PQ}$ and $\overrightarrow{QR}$ are not parallel and the points are not on a line.

16. The upward force on the box is $\mathbf{F_1} + \mathbf{F_2} + \mathbf{F_3} = \langle 20,10,15\rangle + \langle -30,-4,6\rangle + \langle 10,-6,8\rangle = \langle 0,0,29\rangle$ pounds. • The box weighs 29 pounds.

21. [Area] $= \frac{1}{2}\sqrt{6}\sqrt{12} = 3\sqrt{2}$

24. (a) $\dfrac{\mathbf{A}\cdot\mathbf{B}}{|\mathbf{B}|} = \dfrac{23}{\sqrt{30}}$ (b) $\dfrac{\mathbf{A}\cdot\mathbf{B}}{|\mathbf{B}|} = \dfrac{-11}{\sqrt{2}}$

25. (a) $\dfrac{\mathbf{A}\cdot\mathbf{B}}{|\mathbf{B}|^2}\mathbf{B} = \langle \frac{5}{2}, \frac{5}{2}, 5\rangle$ (b) $\dfrac{\mathbf{A}\cdot\mathbf{B}}{|\mathbf{B}|^2}\mathbf{B} = \langle 0, -\frac{24}{37}, -\frac{4}{37}\rangle$

26. (a) $\cos\alpha = \dfrac{3}{\sqrt{22}}$, $\cos\beta = -\dfrac{2}{\sqrt{22}}$, $\cos\gamma = -\dfrac{3}{\sqrt{22}}$ •
$\alpha = \cos^{-1}(\dfrac{3}{\sqrt{22}}) \doteq 0.877$, $\beta = \cos^{-1}(-\dfrac{2}{\sqrt{22}}) \doteq 2.011$, $\gamma = \cos^{-1}(-\dfrac{3}{\sqrt{22}}) \doteq 2.265$
(b) $\cos\alpha = \dfrac{1}{\sqrt{21}}$, $\cos\beta = -\dfrac{2}{\sqrt{21}}$, $\cos\gamma = \dfrac{4}{\sqrt{21}}$ •
$\alpha = \cos^{-1}(\dfrac{1}{\sqrt{21}}) \doteq 1.351$, $\beta = \cos^{-1}(-\dfrac{2}{\sqrt{21}}) \doteq 2.022$ $\gamma = \cos^{-1}(\dfrac{4}{\sqrt{21}}) \doteq 0.510$

Tune-up exercises 10.3

T1. Set $P = (3,0,-2)$ and $\mathbf{v} = \langle 2,0,-5\rangle$, and $R = (x,y,z)$. • $\overrightarrow{OR} = \overrightarrow{OP} + t\mathbf{v} = \langle 3+2t, 0+0t, -2-5t\rangle$ • $x = 3+2t$, $y = 0$. $z = -2-5t$

T2. Set $P = (-6,4,2)$, $Q = (3,5,-1)$, $\mathbf{v} = \overrightarrow{PQ} = \langle 3-(-6), 5-4, -1-2\rangle = \langle 9,1,-3\rangle$, and $R = (x,y,z)$. • $\overrightarrow{OR} = \overrightarrow{OP} + t\mathbf{v} = \langle -6,4,2\rangle + t\langle 9,1,-3\rangle = \langle -6+9t, 4+t, 2-3t\rangle$ • $x = -6+9t$, $y = 4+t$. $z = 2-3t$ Check: P is at $t=0$ and Q is at $t=1$.

T3. Set $P = (1,3,4)$ and $\mathbf{n} = \langle 2,-4,1\rangle$ • For $Q = (x,y,z)$ on the plane: $\overrightarrow{PQ} = \langle x-1, y-3, z-4\rangle$ and $\mathbf{n}\cdot\overrightarrow{PQ} = 0$ • $2(x-1) - 4(y-3) + (z-4) = 0$

T4. $\mathbf{v} = \langle 4,-5,6\rangle$ is perpendicular to the plane and parallel to the line. • With $P = (0,4,0)$ and $R = (x,y,z) : \overrightarrow{OR} = \overrightarrow{OP} + t\mathbf{v} = \langle 0,4,0\rangle + t\langle 4,-5,6\rangle = \langle 0+4t, 4-5t, 0+6t\rangle$ • $x = 4t$, $y = 4-5t$. $z = 6t$

T5. $\mathbf{n} = \langle -5,1,2\rangle$ is parallel to the line perpendicular to the plane. • With $P = (1,2,3)$ and $Q = (x,y,z)$ a point on the plane: $\mathbf{n}\cdot\overrightarrow{PQ} = 0$ • $-5(x-1) + (y-2) + 2(z-3) = 0$

T6. The planes parallel to $6x = 5y$ are $6x - 5y = k$. • Set $x = 1$, $y = 1$, $z = 5 : 6(1) - 5(1) = k$ • $k = 1$ • $6x - 5y = 1$.

T7. (a) $\begin{vmatrix} 5 & 0 & 1 \\ 0 & 4 & -2 \\ 1 & 2 & 3 \end{vmatrix} = 5\begin{vmatrix} 4 & -2 \\ 2 & 3 \end{vmatrix} - 0\begin{vmatrix} 0 & -2 \\ 1 & 3 \end{vmatrix} + 1\begin{vmatrix} 0 & 4 \\ 1 & 2 \end{vmatrix}$
$= 5[(4)(3) - (-2)(2)] - 0[0(3) - (-2)(1)] + [0(2) - 4(1)] = 5(16) - 0 + -4 = 76$
(b) $\begin{vmatrix} 1 & 1 & 2 \\ 2 & 1 & 0 \\ 1 & 3 & 1 \end{vmatrix} = 1\begin{vmatrix} 1 & 0 \\ 3 & 1 \end{vmatrix} - 1\begin{vmatrix} 2 & 0 \\ 1 & 1 \end{vmatrix} + 2\begin{vmatrix} 2 & 1 \\ 1 & 3 \end{vmatrix}$
$= [(1)(1) - (0)(3)] - [(2)(1) - (0)(1)] + 2[(2)(3) - (1)(1)] = 1 - 2 + 2(5) = 9$

T8. $\mathbf{A}\times\mathbf{B} = \begin{vmatrix} \mathbf{i} & \mathbf{j} & \mathbf{k} \\ 1 & 3 & 7 \\ 6 & -2 & 0 \end{vmatrix} = \begin{vmatrix} 3 & 7 \\ -2 & 0 \end{vmatrix}\mathbf{i} - \begin{vmatrix} 1 & 7 \\ 6 & 0 \end{vmatrix}\mathbf{j} + \begin{vmatrix} 1 & 3 \\ 6 & -2 \end{vmatrix}\mathbf{k} = 14\mathbf{i} + 42\mathbf{j} - 20\mathbf{k}$ •
Partial check: $\langle 1,3,7\rangle \cdot \langle 14,42,-20\rangle = 1(14) + 3(42) + 7(-20) = 0$
$\langle 6,-2,0\rangle \cdot \langle 14,42,-20\rangle = 6(14) - 2(42) + 0(-20) = 0$

Answers 10.3

T9. $\mathbf{v} = \mathbf{A} \times \mathbf{B} = \begin{vmatrix} \mathbf{i} & \mathbf{j} & \mathbf{k} \\ 2 & -1 & 1 \\ 3 & 2 & -2 \end{vmatrix} = \begin{vmatrix} -1 & 1 \\ 2 & -2 \end{vmatrix} \mathbf{i} - \begin{vmatrix} 2 & 1 \\ 3 & -2 \end{vmatrix} \mathbf{j} + \begin{vmatrix} 2 & -1 \\ 3 & 2 \end{vmatrix} \mathbf{k} = 0\mathbf{i} + 7\mathbf{j} + 7\mathbf{k} = \langle 0, 7, 7 \rangle$ •
$P = (1, 0, 3)$ • $R = (x, y, z)$ • $\overrightarrow{OR} = \overrightarrow{OP} + t\mathbf{v} = \langle 1 + 0t, 0 + 7t, 3 + 7t \rangle$ • $x = 1$, $y = 7t$, $z = 3 + 7t$

T10. $\mathbf{n}_1 = \langle 1, -2, 3 \rangle$ is normal to $x - 2y + 3z = 0$ and $\mathbf{n}_2 = \langle 4, -5, -1 \rangle$ is normal to $4x - 5y - z = 0$ •
$\mathbf{v} = \mathbf{n}_1 \times \mathbf{n}_2$ is parallel to the line. • $\mathbf{v} = \begin{vmatrix} \mathbf{i} & \mathbf{j} & \mathbf{k} \\ 1 & -2 & 3 \\ 4 & -5 & -1 \end{vmatrix} = \begin{vmatrix} -2 & 3 \\ -5 & -1 \end{vmatrix} \mathbf{i} - \begin{vmatrix} 1 & 3 \\ 4 & -1 \end{vmatrix} \mathbf{j} + \begin{vmatrix} 1 & -2 \\ 4 & -5 \end{vmatrix} \mathbf{k}$
$= 17\mathbf{i} + 13\mathbf{j} + 3\mathbf{k} = \langle 17, 13, 3 \rangle$ • $P = (3, 2, 0)$ • $R = (x, y, z)$ •
$\overrightarrow{OR} = \overrightarrow{OP} + t\mathbf{v} = \langle 3 + 17t, 2 + 13t, 0 + 3t \rangle$ • $x = 3 + 17t$, $y = 2 + 13t$, $z = 3t$

T11. Set $P = (1, 2, 3)$, $Q = (4, 3, 2)$ and $R = (0, 0, 1)$ • $\overrightarrow{PQ} = \langle 3, 1, -1 \rangle$ and $\overrightarrow{PR} = \langle -1, -2, -2 \rangle$ form adjacent sides of the parallelogram. •
$\overrightarrow{PQ} \times \overrightarrow{PR} = \begin{vmatrix} \mathbf{i} & \mathbf{j} & \mathbf{k} \\ 3 & 1 & -1 \\ -1 & -2 & -2 \end{vmatrix} = \begin{vmatrix} 1 & -1 \\ -2 & -2 \end{vmatrix} \mathbf{i} - \begin{vmatrix} 3 & -1 \\ -1 & -2 \end{vmatrix} \mathbf{j} + \begin{vmatrix} 3 & 1 \\ -1 & -2 \end{vmatrix} \mathbf{k}$
$= -4\mathbf{i} + 7\mathbf{j} - 5\mathbf{k} = \langle -4, 7, -5 \rangle$ • [Area] $= |\overrightarrow{PQ} \times \overrightarrow{PR}| = \sqrt{4^2 + 7^2 + 5^2} = 3\sqrt{10}$

T12. $\mathbf{A} \cdot \mathbf{B} \times \mathbf{C} = \begin{vmatrix} 1 & 3 & 2 \\ -1 & 0 & 4 \\ -2 & 1 & 5 \end{vmatrix} = 1 \begin{vmatrix} 0 & 4 \\ 1 & 5 \end{vmatrix} - 3 \begin{vmatrix} -1 & 4 \\ -2 & 5 \end{vmatrix} + 2 \begin{vmatrix} -1 & 0 \\ -2 & 1 \end{vmatrix}$
$= 1[(0)(5) - (4)(1)] - 3[(-1)(5) - (4)(-2)] + 2[(-1)(1) - (0)(-2)] = -4 - 9 - 2 = -15$

T13. Figure T13 • $\overrightarrow{PQ} = \langle 4 - 1, 3 - 0, 1 - 2 \rangle = \langle 3, 3, -1 \rangle$ • $\overrightarrow{PR} = \langle 5 - 1, 6 - 0, 7 - 2 \rangle = \langle 4, 6, 5 \rangle$ •
$\overrightarrow{PS} = \langle 3 - 1, 2 - 0, 1 - 2 \rangle = \langle 2, 2, -1 \rangle$ • $\overrightarrow{OT} = \overrightarrow{OP} + \overrightarrow{PS} + \overrightarrow{PQ} + \overrightarrow{PR}$
$= \langle 1, 0, 2 \rangle + \langle 2, 2, -1 \rangle + \langle 3, 3, -1 \rangle + \langle 4, 6, 5 \rangle = \langle 10, 11, 5 \rangle$ • $T = (10, 11, 5)$

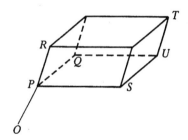

Figure T13

Problems 10.3

1. (a) $\begin{vmatrix} 1 & 3 & 0 \\ 2 & -1 & 1 \\ 0 & 4 & 5 \end{vmatrix} = 1 \begin{vmatrix} -1 & 1 \\ 4 & 5 \end{vmatrix} - 3 \begin{vmatrix} 2 & 1 \\ 0 & 5 \end{vmatrix} + 0 \begin{vmatrix} 2 & -1 \\ 0 & 4 \end{vmatrix}$
$= 1[(-1)(5) - (1)(4)] - 3[(2)(5) - (1)(0)] + 0[(2)(4) - (-1)(0)] = 1(-9) - 30 = -39$

(b) $\begin{vmatrix} 5 & 0 & 1 \\ 0 & 4 & -2 \\ 1 & 2 & 3 \end{vmatrix} = 76$ (c) $\begin{vmatrix} 5 & 0 & 0 \\ 0 & 6 & 4 \\ -3 & 2 & 8 \end{vmatrix} = 200$ (e) $\begin{vmatrix} -1 & 1 & 0 \\ 0 & 2 & 1 \\ 3 & 3 & 3 \end{vmatrix} = 0$

2. (a) $\begin{vmatrix} \mathbf{i} & \mathbf{j} & \mathbf{k} \\ 1 & 4 & -6 \\ 3 & 2 & -4 \end{vmatrix} = \begin{vmatrix} 4 & -6 \\ 2 & -4 \end{vmatrix} \mathbf{i} - \begin{vmatrix} 1 & -6 \\ 3 & -4 \end{vmatrix} \mathbf{j} + \begin{vmatrix} 1 & 4 \\ 3 & 2 \end{vmatrix} \mathbf{k}$
$= [(4)(-4) - (-6)(2)]\mathbf{i} - [(1)(-4) - (-6)(3)]\mathbf{j} + [(1)(2) - (4)(3)]\mathbf{k} = -4\mathbf{i} - 14\mathbf{j} - 10\mathbf{k}$

(b) $\begin{vmatrix} \mathbf{i} & \mathbf{j} & \mathbf{k} \\ 3 & 1 & -2 \\ 1 & 4 & 0 \end{vmatrix} = 8\mathbf{i} - 2\mathbf{j} + 11\mathbf{k}$

(c) $\begin{vmatrix} \mathbf{i} & \mathbf{j} & \mathbf{k} \\ 2 & -1 & 6 \\ 1 & 3 & -2 \end{vmatrix} = -16\mathbf{i} + 10\mathbf{j} + 7\mathbf{k}$

(e) $\begin{vmatrix} \mathbf{i} & \mathbf{j} & \mathbf{k} \\ 1 & 1 & -1 \\ 6 & 0 & -7 \end{vmatrix} = \begin{vmatrix} 1 & -1 \\ 0 & -7 \end{vmatrix} \mathbf{i} - \begin{vmatrix} 1 & -1 \\ 6 & -7 \end{vmatrix} \mathbf{j} + \begin{vmatrix} 1 & 1 \\ 6 & 0 \end{vmatrix} \mathbf{k}$
$= [(1)(-7) - (-1)(0)]\mathbf{i} - [(1)(-7) - (-1)(6)]\mathbf{j} + [(1)(0) - (1)(6)]\mathbf{k} = -7\mathbf{i} + \mathbf{j} - 6\mathbf{k}$

3. (a) True (b) False (c) True (d) False (e) False (f) True (g) False (h) True (i) False (j) True (k) True (ℓ) True

4. (a) Set $P = (2,4,3)$ and $\mathbf{v} = \langle 4,0,-7 \rangle$. • For $R = (x,y,z)$ on the line: $\overrightarrow{OR} = \overrightarrow{OP} + t\mathbf{v} = \langle 2,4,3 \rangle + t\langle 4,0,-7 \rangle = \langle 2+4t, 4+0t, 3-7t \rangle$ • $x = 2+4t$, $y = 4$, $z = 3-7t$
(b) $\mathbf{v} = \langle 6,-4,0 \rangle$ is perpendicular to the plane and parallel to the line. • Set $P = (3,-4,5)$ • For $R = (x,y,z)$ on the line: $\overrightarrow{OR} = \overrightarrow{OP} + t\mathbf{v} = \langle 3,-4,5 \rangle + t\langle 6,-4,0 \rangle = \langle 3+6t, -4-4t, 5+0t \rangle$ • $x = 3+6t$, $y = -4-4t$, $z = 5$
(c) $x = 5$, $y = 6 - t$, $z = 7 + 2t$
(e) Set $P = (6,2,3)$ and $Q = (7,0,-10)$ • $\mathbf{v} = \overrightarrow{PQ} = \langle 7-6, 0-2, -10-3 \rangle = \langle 1,-2,-13 \rangle$ is parallel to the line. • For $R = (x,y,z)$ on the line: $\overrightarrow{OR} = \overrightarrow{OP} + t\mathbf{v} = \langle 6,2,3 \rangle + t\langle 1,-2,-13 \rangle = \langle 6+t, 2-2t, 3-13t \rangle$ • $x = 6+t$, $y = 2-2t$, $z = 3-13t$
(f) $x = 2t$, $y = 5 - 5t$, $z = -4 + 7t$

5. (a) Set $P = (3,0,8)$ and $\mathbf{n} = \langle 4,0,5 \rangle$ • For $R = (x.y.z)$ on the plane: $\overrightarrow{PR} = \langle x-3, y-0, z-8 \rangle$ and $\mathbf{n} \cdot \overrightarrow{PR} = 4(x-3) + 0(y-0) + 5(z-8)$ is zero. • $4(x-3) + 5(z-8) = 0$ (b) $(x-10) - 2(y-5) + 3z = 0$

6. (a) $\mathbf{n}_1 = \langle 2,-4,0 \rangle$ is perpendicular to one plane and $\mathbf{n}_2 = \langle 0,3,5 \rangle$ is perpendicular to the other plane. • $\mathbf{n}_1 \times \mathbf{n}_2$ is parallel to both planes and to the line. • Set $P = (6,0,-3)$ •

$$\mathbf{n}_1 \times \mathbf{n}_2 = \begin{vmatrix} \mathbf{i} & \mathbf{j} & \mathbf{k} \\ 2 & -4 & 0 \\ 0 & 3 & 5 \end{vmatrix} = \begin{vmatrix} -4 & 0 \\ 3 & 5 \end{vmatrix} \mathbf{i} - \begin{vmatrix} 2 & 0 \\ 0 & 5 \end{vmatrix} \mathbf{j} + \begin{vmatrix} 2 & -4 \\ 0 & 3 \end{vmatrix} \mathbf{k}$$

$= [(-4)(5) - (3)(0)]\mathbf{i} - [(2)(5) - (0)(0)]\mathbf{j} + [(2)(3) - (-4)(0)]\mathbf{k} = -20\mathbf{i} - 10\mathbf{j} + 6\mathbf{k}$ • Use $\mathbf{v} = \frac{1}{2} \mathbf{n}_1 \times \mathbf{n}_2 = \langle -10, -5, 3 \rangle$ • For $R = (x,y,z)$ on the line: $\overrightarrow{OR} = \overrightarrow{OP} + t\mathbf{v} = \langle 6,0,-3 \rangle + t\langle -10,-5,3 \rangle = \langle 6-10t, -5t, -3+3t \rangle$ • $x = 6-10t$, $y = -5t$, $z = -3+3t$
(b) $-x + (y - 1) + z = 0$
(c) Two planes are perpendicular if their normal vectors are perpendicular. • $\mathbf{n}_1 = \langle 0, 1, -1 \rangle$ is normal to $y - z = 4$ and $\mathbf{n}_2 = \langle 1, 0, 1 \rangle$ is normal to $x + z = 3$. • $\mathbf{n} = (\mathbf{j} - \mathbf{k}) \times (\mathbf{i} + \mathbf{k})$ is normal to the plane we want. • $\mathbf{n} = \begin{vmatrix} \mathbf{i} & \mathbf{j} & \mathbf{k} \\ 0 & 1 & -1 \\ 1 & 0 & 1 \end{vmatrix} = \begin{vmatrix} 1 & -1 \\ 0 & 1 \end{vmatrix} \mathbf{i} - \begin{vmatrix} 0 & -1 \\ 1 & 1 \end{vmatrix} \mathbf{j} + \begin{vmatrix} 0 & 1 \\ 1 & 0 \end{vmatrix} \mathbf{k}$

$= [(1)(1) - (0)(-1)]\mathbf{i} - [(0)(1) - (-1)(1)]\mathbf{j} + [(0)(0) - (1)(1)]\mathbf{k} = \mathbf{i} - \mathbf{j} - \mathbf{k}$ • Set $P = (5,-1,-2)$ • For $R = (x,y,z)$ on the plane: $\overrightarrow{PR} = \langle x-5, y-(-1), z-(-2) \rangle = \langle x-5, y+1, z+2 \rangle$ and $\mathbf{n} \cdot \overrightarrow{PR} = 0$
• $(x-5) - (y+1) - (z+2) = 0$
(e) Set $P = (1,-1,2)$, $Q = (4,-1,5)$ and $R = (1,2,-1)$ • $\overrightarrow{PQ} = \langle 4-1, -1-(-1), 5-2 \rangle = \langle 3,0,3 \rangle$ and $\overrightarrow{PR} = \langle 1-1, 2-(-1), -1-2 \rangle = \langle 0, 3, -3 \rangle$ are parallel to the plane and $\overrightarrow{PQ} \times \overrightarrow{PR}$ is perpendicular to it. • $\overrightarrow{PQ} \times \overrightarrow{PR} = \begin{vmatrix} \mathbf{i} & \mathbf{j} & \mathbf{k} \\ 3 & 0 & 3 \\ 0 & 3 & -3 \end{vmatrix} = \begin{vmatrix} 0 & 3 \\ 3 & -3 \end{vmatrix} \mathbf{i} - \begin{vmatrix} 3 & 3 \\ 0 & -3 \end{vmatrix} \mathbf{j} + \begin{vmatrix} 3 & 0 \\ 0 & 3 \end{vmatrix} \mathbf{k}$

$= [(0)(-3) - (3)(3)]\mathbf{i} - [(3)(-3) - (3)(0)]\mathbf{j} + [(3)(3) - (0)(0)]\mathbf{k} = -9\mathbf{i} + 9\mathbf{j} + 9\mathbf{k}$ • Use $\mathbf{n} = \frac{1}{9} \overrightarrow{PQ} \times \overrightarrow{PR} = \langle -1,1,1 \rangle$ • For $R = (x,y,z)$ on the plane: $\overrightarrow{PR} = \langle x-1, y-(-1), z-2 \rangle = \langle x-1, y+1, z-2 \rangle$ and $\mathbf{n} \cdot \overrightarrow{PR} = 0$ • $-(x-1) + (y+1) + (z-2) = 0$
(g) The planes parallel to $4x - y + 3z = 0$ are $4x - y + 3z = c$ with constant c. • Set $x = 1$, $y = 2$, $z = 3$: $4(1) - 2 + 3(3) = c$ • $c = 11$ • $4x - y + 3z = 11$

7. (a) $\langle 4,3,1 \rangle \times \langle 5,7,2 \rangle = \begin{vmatrix} \mathbf{i} & \mathbf{j} & \mathbf{k} \\ 4 & 3 & 1 \\ 5 & 7 & 2 \end{vmatrix} = \begin{vmatrix} 3 & 1 \\ 7 & 2 \end{vmatrix} \mathbf{i} - \begin{vmatrix} 4 & 1 \\ 5 & 2 \end{vmatrix} \mathbf{j} + \begin{vmatrix} 4 & 3 \\ 5 & 7 \end{vmatrix} \mathbf{k}$

$= [(3)(2) - (1)(7)]\mathbf{i} - [(4)(2) - (1)(5)]\mathbf{j} + [(4)(7) - (3)(5)]\mathbf{k} = -\mathbf{i} - 3\mathbf{j} + 13\mathbf{k}$ is perpendicular to $\langle 4,3,1 \rangle$ and $\langle 5,7,2 \rangle$ • $|\langle -1,-3,13 \rangle| = \sqrt{1^2 + 3^2 + 13^2} = \sqrt{179}$ • Unit vectors $\pm \frac{1}{\sqrt{179}} \langle -1,-3,13 \rangle$
(b) Unit vectors $\pm \frac{1}{\sqrt{474}} \langle 7, 5, 20 \rangle$

8. $\alpha = 0$

9. $2a + b = 1$

Answers 10.3   127

12. $\langle 2, b, c \rangle$ is parallel to $\langle 1, 3, 5 \rangle \iff \langle 2, b, c \rangle = m \langle 1, 3, 5 \rangle$ for some constant $m \iff 2 = m$, $b = 3m$, $c = 5m$ • $b = 6$, $c = 10$

14. (a) $\mathbf{A} \cdot (\mathbf{B} \times \mathbf{C}) = \begin{vmatrix} 3 & 0 & 1 \\ 1 & 2 & -1 \\ 0 & 1 & 6 \end{vmatrix} = 3 \begin{vmatrix} 2 & -1 \\ 1 & 6 \end{vmatrix} - 0 \begin{vmatrix} 1 & -1 \\ 0 & 6 \end{vmatrix} + 1 \begin{vmatrix} 1 & 2 \\ 0 & 1 \end{vmatrix}$
$= 3[(2)(6) - (-1)(1)] - 0[(1)(6) - (-1)(0)] + [(1)(1) - (2)(0)] = 39 + 0 + 1 = 40$
(b) $\mathbf{A} \cdot (\mathbf{B} \times \mathbf{C}) = 12$

15. (a) $\overrightarrow{PQ} = \langle 6-1, 1-2, -1-4 \rangle = \langle 5, -1, -5 \rangle$ • $\overrightarrow{PS} = \langle 3-1, -2-2, 0-4 \rangle = \langle 2, -4, -4 \rangle$
$\overrightarrow{PQ} \times \overrightarrow{PS} = \begin{vmatrix} \mathbf{i} & \mathbf{j} & \mathbf{k} \\ 5 & -1 & -5 \\ 2 & -4 & -4 \end{vmatrix} = \begin{vmatrix} -1 & -5 \\ -4 & -4 \end{vmatrix} \mathbf{i} - \begin{vmatrix} 5 & -5 \\ 2 & -4 \end{vmatrix} \mathbf{j} + \begin{vmatrix} 5 & -1 \\ 2 & -4 \end{vmatrix} \mathbf{k}$
$= [(-1)(-4) - (-5)(-4)]\mathbf{i} - [(5)(-4) - (-5)(2)]\mathbf{j} + [(5)(-4) - (-1)(2)]\mathbf{k} = -16\mathbf{i} + 10\mathbf{j} - 18\mathbf{k}$ •
[Area] $= |\overrightarrow{PQ} \times \overrightarrow{PS}| = \sqrt{16^2 + 10^2 + 18^2} = \sqrt{680}$

16. (a) $\overrightarrow{PQ} = \langle 3-2, 4-1, 5-0 \rangle = \langle 1, 3, 5 \rangle$ • $\overrightarrow{PR} = \langle 6-2, 1-1, 2-0 \rangle = \langle 4, 0, 2 \rangle$ •
$\overrightarrow{PQ} \times \overrightarrow{PR} = \begin{vmatrix} \mathbf{i} & \mathbf{j} & \mathbf{k} \\ 1 & 3 & 5 \\ 4 & 0 & 2 \end{vmatrix} = \begin{vmatrix} 3 & 5 \\ 0 & 2 \end{vmatrix} \mathbf{i} - \begin{vmatrix} 1 & 5 \\ 4 & 2 \end{vmatrix} \mathbf{j} + \begin{vmatrix} 1 & 3 \\ 4 & 0 \end{vmatrix} \mathbf{k}$
$= [(3)(2) - (5)(0)]\mathbf{i} - [(1)(2) - (5)(4)]\mathbf{j} + [(1)(0) - (3)(4)]\mathbf{k} = 6\mathbf{i} + 18\mathbf{j} - 12\mathbf{k}$ •
[Area] $= \frac{1}{2}|\overrightarrow{PQ} \times \overrightarrow{PS}| = \frac{1}{2}\sqrt{6^2 + 18^2 + 12^2} = \sqrt{126}$

17. (a) $\overrightarrow{PQ} = \langle 0-1, 0-1, -6-1 \rangle = \langle -1, -1, -7 \rangle$ • $\overrightarrow{PR} = \langle 8-1, 1-1, 2-1 \rangle = \langle 7, 0, 1 \rangle$ •
$\overrightarrow{PS} = \langle -4-1, 3-1, 1-1 \rangle = \langle -5, 2, 0 \rangle$ • $\overrightarrow{OT} = \overrightarrow{OP} + \overrightarrow{PQ} + \overrightarrow{PR} + \overrightarrow{PS}$
$= \langle 1, 1, 1 \rangle + \langle -1, -1, -7 \rangle + \langle 7, 0, 1 \rangle + \langle -5, 2, 0 \rangle = \langle 2, 2, -5 \rangle$ • $T = (2, 2, -5)$
(b) $\overrightarrow{PQ} \cdot (\overrightarrow{PR} \times \overrightarrow{PS}) = \begin{vmatrix} -1 & -1 & -7 \\ 7 & 0 & 1 \\ -5 & 2 & 0 \end{vmatrix} = -1 \begin{vmatrix} 0 & 1 \\ 2 & 0 \end{vmatrix} - (-1) \begin{vmatrix} 7 & 1 \\ -5 & 0 \end{vmatrix} + (-7) \begin{vmatrix} 7 & 0 \\ -5 & 2 \end{vmatrix}$
$= -[(0)(0) - (1)(2)] + [(7)(0) - (1)(-5)] - 7[(7)(2) - (0)(-5)] = 2 + 5 - 98 = -91$ •
[Volume] $= |\overrightarrow{PQ} \cdot (\overrightarrow{PR} \times \overrightarrow{PS})| = 91$

19. (a) Set $P = (1, 2, 3)$, $Q = (2, 1, 4)$, $R = (6, 0, 2)$, $S = (4, 4, 4)$ • $\overrightarrow{PQ} = \langle 2-1, 1-2, 4-3 \rangle = \langle 1, -1, 1 \rangle$ • $\overrightarrow{PR} = \langle 6-1, 0-2, 2-3 \rangle = \langle 5, -2, -1 \rangle$ • $\overrightarrow{PS} = \langle 4-1, 4-2, 4-3 \rangle = \langle 3, 2, 1 \rangle$ •
$\overrightarrow{PQ} \cdot (\overrightarrow{PR} \times \overrightarrow{PS}) = \begin{vmatrix} 1 & -1 & 1 \\ 5 & -2 & -1 \\ 3 & 2 & 1 \end{vmatrix} = 1 \begin{vmatrix} -2 & -1 \\ 2 & 1 \end{vmatrix} - (-1) \begin{vmatrix} 5 & -1 \\ 3 & 1 \end{vmatrix} + 1 \begin{vmatrix} 5 & -2 \\ 3 & 2 \end{vmatrix}$
$= [(-2)(1) - (-1)(2)] + [(5)(1) - (-1)(3)] + [(5)(2) - (-2)(3)] = 0 + 8 + 16 = 24$ •
[Volume] $= \frac{1}{6}|\overrightarrow{PQ} \cdot (\overrightarrow{PR} \times \overrightarrow{PS})| = 4$
(b) [Volume] $= \frac{1}{6}|\overrightarrow{PQ} \cdot (\overrightarrow{PR} \times \overrightarrow{PS})| = \frac{1}{6}$

20. (a) $\frac{x}{2} + \frac{y}{3} + \frac{z}{5} = 1$ • Figure A20a (b) $\frac{y}{5} + \frac{z}{2} = 1$ • Figure A20b (c) $x = 10$ • Figure A20c

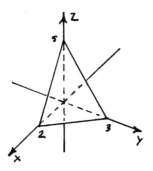

Figure A20a

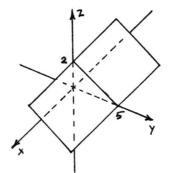

Figure A20b

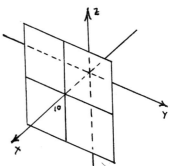

Figure A20c

27. (b) The points are in a plane.

29. (a) Write $x - 3y + 4z - 100 = 0$. • [Distance] $= \dfrac{|0 - 3(1) + 4(0) - 100|}{\sqrt{26}} = \dfrac{103}{\sqrt{26}}$

31. $(0, 0, 5)$ is on the plane $2x + 2y + z = 5$ • [The distance from $(0, 0, 5)$ to $2x + 2y + z - 25 = 0$] $= \dfrac{|2(0) + 2(0) + 5 - 25|}{\sqrt{2^2 + 2^2 + 1^2}} = \dfrac{20}{3}$

32. (a) $\mathbf{n} = \langle 15, 5 \rangle$ • [Distance from the line to $(100, 200)$] $= \dfrac{|15(100) + 5(200) - 47|}{\sqrt{15^2 + 5^2}} = \dfrac{2453}{\sqrt{250}}$
(b) $\mathbf{n} = \langle 4, -3 \rangle$ • [Distance from the line to $(0, 0)$] $= 2$

33. [Distance] $= \dfrac{1}{\sqrt{29}}$

36. (a) $\mathbf{n}_1 = \langle 1, 1, 1 \rangle$ is normal to $x + y + z = 1$ and $\mathbf{n}_2 = \langle 1, -2, -1 \rangle$ is normal to $x - 2y - z = 1$ •
$\mathbf{v} = \mathbf{n}_1 \times \mathbf{n}_2 = \begin{vmatrix} \mathbf{i} & \mathbf{j} & \mathbf{k} \\ 1 & 1 & 1 \\ 1 & -2 & -1 \end{vmatrix} = \begin{vmatrix} 1 & 1 \\ -2 & -1 \end{vmatrix} \mathbf{i} - \begin{vmatrix} 1 & 1 \\ 1 & -1 \end{vmatrix} \mathbf{j} + \begin{vmatrix} 1 & 1 \\ 1 & -2 \end{vmatrix} \mathbf{k}$
$= [(1)(-1) - (1)(-2)]\mathbf{i} - [(1)(-1) - (1)(1)]\mathbf{j} + [(1)(-2) - (1)(1)]\mathbf{k} = \mathbf{i} + 2\mathbf{j} - 3\mathbf{k}$ is parallel to the line of intersection. • To find one point on the line of intersection set $z = 0$ in the equations of the planes: $x + y = 1$, $x - 2y = 1$ • $x = 1$, $y = 0$ • $(1, 0, 0)$ is on the intersection. • $x = 1 + t$, $y = 2t$, $z = -3t$
(b) $x = 2t$, $y = -3 + 4t$, $z = -\frac{19}{4} + 3t$

38. Solve $2t = 4 + s$, $3 - t = 1 + 2s$, $-2 + 4t = 6$ • $t = 2$, $s = 0$ • The intersection is $(4, 1, 6)$

41. (a) $\mathbf{v}_1 = \langle -1, 4, 2 \rangle$ is parallel to L_1 and $\mathbf{v}_2 = \langle 1, 0, 2 \rangle$ is parallel to L_2 •
$\mathbf{v}_1 \times \mathbf{v}_2 = \begin{vmatrix} \mathbf{i} & \mathbf{j} & \mathbf{k} \\ -1 & 4 & 2 \\ 1 & 0 & 2 \end{vmatrix} = \begin{vmatrix} 4 & 2 \\ 0 & 2 \end{vmatrix} \mathbf{i} - \begin{vmatrix} -1 & 2 \\ 1 & 2 \end{vmatrix} \mathbf{j} + \begin{vmatrix} -1 & 4 \\ 1 & 0 \end{vmatrix} \mathbf{k}$
$= [(4)(2) - (2)(0)]\mathbf{i} - [(-1)(2) - (2)(1)]\mathbf{j} + [(-1)(0) - (4)(1)]\mathbf{k} = 8\mathbf{i} + 4\mathbf{j} - 4\mathbf{k}$ is normal to the parallel planes containing the lines. Use $\mathbf{n} = \langle 2, 1, -1 \rangle$ • $P = (2, 3, 0)$ is on L_1 and $Q = (-1, 2, -1)$ is on L_2
• [Distance] $= \dfrac{|\mathbf{n} \cdot \overrightarrow{PQ}|}{|\mathbf{n}|} = \dfrac{|\langle 2, 1, -1 \rangle \cdot \langle -3, -1, -1 \rangle|}{\sqrt{2^2 + 1^2 + 1^2}} = \dfrac{|(-3)(2) + (-1)(1) + (-1)(-1)|}{\sqrt{6}} = \dfrac{6}{\sqrt{6}} = \sqrt{6}$

Tune-up exercises 10.4

T1. (a) For $0 < t < 1$: $\dfrac{dx}{dt} = \dfrac{9-1}{1-0} = 8$ • $\dfrac{dy}{dt} = \dfrac{11-3}{1-0} = 8$ • $\mathbf{v} = \langle 8, 8 \rangle$ meters per minute •
[Speed] $= |\langle 8, 8 \rangle| = 8\sqrt{2}$ meters per minute
For $1 < t < 2$: $\dfrac{dx}{dt} = \dfrac{13-9}{2-1} = 4$ • $\dfrac{dy}{dt} = \dfrac{7-11}{2-1} = -4$ • $\mathbf{v} = \langle 4, -4 \rangle$ meters per minute •
[Speed] $= |\langle 4, -4 \rangle| = 4\sqrt{2}$ meters per minute
For $2 < t < 3$: $\dfrac{dx}{dt} = \dfrac{10-13}{3-2} = -3$ • $\dfrac{dy}{dt} = \dfrac{4-7}{3-2} = -3$ • $\mathbf{v} = \langle -3, -3 \rangle$ meters per minute •
[Speed] $= |\langle -3, -3 \rangle| = 3\sqrt{2}$ meters per minute • Figure T1

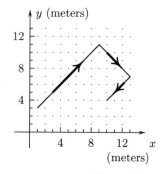

Figure T1

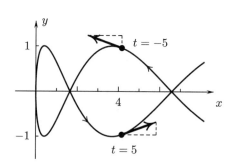

Figure T4

Answers 10.4

T2. $\lim_{t \to 1} \langle 3t^2, \ln t, \cos(\pi t) \rangle = \langle \lim_{t \to 1}(3t^2), \lim_{t \to 1}(\ln t), \lim_{t \to 1}[\cos(\pi t)] \rangle = \langle 3, \ln(1), \cos(\pi) \rangle = \langle 3, 0, -1 \rangle$

T3. $\dfrac{d}{dt}(te^t\,\mathbf{i} - \tan^{-1}t\,\mathbf{j}) = \dfrac{d}{dt}(te^t)\,\mathbf{i} - \dfrac{d}{dt}(\tan^{-1}t)\,\mathbf{j} = [t\dfrac{d}{dt}(e^t) + e^t\dfrac{d}{dt}(t)]\,\mathbf{i} - \dfrac{1}{t^2+1}\,\mathbf{j}$
$= (te^t + e^t)\,\mathbf{i} - \dfrac{1}{t^2+1}\,\mathbf{j}$

T4. (a) $\mathbf{R}(t) = \langle \tfrac{1}{6}t^2, \sin t \rangle$ • $\mathbf{v}(t) = \dfrac{d\mathbf{R}}{dt} = \langle \tfrac{1}{3}t, \cos t \rangle$ • $\mathbf{v}(-5) = \langle -\tfrac{5}{3}, \cos(-5) \rangle, \mathbf{v}(5) = \langle \tfrac{5}{3}, \cos(5) \rangle$ •
(b) $\mathbf{R}(-5) = \langle \tfrac{1}{6}(-5)^2, \sin(-5) \rangle \doteq \langle 4.17, -0.96 \rangle$ • $\mathbf{v}(-5) \doteq \langle -1.67, 0.28 \rangle$ •
$\mathbf{R}(5) = \langle \tfrac{1}{6}(5)^2, \sin(5) \rangle \doteq \langle 4.17, 0.96 \rangle$ • $\mathbf{v}(5) \doteq \langle 1.67, 0.28 \rangle$ • Figure T4

T5. (a) Figure T5a

t	0	1	2	3	4	5	6
$x \approx$	-2.1	-3	-2.1	0	2.1	3	2.1
$y \approx$	2	1	-1	-2	-1	1	2

(b) Figure T5b: $\dfrac{dx}{dt}(2) \approx \dfrac{0.5 - (-2.1)}{3 - 2} = 1.6$ • Figure T5c: $\dfrac{dy}{dt}(2) \approx \dfrac{-2.8 - (-1)}{3 - 2} = -1.8$ •
$\mathbf{v}(2) \approx \langle 1.6, -1.8 \rangle$ • Figure T5a

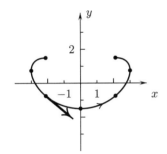

Figure T5a

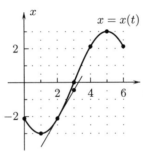

Figure T5b

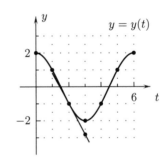

Figure T5c

T6. [Velocity] $= \mathbf{v}(t) = \dfrac{d}{dt}\langle 300t, -400t, 3 \rangle = \langle 300, -400, 0 \rangle$ miles per hour •
[Speed] $= |\langle 300, -400, 0 \rangle| = 500$ miles per hour

T7. $x'(t) = \dfrac{d}{dt}(3\cos t) = -3\sin t$ • $y'(t) = \dfrac{d}{dt}(7\sin t) = 7\cos t$ •
[Length] $= \displaystyle\int_0^{2\pi} \sqrt{[x'(t)]^2 + [y'(t)]^2}\,dt = \int_0^{2\pi} \sqrt{(3\sin t)^2 + (7\cos t)^2}\,dt \approx 32.69$

T8 $\overrightarrow{OR} = \overrightarrow{OP} + \overrightarrow{PQ} + \overrightarrow{QR}$ •
[Velocity of the point R] $= \dfrac{d}{dt}(\overrightarrow{OR}) = \dfrac{d}{dt}(\overrightarrow{OP}) + \dfrac{d}{dt}(\overrightarrow{PQ}) + \dfrac{d}{dt}(\overrightarrow{QR})$
$= \langle 7, 3, 1 \rangle + \langle 2, 1, 3 \rangle + \langle 1, 1, 1 \rangle = \langle 10, 5, 5 \rangle$ yards per minute

Problems 10.4

1. Figure A1 with the points plotted in the following table

t	1	2	3	4	5	6	7	8
x	60	60	10	10	50	50	20	20
y	10	60	60	20	20	50	50	30

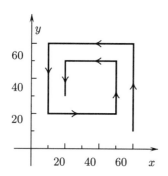

Figure A1

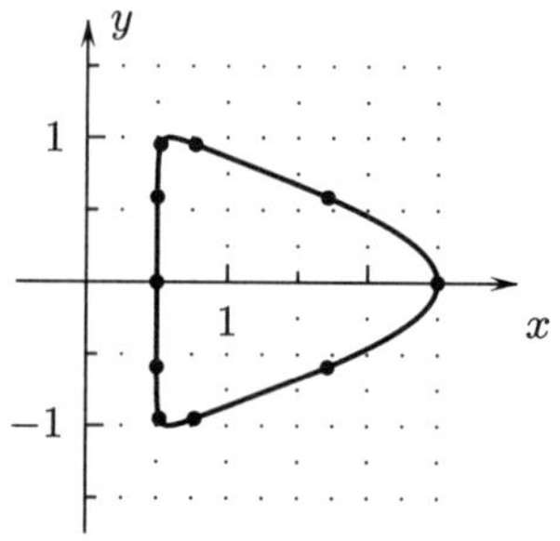

Figure A3

3. Figure A3
5. (a) $\langle 6\cos(\frac{1}{3}\pi), 4\sin(\frac{1}{4}\pi)\rangle = \langle 3, 2\sqrt{2}\rangle$ (b) $\langle 0, 0, 5\rangle$
6. (a) $\langle t\cos t + \sin t, 2te^{t^2}\rangle$ (b) $t(t^2+4)^{-1/2}\mathbf{i} + 2e^{2t}\mathbf{j}$
7. (a) $\left\langle \dfrac{2}{2t+1}, \dfrac{1}{t^2+1}, \dfrac{1}{\sqrt{1-t^2}} \right\rangle$ (b) $\langle \sec t \tan t, -\csc t \cot t\rangle$
8. $x(-2) = -8, y(-2) = 0$ • $\mathbf{v}(-2) = \langle 0, 5\rangle$ •
   $x(0) = 0, y(0) = 9$ • $\mathbf{v}(0) = \langle 6, 0\rangle$ •
   $x(1) = \frac{11}{2}, y(1) = 6$ • $\mathbf{v}(1) = \langle \frac{9}{2}, -\frac{11}{2}\rangle$ • Figure A8

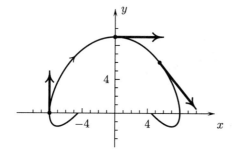

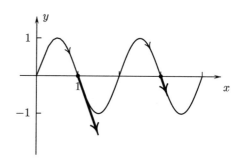

Figure A8

Figure A11

11. (a) Figure A11 (b) $x(1) = 1, y(1) = 0$ • $\mathbf{v}(1) = \langle \frac{1}{2}, -\frac{1}{2}\pi\rangle$ •
    $x(9) = 3, y(9) = 0$ • $\mathbf{v}(9) = \langle \frac{1}{6}, -\frac{1}{6}\pi\rangle$ • Figure A11 (c) $|\mathbf{v}(1)| = \frac{1}{2}\sqrt{1+\pi^2}$ •
    $|\mathbf{v}(9)| = \frac{1}{6}\sqrt{1+\pi^2}$ • The snail is moving three times as fast at $t = 1$ than at $t = 9$ in the same direction.

Answers 10.4

13. (a) Figure A13 (b) The curve is an ellipse oriented clockwise. (c) The x-intercept is 4 at $t = \frac{1}{2}\pi$.
• The y-intercept is -3 at $t = \pi$.

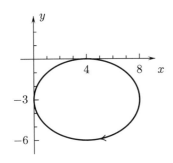

Figure A13

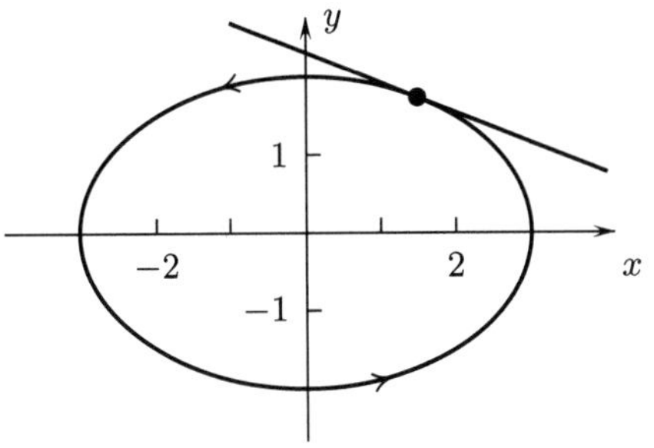

Figure A14

14. $\mathbf{R}(t) = \langle 3\cos t, 2\sin t\rangle$ • $\mathbf{v} = \langle -3\sin t, 2\cos t\rangle$ • $\mathbf{R}(\frac{1}{3}\pi) = \langle \frac{3}{2}, \sqrt{3}\rangle$ • $\mathbf{v}(\frac{1}{3}\pi) = \langle -\frac{3}{2}\sqrt{3}, 1\rangle$ •
The tangent line passes through $(\frac{3}{2}, \sqrt{3})$ and has slope $(-\frac{3}{2}\sqrt{3})^{-1} = -\frac{2}{9}\sqrt{3}$ •
Tangent line: $y = \sqrt{3} - \frac{2}{9}\sqrt{3}(x - \frac{3}{2})$ • Figure A14

16. $\frac{1}{3}(5^{3/2} - 1)$

17. (a) $x'(t) = 4t^3 - 2t, y'(t) = 5t^4 - 3t^2$ • [Length] $= \int_{-2}^{2}\sqrt{(4t^3 - 2t)^2 + (5t^4 - 3t^2)^2}\, dt \approx 55.09$

(b) [Length] $= \int_{0}^{10}\sqrt{\cos^2 t + (te^t + e^t)^2 + (3t^2 - 3)^2}\, dt \approx 220{,}281.85$

22. (a) The curve is a spiral because the projection $(x, y, 0) = (\cos t, \sin t, 0)$ on the xy-plane of a point on the curve goes around the circle $x^2 + y^2 = 1$ while the z-coordinate t of the point increases.
(b) [Velocity] $= \mathbf{v}(t) = \langle -\sin t, \cos t, \frac{1}{3}\rangle$ meters per second • [Speed] $= |\mathbf{v}(t)| = \frac{1}{3}\sqrt{10}$ meters per second
(c) $2\pi\sqrt{10}$ meters

24. (a) $\frac{4}{x^2} + \frac{4}{y^2} = \cos^2 t + \sin^2 t = 1$ (b) $(2\sqrt{2}, -2\sqrt{2})$ (c) 4 meters per minute

25. (a) Tangent line at $(1, 1)$: $y = x$ • Tangent line at $(1, -1)$: $y = -x$ (b) [Speed at the origin] $= 0$

26. (a) $x'(t) = 3t^2 - 1, y'(t) = 2t - 16t^3$ • $x(1) = 0, y(1) = -3$ • [Slope at $t = 1$] $= y'(1)/x'(1) = -7$
• Tangent line: $y = -3 - 7x$ (b) Tangent line: $y = 0$ (c) Tangent line: $y = -\sqrt{2} - \frac{2}{3}(x - \frac{1}{2}\sqrt{2})$

27. (a) Tangent line: $x = t, y = 2t, z = 3t$

28. (a) Figure A28 (b) [Minimum speed] ≈ 1.1744 at $x \approx 0.1498, y \approx 0.7988$

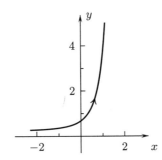

Figure A28

132　　　　　　　　　　　　　　　　　　　　　　　　　　　　Chapter 10: Vectors, Curves, and Surfaces

30. [Velocity of the car relative to the ground] $= \langle 60, 0 \rangle$ miles per hour • [Velocity of the ball relative to the car] $= \langle 0, 10 \rangle$ miles per hour • [Velocity of the ball relative to the ground] $= \langle 60, 0 \rangle + \langle 0, 10 \rangle = \langle 60, 10 \rangle$ miles per hour • [Speed] $= |\langle 60, 10 \rangle| = 10\sqrt{37}$ miles per hour

31. [Velocity of the ball relative to the ground] $= \langle \frac{49}{2}\sqrt{2}, \frac{51}{2}\sqrt{2} \rangle$ miles per hour

34. Red ant: $\mathbf{v}(\frac{3}{4}\pi) = \langle -\frac{1}{2}\sqrt{2}, -\frac{1}{2}\sqrt{2} \rangle$ • Figure A34a • Black ant: $\mathbf{v}(\frac{3}{4}\pi) = \langle -\frac{3}{2}\sqrt{2}, -\frac{3}{2}\sqrt{2} \rangle$ • Figure A34b • The black ant is traveling three times as fast as the red ant in the same direction.

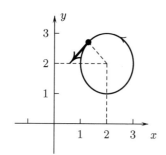
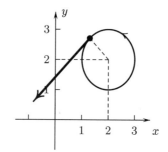

Figure A34a　　　　　　　　　　　　　　　Figure A34b

45. (a) $b = 8\pi$　(b) $\mathbf{v} = \langle 6, 7 \rangle$ when the point goes through the origin from the left (at $t = 0$) and $\mathbf{v} = \langle -6, 7 \rangle$ when the point goes through the origin from the right (at $t = 4\pi$).

49. Solve $t^3 + 6 = u^2 - 3, 3t + 4 = 3u - 5$ • $u = t + 3$ • $t = 0, 3, -2$ • Intersections: $(6, 4), (33, 13), (-2, -2)$

50. (a) The planes do not collide. (b) Their paths intersect at $(300, 1680, 560)$.

52. (a) $\dfrac{dx}{dt} = -e^{-t} - 3t^2, \dfrac{dy}{dt} = \cos t$ • $\dfrac{dy}{dx} = \dfrac{dy/dt}{dx/dt} = \dfrac{-\cos t}{e^{-t} + 3t^2}$

$\dfrac{d^2y}{dx^2} = \dfrac{1}{dx/dt} \dfrac{d}{dt}\left(\dfrac{-\cos t}{e^{-t} + 3t^2}\right) = \dfrac{-(e^{-t} + 3t^2)\sin t + (e^{-t} - 6t)\cos t}{(e^{-t} + 3t^2)^3}$

(b) $\dfrac{dy}{dx} = \dfrac{5t^4}{\ln t + 1}$ • $\dfrac{d^2y}{dx^2} = \dfrac{15t^3 + 20t^3 \ln t}{(\ln t + 1)^3}$

53. $x = k, y = \dfrac{1}{k^2 + 1}$ • $y = \dfrac{1}{x^2 + 1}$

57. [Maximum x] $= 3$ • [Minimum x] $= -\frac{3}{2}$ • [Maximum y] $= \frac{3}{2}\sqrt{3}$ • [Minimum y] $= -\frac{3}{2}\sqrt{3}$

58. [Slope] $= \frac{2}{7}$

Tune-up exercises 10.5

T1. By components: $\dfrac{dx}{dt} = \dfrac{d}{dt}(t^5) = 5t^4$ • $\dfrac{d^2x}{dt^2} = \dfrac{d}{dt}(5t^4) = 20t^3$ •

$\dfrac{dy}{dt} = \dfrac{d}{dt}(-t^7) = -7t^6$ • $\dfrac{d^2y}{dt^2} = \dfrac{d}{dt}(-7t^6) = -42t^5$ • $\mathbf{a}(t) = \langle 20t^3, -42t^5 \rangle$

T2. With vector notation: $\displaystyle\int \langle t^3, -t^4, t^5 \rangle \, dt = \left\langle \int t^3 \, dt, \int (-t^4) \, dt, \int t^5 \, dt \right\rangle$

$= \langle \frac{1}{4}t^4, -\frac{1}{5}t^5, \frac{1}{6}t^6 \rangle + \mathbf{C}$ with a constant vector $\mathbf{C} = \langle C_1, C_2, C_3 \rangle$

T3. The units (centimeters, grams, dynes, and seconds) are compatible for Newton's Law. •

$m = 5$ grams • $\mathbf{a}(t) = \dfrac{1}{m}\mathbf{F} = \frac{1}{5}\langle 10t, 5t^2 \rangle = \langle 2t, t^2 \rangle$ centimeters per second^2

$\mathbf{v}(t) = \displaystyle\int \mathbf{a}(t) \, dt = \int \langle 2t, t^2 \rangle \, dt = \langle t^2 + C_1, \frac{1}{3}t^3 + C_2 \rangle$ • $\mathbf{v}(0) = \langle C_1, C_2 \rangle$ is given as $\langle 2, 1 \rangle$. •

$C_1 = 2, C_2 = 1$, and $\mathbf{v}(t) = \langle t^2 + 2, \frac{1}{3}t^3 + 1 \rangle$ centimeters per second •

$\mathbf{R}(t) = \displaystyle\int \mathbf{v}(t) \, dt = \int \langle t^2 + 2, \frac{1}{3}t^3 + 1 \rangle = \langle \frac{1}{3}t^3 + 2t + C_3, \frac{1}{12}t^4 + t + C_4 \rangle$ •

$\mathbf{R}(0) = \langle C_3, C_4 \rangle$ is given as $\langle 1, 0 \rangle$. • $C_3 = 1, C_4 = 0$ • $\mathbf{R}(t) = \langle \frac{1}{3}t^3 + 2t + 1, \frac{1}{12}t^4 + t \rangle$ centimeters

Answers 10.5

T4. (a) This circle has radius $\rho = 3$ and is oriented counterclockwise, so its curvature is $\kappa = 1/\rho = \frac{1}{3}$.
(a) This circle has radius $\rho = \frac{1}{2}$ and is oriented clockwise, so its curvature is $\kappa = -1/\rho = -2$.

T5. (a) $y = \ln x$ with x as parameter • $\frac{dy}{dx} = x^{-1}$, $\frac{d^2y}{dx^2} = -x^{-2}$ • $\frac{dy}{dx}(1) = 1$ • $\langle 1, 1 \rangle$ is tangent to the graph at $x = 1$ and points in the direction of the curve's orientation. • $\mathbf{T}(1) = \frac{\langle 1, 1 \rangle}{|\langle 1, 1 \rangle|} = \frac{\langle 1, 1 \rangle}{\sqrt{2}}$
• $\mathbf{N}(1) = \frac{\langle -1, 1 \rangle}{\sqrt{2}}$ • Figure T5a

(b) $\kappa(1) = \frac{y''(1)}{\{1 + [y'(1)]^2\}^{3/2}} = \frac{-1}{2^{3/2}}$

(c) $\rho(1) = \frac{1}{|\kappa(1)|} = 2^{3/2}$

(d) With C the center of curvature: $\overrightarrow{OC} = \mathbf{R}(1) + \frac{\mathbf{N}(1)}{\kappa(1)} = \langle 1, 0 \rangle - 2^{3/2} \frac{\langle -1, 1 \rangle}{\sqrt{2}} = \langle 3, -2 \rangle$ •
$C = (3, -2)$ • Figure T5b

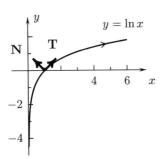

Figure T5a

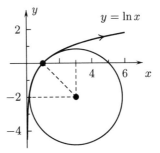

Figure T5b

T6. (a) $\mathbf{a} = -7\mathbf{T} + 18\mathbf{N} \implies \frac{d^2s}{dt^2} = -7$ and $\kappa \left(\frac{ds}{dt}\right)^2 = 18$ • $\kappa = 2 \implies \frac{ds}{dt} = 3$ •
The car is traveling 3 meters per second.
(b) The car is slowing down 7 meters per second^2.

T7. $\kappa = -\frac{1}{4}$ at P • $\mathbf{a} = s''\mathbf{T} + \kappa(s')^2\mathbf{N} = 5\mathbf{T} - \frac{1}{4}(6^2)\mathbf{N} = 5\mathbf{T} - 9\mathbf{N}$ • $\mathbf{T}$ points down to the right, so $\mathbf{N}$ points up to the right. • $\mathbf{a}$ forms the diagonal of the rectangle with sides formed by $5\mathbf{T}$ and $-9\mathbf{N}$ in Figure T7.

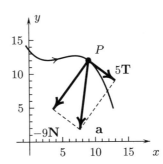

Figure T7

Problems 10.5

1. (a) $\mathbf{R}(t) = \langle te^t, -\sin t\rangle$ • By components: $x = te^t$, $\dfrac{dx}{dt} = t\dfrac{d}{dt}(e^t) + e^t\dfrac{d}{dt}(t) = te^t + e^t$,
$\dfrac{d^2x}{dt^2} = \dfrac{d}{dt}(te^t + e^t) = te^t + 2e^t$ • $y = -\sin t$, $\dfrac{dy}{dt} = -\cot t$, $\dfrac{d^2y}{dt^2} = \sin t$ •
$\mathbf{R}'(t) = \langle te^t + e^t, -\cos t\rangle$ • $\mathbf{R}''(t) = \langle te^t + 2e^t, \sin t\rangle$
   (b) $\mathbf{R}'(t) = \left\langle \dfrac{1}{t}, \dfrac{3}{1+9t^2}\right\rangle$ • $\mathbf{R}''(t) = \left\langle \dfrac{-1}{t^2}, \dfrac{-54t}{(1+9t^2)^2}\right\rangle$
   (c) $\mathbf{R}'(t) = \langle -3\sin(3t), 3\cos(3t), 1\rangle$ • $\mathbf{R}''(t) = \langle -9\cos(3t), -9\sin(3t), 0\rangle$

2. (a) $\mathbf{R}(t) = \langle \sin(2t), 3\cos(4t)\rangle$ • $\mathbf{v}(t) = \mathbf{R}'(t) = \langle \cos(2t)\dfrac{d}{dt}(2t), -3\sin(4t)\dfrac{d}{dt}(4t)\rangle$
$= \langle 2\cos(2t), -12\sin(4t)\rangle$ millimeters per minute •
[Speed] $= |\mathbf{v}(t)| = \sqrt{[2\cos(2t)]^2 + [12\sin(4t)]^2}$ millimeters per minute •
$\mathbf{a}(t) = \mathbf{R}''(t) = \langle -4\sin(2t), -48\cos(4t)\rangle$ millimeters per minute^2
   (b) $\mathbf{v} = \langle 6t, \cos t, -2\cos t \sin t\rangle$ feet per second •
[Speed] $= \sqrt{36t^2 + \cos^2 t + 4\cos^2 t \sin^2 t}$ feet per second^2
$\mathbf{a} = \langle 6, -\sin t, 2\sin^2 t - 2\cos^2 t\rangle$ feet per second

3. (a) $\displaystyle\int \langle \sin t, \cos t, e^t\rangle\,dt = \left\langle \int \sin t\,dt, \int \cos t\,dt, \int e^t\,dt\right\rangle = \langle -\cos t, \sin t, e^t\rangle + \mathbf{C}$
   (b) $\langle \tfrac{1}{2}\sin(t^2), \tfrac{1}{3}\sin(t^3)\rangle + \mathbf{C}$

4. (a) $\mathbf{R}(t) = \displaystyle\int \langle t, -t^2\rangle\,dt = \langle \tfrac{1}{2}t^2 + C_1, -\tfrac{1}{3}t^3 + C_2\rangle$ • $\mathbf{R}(1) = \langle \tfrac{1}{2} + C_1, -\tfrac{1}{3} + C_2\rangle$ equals $\langle 1, 0\rangle$ •
$C_1 = \tfrac{1}{2}, C_2 = \tfrac{1}{3}$ • $\mathbf{R}(t) = \langle \tfrac{1}{2}t^2 + \tfrac{1}{2}, -\tfrac{1}{3}t^3 + \tfrac{1}{3}\rangle$
   (b) $\mathbf{R}(t) = \langle -\tfrac{1}{3}\cos(3t) + \tfrac{13}{3}, \tfrac{1}{4}\sin(4t) + 5\rangle$

5. (a) $\mathbf{v}(t) = \displaystyle\int \langle t^4, t^3, t^2\rangle\,dt = \langle \tfrac{1}{5}t^5 + C_1, \tfrac{1}{4}t^4 + C_2, \tfrac{1}{3}t^3 + C_3\rangle$ • $\mathbf{v}(0) = \langle C_1, C_2, C_3\rangle$ is given as
$\langle 1, 2, 3\rangle$ • $C_1 = 1, C_2 = 2, C_3 = 3$ and $\mathbf{v}(t) = \langle \tfrac{1}{5}t^5 + 1, \tfrac{1}{4}t^4 + 2, \tfrac{1}{3}t^3 + 3\rangle$ •
$\mathbf{R}(t) = \displaystyle\int \langle \tfrac{1}{5}t^5 + 1, \tfrac{1}{4}t^4 + 2, \tfrac{1}{3}t^3 + 3\rangle\,dt = \langle \tfrac{1}{30}t^6 + t + C_4, \tfrac{1}{20}t^5 + 2t + C_5, \tfrac{1}{12}t^4 + 3t + C_6\rangle$ •
$\mathbf{R}(0) = \langle C_4, C_5, C_6\rangle$ is given as $\langle 7, 6, 5\rangle$ • $C_4 = 7, C_5 = 6, C_6 = 5$ •
$\mathbf{R}(t) = \langle \tfrac{1}{30}t^6 + t + 7, \tfrac{1}{20}t^5 + 2t + 6, \tfrac{1}{12}t^4 + 3t + 5\rangle$
   (b) $\mathbf{v}(t) = \langle -\tfrac{1}{2}\cos(2t) + \tfrac{1}{2}, 5, \tfrac{1}{2}\cos(2t) - \tfrac{1}{2}\rangle$ • $\mathbf{R}(t) = \langle -\tfrac{1}{4}\sin(2t) + \tfrac{1}{2}t + 5, 5t, \tfrac{1}{4}\sin(2t) - \tfrac{1}{2}t\rangle$

6. $\mathbf{a}(t) = \langle 0, 0, -9.8\rangle$ meters per second^2 • $\mathbf{v}(t) = \langle C_1, C_2, -9.8t + C_3\rangle$ •
$\mathbf{R}(t) = \langle C_1 t + C_4, C_2 t + C_5, -4.9t^2 + C_3 t + C_6\rangle$ • $\mathbf{R}(0) = \langle C_4, C_5, C_6\rangle$ is given as $\langle 0, 0, 0\rangle$. •
$C_4 = 0, C_5 = 0, C_6 = 0$, and $\mathbf{R}(t) = \langle C_1 t, C_2 t, -4.9t^2 + C_3 t\rangle$ •
At the highest point, $dz/dt = -9.8t + C_3$ is 0, $t = C_3/9.8$, and $z = (C_3)^2/19.6$. •
$(C_3)^2/19.6 = 490 \implies C_3 = 98$ and $t = C_3/9.8 = 10$ • $x(10) = 250 \implies C_1 = 25$ •
$y(10) = 0 \implies C_2 = 0$ • $\mathbf{R}(t) = \langle 25t, 0, -4.9t^2 + 98t\rangle$ meters

8. $\mathbf{R}(t) = \langle \tfrac{4}{5}t^{5/2} + 3t + \tfrac{31}{5}, -\ln t - 10\rangle$

9. $x = t^2 - 3\sin t + 5$, $y = \tfrac{1}{36}t^4 + \tfrac{1}{3}\cos t - \tfrac{19}{3}$

14. (a) $\mathbf{v}(x) = \langle 1, \tfrac{3}{4}x^2\rangle$ • $\mathbf{v}(1) = \langle 1, \tfrac{3}{4}\rangle$ • $\mathbf{T} = \langle \tfrac{4}{5}, \tfrac{3}{5}\rangle$ • $\mathbf{N} = \langle -\tfrac{3}{5}, \tfrac{4}{5}\rangle$
    (b) $\mathbf{T} = \dfrac{\langle -1, 1\rangle}{\sqrt{2}}$ • $\mathbf{N} = \dfrac{\langle -1, -1\rangle}{\sqrt{2}}$

16. $\kappa \approx -2$ at P • $\kappa \approx 1$ at Q • $\kappa \approx -\tfrac{2}{3}$ at R • Figure A16

18. $\kappa = \tfrac{5}{9}$ • $\rho = \tfrac{9}{5}$ • Figure A18

19. $\kappa = -2^{-3/2}$ • $\rho = 2^{3/2}$ • [Center of curvature] $= (\tfrac{7}{3}, -\tfrac{3}{2})$ • Figure A19

Answers 10.5

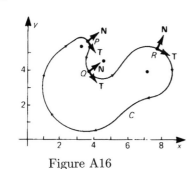

Figure A16

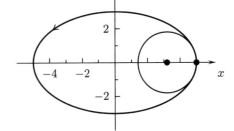

Figure A18

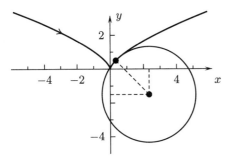

Figure A19

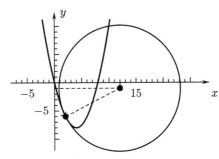

Figure A20

20. [Center of curvature] $= (12, -1)$ • Figure A20

26. $m = 96/32 = 3$ slugs • For $y = e^{-x}$: $y' = -e^{-x}, y'' = e^{-x}$ •
At $x = 0$: $y' = -1$, $\mathbf{T} = \dfrac{\langle 1, -1\rangle}{\sqrt{2}}, \mathbf{N} = \dfrac{\langle 1, 1\rangle}{\sqrt{2}}, y'' = 1, \kappa = 2^{-3/2}$ •
$\mathbf{a} = \sqrt{8}\,\mathbf{T} + 2^{-3/2}(6^2)\mathbf{N} = \langle 11, 7\rangle$ feet per second^2 • $\mathbf{F} = m\mathbf{a} = \langle 33, 21\rangle$ pounds

27. The object's speed is 8 centimeters per second and it is slowing down 4 centimeters per second^2.

30. The object's speed is $\sqrt{\dfrac{45}{2}}$ feet per second and it is speeding up 3 feet per second^2.

33. (a) $\rho \approx \dfrac{3}{4}$ and $\kappa \approx -\dfrac{4}{3}$
(b) $\mathbf{T} = \dfrac{\langle 1, 1\rangle}{\sqrt{2}}, \mathbf{N} = \dfrac{\langle -1, 1\rangle}{\sqrt{2}}$ • $\dfrac{1}{\sqrt{2}}\langle 5, -3\rangle \approx \dfrac{1}{\sqrt{2}}\left[\dfrac{d^2s}{dt^2}\langle 1,1\rangle - \dfrac{4}{3}\left(\dfrac{ds}{dt}\right)^2\langle -1,1\rangle\right]$
$\dfrac{d^2s}{dt^2} + \dfrac{4}{3}\left(\dfrac{ds}{dt}\right)^2 \approx 5$ and $\dfrac{d^2s}{dt^2} - \dfrac{4}{3}\left(\dfrac{ds}{dt}\right)^2 \approx -3$ • $\dfrac{d^2s}{dt^2} \approx 1$ and $\left(\dfrac{ds}{dt}\right)^2 \approx 3$ •
The car's speed is approximately $\sqrt{3}$ feet per second and it is speeding up approximately 1 foot per second^2.

34. The car's speed is approximately 2 feet per second and it is slowing down approximately 1 foot per second^2.

37. The path crosses the x-axis at $x = 15$ meters.

38. $50\sqrt{626}$ meters

45. (a) $\mathbf{R}(t) = \langle 40t - 3t^2 + 32, 50t - \tfrac{1}{2}t^2 + 2\rangle$ meters • $\mathbf{F}(t) = \langle -12000, -2000\rangle$ newtons

62. (a) The ball hits the ground at time $t^* = \dfrac{15}{32}\sqrt{2} + \dfrac{1}{32}\sqrt{642}$ at a distance $(15\sqrt{2})t^* \doteq 30.9$ feet from where it was thrown, so it does not reach the building.
(b) It would not reach the building with air resistance because it would not go as far as without air resistance.

CHAPTER 11
DERIVATIVES WITH TWO OR MORE VARIABLES

Many mathematical models involve functions of two or more variables. The elevation of the points on a mountain, for example, is a function of two horizontal coordinates; the density of the earth at points in its interior is a function of three coordinates; the pressure in a gas-filled balloon is a function of its temperature and volume; and if a store sells fifty items, its profit might be studied as a function of the amounts of the fifty items that it sells. This chapter deals with the differential calculus of such functions. We study graphs, level curves, limits, and continuity of functions with two variables in Section 11.1 and their PARTIAL DERIVATIVES in Section 11.2. DIRECTIONAL DERIVATIVES, tangent planes, and GRADIENT VECTORS of functions with two variables are discussed in Section 11.3. Section 11.4 deals with derivatives of functions with three or more variables, and Section 11.5 covers the First- and Second-Derivative Tests for local maxima and minima and the method of Lagrange multipliers for finding maxima and minima with the variables constrained to be on curves and surfaces.

Section 11.1
Functions of two variables

OVERVIEW: *In this section we discuss domains, ranges, graphs, level curves, limits, and continuity of functions with two variables.*

Topics:

- *The domain, range, and graph of* **f(x, y)**
- *Fixing* **x** *or* **y**: *vertical cross sections of graphs*
- *Other vertical cross sections*
- *Horizontal cross sections of graphs and level curves*
- *Open and closed sets in a plane*
- *Limits and continuity of functions with two variables*

The domain, range, and graph of f(x, y)

The definitions and notation for functions with two variables are similar to those for one variable.

Definition 1 *A* FUNCTION *f of the two variables x and y is a rule that assigns a number, denoted $f(x,y)$, to each point (x,y) in a portion or all of the xy-plane. $f(x,y)$ is called the* VALUE *of the function at (x,y), and the set of points where the function is defined is called its* DOMAIN. *The* RANGE *of the function is the set of its possible values.*

We often use the symbol $f(x, y)$ as another name of the function f to show that its variables are x and y. Also, if a function $f(x, y)$ is given by a formula, we assume that its domain consists of all points (x, y) for which the formula makes sense, unless a different domain is specified.

Example 1 (a) What is the domain of $f(x, y) = x^2 + y^2$? (b) What are the values $f(2, 3)$ and $f(-2, -3)$ of this function at $(2, 3)$ and $(-2, -3)$? (c) What is its range?

SOLUTION (a) Because the expression $x^2 + y^2$ is defined for all x and y, the domain of f is the entire xy-plane.
(b) $f(2, 3) = 2^2 + 3^2 = 13$ and $f(-2, -3) = (-2)^2 + (-3)^2 = 13$.

♦ SUGGESTIONS TO INSTRUCTORS: The main topics in this section can be divided in two parts which can be covered in three or more class meetings.
Part 1 (Functions of two variables, graphs and their cross sections, level curves; Tune-up Exercises T1–T12; Problems 1–40, 46–48, 50) Have students begin Questions 1–6 before class. Question 4 with the analysis of $z = 2xy$ that follows it and Problems 2, 5, 6, 7, 9, 12, 18, 26, 27, and 40 are good for in-class discussion.
Part 2 (Open and closed sets, limits, and continuity; Tune-up Exercise T13; Problems 41–45, 49) Have students begin Question 7 before class. Problems 42, 43, 49 are good for in-class discussion.

(c) The values $x^2 + y^2$ of the function are all ≥ 0 and it has every nonnegative number z as its value at at least one point (x,y). For instance, $x^2 + y^2 = z$ at $(\sqrt{z}, 0)$. Therefore, the range of f is the closed infinite interval $[0, \infty)$. $\square$

Question 1 (a) What are $g(3,1)$ and $g(1,3)$ for $g(x,y) = y^2 - x^2$? (b) What are the domain and range of this function?

Recall that the graph of a function $f(x)$ of one variable is the curve $y = f(x)$ in an xy-plane consisting of the points (x,y) with x in the domain of the function and $y = f(x)$. The graph of a function of two variables is a surface in three-dimensional space.

Definition 2 *The graph of $f(x,y)$ is the surface $z = f(x,y)$ formed by the points (x,y,z) in xyz-space with (x,y) in the domain of the function and $z = f(x,y)$ (Figure 1).*

For a point (x,y) in the domain of the function, its value $f(x,y)$ at (x,y) is determined by moving vertically (parallel to the z-axis) from (x,y) in the xy-plane to the graph and then horizontally (parallel to the xy-plane) to $f(x,y)$ on the z-axis, as is shown in Figure 1.

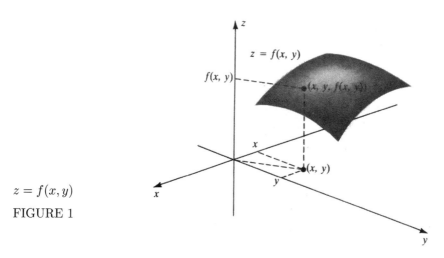

$z = f(x,y)$

FIGURE 1

Fixing x or y: vertical cross sections of graphs

The first step in studying the graph $z = f(x,y)$ of a function of two variables is to study the graphs of the functions of one variable that are obtained by holding x or y constant. To understand this process, we have to look at the geometric meaning of setting x or y equal to a constant in xyz-space.

The meaning of the equation $x = c$, with c a constant, depends on the context in which it is used. If we are dealing with points on an x-axis, then the equation $x = c$ denotes the set[†]

$$\{x : x = c\}$$

consisting of the one point with x-coordinate c (Figure 2). If, however, we are talking about points in an xy-plane, then the equation $x = c$ denotes the vertical line

$$\{(x,y) : x = c\}$$

in Figure 3 consisting of all points (x,y) with $x = c$. Notice that this line is perpendicular to the x-axis and intersects it at $x = c$.

[†]We are using here SET-BUILDER NOTATION $\{P : Q\}$ for the set of points P such that condition Q is satisfied.

11.1 Functions of two variables

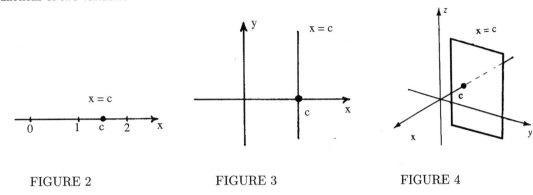

FIGURE 2   FIGURE 3   FIGURE 4

When we are dealing with xyz-space, the equation $x = c$ denotes the vertical plane

$$\{(x, y, z) : x = c\}$$

consisting of all points (x, y, z) with $x = c$. This is the plane perpendicular to the x-axis (parallel to the yz-plane) that intersects the x-axis at $x = c$. With the axes oriented as in Figure 4, this plane is in front of the yz-plane if c is positive and behind the yz-plane if c is negative.

Simiarly, $y = c$ denotes the point $\{y : y = c\}$ on a y-axis (Figure 5), denotes the horizontal line $\{(x, y) : y = c\}$ perpendicular to the y-axis at $y = c$ in an xy-plane (Figure 6), and denotes the vertical plane $\{(x, y, z) : y = c\}$ perpendicular to the y-axis (parallel to the xz-plane) at $y = c$ in xyz-space. With axes oriented as in Figure 7, this plane is to the right of the xz-plane for positive c and to the left of the xz-plane for negative c.

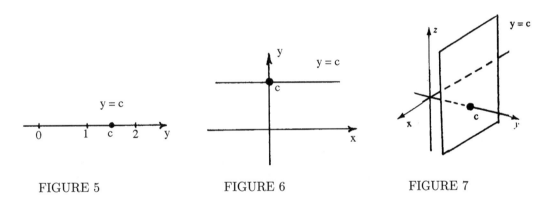

FIGURE 5   FIGURE 6   FIGURE 7

Question 2 (a) Does increasing c move the plane $x = c$ to the front or to the back if the axes are oriented as in Figure 4? (b) Does increasing c move the plane $y = c$ to the left or to the right if the axes are oriented as in Figure 7?

The surfaces $z = x^2 + y^2$ and $z = x^2 - y^2$

Now we will see how a surface can be analyzed by using one-variable techniques to study its cross sections in vertical planes $y = c$ and $x = c$.

Example 2   Figure 8 shows the surface $z = x^2 + y^2$ and Figure 9 some of its vertical cross sections. Describe **(a)** the cross sections in the planes $y = c$ perpendicular to the y-axis and **(b)** the cross sections in the planes $x = c$ perpendicular to the x-axis.

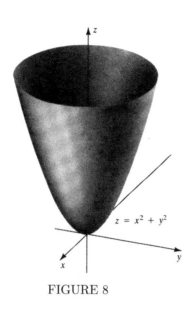

FIGURE 8   FIGURE 9

SOLUTION   **(a)** The intersection of the surface $z = x^2 + y^2$ with the plane $y = c$ is determined by the simultaneous equations,

$$\begin{cases} z = x^2 + y^2 \\ y = c. \end{cases}$$

Replacing y by c in the first equation yields the equivalent pair of equations

$$\begin{cases} z = x^2 + c^2 \\ y = c. \end{cases} \tag{1}$$

These show that the intersection is a parabola in the plane $y = c$ that opens upward and whose vertex is at the origin if $c = 0$ and is c^2 units above the xy-plane if $c \neq 0$. It has the shape of the curve $z = x^2 + c^2$ in the xz-plane of Figure 10.

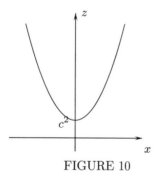   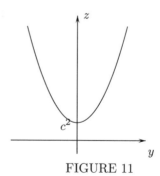

FIGURE 10   FIGURE 11

11.1 Functions of two variables

(b) The intersection of the surface $z = x^2 + y^2$ with the plane $x = c$ is determined by the simultaneous equations

$$\begin{cases} z = x^2 + y^2 \\ x = c \end{cases}$$

which are equivalent to

$$\begin{cases} z = c^2 + y^2 \\ x = c. \end{cases} \quad (2)$$

Consequently, the intersection is a parabola in the plane $x = c$ that opens upward and whose vertex is at the origin of $c = 0$ and is c^2 units above the xy-plane if $c \neq 0$. It has the shape of the curve $z = c^2 + y^2$ in the yz-plane of Figure 11. □

Notice how the parabolas on the surface in Figure 9 vary as c changes. The parabolas (1) in the planes $y = c$ perpendicular to the y-axis are on the left side of the surface for negative c, move down and to the right as c increases to 0, touch the origin for $c = 0$, and move up and to the right as c increases through positive values. Similarly, the parabolas (2) in the planes $x = c$ perpendicular to the x-axis are at the back of the surface in Figure 9 for negative c, move down and forward as c increases to 0, touch the origin for $c = 0$, and move up and forward as c increases through positive values.

Question 3 Explain the shape of the surface $z = y^2 - x^2$ in Figure 12 by analyzing **(a)** its cross sections perpendicular to the y-axis, and **(b)** its cross sections perpendicular to the x-axes that are shown in Figure 13.

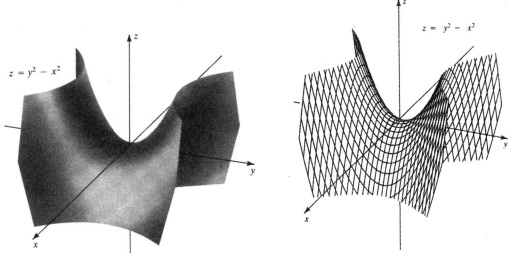

FIGURE 12   FIGURE 13

Example 3 Explain the shape of the surface $z = y - \frac{1}{12}y^3 - \frac{1}{4}x^2$ in Figure 14 by analyzing **(a)** its cross sections in the planes $x = c$, perpendicular to the x-axis, and **(b)** its cross sections in the planes $y = c$, perpendicular to the y-axis.

$z = y - \frac{1}{12}y^3 - \frac{1}{4}x^2$

FIGURE 14

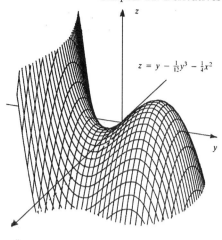

SOLUTION (a) The cross section of the surface in the plane $x = c$ has the equations

$$\begin{cases} z = y - \frac{1}{12}y^3 - \frac{1}{4}x^2 \\ x = c \end{cases}$$

which are equivalent to

$$\begin{cases} z = y - \frac{1}{12}y^3 - \frac{1}{4}c^2 \\ x = c. \end{cases} \qquad (3)$$

We first consider the case of $c = 0$, where the cross section **(3)** is in the yz-plane and is the curve $z = y - \frac{1}{12}y^3$ shown in Figure 15. This curve is symmetric about the origin and passes through it. The highest point on this curve for positive y is at $y = 2$ and its lowest point for negative y is at $y = -2$ because the derivative $\dfrac{d}{dy}(y - \frac{1}{12}y^3) = 1 - \frac{1}{4}y^2$ is negative for $y < -2$, positive for $-2 < y < 2$ and negative for $y > 2$.

Changing c in **(3)** from zero to a nonzero value moves the cross section in Figure 14 to the front or back and lowers it by the amount $\frac{1}{4}c^2$. This gives the surface in Figure 14 its boot-like shape.

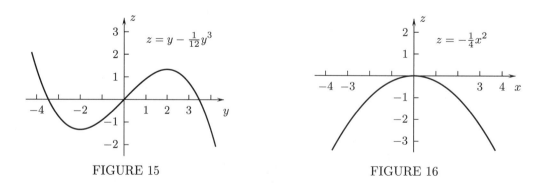

FIGURE 15   FIGURE 16

(b) To analyze the surface by studying the cross sections in the planes perpendicular to the y-axis, we start with the equations for the cross section in the plane $y = c$:

$$\begin{cases} z = y - \frac{1}{12}y^3 - \frac{1}{4}x^2 \\ y = c. \end{cases}$$

11.1 Functions of two variables

These are equivalent to

$$\begin{cases} z = c - \tfrac{1}{12}c^3 - \tfrac{1}{4}x^2 \\ y = c. \end{cases}$$

For $c = 0$, this cross section is in the xz-plane and is the parabola $z = -\tfrac{1}{12}y^2$ in Figure 16. It passes through the origin and opens downward.

Changing c from zero to a nonzero value moves the cross section to the right or left and raises or lowers it depending whether $c - \tfrac{1}{12}c^3$ is positive or negative. As c increases from negative values to zero, the parabola moves down and to the right. Then as c increases through positive values, the parabola moves up and to the right and then down and to the right. (Another way to visualize this result is to imagine that the cross section in the xz-plane in Figure 15 is formed of wire and that wire hoops in the shape of the parabola of Figure 16 are hung from it in perpendicular planes.) □

Example 4   The surface $z = -\tfrac{1}{8}y^3$ in Figure 17 is the graph of the function $h(x,y) = -\tfrac{1}{8}y^3$. Explain its shape.

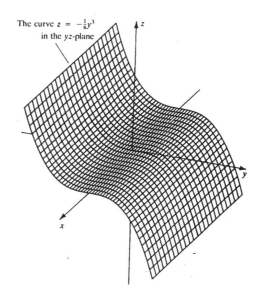

FIGURE 17

SOLUTION   The surface $z = -\tfrac{1}{8}y^3$ is especially easy to analyze because the equation does not involve the variable x. This implies that if one point (x_0, y_0, z_0) with y-coordinate y_0 and z-coordinate z_0 is on it, then the line parallel to the x axis formed by the points (x, y_0, z_0) as x ranges over all numbers is on the surface and the surface consists of lines parallel to the x-axis. In this case, the intersection of the surface with the yz-plane, where $x = 0$, is the curve $z = -\tfrac{1}{8}y^3$, and the surface in Figure 17 consists of this curve and all lines through it parallel to the x-axis. □

The surface $z = 2xy$

It is difficult to visualize the surface $z = 2xy$ by looking only at its cross sections with x or y constant because those cross sections are the lines $z = 2cx, y = c$ and $z = 2cy, x = c$. We can understand the surface better relative to new, rotated axes. We introduce new $x'y'$-axes in the xy-plane with the positive x'-axis at an angle of $\tfrac{1}{4}\pi$ from the positive x-axis, as in Figure 18, and consider a point P as in that drawing with positive xy-coordinates x and y and positive $x'y'$-coordinates $x' = \overline{OQ}$ and $y' = \overline{QP}$.

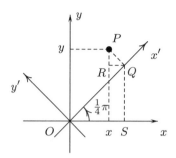

FIGURE 18

Question 4   (a) Use the fact that the right triangles OSQ and PQR in Figure 18 are isosceles to express the lengths $\overline{OS}$ and $\overline{SQ}$ in terms of x' and to express $\overline{RQ}$ and $\overline{RP}$ in terms of y'.   (b) Express x in terms of $\overline{OS}$ and $\overline{RQ}$ and y in terms of $\overline{SQ}$ and $\overline{RP}$.

Combining the results of parts (a) and (b) of Question 4 yields formulas

$$x = \frac{1}{\sqrt{2}}(x' - y'), \ y = \frac{1}{\sqrt{2}}(x' + y') \tag{4}$$

for calculating x and y from x' and y'.[†] They are valid for any location of the point P, and making these substitutions in the equation $z = 2xy$ gives

$$z = 2\left[\frac{1}{\sqrt{2}}(x' - y')\right]\left[\frac{1}{\sqrt{2}}(x' + y')\right].$$

This equation simplifies to $z = (x' - y')(x' + y')$ and then to

$$z = (x')^2 - (y')^2.$$

Figure 19 shows the hyperbolic paraboloid $z = (x')^2 - (y')^2$ in $x'y'z$-space with the positive x'-axis pointing to the right and the positive y'-axis pointing to the back. The calculations above show that the surface is also the graph of $z = 2xy$, relative to x-and y-axes rotated $\frac{1}{4}\pi$ radians clockwise from the x'- and y'-axes, as in Figure 19. Notice that, the x- and y-axes lie on this surface because $z = 2xy$ is zero if x or y is zero; that the surface $z = 2xy$ is above the xy-plane for (x, y) between the positive axes, where x and y are both positive and between the negative axes, where x and y are both negative; and that the surface is below the xy-plane for (x, y) in the other quadrants, where x and y have opposite signs. The surfaces $z = kxy$ with other values of $k \neq 0$ are similar hyperbolic paraboloids.

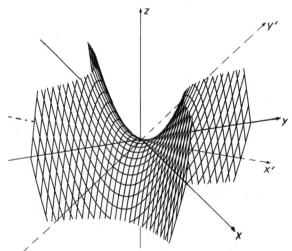

$z = 2xy$
FIGURE 19

[†]The general formulas for rotating axes are discussed in Section 9.1. We require in this section only the special case (4) of rotation by $\frac{1}{4}\pi$ radians to analyze the surfaces $z = kxy$ with nonzero constants k.

Horizontal cross sections and level curves

Another way to determine the shapes of graphs of functions with two variables is to analyze their horizontal cross sections. Horizontal planes, parallel to the xy-plane and perpendicular to the z-axis, have the equations $z = c$ with constant c. Consequently, the horizontal cross sections of the surface $z = x^2 + y^2$ of Figure 8 are given by the equations

$$\begin{cases} z = x^2 + y^2 \\ z = c \end{cases} \quad \text{or} \quad \begin{cases} c = x^2 + y^2 \\ z = c. \end{cases} \tag{5}$$

The ten cross sections for $c = 2, 4, 6, \ldots, 18, 20$ are the circles shown in Figure 20. The cross section for $c = 0$ consists of one point, the origin, and the cross section is empty for $c < 0$

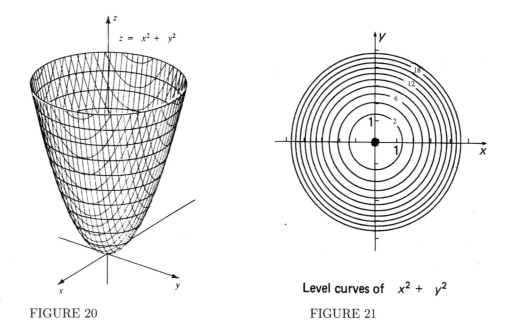

FIGURE 20   Level curves of $x^2 + y^2$ FIGURE 21

If we drop the ten horizontal cross sections in Figure 20 down to the xy-plane, we obtain the ten concentric circles in Figure 21. These and the point at the origin are curves on which $x^2 + y^2$ is constant, and are called LEVEL CURVES or CONTOUR CURVES of the function $z = x^2 + y^2$.

Definition 3 *The level curves (contour curves) of $f(x, y)$ are the curves in the xy-plane where the function is constant. They have the equations $f(x, y) = c$ with constants c.*

The numbers on the level curves in Figure 21 indicate the values of the function on those circles. The function has the values 4, 8, 10, 14, 16, and 20 on the other circles in that drawing.

To visualize the surface in Figure 20 from the level curves, imagine that the xy-plane in Figure 21 is horizontal in xyz-space, that the innermost circle, labeled "2" is lifted 2 units to $z = 2$, the next circle is lifted up to $z = 4$, the circle labeled "6" is lifted to $z = 6$, and so forth. This gives the horizontal cross sections of the surface in Figure 20.

The surface $z = x^2 + y^2$ in Figure 20 is called a CIRCULAR PARABOLOID because its vertical cross sections are parabolas and its horizontal cross sections are circles.

Question 5   Horizontal cross sections of $z = y^2 - x^2$ are shown in Figure 22. When they are moved up or down to the xy-plane, they form the level curves of $y^2 - x^2$ in Figure 23. The curves labeled A, B, and C in Figure 23 are for z-values $A = 2, B = 4$, and $C = 6$. What are the values of D, E, and F?

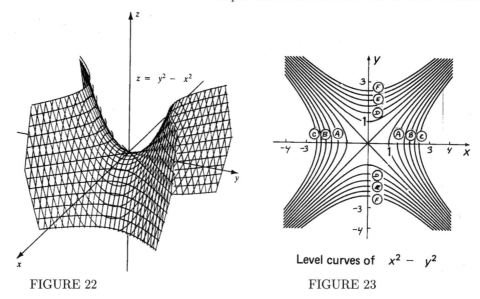

FIGURE 22

Level curves of $x^2 - y^2$
FIGURE 23

The level curves $x^2 - y^2 = c$ in Figure 23 are hyperbolas. Consequently the horizontal cross sections of $z = x^2 - y^2$ in Figure 22 are hyperbolas. Because its vertical cross sections are parabolas, the surface is called a HYPERBOLIC PARABOLOID.

Question 6 Horizontal cross sections of $z = y - \frac{1}{12}y^3 - \frac{1}{4}x^2$ at integer values of z are shown in Figure 24, and the corresponding level curves of the function $g(x,y) = y - \frac{1}{12}y^3 - \frac{1}{4}x^2$ are in Figure 25. Identify the curve(s) where g has the values 0, 1, 5, and -5.

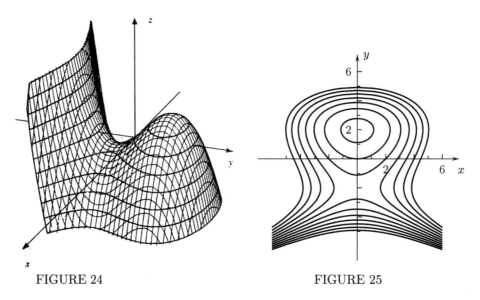

FIGURE 24

FIGURE 25

Open and closed sets in the plane

The definitions of closed and open sets in an xy-plane are analogous to the definitions of closed and open intervals.

Recall that an interval is closed if it includes its endpoints and is open if it does not include its endpoints, and that the closure of an interval is obtained by adding any endpoints it does not already include and the interior is obtained by removing any endpoints. Figure 26 shows, for example, the interval $(-1, 3]$ (which is neither closed nor open), its closure $[-1, 3]$, and its interior $(-1, 3)$.

11.1 Functions of two variables

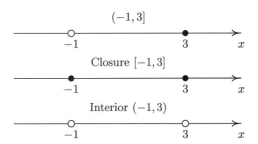

FIGURE 26

Regions in xy-planes do not have endpoints. Instead they have BOUNDARIES and they are closed or open, depending on whether they include or do not include their boundaries.

Definition 4 (a) The point (x_0, y_0) is in the BOUNDARY of a set R in an xy-plane if for every positive number r, the disk $\{(x, y) : (x - x_0)^2 + (y - y_0)^2 < r^2\}$ of radius r and centered at (x_0, y_0) contains at least one point in R and at least one point not in R.
(b) R is CLOSED if it includes all of its boundary and is OPEN if it does not include any of its boundary.
(c) The CLOSURE of R is the closed region obtained by adding any boundary points that are not already in it. The INTERIOR of R is the open region obtained by removing any boundary points that are in it.

Example 5   Figure 27 shows the square $R = \{(x, y) : 1 \leq x < 4 \text{ and } 1 \leq y < 4\}$. The left side and bottom of the square are drawn with solid lines to show that those points are included in R. The top and right sides are drawn with broken lines to indicate that those points are not in R. What are the boundary, closure, and interior of R?

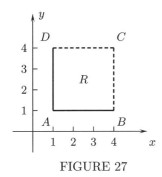
FIGURE 27

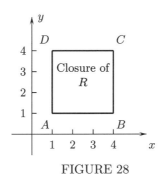
FIGURE 28

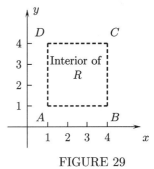
FIGURE 29

SOLUTION   The boundary of R consists of its four sides AB, BC, CD, and DA; its closure is the closed square $\{(x, y) : 1 \leq x \leq 4, 1 \leq y \leq 4\}$ in Figure 28; and its interior is the open square $\{(x, y) : 1 < x < 4, 1 < y < 4\}$ in Figure 29. □

Limits

In studying functions of one variable we used one- and two-sided limits to determine whether functions were continuous at points and in intervals. We cannot talk of two-sided or one-sided limits of functions of two variables. Instead we consider LIMITS ALONG ALL PATHS and LIMITS IN A SET:

Definition 5 (a) *Suppose there is a disk of positive radius centered at (x_0, y_0) all of which, except possibly the point (x_0, y_0), is contained in the domain of $f(x, y)$. Then the limit of $f(x, y)$ as (x, y) approaches (x_0, y_0) along all paths is L and we write*

$$\lim_{(x,y) \to (x_0, y_0)} f(x, y) = L \tag{6}$$

if $f(x, y) \to L$ as (x, y) approaches (x_0, y_0) along all paths that lie in the domain of f and do not contain the point (x_0, y_0) (Figure 30).

(b) *Suppose that $f(x, y)$ is defined in the set D and that (x_0, y_0) is in its closure. Then the limit of $f(x, y)$ as (x, y) approaches (x_0, y_0) in D is L, and we write*

$$\lim_{\substack{(x,y) \to (x_0, y_0) \\ (x,y) \text{ in } D}} f(x, y) = L \tag{7}$$

if $f(x, y) \to L$ as (x, y) approaches (x_0, y_0) along all paths that lie in D and do not contain the point (x_0, y_0) (Figure 31).

In this definition L can be a number, ∞ or $-\infty$.

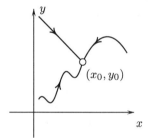

Paths approaching (x_0, y_0)

FIGURE 30

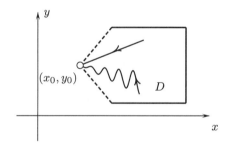

Paths in D approaching (x_0, y_0)

FIGURE 31

Example 6 What is $\lim_{(x,y) \to (3,2)} (x^2 + y^2)$?

SOLUTION Because x^2 is continuous for all x and y^2 is continuous for all y, $x^2 \to 3^2 = 9$ as $x \to 3$ and $y^2 \to 2^2 = 4$ as $y \to 2$. Therefore $x^2 + y^2 \to 9 + 4 = 13$ as $(x, y) \to (3, 2)$ and $\lim_{(x,y) \to (3,2)} (x^2 + y^2) = 13.$ □

Example 7 What are $\lim_{\substack{(x,y) \to (x,0) \\ y > 0}} \dfrac{1}{y}$ and $\lim_{\substack{(x,y) \to (x,0)) \\ y < 0}} \dfrac{1}{y}$?

SOLUTION Since $\dfrac{1}{y} \to \infty$ as $y \to 0^+$ and $\dfrac{1}{y} \to -\infty$ as $y \to 0^-$, we have for any x,

$$\lim_{\substack{(x,y) \to (x,0) \\ y > 0}} \frac{1}{y} = \infty \quad \text{and} \quad \lim_{\substack{(x,y) \to (x,0) \\ y < 0}} \frac{1}{y} = -\infty. \quad \square$$

11.1 Functions of two variables

The results of Example 7 are illustrated in Figure 32, which shows the graph of the function of (x, y) given by $z = 1/y$. The cross section of the graph in planes $x = c$ have the shape of the hyperbola $z = 1/y$ and the cross sections in planes $y = c$ are horizontal lines, and you can see from the drawing that $z \to \infty$ as (x, y) approaches any point on the y-axis on a path with $y > 0$, and that $z \to -\infty$ as (x, y) approaches any point on the y-axis on any path with $y < 0$.

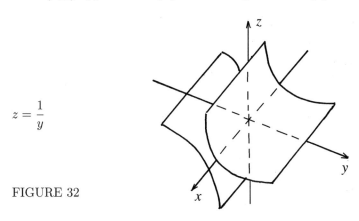

$z = \dfrac{1}{y}$

FIGURE 32

Continuity at points and in regions

Recall that a function $F(x)$ of one variable is continuous at a point x_0 if x_0 is in the interior of an interval where $F(x)$ is defined and if $\lim\limits_{x \to x_0} F(x) = F(x_0)$ and that $F(x)$ is continuous in an interval provided that (i) it is continuous at each point in the interior of the interval, (ii) $\lim\limits_{x \to a^+} F(x) = F(a)$ if the interval includes a left endpoint a, and (iii) $\lim\limits_{x \to b^-} F(x) = F(b)$ if the interval includes a right endpoint b. The function $F(x)$ of Figure 33, for instance, is continuous at all x with $-2 < x < -1, -1 < x < 1$, and $1 < x < 2$ and is continuous in the intervals $[-2, -1], (-1, 1)$, and $[1, 2]$.

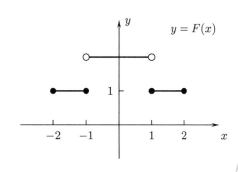

FIGURE 33

The definitions of continuity at a point and in a region for a function of two variables are similar:

Definition 6 (a) $f(x, y)$ is continuous at (x_0, y_0) if (x_0, y_0) is in the interior of the domain of $f(x, y)$ and $\lim\limits_{(x,y) \to (x_0, y_0)} f(x, y) = f(x_0, y_0)$.

(b) $f(x, y)$ is continuous in a set D if $\lim\limits_{\substack{(x, y) \to (x_0, y_0) \\ (x, y) \text{ in } D}} f(x, y) = f(x_0, y_0)$ for all (x_0, y_0) in D.

Example 8 The function $P(x,y)$ is defined for $\sqrt{x^2+y^2} \le 2$ by

$$P(x,y) = \begin{cases} 2 & \text{for} \quad \sqrt{x^2+y^2} < 1 \\ 1 & \text{for} \quad 1 \le \sqrt{x^2+y^2} \le 2. \end{cases}$$

Its graph in Figure 34 consists of a circle of radius 1 at $z = 2$ and a ring with inner radius 1 and outer radius 2 at $z = 1$. **(a)** At what points is $P(x,y)$ continuous? **(b)** What are the most extensive regions in which $P(x,y)$ is continuous?

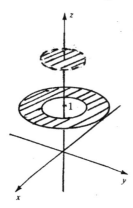

FIGURE 34

SOLUTION **(a)** $P(x,y)$ is continuous at all (x,y) in the open circle $\sqrt{x^2+y^2} < 1$ and in the open ring $\sqrt{x^2+y^2} < 2$.
(b) The most extensive regions in which $P(x,y)$ is continuous are the open circle $\{(x,y) : \sqrt{x^2+y^2} < 1\}$ and the closed ring $\{(x,y) : 1 \le \sqrt{x^2+y^2} \le 2\}$. □

Continuity of functions given by single formulas

Recall from Theorem 1 of Section 2.2 that a function of one variable that is given by a single formula formed by taking sums, differences, products, quotients, and compositions of the power functions x^n, the absolute-value function, exponential functions, logarithms, trigonometric functions, and inverse trigonometric functions is continuous in its domain. The analogous statememt is true for functions of two variables:

Theorem 1 *A function $f(x,y)$ that is given by a single formula formed by taking sums, differences, products, quotients, and compositions of power functions, the absolute-value function, exponential functions, logarithms, trigonometric functions, and inverse trigonometric functions of x and y is continuous in its domain.*

This Theorem, like Theorem 1 of Section 2.2, can be used to find limits.

Example 9 What is $\lim\limits_{(x,y) \to (3,0)} e^{x+\cos y}$.

SOLUTION Because $f(x,y) = e^{x+\cos y}$ is given by a single formula and is defined for all (x,y), it is continuous in the entire xy-plane and

$$\lim_{(x,y)\to(3,0)} e^{x+\cos y} = \left[e^{x+\cos y}\right]_{x=3,y=0} = e^{3+\cos(0)} = e^4. \quad \square$$

Question 7 The graph of $z = \sqrt{2-x^2-y^2}$ is the hemispherical surface in Figure 35. **(a)** What is the domain of this function? **(b)** What is its limit as $(x,y) \to (1,1)$ in its domain?

11.1 Functions of two variables

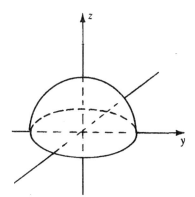

$z = \sqrt{2 - x^2 - y^2}$

FIGURE 35

$\epsilon\delta$- and $Z\delta$-definitions

Definition 5 of limits can not be used directly because we cannot check the limits along all possible paths. Instead, the theory of limits of functions with two variables is based on the following $\epsilon\delta$-definitions, which are used to derive results, such as Theorem 1 above, for studying such limits.

Definition 7 (a) $\lim\limits_{\substack{(x,y) \to (x_0, y_0) \\ (x,y) \text{ in } D}} f(x,y) = L$ with a number L provided that for each $\epsilon > 0$ there is a $\delta > 0$ such that $|f(x,y) - L| < \epsilon$ for $0 < \sqrt{(x-x_0)^2 + (y-y_0)^2} < \delta$.

(b) $\lim\limits_{\substack{(x,y) \to (x_0, y_0) \\ (x,y) \text{ in } D}} f(x,y) = L$ with a number L provided (x_0, y_0) is in the closure of D and for each $\epsilon > 0$ there is a $\delta > 0$ such that $|f(x,y) - L| < \epsilon$ for all (x,y) in D with $0 < \sqrt{(x-x_0)^2 + (y-y_0)^2} < \delta$.

The $Z\delta$-definitions for $L = \pm\infty$ are obtained by replacing the phrase "for each $\epsilon > 0$ there is a $\delta > 0$ such that $|f(x,y) - L| < \epsilon$" in Definition 7 with "for every positive Z there is a $\delta > 0$ such that $f(x,y) > Z$" in the case of $L = \infty$ and with "for every negative Z there is a $\delta > 0$ such that $f(x,y) < Z$" for $L = -\infty$.

Principles and procedures

- In earlier chapters, when you were dealing with functions of one variable, you could rely extensively on your hand-drawn sketches or calculator- or computer-generated graphs to guide your reasoning and check your work. In dealing with functions of two variables, you will generally have to reason more abstractly, because it is difficult to draw good pictures of most surfaces, and drawings of surfaces generated by graphing calculators or computers are frequently difficult to interpret. Moreover, even when you have a good sketch of the graph of a function with two variables, you cannot determine the function's values from it because you cannot tell from the two-dimensional picture where vertical lines intersect the graph.

- To draw a surface, show the portion in an imaginary box with sides parallel to the coordinate planes by drawing the portions of any profiles of the surface in the box and any curves that are obtained by chopping off the surface at the sides of the box. Then add coordinate axes to fit your drawing.

- To draw a circular or elliptic paraboloid $z = a + bx^2 + cx^2$ or $z = a - bx^2 - cx^2$ with $b > 0, c > 0$, as in Figure 36, first draw a horizontal circle to represent the circle or ellipse where the surface is chopped off. Add a parabola for the profile of the surface. Then draw coordinate axes as appropriate.

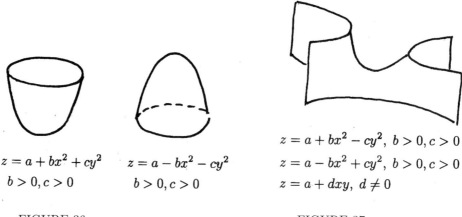

$z = a + bx^2 + cy^2$
$b > 0, c > 0$

$z = a - bx^2 - cy^2$
$b > 0, c > 0$

$z = a + bx^2 - cy^2, \ b > 0, c > 0$
$z = a - bx^2 + cy^2, \ b > 0, c > 0$
$z = a + dxy, \ d \neq 0$

FIGURE 36   FIGURE 37

- To draw a hyperbolic paraboloid $z = a + bx^2 - cx^2$ or $z = a - bx^2 + cx^2$ with $b > 0, c > 0$ or $z = kxy$ with $k \neq 0$, as in Figure 37, first draw part of a parabola that opens upwards as the profile of the surface. Add portions of horizontal hyperbolas where the surface is chopped off at the top and bottom, and draw parabolas or vertical lines to represent where the surface is chopped off at the sides. Then draw coordinate axes as appropriate.

- Many graphs of functions $f(x,y) = G(x)$ or $f(x,y) = G(y)$ that depend on only one of the variables x or y are easy to sketch because their cross sections in one direction are horizontal lines and in the other direction have the shape of the graph of the function G of one variable. For example, to draw the surface $z = a + bx^2, z = a + by^2, z = a - bx^2$, or $z = a - by^2$ with $b > 0$, as in Figure 38, draw parts of vertical parabolas to represent parabolic ends of the surface where it is cut off by vertical planes and parallel lines to represent its profile and where is it chopped off by a horizontal plane.

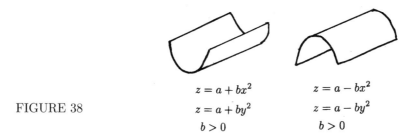

FIGURE 38

$z = a + bx^2$
$z = a + by^2$
$b > 0$

$z = a - bx^2$
$z = a - by^2$
$b > 0$

- Most statements concerning domains, graphs, limits, and continuity of functions of two variables are fairly easy to figure out because they involve the same concepts as corresponding statements about functions of one variable. The statements with two variables, however, are generally considerably more complicated because of the extra variables and have to be read closely to see the ideas behind them.

11.1 Functions of two variables

Tune-Up Exercises 11.1

A Answer provided. **O** Outline of solution provided.

T1. Calculate $f(2,3)$ and $f(-1,10)$ for $f(x,y) = x^2 y^3$.
◆ Type 1, value of a function

T2. (a) Is $g(x,y) = x^3 y$ an increasing or decreasing function of x for $y = 10$? (b) Is $g(x,y)$ from part (a) an increasing or decreasing function of y for $x = -2$?
◆ Type 1, fixing one variable

T3. What is the global minimum of $h(x,y) = (x-y)^2 + 10$ and at what values of x and y does it occur?
◆ Type 1, finding a maximum from a formula

T4. What is the global maximum of $k(x,y) = \dfrac{6}{x^2 + y^2 + 2}$ and at what point (x,y) does it occur?
◆ Type 1, finding a maximum from a formula

T5. (a) Describe the intersections of the graph $z = x^2 + y^2 + 6$ with the planes $x = c, y = c$, and $z = c$ for constants c. (b) Add axes to the surface in Figure 39 so it represents the graph from part (a).
◆ Type 1, horizontal and vertical cross sections

FIGURE 39

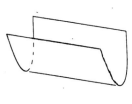
FIGURE 40

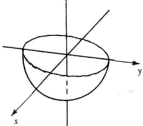
FIGURE 41

T6. (a) Describe the intersections of the graph of the function $f(x,y) = y^2$ with the planes $x = c, y = c$, and $z = c$ for constants c. (b) Copy the surface in Figure 40 and add axes and label them so the drawing represents the graph from part (a).
◆ Type 1, horizontal and vertical cross sections

T7. Figure 41 shows the surface $z = -\sqrt{1 - x^2 - y^2}$ Explain why it is the lower half of the sphere $x^2 + y^2 + z^2 = 1$ of radius 1 with its center at the origin. (b) Draw the graph of the function $g(x,y) = 2 - \sqrt{1 - x^2 - y^2}$.
◆ Type 1, hemispheres

T8. Draw the graph of the function $L(x,y) = -3$.
◆ Type 1, graph of a constant function

T9. (a) What is the domain of $z(x,y) = \dfrac{x}{\sqrt{y}}$? (b) For what values of x and y is the function from part (a) positive? For what values of x and y is it negative? For what values of x and y is it zero?
◆ Type 1, finding the sign of a function from its formula

T10. Draw the level curves $x^2 + y = c$ of $M(x,y) = x^2 + y$ for $c = -2, 0$ and 2.
◆ Type 1, level curves

T11. Draw the level curves $y - \frac{1}{2}x = c$ of $N(x,y) = y - \frac{1}{2}x$ for $c = 0, \pm 1$, and ± 2.
◆ Type 1, level curves

◆The Tune-up Exercises and Problems are classified by type and content. The types are (1) basic, reactive; (2) basic reflective; (3) intermediate, reactive; (4) intermediate, reflective; (5) advanced, reactive; (6) advanced, reflective; and (7) advanced, theoretical.

T12.[A] What are the values of $P(x,y) = x^2 + 4y^2$ on its three level curves in Figure 42?

◆ Type 1, level curves

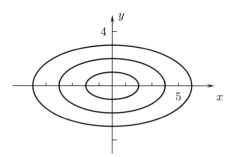

FIGURE 42

T13.[A] (a) What is the limit of $\dfrac{\sin(x-y)}{1+x^2+y^2}$ as $(x,y) \to (3,2)$? (b) Find $\lim\limits_{\substack{(x,y) \to (3,0) \\ y>0}} \dfrac{x^2}{\cos(\sqrt{y})}$.

◆ Type 1, limits

Problems 11.1

[A] Answer provided. [O] Outline of solution provided.

1.[A] (a[O]) What is $f(1,-1)$ for $f(x,y) = x^2 e^{-y} - y^3 e^x$? (b[A]) What is $g(1,3)$ for $g(x,y) = \dfrac{x-y}{x+y}$?
(c) Find the value of $z = \ln(xy+3)$ at $x=4, y=5$. (d) What is $h(3,2)$ if $h(x,y) = x^y + y^x$?

◆ Type 1, values of functions

2.[A] Which is the greatest and which is the least of the numbers $M(1,2), M(2,1)$, and $M(2,2)$ if $M(x,y) = \dfrac{1+\cos x}{2+\sin y}$?

◆ Type 1, values of functions

3.[A] The total resistance $R = R(r_1, r_2)$ of an electrical circuit consisting of resistances of r_1 and r_2 ohms (Figure 43) is determined by the equation $\dfrac{1}{R} = \dfrac{1}{r_1} + \dfrac{1}{r_2}$.[(1)] (a) Show that $R(r_1, r_2) = \dfrac{r_1 r_2}{r_1 + r_2}$.
(b) Does $R(r_1, r_2)$ increase or decrease as r_1 or r_2 increases?

◆ Type 2, rewriting a formula and fixing one variable

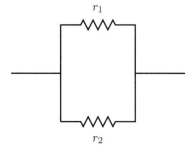

FIGURE 43

[(1)] *CRC Handbook of Chemistry and Physics*, Boca Raton FL: CRC Press, Inc., 1981, p. F-112.

11.1 Functions of two variables

4. When a very small spherical pebble falls under the force of gravity in a deep body of still water, it quickly approaches a constant speed, called its TERMINAL SPEED. By STOKES' LAW the terminal speed is $v(r,\rho) = 21800(\rho - 1)r^2$ centimeters per second if the radius of the pebble is r centimeters and its density is $\rho \geq 1$ grams per cubic centimeter.[2] **(a)** What is the terminal speed of a quartz pebble of density 2.6 grams per cubic centimeter if its radius is 0.01 centimeters? **(b)** Which of two pebbles has the greater terminal speed if they have the same density and one is larger than the other? **(c)** What happens to the pebble if its density is 1 gram per centimeter, the density of water?

♦ Type 2, values of a function

5. The table below gives the equivalent human age $A(t,w)$ of a dog that is t years old and weighs w pounds.[3] **(a)** What does $A(11, 50)$ represent and, based on the table, what is its approximate value? **(b)** What does $A(14, 70)$ represent and what is its approximate value?

♦ Type 2, values of a function

$A(t, w) =$ EQUIVALENT HUMAN AGE

	$t = 6$	$t = 8$	$t = 10$	$t = 12$	$t = 14$	$t = 16$
$w = 20$	40	48	56	64	72	80
$w = 50$	42	51	60	69	78	87
$w = 90$	45	55	66	77	88	99

6.º Describe the level curve $N(x, y) = 1$ of $N(x, y) = \dfrac{x + 2y}{3x + y}$.

♦ Type 2, level curves

7.º What are the values of $L(x, y) = |x| + |y|$ on its three level curves in Figure 44?

♦ Type 2, level curves

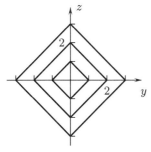

FIGURE 44

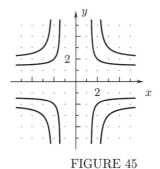

FIGURE 45

8. What are the values of $K(x, y) = \dfrac{1}{x^2} + \dfrac{1}{y^2}$ on its eight level curves in Figure 45?

♦ Type 2, level curves

9.ᴬ Figure 46 shows level curves of the function $F(x, y) = Ax + By + C$. What are the values of the constants A, B, and C?

♦ Type 2, level curves

---

[2]*CRC Handbook of Chemistry and Physics*, Boca Raton FL: CRC Press, Inc., 1981, p. F-115.
[3]Data from "Senior-Care Health Report" by Pfizer, Inc., based on a chart developed by F. Menger, State College, PA.

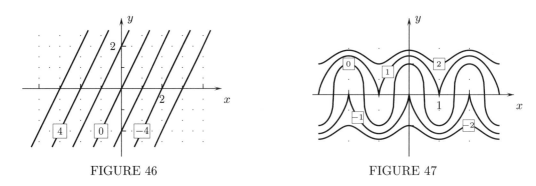

FIGURE 46   FIGURE 47

10. Level curves of $G(x,y) = Ay^3 - \cos(Bx)$ are shown in Figure 47. What are the constants A and B?
◆ Type 2, level curves

11.^A (a) Figures 48 through 50 show the surfaces $z^2 = 2x^2+2y^2$, $z^2 = 2x^2+2y^2-1$, and $z^2 = 2x^2+2y^2+1$. Use these drawing to sketch the graphs of $f(x,y) = \sqrt{2x^2+2y^2}, g(x,y) = \sqrt{2x^2+2y^2-1}$, and $h(x,y) = \sqrt{2x^2+2y^2+1}$. (b) What are the domains and ranges of $f, g,$ and h from part (a)?
◆ Type 2, graphs of functions

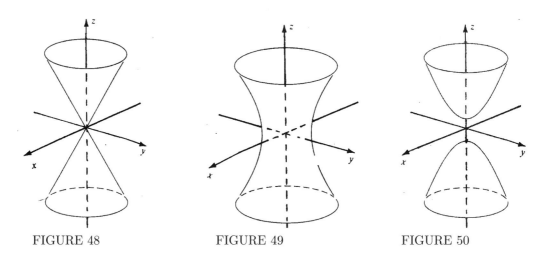

FIGURE 48   FIGURE 49   FIGURE 50

12. Label positive ends of the x- and y-axes in Figure 51 so that the surface has the equation $z = |x|$.
◆ Type 2, graph of a function

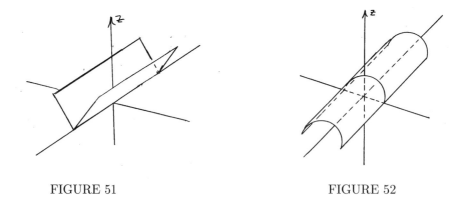

FIGURE 51   FIGURE 52

11.1 Functions of two variables

13. Select positive ends of the x- and y-axes in Figure 52 so that the surface is the graph of $Q(x,y) = \sqrt{1-y^2}$.

♦ Type 2, graph of a function

14. Identify the positive ends of the x- and y-axes in Figure 53 **(a^A)** so that the surface is $z = y^2 - x^2$ and **(b)** so that the surface is $z = x^2 - y^2$.

♦ Type 2, graph of a function

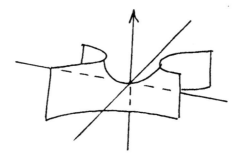

FIGURE 53

15.^{A} Draw **(a)** the graph of $R(x,y) = \sqrt{4-x^2}$. **(b)** Describe the level curves where the function of part (a) has the values $0, 1$, and 2.

♦ Type 2, graph and level curves of a function

16. Draw the surface $z = x^2$ in xyz-space.

♦ Type 2, graph of a function

17 Draw the graph of $H(x,y) = -xy$.

♦ Type 2, graph of a function

18.^{A} Draw and label the level curves of $S(x,y) = y - \sin x$ where it has the values $0, \pm 2, \pm 4$.

♦ Type 2, level curves of a function

19.^{A} Draw and label the level curves of $T(x,y) = \sqrt[3]{x-y}$ where it has the values $0, \pm 1$, and ± 2.

♦ Type 2, level curves of a function

20. Draw the graph of $V(x,y) = -\frac{1}{2}\sqrt{x^2+y^2}$.

♦ Type 2, level curves of a function

21. **(a)** What is the domain of $U(x,y) = \dfrac{x}{y}$? **(b)** Draw the level curves of U where it has the values $0, \pm 1, \pm 2$.

♦ Type 2, domain and level curves of a function

22. Draw level curves **(a^A)** of $V(x,y) = 4x + y^2$ and **(b)** of $W(x,y) = x - y^2$.

♦ Type 2, level curves of a function

23. Draw level curves **(a)** of $A(x,y) = x + 3y$, **(c)** of $B(x,y) = x^2 y$, and **(c)** of $C(x,y) = 2^{x-y}$.

♦ Type 2, level curves of functions

24. Describe and draw the graph of $G(x,y) = -1 - (x+1)^2 - (y-1)^2$.

♦ Type 2, graph of a function

25. Describe some ways in which the surfaces $z = 3 - x^2 - y^2$ and $z = \sqrt{9-x^2-y^2}$ are similar and in which they differ.

♦ Type 2, graphs of functions

26. The graphs of **(a^A)** $\sin y - \frac{1}{9}x^3 + \frac{1}{2}$, **(b^A)** $\sin y$, **(c)** $-\sin x \sin y$, **(d^A)** $\sin^2 y + \frac{1}{2}x^2$, **(e)** $-\frac{1}{9}x^3 \sin y$, and **(f)** $3e^{-x/5} \sin y$ are shown in Figures 54 through 59. Match the functions to their graphs and explain how the shapes of the surfaces are determined by their equations.

♦ Type 2, matching graphs by analyzing vertical cross sections

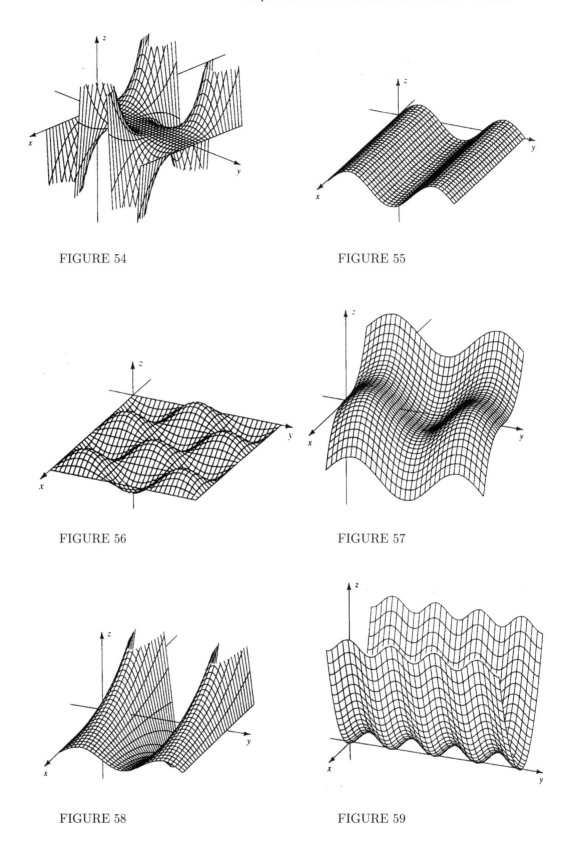

FIGURE 54

FIGURE 55

FIGURE 56

FIGURE 57

FIGURE 58

FIGURE 59

27 Level curves of (**a^A**) $\sin y - \frac{1}{9}x^3 + \frac{1}{2}$, (**b^A**) $\sin y$, (**c**) $-\sin x \sin y$, (**d^A**) $\sin^2 y + \frac{1}{2}x^2$, (**e**) $-\frac{1}{9}x^3 \sin y$, and (**f**) $3e^{-x/5}\sin y$ from Problem 26 are shown in Figures 60 through 65. Match the functions to their level curves by comparing the level curves with the graphs in Figures 54 through 59.

♦ Type 2, matching level curves to graphs

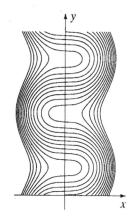

FIGURE 60

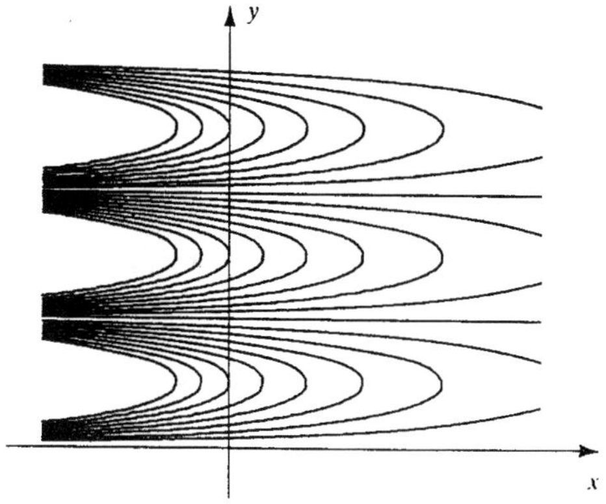

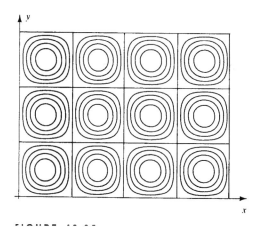

FIGURE 61

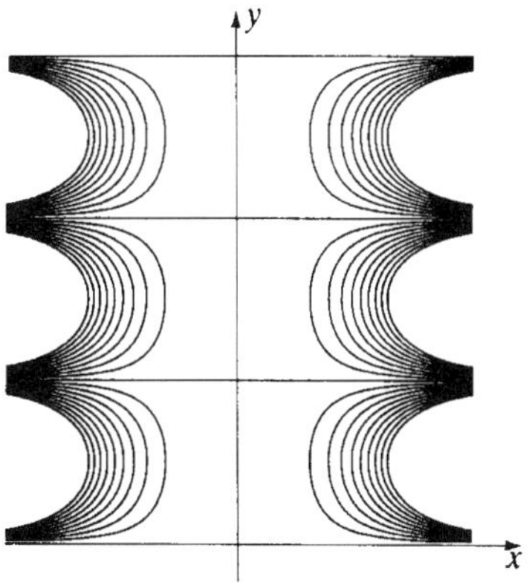

FIGURE 62

FIGURE 63

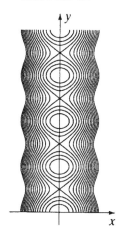

FIGURE 64

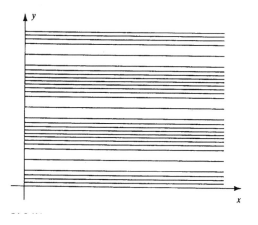

FIGURE 65

28. (a) Explain why the horizontal cross sections of $z = \ln(x^2 + y^2)$ and of $z = \dfrac{1}{\sqrt{x^2 + y^2}}$ are circles.

(b) Match the surfaces in Figures 66 and 67 to their equations in part (a). Explain your choices.

♦ Type 2, matching functions to graphs

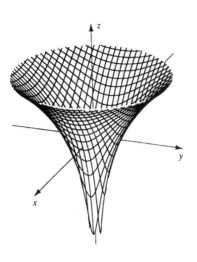

FIGURE 66   FIGURE 67

29. Large earthquakes with epicenters near coastlines in an ocean generate tsunami waves which can travel up to 600 miles per hour in very deep water and can be up to 90 feet high near shores. Figure 68 shows level curves of $T(P)$, where for points P in the Pacific Ocean and on its shores, $T(P)$ is the time it took the tsunami wave created by an earthquake in 1964 to reach P. The dot in the drawing is the epicenter of the earthquake.[4] Approximately how much longer did it take the wave to reach New Zealand than to reach Japan?

♦ Type 2, interpreting level curves

FIGURE 68

30.° Generate the level curves where $p(x, y) = xy + \frac{1}{4}x^3$ has the values $-4, 0,$ and 4 by solving for y and generating the graphs of the resulting functions of x. Use the window $-5 \leq x \leq 5, -7 \leq y \leq 7$. Copy the curves on your paper and label them with the corresponding values of p.

♦ Type 4, generating level curves with a calculator or computer

---

[4] Adapted from A. Strahler and A. Strahler, *Environmental Geoscience*, Santa Babara: Hamilton Publishing Co., 1973.

11.1 Functions of two variables

31. Generate the level curves where $q(x, y) = e^y - \sin(\pi x)$ has the values $0, 1, 2, 3, 4$ by solving for y and generating the graphs of the resulting functions of x. Use the window $-3 \leq x \leq 3, -2 \leq y \leq 2$. Copy the curves on your paper and label them with the corresponding values of q.
♦ Type 4, generating level curves with a calculator or computer

32. Generate the level curves where $r(x, y) = e^y - e^x$ has the values $0, \pm 0.5, \pm 1, \pm 1.5$ by solving for y and generating the graphs of the resulting functions of x. Use a window with equal scales on the axes. Copy the curves on your paper and label those where $r = 0, \pm 1$.
♦ Type 4, generating level curves with a calculator or computer

33. Find the values of a such that the curve with parametric equations $x = a\cos^3 t, y = a\sin^3 t$ are the level curves where $r(x, y) = x^{2/3} + y^{2/3}$ has the values 1, 2, and 3. Then generate the three curves in a window with equal scales on the axes and copy and label them.
♦ Type 4, generating level curves with a calculator or computer

34. A person's BODY-MASS INDEX is the number $I(w, h) = \dfrac{w}{h^2}$, where w is his or her weight, measured in kilograms, and h is his or her height, measured in meters. **(a)** What is your body-mass index? (A kilogram is 2.2 pounds and a meter is 39.37 inches.) **(b)** A study of middle-aged women found that those with a body-mass index of over 29 had twice the risk of death than those whose body-mass index was less than 19. Suppose a woman is 1.5 meters tall and has a body-mass index of 29. How much weight would she have to lose to reduce her body-mass index to 19?
♦ Type 2, using a formula in a narrative problem

35. Sketch the graph of $H(x) = \begin{cases} x^2 + y^2 & \text{for} & x^2 + y^2 < 1 \\ 2 - (x^2 + y^2) & \text{for} & x^2 + y^2 \geq 1 \end{cases}$.
♦ Type 4, the graph of a function

36. **(a)** What is the domain of $\dfrac{\ln y}{x}$? **(b)** Draw some of its level curves.
♦ Type 4, domain and level curves

37. **(a)** What is the domain of $\dfrac{1}{\sqrt{1 - xy}}$? **(b)** What is its global maximum value and at what points does it occur?
♦ Type 4, domain and maximum

38.^{A} **(a)** What is the domain of $\ln(xy)$? **(b)** How are its level curves related to those of the function xy?
♦ Type 4, domain and level curves

39. **(a)** What is the domain of $\sin^{-1}(x^2 + y^2 - 2)$? **(b)** What are its global maximum and minmimum values and at what points do they occur?
♦ Type 4, domain and global maxima and minima

40. **(a)** What is the domain of $f(x, y) = y \csc x$? **(b)** Draw the level curves where it has the values $\pm 1, \pm 2, \pm 3$
♦ Type 4, domain and level curves

41. Find **(a**^{A}**)** $\lim\limits_{(x,y) \to (-5,3)} \dfrac{x^3 y}{x^2 - y}$, **(b**^{A}**)** $\lim\limits_{(x,y) \to (-2,7)} x\sin(xy)$, **(c)** $\lim\limits_{(x,y) \to (0,0)} e^{3x - 2y}$, and
(d) $\lim\limits_{(x,y) \to (1,3)} \ln(x^2 + 3y)$.
♦ Type 1, limits

42.^{A} Use polar coordinates to find the following limits or show that they do not exist:
(a) $\lim\limits_{(x,y) \to (0,0)} \dfrac{xy}{\sqrt{x^2 + y^2}}$, **(b)** $\lim\limits_{(x,y) \to (0,0)} \dfrac{x + y}{x^2 + y^2}$, **(c)** $\lim\limits_{(x,y) \to (0,0)} \dfrac{(x^2 + y^2)^2 + x^3 y^3}{(x^2 + y^2)^2}$.
♦ Type 6, existence of limits using polar coordinates

43. Use polar coordinates to find the value of $\lim\limits_{(x,y) \to (0,0)} \dfrac{y}{x}$ or show that the limit does not exist.
♦ Type 6, existence of a limit using polar coordinates

44. Show that $\displaystyle\lim_{(x,y)\to(0,0)} \frac{x^2 y}{x^4 + y^2}$ does not exist by considering (x, y) that approach $(0, 0)$ along different parabolas.

♦ Type 6, existence of limits using parabolic paths

45. The total area of the base and lateral surface of a right circular cone of height h and with base of radius r is $A(r, h) = \pi r^2 + \pi r \sqrt{r^2 + h^2}$ for positive h and r. **(a)** What is the total area of the base and lateral surface if $r = h$? **(b)** What is the limit of $A(h, r)$ as $h \to 0^+$ for fixed $r > 0$? **(c)** What is the limit of $\dfrac{A(h, r)}{\pi r^2}$ as $r \to \infty$ for fixed $hr > 0$? Give geometric interpretations of the results in parts (b) and (c).

♦ Type 4, interpreting limits

46. Figure 69 shows the graph of the function $f(x, y) = \dfrac{-5y}{x^2 + y^2 + 1}$. **(a)** Generate the vertical cross sections in the planes $y = 0, \pm 1, \pm 2$ on your calculator or computer. Copy and label them in one drawing. **(b)** Follow the instructions of part (a) with the cross sections in the planes $x = 0, \pm 1, \pm 2$. **(c)** Figure 70 shows level curves $\dfrac{-5y}{x^2 + y^2 + 1} = c$ of the function from part (a). Show that one of the level curves is a line and the others are circles. (For $c \neq 0$, set $k = 1/c$ and complete the square.)

♦ Type 4, vertical and horizontal cross sections

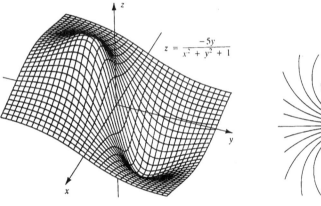

FIGURE 69   FIGURE 70

47.^A Figure 71 shows the graph of $g(x, y) = \dfrac{10 \cos(xy)}{1 + 2y^2}$. Find, without using derivatives, the global maximum of $g(x, y)$ and the values of (x, y) where it occurs.

♦ Type 4, finding a global maximum from a formula

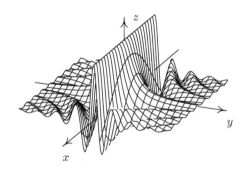 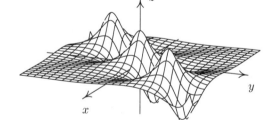

FIGURE 71   FIGURE 72

11.1 Functions of two variables

48. Find, without using derivatives, the global maximum and minumum of $h(x,y) = \dfrac{3\cos(x+y)}{1+(x-y)^2}$. and the values of (x,y) where they occur. The graph of $h(x,y)$ is in Figure 72.

♦ Type 4, finding a global maximum from a formula

49. Figure 73 shows the graph of $h(x,y) = \sin y - \dfrac{1}{y}$. What are **(a)** $\lim\limits_{y \to 0^+} C(x,y)$, $\lim\limits_{y \to 0^-} C(x,y)$, $\lim\limits_{y \to \infty} C(x,y)$, and $\lim\limits_{y \to -\infty} C(x,y)$, for fixed x? **(b)** What are $\lim\limits_{x \to 0} C(x,y)$, $\lim\limits_{x \to \infty} C(x,y)$, and $\lim\limits_{x \to -\infty} C(x,y)$ for fixed nonzero y?

♦ Type 4, limits

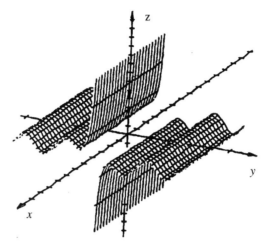

FIGURE 73

50. Match the functions **(a)** $A(x,y) = 2\cos(x+y) - x$, **(b)** $B(x,y) = 2\sin(y^2) + \dfrac{8}{1+x^2+y^2}$, **(c)** $C(x,y) = 4\cos(x+y) + \tfrac{1}{3}(x^2 + y^2)$, and **(d)** $D(x,y) = 2\sin(2\pi y) + x\sin(\pi y)$ to their graphs in Figures 74 through 77 and explain in each case how the shape of the surface is determined by the formula for the function.

♦ Type 4, matching functions to graphs and explaining the shapes

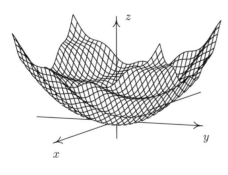

FIGURE 74

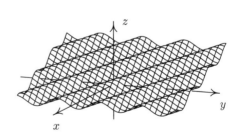

FIGURE 75

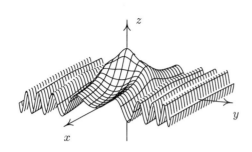
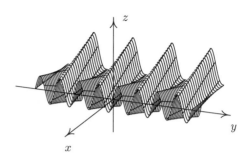

FIGURE 76

FIGURE 77

Section 11.2

Partial derivatives with two variables

OVERVIEW: *The differential calculus of functions of two variables is based on applying rules and formulas for functions of one variable to functions of one variable that are constructed from the functions of two variables. In this section we study* PARTIAL DERIVATIVES, *which are obtained by taking derivatives with respect to one variable while holding the other variable constant. We describe the geometric interpretations of partial derivatives, illustrate how formulas for them can be found with differentiation formulas with one variable, and show how they can be estimated from tables and level curves. Then we discuss higher order derivatives and Chain Rules for differentiating composite functions with two variables.*

Topics:
- *Partial derivatives*
- *Estimating partial derivatives from tables*
- *Estimating partial derivatives from level curves*
- *Derivatives of compositions formed from elementary functions*
- *Derivatives of* **F(g(x,y))**
- *The derivative of* **f(x(t), y(t))**
- *Derivatives of* **f(x(s,t), y(s,t))**
- *Higher-order derivatives*

Partial derivatives

The x-PARTIAL DERIVATIVE or x-DERIVATIVE of a function $f(x,y)$ of two variables is obtained by holding y constant and differentiating with respect to x. This derivative is denoted $\dfrac{\partial f}{\partial x}(x,y)$ or $f_x(x,y)$ and is the (instantaneous) rate of change of $f(x,y)$ with respect to x at (x,y).

Similarly, if we hold x constant and differentiate $f(x,y)$ with respect to y, we obtain the y-PARTIAL DERIVATIVE or y-DERIVATIVE, which is denoted $\dfrac{\partial f}{\partial y}(x,y)$ or $f_y(x,y)$ and is the (instantaneous) rate of change of $f(x,y)$ with respect to y at (x,y). The x- and y-derivatives of $f(x,y)$ are referred to as its FIRST DERIVATIVES or FIRST-ORDER DERIVATIVES.

We use the symbols $\dfrac{\partial}{\partial x}$ and $\dfrac{\partial}{\partial y}$ for the PARTIAL-DIFFERENTIATION OPERATORS that transform a function of two variables into its x- and y-partial derivatives.

Example 1 Find the x- and y-derivatives of $f(x,y) = x^3y - x^2y^5 + x$.

SOLUTION To obtain the x-derivative, we consider y to be a constant and differentiate with respect to x:

$$\frac{\partial f}{\partial x} = \frac{\partial}{\partial x}(x^3y - x^2y^5 + x) = \left[\frac{\partial}{\partial x}(x^3)\right]y - \left[\frac{\partial}{\partial x}(x^2)\right]y^5 + \frac{\partial}{\partial x}(x)$$
$$= 3x^2y - 2xy^5 + 1.$$

♦ SUGGESTIONS TO INSTRUCTORS: This main topics in this section can be divided in two parts to be covered in two or more lectures.
Part 1 (Partial derivatives from formulas and estimating partial derivatives from tables and level curves; Tune-up Exercises T1–T6; Problems 1–21) Have students begin Questions 1 through 6 before class. Problems 1a, 2a, 3, 5, 7, 12, 13, 16 are good for in-class discussions.
Part 2 (The Chain Rule with two variables and higher-order derivatives; Tune-up Exercises T7–T12; Problems 22–44) Have students begin work on Questions 7 through 12 before class. Problems 23, 24, 25a, 26, 30, 31, 41 are good for in-class discussion.

To find the y-derivative, hold x fixed and differentiate with respect to y:

$$\frac{\partial f}{\partial y} = \frac{\partial}{\partial y}(x^3 y - x^2 y^5 + x) = x^3 \left[\frac{\partial}{\partial y}(y)\right] - x^2 \left[\frac{\partial}{\partial y}(y^5)\right] + \frac{\partial}{\partial y}(x)$$
$$= x^3 - 5x^2 y^4. \ \square$$

Question 1 What are the first-order partial derivatives of $F(x, y) = x^3 y^6 + 3x - 4y$?

Example 2 What are $g_x(2, 5)$ and $g_y(2, 5)$ for $g(x, y) = x^2 e^{3y}$?

SOLUTION Differentiating with respect to x with y constant gives

$$g_x(x, y) = \frac{\partial}{\partial x}(x^2 e^{3y}) = 2x e^{3y}.$$

To differentiate with respect to y with x constant, we need the Chain Rule for functions of one variable. We obtain

$$g_y(x, y) = \frac{\partial}{\partial y}(x^2 e^{3y}) = x^2 e^{3y} \frac{\partial}{\partial y}(3y) = 3x^2 e^{3y}.$$

Setting $x = 2$ and $y = 5$ in these formulas gives $g_x(2, 5) = 2(2)e^{3(5)} = 4e^{15}$ and $g_y(2, 5) = 3(2)^2 e^{3(5)} = 12 e^{15}. \ \square$

Example 3 The volume of a right circular cylinder of radius r and height h is equal to the product $V(r, h) = \pi r^2 h$ of its height h and the area πr^2 of its base (Figure 1). What is the rate of change of the volume with respect to the radius and what is its geometric significance?

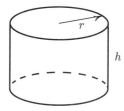

[Volume] $= \pi r^2 h$

[Lateral surface area] $= 2\pi r h$

FIGURE 1

SOLUTION The rate of change of V with respect to r is $\dfrac{\partial V}{\partial r} = \dfrac{\partial}{\partial r}(\pi r^2 h) = 2\pi r h$ and equals the circumference of its base $2\pi r$, multiplied by the height h, which is the area of the lateral surface (the sides) of the cylinder. $\square$

Because the derivative of the volume of the cylinder with respect to its radius is its lateral surface area, the change in the volume when the radius is increased a small amount Δr from r to $r + \Delta r$ is approximately equal to Δr multiplied by the lateral surface area:

$$V(r + \Delta r, h) - V(r, h) \approx \frac{\partial V}{\partial r}(r, h) \Delta r = 2\pi r h \Delta r. \tag{1}$$

To understand this formula, imagine that the lateral surface of the cylinder in Figure 1 is to be painted with a thin layer of paint of thickness Δr. Then **(1)** states that the increase $V(r + \Delta r) - V(r)$ in the volume of the cylinder due to the coat of paint is approximately equal to the lateral surface area $2\pi r h$ of the cylinder, multiplied by the thickness of the paint.

Question 2 (a) What is the rate of change of the volume of a right circular cylinder with respect to its height and what is its geometric significance? (b) How much would the volume of the cylinder in Figure 1 be increased if its top is painted with a coat of paint of thickness Δh?

11.2 Partial derivatives with two variables

Partial derivatives as slopes of tangent lines

When we hold y equal to a constant $y = y_0$, $f(x, y)$ becomes the function $f(x, y_0)$ of x, whose graph is the intersection of the surface $z = f(x, y)$ with the vertical plane $y = y_0$ (Figure 2). The x-derivative $\dfrac{\partial f}{\partial x}(x_0, y_0)$ is the slope in the positive x-direction of the tangent line to this curve at $x = x_0$.

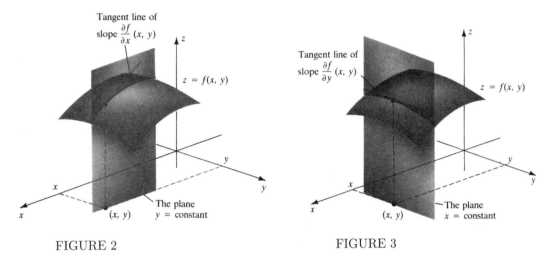

FIGURE 2   FIGURE 3

Similarly, when we hold x equal to a constant x_0, $f(x, y)$ becomes the function $f(x_0, y)$ of y, whose graph is the intersection of the surface with the plane $x = x_0$ (Figure 3), and the y-derivative $\dfrac{\partial f}{\partial y}(x_0, y_0)$ is the slope in the positive y-direction of the tangent line to this curve at $y = y_0$.

Example 4   The MONKEY SADDLE in Figure 4 is the graph of $g(x, y) = \frac{1}{3}y^3 - x^2 y$ **(a)** Where is the plane $x = 2$ in Figure 4?   **(b)** The intersection of the surface with the plane $x = 2$ is one of the curves drawn with a heavy line in Figure 4. Generate this curve in the window $-4 \le y \le 4, -8 \le z \le 8$ and draw it in an yz-plane with its tangent line at $y = 1$.   **(c)** How is the slope of the tangent line related to $g(x, y)$?

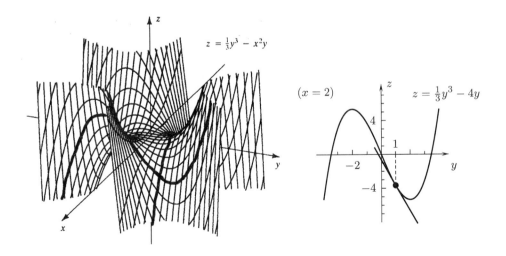

FIGURE 4   FIGURE 5

SOLUTION (a) The plane $x = 2$ is parallel to and in front of the yz-plane in Figure 4.
(b) Because $g(2, y) = \frac{1}{3}y^3 - (2^2)y = \frac{1}{3}y^3 - 4y$, the intersection of the plane $x = 2$ with the graph of g is given by $z = \frac{1}{3}y^3 - 4y, x = 2$. Its graph is in Figure 5. Since

$$\left[\tfrac{1}{3}y^3 - 4y\right]_{y=1} = \tfrac{1}{3}1^3 - 4(1) = -\tfrac{11}{3}$$

and

$$\left[\frac{d}{dy}(\tfrac{1}{3}y^3 - 4y)\right]_{y=1} = \left[y^2 - 4\right]_{y=1} = 1 - 4 = -3$$

the tangent line at $y = 1$ has the equation $z = -\tfrac{11}{3} - 3(y - 1)$.
(c) The slope of the tangent line is the value of the partial derivative

$$\frac{\partial g}{\partial y}(2, 1) = \left[y^2 - x^2\right]_{x=2, y=1} = -3 \text{ at } (2, 1). \quad \square$$

Question 3 (a) Where is the plane $y = 1$ in Figure 4? (b) The intersection of the surface with the plane $y = 1$ is drawn with a heavy line in Figure 4. Generate this curve in the window $-4 \leq x \leq 4, -12 \leq z \leq 6$ and draw it with its tangent line at $x = 2$. (b) How is the slope of the tangent line related to $g(x, y)$?

Estimating partial derivatives from tables

In the next Example we estimate partial derivatives of a function of two variables whose values are given in a table by employing procedures that we used Section 3.3 to estimate derivatives of functions of one variable from tables.

Example 5 The table below is from a study of the effect of exercise on the blood pressure of women. $P(t, E)$ is the average blood pressure, measured in millimeters of mercury (mm Hg), of women of age t years who are exercising at the rate of E watts.[1]
(a) Based on the table, is the average blood pressure of women at rest an increasing or decreasing function of their age? (b) What is the approximate rate of change with respect to age of the average blood pressure of forty-five-year old women who are exercising at the rate of 100 watts?

TABLE 1. $P(t, E)$ (millimeters of mercury)

	$t = 25$	$t = 35$	$t = 45$	$t = 55$	$t = 65$
$E = 150$	178	180	197	209	195
$E = 100$	163	165	181	199	200
$E = 50$	145	149	167	177	181
$E = 0$	122	125	132	140	158

SOLUTION (a) For women at rest ($E = 0$) in the study the average blood pressure was 122 mm Hg for twenty-five-year-olds, 125 mm Hg for thirty-five-year-olds, 132 mm Hg for forty-five-year-olds, etc. The blood pressure increases with age.

---

[1]Data adapted from *Geigy Scientific Tables*, edited by C. Lentner, Vol. 5, Basel, Switzerland: CIBA-GEIGY Limited, 1990, p. 29.

11.2 Partial derivatives with two variables

(b) ONE ANSWER: $\dfrac{\partial P}{\partial t}(45, 100)$ is approximately equal to the average rate of change of $P(t, 100)$ from $t = 45$ to $t = 55$:

$$\frac{\partial P}{\partial t}(45, 100) \approx \frac{P(55, 100) - P(45, 100)}{55 - 45} = \frac{199 - 181}{10}$$
$$= 1.8 \text{ millimeters of mercury per year.}$$

ANOTHER ANSWER: $\dfrac{\partial P}{\partial t}(45, 100)$ is approximately equal to the average rate of change of $P(t, 100)$ from $t = 35$ to $t = 45$:

$$\frac{\partial P}{\partial t}(45, 100) \approx \frac{P(45, 100) - P(35, 100)}{45 - 35} = \frac{181 - 165}{10}$$
$$= 1.6 \text{ millimeters of mercury per year.}$$

A THIRD ANSWER: $\dfrac{\partial P}{\partial t}(45, 100)$ is approximately equal to the average rate of change of $P(t, 100)$ from $t = 35$ to $t = 55$:

$$\frac{\partial P}{\partial t}(45, 100) \approx \frac{P(55, 100) - P(35, 100)}{55 - 35} = \frac{199 - 165}{20}$$
$$= 1.7 \text{ millimeters of mercury per year.}$$

Question 4   **(a)** The blood pressure in Table 1 is an increasing function of the level of exercise E for each age of women with one exception. What is the exception? **(b)** Based on Table 1, what is the approximate value of $\dfrac{\partial P}{\partial E}(45, 100)$?

To estimate first derivatives at points that are between those in a table, we can use average rates of change with nearby points that are in the table. For example, we can estimate $\dfrac{\partial P}{\partial t}(62, 75)$ by using either the average rate of change of $P(t, 50)$ with respect to t from $t = 55$ to $t = 65$,

$$\frac{\partial P}{\partial t}(62, 75) \approx \frac{P(65, 50) - P(55, 50)}{65 - 55} = \frac{181 - 177}{10}$$
$$= 0.4 \text{ millimeters of mercury per year}$$

or by using the average rate of change of $P(t, 100)$ with respect to t from $t = 55$ to $t = 65$,

$$\frac{\partial P}{\partial t}(62, 75) \approx \frac{P(65, 100) - P(55, 100)}{65 - 55} = \frac{200 - 199}{10}$$
$$= 0.1 \text{ millimeters of mercury per year.}$$

Question 5   Based on Table 1, what is the approximate rate of change of $P(t, E)$ with respect to E at $t = 62, E = 75$?

Estimating partial derivatives from level curves

We can estimate first-order partial derivatives of a function from a drawing of its level curves by using average rates of change with values at points on the level curves.

Example 6 Figure 6 shows level curves of the temperature $T(t,h)$ (degrees Fahrenheit) as a function of time t (hours) and the depth h (centimeters) beneath the surface of the ground at O'Neil, Nebraska, from midnight one day ($t = 0$) until midnight the next.[2] What is the approximate rate of change of the temperature with respect to time at 2:00 PM at a point ten centimeters beneath the surface of the ground?

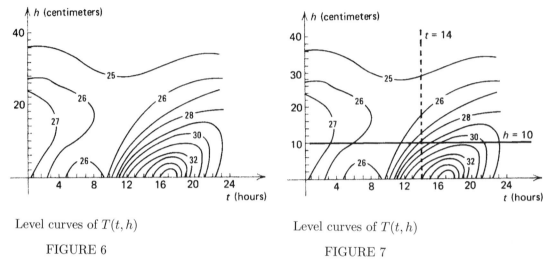

Level curves of $T(t,h)$ — FIGURE 6

Level curves of $T(t,h)$ — FIGURE 7

SOLUTION   Because $t = 14$ at 2:00 PM and $h = 10$ ten centimeters below the surface of the ground, the required rate of change is the t-derivative $\dfrac{\partial T}{\partial t}(t, 10)$ at $t = 14$. To find its approximate value, we draw the horizontal line $h = 10$, as in Figure 7. The point $(14, 10)$ is between the level curves $T = 28$ and $T = 29$ of the temperature, so the change ΔT in temperature from the left curve to the right curve is 1 degree. The horizontal distance Δt along $h = 10$ from the left curve to the right curve is approximately 1 hour. Consequently,

$$\frac{\partial T}{\partial t}(14, 10) \approx \frac{\Delta T}{\Delta t} \approx \frac{1 \text{ degrees}}{1 \text{ hour}} = 1 \text{ degree per hour.} \ \square$$

Question 6 (a) What does $\dfrac{\partial T}{\partial h}(14, 10)$ represent as a rate of change? Give its units. (b) Use values of $T(t,h)$ on the horizontal line $t = 14$ to find the approximate value of $\dfrac{\partial T}{\partial h}(14, 10)$.

---

[2] Data adapted from *Fundamentals of Air Pollution* by S. Williamson, Reading, MA: Addison Wesley, 1973, p. 162.

11.2 Partial derivatives with two variables

Derivatives of compositions formed from elementary functions

If $F(g)$ and $g(x,y)$ are formed by combining sums, products, and quotients with elementary functions of one variable (powers, trigonometric functions, exponential functions, and logarithms), then the composite function $F(g(x,y))$ can be written directly in terms of x and y and the resulting formula can be used to find the partial derivatives $\frac{\partial}{\partial x}[F(g(x,y))]$ and $\frac{\partial}{\partial y}[F(g(x,y))]$.

Example 7 Find the x- and y-derivatives of $F(g(x,y))$ for $F(g) = g^{10}$ and $g(x,y) = x^2y^3 + \sin x$.

SOLUTION Because $F(g(x,y)) = (x^2y^3 + \sin x)^{10}$, we obtain with the Chain Rule for one variable

$$\frac{\partial}{\partial x}[F(g(x,y))] = \frac{\partial}{\partial x}[(x^2y^3 + \sin x)^{10}] = 10(x^2y^3 + \sin x)^9 \frac{\partial}{\partial x}(x^2y^3 + \sin x)$$
$$= 10(x^2y^3 + \sin x)^9(2xy^3 + \cos x)$$
$$\frac{\partial}{\partial y}[F(g(x,y))] = \frac{\partial}{\partial y}[(x^2y^3 + \sin x)^{10}] = 10(x^2y^3 + \sin x)^9 \frac{\partial}{\partial y}(x^2y^3 + \sin x)$$
$$= 10(x^2y^3 + \sin x)^9(3x^2y^2). \square$$

Similarly, if $f(x,y)$, $x(t)$, and $y(t)$ are constructed from elementary functions, then the derivative $\frac{d}{dt}[f(x(t),y(t))]$ of the composite function $\frac{d}{dt}[f(x(t),y(t))]$ can be calculated directly from the formulas.

Question 7 (a) Find a formula in terms of t for $f(x(t),y(t))$ where $f(x,y) = x^5y^6$, $x(t) = e^t$, and $y(t) = \sqrt{t}$. (b) Give a formula for for $\frac{d}{dt}[f(x(t),y(t))]$ with the functions from part (a).

Derivatives of F(g(x,y)) where g has only a letter name

If the function $g(x,y)$ is not given by a formula involving elementary functions of one variable, then we can use the Chain Rule for functions of one variable to find the partial derivatives of composite functions of the form $F(g(x,y))$ where $F(g)$ is a function of one variable.

Example 8 Express the x- and y-derivatives of $F(g(x,y))$ in terms of derivative $\frac{dF}{dg}$ of $F(g)$ and the derivatives $\frac{\partial g}{\partial x}$ and $\frac{\partial g}{\partial x}$ of $g(x,y)$.

SOLUTION We can use the Chain Rule for functions of one variable in this case, because y is held constant when we take the x-derivative and x is held constant when we take the y-derivative:

$$\frac{\partial}{\partial x}[F(g(x,y))] = \frac{dF}{dg}(g(x,y))\frac{\partial g}{\partial x}(x,y)$$
$$\frac{\partial}{\partial y}[F(g(x,y))] = \frac{dF}{dg}(g(x,y))\frac{\partial g}{\partial y}(x,y). \square$$

(2)

Example 9 What are the x- and y-derivatives of $F(g(x,y))$ at $x=5, y=6$ if
$$g(5,6)=10, \frac{dF}{dg}(10)=-7, \frac{\partial g}{\partial x}(5,6)=3, \text{ and } \frac{\partial g}{\partial y}(5,6)=11?$$

SOLUTION Formulas **(2)** yield

$$\left[\frac{\partial}{\partial x}\{F(g(x,y))\}\right]_{x=5,y=6} = \frac{dF}{dg}(g(5,6))\frac{\partial g}{\partial x}(5,6)$$
$$= \frac{dF}{dg}(10)\frac{\partial g}{\partial x}(5,6) = (-7)(3) = -21$$
$$\left[\frac{\partial}{\partial y}\{F(g(x,y))\}\right]_{x=5,y=6} = \frac{dF}{dg}(g(5,6))\frac{\partial g}{\partial y}(5,6) = (-7)(11) = -77. \square$$

A new Chain Rule

We need a new version of the Chain Rule to express the t-derivative of a composite function of the form $f(x(t),y(t))$ in terms of the x- and y-derivatives of $f(x,y)$ and the t-derivatives of $x(t)$ and $y(t)$ if f is only given by its letter name. We suppose that $x(t)$ and $y(t)$ have derivatives at a fixed value t_0 and that $f(x,y)$ has continuous first derivatives in an open disk centered at (x_0, y_0) where $x_0 = x(t_0)$ and $y_0 = y(t_0)$. We consider nonzero Δt so small that $(x_1, y_1) = (x(t_0 + \Delta t), y(t_0 + \Delta t))$ is also in the disk. Then

$$\left[\frac{d}{dt}\{f(x(t),y(t))\}\right]_{t=t_0} = \lim_{\Delta t \to 0} \frac{f(x(t_0+\Delta t), y(t_0+\Delta t)) - f(x(t_0), y(t_0))}{\Delta t}$$
$$= \lim_{\Delta t \to 0} \frac{f(x_1,y_1) - f(x_0,y_0)}{\Delta t}. \tag{3}$$

We express the change $f(x_1, y_1) - f(x_0, y_0)$ in the value of $f(x,y)$ from (x_0, y_0) to (x_1, y_1) as the change in the x-direction from (x_0, y_0) to (x_1, y_0) plus the change in the y-direction from (x_1, y_0) to (x_1, y_1), as indicated in Figure 8:

$$f(x_1,y_1) - f(x_0,y_0) = [f(x_1,y_0) - f(x_0,y_0)] + [f(x_1,y_1) - f(x_1,y_0)]. \tag{4}$$

(Notice that the terms $f(x_1, y_0)$ and $-f(x_1, y_0)$ on the right side of **(4)** cancel to give the left side.)

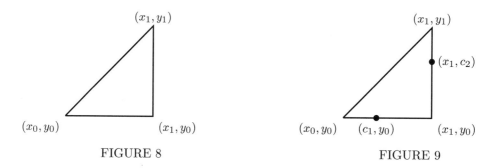

FIGURE 8   FIGURE 9

We can apply the Mean-Value Theorem from Section 3.3 to the expression in the first set of square brackets on the right of **(4)** where y is constant and to the expression in the second set of square brackets where x is constant: We conclude that there is a number c_1 between x_0 and x_1 and a number c_2 between y_0 and y_1 (see Figure 9) such that

$$f(x_1, y_0) - f(x_0, y_0) = \frac{\partial f}{\partial x}(c_1, y_0)(x_1 - x_0)$$
$$f(x_1, y_1) - f(x_1, y_0) = \frac{\partial f}{\partial y}(x_1, c_2)(y_1 - y_0). \tag{5}$$

11.2 Partial derivatives with two variables

Combining equations **(4)** and **(5)** yields

$$f(x_1, y_1) - f(x_0, y_0) = \frac{\partial f}{\partial x}(c_1, y_0)(x_1 - x_0) + \frac{\partial f}{\partial y}(x_1, c_2)(y_1 - y_0). \tag{6}$$

This equation states that the change in the value of f from the lower left corner (x_0, y_0) of the triangle in Figure 9 to the upper right corner (x_1, y_1) equals the rate of change of f with respect to x at the point (c_1, y_0) on the bottom of the triangle, multiplied by the change in x, plus the rate of change of f with respect to y at (x_0, c_2) on the side, multiplied by the change in y.

We divide both sides of **(6)** by Δt to obtain

$$\frac{f(x_1, y_1) - f(x_0, y_0)}{\Delta t} = \frac{\partial f}{\partial x}(c_1, y_0)\left[\frac{x_1 - x_0}{\Delta t}\right] + \frac{\partial f}{\partial y}(x_1, c_2)\left[\frac{y_1 - y_0}{\Delta t}\right]. \tag{7}$$

Question 8   What are the limits of $\dfrac{x_1 - x_0}{\Delta t}$ and $\dfrac{y_1 - y_0}{\Delta t}$ as $\Delta t \to 0$?

Because $x(t)$ and $y(t)$ have derivatives at t_0, they are continuous there and the triangle in Figure 7 collapses to the point $(x_0, y_0) = (x(t_0), y(t_0))$ as $\Delta t \to 0$. Also, because we assume that the partial derivatives of $f(x, y)$ are continuous, the term $\frac{\partial f}{\partial x}(c_1, y_0)$ in **(7)** tends to $\frac{\partial f}{\partial x}(x_0, y_0) = \frac{\partial f}{\partial x}(x(t_0), y(t_0))$ and the term $\frac{\partial f}{\partial y}(x_0, c_2)$ tends to $\frac{\partial f}{\partial y}(x_0, y_0) = \frac{\partial f}{\partial y}(x(t_0), y(t_0))$ as $\Delta t \to 0$. Hence, with the results of Question 8, we have the following result.

Theorem 1   Suppose that $x(t)$ and $y(t)$ have first derivatives at t and that $f(x, y)$ has continuous first-order derivatives in an open disk centered at $(x(t), y(t))$. Then

$$\frac{d}{dt}[f(x(t), y(t))] = \frac{\partial f}{\partial x}(x(t), y(t))\frac{dx}{dt}(t) + \frac{\partial f}{\partial y}(x(t), y(t))\frac{dy}{dt}(t). \tag{8}$$

Formula **(8)** states that the rate of change of $f(x(t), y(t))$ with respect to t equals the rate of change of f with respect to x multiplied by the rate of change of x with respect to t, plus the rate of change of f with respect to y multiplied by the rate of change of y with respect to t. It is perhaps easier to remember in the abbreviated form

$$\frac{d}{dt}(f) = \frac{\partial f}{\partial x}\frac{dx}{dt} + \frac{\partial f}{\partial y}\frac{dy}{dt}$$

where no reference is made to the values of the variables.

We will assume in all calculations of derivatives of the form in Theorem 1 that the functions considered have continuous first derivatives so that the Theorem can be applied—unless information to the contrary is stated specifically.

Example 10   What is the t-derivative of $f(x(t), y(t))$ at $t = 1$ if $x(1) = 2, y(1) = 3$, $\frac{dx}{dt}(1) = -4, \frac{dy}{dt}(1) = 5, \frac{\partial f}{\partial x}(2, 3) = -6$, and $\frac{\partial f}{\partial y}(2, 3) = 7$?

SOLUTION   By **(8)** with $t = 1$

$$\left[\frac{d}{dt}\{f(x(t), y(t))\}\right]_{t=1} = \frac{\partial f}{\partial x}(x(1), y(1))\frac{dx}{dt}(1) + \frac{\partial f}{\partial y}(x(1), y(1))\frac{dy}{dt}(1)$$

$$= \frac{\partial f}{\partial x}(2, 3)\frac{dx}{dt}(1) + \frac{\partial f}{\partial y}(2, 3)\frac{dy}{dt}(1)$$

$$= (-6)(-4) + (7)(5) = 59. \square$$

Question 9 What is the t-derivative of $F(t^2, t^3)$ at $t = 2$ if $\dfrac{\partial F}{\partial x}(4,8) = 10$, and $\dfrac{\partial F}{\partial y}(4,8) = -20$?

Example 11 A small plane uses gasoline at the rate of $R(h,v)$ gallons per hour if it is flying at an elevation of h feet above the ground and its air speed is v knots (nautical miles per hour), where $R(8000, 120) = 7.2$ gallons per hour, $\dfrac{\partial R}{\partial h}(800, 120) = -2 \times 10^{-4}$ gallons per hour per foot, and $\dfrac{\partial R}{\partial v}(8000, 120) = 0.13$ gallons per hour per knot.[3] At a moment when the plane has an altitude of 8000 feet and a speed of 120 knots, its height is increasing 500 feet per minute and it is accelerating 3 knots per minute. At what rate is its rate of gasoline consumption increasing or decreasing at that moment?

SOLUTION At the moment in question the plane's rate of gas consumption R is changing at the rate

$$\frac{d}{dt}(R) = \left[\frac{\partial R}{\partial h} \frac{\text{gallons per hour}}{\text{foot}}\right]\left[\frac{dh}{dt} \frac{\text{feet}}{\text{minute}}\right]$$
$$+ \left[\frac{\partial R}{\partial v} \text{ gallons per hour per knot}\right]\left[\frac{dv}{dt} \frac{\text{knots}}{\text{minute}}\right]$$
$$= \left[-2 \times 10^{-4} \frac{\text{gallons per hour}}{\text{foot}}\right][500 \text{ feet per minute}]$$
$$+ \left[0.13 \frac{\text{gallons per hour}}{\text{knot}}\right]\left[3 \frac{\text{knots}}{\text{minute}}\right]$$
$$= (-2 \times 10^{-4})(500) + (0.13)(3) = 0.29 \frac{\text{gallons per hour}}{\text{minute}}.$$

The plane's rate of fuel consumption is increasing 0.29 gallons per hour per minute. □

Question 10 At what rate is the rate of fuel consumption of the plane from Example 11 increasing or decreasing at a moment when it is flying at an altitude of 8000 feet and a speed of 120 knots, its altitude is decreasing 400 feet per minute and it is decelarating 10 knots per minute?

Derivatives of $f(x(s,t), y(s,t))$

Theorem 1 can be applied to find the s- and t-derivatives of $f(x(s,t), y(s,t))$ because in taking the derivative with respect to s or t, the other variable is constant. We obtain

$$\frac{\partial}{\partial s}[f(x(s,t), y(s,t))] = \frac{\partial f}{\partial x}(x(s,t), y(s,t))\frac{\partial x}{\partial s}(s,t) + \frac{\partial f}{\partial y}(x(s,t), y(s,t))\frac{\partial y}{\partial s}(s,t) \tag{9}$$

$$\frac{\partial}{\partial t}[f(x(s,t), y(s,t))] = \frac{\partial f}{\partial x}(x(s,t), y(s,t))\frac{\partial x}{\partial t}(s,t) + \frac{\partial f}{\partial y}(x(s,t), y(s,t))\frac{\partial y}{\partial t}(s,t). \tag{10}$$

---

[3] Data adapted from *Cessna 172N Information Manual*, Wichita Kansas: Cessna Aircraft Company, 1978, p.5-16.

11.2 Partial derivatives with two variables

Example 12 What are the s- and t-derivatives of $f(st^2, te^s)$ at $s = 0, t = 2$ if $\dfrac{\partial f}{\partial x}(0, 2) = 10$ and $\dfrac{\partial f}{\partial y}(0, 2) = -5$?

SOLUTION   By **(9)** and **(10)**,

$$\frac{\partial}{\partial s}[f(st^2, te^s)] = \frac{\partial f}{\partial x}(st^2, te^s)\frac{\partial}{\partial s}(st^2) + \frac{\partial f}{\partial y}(st^2, te^s)\frac{\partial}{\partial s}(te^s).$$

$$= t^2\frac{\partial f}{\partial x}(st^2, te^s) + te^s\frac{\partial f}{\partial y}(st^2, te^s)$$

$$\frac{\partial}{\partial t}[f(st^2, te^s)] = \frac{\partial f}{\partial x}(st^2, te^s)\frac{\partial}{\partial t}(st^2) + \frac{\partial f}{\partial y}(st^2, te^s)\frac{\partial}{\partial t}(te^s)$$

$$= 2st\frac{\partial f}{\partial x}(st^2, te^s) + e^s\frac{\partial f}{\partial y}(st^2, te^s).$$

We set $s = 0$ and $t = 2$ and use the given values of the derivatives of $f(x, y)$ to obtain

$$\left[\frac{\partial}{\partial s}\{f(st^2, te^s)\}\right]_{s=0, t=2} = 2^2\frac{\partial f}{\partial x}(0, 2) + 2e^0\frac{\partial f}{\partial y}(0, 2) = 4(10) + 2(-5) = 30$$

$$\left[\frac{\partial}{\partial t}\{f(st^2, te^s)\}\right]_{s=0, t=2} = 2(0)(2)\frac{\partial f}{\partial x}(0, 2) + e^0\frac{\partial f}{\partial y}(0, 2) = 0(10) + 1(-5) = -5. \ \square$$

Higher-order partial derivatives

The first-order partial derivatives $f_x = \dfrac{\partial f}{\partial x}$ and $f_y = \dfrac{\partial f}{\partial y}$ of $f(x, y)$ can be differentiated with respect to x and y to obtain the second x-derivative

$$f_{xx} = \frac{\partial^2 f}{\partial x^2} = \frac{\partial}{\partial x}\left(\frac{\partial f}{\partial x}\right)$$

the second y-derivative

$$f_{yy} = \frac{\partial^2 f}{\partial y^2} = \frac{\partial}{\partial y}\left(\frac{\partial f}{\partial y}\right)$$

and the MIXED SECOND DERIVATIVES

$$f_{yx} = \frac{\partial^2 f}{\partial x \partial y} = \frac{\partial}{\partial x}\left(\frac{\partial f}{\partial y}\right) \ \text{and} \ f_{xy} = \frac{\partial^2 f}{\partial y \partial x} = \frac{\partial}{\partial y}\left(\frac{\partial f}{\partial x}\right). \tag{11}$$

In all cases that we will encounter, the mixed second derivatives **(11)** are equal, so this derivative may be calculated either by differentiating first with respect to x and then with respect to y or by differentiating first with respect to y and then with respect to x. This is a consequence of the following theorem, whose proof is outlined in Problem 44.

Theorem 2 If $\dfrac{\partial^2 f}{\partial x \partial y}$ and $\dfrac{\partial^2 f}{\partial y \partial x}$ are defined and continuous in an open disk centered at (x, y), then $\dfrac{\partial^2 f}{\partial x \partial y}(x, y) = \dfrac{\partial^2 f}{\partial y \partial x}(x, y).$

Because the mixed partial derivatives **(11)** are equal, you can calculate them with both orders of differentiation to check your work.

Example 13 What are the second-order derivatives of $f(x,y) = xy^2 + x^3y^5$?

SOLUTION We begin with the first derivatives:

$$f_x = \frac{\partial}{\partial x}(xy^2 + x^3y^5) = y^2 + 3x^2y^5$$

$$f_y = \frac{\partial}{\partial y}(xy^2 + x^3y^5) = 2xy + 5x^3y^4.$$

Then we find the second derivatives:

$$f_{xx} = \frac{\partial}{\partial x}(f_x) = \frac{\partial}{\partial x}(y^2 + 3x^2y^5) = 6xy^5$$

$$f_{xy} = \frac{\partial}{\partial y}(f_x) = \frac{\partial}{\partial y}(y^2 + 3x^2y^5) = 2y + 15x^2y^4$$

$$f_{yx} = \frac{\partial}{\partial x}(f_y) = \frac{\partial}{\partial x}(2xy + 5x^3y^4) = 2y + 15x^2y^4$$

$$f_{yy} = \frac{\partial}{\partial y}(f_y) = \frac{\partial}{\partial y}(2xy + 5x^3y^4) = 2x + 20x^3y^3.$$

Notice that f_{xy} and f_{yx} are equal, as should be the case because these functions are continuous. □

Question 11 Find the second-order derivatives of $g(x,y) = x^4 \sin(3y) + 5x - 6y$.

Question 12 What is K_{yyy} for $K(x,y) = e^{xy}$?

Higher-order derivatives are defined similarly and mixed derivatives can generally be taken in any order for functions given by formulas.

Example 14 Find $\dfrac{\partial^3 H}{\partial^2 x \partial y}$ in three different ways where $H(x,y) = x^3 y^8$.

SOLUTION There are three ways to obtain the answer because we need to take two x-derivatives and one y-derivative in any order. If we take the two x-derivatives first we obtain

$$H_x = \frac{\partial}{\partial x}(x^3y^8) = 3x^2y^8,\ H_{xx} = \frac{\partial}{\partial x}(3x^2y^8) = 6xy^8,\ H_{xxy} = \frac{\partial}{\partial y}(6xy^8) = 48xy^7.$$

We can also take first an x-derivatives, then a y-derivative, and then an x-derivative to have

$$H_x = \frac{\partial}{\partial x}(x^3y^8) = 3x^2y^8,\ H_{xy} = \frac{\partial}{\partial y}(3x^2y^8) = 24x^2y^7,\ H_{xyx} = \frac{\partial}{\partial x}(24x^2y^7) = 48xy^7.$$

Or, we can take the y-derivative first and obtain

$$H_y = \frac{\partial}{\partial y}(x^3y^8) = 8x^3y^7,\ H_{yx} = \frac{\partial}{\partial x}(8x^3y^7) = 24x^2y^7,\ H_{yxx} = \frac{\partial}{\partial x}(24x^2y^7) = 48xy^7.$$

In any case, we obtain the formula $48xy^7$ for the mixed third derivative. □

11.2 Partial derivatives with two variables

Principles and procedures

- Finding partial derivatives of functions given by formulas is generally not much more difficult than finding ordinary derivatives, once you have had enough practice keeping track of the constant variables and once you have developed an orderly way of carrying out your calculations. If you have difficulty obtaining correct answers, write down what is required at each step before you carry it out and avoid doing more than one step at a time your head. For example, to find the x-derivative of $\cos(x^2 y^3)$ write

$$\frac{\partial}{\partial x}[\cos(x^2 y^3)] = -\sin(x^2 y^3)\frac{\partial}{\partial x}(x^2 y^3) = -2xy^3 \sin(x^2 y^3)$$

where you show how the Chain Rule from Chapter 3 is employed, instead of only applying the Chain Rule mentally by writing

$$\frac{\partial}{\partial x}[\cos(x^2 y^3)] = -2xy^3 \sin(x^2 y^3).$$

- Conceptually, partial derivatives with two variables are nothing new: they are just ordinary derivatives (rates of change) of functions with respect to one variable with the other variable held constant. The geometric interpretation of the first-order derivatives of $f(x,y)$ as slopes of tangent lines to cross sections of the graph of f is, however, a new idea that you need to be sure to understand.

- The procedures for estimating first-order partial derivatives of functions with two variables from tables are the same as used in Chapter 3 for estimating first derivatives of functions of one variable because in the case of two variables you look at one row or column of the table at a time.

- Estimating first-order derivatives from level curves is similar to estimating derivatives from tables, except that you use values on level curves instead of numbers in a table.

- The Chain Rule $\dfrac{d}{dt}(f) = \dfrac{\partial f}{\partial x}\dfrac{dx}{dt} + \dfrac{\partial f}{\partial y}\dfrac{dy}{dt}$ for $f(x(t), y(t))$ and the Chain Rules $\dfrac{\partial}{\partial s}(f) = \dfrac{\partial f}{\partial x}\dfrac{\partial x}{\partial s} + \dfrac{\partial f}{\partial y}\dfrac{\partial y}{\partial s}$ and $\dfrac{\partial}{\partial t}(f) = \dfrac{\partial f}{\partial x}\dfrac{\partial x}{\partial t} + \dfrac{\partial f}{\partial y}\dfrac{\partial y}{\partial t}$ for $f(x(s,t), y(s,t))$ are not needed if $f(x,y)$ is given by a formula constructed from elementary functions but are required if f is given only by its letter name.

- Higher-order derivatives of functions $f(x,y)$ given by formulas are, in principle, easy to find, but require careful calculations. When you find mixed second-order or higher-order derivatives, take the derivatives in all possible orders to check your work.

Tune-Up Exercises 11.2[♦]

A Answer provided. **O** Outline of solution provided.

T1.[O] Find (a) $\dfrac{\partial}{\partial x}(xy^5 - 4y^2 + 6x^4 y^7)$ and (b) $\dfrac{\partial}{\partial y}(xy^5 - 4y^2 + 6x^4 y^7)$.

♦ Type 1, partial derivatives with formulas

T2.[O] What is $G_y(x,y)$ if $G(x,y) = x^2 \sin(xy) + y - x$?

♦ Type 1, partial derivatives with formulas

[♦] The Tune-up Exercises and Problems are classified by type and content. The types are (1) basic, reactive; (2) basic reflective; (3) intermediate, reactive; (4) intermediate, reflective; (5) advanced, reactive; (6) advanced, reflective; and (7) advanced, theoretical.

T3.⁰ Use the following table of values of $g(x,y)$ to estimate **(a)** $g_x(2,5)$ and **(b)** $g_y(2,5)$.

♦ Type 1, estimating partial derivatives from a table

	$x=1$	$x=1.5$	$x=2$	$x=2.5$	$x=3$
$y=5.2$	150	160	172	184	195
$y=5.0$	187	200	212	223	235
$y=4.8$	231	242	253	266	278
$y=4.6$	273	283	293	305	316

T4.⁰ Level curves of $G(x,y)$ are shown in Figure 10. Find its approximate x- and y-derivatives at $(3,3)$.

♦ Type 1, estimating partial derivatives from level curves

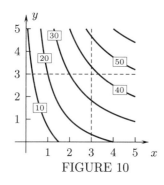
FIGURE 10

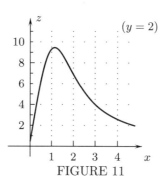
FIGURE 11

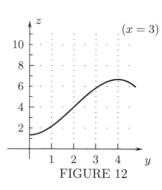
FIGURE 12

T5.⁰ Figure 11 shows the graph of the function $P(x,2)$ of x that is obtained from $P(x,y)$ by setting $y=2$, and Figure 12 shows the graph of the function $P(3,y)$ of y that is obtained from $P(x,y)$ by setting $x=3$. Use the graphs to find the approximate values of $\dfrac{\partial P}{\partial x}(3,2)$ and $\dfrac{\partial P}{\partial y}(3,2)$.

♦ Type 1, estimating partial derivatives from graphs of cross sections

T6.⁰ **(a)** Draw in an xz-plane the intersection of the plane $y=2$ with the graph of $H(x,y) = \tfrac{1}{2}y^2 - x^2$ and the tangent line to the curve whose slope in the positive x-direction is $\dfrac{\partial H}{\partial x}(1,2)$. **(b)** Draw in a yz-plane the intersection of the plane $x=1$ with the graph of $H(x,y)$ and the tangent line to the curve whose slope in the positive y-direction is $\dfrac{\partial H}{\partial y}(1,2)$.

♦ Type 1, relating partial derivatives to graphs of cross sections

T7.⁰ Express $f(x(t), y(t))$ in terms of t where $f(x,y) = x\sin(xy)$, $x(t) = t^5$, and $y(t) = t^{-3}$.

♦ Type 1, simplifying the formula for a composite function

T8.⁰ Express the x- and y-derivatives of $W(x^3 y^5)$ in terms of x, y, and $W'(t) = \dfrac{dW}{dt}(t)$.

♦ Type 3, the one-variable Chain Rule used with two variables

T9.⁰ What is $\dfrac{d}{dt}[F(x(t), y(t))]$ at $t=0$ if $x(0)=3$, $y(0)=7$, $\dfrac{dx}{dt}(0) = -4$, $\dfrac{dy}{dt}(0) = 6$, $\dfrac{\partial F}{\partial x}(3,7) = 8$, and $\dfrac{\partial F}{\partial y}(3,7) = 2$?

♦ Type 5, the two-variable Chain Rule

T10.⁰ Find the u- and v-derivatives of $f(x(u,v), y(u,v))$ with $f(x,y) = 3x - 4y + 7$, $x = 2u - 5v$, and $y = 3u + 8v$ **(a)** by using the Chain Rule $\dfrac{\partial}{\partial u}(f) = \dfrac{\partial f}{\partial x}\dfrac{\partial x}{\partial u} + \dfrac{\partial f}{\partial y}\dfrac{\partial y}{\partial u}$, $\dfrac{\partial}{\partial v}(f) = \dfrac{\partial f}{\partial x}\dfrac{\partial x}{\partial v} + \dfrac{\partial f}{\partial y}\dfrac{\partial y}{\partial v}$ and **(b)** by first writing $f(x(u,v), y(u,v))$ in terms of u, and v.

♦ Type 5, the two-variable Chain Rule

11.2 Partial derivatives with two variables

T11.^{O} What is $\dfrac{\partial U}{\partial t}(2,6)$ if $U(s,t) = V(x(s,t), y(s,t))$, $x(2,6) = 5, y(2,6) = -7, \dfrac{\partial x}{\partial s}(2,6) = 4$, $\dfrac{\partial x}{\partial t}(2,6) = 8, \dfrac{\partial y}{\partial s}(2,6) = -1, \dfrac{\partial y}{\partial t}(2,6) = 10, \dfrac{\partial V}{\partial x}(5,-7) = 11$, and $\dfrac{\partial V}{\partial y}(5,-7) = 12$?

♦ Type 5, the two-variable Chain Rule

T12.^{O} Find the second-order derivatives of $R(x,y) = x^{1/2}y^{1/3}$.

♦ Type 1, second-order derivatives

Problems 11.2

^{A}Answer provided. ^{O}Outline of solution provided.

1. Find **(a^A)** $\dfrac{\partial}{\partial x}(x^3y^2 - x + y)$, **(b^A)** $\dfrac{\partial}{\partial y}(x^2e^{3y} + y^2e^{3x})$, and **(c)** $\dfrac{\partial}{\partial x}(xe^y + 6x^2 - y)$

   ♦ Type 1, partial derivatives with formulas

2. Find the x- and y-derivatives of **(a^A)** $F(x,y) = \sin(x^2y^4)$, **(b^A)** $G(x,y) = \ln(1-xy)$, **(c)** $H(x,y) = (x^2+x+1)(y^2+y-3)$, **(d)** $P(x,y) = e^{u^2}\cos(v^2)$, and **(e)** $Q(x,y) = x^{1/2}y^{1/4} + x^2y^4$.

   ♦ Type 1, partial derivatives with formulas

3.^{A} The volume of a right circular cone of height h meters and with a base of radius r meters is $V = \frac{1}{3}\pi r^2 h$ cubic meters. What is the radius of change of the volume with respect to the radius?

   ♦ Type 1, partial derivatives with formulas

4. If a constant current of I amperes flows through a circuit with a resistance of 100 ohms for t seconds, it will produce $H(I,t) = 23.9I^2t$ calories of heat.$^{(4)}$ **(a)** Is $H(I,t)$ a linear function of I for fixed t or a linear function of t for fixed I? **(b)** What is the rate of change of the heat production with respect to I at $t = 5, I = 10$? Give the units.

   ♦ Type 1, partial derivatives with formulas

5. If a gas has density ρ_0 grams per cubic centimeter at $0°C$ and pressure of one atmosphere, then its density at $T°C$ and pressure P atmospheres is $\rho(T,P) = \dfrac{\rho_0(1 + \frac{1}{273}T)}{P}$ grams per cubic centimeter.$^{(5)}$ **(a)** Find formulas for $\dfrac{\partial \rho}{\partial T}$ and $\dfrac{\partial \rho}{\partial P}$ in terms of T, P, and the parameter ρ_0. Give their units. **(b)** One of the derivatives in part (a) is positive and the other is negative for $T > -273$ and positive P. Explain in terms of gasses why this could be expected.

   ♦ Type 1, partial derivatives with formulas

6.^{A} The table below gives the volume $V(p,T)$ (cubic feet) of a pound of a p-percent solution of sulfuric acid in water that is at a temperature of $T°C$.$^{(6)}$ **(a)** Does the volume of a pound of a solution increase or decrease as the temperature increases? **(b)** Suppose that two solutions are at the same temperature but one contains a greater concentration of sulfuric acid. Which has the greater volume? **(c)** What rate of change is represented by $\dfrac{\partial V}{\partial p}(15,80)$ and what is its approximate value? **(d)** What is the approximate value of $\dfrac{\partial V}{\partial T}(p,T)$ for $10 \leq p \leq 20$ and $60 \leq T \leq 100$?

   ♦ Type 1, estimating partial derivatives from a table

	$p = 10\%$	$p = 15\%$	$p = 20\%$	$p = 25\%$
$T = 100°C$	0.0157	0.0152	0.0147	0.0143
$T = 80°C$	0.0155	0.0150	0.0145	0.0141
$T = 60°C$	0.0153	0.0148	0.0143	0.0139
$T = 40°C$	0.0151	0.0146	0.0142	0.0138

$^{(4)}$Data adapted from *CRS Handbook of Chemistry and Physics*, 62nd edition, Boca Raton, FL: CRC Press, Inc., 1981, p. F-98.

$^{(5)}$Ibid, p. F-94

$^{(6)}$Data adapted from *Handbook of Engineering Materials* by F. Miner and J. Seastone, New York, NY: John Wiley & Sons, 1955, p. 3-407

7. The next table gives the wind chill $W(T, v)$ (degrees Fahrenheit) as a function of the Fahrenheit temperature T and the velocity of the wind v, measured in miles per hour for five temperatures below the freezing point of water ($32°$F) and four wind speeds. $W(T, v)$ is the constant temperature which with no wind has the same cooling effect as temperature $T°$F in a wind of constant velocity v miles per hour. (a) Is $W(T, v)$ an increasing or a decreasing function of T for fixed v and why is this plausible? (b) What is the approximate rate of change of $W(T, v)$ with respect to T if the temperature is $0°$F and the wind velocity is 20 miles per hour? (c) Is wind chill $W(T, v)$ an increasing or a decreasing function of v for fixed T, and why? (d) What is the approximate rate of change of the wind chill with respect to the velocity of the wind when the temperature is $0°$F and the wind velocity is 20 miles per hour?

♦ Type 1, estimating partial derivatives from a table

	$T = -20$	$T = -10$	$T = 0$	$T = 10$	$T = 20$
$v = 30$	-79	-64	-49	-33	-18
$v = 20$	-67	-53	-39	-24	-10
$v = 10$	-46	-34	-22	-9	3
$v = 0$	-20	-10	0	10	20

8. The next table gives the amount of food $F(w, t)$ (pounds) required each day by a horse that weighs w pounds and is ridden t hours a day. (a) Give approximate values of $\dfrac{\partial F}{\partial w}(1000, 4)$ and $\dfrac{\partial F}{\partial t}(1000, 4)$. (a) Why, in terms of horses, is $F(w, t)$ an increasing function of w for fixed t and an increasing function of t for fixed w? (c) Is it possible, based on the data in the table, that $F(w, t)$ is a linear function of w for one of the values of t in the table or that $F(w, t)$ is a linear function of t for one of the values of w in the table? Explain.

♦ Type 1, estimating partial derivatives from a table

	$w = 800$	$w = 900$	$w = 1000$	$w = 1100$	$w = 1200$
$t = 6$	18.7	20.5	22.2	23.8	25.4
$t = 4$	17.9	19.6	21.2	22.8	24.3
$t = 2$	16.9	18.5	20.1	21.5	23.0
$t = 0$	12.9	14.1	15.3	16.4	17.5

9.° Use the level curves of $K(x, y)$ in Figure 13 to give its approximate x- and y-derivatives at $(6, 2)$.

♦ Type 1, estimating partial derivatives from level curves

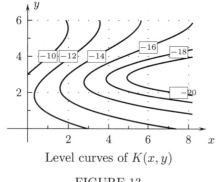

Level curves of $K(x, y)$

FIGURE 13

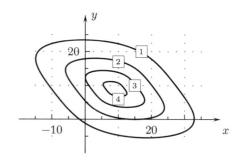

Level curves of $L(x, y)$

FIGURE 14

11.2 Partial derivatives with two variables

10. Based on the level curves of $L(x,y)$ in Figure 14, what are the approximate values of $\dfrac{\partial L}{\partial x}(20,10)$ and $\dfrac{\partial L}{\partial y}(20,10)$?

♦ Type 1, estimating partial derivatives from level curves

11. The x- and y-derivatives of the function $h(x,y)$ of Figure 15, are constant. What are their approximate values?

♦ Type 1, finding partial derivatives from level curves

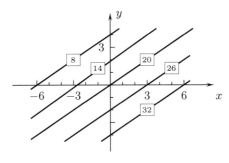

Level curves of $h(x,y)$

FIGURE 15

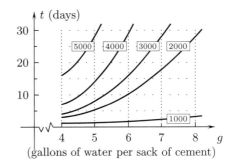

Level curves of S

FIGURE 16

12.^A Figure 16 gives level curves of the compressive strength $S(g,t)$ (pounds per square inch) of portland concrete that is made with g gallons of water per sack of cement and that has cured t days.[9] What are the approximate values of $\dfrac{\partial S}{\partial g}(6,14)$ and $\dfrac{\partial S}{\partial t}(6,14)$?

♦ Type 1, estimating partial derivatives from level curves

13. Figure 17 gives level curves of the amount of solar radiation $R(t,L)$ (calories) received by a horizontal plate one centimeter square during a cloudless day at a latitude of L (degrees) and at time t (month) of the year. What are the approximate values of **(a)** R, **(b)** R_L, and **(c)** R_t at May 1 and a latitude of $40°$? **(d)** Why, based on the seasons, is $R(6,60)$ greater than $R(6,-60)$ and $R(1,60)$? **(d)** Why are $R(1,80)$ and $R(6,-80)$ zero? **(e)** Where and when is R greatest?

♦ Type 1, estimating partial derivatives from level curves

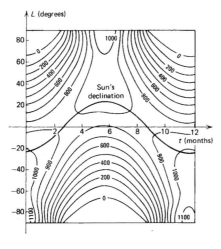

FIGURE 17

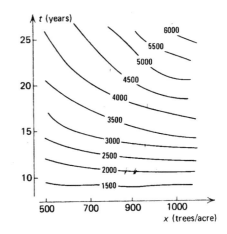

FIGURE 18

---

[9] Data adapted from *Handbook Of Engineering Materials*, Ibid., p. 4-14

14. Figure 18 shows level curves of the amount of the yield $Y(x,t)$ (cubic feet per acre) from a pine plantation with x trees per acre that are harvested t years after planting. **(a)** Determine without doing any calculations whether $\dfrac{\partial Y}{\partial t}(700, 20)$ is less than or greater than $\dfrac{\partial Y}{\partial t}(1000, 15)$. Explain your reasoning and describe what this indicates about the plantation. **(b)** Is it better to have 600 trees per acre or 1000 trees per acre if your only goal is to maximize the yield?

♦ Type 1, estimating partial derivatives from level curves

15. The FETCH of the wind at a point on a body of water is the distance that the wind has blown over water before it reaches the point. The next table gives the height $h(v, f)$ (feet) of waves as a function of the velocity v (knots) and of the fetch f (nautical miles). (Knots are nautical miles per hour). **(a)** Based on the table, what is $\dfrac{\partial h}{\partial f}(10, f)$ for all f? What does this say about the waves? **(b)** Based on the table, is $\dfrac{\partial h}{\partial v}(40, f)$ an increasing or a decreasing function of f? What does this say about the waves? **(c)** What do you think would happen to $h(v, f)$ as $f \to \infty$ for each fixed v and why?

♦ Type 1, estimating partial derivatives from a table

	$v = 10$	$v = 20$	$v = 30$	$v = 40$
$f = 1000$	3	8	18	50
$f = 500$	3	8	18	47.5
$f = 200$	3	7.5	17	39.5
$f = 100$	3	7	14.5	31
$f = 50$	3	6	12	22.5

16. Find the approximate maximum and minimum values of $\dfrac{\partial W}{\partial y}(x, y)$ for $0 \leq x \leq 5, 1 \leq y \leq 4$, where $W(x, y)$ is the the function whose level curves are shown in Figure 19.

♦ Type 3, estimating partial derivatives from level curves

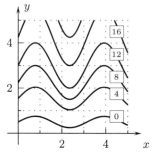

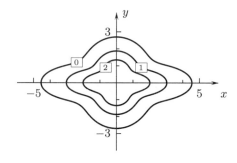

FIGURE 19

FIGURE 20

17. **(a)** Describe the shape of the graph of the function $Z(x, y)$ whose level curves are shown in Figure 20. **(b)** Describe how $\dfrac{\partial Z}{\partial x}(x, 0)$ changes as x increases from -5 to 5 and how $\dfrac{\partial Z}{\partial y}(0, y)$ changes as y increases from -3 to 3 **(c)** $Z(x, y)$ is said to be an even function of x for each y and an even function of y for each x because $Z(-x, y) = Z(x, y)$ and $Z(x, -y) = Z(x, y)$ for all (x, y). Are $\dfrac{\partial Z}{\partial x}(0, y)$ and $\dfrac{\partial Z}{\partial y}(x, 0)$ even or odd functions of y and x, respectively?

♦ Type 4, analyzing level curves

11.2 Partial derivatives with two variables

18.[A] Figure 21 shows the graph of $p(x,y) = \sin y - \frac{1}{9}x^3$. **(a)** Draw the intersection of the surface with the plane $x = 0$ and its tangent line whose slope is $\frac{\partial p}{\partial y}(0, \frac{3}{4}\pi)$. **(b)** Draw the intersection of the surface with the plane $y = \frac{3}{4}\pi$ and its tangent line whose slope is $\frac{\partial p}{\partial x}(0, \frac{3}{4}\pi)$.

♦ Type 2, drawing vertical cross sections

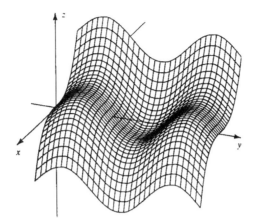

FIGURE 21

19.[A] **(a)** Generate the intersection of the graph of $f(x,y) = x^2 e^{(1-y^2)/2}$ with the plane $y = -1$. Use the window $-4 \le x \le 4, -2 \le z \le 10$. Then draw the curve with the tangent line whose slope is $\frac{\partial f}{\partial x}(2,-1)$. **(b)** Generate the intersection of the graph of $f(x,y)$ with the plane $x = 2$, using the window $-4 \le y \le 4, -2 \le z \le 10$. Then draw the curve with the tangent line whose slope is $\frac{\partial f}{\partial y}(2,-1)$.

♦ Type 2, generating vertical cross sections and tangent lines with a calculator or computer

20. **(a)** Draw the intersection of the graph of $g(x,y) = x^4 - xy^2 + y^3$ with the plane $y = -1$ and the tangent line whose slope is $\frac{\partial g}{\partial x}(1,-1)$. **(b)** Draw the intersection of the graph of $g(x,y)$ with the plane $x = 1$ with the tangent line whose slope is $\frac{\partial g}{\partial y}(1,-1)$.

♦ Type 2, drawing vertical cross sections and tangent lines

21. The volume of water (liters) in the body of a person who weighs w kilograms and is h centimeters high can be predicted with the formula $V(w,h) = 0.135 w^{2/3} h^{1/2}$.[13] Suppose that a man is 169 centimeters high and that at his current weight, the volume of water in his body would increase by approximately $0.03 \Delta w$ if his weight increased by a small amount Δw. Based on the formula, how much does he weigh?

♦ Type 4, interpreting a partial derivative

22.[O] Wool contains 17.5% water (measured by weight) when the ambient temperature is 96°F and the relative humidity is 75%. Moreover, at this temperature and humidity the rates of change with respect of the percent of water in wool with respect to the temperature and relative humidity are -0.06 percent per degree and 0.025 percent per percent, respectively.[14] **(a)** What is the approximate percent of water in wool when the temperature is 96°F and the relative humidity is 73%? **(b)** What is the approximate percent of water in wool when the temperature is 98°F and the relative humidity is 75%? **(c)** At what rate is the percent of water in the wool increasing or decreasing if the ambient temperature is 96°F and is increasing 3 degrees per hour while the relative humidity is 75% and is increasing 0.4% per hour?

♦ Type 5, application of the two-variable Chain Rule

---

[13] Data adapted from *Report of the Task Group on Reference Man*, International Commission on Radiological Protection, TarryTown, NY: Elsevier Science, Inc., 1975, p. 28.

[14] Data adapted from *Handbook of Engineering Materials*, Ibid. p. 3-167.

23. At a temperature of 40°C and pressure of 2 atmospheres, the density of hydrogen gas is 1.5×10^{-4} grams per cubic centimter, the rate of change of the density with respect to temperature is -5×10^{-7} grams per cubic centimeter per degree, and the rate of change of the density with respect to atmospheric is 7.7×10^{-5} grams per cubic centimeter per atmosphere.[15] **(a)** What is the approximate density of hydrogen when the temperature is 50°C and the pressure is 2 atmospheres? **(b)** What is the approximate density of hydrogen when the temperature is 40°F and the pressure is 1.8 atmospheres? **(c)** At what rate is the density of hydrogen increasing or decreasing if the temperature is 40°C and is decreasing 0.2 degrees per hour while the pressure is 2 atmospheres and is decreasing 0.06 atmospheres per hour?

♦ Type 5, application of the two-variable Chain Rule

24. A company sells two products, A and B. When it has sold 2000 units of product A and 3000 units of product B, it earns $3 per unit profit on the sale of item A and $4.50 per unit profit on the sale of item B. At what rate is its total profit from the two items increasing at a time when it has sold 2000 units of product A and 3000 units of product B and it is selling item A at the rate of 200 units per month and item B at the rate of 250 units per month?

♦ Type 5, application of the two-variable Chain Rule

25. **(a^A)** Express $F(t) = f(x(t), y(t))$ by a formula in terms of t for $f(x, y) = x\sin(xy), x(t) = t^5$, and $y(t) = t^{-3}$. **(b^A)** Express $P(u,v) = f(x(u,v), y(u,v))$ by a formula in terms of u and v where $f(x,y) = x\ln(xy+1), x(u,v) = ue^v$, and $y(u,v) = u^2 v^3$. **(c)** Express $Q(u,v) = f(x(u,v), y(u,v))$ by a formula in terms of u and v where $f(x,y) = \sqrt{x^2 + y^2}, x(u,v) = u^2 v^2$, and $y(u,v) = 2u - 3v$.

♦ Type 1, simplifying formulas for composite functions

26.º Find the values of **(a)** $g(2)$ and **(b)** $\dfrac{dg}{dt}(2)$ for $g(t) = f(t^3, t^4)$, where $f(8, 16) = 3, \dfrac{\partial f}{\partial x}(8, 16) = 5$, and $\dfrac{\partial f}{\partial y}(8, 16) = -7$.

♦ Type 5, a two-variable Chain Rule

27.^A What are **(a)** $h(3)$ and **(b)** $\dfrac{dh}{dt}(3)$ for $h(t) = g(t^3 - 5t, 11t - 1)$ if $g(12, 32) = 0, g_x(12, 32) = -3$, and $g_y(12, 32) = 2$?

♦ Type 5, a two-variable Chain Rule

28. What are **(a)** $K(0)$ and **(b)** $K'(0)$ for $K(t) = L(\sin t, \cos t)$ if $L(0, 1) = 50, L_x(0, 1) = 10$, and $L_y(0, 1) = -7$?

♦ Type 5, a two-variable Chain Rule

29. Find **(a)** $Z(0)$ and **(b)** $Z'(0)$ for $Z(t) = W(\ln(1+t), e^{5t})$, where $W(0, 1) = -3, W_x(0, 1) = 2$, and $W_y(0, 1) = 3$.

♦ Type 5, a two-variable Chain Rule

30.º What are **(a)** $\dfrac{\partial T}{\partial x}(1, 2)$ and **(b)** $\dfrac{\partial T}{\partial y}(1, 2)$ for $T(x, y) = U(v(x, y))$ if $v(1, 2) = 3, v_x(1, 2) = 5$, $v_y(1, 2) = 7$, and $U'(3) = 9$?

♦ Type 5, a two-variable Chain Rule

31.^A Find **(a)** $J_x(1, 1)$ and **(b)** $J_y(1, 1)$ for $J(x, y) = \sqrt{K(x, y)}$, where $K(1, 1) = 9, K_x(1, 1) = 25$, and $K_y(1, 1) = 16$.

♦ Type 3, a one-variable Chain Rule used with two variables

32. Calculate **(a)** $W_u(5, 10)$ and **(b)** $W_v(5, 10)$ for $W(u, v) = u\ln[H(u, v)]$, where $H(5, 10) = e$, $H_u(5, 10) = 6$, and $H_v(5, 10) = -7$.

♦ Type 5, a two-variable Chain Rule with the Product Rule

33.º Give the values of **(a)** $F(0, 2)$, **(b)** $F_u(0, 2)$ and **(c)** $F_v(0, 2)$ where $F(u, v) = f(v\sin u, u\sin v)$ with $f(0, 0) = 4, f_x(0, 0) = 10$, and $f_y(0, 0) = 2$.

♦ Type 5, a two-variable Chain Rule

34.^A What are **(a)** $G(1, 2)$, **(b)** $G_u(1, 2)$ and **(c)** $G_v(1, 2)$ where $G(u, v) = g(u^2 v^3, 3u + 7v)$ with $g(8, 17) = 10, g_x(8, 17) = 5$, and $g_y(8, 17) = -1$.

♦ Type 5, a two-variable Chain Rule

---

[15] Data adapted from *CRC Handbook of Tables for Applied Engineering Science*, 2nd Edition, R. Bolz and G. Tuve, editors, Boca Raton, FL: CRC Press, 1973, . pp. 65–67.

11.2 Partial derivatives with two variables

35.[O] Find **(a)** $\dfrac{\partial}{\partial x}[P(q(x,y))]$ and **(b)** $\dfrac{\partial}{\partial y}[P(q(x,y))]$ at $x = 3, y = 7$ where $q(3,7) = 9, q_x(3,7) = 6, q_y(3,7) = 10$, and $P'(9) = -11$.

♦ Type 4, the one-variable Chain Rule used with two variables

36.[A] What are **(a)** $P(3,2)$, **(b)** $P_u(3,2)$, and **(c)** $P_v(3,2)$ for $P(u,v) = R(x(u,v), y(u,v))$ where $x(3,2) = 1, y(3,2) = 0, x_u(3,2) = 5, x_v(3,2) = 6, y_u(3,2) = 7, y_v(3,2) = 4, R(1,0) = 8, R_x(1,0) = 9$ and $R_y(1,0) = 10$?

♦ Type 5, a two-variable Chain Rule

37. Find **(a)** $W(1,10)$, **(b)** $W_u(11,10)$, and **(c)** $W_v(11,10)$ for $W(u,v) = Z(x(u,v), y(u,v))$ where $x(11,10) = 6, y(1,10) = 8, x_u(11,10) = 7, x_v(11,10) = 6, y_u(11,10) = 5, y_v(11,10) = 4, Z(9,8) = 3, Z_x(9,8) = 2$ and $Z_y(9,8) = 1$.

♦ Type 5, a two-variable Chain Rule

38. Find the second-order derivatives of **(a**[O]**)** $f(x,y) = x^4 y^5$, **(b**[A]**)** $k(x,y) = \ln(2x - 3y)$, **(c**[O]**)** $M(x,y) = (1 + xy)^{10}$, and **(d)** $N(x,y) = \tan^{-1}(xy)$.

♦ Type 1, higher-order derivatives

39. Find **(a**[A]**)** $\dfrac{\partial f^3}{\partial x^2 \partial y}$ for $f(x,y) = x^3 y^4$ and **(b)** $\dfrac{\partial g^3}{\partial x^2 \partial y}$ for $g(x,y) = y^3 e^{-4x}$.

♦ Type 1, higher-order derivatives

40. A farmer will earn $P(x,t)$ dollars profit if he takes x pounds of cattle to market t weeks after the market opens for the season. The total weight of his herd is a function $x = x(t)$ of the time. Express the rate of change of his profit with respect to time in terms of the derivatives of $P(x,t)$ and $x(t)$ with appropriate units.

♦ Type 6, a two-variable Chain Rule in an unusual context

41. The temperature $T(x,y)°$F at the point (x,y) on a metal plate does not change with time, and an ant crosing the plate is at $x = t^2, y = 4t + 1$ feet at time t minutes. What is the temperature at the ant's position at $t = 3$ and what is the rate of change of the temperature at the ant's position with respect to time at $t = 3$ if $T(9,13) = 50$ degrees, $T_x(9,13) = 5$ degrees per foot, and $T_y(9,13) = -1$ degree per foot.

♦ Type 5, a two-variable Chain Rule

42. A metal rod extends from $x = 0$ to $x = \pi$ on an x-axis. Its temperature at time $t \geq 0$ and at point x is $T(x,t) = \sin(x)e^{-t} + \sin(2x)e^{-4t} + \sin(4x)e^{-16t}$. **(a)** What is the temperature at the ends of the rod? **(b)** Generate together in an xy-plane the graphs of the temperature for $t = 0, t = 0.03$ and $t = 0.06$. Use the window $0 \leq x \leq \pi, -1.5 \leq y \leq 2.5$. You should obtain the curves in Figures 22 through 24 in one drawing. Based on these curves, does the temperature appear to be an increasing or a decreasing function of t where its graph as a function of x is concave up? Does the temperature appear to be an increasing or a decreasing function of t where its graph as a function of x is concave down? **(c)** Show that the temperature satisfies the HEAT-FLOW EQUATION $\dfrac{\partial T}{\partial t} = \dfrac{\partial^2 T}{\partial x^2}$. **(d)** Use the heat-flow equation to verify the predictions from part (b).

♦ Type 4, partial derivatives

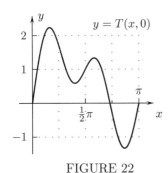

FIGURE 22

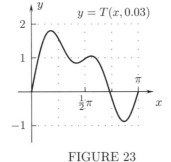

FIGURE 23

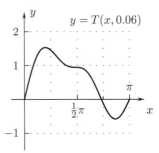

FIGURE 24

43. The ends of a vibrating violin string are fastened to the x-axis at $x = 0$ to $x = \pi$. and at time t it has the shape of the graph of $S(x,t) = \sin(3x)\cos(3t)$. **(a)** Show that the string has zero vertical velocity at $t = 0$ **(b)** Generate together in an xy-plane the graphs of the shape of the string for $t = 0, t = 0.2$ and $t = 0.4$. Use the window $0 \leq x \leq \pi, -1.5 \leq y \leq 1.5$. You should obtain the curves in Figures 24 through 26 in one drawing. Based on these curves, does the string appear to be accelerating upward or downward where it is concave up? Does the string appear to be accelerating upward or downward where it is concave down? **(c)** Show that the function $S(x,t)$ satisfies the WAVE EQUATION $\dfrac{\partial^2 S}{\partial^2 t} = \dfrac{\partial^2 S}{\partial x^2}$. **(d)** Use the wave equation to verify the predictions from part (b). **(e)** Describe the motion of the string as t increases.

♦ Type 4, partial derivatives

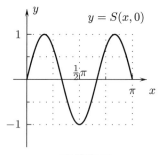
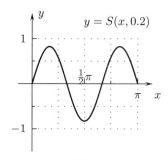
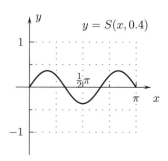

FIGURE 21 FIGURE 22 FIGURE 23

44. (EQUALITY OF MIXED PARTIAL DERIVATIVES) Show that $f_{xy}(x_0, y_0) = f_{yx}(x_0, y_0)$ for any function $f(x,y)$ that has continuous second-order derivatives in an open disk centered at (x_0, y_0) by the following argument: Use the Mean-Value Theorem for functions of one variable four times to show that for any point (x, y) in the circle, the quantity $f(x, y) + f(x_0, y_0) - f(x, y_0) - f(x_0, y)$ can be expressed either in the form $f_{xy}(c_1, c_2)(x - x_0)(y - y_0)$ or $f_{yx}(c_3, c_4)(x - x_0)(y - y_0)$, where the points (c_1, c_2) and (c_3, c_4) are in a rectangle with (x_0, y_0) and (x, y) as opposite vertices. This shows that $f_{xy}(c_1, c_2) = f_{yx}(c_3, c_4)$. Complete the derivation by having (x, y) tend to (x_0, y_0).

♦ Type 7, a proof

Section 11.3

Directional derivatives and tangent planes

OVERVIEW: *The partial derivatives $f_x(x_0, y_0)$ and $f_y(x_0, y_0)$ are the rates of change of $f(x,y)$ at (x_0, y_0) in the positive x- and y-directions. Rates of change in other directions are given by* DIRECTIONAL DERIVATIVES. *We open this section by defining directional derivatives and then use the Chain Rule from the last section to derive a formula for their values in terms of partial derivatives. Then we study* LINEAR FUNCTIONS *of two variables and use a formula for the graph of a linear function to obtain an equation of a* TANGENT PLANE *to the graph of a general function of two variables. Next, we study* GRADIENT VECTORS, *which are used to find maximum and minimum directional derivatives, and at the end of the section we discuss* DIFFERENTIALS *and their use in estimating errors.*

Topics:

- **Directional derivatives**
- **Linear functions**
- **Tangent planes**
- **The gradient vector**
- **Differentials and error estimation**

Directional derivatives

To find the derivative of $f(x,y)$ at (x_0, y_0) in the direction of the unit vector $\mathbf{u} = \langle u_1, u_2 \rangle$ in the xy-plane, we introduce an s-axis with its origin at (x_0, y_0) and its positive direction in the direction of $\mathbf{u}$ as in Figure 1. Then the point at s on the s-axis has xy-coordinates $(x_0 + su_1, y_0 + su_2)$, and the value of $f(x,y)$ at the point s on the s-axis is

$$F(s) = f(x_0 + su_1, y_0 + su_2). \tag{1}$$

We call $F(s)$ the CROSS SECTION through (x_0, y_0) of $f(x,y)$ in the direction of $\mathbf{u}$.

The graph of $F(s)$ is the intersection of the surface $f(x,y)$ with the vertical plane through the s-axis in the direction of $\mathbf{u}$ (Figure 2). We introduce a vertical w-axis with its origin at the point (x_0, y_0) (the origin on the s-axis) as in Figure 2. Then $w = F(s)$ is a curve in the sw-plane.

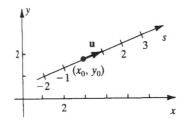

$$x = x_0 + su_1$$
$$y = y_0 + su_2$$

FIGURE 1

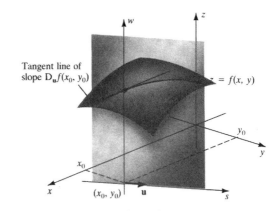

Tangent line of slope $D_{\mathbf{u}}f(x_0, y_0)$

FIGURE 2

♦ SUGGESTIONS TO INSTRUCTORS: The main topics in this section can be divided into two parts to be covered in two or more class meetings.
Part 1 (Directional derivatives, linear functions, and tangent planes; Tune-up Exercises T1–T6; Problems 1–24) Have students begin work on Questions 1 through 8 before class. Problems 1, 5, 7, 13, 14, 15, 16 are good for in-class discussions.
Part 2 (The gradient vector and differentials; Tune-up Exercises T7–T11; Problems 25-49) Have students begin Question 11 and Tune-up Exercises T7, T8, and T9 before class. Problems 25b, 26, 27, 34, 41 are good for in-class discussions.

The derivative $F'(0)$ of $F(s)$ at $s = 0$ is the slope of the tangent line to this curve in the positive s-direction at the point $(x_0, y_0, f(x_0, y_0))$. We define this slope to be the derivative of $f(x, y)$ at (x_0, y_0) in the direction of **u** and denote it by the symbol $D_{\mathbf{u}} f(x_0, y_0)$.

Example 1 Figure 3 shows level curves of $f(x, y) = 4xy - \frac{1}{4}x^4 - \frac{1}{4}y^4$ and an s-axis with its origin at the point $(1, 1)$ and its positive side pointing in the direction of the unit vector $\mathbf{u} = \langle 1/\sqrt{2}, 1/\sqrt{2} \rangle$ at an angle of $\frac{1}{4}\pi$ radians from the positive x-direction. **(a)** Give a formula in terms of s for the cross section $F(s)$ of $f(x, y)$ determined by this s-axis and generate it in a sw-plane using the window $-9 \leq s \leq 5, -5 \leq w \leq 20$. **(b)** Use the formula from part (a) to find the derivative $D_{\mathbf{u}} f(1, 1)$ of $f(x, y)$ at $(1, 1)$ in the direction of **u**.

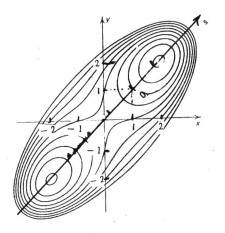
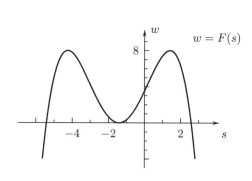

Level curves of $f(x, y)$

FIGURE 3

The graph of $w = F(s)$

FIGURE 4

SOLUTION **(a)** By **(1)** with $(x_0, y_0) = (1, 1)$ and $\langle u_1, u_2 \rangle = \langle 1/\sqrt{2}, 1/\sqrt{2} \rangle$, the s-axis with $s = 0$ at the point $(1, 1)$ has the parametric equations

$$x = 1 + \frac{s}{\sqrt{2}}, y = 1 + \frac{s}{\sqrt{2}}. \tag{2}$$

Substituting formulas **(2)** for x and y in the formula $f(x, y) = 4xy - \frac{1}{4}x^4 - \frac{1}{4}y^4$ gives

$$F(s) = 4\left(1 + \frac{s}{\sqrt{2}}\right)\left(1 + \frac{s}{\sqrt{2}}\right) - \frac{1}{4}\left(1 + \frac{s}{\sqrt{2}}\right)^4 - \frac{1}{4}\left(1 + \frac{s}{\sqrt{2}}\right)^4$$

$$= 4\left(1 + \frac{s}{\sqrt{2}}\right)^2 - \frac{1}{2}\left(1 + \frac{s}{\sqrt{2}}\right)^4.$$

The graph of this function is shown in the sw-plane of Figure 4.

(b) $D_{\mathbf{u}} f(1, 1)$ is the derivative of $F(s)$ from part (a) with respect to s at $s = 0$. The Chain Rule for functions of one variable gives

$$F'(s) = \frac{d}{ds}\left[4\left(1 + \frac{s}{\sqrt{2}}\right)^2 - \frac{1}{2}\left(1 + \frac{s}{\sqrt{2}}\right)^4\right]$$

$$= 8\left(1 + \frac{s}{\sqrt{2}}\right)\frac{d}{ds}\left(1 + \frac{s}{\sqrt{2}}\right) - 2\left(1 + \frac{s}{\sqrt{2}}\right)^3 \frac{d}{ds}\left(1 + \frac{s}{\sqrt{2}}\right)$$

$$= \frac{8}{\sqrt{2}}\left(1 + \frac{s}{\sqrt{2}}\right) - \frac{2}{\sqrt{2}}\left(1 + \frac{s}{\sqrt{2}}\right)^3.$$

We set $s = 0$ to determine that $D_{\mathbf{u}} f(1, 1) = F'(0) = \dfrac{8}{\sqrt{2}} - \dfrac{2}{\sqrt{2}} = \dfrac{6}{\sqrt{2}} = 3\sqrt{2}.$ □

11.3 Directional derivatives and tangent planes

Figure 5 shows the tangent line of slope $3\sqrt{2}$ at $s = 0$ to the graph of $F(s)$ from Example 1, and Figure 6 shows graph of $f(x,y)$ and its intersection with the vertical plane $y = x$ which has the shape of the graph of the function $F(s)$. The point labeled P in Figure 6 is the point on the surface at $x = 1, y = 1$ and the directional derivative $D_{\mathbf{u}}f(1,1) = 3\sqrt{2}$ is the slope of the cross section at that point.

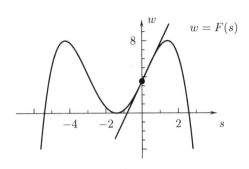
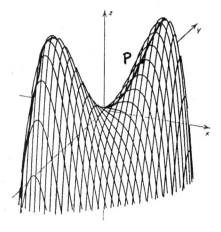

FIGURE 5 　　　　　　　　　　　　　　　FIGURE 6

Question 1 Figure 7 shows the graph of the function $g(x,y) = e^y - 1$ and its intersection with the vertical plane through the origin in the direction of the unit vector $\mathbf{u} = \langle \frac{1}{2}, \frac{1}{2}\sqrt{3} \rangle$ at an angle $\frac{1}{3}\pi$ from the positive x-axis in the xy-plane. The axes are oriented with the x-axis pointing to the right and the y-axis points back in Figure 7 to improve the drawing of the surface. **(a)** Find a formula for the cross section $F(s)$ and generate its graph in an sw-plane with the window $-3 \leq s \leq 2, -3 \leq w \leq 5$. **(b)** Use the formula from part (a) to find the directional derivative $D_{\mathbf{u}}g(0,0) = F'(0)$ and draw the tangent line with this slope with the graph of $F(s)$.

FIGURE 7
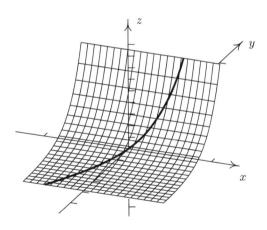

Finding directional derivatives from partial derivatives

Instead of determining directional derivatives of $f(x,y)$ by finding formulas for the cross sections $F(s)$ as we did in Example 1 and Question 1 above, we can use a general formula for the directional derivative $D_{\mathbf{u}}f(x_0, y_0)$ in terms of the partial derivatives $\dfrac{\partial f}{\partial x}(x_0, y_0)$ and $\dfrac{\partial f}{\partial y}(x_0, y_0)$ and the components of $\mathbf{u} = \langle u_1, u_2 \rangle$. We will obtain the necessary formula by using the Chain Rule with two variables from the last section.

We assume that $f(x, y)$ has continuous first-order derivatives in an open disk centered at (x_0, y_0) so that the Chain Rule can be applied to $F(s)$ defined by **(1)**. We obtain

$$F'(s) = \frac{d}{dx}[f(x_0 + u_1 s, y_0 + u_2 s)]$$
$$= \frac{\partial f}{\partial x}(x_0 + u_1 s, y_0 + u_2 s)\frac{d}{ds}(x_0 + u_1 s) + \frac{\partial f}{\partial y}(x_0 + u_1 s, y_0 + u_2 s)\frac{d}{ds}(y_0 + u_2 s)$$
$$= \frac{\partial f}{\partial x}(x_0 + u_1 s, y_0 + u_2 s)u_1 + \frac{\partial f}{\partial y}(x_0 + u_1 s, y_0 + u_2 s)u_2.$$

Then setting $s = 0$ gives

$$F'(0) = \frac{\partial f}{\partial x}(x_0, y_0)u_1 + \frac{\partial f}{\partial y}(x_0, y_0)u_2.$$

Since $F'(0)$ is the directional derivative of f, we obtain the following result:

Theorem 1 *Suppose that $f(x,y)$ has continuous first-order partial derivatives in an open disk centered at (x_0, y_0). Then for any unit vector $\mathbf{u} = \langle u_1, u_2 \rangle$, the (directional) derivative of f at (x_0, y_0) in the direction of $\mathbf{u}$ is*

$$D_{\mathbf{u}}f(x_0, y_0) = \frac{\partial f}{\partial x}(x_0, y_0)u_1 + \frac{\partial f}{\partial y}(x_0, y_0)u_2. \tag{3}$$

Remember formula **(3)** in words: The directional derivative of f in the direction of $\mathbf{u}$ equals the x-derivative of f multiplied by the x-component of $\mathbf{u}$, plus the y-derivative of f multiplied by the y-component of $\mathbf{u}$.

Example 2 What is the derivative of $x^2 y^5$ at $(3, 1)$ in the direction toward $(4, -3)$?

SOLUTION For $f = x^2 y^5$, we have $f_x = 2xy^5$ and $f_y = 5x^2 y^4$, so that $f_x(3,1) = 6$, and $f_y(3,1) = 45$.

To find the unit vector in the direction from $P = (3,1)$ to $Q = (4,-3)$, we first find the displacement vector $\overrightarrow{PQ} = \langle 4-3, -3-1 \rangle = \langle 1, -4 \rangle$ from P to Q and calculate its length $|\overrightarrow{PQ}| = \sqrt{1^2 + (-4)^2} = \sqrt{17}$. Then

$$\mathbf{u} = \langle u_1, u_2 \rangle = \frac{\overrightarrow{PQ}}{|\overrightarrow{PQ}|} = \frac{\langle 1, -4 \rangle}{\sqrt{17}}.$$

Therefore, $u_1 = \dfrac{1}{\sqrt{17}}$ and $u_2 = \dfrac{-4}{\sqrt{17}}$, so that

$$D_{\mathbf{u}}f(3,1) = f_x(3,1)u_1 + f_y(3,1)u_2$$
$$= 6\left(\frac{1}{\sqrt{17}}\right) + 45\left(\frac{-4}{\sqrt{17}}\right) = -\frac{174}{\sqrt{17}}. \ \square$$

Question 2 What is the derivative of $x^2 y^5$ at $(3,1)$ in the direction of the unit vector $\mathbf{u} = \langle \frac{3}{5}, \frac{4}{5} \rangle$?

Question 3 How are the directional derivatives **(a)** in the positive x-direction, **(b)** in the positive y-direction, **(c)** in the negative x-direction, and **(d)** in the negative y-direction related to the x- and y-derivatives?

11.3 Directional derivatives and tangent planes

Using angles of inclination
If the direction of a directional derivative is described by giving the angle θ of inclination of the unit vector **u**, then we can use the expression

$$\mathbf{u} = \langle \cos\theta, \sin\theta \rangle \tag{4}$$

for **u** in terms of θ to calculate the directional derivative (Figure 8).

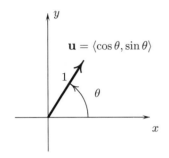
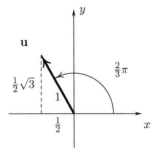

FIGURE 8   FIGURE 9

Example 3   What is the derivative of $h(x,y) = e^{xy}$ at $(2,3)$ in the direction at an angle of $\tfrac{2}{3}\pi$ from the positive x-direction?

SOLUTION   The partial derivatives are $\dfrac{\partial h}{\partial x} = e^{xy}\dfrac{\partial}{\partial x}(xy) = ye^{xy}$ and $\dfrac{\partial h}{\partial y} = e^{xy}\dfrac{\partial}{\partial y}(xy) = xe^{xy}$, whose values at $(2,3)$ are $\dfrac{\partial h}{\partial x}(2,3) = 3e^6$ and $\dfrac{\partial h}{\partial y}(2,3) = 2e^6$.

The unit vector **u** with angle of inclination $\tfrac{2}{3}\pi$ forms the hypotenuse of the $30°$-$60°$-right triangle in Figure 9 whose base is $\tfrac{1}{2}$ and height is $\tfrac{1}{2}\sqrt{3}$. Therefore, $\mathbf{u} = \langle u_1, u_2 \rangle$ with $u_1 = -\tfrac{1}{2}$ and $u_2 = \tfrac{1}{2}\sqrt{3}$, and

$$D_\mathbf{u} h(2,3) = f_x(2,3)u_1 + f_y(2,3)u_2$$
$$= 3e^6\left(-\tfrac{1}{2}\right) + 2e^6\left(\tfrac{1}{2}\sqrt{3}\right) = \left(-\tfrac{3}{2} + \sqrt{3}\right)e^6. \; \square$$

Finding approximate directional derivatives from level curves
We can estimate directional derivatives from level curves just as we estimated x- and y-derivatives from level curves in the last section.

Example 4   Figure 10 shows level curves of a temperature reading $T(x,y)°C$ of the surface of the ocean off the west coast of the United States.[1] **(a)** Express the rate of change toward the northeast of the temperature at point P in the drawing as a directional derivative. **(b)** Find the approximate value of this rate of change.

SOLUTION   **(a)** If we suppose that the point P has coordinates $(1240, 1000)$ and that the unit vector pointing toward the northeast is **u**, then the rate of change of the temperature toward the northeast at P is $D_\mathbf{u} T(1240, 1000)$.

(b) We draw an s-axis in the direction of **u** with its origin at P and with the same units as used on the x- and y-axes (Figure 11). This axis crosses the level curve $T = 18°C$ at a point just below P and crosses the level curve $T = 17°C$ at a point just above it. The change in the temperature from the lower to the upper point is $\Delta T = 17° - 18° = -1°$, and s increases by approximately $\Delta s = 200$ miles from the

---

[1] Data adapted from *Zoogeography of the Sea* by S. Elkman, London: Sidgwich and Jackson, 1953, p. 144.

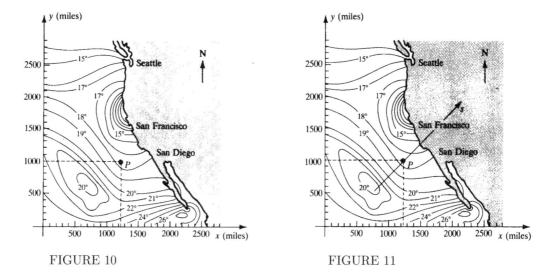

FIGURE 10   FIGURE 11

lower point to the upper point. Consequently, the rate of change of T at P in the direction of the positive s-axis is approximately $\dfrac{\Delta T}{\Delta s} = \dfrac{-1}{200} = -0.005$ degrees per mile. □

Question 4   What is the approximate rate of change of the temperature toward the southwest at P in Figure 10?

Linear functions of two variables

Recall that a function $F(x)$ of one variable is linear if its graph in an xy-plane is a line and that in this case its derivative is constant and equals the slope of the line. In studying such functions in earlier chapters, we frequently used either the slope-intercept equation

$$y = mx + b \tag{5}$$

for the line, where m is the slope m and b the y-intercept of the line (Figure 12), or the point-slope equation

$$y = y_0 + m(x - x_0) \tag{6}$$

where m is the slope and (x_0, y_0) a point on the line (Figure 13).

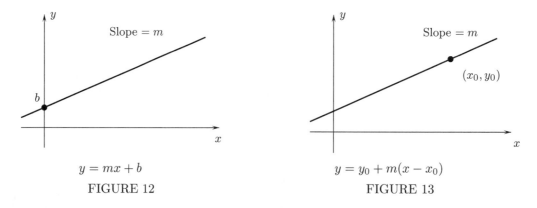

$y = mx + b$ $y = y_0 + m(x - x_0)$
FIGURE 12   FIGURE 13

A function $f(x, y)$ of two variables is LINEAR if its graph in xyz-space is a plane. We found equations of planes in Section 10.3 in terms of their normal vectors. Here we will need equations for planes in terms of the slopes of their cross sections in the x- and y-directions.

11.3 Directional derivatives and tangent planes

Suppose first that the z-intercept of a plane is b, that the slope of its vertical cross sections in the x-direction is m_1, and that the slope of its vertical cross sections in the y-direction is m_2. Figure 14 shows the portion of such a plane over a rectangle with one corner at the origin in the xy-plane and the opposite corner at (x, y), in a case where x, y, b, m_1, and m_2 are positive. To obtain an equation for the plane, we need to find the z-coordinate of the point R on it in terms of x and y.

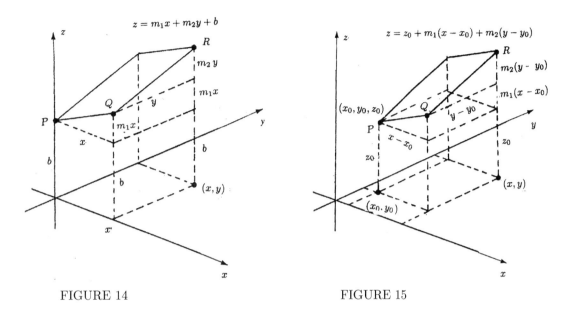

FIGURE 14   FIGURE 15

The z-coordinate of P is the z-intercept b, and because the line PQ has slope m_1, the z-coordinate of Q is $b + m_1 x$. Then, because the line QR has slope m_2, the z-coordinate of R is $b + m_1 x + m_2 y$. This gives what we will call the SLOPES-INTERCEPT EQUATION of the plane with slope m_1 in the positive x-direction, slope m_2 in the positive y-direction, and z-intercept b:

$$z = m_1 x + m_2 y + b. \tag{7}$$

Example 5 Give an equation of the plane with slope -6 in the positive x-direction, slope 7 in the positive y-direction, and z-intercept 10.

SOLUTION By **(7)** with $m_1 = -6, m_2 = 7$, and $b = 10$, the plane has the equation $z = -6x + 7y + 10$. □

The result of the next Question will enable us to give an equation of a plane in terms of the slopes m_1 and m_2 and coordinates of a point (x_0, y_0, z_0) on it:

Question 5 The point P on the plane in Figure 15 is (x_0, y_0, z_0), the vertical cross sections of the plane in the positive x-direction have slope m_1, and the vertical cross sections of the plane in the positive y-direction have slope m_2. **(a)** What is the z-coordinate of the point Q? **(b)** What is the z-coordinate of R?

The result of Question 5 gives what we will call the POINT-SLOPES EQUATION of the plane through (x_0, y_0, z_0) with slope m_1 in the positive x-direction and slope m_2 in the positive y-direction:

$$z = z_0 + m_1(x - x_0) + m_2(y - y_0). \tag{8}$$

Question 6 Give an equation of the plane through the point $(1, 2, 3)$ with slope 4 in the positive x-direction and slope -5 in the positive y-direction.

As we would expect, the x-derivative of the linear function $f(x,y) = m_1 x + m_2 y + b$ equals the constant slope m_1 of all of the vertical cross sections of its graph in the positive x-direction, and its y-derivative equals the constant slope m_2 of all of the vertical cross sections of its graph in the positive y-direction:

$$\frac{\partial f}{\partial x} = \frac{\partial}{\partial x}(m_1 x + m_2 y + b) = m_1 \quad \text{and} \quad \frac{\partial f}{\partial y} = \frac{\partial}{\partial y}(m_1 x + m_2 y + b) = m_2.$$

A similar calculaton yields the same conclusion if the function is given in the point-slopes form $f(x,y) = z_0 + m_1(x - x_0) + m_2(y - y_0)$.

Example 6   The table below gives values of a linear function $g(x,y)$. **(a)** What are its x- and y-derivatives? **(b)** What is the z-intercept of its graph? **(c)** Give a formula for $g(x,y)$ in terms of x and y.

VALUES OF $g(x,y)$

	$x = -4$	$x = -2$	$x = 0$	$x = 2$	$x = 4$
$y = 4$	0	4	8	12	16
$y = 2$	6	10	14	18	22
$y = 0$	12	16	20	24	28
$y = -2$	18	22	26	30	34
$y = -4$	24	28	32	36	40

SOLUTION   **(a)** Because g is linear, its x-derivative is constant and equals the slope of each its cross sections in the positive x-direction. Similarly, its y-derivative is constant and equals the slope of each of its cross sections in the positive y-direction. Therefore, we can use any two values in any row to calculate the x-derivative and any two values in any column to find the y-derivative.

With $g(4,4) = 16$ and $g(2,4) = 12$ we obtain

$$\frac{\partial g}{\partial x} = \frac{g(4,4) - g(2,4)}{4 - 2} = \frac{16 - 12}{2} = \tfrac{4}{2} = 2$$

Similarly, the values $g(4,4) = 16$ and $g(4,2) = 22$ give

$$\frac{\partial g}{\partial y} = \frac{g(4,4) - g(4,2)}{4 - 2} = \frac{16 - 22}{2} = -\tfrac{6}{2} = -3.$$

(b) The z-intercept of the graph of g is its value $z = 20$ at $x = 0, y = 0$.

(c) By the slopes-intercept equation (7) with $m_1 = 2, m_2 = -3$, and $b = 20$, the graph of g has the equation $z = 2x - 3y + 20$ and $g(x,y) = 2x - 3y + 20$. □

Level curves of linear functions

The level curves of a linear function $f(x,y) = m_1 x + m_2 y + b$ are horizontal cross sections of the plane $z = m_1 x + m_2 y + b$ and are parallel lines, which are equally spaced if, as in Figure 16, the values of the function that are represented are equally spaced.

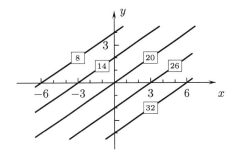

FIGURE 16

Question 7 Find a formula for the linear function $h(x,y)$ whose level curves are shown in Figure 16.

Zooming in on level curves of a nonlinear f(x, y)

Recall from Section 3.3 that if a function $f(x)$ of one variable has a derivative at x_0 and the graph $y = f(x)$ is generated by a calculator or computer in a small enough window containing the point $(x_0, f(x_0))$, the displayed portion of the graph will look like a line. This occurs because the graph is closely approximated by the tangent line at that point.

What happens if we zoom in on the graph $z = f(x,y)$ of a function of two variables? If the function has continuous first derivatives, the viewed portion of the graph will look like a plane if the viewing window is sufficiently small. This occurs because, as will see below, the surface has a TANGENT PLANE at each point which closely approximates the surface near the point of tangency.

The fact that graphs of functions of two variables with continuous first derivatives look like planes when viewed in small windows is illustrated by the level curves of $K(x,y) = 3x^2 y^3 + x$ in Figures 17 through 19. Figure 17 shows a square window $0.6 \leq x, y \leq 1.4$ of width 0.8. The function has the value 1 on the level curve at the lower left, 2 on the next level curve, 3 on the next, 4 on the level curve through $(1,1)$, up to 16 on the level curve at the upper right. This portion of the graph does not look much like a plane, as is evidenced by the curved shapes of the level curves.

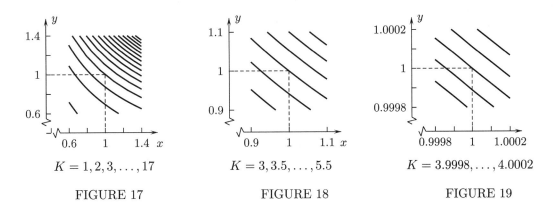

FIGURE 17   FIGURE 18   FIGURE 19

The level curves are closer to being straight lines in the square window $0.9 \leq x, y \leq 1.1$ of width 0.2 in Figure 18, and appear even more like straight lines in the square window of width 0.0004 in Figure 19. Accordingly, the portion of the graph of $K(x,y)$ corresponding to Figure 18 is close to being a plane and the portion corresponding to Figure 19 is even closer to being a plane.

Equations of tangent planes

If $f(x,y)$ has continuous first derivatives in a circle centered at (x_0, y_0), then the linear function

$$L(x,y) = f(x_0, y_0) + f_x(x_0, y_0)(x - x_0) + f_y(x_0, y_0)(y - y_0) \tag{9}$$

has the same value and the same first derivatives at (x_0, y_0) as does $f(x, y)$ and consequently, has the same directional derivatives in all directions at (x_0, y_0) as does $f(x, y)$. This implies that all of the tangent lines to vertical cross sections of the graph of f through $x = x_0, y = y_0$ (Figure 20) are in the plane that is the graph of L (Figure 21). We define that plane to be the tangent plane to the graph of $f(x, y)$ at $x = x_0, y = y_0$:

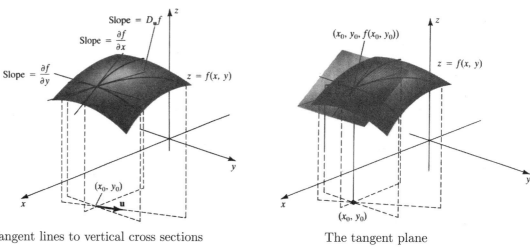

Tangent lines to vertical cross sections

FIGURE 20

The tangent plane

FIGURE 21

Definition 1 If $f(x, y)$ has continuous first-order derivatives in an open disk centered at (x_0, y_0) then the tangent plane to its graph at $x = x_0, y = y_0$ is

$$z = f(x_0, y_0) + f_x(x_0, y_0)(x - x_0) + f_y(x_0, y_0)(y - y_0). \tag{10}$$

Formula (10) can also be applied with somewhat more generality. In deriving the Chain Rule with two variables (Theorem 1) in Section 11.2, we showed that if (x, y) is in a open disk centered at (x_0, y_0) where f has continuous first derivatives, then

$$f(x, y) = f(x_0, y_0) + f_x(c_1, y_0)(x - x_0) + f_y(x, c_2)(y - y_0)$$

where (c_1, y_0) and (x, c_2) are in the rectangle with corners (x_0, y_0) and (x, y). Because f_x and f_y are continuous and $f_x(c_1, y_0)$ tends to $f_x(x_0, y_0)$ and $f_y(x, c_2)$ tends to $f_y(x_0, y_0)$ as (x, y) tends to (x_0, y_0), the last equation gives

$$\begin{aligned} f(x, y) &= f(x_0, y_0) + [f_x(x_0, y_0) + \epsilon_1](x - x_0) + [f_y(x_0, y_0) + \epsilon_2](y - y_0) \\ &= L(x, y) + \epsilon_1(x - x_0) + \epsilon_2(y - y_0) \end{aligned} \tag{11}$$

where L is the linear function (10) and ϵ_1 and ϵ_2 are functions that tend to zero as $(x, y) \to (x_0, y_0)$.

A function $f(x, y)$ that can be written in the form (11) with $L(x, y)$ linear and $\epsilon_1(x, y)$ and $\epsilon_2(x, y)$ functions that tend to zero as $(x, y) \to (x_0, y_0)$ is said to be DIFFERENTIABLE at (x_0, y_0). Formula (11) for the tangent plane can be used if $f(x, y)$ is differentiable at $x = x_0, y = y_0$, even if it does not have continuous first-order derivatives in an open disk centered at (x_0, y_0), as we required in Definition 1.

11.3 Directional derivatives and tangent planes

Example 7 (a) Give an equation of the tangent plane to the graph of $f(x,y) = x^3 y^4$ at $x=1, y=2$. (b) How closely is $f(x,y)$ approximated at $(1.01, 2.01)$ by the linear function $L(x,y)$ whose graph is the tangent plane? (c) How closely is $f(x,y)$ approximated by $L(x,y)$ at $(1.001, 2.001)$?

SOLUTION (a) We find the partial derivatives

$$\frac{\partial f}{\partial x} = \frac{\partial}{\partial x}(x^3 y^4) = 3x^2 y^4 \text{ and } \frac{\partial f}{\partial y} = \frac{\partial}{\partial y}(x^3 y^4) = 4x^3 y^3.$$

Equation (**10**) with $x_0 = 1, y_0 = 2, f(1,2) = (1)^3(2)^4 = 16, f_x(1,2) = 3(1)^2(2)^4 = 48$, and $f_y(1,2) = 4(1)^3(2)^3 = 32$ gives the equation

$$z = 16 + 48(x-1) + 32(y-2) \tag{12}$$

for the tangent plane.

(b) The linear approximation $L(x,y) = 16 + 48(x-1) + 32(y-2)$ of $f(x,y)$ near $(1,2)$ is the function whose graph is the tangent plane (**12**). The error made in using $L(x,y)$ in place of $f(x,y)$ at $(1.01, 2.01)$ is

$$|f(1.01, 2.01) - L(1.01, 2.01)|$$
$$= (1.01)^3 (2.01)^4 - [16 + 48(0.01) + 32(0.01)] \doteq 1.7 \times 10^{-2}.$$

(c) The error at $(1.001, 2.001)$ is

$$|f(1.001, 2.001) - L(1.001, 2.001)|$$
$$= (1.001)^3 (2.001)^4 - [16 + 48(0.001) + 32(0.001)] \doteq 1.7 \times 10^{-4}. \square$$

Notice that the distance from $(1.001, 2.001)$ to $(1,2)$ in part (c) of Example 7 is one-tenth the distance from $(1.01, 2.01)$ to $(1,2)$ in part (b), whereas the error in part (c) is one-hundredth the error in part (b).

Question 8 Give an equation of the tangent plane to $z = 2\sin x + 3e^y + 4$ at $x=0, y=0$.

Writing equation (**10**) in the form

$$f(x_0, y_0) + f_x(x_0, y_0)(x - x_0) + f_y(x_0, y_0)(y - y_0) - z = 0$$

where the coefficients of x, y, and z are $f_x(x_0, y_0), f_y(x_0, y_0)$, and -1, respectively, shows that the vector

$$\mathbf{n} = \langle f_x(x_0, y_0), f_y(x_0, y_0), -1 \rangle \tag{13}$$

in space is normal to the tangent plane. The vector $\mathbf{n} = \langle 48, 32, -1 \rangle$, for example, is normal to the tangent plane (**12**) of Example 7.

The gradient vector

The formula

$$D_{\mathbf{u}} f(x_0, y_0) = \frac{\partial f}{\partial x}(x_0, y_0) u_1 + \frac{\partial f}{\partial y}(x_0, y_0) u_2 \tag{14}$$

from Theorem 1 for the derivative of f at (x_0, y_0) in the direction of the unit vector $\mathbf{u} = \langle u_1, u_2 \rangle$ has the form of the dot product of $\mathbf{u}$ with the vector $\langle f_x, f_y \rangle$ at (x_0, y_0). This leads us to define the latter vector to be the GRADIENT VECTOR of f, which is denoted ∇f.[†]

[†]The symbol ∇ is called "nabla" or "del."

Definition 2 The gradient vector of $f(x,y)$ at (x_0, y_0) is

$$\nabla f(x_0, y_0) = \langle f_x(x_0, y_0), f_y(x_0, y_0)\rangle. \tag{15}$$

The gradient vector **(15)** is drawn as an arrow with its base at (x_0, y_0). Because its length is a derivative (a rate of change), its length can be measured with any convenient scale. We will, however, use the scales on the coordinate axes whenever possible.

Example 8 Draw $\nabla f(1,1), \nabla f(-1,2)$, and $\nabla f(-2,-1)$ for $f(x,y) = x^2 y$. Use the scale on the x- and y-axes to measure the lengths of the arrows.

SOLUTION We calculate $\nabla f = \left\langle \dfrac{\partial}{\partial x}(x^2 y), \dfrac{\partial}{\partial y}(x^2 y) \right\rangle = \langle 2xy, x^2 \rangle$, and then $\nabla f(1,1) = \langle 2, 1\rangle, \nabla f(-1, 2) = \langle -4, 1\rangle$, and $\nabla f(-2,-1) = \langle 4, 4\rangle$. These vectors are drawn in Figure 22. □

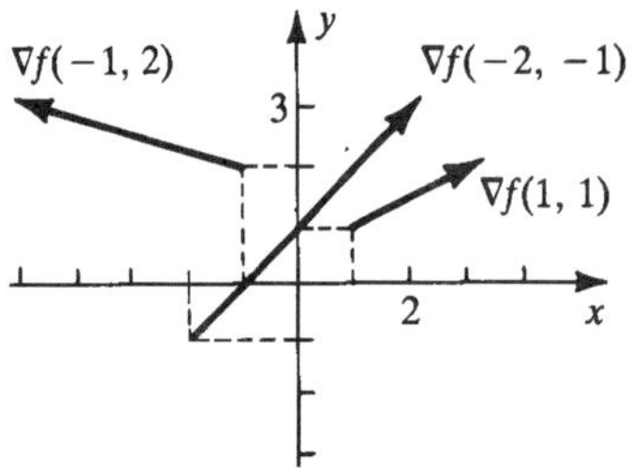

FIGURE 22

With Definition 2, formula **(3)** for the directional derivative becomes

$$D_{\mathbf{u}} f(x_0, y_0) = \nabla f(x_0, y_0) \cdot \mathbf{u}. \tag{16}$$

This representation is useful because we know from Section 10.1 that the dot product $\mathbf{A} \cdot \mathbf{B}$ of two nonzero vectors equals the product $|\mathbf{A}||\mathbf{B}|\cos\theta$ of their lengths and the cosine of an angle between them. Because $|\mathbf{u}| = 1$, we obtain the following theorem:

Theorem 2 If $f(x, y)$ has continuous first-order derivatives in an open disk centered at (x_0, y_0) and $\nabla f(x_0, y_0)$ is not the zero vector, then for any unit vector $\mathbf{u}$,

$$D_{\mathbf{u}} f(x_0, y_0) = |\nabla f(x_0, y_0)| \cos\theta \tag{17}$$

where θ is an angle between ∇f and $\mathbf{u}$ (Figure 23). If $\nabla f(x_0, y_0)$ is the zero vector, then $D_{\mathbf{u}}(x_0, y_0) = 0$ for all unit vectors $\mathbf{u}$.

FIGURE 23

11.3 Directional derivatives and tangent planes

Look closely at formula **(17)**. If the point (x_0, y_0) is fixed, then $|\nabla f(x_0, y_0)|$ is a positive constant and as θ varies, $\cos \theta$ varies between 1 when $\nabla f(x_0, y_0)$ and **u** have the same direction and -1 when $\nabla f(x_0, y_0)$ and **u** have opposite directions and θ is a straight angle. Also $\cos \theta$ is zero when $\nabla f(x_0, y_0)$ and **u** are perpendicular and θ is a right angle. This establishes the next result.

Theorem 3 Suppose that $f(x, y)$ has continuous first-order derivatives in an open disk containing (x_0, y_0) and that $\nabla f(x_0, y_0)$ is not zero. Then **(a)** the maximum directional derivative of f at (x_0, y_0) is $|\nabla f(x_0, y_0)|$ and occurs for **u** with the same direction as $\nabla f(x_0, y_0)$, **(b)** the minimum directional derivative of f at (x_0, y_0) is $-|\nabla f(x_0, y_0)|$ and occurs for **u** with the opposite direction as $\nabla f(x_0, y_0)$, and **(c)** the directional derivative of f at (x_0, y_0) is zero for **u** with either of the two directions perpendicular to $\nabla f(x_0, y_0)$.

Example 9 What is the maximum directional derivative of $h(x, y) = y^2 e^{2x}$ at $(2, -1)$ and in the direction of what unit vector does it occur?

SOLUTION Because $\nabla h = \left\langle \dfrac{\partial}{\partial x}(y^2 e^{2x}), \dfrac{\partial}{\partial y}(y^2 e^{2x}) \right\rangle = \langle 2y^2 e^{2x}, 2y e^{2x} \rangle$, we have $\nabla h(2, -1) = \langle 2e^4, -2e^4 \rangle$. The maximum directional derivative is $|\nabla h(2, -1)| = |\langle 2e^4, -2e^4 \rangle| = \sqrt{(2e^4)^2 + (-2e^4)} = \sqrt{8}\, e^4$. It occurs in the direction of the unit vector

$$\mathbf{u} = \frac{\nabla h(2, -1)}{|\nabla h(2, -1)|} = \frac{\langle 2e^4, -2e^4 \rangle}{|\langle 2e^4, -2e^4 \rangle|} = \frac{\langle 1, -1 \rangle}{\sqrt{2}}. \quad \square$$

Question 9 What is the minimum directional derivative at $(2, -1)$ of the function $h(x, y)$ from Example 9 and what is the unit vector in that direction?

Example 10 Give the two unit vectors **u** such that the function $h(x, y)$ of Example 9 has zero derivatives at $(2, -1)$ in the direction of **u**.

SOLUTION The derivative is zero in the two directions perpendicular to the unit vector $\dfrac{\langle 1, -1 \rangle}{\sqrt{2}}$ with the direction of the gradient. Interchanging the components and multiplying one or the other of the components by -1 gives the perpendicular unit vectors. The directional derivative is zero in the directions of $\mathbf{u} = \dfrac{\langle -1, -1 \rangle}{\sqrt{2}}$ and $\mathbf{u} = \dfrac{\langle 1, 1 \rangle}{\sqrt{2}}$. $\square$

Gradient vectors and level curves

Not all level curves of function $f(x, y)$ with continuous first derivatives are what we would normally call a "curve." The level curve $x^2 + y^2 = 0$ of $f = x^2 + y^2$, for example, consists of the single point $(0, 0)$ in Figure 24. Also, even in cases where the level curve is what we normally would call a "curve," it may not have tangent lines at all points. The level curve $x^2 - y^3 = 0$ of $g = x^2 - y^3$, for example, is the curve $y = x^{2/3}$ in Figure 25, which has a cusp and no tangent line at the origin. Such examples share a common property, however, which is explored in the next Question.

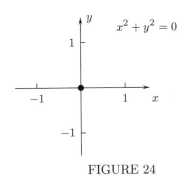

FIGURE 24

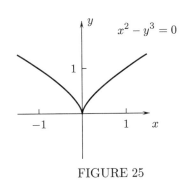

FIGURE 25

Question 10 Find the gradient vectors **(a)** of $f = x^2 + y^2$ and **(b)** of $g = x^2 - y^3$ at $(0,0)$.

The results of Question 10 serve as background for the next theorem, which is established in advanced calculus classes.

Theorem 4 (The Implicit-Function Theorem) If $f(x,y)$ has continuous first-order derivatives in an open disk centered at (x_0, y_0) and $\nabla f(x_0, y_0)$ is not zero, then the portion of the level curve of f through (x_0, y_0) in a (possibly smaller) open disk consists of a parametrized curve with a nonzero velocity vector and therefore a tangent line at (x_0, y_0).

Theorem 4 cannot be applied at the origin to the level curves of Figures 24 and 25 because, as you saw in Question 10, the gradients are zero at that point.

Suppose that the conditions of Theorem 4 are satisfied and the parametric equations of the level curve $f = c$ are $x = x(t), y = y(t)$ with $x(t_0) = x_0$ and $y(t_0) = y_0$. Then for t near t_0 the composite function $f(x(t), y(t))$ has the constant value c, and its derivative

$$\frac{d}{dt}[f(x(t), y(t))] = f_x(x(t), y(t))\frac{dx}{dt}(t) + f_y(x(t), y(t))\frac{dy}{dt}(t)$$

is zero. Setting $t = t_0$ gives

$$0 = f_x(x_0, y_0)\frac{dx}{dt}(t_0) + f_y(x_0, y)\frac{dy}{dt}(t_0) = \nabla f(x_0, y_0) \cdot \langle \frac{dx}{dt}(t_0), \frac{dy}{dt}(t_0) \rangle.$$

Since $\langle \frac{dx}{dt}(t_0), \frac{dy}{dt}(t_0) \rangle$ is a nonzero velocity vector to the level curve and is tangent to it, $\nabla f(x_0, y_0)$ is perpendicular to the level curve and we have the next result.

Theorem 5 If $f(x,y)$ has continuous first derivatives in an open disk centered at (x_0, y_0) and $\nabla f(x_0, y_0)$ is not zero, then $\nabla f(x_0, y_0)$ is perpendicular to the level curve of f through (x_0, y_0) (Figure 26).

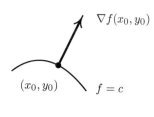

FIGURE 26

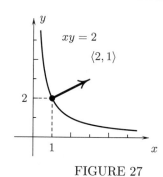

FIGURE 27

Example 11 Draw the gradient of xy at $(1,2)$ and the level curve of xy through that point.

SOLUTION Because $\nabla(xy) = \langle \frac{\partial}{\partial x}(xy), \frac{\partial}{\partial y}(xy) \rangle = \langle y, x \rangle$, the gradient at $(1,2)$ is $\langle 2, 1 \rangle$. Also, because $xy = 2$ at $(1,2)$, the level curve is $xy = 2$ or $y = 2/x$ (Figure 27). Notice that the gradient vector is perpendicular to the level curve. □

Question 11 Draw $\nabla f(-3, 1)$ and the level curve of f through $(-3, 1)$ for $f(x,y) = xy$.

Estimating gradient vectors from level curves

To estimate the gradient of a function from its level curves, we could estimate the x- and y-derivatives. Here we want to emphasize the gradient vector's geometric properties, so we will instead estimate the length and direction of the gradient vector directly from the level curves.

Example 12 Level curves of a function f are shown in Figure 28. Find the approximate length and direction of $\nabla f(3, 2)$ and then draw it with the level curves, using the scales on the axes to measure the length of the arrow.

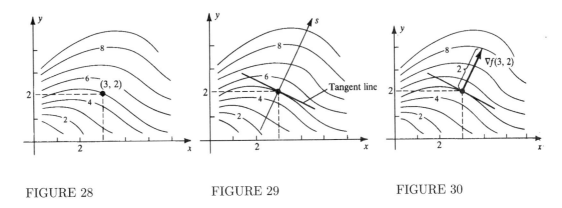

FIGURE 28   FIGURE 29   FIGURE 30

SOLUTION   We draw an approximate tangent line at $(3, 2)$ to the level curve of f through that point and a perpendicular s-axis with its positive side in the direction in which f increases, as in Figure 29. By Theorem 5, $\nabla f(3,2)$ points in the direction of the positive s-axis and by Theorem 3, its length is the rate of change of f at $(3, 2)$ in that direction. The change in f from the level curve $f = 5$ at $(3, 2)$ to the level curve $f = 6$ above it is $\Delta f = 6 - 5 = 1$, and the distance between the level curves along the s-axis is $\Delta s \approx \frac{1}{2}$. Therefore, for $\mathbf{u}$ in the positive s-direction, $D_{\mathbf{u}}(3,2) \approx \dfrac{\Delta f}{\Delta s} = \dfrac{1}{\frac{1}{2}} = 2$.

Since $D_{\mathbf{u}} f(3,2) = |\nabla f(3,2)|$ for this vector $\mathbf{u}$, we draw $\nabla f(3,2)$ as an arrow of length 2 pointing in the direction of the positive s-axis (Figure 30). □

Differentials and error estimates

Recall that the differentials dx and dy to a plane curve at a point where it has a tangent line are corresponding changes in x and y along that line. Similarly, the differentials dx, dy, and dz at a point on a surface where it has a tangent plane are corresponding changes in x, y, and z along the tangent plane. In the case where the surface is the graph of a function $f(x, y)$, we write df for dz and call it the TOTAL DIFFERENTIAL of f:

Definition 3 *If $f(x, y)$ has continuous first-order derivatives in an open disk centered at (x_0, y_0), then the* TOTAL DIFFERENTIAL *of f at (x_0, y_0) is a function df of the differentials dx and dy, given by*

$$df = \frac{\partial f}{\partial x}(x_0, y_0)\, dx + \frac{\partial f}{\partial y}(x_0, y_0)\, dy. \tag{18}$$

The total differential df (18) is the approximate change in f from its value at (x_0, y_0) to its value at $(x_0 + dx, y_0 + dy)$:

$$df \approx f(x_0 + dx, y_0 + dy) - f(x_0, y_0). \tag{19}$$

This notation is convenient in error analysis, as is illustrated in the next Example.

Example 13   The radius of a right circular cylinder is measured to be $r = 10 \pm 0.01$ centimeters (meaning that it is measured to be 10 centimeters with an error ≤ 0.01) and its height is measured to be $h = 15 \pm 0.005$ centimeters (meaning that it is measured to be 15 centimeters with an error ≤ 0.005). Use differentials to estimate the maximum possible error in the volume of the cylinder that is calculated using $r = 10$ and $h = 15$.

SOLUTION   The volume of a right circular cylinder of radius r and height h is $V = \pi r^2 h$, for which $V_r = \dfrac{\partial}{\partial r}(\pi r^2 h) = 2\pi r h$ and $V_h = \dfrac{\partial}{\partial h}(\pi r^2 h) = \pi r^2$. At $r = 10, h = 15$, these derivatives have the values $V_r = 300\pi$ and $V_h = 100\pi$, and consequently,

$$dV = 300\pi \, dr + 100\pi \, dh.$$

If the (unknown) exact values of the radius and height are $r = 10 + dr$ and $h = 15 + dh$, then $|dr| \leq 0.01$ and $|dh| \leq 0.005$, and the error in the calculated volume is approximately

$$|dV| = |300\pi \, dr + 100\pi \, dh| \leq 300\pi \, |dr| + 100\pi \, |dh|$$
$$\leq 300\pi(0.01) + 100\pi(0.005) = 3.5\pi \doteq 11 \text{ cubic centimeters}.$$

The maximum possible error is approximately 11 cubic centimeters. □

Principles and procedures

- You can think of the directional derivative $D_\mathbf{u}(x_0, y_0)$ either as the (instantaneous) rate of change of f at (x_0, y_0) in the direction of the unit vector $\mathbf{u}$ or as the slope in the $\mathbf{u}$-direction of the tangent line at $(x_0, y_0, f(x_0, y_0))$ to the cross section of the function in the vertical plane determined by $\mathbf{u}$.

- The directional derivative $D_\mathbf{u}(x_0, y_0)$ can be found as the s-derivative at $s = 0$ of the cross section $F(s) = f(x_0 + su_1, y_0 + su_2)$, but if you do not need a formula for the cross section, it is easier to use the formula $D_\mathbf{u}(x_0, y_0) = f_x(x_0, y_0)u_1 + f_y(x_0, y_0)u_2$ from Theorem 1.

- You can find the approximate directional derivatives at a point P of a function $f(x, y)$ given by level curves by drawing an s-axis through P pointing in the given direction. Then pick points Q and R on the s-axis near P where you can determine or estimate the values of f, with R in the positive s-direction from Q. Then the directional derivative is approximately equal to $\dfrac{\Delta F}{\Delta s}$, where ΔF is the approximate change in F from Q to R and Δs is the approximate (positive) distance between Q and R.

- Because the graphs of linear functions $L(x, y)$ are planes, they have constant x- and y-derivatives, all vertical cross sections of their graphs are parallel lines, and their level curves are lines.

- Remember the slopes-intercept equation $z = m_1 x + m_2 y + b$ of the plane with slope m_1 in the positive x-direction, slope m_1 in the positive y-direction, and z-intercept b and the point-slopes equation $z = z_0 + m_1(x - x_0) + m_2(y - y_0)$ of the plane passing through (x_0, y_0, z_0) with slope m_1 in the positive x-direction and slope m_2 in the positive y-direction as generalizations of the slope-intercept and point-slope equations of lines.

- Remember the equation $z = f(x_0, y_0) + f_x(x_0, y_0)(x - x_0) + f_y(x_0, y_0)(y - y_0)$ of the tangent plane to $z = f(x, y)$ at $x = x_0, y = y_0$ as an application of the point-slopes equation.

- Know both the description of the gradient vector ∇f in terms of its components $\langle f_x, f_y \rangle$ and its description in Theorems 3 and 5 as the vector whose length is the greatest directional derivative and whose direction is perpendicular to the level curve and points in the direction of increasing f.

- To estimate a gradient vector at a point from level curves, choose its direction to be perpendicular to the level curve through the point and its length to be the approximate directional derivative in that direction (the maximum directional derivative at the point).

11.3 Directional derivatives and tangent planes

- The formula $df = f_x(x_0, y_0)\,dx + f_y(x_0, y_0)\,dy$ for the total differential comes from writing equation **(11)** of a tangent plane in the form
$$z - f(x_0, y_0) = f_x(x_0, y_0)(x - x_0) + f_y(x_0, y_0)(y - y_0)(x - x_0)$$
where $dx = x - x_0, dy = y - y_0$, and $df = z - f(x_0, y_0)$ are corresponding changes in x, y, and z on the tangent plane.

Tune-Up Exercises 11.3♦

^AAnswer provided. ^OOutline of solution provided.

T1.^O (a) Find an equation for the cross section $F(s)$ of $f(x,y) = x + \sin y + 2$ through the point $x = 0, y = 0$ and in the direction of the unit vector $\mathbf{u} = \langle \frac{3}{5}, \frac{4}{5}\rangle$. (b) Use the formula from part (a) to find $D_{\mathbf{u}}f(0,0)$. (c) Generate in an sw-plane the graph of $F(s)$ and the tangent line whose slope is $D_{\mathbf{u}}f(0,0)$ and copy them on your paper. Use the window $-8 \le s \le 8, -3 \le w \le 6$.
♦ Type 1, generating a cross section and its tangent line

T2.^O Find the derivative of $g(x,y) = x^2 y^3$ at $(1, -1)$ in the direction toward the point $(2,2)$.
♦ Type 1, a directional derivative

T3.^O Level curves of a function $K(x, y)$ are shown in Figure 31. Find the approximate derivative of K at $(5, 5)$ in the direction toward the origin.
♦ Type 1, estimating a directional derivative from level curves

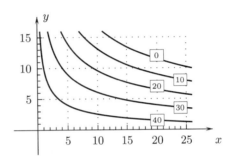

Level curves of $K(x, y)$
FIGURE 31

T4.^O Give an equation of the tangent plane to the graph of $z = x^2 y^{-3}$ at $(2, 1)$.
♦ Type 1, equation of a tangent plane

T5.^O Give a formula for the linear function $P(x, y)$ whose values are given in the following table.
♦ Type 1, finding an equation of a linear function

	$x = -10$	$x = -5$	$x = 0$	$x = 5$	$x = 10$
$y = 10$	-105	-95	-85	-75	-65
$y = 5$	-55	-45	-35	-25	-15
$y = 0$	-5	5	15	25	35
$y = -5$	45	55	65	75	85

T6.^O Draw a few level curves of $Q(x, y) = 600x - 200y + 400$.
♦ Type 1, level curves of a linear function

T7.^O What is the gradient vector of $f = 2x - \ln y$ at $(0, 3)$?
♦ Type 1, a gradient vector

♦The Tune-up Exercises and Problems are classified by type and content. The types are (1) basic, reactive; (2) basic reflective; (3) intermediate, reactive; (4) intermediate, reflective; (5) advanced, reactive; (6) advanced, reflective; and (7) advanced, theoretical.

204   Chapter 11: Derivatives of Functions of Two or More Variables

T8.⁰   Draw the approximate gradient vector $\nabla G(-1,2)$ and estimate its components, where G is the function whose level curves are in Figure 32.

♦ Type 3, estimating a gradient vector from level curves

FIGURE 32

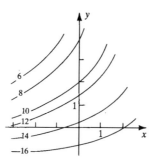

T9.⁰   What is the maximum directional derivative of $f = x\sin y$ at $(5, \tfrac{1}{3}\pi)$ and what is the unit vector in that direction?

♦ Type 3, finding a maximum directional derivative

T10.⁰   The directional derivative of $g = x^2 - 3y^3$ at $(3,2)$ is zero in two directions. Give the unit vectors in those directions.

♦ Type 3, finding zero directional derivatives

T11.⁰   The quantity x is measured to be 1 with an error no greater than 3×10^{-4} and the quantity y is measured to be -1 with an error no greater than 5×10^{-4}. The numbers $x = 1$ and $y = -1$ are then used to calculate $Z(x,y) = x^2 y^3$. What is the approximate maximum possible error in the calculated value of Z?

♦ Type 3, differentials

Problems 11.3
ᴬAnswer provided. ⁰Outline of solution provided.

1.⁰   (a) Draw in an *sw*-plane the graph of the cross section $F(s)$ of $f(x,y) = x^2 - y^2$ through $(2,1)$ in the direction of the unit vector $\mathbf{u} = \dfrac{\langle 1,-2 \rangle}{\sqrt{5}}$. Set $s = 0$ at $(0,0)$. (b) Use the formula for $F(s)$ from part (a) to find $D_\mathbf{u} f(2,1)$ and draw the tangent line with the slope with the graph of F.

♦ Type 1, drawing a cross section and its tangent line

2.⁰   (a) Draw in an *sw*-plane the graph of the cross section $F(s)$ of $g(x,y) = xe^y$ through $(0,0)$ in the direction of the unit vector $\mathbf{u} = \dfrac{\langle 3,1 \rangle}{\sqrt{10}}$. Set $s = 0$ at $(0,0)$. (b) Use the formula for $F(s)$ from part (a) to find $D_\mathbf{u} g(0,0)$ and draw the tangent line with this slope with the graph of F.

♦ Type 1, drawing a cross section and its tangent line

3.   (a) Draw in an *sw*-plane the graph of the cross section $F(s)$ of $h(x,y) = xy^2$ through $(2,2)$ in the direction of the unit vector $\mathbf{u} = \dfrac{\langle 1,1 \rangle}{\sqrt{2}}$. Set $s = 0$ at $(2,2)$. (b) Use the formula for $F(s)$ from part (a) to find $D_\mathbf{u} h(2,2)$ and draw the tangent line with the slope with the graph of F.

♦ Type 1, drawing a cross section and its tangent line

4.   (a) Draw the graph of the cross section $F(s)$ of $k(x,y) = 2x + 3y - 4$ through $(1,2)$ in the direction of the unit vector $\mathbf{u} = \left\langle \tfrac{5}{13}, \tfrac{12}{13} \right\rangle$. Set $s = 0$ at $(1,2)$. (b) How is the graph from part (a) related to the number $D_\mathbf{u} k(1,2)$?

♦ Type 1, drawing a cross section and its tangent line

5.ᴬ   Figure 33 shows level curves of the depth $D(x,y)$ (feet) of the ocean in the Monterey Canyon off the coast of California.⁽²⁾ (a) Sketch the graph of the cross section $F(s)$ of $h(x,y)$ determined by the *s*-axis in the drawing. (b) What is the approximate rate of change of the depth with respect to distance at P in the direction of the *s*-axis?

♦ Type 1, generating a cross section and its tangent line

⁽²⁾Adapted from *Submarine Canyons and Other Sea Valleys* by F. Shepard and R. Dill, Skokie, IL: Rand McNally, 1966, p. 82

11.3 Directional derivatives and tangent planes

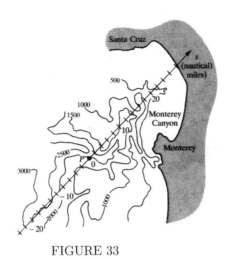

FIGURE 33

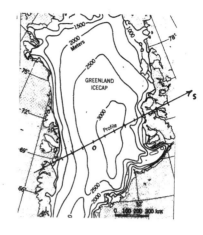

FIGURE 34

6. Figure 34 shows level curves of the elevation $h(x,y)$ of the Greenland icecap above sea level.[3] **(a)** Sketch the graph of the cross section $F(s)$ of $h(x,y)$ determined by the s-axis in the drawing. The scale on the s-axis is given in kilometers. **(b)** Suppose that **u** is the unit vector in the positive s-direction. Based on the level curves, what is the approximate maximum value of $D_{\mathbf{u}}h$ on the s-axis?

♦ Type 3, estimating a directional derivative from level curves

7.[O] What is the derivative of $f = x^2y - xy^3$ at $(3,-2)$ in the direction toward $(5,6)$?

♦ Type 1, a directional derivative

8.[A] What is the derivative of $g = \sin(xy)$ at $(\tfrac{3}{4}, \pi)$ in the direction of the unit vector $\mathbf{u} = (-\mathbf{i}+2\mathbf{j})/\sqrt{5}$?

♦ Type 1, a directional derivative

9. What is the derivative of $h = x^2 e^{2y}$ at $(4,3)$ in the direction of the (nonunit) vector $2\mathbf{i} - 3\mathbf{j}$?

♦ Type 1, a directional derivative

10. What is the derivative of $k = \ln(x^2-y^2)$ at $(4,1)$ in the direction toward $(4,-5)$?

♦ Type 1, a directional derivative

11.[A] Figure 35 shows level curves of $g(x,y)$. Find the approximate derivative of $g(x,y)$ at $(-1,1)$ in the direction toward $(0,-2)$.

♦ Type 3, estimating a directional derivative from level curves

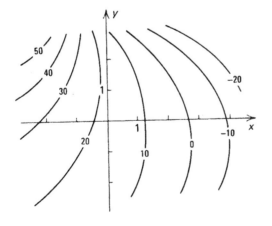

FIGURE 35

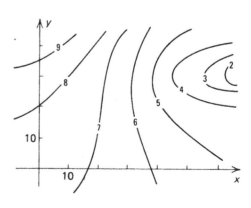

FIGURE 36

---

[3] Adapted from *Introduction to Physical Geography* by A. Strahler, New York, NY: John Wiley & Sons, 1970, p. 343

12. Figure 36 shows level curves of $h(x,y)$. What is the approximate derivative of $h(x,y)$ at $(50,30)$ in the direction of the vector $\langle -5, -2 \rangle$?

    ♦ Type 3, estimating a directional derivative from level curves

13.$^\text{A}$ Give a formula for the linear function $L(x,y)$ such that $\dfrac{\partial L}{\partial x} = 6, \dfrac{\partial L}{\partial y} = -3$, and $L(0,0) = 7$.

    ♦ Type 2, finding a formula for a linear function

14. Give a formula for the linear function $M(x,y)$ with the values $M(1,1) = 10, M(6,1) = 30$, and $M(1,6) = 0$.

    ♦ Type 2, finding a formula for a linear function

15. Values of a function $N(x,y)$ are given in the next table. Is it possible that N is linear? If so, give a formula for such a function and if not explain why not

    ♦ Type 3, determining if a function is linear

	$x = -5$	$x = 0$	$x = 5$
$y = 10$	50	305	10
$y = 5$	40	25	0

16. Give equations (a$^\text{O}$) of the tangent plane to $z = x^2 y^{-3}$ at $x = 2, y = 1$ (b$^\text{A}$) of the tangent plane to $z = xe^{-y}$ at $x = 3, y = 0$, and (c) of the tangent plane to $z = x^3 y - y^{1/2}$ at $x = 2, y = 4$.

    ♦ Type 1, equations of tangent planes

17. Give nonzero normal vectors (a$^\text{O}$) to the tangent plane of Problem 16(a), (b$^\text{A}$) to the tangent plane of Problem 16(b), and (c) to the tangent plane of Problem 16(c).

    ♦ Type 1, normal vectors

18. Give equations (a$^\text{O}$) of the tangent plane to $z = e^{2x} \sin y$ at $x = 2, y = \frac{1}{6}\pi$ (b$^\text{A}$) of the tangent plane to $z = \frac{1}{2}\ln(x^2 + y^2)$ at $x = 3, y = 4$, and (c) of the tangent plane to $z = \sin x \cos y$ at $x = \frac{1}{4}\pi, y = \frac{2}{3}\pi$.

    ♦ Type 1, equations of tangent planes

19. Give nonzero normal vectors (a$^\text{O}$) to the tangent plane of Problem 18(a), (b$^\text{A}$) to the tangent plane of Problem 18(b), and (c) to the tangent plane of Problem 18(c).

    ♦ Type 1, normal vectors

20. Give an equation of the tangent plane to $z = 4 + 6x - 2y$ at $x = 1, y = 5$.

    ♦ Type 1, equations of tangent planes

21. The tangent plane to the graph of $W(x,y)$ at $x = 3, y = 6$ is $z = 14 - 12(x - 3) + 7(y - 6)$. What are $W(3,6), W_x(3,6)$, and $W_y(3,6)$?

    ♦ Type 4, using an equation of a tangent plane

22.$^\text{A}$ (a) Give an equation of the tangent plane to the graph of $f(x,y) = xy$ at $x = 3, y = 5$.
    (b) Let $L(x,y)$ be the linear function whose graph is the tangent plane from part (a). How closely does $L(3.01, 5.01)$ approximate $f(3.01, 5.01)$? (c) How closely does $L(3.001, 5.001)$ approximate $f(3.001, 5.001)$?

    ♦ Type 2, equations of tangent planes

23.$^\text{A}$ (a) Give a formula for the linear function $L(x,y)$ whose graph is the tangent plane to the graph of $g(x,y) = \sin(\pi x) + \cos(\pi y)$ at $x = 1, y = 1$. (b) How closely does $L(1.02, 1.03)$ approximate $g(1.02, 1.03)$?

    ♦ Type 2, equations of tangent planes

24. (a) Give a formula for the linear function $L(x,y)$ whose graph is the tangent plane to the graph of $h(x,y) = \sqrt{x^2 + y^2}$ at $x = 3, y = 4$. (b) How closely does $L(4,5)$ approximate $h(4,5)$?

    ♦ Type 2, equations of tangent planes

25. Find (a$^\text{O}$) the gradient of $f(x,y) = \ln(xy)$ at $(5,10)$, (b$^\text{A}$) the gradient of $g(x,y) = x^5 y^{20}$ at $(-1,1)$, (c) the gradient of $h(x,y) = x^3 y^2 - y^3 x^2$ at $(2,-3)$, and (d) the gradient of $k(x,y) = (x^2 - y^2)^{3/2}$ at $(5,4)$.

    ♦ Type 1, gradient vectors

11.3 Directional derivatives and tangent planes

26.º If the gradient of $P(x,y)$ is $\langle 4,2 \rangle$ at $(0,0)$, what is the derivative of $P(x,y)$ at $(0,0)$ in the direction toward $(2,2)$?

♦ Type 4, using a gradient vector

27.º What is the minimum directional derivative of $x^5 e^{4y}$ at $(1,0)$?

♦ Type 3, a minimum directional derivative

28.^{A} Give unit vectors in the directions in which the directional derivative of $x + \sin(5y)$ at $(2,0)$ are zero.

♦ Type 3, zero directional derivatives

29.º Draw the level curves of $f = y - \frac{1}{4}x^2$ through the points $(0,0), (-1,-2)$, and $(-1,3)$ and ∇f at those points.

♦ Type 3, drawing gradient vectors and level curves

30.^{A} Draw the level curves of $g = \frac{1}{4}x^2 + \frac{1}{4}y^2$ through the points $(1,1), (1,-2)$, and $(-3,-1)$ and ∇g at those points.

♦ Type 3, drawing gradient vectors and level curves

31. Draw the level curve $y + \sin x = 2$ of $k(x,y) = y + \sin x$ and $\nabla k(x,y)$ at several points on it. Describe how ∇k varies along the curve.

♦ Type 3, drawing gradient vectors and level curves

32. Draw the level curve of $L(x,y) = \ln(2y - x)$ through $(2,2)$ and $\nabla L(x,y)$ at several points on it. Describe how ∇L varies along the curve.

♦ Type 3, drawing gradient vectors and level curves

33. Draw the level curves of $h(x,y) = \frac{1}{2}x^2 y$ through the two points where $\nabla h(x,y) = \langle 2, \frac{1}{2} \rangle$ and $\nabla h(x,y)$ at those points.

♦ Type 3, drawing gradient vectors and level curves

34.º Find the approximate x- and y-components of $\nabla f(2,1)$ for the function $f(x,y)$ whose level curves are shown in Figure 37.

♦ Type 3, estimating gradient vectors from level curves

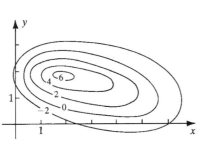
FIGURE 37

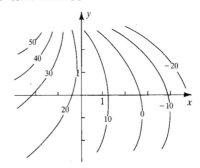
FIGURE 38

35.^{A} Find the approximate x- and y-components of $\nabla g(2,2)$ for the function $g(x,y)$ whose level curves are shown in Figure 38.

♦ Type 3, estimating gradient vectors from level curves

36. Find the approximate x- and y-components of $\nabla h(30,30)$ for the function $h(x,y)$ whose level curves are shown in Figure 39.

♦ Type 3, estimating gradient vectors from level curves

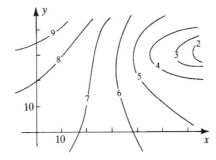
FIGURE 39

37. Figure 40 shows level curves of the elevation $h(x,y)$ above sea level in Antarctica.[4] **(a)** What are the approximate components of ∇h at the south pole? **(b)** Why does ∇h point north at the south pole?

♦ Type 4, estimating gradient vectors from level curves

FIGURE 40   FIGURE 41

38. At most points on the earth magnets do not point exactly toward the north pole. Figure 41 gives level curves of the magnetic declination D, which is the angle (degrees) between a vector pointing toward the north pole and one pointing in the direction given by a compass, with positive values where compasses points west of true north and negative values where compasses point east of true north.[5] **(a)** By approximately how much does D vary in North America? **(b)** What is the approximate maximum directional derivative of D with respect to distance at the most eastern point on the equator in South America? The diameter of the earth is approximately 8000 miles.

♦ Type 4, estimating a maximum directional derivative from level curves

39. An ant is on a metal plate whose temperature at (x,y) is $3x^2y - y^3$ degrees Celsius. When he is at the point $(5,1)$, he is anxious to move in the direction in which the temperature drops the most rapidly. Give the unit vector in that direction.

♦ Type 3, a negative directional derivative

40. **(a^O)** Give the two unit vectors normal to the curve $xy^3 + 6x^2y = -7$ at $(1,-1)$. **(b^A)** Give the two unit vectors normal to the curve $x - y^2 = 0$ at $(4,2)$. **(c)** Give the two unit vectors normal to the curve $e^{x-y^2} = 1$ at $(4,2)$.

♦ Type 3, normal vectors to level curves

41. Figure 42 shows ∇F at $(2,1), (3,3)$, and $(5,-1)$ with the lengths of the arrows measured by the scales on the axes. Give approximate values of **(a^O)** $F_x(5,-1)$, **(b^A)** $F_y(3,3)$, **(c^A)** $F_x(2,1)$, **(d)** $F_y(2,1)$, **(e^A)** $D_{\mathbf{u}}F(2,1)$ with $\mathbf{u} = \langle 1,1\rangle/\sqrt{2}$, and **(f)** $D_{\mathbf{u}}F(3,3)$ with $\mathbf{u} = \langle 1,-3\rangle/\sqrt{10}$.

♦ Type 3, estimating directional derivatives from a sketch of the gradient

FIGURE 42

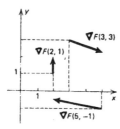

---

[4] Strahler p. 344
[5] Strahler p. 432

11.3 Directional derivatives and tangent planes

42. Figure 43 shows level curves of the depth $f(x,y)$ (meters) of the ocean in the Monterey Canyons off the coast of California. **(a)** Approximately how deep is the ocean at the point P? **(b)** Give six approximate values of s for points on the s-axis where the derivative of f in the positive s-direction is zero.

♦ Type 3, estimating directional derivatives from level curves

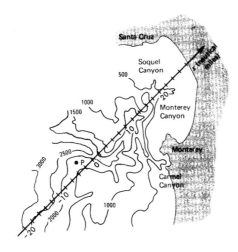

FIGURE 43   FIGURE 44

43. Excessive algae in a lake can give rise to levels of dissolved oxygen that are toxic to fish. Figure 44 shows level curves of the oxygen concentration $C(x,y)$ (parts per million) in the bottom waters of Lake Erie in 1960. **(a)** What is the approximate rate of change of $C(x,y)$ with respect to distance at the point P in the direction toward Cleveland? **(b)** Give approximate components of ∇C at the point Q.

♦ Type 4, estimating gradient vectors from level curves

44. Suppose that $f(x,y)$ has continuous first-order derivatives in an open disk containing (x_0, y_0), that $\nabla f(x_0, y_0) \neq \mathbf{0}$, and that $F(t)$ has a nonzero derivative at $t = f(x_0, y_0)$. Set $g(x,y) = F(f(x,y))$. **(a)** Show that $\nabla g(x_0, y_0)$ and $\nabla f(x_0, y_0)$ are parallel. **(b)** How are the level curves of $f(x,y)$ and $g(x,y)$ related?

♦ Type 6, gradient vectors of composite functions

45. Suppose that $G(x,y)$ has continuous first-order derivatives in an open disk containing the origin, that $G(0,0) = 16$, and $\nabla G(0,0) = \mathbf{i} - 5\mathbf{j}$. Set $H(x,y) = \sqrt{G(x,y)}$. What is $\nabla H(0,0)$?

♦ Type 4, finding a gradient vector from directional derivatives

46. **(a)** Find constants A and B such that $\nabla g(2,2) = \langle -2, 1 \rangle$ for $g(x,y) = Ax + By - \sin(\pi x)$.
(b) Generate the level curve of $g(x,y)$ through $(2,2)$ draw it with $\nabla g(2,2)$ on your paper.

♦ Type 4, finding parameters to match a gradient vector

47. **(a)** Find constants A and B such that $\nabla f(1,4) = \langle 3, 1 \rangle$ for $f(x,y) = Axy + Bx^2 + 5x$.
(b) Generate the level curve of $f(x,y)$ through $(1,4)$ draw it with $\nabla f(1,4)$ on your paper.

♦ Type 4, finding parameters to match a gradient vector

48. If $D_\mathbf{u} W(5,10) = -17$ for $\mathbf{u} = \langle \frac{12}{13}, -\frac{5}{13} \rangle$ and $D_\mathbf{u} W(5,10) = 13\sqrt{2}$ for $\mathbf{u} = \langle -1, 1 \rangle / \sqrt{2}$, what are the x- and y-derivatives of $W(x,y)$ at $(5,10)$?

♦ Type 4, finding partial derivatives from directional derivatives

49. Find the unit vectors $\mathbf{u}$ such that $D_\mathbf{u} f(x_0, y_0) = 1$ if $\nabla f(x_0, y_0) = \mathbf{i} + \sqrt{3}\,\mathbf{j}$.

♦ Type 4, finding directions from directional derivatives

Section 11.4

Derivatives with three or more variables

OVERVIEW: *In this section we modify definitions and results from Sections 11.1 through 11.3 so they apply to functions with three or more variables. We first discuss level surfaces, partial derivatives, the Chain Rule, directional derivatives, gradient vectors, and differentials of functions of three variables. Then we look briefly at functions with four or more variables defined on subsets of n-dimensional* EUCLIDEAN SPACES $\Re^n$ *with* $n \geq 4$.

Topics:

- *Functions of three variables and their level surfaces*
- *Quadric surfaces*
- *Partial derivatives, directional derivatives and gradients with three variables*
- *Tangent planes to level surfaces in xyz-space*
- *n-dimensional Euclidean space and functions with four or more variables*

Functions with three variables

A FUNCTION f with three variables is a rule that assigns a number $f(x, y, z)$ to each point (x, y, z) in its domain. The domain is a portion or all of xyz-space.

Example 1   The volume of a frustum of a right circular cone with base of radius R, top of radius r and height h (Figure 1) is $V(R, r, h) = \frac{1}{3}\pi(R^2 + rR + r^2)h$. **(a)** What does $V(6, 4, 9)$ represent and what is its value if lengths are measured in meters? **(b)** What does $V(R, 0, h)$ represent?

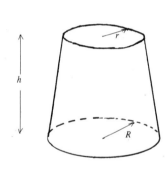

FIGURE 1   FIGURE 2

SOLUTION   **(a)** $V(9, 6, 4)$ is the volume of a frustum with base of radius 9 meters, top of radius 6 meters, and height 4 meters. $V(9, 6, 4) = \frac{1}{3}\pi(9^2 + 6(9) + 6^2)(4) = 228\pi$ cubic meters.

(b) $V(R, 0, h) = \frac{1}{3}\pi R^2 h$ is the volume of a right circular cone whose base has radius R and whose height is h (Figure 2). □

Question 1   What does $V(R, R, h)$ represent for the function V of Example 1?

We cannot draw or visualize the graph of a function of three variables because it would be in four-dimensions. Instead we consider only its level surfaces.

♦ SUGGESTIONS TO INSTRUCTORS: The main topics in this section are extensions of concepts and results from Sections 11.1 through 11.3 to the case of three variables. Partial derivatives of functions with more than three variables are also discussed. Have students begin Questions 1 through 6 and Tune-up Exercises T1, T2, and T4 before class. Problems 2, 4a, 6, 10, and 14 are good for in-class discussions.

Defintion 1 *A level surface of a function $f(x,y,z)$ is the set of points $f(x,y,z) = c$ consisting of all points (x,y,z) in the domain of the function where it has the value c for some constant c.*

Example 2   Describe the level surfaces of $f(x,y,z) = x^2 + y^2 + z^2$.

SOLUTION   Because $\sqrt{x^2+y^2+z^2}$ is the distance from (x,y,z) to the origin $(0,0,0)$, the level surface $f = c$ consists of the points that are a distance $\sqrt{c}$ from the origin. The level surface is the sphere of radius $\sqrt{c}$ centered at the origin if $c > 0$ (Figure 3), consists of a single point (the origin) if $c = 0$, and is empty (contains no points) if $c < 0$. □

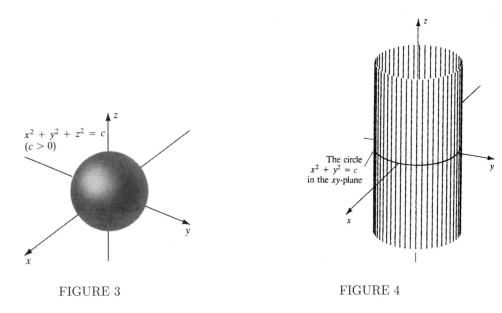

FIGURE 3   FIGURE 4

Example 3   Describe the level surfaces of $g(x,y,z) = x^2 + y^2$.

SOLUTION   Because the variable z does not appear in the equations $x^2 + y^2 = c$ of level surfaces of g, the surfaces consist of vertical lines, parallel to the z-axis. If the constant c is positive, then the surface is the cylinder in Figure 4 formed by the vertical lines through the circle $x^2 + y^2 = c$ of radius $\sqrt{c}$ in the xy-plane. The level surface $x^2 + y^2 = 0$ with $c = 0$ is the z-axis and the surface $x^2 + y^2 = c$ is empty if $c < 0$. □

Quadric surfaces

As we saw in Chapter 9, level curves

$$Ax^2 + Bxy + Cy^2 + Dx + Ey + F = c$$

of second-degree polynomials in the two variables x and y are either conic sections or degenerate conic sections. Level surfaces

$$Ax^2 + By^2 + Cz^2 + Dxy + Exz + Fyz + Gx + Hy + Iz + J = c \tag{1}$$

of second-degree polynomials with three variables are called QUADRIC SURFACES. The sphere in Figure 3 and the cylinder in Figure 4 are quadric surfaces, as is the surface $\frac{1}{9}x^2 + \frac{1}{16}y^2 + z^2 = 1$ in Figure 5. This last surface is called an ELLIPSOID because all of its planar cross sections are ellipses.

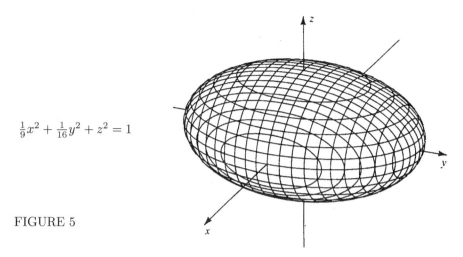

$\frac{1}{9}x^2 + \frac{1}{16}y^2 + z^2 = 1$

FIGURE 5

Question 2 What are the x-, y-, and z-intercepts of the ellipsoid in Figure 5?

The paraboloids
$$z = Ax^2 + Bxy + Cy^2 + Dx + Ey + F \qquad (2)$$
that we studied in Section 11.1 are also quadric surfaces. Recall that the vertical cross sections of these surfaces are either parabolas or lines, and that such a surface is an elliptical paraboloid if its horizontal cross sections are ellipses and a hyperbolic paraboloid if its horizontal cross sections are hyperbolas. We consider three other types of quadric surfaces in the next Question.

Question 3 What are the x-, y-, and z-intercepts **(a)** of the surface $\frac{1}{3}z^2 = x^2 + y^2$ in Figure 6, **(b)** of the surface $\frac{1}{3}z^2 = x^2 + y^2 + 1$ in Figure 7, and **(c)** of the surface $\frac{1}{3}z^2 = x^2 + y^2 - 1$ in Figure 8?

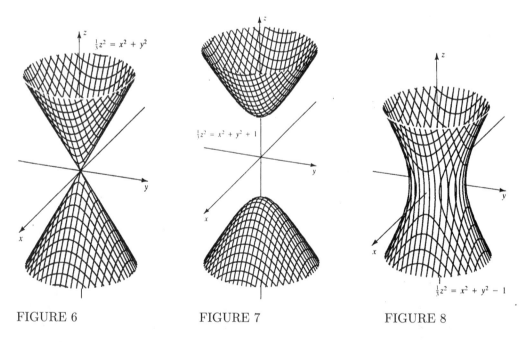

FIGURE 6   FIGURE 7   FIGURE 8

Setting z equal to a constant in any of the equations of the surfaces in Figures 6 through 8 gives an equation of the form $x^2 + y^2 = k$ with a constant k. Accordingly, all of the horizontal cross sections of these surfaces are either circles (if $k > 0$), points (if $k = 0$), or empty (if $k < 0$), and we can determine the shape of each surface by studying its cross section in a vertical plane. In particular, we can set $x = 0$ to obtain equations of the cross sections in the yz-plane.

11.4 Derivatives with three or more variables

When we set $x = 0$ in the equation $\frac{1}{3}z^2 = x^2 + y^2$ of the surface in Figure 6, we obtain $\frac{1}{3}z^2 = y^2$, which is equivalent to $z = \pm\sqrt{3}\,y$. Consequently, the cross section consists of the lines in Figure 9 and the surface in Figure 6 is a double circular cone.

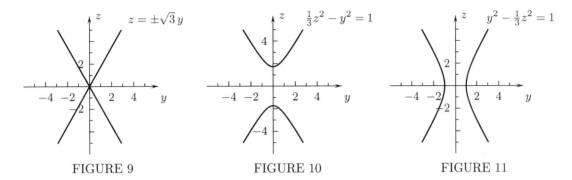

FIGURE 9 FIGURE 10 FIGURE 11

Setting $x = 0$ in the equation $\frac{1}{3}z^2 = x^2 + y^2 + 1$ of the surface in Figure 7 gives the equation of the hyperbola $\frac{1}{3}z^2 - y^2 = 1$ in Figure 10, and setting $x = 0$ in the equation $\frac{1}{3}z^2 = x^2 + y^2 - 1$ of the surface in Figure 7, gives the equation of the hyperbola $y^2 - \frac{1}{3}z^2 = 1$ in Figure 11. These curves, combined with the fact that the horizontal cross sections of the surfaces are circles, determine the shapes of the surfaces.

The quadric surface in Figure 7 is called a HYPERBOLOID OF TWO SHEETS and the surface in Figure 8 is called a HYPERBOLOID OF ONE SHEET.

Example 4 Describe the level surfaces of $g(x, y, z) = \frac{1}{3}z^2 - x^2 - y^2$.

SOLUTION The level surface $\frac{1}{3}z^2 - x^2 - y^2 = 0$ is the double cone $\frac{1}{3}z^2 = x^2 + y^2$ in Figure 6, $\frac{1}{3}z^2 - x^2 - y^2 = 1$ is the hyperboloid of two sheest $\frac{1}{3}z^2 = x^2 + y^2 + 1$ in Figure 7, and the level surface $\frac{1}{3}z^2 - x^2 - y^2 = -1$ is the hyperboloid of one sheet $\frac{1}{3}z^2 = x^2 + y^2 - 1$ in Figure 8. These are shown together in Figure 12. Calculations similar to those made above would show that the level surfaces for $c > 0$ are hyperboloids of two sheets inside the double cone whose pieces move away from the xy-plane as c increases and that the level surfaces for $c < 0$ are hyperboloids of one sheet that surround the double cone and expand as c becomes more negative. □

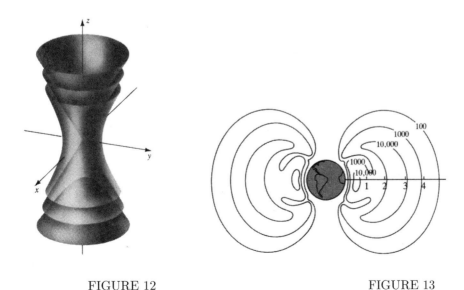

FIGURE 12 FIGURE 13

Example 5 Figure 13 is a cross-sectional view of level surfaces of the Van Allen belts of cosmic radiation that surround the earth. (To visualize the level surfaces imagine that the curves as drawn are rotated around the north-south axis of the earth.) The radiation is measured in counts per second and the scale shows the distance in earth radii ($\approx$ 4000 miles). At approximately what distances from the equator is the radiation the greatest?

SOLUTION Based on the drawing, the radiation is greatest at approximately 0.5 and 2.5 earth radii from the equator. $\square$.

Boundary points, closed and open sets, limits, and continuity

The definitions of boundary points, closed and open sets, limits, and continuity with three variables are almost the same as with two variables. A point (x_0, y_0, z_0) is in the BOUNDARY of a subset D of xyz-space if every ball $\{(x,y,z) : (x-x_0)^2 + (y-y_0)^2 + (z-z_0)^2 < r\}$ centered at (x_0, y_0, z_0), no matter how small, contains at least one point in D and one point not in D. A set is CLOSED if it includes all of its boundary points and is OPEN if it does not include any of its boundary points. The CLOSURE of a set is obtained by adding to it any boundary points that are not already in it, and the INTERIOR of the set is obtained by removing any boundary points that are in it.

We say that the LIMIT of $f(x, y, z)$ as (x, y, z) tends to (x_0, y_0, z_0) is L with L a number, ∞, or $-\infty$ if $f(x, y, z)$ tends to L as (x, y, z) approaches (x_0, y_0, z_0) on all possible paths, and that the LIMIT IN D of $f(x, y, z)$ as (x, y, z) tends to (x_0, y_0, z_0) is L if $f(x, y, z)$ tends to L as (x, y, z) approaches (x_0, y_0, z_0) on all possible paths

The function $f(x, y, z)$ is CONTINUOUS AT (x_0, y_0, z_0) if (x_0, y_0, z_0) is in the interior of the domain of f and $f(x_0, y_0, z_0)$ is the limit of $f(x, y, z)$ as (x, y, z) tends to (x_0, y_0, z_0). The function $f(x, y, z)$ is CONTINUOUS IN A SET D if $f(x_0, y_0, z_0)$ is the limit of $f(x, y, z)$ as (x, y, z) tends to (x_0, y_0, z_0) in D.

Partial derivatives

Partial derivatives of $f(x, y, z)$ are obtained by holding two of the variables constant and differentiating with respect to the third.

Example 6 What are the first-order partial derivatives of $f = x^2 y^3 z^4$?

SOLUTION The derivatives are $f_x = \dfrac{\partial}{\partial x}(x^2 y^3 z^4) = 2xy^3 z^4$, $f_y = \dfrac{\partial}{\partial y}(x^2 y^3 z^4) = 3x^2 y^2 z^4$, and $f_z = \dfrac{\partial}{\partial z}(x^2 y^3 z^4) = 4x^2 y^3 z^3$. $\square$

Example 7 A rectangular box has width x, depth y, and height z feet and consequently volume $V = xyz$ cubic feet (Figure 14). **(a)** Find formulas for $\dfrac{\partial V}{\partial x}, \dfrac{\partial V}{\partial y}$, and $\dfrac{\partial V}{\partial z}$. **(b)** Give geometric explanations of the formulas from part (a).

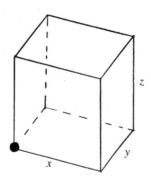

FIGURE 14

11.4 Derivatives with three or more variables

SOLUTION   (a) $\dfrac{\partial V}{\partial x} = \dfrac{\partial}{\partial x}(xyz) = yz$, $\dfrac{\partial V}{\partial y} = \dfrac{\partial}{\partial y}(xyz) = xz$, and $\dfrac{\partial V}{\partial z} = \dfrac{\partial}{\partial z}(xyz) = xy$.

(b) Imagine that the left, lower, front corner of the box is fixed, as suggested by the dot in Figure 14. Then as x increases, the right side moves to the right, and the rate of change of the volume with respect to x is equal to the area yz of the right side. Similarly, as y increases, the back moves back, and the rate of change of the volume with respect to y is equal to the area xz of the back; and, as z increases, the top moves up, and the rate of change of the volume with respect to z equals the area xy of the top. □

As with functions of two variables mixed partial derivatives with three variables can be calculated in any order, provided that the resulting derivatives are continuous.

Question 4   Find a formula for h_{yz} in two ways where $h(x,y,z) = e^x \sin y \cos z$.

Chain Rules

The basic Chain Rule for functions of three variables is

$$\dfrac{d}{dt}[f(x(t),y(t),z(t))] = \dfrac{\partial f}{\partial x}(x(t),y(t),z(t))\dfrac{dx}{dt}(t) + \dfrac{\partial f}{\partial y}(x(t),y(t),z(t))\dfrac{dy}{dt}(t)$$
$$+ \dfrac{\partial f}{\partial z}(x(t),y(t),z(t))\dfrac{dz}{dt}(t). \quad (3)$$

It holds provided that the first derivatives of $x(t), y(t),$ and $z(t)$ exist and that the first-order partial derivatives of f are continuous in an open ball centered at $(x(t), y(t), z(t))$.† Without the references to the values of the variables, it takes the more concise form

$$\dfrac{d}{dt}(f) = \dfrac{\partial f}{\partial x}\dfrac{dx}{dt} + \dfrac{\partial f}{\partial y}\dfrac{dy}{dt} + \dfrac{\partial f}{\partial z}\dfrac{dz}{dt}. \quad (4)$$

Example 8   What is the t-derivative of $f(t^7, t^8, t^9)$ at $t=1$ if $f_x(1,1,1) = 4$, $f_y(1,1,1) = 5$, and $f_z(1,1,1) = 6$?

SOLUTION   By the Chain Rule (3) with $x(t) = t^7, y(t) = t^8,$ and $z(t) = t^9$

$$\dfrac{d}{dt}[f(t^7,t^8,t^9)] = f_x(t^7,t^8,t^9)\dfrac{d}{dt}(t^7) + f_y(t^7,t^8,t^9)\dfrac{d}{dt}(t^8) + f_z(t^7,t^8,t^9)\dfrac{d}{dt}(t^9)$$
$$= f_x(t^7,t^8,t^9)(7t^6) + f_y(t^7,t^8,t^9)(8t^7) + f_z(t^7,t^8,t^9)(9t^8).$$

Setting $t=1$ gives

$$\left[\dfrac{d}{dt}[f(t^7,t^8,t^9)]\right]_{t=1} = f_x(1,1,1)(7) + f_y(1,1,1)(8) + f_z(1,1,1)(9)$$
$$= 4(7) + 5(8) + 6(9) = 122. \ \square$$

Question 5   Express $\dfrac{\partial R}{\partial v}(5,10)$ for $R(u,v) = Q(x(u,v), y(u,v), z(u,v))$ in terms of derivatives of $Q(x,y,z), x(u,v), y(u,v),$ and $z(u,v)$.

---

†As with the Chain Rule for functions with two variables, we will assume that these conditions are satisfied in all applications of the Chain-Rule with more variables.

Directional derivatives and gradient vectors

By analogy with the definition of directional derivatives with two variables, we define the derivative $D_{\mathbf{u}}f(x_0, y_0, z_0)$ of $f(x, y, z)$ at (x_0, y_0, z_0) in the direction of the unit vector $\mathbf{u} = \langle u_1, u_2, u_3\rangle$ by introducing an s-axis with its origin at (x_0, y_0, z_0), with its positive side pointing in the direction of $\mathbf{u}$, and with the same scale as on the coordinate axes (Figure 15). Because the point at s on the s-axis has coordinates $x = x_0 + u_1 s, y = y_0 + u_2 s, z = z_0 + u_3 s$ we define the CROSS SECTION of $f(x, y, z)$ corresponding to this s-axis to be

$$F(s) = f(x_0 + u_1 s, y_0 + u_2 s, z_0 + u_3 s).$$

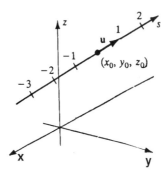

FIGURE 15

The directional derivative of f at (x_0, y_0, z_0) in the direction of the unit vector $\mathbf{u}$ is the derivative of $F(s)$ at $t = 0$. The Chain Rule **(3)** gives

$$F'(s) = \frac{d}{ds}[f(x_0 + u_1 s, y_0 + u_2 s, z_0 + u_3 s)]$$
$$= \frac{\partial f}{\partial x}\frac{d}{ds}(x_0 + u_1 s) + \frac{\partial f}{\partial y}\frac{d}{ds}(y_0 + u_2 s) + \frac{\partial f}{\partial z}\frac{d}{ds}(z_0 + u_3 s)$$
$$= \frac{\partial f}{\partial x}u_1 + \frac{\partial f}{\partial y}u_2 + \frac{\partial f}{\partial z}u_3$$

with the derivatives of f evaluated at $(x_0 + u_1 s, y_0 + u_2 s, z_0 + u_3 s)$. We set $s = 0$ and obtain the following result:

Theorem 1 *If $f(x, y, z)$ has continuous first-order derivatives in a ball centered at (x_0, y_0, z_0), then the derivative of f at (x_0, y_0, z_0) in the direction of the unit vector $\mathbf{u} = \langle u_1, u_2, u_3\rangle$ is*

$$D_{\mathbf{u}}f = \frac{\partial f}{\partial x}u_1 + \frac{\partial f}{\partial y}u_2 + \frac{\partial f}{\partial z}u_3 \tag{5}$$

with the derivatives of f evaluated at (x_0, y_0, z_0).

Example 9   What is the derivative of $g(x, y, z) = x^3 y^2 z$ at $(-1, 3, 2)$ in the direction toward $(4, 2, 6)$?

SOLUTION   The displacement vector from $P = (-1, 3, 2)$ to $Q = (4, 2, 6)$ is

$$\overrightarrow{PQ} = \langle 4 - (-1), 2 - 3, 6 - 2\rangle = \langle 5, -1, 4\rangle$$

and its length is $|\langle 5, -1, 4\rangle| = \sqrt{5^2 + 1^2 + 4^2} = \sqrt{42}$, so the unit vector in the required direction is $\mathbf{u} = \langle 5, -1, 4\rangle/\sqrt{42}$. Since $g_x = 3x^2 y^2 z, g_y = 2x^3 yz$, and $g_z = x^3 y^2$, we have $g_x(-1, 3, 2) = 3(-1)^2(3)^2(2) = 54, g_y(-1, 3, 2) = 2(-1)^3(3)(2) = -12$, and $g_z(-1, 3, 2) = (-1)^3(3)^2 = -9$. By **(5)**

11.4 Derivatives with three or more variables

$$D_{\mathbf{u}}g(-1,3,2) = g_x(-1,3,2)u_1 + g_y(-1,3,2)u_2 + g_z(-1,3,2)u_3$$
$$= 54\left(\frac{5}{\sqrt{42}}\right) + (-12)\left(\frac{-1}{\sqrt{42}}\right) + (-9)\left(\frac{4}{\sqrt{42}}\right) = \frac{246}{\sqrt{42}}. \square$$

As in the case of two variables, formula (5) leads us to define the GRADIENT VECTOR of f by

$$\nabla f = \left\langle \frac{\partial f}{\partial x}, \frac{\partial f}{\partial y}, \frac{\partial f}{\partial z} \right\rangle \tag{6}$$

so that (5) gives $D_{\mathbf{u}}f = \nabla f \cdot \mathbf{u}$. If ∇f is not zero at (x_0, y_0, z_0), then the last equation gives $D_{\mathbf{u}}f = |\nabla f||\mathbf{u}|\cos\theta$ with θ an angle between ∇f and $\mathbf{u}$. This establishes the next theorem.

Theorem 2 If $f(x,y,z)$ has continuous first-order derivatives in a ball centered at (x_0, y_0, z_0) then for unit vectors $\mathbf{u} = \langle u_1, u_2, u_3 \rangle$

$$D_{\mathbf{u}}f(x_0, y_0, z_0) = \nabla f(x_0, y_0, z_0) \cdot \mathbf{u}. \tag{7}$$

Moreover, if $\nabla f(x_0, y_0, z_0)$ is not zero, then

$$D_{\mathbf{u}}f(x_0, y_0, z_0) = |\nabla f(x_0, y_0, z_0)|\cos\theta \tag{8}$$

where θ is an angle between $\nabla f(x_0, y_0, z_0)$ and u. In particular, the maximum directional derivative of f at $f(x_0, y_0, z_0)$ is $|\nabla f(x_0, y_0, z_0)|$ and occurs for $\mathbf{u}$ in the direction of $\nabla f(x_0, y_0, z_0)$; the minimum directional derivative of f at $f(x_0, y_0, z_0)$ is $-|\nabla f(x_0, y_0, z_0)|$ and occurs for $\mathbf{u}$ in the direction of $-\nabla f(x_0, y_0, z_0)$; and the directional derivative is zero for $\mathbf{u}$ perpendicular to $\nabla f(x_0, y_0, z_0)$.

Example 10 (a) What is the gradient vector of $g(x,y,z) = x^3y^2z$ at $(-1,3,2)$ for the function g of Example 9? (b) What is the maximum directional derivative of g at $(-1,3,2)$ and what is the unit vector in that direction?

SOLUTION (a) The gradient vector is $\nabla g(-1,3,2) = \langle g_x(-1,3,2), g_y(-1,3,2), g_z(-1,3,2) \rangle = \langle 54, -12, -9 \rangle$.

(b) By Theorem 2 the greatest directional derivative of g at $(-1,3,2)$ is $|\nabla g(-1,3,2)| = |\langle 54, -12, -9 \rangle| = \sqrt{54^2 + 12^2 + 9^2} = \sqrt{3141}$ and it occurs in the direction of $\mathbf{u} = \nabla g(-1,3,2)/|\nabla g(-1,3,2)| = \langle 54, -12, -9 \rangle/\sqrt{3141}$. $\square$

Question 6 What is the minimum directional derivative at $(-1,3,2)$ of the function g of Example 10 and what is the unit vector in that direction?

Gradient vectors and level surfaces

Just as not all level curves of functions of two variables are what we would normally call "curves," not all level surfaces of functions with three variables are what we would call "surfaces." However, we get what we would normally call a surface through any point where the function $f(x,y,z)$ has a nonzero gradient, as is shown by the following theorem, which is proved in advanced calculus classes.

Theorem 3 (The Implicit Function Theorem) If $f(x,y,z)$ has continuous first derivatives in an open ball centered at (x_0, y_0, z_0) and $\nabla f(x_0, y_0, z_0)$ is not zero, then the portion in an open ball centered at (x_0, y_0, z_0) of the level surface of f through (x_0, y_0, z_0) is a surface with a tangent plane at (x_0, y_0, z_0). Moreover, $\nabla f(x_0, y_0, z_0)$ is a normal vector to the surface (Figure 16), so the tangent plane has the equation

$$f_x(x_0, y_0, z_0)(x - x_0) + f_y(x_0, y_0, z_0)(y - y_0) + f_z(x_0, y_0, z_0)(z - z_0). \tag{9}$$

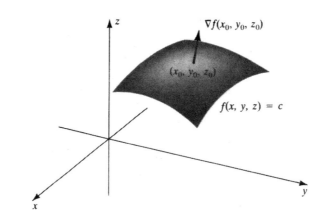

FIGURE 16

Example 11 Give an equation of the tangent plane to the hyperboloid of two sheets $\frac{1}{3}z^2 - x^2 - y^2 = 1$ of Figure 7 at the point $(1,1,3)$.

SOLUTION We set $f = \frac{1}{3}z^2 - x^2 - y^2$ so that the hyperboloid is the level surface $f = 1$. Then $\nabla f = \langle -2x, -2y, \frac{2}{3}z\rangle$ and $\nabla f(1,1,3) = \langle -2, -2, 2\rangle$ is a normal vector to the level surface at $(1,1,3)$. The tangent plane at that point has the equation $-2(x-1) - 2(y-1) + 2(z-3) = 0$ or $(x-1) + (y-1) - (z-3) = 0$. □

Linear approximations and differentials

If $f(x,y,z)$ has continuous first-order partial derivatives in an open ball centered at (x_0, y_0, z_0), then the linear function that best approximates $f(x,y,z)$ near (x_0, y_0, z_0) is

$$L(x,y,z) = f(x_0,y_0,z_0) + \frac{\partial f}{\partial x}(x_0,y_0,z_0)(x-x_0) + \frac{\partial f}{\partial y}(x_0,y_0,z_0)(y-y_0) \\ + \frac{\partial f}{\partial z}(x_0,y_0,z_0)(z-z_0). \tag{10}$$

This formula with differential notation reads

$$df = \frac{\partial f}{\partial x}(x_0,y_0,z_0)dx + \frac{\partial f}{\partial y}(x_0,y_0,z_0)dy + \frac{\partial f}{\partial z}(x_0,y_0,z_0)dz. \tag{11}$$

Here df is the TOTAL DIFFERENTIAL of f at (x_0, y_0, z_0), which equals the change in the linear function (10) from its value at (x_0, y_0, z_0) to its value at $(x_0 + dx, y_0 + dy, z_0 + dz)$ and, for small dx, dy, and dz, is approximately equal to the corresponding change in the value of f.

Example 12 The sides of a rectangular box are measured to be $x = 10 \pm 0.001$ meters, $y = 15 \pm 0.002$ meters, and $z = 20 \pm 0.003$ meters. The numbers $x = 10, y = 15$, and $z = 20$ are used to calculate the volume $V = xyz = 3000$ cubic meters. Use differentials to estimate the maximum possible error in this value.

SOLUTION Since $V_x = yz, V_y = xz$, and $V_z = xy$ have the values $V_x(10,15,20) = 300$, $V_y(10,15,20) = 200$, and $V_z(10,15,20) = 150$, the total differential of V at $(10,15,20)$ is

$$dV = 300\,dx + 200\,dy + 150\,dz.$$

For $|dx| \leq 0.001, |dy| \leq 0.002$, and $|dz| \leq 0.003$

$$|dV| \leq 300(0.001) + 200(0.002) + 150(0.003) = 1.15 \text{ cubic meters}.$$

and the maximum possible error is aproximately 1.15 cubic meters □

Functions with more than three variables

Most results concerning functions of two and three variables carry over to functions with more variables when we provide for the extra variables in definitions, theorems, and descriptions of procedures. The main difference is that we cannot visualize geometric interpretations of results and procedures involving more than three variables.

Nevertheless, we can use our geometric intuition to guide us in dealing with the higher dimensional cases by visualizing the analogous calculations and relationships in three-dimensional space. For this purpose, we use terms such as "point," "plane," "surface," and "space" in dicussions of the corresponding mathematical entities in higher dimensions.

We define a POINT in n-dimensional space for an arbitrary positive integer n to be an ordered set $P = (x_1, x_2, x_3, \ldots, x_n)$ of n real numbers and call the numbers the COORDINATES of P. The set of all such points is called n-DIMENSIONAL EUCLIDEAN SPACE and is denoted $\Re^n$.

A real-valued FUNCTION f of n variables is a rule that assigns a number denoted $f(x_1, x_2, x_3, \ldots, x_n)$ to each point in the domain of the function. The definitions of limits, continuity, and derivatives of such functions are similar to those for functions of two or three variables. We will consider only their partial derivatives, which are obtained by differentiating with respect to one variable while holding the others constant. In all cases we will encounter their higher order mixed partial derivatives can be calculated in any order.

Example 13 What is $\dfrac{\partial f^4}{\partial x \partial y \partial z \partial w}$ for $f = x^2 y^3 z^4 w^5$?

SOLUTION $\dfrac{\partial f}{\partial x} = \dfrac{\partial}{\partial x}(x^2 y^3 z^4 w^5) = 2xy^3 z^4 w^5$, $\dfrac{\partial^2 f}{\partial x \partial y} = \dfrac{\partial}{\partial y}(2xy^3 z^4 w^5) = 6xy^2 z^4 w^5$, $\dfrac{\partial^3 f}{\partial x \partial y \partial z} = \dfrac{\partial}{\partial z}(6xy^2 z^4 w^5) = 24xy^2 z^3 w^5$, and $\dfrac{\partial^4 f}{\partial x \partial y \partial z \partial w} = \dfrac{\partial}{\partial w}(24xy^2 z^3 w^5) = 120xy^2 z^3 w^4$. □

Principles and procedures

- The definitions and techniques for dealing with derivatives of functions with three variables are the same as with two variables once allowance is made for the third variable. The main difference is that with three variables we study level surfaces rather than level curves and the gradients are vectors in space rather than in a plane. Also, we do not consider graphs of functions with three variables because they are surfaces in four-dimensional space

- The equation $f_x(P)(x - x_0) + f_y(P)(y - y_0) + f_z(P)(z - z_0) = 0$ for the tangent plane to the level surface of f through $P = (x_0, y_0, z_0)$ is obtained from the fact that $\nabla f(P) = \langle f_x(P), f_y(P), f_z(P) \rangle$ is normal to the level surface.

Tune-Up Exercises 11.3♦

^AAnswer provided. ^OOutline of solution provided.

T1.^O (a) What is the domain of $f(x,y,z) = \sqrt{25 - x^2 - y^2 - z^2}$? (b) Describe the level surface $f = 4$.
♦ Type 1, a domain and level surface

T2.^O Give a formula for R_{xy} where $R(x,y,z) = (x + 2y + 3z)^{10}$?
♦ Type 1, a second-order derivative from a formula

T3.^O What is the t-derivative of $G(x(t), y(t), z(t))$ at $t = 10$ if $x(10) = 1, y(10) = 2, z(10) = 3$, $x'(10) = 8, y'(10) = 7, z'(10) = 6, G_x(1,2,3) = 50, G_y(1,2,3) = 60$, and $G_z(1,2,3) = 70$?
♦ Type 3, the three-variable Chain Rule

♦The Tune-up Exercises and Problems are classified by type and content. The types are (1) basic, reactive; (2) basic reflective; (3) intermediate, reactive; (4) intermediate, reflective; (5) advanced, reactive; (6) advanced, reflective; and (7) advanced, theoretical.

T4.º (a) What is ∇H for $H(x,y,z) = x^2 + y^3 + z^4$? (b) What is the derivative of H at $(3,2,1)$ in the direction toward $(4,4,3)$ for the function H of part (a)? (c) What is the maximum directional derivative of H at $(3,2,1)$ and what is the unit vector in that direction?
♦ Type 3, a gradient, directional derivative, and maximum directional derivative

T5.º Give a nonzero normal vector and an equation of the tangent plane at $(3,2,1)$ on the level surface $H = 18$ for the function H of Exercise T4.
♦ Type 1, a tangent plane and normal vector to a level surface

T6.º Find $g_{xyzw}(x,y,z,w)$ where $g(x,y,z,w) = (\sin x)(\cos y)(\ln z)(e^{2w})$.
♦ Type 1, a fourth-order derivative

Problems 11.4

^AAnswer provided. ºOutline of solution provided.

1.^A For a yacht to be in the R–5.5 class at the Olympic Games, its rating $R(L, V, A)$ $= 0.9\left(\dfrac{L\sqrt{A}}{12\sqrt[3]{V}} + \tfrac{1}{4}(L + \sqrt{A})\right)$ must be less than 5.5, where L (meters) is the length of the yacht, A (square meters) is the surface area of its sails, and V (cubic meters) is the volume of water it displaces.[1] Would your yacht qualify for the R–5.5 class if it is 10 meters long, has 36 square meters of sail, and displaces 8 cubic meters of water?
♦ Type 2, using a value of a function

2. The maximum rate r (liters per minute) at which men 55 years old or older can take in oxygen under strenuous exercise can be predicted with the formula $r = 3 - 0.05A + 0.003m + 0.01h$ where A is the man's age, m is his weight in kilograms, and h is his height in meters.[2] Based on this formula, how much weight would a man have to gain every year to maintain the same maximum rate of oxygen intake?
♦ Type 2, using a value of a function

3. The National Football League rating of quarterbacks is obtained from the formula $R = \tfrac{5}{6}P + \tfrac{25}{6}A + \tfrac{10}{3}T - \tfrac{25}{6}I + \tfrac{25}{12}$, where P is the percentage of passes that the quarterback has completed, A is the average number of yards gained per pass, T is the percentage of attempted passes that score touchdowns, and I is the percentage of attempted passes that are intercepted.[3] Which has the greater effect on a quarterback's rating, increasing the percentage of attempted passes that score touchdowns by one percent or decreasing the percentage of attempted passes that are intercepted by one percent?
♦ Type 4, using values of a function

4. Find the first-order partial derivatives of (**a**^A) $f = x^{1/2}y^{1/4} + y^{1/5}z^{1/6}$, (**b**^A) $g = \ln(x + 3y - 4z)$, (**c**) $T = u^{1/3}v^{1/4}w^{1/5}$, and (**d**) $h = x\sin y - y\cos z$.
♦ Type 1, partial derivatives

5. What are the first-order partial derivatives of (**a**º) $f = x^2y^3(1+xyz)^2$, (**b**^A) $g = xy\tan^{-1}(yz)$, (**c**) $h = \dfrac{xyz}{x+y+z}$, and (**d**) $k = xye^{x^2z}$?
♦ Type 1, partial derivatives

6.^A Find M_{xy}, M_{yz}, and M_{xz} for $M(x,y,z) = x^2 e^{-3y}\sin(4z)$.
♦ Type 3, partial derivatives

7. What are the second-order partial derivatives of $N(x,y,z) = x^2y - y^3z^2 + xz$?
♦ Type 3, partial derivatives

8. What is H_{xxyyzz} for $H(x,y,z) = e^z \sin x \cos y$?
♦ Type 3, a partial derivative

9.º (a) What is the gradient of $f(x,y,z) = x\sin(yz^2)$ at $(3,2,1)$? (b) What is the derivative of $f(x,y,z) = x\sin(yz^2)$ at $(3,2,1)$ in the direction toward the origin?
♦ Type 3, a gradient and directional derivative

---

[1] *Sailing Theory and Practice* by C. Marchaj, New York, NY: Dodd, Mead, and Company, 1964, p. 75.

[2] Data adapted from *Geigy Scientific Tables*, edited by C. Lentner, Basel, Switzerland: CIBA-GEIGY Limite, 1990, p. 210.

[3] Data adapted from *The Times Picayune* New Orleans LA: The Times Picayune, Inc., November 22, 1998.

11.4 Derivatives with three or more variables

10. (a^A) A silo has the shape of a right circular cylinder of radius r and height H with a right circular cone of radius r and height h on top of it. Give a formula for the volume V of the silo. **(b)** What are the rates of change of V with respect to r, H, and h? **(c)** The exterior surface area of the silo is $A = 2\pi r H + \pi r \sqrt{h^2 + r^2}$. What are the rates of change of A with respect to r, H, and h?
♦ Type 1, partial derivatives

11.^{A} If the sides of lengths a and b in a triangle form an angle θ, then, by the Law of Cosines, the third side has length $c = \sqrt{a^2 + b^2 - 2ab\cos\theta}$. What are the rates of change of c with respect to a, b, and θ?
♦ Type 4, partial derivatives

12. If two thin lenses with focal lengths f_1 and f_2 are separated a distance $d < f_1 + f_2$, then together they have the same effect as one lens of focal length $F = \dfrac{f_1 f_2}{f_1 + f_2 - d}$.$^{(4)}$ What are the derivatives of F with respect to f_1, f_2, and d?
♦ Type 4, partial derivatives

13. (a^O) What is the derivative of $f = x^4 y^{-2} z^3$ at $(1, 2, 3)$ in the direction of the (nonunit) vector $3\mathbf{i} - 4\mathbf{j} - 12\mathbf{k}$? **(b^A)** Find the derivative of $g = e^{3x - 2y + z}$ at $(0, 0, 0)$ in the direction toward $(1, -1, 1)$. **(c)** Find the derivative of $h = \ln(xyz)$ at $(5, 2, 6)$ in the direction toward $(4, 3, 7)$.
♦ Type 3, directional derivatives

14. (a^A) What is the maximum directional derivative of $f = e^{xyz}$ at $(2, 3, 4)$? **(b)** What is the minimum directional derivative of $g = xy\sin z$ at $(5, 10, \frac{1}{6}\pi)$?
♦ Type 3, maximum and minimum directional derivatives

15. Give equations for the tangent planes **(a^O)** to $3xy^2 + 2yz^3 + xz = 93$ at $(-1, 2, 3)$, **(b^A)** to $x\ln(xz) = 0$ at $(5, 10, \frac{1}{5})$, **(c)** to $\sqrt{x} + \sqrt{y} + \sqrt{z} = 6$ at $(1, 4, 9)$, and **(d)** to $x^3 + y^4 - z^5 = 10$ at $(2, -1, -1)$.
♦ Type 3, equations of tangent planes

16. (a^A) Express $F(t) = f(x(t), y(t), z(t))$ in terms of t where $f(x, y, z) = x^3 y^5 z^4$, $x(t) = t^{-1}$, $y(t) = t^3$, and $z(t) = -2t$. **(b)** Express $R(s, t) = V(p(s, t), q(s, t), r(s, t))$ in terms of s and t for $V(p, q, r) = p\cos(qr)$, $p(s, t) = st$, $q(s, t) = s^2 t^{-2}$, and $r(s, t) = s^2 + t^2$.
♦ Type 3, simplifying formulas for composite functions

17.º Give the values of $P(2)$ and $P'(2)$ where $P(t) = f(t, t^2, t^3)$ and $f(x, y, z)$ satisfies $f(2, 4, 8) = 3, f_x(2, 4, 8) = 4, f_y(2, 4, 8) = 5$, and $f_z(2, 4, 8) = -6$.
♦ Type 4, the three-variable Chain Rule

18.^{A} What is $G_p(2, 3)$ if $G(p, q) = H(p^2 q, p - q, pq^2)$ and $H(x, y, z)$ satisfies $H_x(12, -1, 18) = 5$, $H_y(12, -1, 18) = 10$, and $H_z(12, -1, 18) = 20$?
♦ Type 4, the three-variable Chain Rule

19.^{A} What are the first-order derivatives of $G(x, y, z) = H(L(x, y, z))$ at $(-5, 0, 5)$ if $L(-5, 0, 5) = 0$, $L_x(-5, 0, 5) = 4, L_y(-5, 0, 5) = 8, L_z(-5, 0, 5) = 10, H'(0) = 10, H'(4) = 16$, and $H'(8) = 12$?
♦ Type 4, the three-variable Chain Rule

20. Find the first-order derivatives of $Z(A(x, y, z))$ at $(0, 10, 20)$ where $A_x(0, 10, 20) = 5$, $A_y(0, 10, 20) = 6, A_z(0, 10, 20) = 7, Z'(5) = -1, Z'(6) = -5, ZX'(7) = 4$, and $Z'(100) = 2$.
♦ Type 4, the one-variable Chain Rule used with three variables

21.^{A} What is $B_x(3, 2, 1)$ if $B(x, y, z) = \sin[\theta(x, y, z)], \theta(3, 2, 1) = \frac{1}{6}\pi, \theta_x(3, 2, 1) = \frac{1}{2}\pi, \theta_y(3, 2, 1) = \frac{3}{4}\pi$, and $\theta_z(3, 2, 1) = \frac{2}{3}\pi$?
♦ Type 4, the one-variable Chain Rule used with three variables

22. What is $g_y(1, 3)$ if $g(x, y) = H(u(x, y), v(x, y), w(x, y)), u(1, 3) = 4, v(1, 3) = 5, w(1, 2) = 6$, $u_y(1, 3) = 2, v_y(1, 3) = 4, w_y(1, 3) = 6, H_u(4, 5, 6) = -7, H_v(4, 5, 6) = 10$, and $H + w(4, 5, 6) = -4$?
♦ Type 4, the three-variable Chain Rule

23.^{A} Find $G(1, 2, 3), G_x(1, 2, 3), G_y(1, 2, 3)$ and $G_z(1, 2, 3)$ where $G(x, y, z) = f(x)[f(y)]^2 + [f(z)]^3$, $f(1) = 4, f(2) = 3, f(3) = -1, f'(1) = 2, f'(2) = -5$, and $f'(3) = 10$.
♦ Type 4, the one-variable Chain Rule used with three variables

---

$^{(4)}$Adapted from *CRC Handbook of Chemistry and Physics*, edited by R. Weast, Boca Raton, FL: CRC Pres, Inc., 1981, p. F-103.

24. Find $A(1, 10, 100)$, $A_u(1, 10, 100)$, $A_v(1, 10, 100)$ and $A_w(1, 10, 100)$ where $A(u, v, w) = u^2 v^2 [B(w)]^{-1}$, $B(100) = 1000$, and $B'(100) = 30$.

♦ Type 4, the one-variable Chain Rule used with three variables

25. Express the first-order derivatives of $K(x, y, z) = A(x^2) + [A(y)]^3 + A(z^4)$ in terms of A and A'.

♦ Type 4, the one-variable Chain Rule used with three variables

26.[A] **(a)** What is the domain of $f(x, y, z) = \dfrac{2z}{\sqrt{x^2 + y^2}}$? **(b)** Describe and draw the level surfaces $f = 1$ and $f = -2$.

♦ Type 4, a domain and level surfaces

27. **(a)** Find the domain of $g(x, y, z) = (z - x^2 - y^2)^{1/4}$. **(b)** Describe and draw the level surface $h = \frac{1}{2}$.

♦ Type 4, a domain and level surface

28. **(a)** What is the domain of $h(x, y, z) = \dfrac{x^2 + y^2 + z^2}{x^2 + y^2 + z^2 - 1}$? **(b)** Describe the level surface $h = \frac{4}{3}$.

♦ Type 4, a domain and level surface

29. Draw the level surfaces $F = -2$, $F = 0$, and $F = 2$ of $F(x, y, z) = x^2 + y^2 - z$.

♦ Type 4, level surfaces

30. Find the sixth derivative $\dfrac{\partial^6}{\partial x^2 \partial y^2 \partial z^2}(x^{10} y^{20} z^{30})$.

♦ Type 4, a sixth-order derivative

31. In a mathematical model of a vibrating drum head, we suppose that the drum has the shape of the surface $z = f(x, y, t)$ at each fixed time t and the function $f(x, y, t)$ satisfies the WAVE EQUATION $f_{xx} + f_{yy} = k^{-2} f_{tt}$. Show that $f(x, y, t) = \sin x \sin y \cos(k\sqrt{2}\,t)$ satisfies this differential equation.

♦ Type 5, a partial-differential equation

32. **(a°)** Find the total differential of $f(x, y, z) = x^2 y^3 e^{4z}$ at $(2, -1, 0)$. **(b[A])** What is the total differential of $g(x, y, z) = x^2 e^{3y - 4z}$ at $(10, 3, 1)$?

♦ Type 3, total differentials

33. Find the total differentials at (x, y, z) **(a°)** of $g(x, y, z) = x^2 e^{3y - 4z}$, **(b[A])** of $Q(x, y, z) = xy(1 + e^{yz})$, and **(c)** of $K(x, y, z) = (x^2 + y^3 - z^4)^{1/5}$.

♦ Type 3, total differentials

34. Suppose that a sound is emitted at frequency f (cycles per second) from a source that is moving with velocity v meters per second in the positive direction on an x-axis; that the wind is blowing parallel to the x-axis with velocity w meters per second; and that an observer is moving on the s-axis with velocity V meters per second. In this case the observer hears the sound as if it had frequency $F = \left[\dfrac{c + w - V}{c + w - v}\right] f$. (This change in the frequency is called the DOPPLER EFFECT.) Here the constant c is the speed of sound. Find the derivatives of F with respect to v, w, V, and f and give their units.

♦ Type 4, using total differentials

35.[A] The work done by a constant force of magnitude F on an object that moves a distance s along a straight line at an angle θ with the direction of the force is $W = Fs \cos \theta$. The force F is measured to be 10 pounds with an error no greater than 0.1 pounds; the distance s is measured to be 100 feet with an error no greater than one inch; and the angle is measured to be $30°$ with an error no greater than one degree. Use differentials to estimate the maximum possible error in the work if the measured values are used to calculate it. (Be careful with the units.)

♦ Type 4, using total differentials

36.[A] A steel rod L inches long with a rectangular cross section of width w inches and height h inches will stretch $S = \dfrac{(3.6 \times 10^{-8}) LF}{wh}$ inches if one end is held fixed and a force of F pounds is applied to the other end. **(a)** How much would such a rod stretch under a force of 1500 pounds if the rod is 60 inches long, 3 inches wide, and 6 inches high? **(b)** Use differentials to estimate the maximum possible error in the number from part (a) if an error up to 0.05 pound might be made in measuring the force, an error up to 0.1 inch might be made in measuring the length, and errors up to 0.005 inches might be made in measuring the width and the height.

♦ Type 4, using total differentials

11.4 Derivatives with three or more variables

37. The height h of the fustrum of a right circular cone is measured to be 3 inches with an error no greater than 0.01 inches. The radius R of its base and the radius r of its top are measured to be 10 inches and 5 inches, with errors no greater than 0.03 inches in each measurement. Use differentials to estimate the maximum possible error in the computed volume $V = \frac{1}{3}\pi h(R^2 + Rr + r^2)$ of the frustum.

♦ Type 4, using total differentials

38. What is the domain of $k(x,y,z) = \sin^{-1}(x^2 + y^2 + z^2)$ and what is the level surface $k = \frac{1}{4}\pi$?

♦ Type 4, a domain and level surface

39. Find the domain of $L(x,y,z) = \ln x + \ln y - \ln z$ and describe the level surface $L = 0$.

♦ Type 4, a domain and level surface

40. Show that the tangent plane to the quadric surface $Ax^2 + By^2 + Cz^2 + D = 0$ at the point (x_0, y_0, z_0) on it has the equation $Ax_0 x + By_0 y + Cz_0 z + D = 0$.

♦ Type 6, a tangent plane

41. (a^A) Give parametric equations of the normal line to the surface $x^2 y^3 z^4 + xyz = 2$ at $(2,1,-1)$.
(b) Give parametric equations of the normal line to the hyperboloid of two sheets $z^2 - 4x^2 - 9y^2 = 11$ at $(2,1,6)$.

♦ Type 6, parametric equations of normal lines

42.^{A} What is the y-derivative of $J(u,v,w,x,y) = u^2 v^2 w K(x,y)$ at $(u,v,w,x,y) = (5,-4,3,-2,1)$ if the y-derivative of $K(x,y)$ is 30 at $x = -2, y = 1$?

♦ Type 3, a partial derivative with five variables

43. What are the first-order partial derivatives of $S(K,L,M,N) = \sin(R(K,J,M,N))$ at $(1,2,3,4)$ if $R(1,2,3,4) = \frac{1}{3}\pi$, $R_K(1,2,3,4) = 5$, $R_L(1,2,3,4) = -6$, $R_M(1,2,3,4) = 10$, and $R_N(1,2,3,4) = 4$?

♦ Type 3, a partial derivatives with four variables

44. What is $\dfrac{\partial F}{\partial x}(1,1,1,1)$ if $F(x,y,z,w) = \sqrt{G(x,y,z,w)}$, $G(1,1,1,1) = 16$, and $\dfrac{\partial G}{\partial x}(1,1,1,1) = 14$.

♦ Type 3, a partial derivatives with four variables

45. Find the set of points in xyz-space where $\nabla(ye^{xz}) = \langle 0,1,2 \rangle$.

♦ Type 4, finding where a gradient is a given vector

46. Sketch the quadric surfaces (a) $x^2 + z^2 = 1$, (b) $y = x^2$, (c) $x^2 + 4y^2 = 4$, and (d) $z^2 = x^2 - y^2$ in xyz-space.

♦ Type 4, quadric surfaces

Section 11.5

Maxima and minima with two or three variables

OVERVIEW: *In this section we discuss the First-Derivative Test and the method of Lagrange multipliers for finding local maxima and minima of functions of in two- and three-dimensional regions and on curves in coordinate planes and surfaces in space, and we present the Second-Derivative Test for local maxima and minima with two variables.*

Topics:

- *Maxima and minima with two variables*
- *Local maxima and minima and the First-Derivative Test*
- *The Second-Derivative Test with two variables*
- *Lagrange multipliers*

Maxima and minima with two variables

Recall from Chapter 3 that in finding the maximum and minimum values in an interval of a continuous function $f(x)$ of one variable, we generally first found the critical points in the interval—the values of x where the derivative $f'(x)$ was zero or did not exist. These were the points where the function might have local maxima or minima. Then we determined the open intervals where the function was increasing and decreasing by determining where the derivative was positive and negative, and, if appropriate, compared values of the function at the critical points and endpoints of the interval.

The situation is more complicated when we have two variables, as is illustrated in the next Question.

Question 1   The circles centered at the origin that are drawn with fine lines in Figure 1 are the level curves where $f(x,y) = x^2 + y^2$ has the values 6, 12, 18, and 24. The dot at the origin represents the point where $f(x,y) = 0$. The circle K with the equation $(x-1)^2 + (y-1)^2 = 8$ is drawn with a heavier line. What are the minimum and maximum values of $f(x,y)$ in the closed disk $R = \{(x,y) : (x-1)^2 + (y-1)^2 \leq 8\}$ consisting of the circle K and the region inside it? At what points do the maximum and minimum values occur?

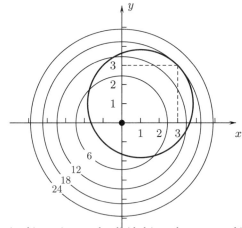

Level curves of
$f(x,y) = x^2 + y^2$
and the circle K

FIGURE 1

♦ SUGGESTIONS TO INSTRUCTORS: The main topics in this section can be divided into three parts which can be covered in four or more class meetings.

Part 1 (The First-Derivative Test with two or three variables; Tune-up Exercises T1–T4; Problems 1–13, 37, 39, 45–51, 53, 54, 57) Have students begin work on Questions 1 through 7 before class. Problems 1, 3, 5, 9, 11, 45, 47 are good for in-class discussion.

Part 2 (The Second-Derivative Test with two variables; Tune-up Exercise T5; Problems 14–19, 34–36, 41–44) Have students start work on Questions 8 and 9 before class. Problems 14, 16, 18, 19 are good for in-class discussion.

Part 3 (Lagrange multipliers with two or three variables; Tune-up Exercise T6; Problems 21-33, 52, 55, 56, 58–62) Have students begin work on Questions 10 and 11 before class. Figures 23–25, Problems 21 (with a drawing of the ellipse and level lines), 26, 32 are good for in-class discussion.

11.5 Maxima and minima with two or three variables

Notice that the minimum of the function in Question 1 in the disk R occurs at the origin, which is in the interior of the set, and the maximum occurs on the boundary of the disk. As we will see below, the minimum can be found using the First-Derivative Test with two variables and the maximum with the method of Lagrange multipliers.

Local maxima and minima and the First-Derivative Test

The function $f(x, y) = x^2 + y^2$ from Question 1 has a LOCAL MINIMUM of 0 at $(0, 0)$, according to the following general definition:

Definition 1 (a) *A function $f(x, y)$ has a* LOCAL MAXIMUM *at (x_0, y_0) if f is defined in an open disk centered at (x_0, y_0) and $f(x, y) \leq f(x_0, y_0)$ for all points (x, y) in the disk. The function has a* LOCAL MINIMUM *at (x_0, y_0) if $f(x, y) \geq f(x_0, y_0)$ for all points (x, y, z) in such a disk.*
(b) *A function $f(x, y, z)$ has a* LOCAL MAXIMUM *at (x_0, y_0, z_0) if f is defined in an open ball centered at (x_0, y_0, z_0) and $f(x, y, z) \leq f(x_0, y_0, z_0)$ for all points (x, y, z) in the ball. The function has a* LOCAL MINIMUM *at (x_0, y_0, z_0) if $f(x, y, z) \geq f(x_0, y_0, z_0)$ for all points (x, y, z) in such a ball.*

The graph of the function $f = x^2 + y^2$ of Question 1 is the circular paraboloid $z = x^2 + y^2$ in Figure 2. The function has a local minimum of 0 at $(0, 0)$. This local minimum is also a GLOBAL MINIMUM because it is the least value of f for all (x, y). In particular, this is the minimum value in the disk R of Figure 1, as you saw in Question 1.

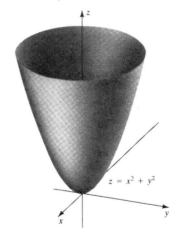

FIGURE 2

Question 2 Give an equation of the tangent plane at $x = 0, y = 0$ to the graph of $f(x, y) = x^2 + y^2$ in Figure 2.

Recall from Section 3.2 that if a function $f(x)$ of one variable has a local maximum or a local miniumum at $x = x_0$, then x_0 is a critical point of $f(x)$, meaning that either the graph $y = f(x)$ does not have a tangent line at $x = x_0$ or, as is illustrated in Figures 3 and 4, the tangent line is horizontal.

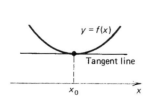

A local minimum
FIGURE 3

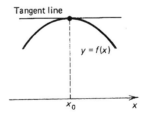

A local maximum
FIGURE 4

CRITICAL POINTS of functions with two or three variables are defined similarly:

Definition 2 (a) *The point (x_0, y_0) is a* CRITICAL POINT *of $f(x,y)$ if $f(x,y)$ is defined in an open disk centered at (x_0, y_0) and each of the partial derivatives $f_x(x_0, y_0)$ and $f_y(x_0, y_0)$ either does not exist or is zero.*
(b) *The point (x_0, y_0, z_0) is a* CRITICAL POINT *of $f(x, y, z)$ if $f(x, y, z)$ is defined in an open ball centered at (x_0, y_0, z_0) and each of the partial derivatives $f_x(x_0, y_0, z_0), f_y(x_0, y_0, z_0)$ and $f_z(x_0, y_0, z_0)$ either does not exist or is zero.*

We will show in the next Theorem that if a function $f(x,y)$ of two variables has a local maximum or local minimum at (x_0, y_0), then (x_0, y_0) is a critical point of $f(x,y)$ so that either the graph $z = f(x,y)$ does not have a tangent plane at that point or the tangent plane is horizontal. The latter case is illustrated by the result of Question 2: the function $f(x,y) = x^2 + y^2$ of Figure 2 has a local minimum at $x = 0, y = 0$ and the tangent plane to its graph at that point is the xy-plane, which is horizontal.

Theorem 1 *If a function f of two or three variables has a local maximum or minimum at a point P, then P is a critical point of f.*

To establish this theorem for functions with two variables, we suppose that $f(x,y)$ has a local maximum or minimum at (x_0, y_0). Then the function $f(x, y_0)$ of the one variable x has a local maximum or minimum at $x = x_0$, so that by Theorem 1 of Section 3.1, the x-derivative $f_x(x_0, y_0)$ of $f(x, y_0)$ at $x = x_0$ either is zero or does not exist. Moreover, the function $f(x_0, y)$ of y has a local maximum or minimum at $y = y_0$, so that the y-derivative $f_y(x_0, y_0)$ of $f(x_0, y)$ at $y = y_0$ either is zero or does not exist. Consequently, (x_0, y_0) is a critical point of $f(x, y)$.

Theorem 1 can be obtained for $f(x, y, z)$ by similar reasoning using the function $f(x, y_0, z_0)$ of x, the function $f(x_0, y, z_0)$ of y, and the functiion $f(x_0, y_0, z)$ of z.

Example 1 The function $g(x,y) = x^2 + y^2 - 8x - 2y + 18$ has a global minimum. Find it.

SOLUTION Because $g(x,y)$ is defined for all (x,y), the global minimum is a local minimum and we can apply Theorem 1 to locate it. We first find the partial derivatives

$$\frac{\partial g}{\partial x} = \frac{\partial}{\partial x}(x^2 + y^2 - 8x - 2y + 18) = 2x - 8 = 2(x - 4)$$
$$\frac{\partial g}{\partial y} = \frac{\partial}{\partial y}(x^2 + y^2 - 8x - 2y + 18) = 2y - 2 = 2(y - 1).$$

These derivatives are both zero if $x = 4$ and $y = 1$, so the one critical point is $(4, 1)$. We know that g has a global minimum, so it must occur at the critical point. The global minimum is $g(4,1) = 4^2 + 1^2 - 8(4) - 2(1) + 18 = 1$. □

The graph of $g(x, y)$ from Example 1 is the circular paraboloid in Figure 5. Its lowest point is $(4, 1, 1)$ because the global minimum of $g(x, y)$ is $g(4, 1) = 1$ at $x = 4, y = 1$.

FIGURE 5

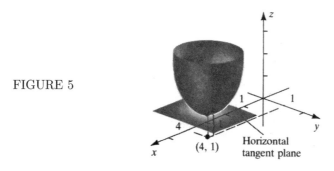

11.5 Maxima and minima with two or three variables

Because $g(x,y)$ in Example 1 is given by a second-degree polynomial with no xy-term, we can also find its global minimum by completing squares. We write

$$g(x,y) = x^2 + y^2 - 8x + 2y = (x^2 - 8x) + (y^2 - 2y) + 18$$
$$= (x^2 - 8y + 16) + (y^2 - 2y + 1) + 18 - 16 - 1$$
$$= (x-4)^2 + (y-1)^2 + 1.$$

Since the sum $(x-4)^2 + (y-1)^2$ is zero at $(x,y) = (4,1)$ and positive elsewhere, this formula shows that $g(x,y)$ has an absolute minimum of 1 at $(4,1)$, as we determined in Example 1.

Question 3 (a) Find without using calculus the global maximum of $h(x,y) = 3e^{-x^2-y^2}$ whose graph is shown in Figure 6. (b) Apply Theorem 1 to this function.

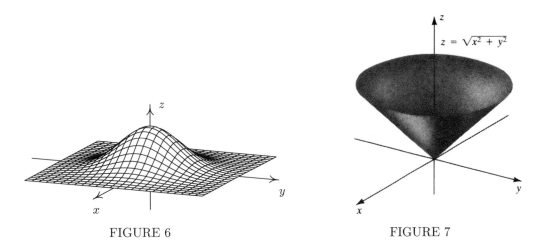

FIGURE 6 FIGURE 7

Question 4 Figure 7 shows the cone $z = \sqrt{x^2+y^2}$ that is the graph of $k(x,y) = \sqrt{x^2+y^2}$. (a) What is the global minimum of $k(x,y)$? (b) Show by considering the functions $k(x,0)$ and $k(0,y)$ that $k_x(0,0)$ and $k_y(0,0)$ do not exist. How do these facts relate to Theorem 1?

Example 2 The function $M(x,y) = \dfrac{-5y}{x^2+y^2+1}$, whose graph is shown in Figure 8, has a global maximum and a global minimum. Find them.

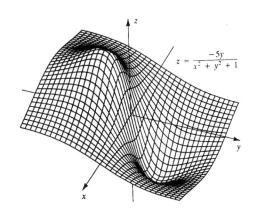

FIGURE 8

SOLUTION We find the first derivatives, using the Chain Rule for the x-derivative and the Quotient Rule for the y-derivative:

$$\frac{\partial M}{\partial x} = \frac{\partial}{\partial x}\left[\frac{-5y}{x^2+y^2+1}\right] = \frac{\partial}{\partial x}[-5y(x^2+y^2+1)^{-1}]$$

$$= 5y(x^2+y^2+1)^{-2}\frac{\partial}{\partial x}(x^2+y^2+1) = \frac{10xy}{(x^2+y^2+1)^2}$$

$$\frac{\partial M}{\partial y} = \frac{\partial}{\partial y}\left[\frac{-5y}{x^2+y^2+1}\right] = \frac{(x^2+y^2+1)\frac{\partial}{\partial y}(-5y) - (-5y)\frac{\partial}{\partial y}(x^2+y^2+1)}{(x^2+y^2+1)^2}$$

$$= \frac{-5(x^2+y^2+1)+5y(2y)}{(x^2+y^2+1)^2} = \frac{5(y^2-x^2-1)}{(x^2+y^2+1)^2}.$$

The x-derivative is zero if $xy = 0$ and the y-derivative is zero if $y^2 - x^2 - 1 = 0$, so the critical points are the solutions (x,y) of the system of equations

$$\begin{cases} xy = 0 \\ y^2 - x^2 - 1 = 0. \end{cases} \quad (1)$$

We begin with the simpler, first equation $xy = 0$ in **(1)**. For it to be satisfied, either x or y must be zero. We study the second equation in **(1)** in these two cases separately.

For $y = 0$, the second equation becomes $-x^2 - 1 = 0$, which has no solutions. For $x = 0$, the second equation reads $y^2 - 1 = 0$ and has the solutions $y = \pm 1$. The critical points are therefore $(0,1)$ and $(0,-1)$.

Because $M(x,y)$ is defined for all (x,y), its global maximum and minimum must be local maximum and minimum and occur at the critical points. Then since $M(0,1) = \frac{-5(1)}{1^2+0^2+1} = -\frac{5}{2}$ and $M(0,-1) = \frac{-5(-1)}{1^2+0^2+1} = \frac{5}{2}$, the global maximum is $\frac{5}{2}$ at $(0,-1)$ and the global minimum is $-\frac{5}{2}$ at $(0,1)$. Notice that these conclusions are consistent with the drawing of the graph of $M(x,y)$ in Figure 8. □

Narrative problems

In the next illustration of the First-Derivative Test, we will find the minimum of a quantity that depends on three variables, subject to a side condition. We use the side condition to express one of the three variables in terms of the other two and then use the First-Derivative Test to find the minimum of the resulting function of two variables.

Suppose that a large number of rectangular boxes the same shape and no top are to be manufactured so that each has a volume of 6 cubic feet and that the boxes are to be made from material that costs 6 dollars per square foot for the bottoms, 2 dollars per square foot for the fronts and backs, and 1 dollar per square foot for the sides. We want to find the dimensions that minimize the cost of manufacturing each box.

We denote the width of the front and back of the box by x, the length of the sides by y, and the height by z, all measured in feet, as in Figure 9. Then the area of the bottom is xy, the combined area of the front and back is $2xz$, and the combined area of the sides is $2yz$, all measured in square feet.

11.5 Maxima and minima with two or three variables

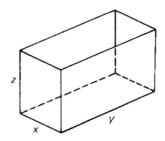

FIGURE 9

Question 5 Give a formula for the cost of the box in terms of x, y, and z.

Example 3 Find the dimensions that minimize the cost of the box with the requirement that its volume be 6 cubic feet.

SOLUTION Because the volume of the box is xyz and is to be 6 cubic feet, we must have $xyz = 6$ or $z = \dfrac{6}{xy}$. When we substitute this into the formula for the cost from Question 4, we obtain the following formula for the cost of the box in terms of x and y:

$$C(x,y) = 6xy + 4xz + 2yz$$
$$= 6xy + 4x\left(\frac{6}{xy}\right) + 2y\left(\frac{6}{xy}\right) = 6xy + \frac{24}{y} + \frac{12}{x}. \tag{2}$$

The first-order derivatives of this function are

$$C_x = \frac{\partial}{\partial x}[6xy + 24y^{-1} + 12x^{-1}] = 6y - 12x^{-2} = 6(y - 2x^{-2})$$
$$C_y = \frac{\partial}{\partial y}[6xy + 24y^{-1} + 12x^{-1}] = 6x - 24y^{-2} = 6(x - 4y^{-2}). \tag{3}$$

The critical points of $C(x,y)$ are the solutions of the equations $C_x = 0, C_y = 0$, which, because of (3) are equivalent to

$$\begin{cases} y - \dfrac{2}{x^2} = 0 \\ x - \dfrac{4}{y^2} = 0. \end{cases} \tag{4}$$

The first of equations (4) gives $y = \dfrac{2}{x^2}$, which, when substituted into the second equation, yields $x = \dfrac{4}{(2/x^2)^2}$ or $x = x^4$. We write this in the form $x - x^4 = 0$ and then $x(1 - x^3) = 0$. The last equation has solutions $x = 0$ and $x = 1$, but $x = 0$ cannot be used in (4), so the only solution is with $x = 1$. Then the first of equations (4) implies that $y = 2$ and the only critical point of $C(x,y)$ is $(1,2)$.

Formula (2) shows that $C(x,y)$ is very large for very small x, for very small y, and if x is large and y is not too small, and if y is large and x is not too small.[†] Consequently, $C(x,y)$ has a minimum value for $x > 0, y > 0$, which occurs at the critical point $(1,2)$. The cost is a mimumum for $x = 1, y = 2$, for which $z = 6/(xy) = 3$. The box should be manufactured to be 1 foot wide, 2 feet long, and 3 feet high. □

Question 6 What is the minimum cost of the box in Example 3?

[†]We might also reason as follows: For $0 < x < \frac{1}{10}$, the third term, $12/x$, in (2) is greater than 120; for $0 < y < \frac{1}{10}$, the second term, $24/y$, is greater than 240; for $x > 10, y \geq \frac{1}{10}$ and for $x \geq \frac{1}{10}, y > 10$ the first term, $6xy$, is ≥ 60. Consequently, $C(x,y) > 60$ outside the rectangle $\frac{1}{10} \leq x \leq 10, \frac{1}{10} \leq y \leq 10$ and the value 36 at (1,2) has to be the minimum. We conclude that the cost is a minimum at the critical point $(1,4)$.

In the next Question we apply Theorem 1 to a function of three variables.

Question 7 The function $N(x, y, z) = x^2 e^{-x} + 4y - y^4 - z^2$ has a global maximum. What is it?

The Second-Derivative Test with two variables

As we saw in Chapter 3, we can often determine whether a function $f(x)$ of one variable has a local maximum or minimum at a point $x = x_0$ where its first derivative is zero and its tangent line is horizontal by calculating its second derivative at that point. If $f''(x_0)$ is positive, then the graph of $f(x)$ is concave up at $x = x_0$ and is above the tangent line for x near x_0, so the function has a local minimum at that point (Figure 10). If $f''(x_0)$ is negative, then the graph of $f(x)$ is concave down at $x = x_0$ and below the tangent line for x near x_0 and the function has a local minimum there (Figure 11). If $f''(x_0)$ is zero, then this Second-Derivative Test fails and we have to analyze the function further to determine if it has a local maximum or minimum at the point.

FIGURE 10　　　　　　　　　　　　FIGURE 11

To deal with the case of two variables, we suppose that $f(x, y)$ has continuous second-order derivatives in an open disk containing the point (x_0, y_0) and that $f_x(x_0, y_0)$ and $f_y(x_0, y_0)$ are zero, so that (x_0, y_0) is a critical point of f. To determine whether $f(x, y)$ has a local maximum or minimum at (x_0, y_0), we will approximate it by the second-degree polynomial

$$P(x, y) = \tfrac{1}{2}A(x - x_0)^2 + B(x - x_0)(y - y_0) + \tfrac{1}{2}C(y - y_0)^2 + f(x_0, y_0) \tag{5}$$

where

$$A = f_{xx}(x_0, y_0), \ B = f_{xy}(x_0, y_0), \ C = f_{yy}(x_0, y_0). \tag{6}$$

Question 8 Show that the polynomial $P(x, y)$ defined by **(5)** and **(6)** has the same value and the same first- and second-order derivatives as $f(x, y)$ at (x_0, y_0).

Because $P(x, y)$ has the same value and the same first- and second-order derivatives as $f(x, y)$ at (x_0, y_0), it is the second-degree polynomial that best approximates $f(x, y)$ near (x_0, y_0).

If A, B, and C are all zero, then $P(x, y)$, given by **(5)**, is constant and its graph is a horizontal plane. Otherwise, all of the vertical cross sections of its graph are either parabolas or straight lines, depending on the values of A, B, and C. The graph $z = P(x, y)$ is one of three types that is determined by the shape of its horizontal cross sections. The graph is an elliptic parabololoid with a bowl-like shape (Figure 12) if its horizontal cross sections are ellipses, a hyperbolic paraboloid with a saddle shape (Figure 13) if its horizontal cross sections are hyperbolas, and a parabolic ridge or trough (Figure 14) if its horizontal cross sections are lines.

11.5 Maxima and minima with two or three variables

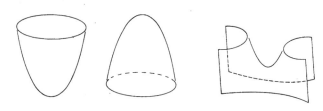
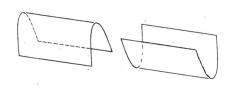

Elliptic paraboloids
FIGURE 12

Hyperbolic paraboloid
FIGURE 13

Parabololic ridge and trough
FIGURE 14

The intersections of the graph of P with the horizontal planes $z = c$ have the equations

$$\tfrac{1}{2}A(x - x_0)^2 + B(x - x_0)(y - y_0) + \tfrac{1}{2}C(y - y_0)^2 + f(x_0, y_0) = c, z = c \qquad (7)$$

with constants c. By Theorem 1 of Section 9.2 they are ellipses if $AC - B^2$ is positive, they are hyperbolas if $AC - B^2$ is negative, and they are parabolas if $AC - B^2$ is zero.

If $AC - B^2$ is positive, then A and B are both positive or both negative. If $AC - B^2, A$, and B are all positive, then the graph of P is an elliptic paraboloid that opens upward (Figure 15). The graph of P approximates the graph of f so closely that all of the vertical cross sections of the graph of f through $x = x_0, y = y_0$ are concave up at $(x_0, y_0, f(x_0, y_0))$ (Figure 16), and f has a local minimum at (x_0, y_0).

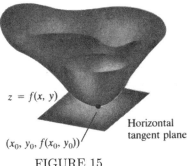
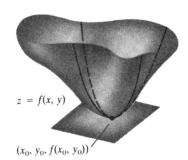

FIGURE 15

FIGURE 16

If $AC - B^2$ is positive and A and C are negative, then the approximating surface is an elliptic paraboloid that opens downward (Figure 17), all of the vertical cross sections of the graph of f through $x = x_0, y = y_0$ are concave down at $(x_0, y_0, f(x_0, y_0))$ (Figure 18), and f has a local maximum at (x_0, y_0).

FIGURE 17

FIGURE 18

If $AC - B^2$ is negative, then the approximating surface is a hyperbolic paraboloid (Figure 19). Some of the vertical cross sections of the graph of f are concave up and others are concave down at $(x_0, y_0, f(x_0, y_0))$ (Figure 20), and f has neither a local maximum nor a local minimum at (x_0, y_0). In this case f is said to have a SADDLE POINT at (x_0, y_0).

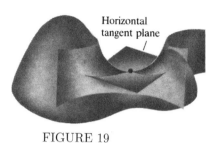

FIGURE 19　　　　　　　　　　FIGURE 20

In the one remaining case, where $AC - B^2 = 0$ and the approximating surface is a parabolic trough or ridge, we cannot tell from its second derivatives at (x_0, y_0) alone whether f has a local maximum or a local minimum at that point

We state these results in the next Theorem.

Theorem 2 (The Second-Derivative Test with two variables) *Suppose that $f(x, y)$ has continuous second-order derivatives in an open disk centered at (x_0, y_0) and that (x_0, y_0) is a critical point of f. Set $A = f_{xx}(x_0, y_0), B = f_{xy}(x_0, y_0)$, and $C = f_{yy}(x_0, y_0)$* **(a)** *If $AC - B^2, A$, and C are all positive, then f has a local minimum at (x_0, y_0).* **(b)** *If $AC - B^2$ is positive and A and C are negative, then f has a local maximum at (x_0, y_0).* **(c)** *If $AC - B^2$ is negative, then f has a saddle point at (x_0, y_0).*

If $AC - B^2$ is zero under the conditions of Theorem 2, then this Second Derivative Test fails and we would have to analyze the function more closely to determine its behavior near (x_0, y_0).

The number $AC - B^2 = f_{xx}(x_0, y_0)f_{yy}(x_0, y_0) - [f_{xy}(x_0, y_0)]^2$ is called the DISCRIMINANT of f at (x_0, y_0). It is often given as the determinant

$$\begin{vmatrix} f_{xx} & f_{xy} \\ f_{xy} & f_{yy} \end{vmatrix} = f_{xx}f_{yy} - (f_{xy})^2.$$

A more complete proof of Theorem 2 is given at the end of this section.

Example 4　Figure 21 shows the graph of $f = -x^4 - y^4 - 4xy + \frac{1}{16}$ and Figure 22 shows its level curves. Find its critical points and use the Second-Derivative Test to classify them.

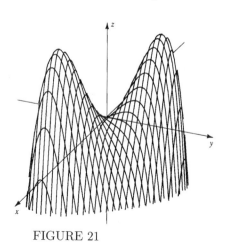
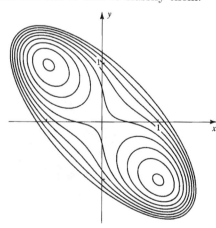

FIGURE 21　　　　　　　　　　FIGURE 22

11.5 Maxima and minima with two or three variables

SOLUTION   We find the first-orer derivatives:

$$f_x = \frac{\partial}{\partial x}(-x^4 - y^4 - 4xy + \tfrac{1}{16}) = -4x^3 - 4y = -4(x^3 + y)$$
$$f_y = \frac{\partial}{\partial y}(-x^4 - y^4 - 4xy + \tfrac{1}{16}) = -4y^3 - 4x = -4(y^3 + x).$$

(8)

Setting $f_x = 0$, and $f_y = 0$ gives the equations $y = -x^3, x = -y^3$ for the critical points. We substitute the first of these equations into the second to obtain the equation $x = -(-x^3)^3$ or $x - x^9 = 0$ for x. Factoring it as $x(1 - x^8) = 0$ shows that it has the three solutions $x = 0, 1, -1$. Then with the equation $y = -x^3$, we see that the critical points are $(0,0), (1,-1)$, and $(-1,1)$.

From formulas (8) we obtain

$$f_{xx} = \frac{\partial}{\partial x}(-4x^3 - 4y) = -12x^2$$
$$f_{xy} = \frac{\partial}{\partial y}(-4x^3 - 4y) = -4$$
$$f_{yy} = \frac{\partial}{\partial y}(-4y^3 - 4x) = -12y^2.$$

We then make the following table of values of $A = f_{xx}, B = f_{xy}, C = f_{yy}$, and $AC - B^2$ at the critical points:

Critical point	$A = f_{xx}$,	$B = f_{xy}$	$C = f_{yy}$	$AC - B^2$
$(0,0)$	0	-4	0	$0(0) - (-4)^2 = -16$
$(1,-1)$	-12	-4	-12	$-12(-12) - 4^2 = 128$
$(-1,1)$	-12	-4	-12	$-12(-12) - 4^2 = 128$

The function $f(x, y)$ has a saddle point at $x = 0, y = 0$ because the discriminant $AC - B^2$ is negative there. The function has local maxima at $x = 1, y = -1$ and at $x = -1, y = 1$ because $AC - B^2$ is positive and A and C are negative at those points. These properties of the function can be seen from its graph and level curves in Figures 21 and 22. □

Question 9 Apply the Second-Derivative Test to (a) $f(x,y) = x^2 + y^2$ from Question 1, (b) $g(x,y) = -x^2 - y^2$, (c) $h(x,y) = x^2 - y^2$, and (d) $k(x,y) = xy$.

You can use the functions in Question 9 to recall the signs in the Second-Derivative Test.

Maxima and minima on curves

Imagine that the curve in Figure 23 is a mirror and that a viewer at point F_2 is looking at the image in the mirror of an object at point F_1. We want to find the location P of the image in the mirror. According to FERMAT'S PRINCIPLE from physics, P will be the point on the mirror such that the total distance

$$f(P) = \overline{PF_1} + \overline{PF_2} \tag{9}$$

that the light travels from the object to the viewer is a minimum.

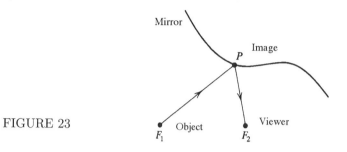

FIGURE 23

For each number c that is greater than the distance $\overline{F_1 F_2}$ between the object and the viewer, the level curve $\overline{PF_1} + \overline{PF_2} = c$ of the distance $f(P)$ is an ellipse. Figure 24 shows eight of these ellipses.[†] The value of $f(P)$ is greater on larger ellipses, so the minimum value of $f(P)$ for P on the mirror can be determined by finding the smallest ellipse that touches the mirror and P is the point where that ellipse touches the mirror, as is shown in Figure 24.

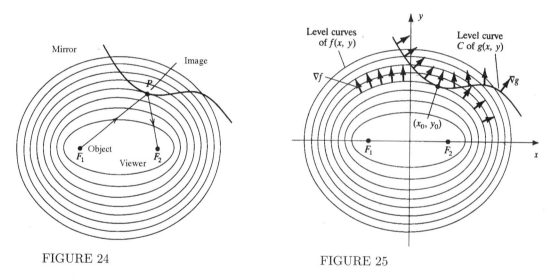

FIGURE 24　　　　　　　　　　FIGURE 25

The key to finding the point P is that the smallest ellipse that touches the mirror in Figure 24 is tangent to the mirror at P. To see why this is the case, imagine a slightly larger ellipse. It would intersect the mirror at two nearby points, and a line through these points would be a secant line to the mirror and to the ellpise. As the imagined ellipse shrinks down to the smallest ellipse that touches the mirror, the secant line becomes the tangent line to the mirror and to the ellipse at P, so both curves have the same tangent line at that point.

To express this idea in a form we can deal with, we introduce xy-axes, as in Figure 25, and let $f(x, y)$ be the sum **(9)** of the distances from (x, y) to F_1 and F_2. We refer to the mirror as the curve C and assume that it is a level curve $g(x, y) = c$ of another function with a nonzero gradient vector. We call C the CONSTRAINT CURVE.

[†]The ellipse $\overline{PF_1} + \overline{PF_2} = c$ could be drawn by pinning the ends of a string of length c at F_1 and F_2 and running a pencil point around inside the taut string.

11.5 Maxima and minima with two or three variables

At $P = (x_0, y_0)$ where the smallest ellipse and C intersect, $\nabla f(x_0, y_0)$ is perpendicular to the ellipse, which is its level curve, and $\nabla g(x_0, y_0)$ is perpendicular to C, which is its level curve (Figure 26). Since the curves are tangent at (x_0, y_0), the two gradient vectors are parallel, and there is a number λ such that $\nabla f(x_0, y_0) = \lambda \nabla g(x_0, y_0)$. The number λ is called a LAGRANGE MULTIPLIER. Similar reasoning could be applied to the maximum or minimum of any $f(x, y)$ on any level curve $g(x, y) = c$ and leads to the next result.

Theorem 3 *Suppose that C is a level curve of $g(x, y)$, that $\nabla g(x, y)$ is not zero on C, and that $g(x, y)$ and $f(x, y)$ have continuous first derivatives near C. Then, if $f(x, y)$ has a local maximum or a local minimum for (x, y) on C, then that maximum or minimum occurs at a point where*

$$\nabla f(x_0, y_0) = \lambda \nabla g(x_0, y_0) \tag{10}$$

for some number λ.

The local maximum or minimum of $f(x, y)$ with (x, y) restricted to the constraint curve might occur because $f(x, y)$ as a function in an open disk containing (x_0, y_0) has a local maximum or minimum at (x_0, y_0). In this case $\nabla f(x_0, y_0)$ is the zero vector and λ is zero. We describe a more complete proof of this Theorem at the end of the section.

Example 5   Use Lagrange multipliers to find the rectangle of perimeter 12 that has the smallest area.

SOLUTION   We let x denote the width and y the height of the rectangle (Figure 26). Then its area is xy and its perimeter is $2x + 2y$, so we want the maximum of xy subject to the constraint

$$2x + 2y = 12. \tag{11}$$

FIGURE 26

To apply Theorem 3, we set $f = xy$ and $g = 2x + 2y$. Then $\nabla f = \langle y, x \rangle$ and $\nabla g = \langle 2, 2 \rangle$, and the maximum occurs where

$$\langle y, x \rangle = \lambda \langle 2, 2 \rangle$$

with some number λ. This vector equation can be written as the two numerical equations

$$y = 2\lambda, \quad x = 2\lambda. \tag{12}$$

As in many Lagrange multiplier problems with two variables, there are two basic approaches. We can find λ by substituting (12) in (11) to obtain $2(2\lambda) + 2(2\lambda) = 12$ or $8\lambda = 12$ and then $\lambda = \frac{3}{2}$. Then equations (12) give $x = 2(\frac{3}{2}) = 3, y = 2(\frac{3}{2}) = 3$ as the solution.

With the other approach, we eliminate λ from (12) to obtain $y = x$ and then (11) gives $4x = 12$. We obtain again the solution $x = 3, y = 3$. Both procedures show that the rectangle with perimeter 12 and maximum area is a square. $\square$

The constraint curve in Example 5 is the line $2x + 2y = 12$ and the level curves of the area xy are the hyperbolas $xy = k$ in Figure 27. The area is a maximum at $x = 3, y = 3$, where the level curve is tangent to the constraint curve.

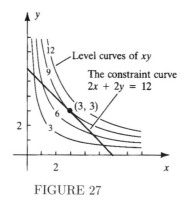

FIGURE 27

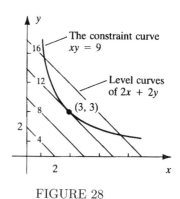

FIGURE 28

In many cases the maximum value of $f(x, y)$ on a level curve of $g(x, y)$ occurs at the same point as the minimum of $g(x, y)$ on a level curve of $f(x, y)$. This happens because in both cases the maximum or minimum occurs at a point (x, y) where $\nabla f(x, y)$ and $\nabla g(x, y)$ are parallel.

Suppose, for example, we are to find the rectangle of area 9 with minimum perimeter. The level curves of the perimeter $2x + 2y$ are the lines in Figure 28 and the constraint curve in this case is the hyperbola $xy = 9$.

Question 10 Find the minimum of $2x + 2y$ on $xy = 9$ by the method of Lagrange multipliers.

The rectangle of area 9 and minimum perimeter is the square of width 3, and the minimum perimeter occurs at $x = 3, y = 3$ in Figure 28 where the level curve of the perimeter is tangent to the constraint curve.

This, like many Lagrange multiplier problems, can also be solved with techniques of one-variable calculus that we studied in Chapter 3.

Question 11 Solve the problem in Question 10 by finding the minimum of a function of one variable.

In the next Example, we use Lagrange multipliers to find the maximum in Question 1.

Example 6 Find the maximum of $f(x, y) = x^2 + y^2$ in the disk $R : (x-1)^2 + (y-1)^2 \leq 8$ of Figure 1.

SOLUTION We saw that the only critical point of f is the origin, where f has a local and global minimum. The maximum value of f must, therefore, occur on the circle $K : (x-1)^2 + (y-1)^2 = 8$. We set $g(x, y) = (x-1)^2 + (y-1)^2$. Then $\nabla f = \langle 2x, 2y \rangle$ and $\nabla g = \langle 2(x-1), 2(y-1) \rangle$, so that at the point on K where f is a maximum

$$\langle 2x, 2y \rangle = \lambda \langle 2(x-1), 2(y-1) \rangle \tag{13}$$

with a constant λ. This gives the two equations $2x = 2\lambda(x-1), 2y = 2\lambda(y-1)$, which imply that $(\lambda - 1)x = \lambda, (\lambda - 1)y = \lambda$ and then $x = \dfrac{\lambda}{\lambda - 1}, x = \dfrac{\lambda}{\lambda - 1}$.

Instead of substituting the last formulas into the equation $(x-1)^2 + (y-1)^2 = 8$ for the circle K and solving for λ, we note that these formulas imply that $y = x$, so that $(x-1)^2 + (x-1)^2 = 8$. This shows that $(x-1)^2 = 4$, so that $x - 1 = \pm 2$ and $x = 3$ or $x = 1$. The points on the circle where **(13)** is satisfied are $(1, 1)$, where $f(1, 1) = 1^2 + 1^2 = 2$, and $(3, 3)$, where $f(3, 3) = 3^2 + 3^2 = 18$. The maximum value of f is 18, as you probably found in answering Question 1. □

11.5 Maxima and minima with two or three variables

Example 7 Figure 29 shows level curves of the yield of corn, measured in thousand of pounds per acre, that a farmer will obtain if he applies x acre-feet of irrigation water and y pounds of fertilizer per acre during the growing season.[†] Suppose that the water costs $60 per acre-foot, the fertilizer costs 9 dollars per pound and the farmer has $180 to invest per acre for water and fertilizer. How much water and how much fertilizer should he buy to maximize his yield?

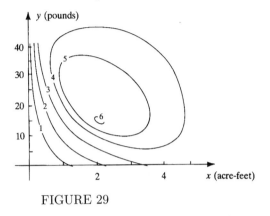
FIGURE 29

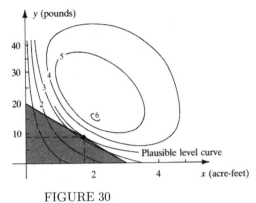
FIGURE 30

SOLUTION It would cost $C(x, y) = 60x + 9y$ dollars for x acre-feet of water and y-pounds of fertilizer, so x and y must satisfy $x \geq 0, y \geq 0, 60x + 9y \leq 180$ and (x, y) must be in the shaded triangle of Figure 30. As the drawing shows, the maximum yield occurs on the boundary line $60x + 9y = 180$. To estimate where it occurs, we draw a plausible level curve of the yield tangent to the line, as in Figure 30, and use the coordinates of the point of tangency. The farmer should buy approximately 1.75 acre-feet of water and 8.3 pounds of fertilizer for each acre. □

Lagrange multipliers with three variables

The Method of Lagrange Mulitpliers for functions of three variables is based on the principle that a maximum or minimum of $f(x, y, z)$ restricted to a level surface of $g(x, y, z)$ at a point where ∇g is not zero must occur either at a critical point of f or at a point where the level surfaces of f and g are tangent and, consequently, ∇f and ∇g are parallel.

Theorem 4 *Suppose that Σ is a level surface of $g(x, y, z)$, that $\nabla g(x, y, z)$ is not zero on Σ, and that $f(x, y, z)$ and $g(x, y, z)$ have continuous first derivatives in an open set containing Σ. If f, restricted to Σ, has a local maximum or local minimum on Σ, then that extreme value occurs at a point (x_0, y_0, z_0) where, for some number λ*

$$\nabla f(x_0, y_0, z_0) = \lambda \nabla g(x_0, y_0, z_0). \tag{14}$$

The Lagrange multiplier λ is zero in (14) if $f(x, y, z)$, considered in an open ball containing (x_0, y_0, z_0) has a local maximum or minimum at (x_0, y_0, z_0).

Example 8 Find the maximum and minimum values of $f = 6x + 3y + 2z - 5$ on the ellipsoid $4x^2 + 2y^2 + z^2 = 70$.

SOLUTION We set $g = 4x^2 + 2y^2 + z^2$, so the ellipsoid is the level surface $g = 70$. We find

$$\nabla f = \langle \frac{\partial}{\partial x}, \frac{\partial}{\partial y}, \frac{\partial}{\partial z} \rangle (6x + 3y + 2z - 5) = \langle 6, 3, 2 \rangle$$

$$\nabla g = \langle \frac{\partial}{\partial x}, \frac{\partial}{\partial y}, \frac{\partial}{\partial z} \rangle (4x^2 + 2y^2 + z^2) = \langle 8x, 4y, 2z \rangle$$

[†] An acre foot of water would cover an acre one-foot deep.

so the condition $\nabla f = \lambda \nabla g$ reads $\langle 6, 3, 2 \rangle = \lambda \langle 8x, 4y, 2z \rangle$ and gives the three equations $6 = 8\lambda x$, $3 = 4\lambda y$, $2 = 2\lambda z$. These imply that

$$x = \frac{3}{4\lambda}, \ y = \frac{3}{4\lambda}, \ z = \frac{1}{\lambda} \tag{15}$$

so the condition $4x^2 + 2y^2 + z^2 = 70$ gives

$$4\left(\frac{3}{4\lambda}\right)^2 + 2\left(\frac{3}{4\lambda}\right)^2 + \left(\frac{1}{\lambda}\right)^2 = 70.$$

We simplify this equation to $\frac{70}{16}\lambda^{-2} = 70$ and then $\lambda^2 = \frac{1}{16}$, so that $\lambda = \pm\frac{1}{4}$.

With $\lambda = \frac{1}{4}$, equations **(15)** give $x = 3, y = 3, z = 4$, and with $\lambda = -\frac{1}{4}$, we obtain $x = -3, y = -3, z = -4$. The maximum and minimum of f on Σ must occur at $(3, 3, 4)$ or $(-3, -3, -4)$. We calculate $f(3, 3, 4) = 6(3) + 3(3) + 2(4) - 5 = 30$ and $f(-3, -3, -4) = 6(-3) + 3(-3) + 2(-4) - 5 = -40$. The maximum is 30 and the minimum is -40. $\square$

More on the proof of Theorem 2

Suppose that $f(x, y)$ has continuous second-order derivatives in an open disk containing (x_0, y_0) and that $f_x(x_0, y_0) = 0$ and $f_y(x_0, y_0) = 0$, so that (x_0, y_0) is a critical point of f. We use the parametric equations $x = x_0 + t, y = y_0 + mt$ of the line through (x_0, y_0) with slope m to define the cross section of $f(x, y)$ along that line:

$$F(t) = f(x_0 + t, y_0 + mt). \tag{16}$$

By the Chain Rule with two variables, $F'(t) = f_x(x_0 + t, y_0 + mt) + mf_y(x_0 + t, y_0 + mt)$ and

$$F''(t) = f_{xx}(x_0 + t, y_0 + mt) + 2mf_{xy}(x_0 + t, y_0 + mt) + m^2 f_{yy}(x_0 + t, y_0 + mt).$$

so that at $t = 0$, which corresponds to the point (x_0, y_0),

$$F''(0) = A + 2Bm + Cm^2 \tag{17}$$

with $A = f_{xx}(x_0, y_0), B = f_{xy}(x_0, y_0)$, and $C = f_{yy}(x_0, y_0)$, as in the statement of Theorem 2.

Suppose first that $AC - B^2 < 0$. Then by the Quadratic Formula, the equation $F''(0) = 0$ with $F''(0)$ given by **(17)**, has two distinct solutions

$$m = \frac{-B \pm \sqrt{B^2 - AC}}{2C}. \tag{18}$$

This implies that $F''(0)$ is positive for some values of m and negative for other values of m. Consequently the graph of the cross section **(16)** is concave up at (x_0, y_0) in some directions and concave down in others, so that $f(x, y)$ has a saddle point at (x_0, y_0), as stated in part (c) of Theorem 2.

Next, suppose that $AC - B^2$ is positive. Then the solutions **(18)** of $F''(0) = 0$ are not real numbers and $F''(0)$ is either positive for all values of m or negative for all values of m. If A and C are also positive, then $F''(0)$ is positive for all m as is the second derivative $C = f_{yy}(x_0, y_0)$ in the y-direction, so all vertical cross sections of the graph of f are concave up at (x_0, y_0) and f has a local minimum at that point. This establishes part (a) of the Theorem. Part (b) concerning local maxima can be established by similar reasoning.

11.5 Maxima and minima with two or three variables

More on the proof of Theorem 3

Suppose that $f(x,y)$ and $g(x,y)$ have continuous first-order derivatives in an open disk containing (x_0, y_0), that the curve $g(x,y) = c$ passs through (x_0, y_0), that $\nabla g(x_0, y_0) \neq \mathbf{0}$, and that $f(x,y)$ with (x,y) constrained to the surface $C: g(x,y) = c$ has a local maximum or minimum at (x_0, y_0).

By the Implicit Function Theorem of Section 11.3, a portion of the curve C near (x_0, y_0) has parametric equations $C: x = x(t), y = y(t)$ with $(x(0), y(0)) = (x_0, y_0)$ and a nonzero velocity vector $\mathbf{v}(0) = \langle x'(0), y'(0) \rangle$ at that point. Then the function $F(t) = f(x(t), y(y))$ has a local maximum or minimum at $t = 0$, so that by the First Derivative Test for functions of one variable, the derivative

$$F'(0) = f_x(x_0, y_0)x'(0) + f_y(x_0, y_0)y'(0) = \nabla f(x_0, y_0) \cdot \mathbf{v}(0)$$

is zero. Because $\mathbf{v}(0)$ is tangent to C at (x_0, y_0), $\nabla f(x_0, y_0)$ is perpendicular to C at (x_0, y_0) and consequently $\nabla f(x_0, y_0)$ is parallel to $\nabla g(x_0, y_0)$. This implies that $\nabla f(x_0, y_0) = \lambda \nabla g(x_0, y_0)$ for some number λ, as is stated in Theorem 3.

Principles and procedures

- For a function $f(x,y)$ with continuous first-order derivatives, the First Derivative Test (Theorem 1) states that any local maximum or minimum of $f(x,y)$ has to occur at a point where its graph has a horizontal tangent plane, since a critical point of such a function is a point where f_x and f_y are zero.

- In solving the pair of equations $\begin{cases} f_x(x,y) = 0 \\ f_y(x,y) = 0 \end{cases}$ to find critical points of f, it is generally a good idea to study the simpler equation first. If it yields one value of x, one value of y, or a relation between x and y, subsitute this information in the other of equations $f_x = 0$ and $f_y = 0$ and solve. If one equation yields more than one value or relation, use this information as separate cases in the other equation.

- Some narrative problems yield directly functions of two variables that are to be maximimized or minimized. In other cases (see Example 3) a function of three variables is given with a constraint that can be used to reduce the number of variables to two, leaving a maximum/minimum problem in an open set to which the First Derivative Test can be applied.

- It is often difficult to prove that there is a maximum or minimum at a critical point. Accordingly, we often do not require that this be verified in applications.

- To recall the conditions on $A = f_{xx}, B = f_{xy}$, and $C = f_{yy}$ in the Second Derivative Test, calculate A, B, and C for (i) $f = x^2 + y^2$, which has a local minimum at $(0,0)$, for (ii) $g = -x^2 - y^2$, which has a local maximum at $(0,0)$, and for (iii) $h = x^2 - y^2$ and (iv) $k = xy$, which have saddle points at $(0,0)$, as in Question 9.

- In the Method of Lagrange Multipliers with two variables, a local maximum or minimum of $f(x,y)$ with (x,y) restricted to a constraint curve $g(x,y) = c$ is found by solving the one vector equation and one scalar equation $\nabla f(x,y) = \lambda \nabla g(x,y).g(x,y) = c$, which are equivalent to the three scalar equations $f_x(x,y) = \lambda g_x(x,y), f_y(x,y) = g_y(x,y), g(x,y) = c$ with the three unknowns x, y, and λ. In the case of three variables, we obtain four equations in the four unknowns x, y, z, and λ. This analysis provides a geometric understanding of maxima and minima with constraints which is important in physics and economics.

- Some problems with two variables that can be solved by Lagrange multipliers can also be solved by using the constraint $g(x,y) = c$ to obtain a formula either for x in terms of y or of y in terms of x, which, when substituted into the formula of $f(x,y)$ gives a maximum/minimum problem with one variable that can be solved by the techniques of Chapter 3.

- In some cases the maximum or minimum of f on the constraint curve or surface $g = c$ occurs at a point P not because of the curve or surface but because f has a local maximum or minimum at P as a function not restricted to the surface . Then P is a critical point of f, $\nabla f = 0$ at P, and the Lagrange multiplier is zero in the equation $\nabla f = \lambda \nabla g$.

Tune-Up Exercises 11.5♦

A Answer provided. **O** Outline of solution provided.

T1.^{O} What is the critical point of $f(x, y) = x^2 - xy + y$?

♦ Type 3, a critical point

T2.^{O} What is the critical point of $g(x, y) = e^{xy}$?

♦ Type 3, a critical points

T3.^{O} Find the critical points of $h(x, y) = 2xy^3 - x^2 + 6y$.

♦ Type 3, critical points

T4.^{O} The function $g(x, y) = 4x - x^4 + 6y - y^2$ has a global maximum. What is it?

♦ Type 3, the First-Derivative Test

T5.^{O} Find the critical point of $h(x, y) = x^4 + 4x - xy$ and use the Second-Derivative Test to classify it.

♦ Type 3, the Second-Derivative Test

T6.^{O} Find the absolute minimum of $f(x, y) = x^2 + y^2$ on the line $3x + y = 20$ **(a)** by the Method of Lagrange Multipliers with the function $g(x, y) = 3x + y$ and **(b)** By using calculus with one variable.

♦ Type 3, Lagrange multipliers and conversion to one variable

Problems 11.5

A Answer provided. **O** Outline of solution provided.

1.^{O} Find the one critical point of $f = x^4 + 32x - y^3$.

♦ Type 3, a critical point

2.^{O} The function $g(x, y) = x^2 + \sin y$ has an infinite number of critical points. What are they?

♦ Type 3, critical points

3.º Find the four critical points of $h(x, y) = 3xy^2 + x^3 - 3x$.

♦ Type 3, critical points

4.   What are the critical points of $f(x, y) = 3xy - x^3 + y^3$?

♦ Type 3, critical points

5.^{A} Find the six critical points of $g(x, y) = x^3 + y^4 - 36y^2 - 12x$.

♦ Type 3, critical points

6.   The function $h(x, y) = 2y^2 + x^4 - 8xy$ has a global minimum. What is it?

♦ Type 3, the First-Derivative Test

7.^{A} Find the global maximum of $k(x, y) = e^{-x^2 - y^2 + 4x - 6y}$.

♦ Type 3, the First-Derivative Test

8. $G(x, y) = 2x^2 - x^4 + 10y - y^2$ has a global maximum. What is it?

♦ Type 3, the First-Derivative Test

9.   Describe the critical points of $H(x, y) = x^2 + 6xy + 9y^2$.

♦ Type 4, critical points

10. **(a**^{A}**)** Find all the critical points of $p(x, y) = x^2$. **(b)** Draw the graph of $p(x, y)$ and give a geometric explanation of the result of part (a).

♦ Type 4, critical points

11. Show, without using derivatives that $K(x, y) = x^2 + e^{-y^2}$ has a global minimum. Then show that the function has a critical point where the global minimum occurs.

♦ Type 3, a global minimum

12.º Find the point on the half-hyperboloid $z = \sqrt{x^2 + y^2 + 3}$ that is closest to the point $(6, 4, 0)$. (Minimize the square $f(x, y)$ of the distance between $(6, 4, 0)$ and the point on the surface with x-coordinate x and y-coordinate y.)

♦ Type 4, a global minimum

♦The Tune-up Exercises and Problems are classified by type and content. The types are (1) basic, reactive; (2) basic reflective; (3) intermediate, reactive; (4) intermediate, reflective; (5) advanced, reactive; (6) advanced, reflective; and (7) advanced, theoretical.

11.5 Maxima and minima with two or three variables

13. A rectangular box of volume 24 cubic feet is to be constructed with material that costs $1.50 per square foot for the sides, $2.25 per square foot for the front and back, and $3 per square foot for the top and bottom. (**a^A**) Give a formula for the cost $C(x,y)$ of the box in terms of the width x and depth y of its base. (**b**) What dimensions would minimize the cost of the box?

♦ Type 4, a minimum narrative problem

14.º Find the critical points of $f = x^3y - 12xy$ and use the Second-Derivative Test to classify them

♦ Type 3, the Second-Derivative Test

15.^A Find the critical points of $g = x \sin y$ and use the Second-Derivative Test to classify them

♦ Type 3, the Second-Derivative Test

16. Show that $h = 4x^5 - 20xy - 5y^4$ has one saddle point and one local maximum.

♦ Type 3, the Second-Derivative Test

17. Use the Second-Derivative Test to classify the critical point of $k = e^{xy}$.

♦ Type 3, the Second-Derivative Test

18. The graphs of (**a^O**) $f = x^3 - y^3 - 3xy + \frac{1}{8}$, (**b^A**) $f = \frac{3}{2}y - \frac{1}{2}y^3 - x^2y + \frac{1}{16}$, (**c**) $f = y^4 - 2y^2 + x^2 - \frac{17}{16}$, (**d^O**) $f = -2x^3 - 3y^4 + 6xy^2 + \frac{1}{16}$, (**e**) $f = y^4 - x^4 - 2y^2 + 2x^2 + \frac{1}{16}$, and (**f**) $f = -y^4 - 4xy - 2x^2 + \frac{1}{16}$ are shown in Figures 31 through 36. Find and classify their critical points and then match the functions to their graphs.

♦ Type 3, the Second-Derivative Test

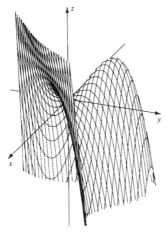

FIGURE 31

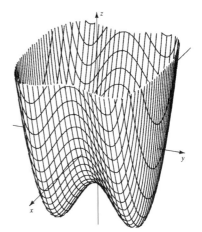

FIGURE 32

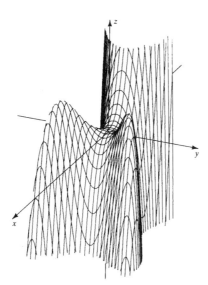

FIGURE 33

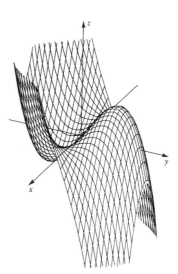

FIGURE 34

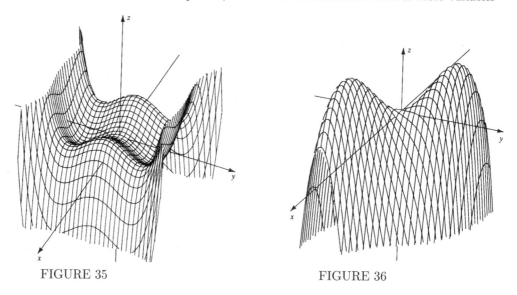

FIGURE 35

FIGURE 36

19. Level curves of the six functions of Problem 18 are shown in Figures 37 through 42. Match the functions to their level curves.

♦ Type 3, matching level surfaces to types of critical points

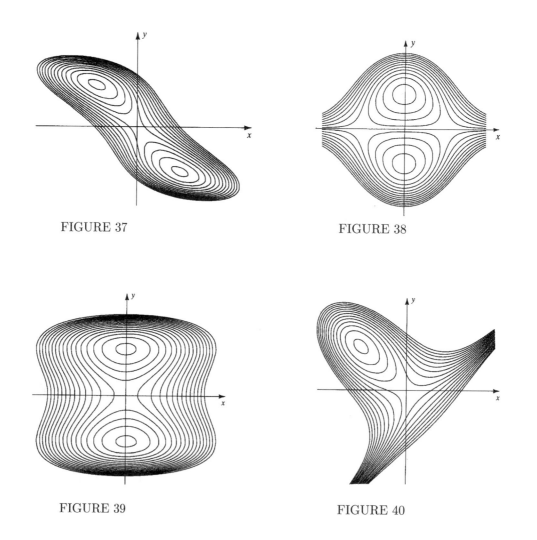

FIGURE 37

FIGURE 38

FIGURE 39

FIGURE 40

11.5 Maxima and minima with two or three variables

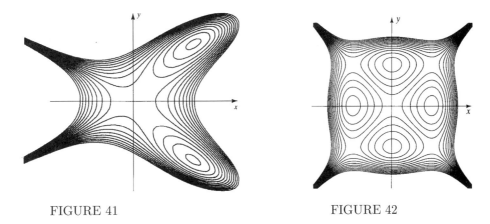

FIGURE 41

FIGURE 42

20. Figure 43 shows level curves in the Eastern Hemisphere of temperature range T, calculated by subtracting the average January temperature from the average June temperature. **(a)** What is the approximate absolute maximum of T in the Eastern Hemisphere? **(b)** What are the approximate local maxima of T in Africa?

♦ Type 3, reading level curves

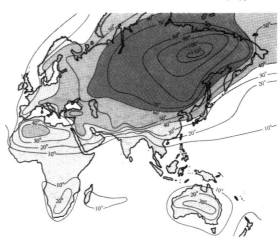

FIGURE 43

In Problems 21 through 26 use Lagrange Multipliers to find the maximum and/or minimum values and the points where they occur.

21.ᴼ The maximum and minimum of $f = 2x + y$ for $x^2 + 2y^2 = 18$
♦ Type 3, Lagrange multipliers

22.ᴬ The maximum and minimum of $f = x + 4y - 2$ for $2x^2 + y^2 = \frac{33}{2}$
♦ Type 3, Lagrange multipliers

23. The maximum and minimum of $f = 4x + y - 2$ for $x^2 + 2y^2 = 66$
♦ Type 3, Lagrange multipliers

24.ᴼ The maximum and minimum of $f = x + 5z$ for $2x^2 + 3y^2 + 5z^2 = 198$
♦ Type 3, Lagrange multipliers

25. The maximum and minimum of $f = 5 - 6x - 3y - 2z$ for $4x^2 + 2y^2 + z^2 = 70$
♦ Type 3, Lagrange multipliers

26.ᴼ The maximum and minimum of $f = xy$ for $4x^2 + 9y^2 = 72$
♦ Type 3, Lagrange multipliers

27.ᴬ The maximum and minimum of $f = x^2 + 2x + y^2$ for $3x^2 + 2y^2 = 48$
♦ Type 3, Lagrange multipliers

28. The maximum and minimum of $f = x^2 y$ for $x^2 + 8y^2 = 24$
♦ Type 3, Lagrange multipliers

29. The minimum of $f = 2x^2 + 3y^2$ for $3x + 4y = 59$
♦ Type 3, Lagrange multipliers

30. The minimum of $f = 3x^2 + y^2$ for $y - 3x = 8$
♦ Type 3, Lagrange multipliers

31.º What are the maximum and minimum of $f = x^2y^2$ for $x^2 + 4y^2 \leq 24$?

♦ Type 4, Lagrange multipliers and the First-Derivative Test

32.^{A} What are the maximum and minimum of $f = x^2 + 2x + y^2$ for $x^2 + y^2 \leq 4$?

♦ Type 4, Lagrange multipliers and the First-Derivative Test

33. Does $f(x, y, z) = 2x^2 + 3y^2 + z^2$ have a maximum or a minimum on the plane $x + 2y + z = 17$? Find the maximum or minimum and where it occurs.

♦ Type 4, Lagrange multipliers with three variables

34. Use the Second-Derivative Test to classify the critical points of **(a**^{O}**)** $f = 4x^2y + 3xy^2 - 12xy$, **(b**^{A}**)** $g = xye^{-x-y}$, **(c)** $h = 4x^5 - 20xy - 5y^4$, and **(d)** $k = x^3y - 9xy$.

♦ Type 4, the Second-Derivative Test

35. **(a**^{A}**)** Find the critical points of $g = \cos(x - y)$. **(b)** At which critical points does g have a global maximum and at which does it have a global minimum. **(c)** Use the Second-Derivative Test to classify the critical points.

♦ Type 4, the Second-Derivative Test

36. Figure 44 shows level curves of $h = -\frac{1}{2}x^2y - x^2 - y^2 - 2y + \frac{1}{16}$ at levels $z = k/4$ with integers k. Find and label the curves where h has the values $0, \pm 1, \pm 2$ by analyzing its critical points.

♦ Type 4, the Second-Derivative Test

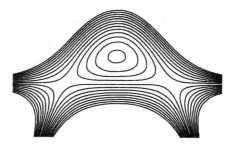

FIGURE 44

37. The function $K = \dfrac{xy}{1 + x^4 + y^4}$ has a global maximum and a global minimum. Find them.

♦ Type 4, the First-Derivative Test

38. The function $M = 4x^4 + y^3 + z^2 - x^2 - 3y - 4z$ has a minimum value for $x \geq 0, y \geq 0, z \geq 0$. What is it?

♦ Type 4, the First-Derivative Test

39. Find and classify the critical points of $f(x, y) = F(x) + G(y)$ where the graphs of $F(x)$ and $G(y)$ are in Figures 45 and 46. The only critical points of $F(x)$ are $x = \pm 1$ and the only critical point of $G(y)$ is $y = 0$.

♦ Type 4, the Second-Derivative Test

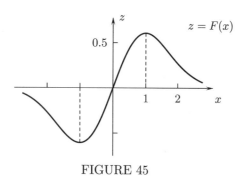
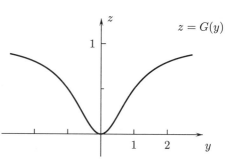

FIGURE 45   FIGURE 46

40. Find the maximum and minimum values of $f = 3x^2y^2 + 2x^3 + 6y$ on the square $0 \leq x \leq 2, 0 \leq y \leq 2$.

♦ Type 4, the First-Derivative Test

11.5 Maxima and minima with two or three variables

Use a calculator or computer algorithm for finding approximate solutions of equations with one variable to find the approximate critical points of the function in Problems 41 through 44. Then use the Second Derivative Test to classify the critical points.

41.º $f = 4x - \frac{1}{2}x^2 - \frac{1}{4}x^4 + 4y^2 + y - y^4$
♦ Type 5, the First and Second-Derivative Tests

43. $f = y^3 - 3y + \frac{1}{600}xe^{x^2} - x$
♦ Type 5, the First and Second-Derivative Tests

42.^{A} $f = x^4 - 4xy^2 - 4y^2 + \frac{8}{3}y^3$
♦ Type 5, the First and Second-Derivative Tests

44. $f = 4xy + y^2 - x^4 - y^4$
♦ Type 5, the First and Second-Derivative Tests

45.^{A} Find the four points on the surface $z = \dfrac{1}{xy}$ that are closest to the origin $(0,0,0)$ by finding the minimum of the square of the distance from $(x, y, (xy)^{-1})$ on the surface to the origin.
♦ Type 5, the First-Derivative Test

46. Use a calculator or computer algorithm for finding approximate solutions of equations with one variable to find the approximate coordinates of the point on the surface $z = xy$ that is closest to $(1, 2, 0)$. (See Problem 45.)
♦ Type 5, the First-Derivative Test

47. Find the dimensions of the rectangular box of maximum volume whose base is the rectangle with corners $(\pm x, \pm y)$ in the xy-plane and whose upper corners are on the elliptic paraboloid $z = 4 - 9x^2 - y^2$.
♦ Type 4, a maximum narrative problem

48. Find the vertices of the tetrahedron(s) of least volume whose sides are formed by the coordinate planes and by a plane with positive intercepts that passes through the point $(1, 2, 3)$.
♦ Type 4, a minimum narrative problem

49. Suppose it costs $C(x)$ dollars to produce x units of a product, that x units sold in one market bring in a revenue of $R_1(x)$ dollars, and that x units sold in a second market bring in $R_2(x)$ dollars. **(a)** Explain why, if x units are sold in the first market and y units in the second market, then the profit from both markets is $C(x + y) - R_1(x) - R_2(y)$ dollars. **(b)** Show that if the profit is maximized by selling x units in the first market and y units in the second, then $R_1'(x) = R_2'(y)$.
♦ Type 6, a maimum narrative problem

50. Find the point (x, y) such that the sum of the squares of its distances from $(x_1, y_1, z_1), (x_2, y_2, z_2)$, and (x_3, y_3, z_3) is a minimum.
♦ Type 4, a minimum narrative problem

51. Use the First Derivative Test to find the point on the plane $z = 2x + 3y$ that is closest to the point $(4, 2, 0)$.
♦ Type 4, a minimum narrative problem

52. Use a sketch of the level curves of $f(x, y, z) = x^2 + 2x + y^2$ to determine its maximum value on the square $|x| + |y| = 4$ and the method of Lagrange multipliers to find its minimum value on the square. Why can Lagrange multipliers not be used to find the maximum?.
♦ Type 4, Lagrange multipliers interpreted

53. The function $f(x, y, z) = \dfrac{4}{x} + \dfrac{1}{y} + \dfrac{1}{z} + xyz$ has a minimum for $x > 0, y > 0, z > 0$. What is it?
♦ Type 4, a global minimum with three variables

54. Find the point on the elliptical paraboloid $z = x^2 + 2x + 2y^2 - 6$ that is closest to the origin.
♦ Type 4, a global minimum with three variables

55. Show that the Second-Derivative Test fails at the critical points of **(a)** $f = (x - y)^4 + (x + y + 2)^2$, **(b)** $g = 1 - x^4 - y^4$ and **(c)** $h = x^3 + (x - y)^2$. Then show that f has a local and global minimum at its critical point, g has a local and global maximum at its critical point, and h has neither a local maximum nor a local minimum at its critical point.
♦ Type 4, the Second-Derivative Test analyzed

56. **(a)** Show by sketching the level curves of $f = x - y$ and the constraint curve $C : y - x - x^5 = 2$ that the method of Lagrange multipliers yields a point where f does not have a local maximum or minimum on C. **(b)** Show by sketching the level curves of $f = e^y$ and the constraint curve $C : y = 3x - x^3$ that the method of Lagrange multipliers yields points where f has a local maximum or minimum on C but not a global maximum or minimum.
♦ Type 4, Lagrange multipliers analyzed

57. A rectangular box with corners at $(\pm x, \pm y, \pm z)$ is inscribed in the ellipsoid $z^2 = 1 - x^2 - 4y^2$. What choices of $x, y,$ and z would maximize the total length of its twelve edges?

♦ Type 4, a maximum narrative problem

58. Find the maximum of $f(x, y, z) = \sin x \sin y \sin z$ for $x, y,$ and z the interior angles of a triangle.

♦ Type 6, a Lagrange multipliers narrative problem

59. Show that the rectangular box of maximum volume that can be inscribed in a sphere is a cube.

♦ Type 6, a Lagrange multipliers narrative problem

60. Find the maximum volume of a rectangular box with sides parallel to the coordinate planes in xyz-space, with one corner at the origin, and with the diagonally opposite corner on the plane $\dfrac{x}{a} + \dfrac{y}{b} + \dfrac{z}{c} = 1$ with positive constants $a, b,$ and c.

♦ Type 6, a Lagrange multipliers narrative problem

61. Find the three positive numbers $x, y,$ and z which maximize the quantity $x^3 y^4 z^5$ under the condition that $x + y + z = 1$.

♦ Type 6, a Lagrange multipliers narrative problem

62. Prove that $\sqrt[3]{xyz} \leq \frac{1}{3}(x + y + z)$ for nonnegative $x, y,$ and z, by finding the minimum of $x + y + z$ for $z = k/(xy)$ with a positive parameter k.

♦ Type 6, a Lagrange multipliers narrative problem

RESPONSES TO QUESTIONS IN CHAPTER 11
Responses 11.1

Response 1 (a) $g(3,1) = 1^2 - 3^2 = -8$ • $g(1,3) = 3^2 - 1^2 = 8$
(b) The domain of g is the entire xy-plane. • Because $g(x,0) = -x^2$ takes on all nnegative values and $g(0,y) = y^2$ assumes all nonnegative values, the range of g is the infinite interval $(-\infty, \infty)$.

Response 2 (a) With the axes oriented as in Figure 4 of Section 11.1, increasing c moves the plane $x = c$ to the front.
(b) With the axes oriented as in Figure 7, increasing c moves the plane $y = c$ to the right.

Response 3 (a) The cross section of $z = y^2 - x^2$ in the plane $y = c$ has the simultaneous equations

$$\begin{cases} z = y^2 - x^2 \\ y = c \end{cases} \quad \text{or} \quad \begin{cases} z = c^2 - x^2 \\ y = c. \end{cases}$$

One description: Each of these cross sections is a parabola that opens downward. • If c is negative, it is on the left of the surface in Figure 13 and its vertex is above the xy-plane. • As c increases to 0, the parabola moves down and to the right and for $c = 0$ its vertex is at the origin. • As c increases through positive values, the parabola moves up and to the right.

(b) The cross section of $z = y^2 - x^2$ in the plane $x = c$ has the simultaneous equations

$$\begin{cases} z = y^2 - x^2 \\ x = c \end{cases} \quad \text{or} \quad \begin{cases} z = y^2 - c^2 \\ x = c. \end{cases}$$

One description: Each of these cross sections is a parabola that opens upward. • If c is negative, it is on the back of the surface in Figure 13 and its vertex is below the xy-plane. • As c increases to 0, the parabola moves up and to the front, and for $c = 0$ its vertex is at the origin. • As c increases through positive values, the parabola moves down and to the front.

Response 4 (a) $\overline{OS}$ and $\overline{SQ}$ are both equal to $\overline{OQ}/\sqrt{2}$, which equals $x'/\sqrt{2}$. • $\overline{RQ}$ and $\overline{SQ}$ are both equal to $\overline{QP}/\sqrt{2}$, which equals $y'/\sqrt{2}$.
(b) $x = \overline{OS} - \overline{RQ}$ • $y = \overline{SQ} + \overline{RP}$.

Response 5 $D = -2, E = -4, F = -6$ because the surface drops down at the front and back

Response 6 One approach: Look at the values of g on the y-axis, where $x = 0$ and $g(0,y) = y - \frac{1}{12}y^3$. • $g(0,0) = 0$, so the level curve through the origin is $g = 0$. • $g(0,2) \doteq 1.33$, so the level curve around $(0,2)$ is $g = 1$. • $g(0,-1) \doteq -0.92$, so the curve $g = -1$ intersects the y-axis just above $(0,-1)$. • The level curves are at integer values of g, so counting down shows that $g = -5$ is the outer curve at the top and counting up shows that other parts of $g = 0$ and $g = 1$ are at the bottom and $g = 5$ is the outermost curve at the bottom, as shown in Figure R6

Response 7 (a) $\sqrt{2 - x^2 - y^2}$ is defined for $x^2 + y^2 \leq 2$.
(b) $\sqrt{2 - x^2 - y^2} \to 0$ as $(x,y) \to (1,1)$ in the domain of the function.

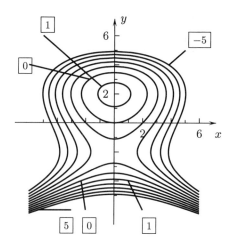

Figure R6

Responses 11.2

Response 1
$F_x(x,y) = \dfrac{\partial}{\partial x}(x^3 y^6 + 3x - 4y) = \left[\dfrac{\partial}{\partial x}(x^3)\right] y^6 + 3\dfrac{\partial}{\partial x}(x) - 4\dfrac{\partial}{\partial x}(y)$
$= 3x^2 y^6 + 3(1) - 4(0) = 3x^2 y^6 + 3$ •

$F_y(x,y) = \dfrac{\partial}{\partial y}(x^3 y^6 + 3x - 4y) = x^3 \dfrac{\partial}{\partial y}(y^6) + 3\dfrac{\partial}{\partial y}(x) - 4\dfrac{\partial}{\partial y}(y)$
$= x^3(6y^5) + 3(0) - 4(1) = 6x^3 y^5 - 4$

Response 2 (a) The rate of change of V with respect to h is $\dfrac{\partial V}{\partial h} = \dfrac{\partial}{\partial h}(\pi r^2 h) = \pi r^2$ and equals the area of the top of the cylinder.
(b) The increase in the volume would be $V(r, h+\Delta h) - V(r,h) = \pi r^2(h+\Delta h) - \pi r^2 h = \pi r^2 \Delta h$ and equals the area of the top multiplied by the thickness of the paint.

Response 3 (a) The plane $y=1$ is parallel to and to the right of the xz-plane in Figure 4.
(b) Because $g(x,1) = \frac{1}{3}1^3 - x^2 = \frac{1}{3} - x^2$, the intersection of the plane $y=1$ with the graph of g is given by $z = \frac{1}{3} - x^2, y = 1$. • Figure R3 •

$\left[\frac{1}{3} - x^2\right]_{x=2} = \frac{1}{3} - 4 = -\frac{11}{3}$ and $\left[\dfrac{d}{dx}(\frac{1}{3} - x^2)\right]_{x=2} = \left[-2x\right]_{x=2} = -4$ •

The tangent line in Figure R3 has the equation $z = -\frac{11}{3} - 4(x-2)$.
(c) The slope of the tangent line is the value of the partial derivative
$\dfrac{\partial g}{\partial x}(2,1) = \left[-2xy\right]_{x=2, y=1} = -4$ at $(2,1)$.

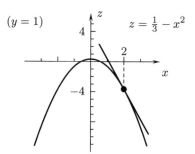

Figure R3

Response 4 (a) $P(t, E)$ is a increasing function of E for fixed t in the table except that $P(65, 150) = 195$ is less than $P(65, 100) = 200$.

(b) ONE ANSWER: $\dfrac{\partial P}{\partial E}(45, 100) \approx \dfrac{P(45, 150) - P(45, 100)}{150 - 100} = \dfrac{197 - 181}{50}$
$= 0.32$ millimeters of mercury per watt •

(b) ANOTHER ANSWER: $\dfrac{\partial P}{\partial E}(45, 100) \approx \dfrac{P(45, 100) - P(45, 50)}{100 - 50} = \dfrac{181 - 167}{50}$
$= 0.28$ millimeters of mercury per watt •

(b) A THIRD ANSWER: $\dfrac{\partial P}{\partial E}(45, 100) \approx \dfrac{P(45, 150) - P(45, 50)}{150 - 50} = \dfrac{197 - 167}{100}$
$= 0.30$ millimeters of mercury per watt

Response 5 ONE ANSWER: $\dfrac{\partial P}{\partial E}(62, 75) \approx \dfrac{P(55, 100) - P(55, 50)}{100 - 50} = \dfrac{199 - 177}{50}$
$= 0.44$ millimeters of mercury per watt •

(b) ANOTHER ANSWER: $\dfrac{\partial P}{\partial E}(62, 75) \approx \dfrac{P(65, 100) - P(65, 50)}{100 - 50} = \dfrac{200 - 181}{50}$
$= 0.38$ millimeters of mercury per watt

Response 6 (a) $\dfrac{\partial T}{\partial h}(14, 10)$ is the (instantaneaus) rate of change of the temperature with respect to depth at 2:00 PM and at a depth of 10 centimeters below the surface of the ground and is measured in degrees Fahrenheit per centimeter.

(b) Along the vertical line $t = 14$ in Figure R5, the distance between the level curves $T = 28$ above and $T = 29$ below the point $h = 10, t = 14$ is approximately 2 centimeters, measured on the h-axis. • The temperature changes $\Delta T = -1$ degree as h increases $\Delta h = 2$ centimeters. •

$\dfrac{\partial T}{\partial h}(14, 10) \approx \dfrac{\Delta T}{\Delta h} \approx \dfrac{-1 \text{ degree}}{2 \text{ centimeters}} = -\tfrac{1}{2}$ degree per centimeter.

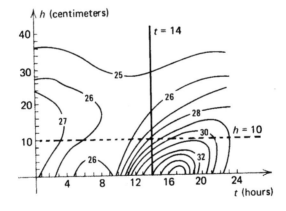

FIGURE R6

Response 7 (a) $f(x(t), y(t)) = [x(t)]^5 [y(t)]^6 = (e^t)^5 (t^{1/2})^6 = e^{5t} t^3$

(b) By the Product Rule:
$\dfrac{d}{dt}[f(x(t), y(t))] = \dfrac{d}{dt}(e^{5t} t^3) = e^{5t} \dfrac{d}{dt}(t^3) + t^3 \dfrac{d}{dt}(e^{5t}) = 3t^2 e^{5t} + t^3 e^{5t} \dfrac{d}{dt}(5t)$
$= 3t^2 e^{5t} + 5t^3 e^{5t}$

Response 8 $\lim\limits_{\Delta t \to 0} \dfrac{x_1 - x_0}{\Delta t} = \lim\limits_{\Delta t \to 0} \dfrac{x(t_0 + \Delta t_0) - x(t_0)}{\Delta t} = \dfrac{dx}{dt}(t_0)$ •

$\lim\limits_{\Delta t \to 0} \dfrac{y_1 - y_0}{\Delta t} = \lim\limits_{\Delta t \to 0} \dfrac{y(t_0 + \Delta t) - y(t_0)}{\Delta t} = \dfrac{dy}{dt}(0)$

Response 9 $\dfrac{d}{dt}[F(t^2, t^3)] = \dfrac{\partial F}{\partial x}(t^2, t^3)\dfrac{d}{dt}(t^2) + \dfrac{\partial F}{\partial y}(t^2, t^3)\dfrac{d}{dt}(t^3)$

$= 2t\dfrac{\partial F}{\partial x}(t^2, t^3) + 3t^2\dfrac{\partial F}{\partial y}(t^2, t^3)$ •

$\left[\dfrac{d}{dt}\{F(t^2, t^3)\}\right]_{t=2} = 2(2)\dfrac{\partial F}{\partial x}(2^2, 2^3) + 3(2)^2\dfrac{\partial F}{\partial y}(2^2, 2^3)$

$= 4\dfrac{\partial F}{\partial x}(4, 8) + 12\dfrac{\partial F}{\partial y}(4, 8) = 4(10) + 12(-20) = -200$

Response 10 At the time under consideration,

$\dfrac{d}{dt}(R) = \left[-2 \times 20^{-4}\ \dfrac{\text{gallons per hour}}{\text{foot}}\right][-400\ \text{feet per minute}]$

$+ \left[0.13\ \dfrac{\text{gallons per hour}}{\text{knot}}\right]\left[-10\ \dfrac{\text{knots}}{\text{minute}}\right]$

$= (-2 \times 10^{-4})(-400) + (0.13)(-10) = -1.22\ \dfrac{\text{gallons per hour}}{\text{minute}}$ •

The plane's rate of fuel consumption is decreasing 1.22 gallons per hour per minute.

Response 11 $g_x = \dfrac{\partial}{\partial x}[x^4 \sin(3y) + 5x - 6y] = 4x^3 \sin(3y) + 5$ •

$g_y = \dfrac{\partial}{\partial y}[x^4 \sin(3y) + 5x - 6y] = x^4 \cos(3y)\dfrac{\partial}{\partial y}(3y) - 6 = 3x^4 \cos(3y) - 6$ •

$g_{xx} = \dfrac{\partial}{\partial x}[4x^3 \sin(3y) + 5] = 12x^2 \sin(3y)$ •

$g_{xy} = \dfrac{\partial}{\partial y}[4x^3 \sin(3y) + 5] = 4x^3 \cos(3y)\dfrac{\partial}{\partial y}(3y) = 12x^3 \cos(3y)$ •

$g_{yx} = \dfrac{\partial}{\partial x}[3x^4 \cos(3y) - 6] = 12x^3 \cos(3y)$ •

$g_{yy} = \dfrac{\partial}{\partial y}[3x^4 \cos(3y) - 6] = -3x^4 \sin(3y)\dfrac{\partial}{\partial y}(3y) = -9x^4 \sin(3y)$

Response 12 $K_y = \dfrac{\partial}{\partial y}(e^{xy}) = e^{xy}\dfrac{\partial}{\partial y}(xy) = xe^{xy}$ •

$K_{yy} = \dfrac{\partial}{\partial y}(xe^{xy}) = xe^{xy}\dfrac{\partial}{\partial y}(xy) = x^2 e^{xy}$ •

$K_{yyy} = \dfrac{\partial}{\partial y}(x^2) = x^2 e^{xy}\dfrac{\partial}{\partial y}(xy) = x^3 e^{xy}$

Responses 11.3

Response 1 (a) $g(x, y) = e^y - 1$ • $(x_0, y_0) = (0, 0)$ • $\mathbf{u} = \langle \tfrac{1}{2}, \tfrac{1}{2}\sqrt{3}\rangle$ • $x = 0 + \tfrac{1}{2}s, y = 0 + \tfrac{1}{2}\sqrt{3}s$ • $F(s) = e^{s\sqrt{3}/2} - 1$ • Figure R1a

(b) $F'(s) = \dfrac{d}{ds}(e^{s\sqrt{3}/2} - 1) = e^{s\sqrt{3}/2}\dfrac{d}{ds}(s\sqrt{3}/2) = \tfrac{1}{2}\sqrt{3}e^{s\sqrt{3}/2}$ •

$D_{\mathbf{u}}g(0, 0) = F'(0) = \tfrac{1}{2}\sqrt{3}$ • $g(0, 0) = F(0) = 0$ •

Tangent line: $w = \tfrac{1}{2}\sqrt{3}s$ • Figure R1b

Response 2 Use $f_x(3, 1) = 6$, and $f_y(3, 1) = 45$ from Example 2. • For $\mathbf{u} = \langle \tfrac{3}{5}, \tfrac{4}{5}\rangle$:

$D_{\mathbf{u}}f(3, 1) = f_x(3, 1)u_1 + f_y(3, 1)u_2 = 6(\tfrac{3}{5}) + 45(\tfrac{4}{5}) = 39.6$

Response 3 (a) The derivative in the positive x-direction is the x-derivative.
(b) The derivative in the positive y-direction is the y-derivative.
(c) The derivative in the negative x-direction is the negative of the x-derivative.
(d) The derivative in the negative y-direction is the negative of the y-derivative.

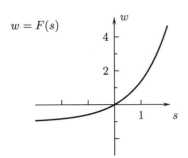

Figure R1a

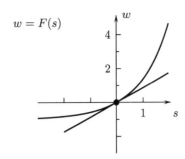

Figure R1b

Response 4 The rate of change toward the southeast is the negative of the rate of change toward the northwest from Example 4, so it is approximately 0.005 degrees per mile.

Response 5 (a) The slope of PQ in Figure 15 is m_1 and the run from P to Q is $x - x_0$, so the rise from P to Q is $m_1(x - x_0)$. • The z-coordinate of P is z_0, so the z-coordinate of Q is $z_0 + m_1(x - x_0)$.
(b) The slope of QR is m_2 and the run from Q to R is $y - y_0$, so the rise from Q to R is $m_2(y - y_0)$. • Because the z-coordinate of Q is $z_0 + m_1(x - x_0)$, the z-coordinate of R is $z_0 + m_1(x - x_0) + m_2(y - y_0)$.

Response 6 Use equation (8) with $x_0 = 1, y_0 = 2, z_0 = 3, m_1 = 4$, and $m_2 = -5$:
$z = 3 + 4(x - 1) - 5(y - 2)$

Response 7 Because we are told that h is linear, we can find its x-derivative from any two values on any horizontal line and its y-derivative from any two values on any vertical line. • One solution: $h(x, 0)$ increases by $\Delta h = 26 - 20 = 6$ as x changes by $\Delta x = 3 - 0$ from $x = 0$ to $x = 3$ on the x-axis, so $\dfrac{\partial h}{\partial x} = \dfrac{\Delta z}{\Delta x} = \dfrac{6}{3} = 2$ • $h(0, y)$ changes by $\Delta h = 14 - 20 = -6$ as y increases by $\Delta y = 2 - 0$ from $y = 0$ to $y = 2$ on the y-axis, so $\dfrac{\partial h}{\partial y} = \dfrac{\Delta z}{\Delta y} = \dfrac{-6}{2} = -3$ • $h(0, 0) = 20$ •
Use the slopes-intercept equation with $m_1 = 2, m_2 = -3$, and $b = 20$:
$h(x, y) = 2x - 3y + 20$ (h is the same as g from Example 6.)

Response 8 Set $f = 2\sin x + 3e^y + 4$. • $f_x = 2\cos x$ • $f_y = 3e^y$ •
$f(0, 0) = 2\sin(0) + 3e^0 + 4 = 7$ • $f_x(0, 0) = 2\cos(0) = 2$ • $f_y(0, 0) = 3e^0 = 3$ •
Tangent plane: $z = 7 + 2x + 3y$

Response 9 The minimum directional derivative is $-|\nabla h(2, -1)| = -\sqrt{8e^4}$. •
The minimum directional derivative is in the direction of $\mathbf{u} = \dfrac{\langle -1, 1 \rangle}{\sqrt{2}}$.

Response 10 (a) $\nabla f = \nabla(x^2 + y^2) = \langle 2x, 2y \rangle$ • $[\nabla f]_{x=0, y=0} = \langle 0, 0 \rangle$
(b) $\nabla g = \nabla(x^2 - y^3) = \langle 2x, -3y^2 \rangle$ • $[\nabla g]_{x=0, y=0} = \langle 0, 0 \rangle$

Response 11 $\nabla(xy) = \langle y, x \rangle$ equals $\langle 1, -3 \rangle$ at $(-3, 1)$ • $xy = -3$ at $(-3, 1)$ • The level curve is $xy = -3$ or $y = -3/x$ (Figure R11). □

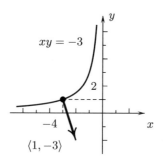

Figure R11

Responses 11.4

Response 1 $V(R, R, h) = \pi R^2 h$ is the volume of a right circular cylinder of radius R and height h (Figure R1)

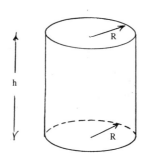

Figure R1

CylinderMay8.00a scaled 400

Response 2   Set $y = 0$ and $z = 0$ to find the x-intercepts: $\frac{1}{9}x^2 = 1$ • $x = \pm 3$ •
Set $x = 0$ and $z = 0$ to find the y-intercepts: $\frac{1}{16}y^2 = 1$ • $y = \pm 4$ •
Set $x = 0$ and $y = 0$ to find the z-intercepts: $z^2 = 1$ • $z = \pm 1$

Response 3   (a) Set $y = 0$ and $z = 0$ in $\frac{1}{3}z^2 = x^2 + y^2$ • $x^2 = 0$ • The x-intercept is 0.
Set $x = 0$ and $z = 0$ in $\frac{1}{3}z^2 = x^2 + y^2$ • $y^2 = 0$ • The y-intercept is 0.
Set $x = 0$ and $y = 0$ in $\frac{1}{3}z^2 = x^2 + y^2$ • $\frac{1}{3}z^2 = 0$ • The z-intercept is 0.
(b) Set $y = 0$ and $z = 0$ in $\frac{1}{3}z^2 = x^2 + y^2 + 1$ • $x^2 + 1 = 0$ •
There are no x-intercepts.
Set $x = 0$ and $z = 0$ in $\frac{1}{3}z^2 = x^2 + y^2 + 1$ • $y^2 + 1 = 0$ •
There are no y-intercepts.
Set $x = 0$ and $y = 0$ in $\frac{1}{3}z^2 = x^2 + y^2 + 1$ • $\frac{1}{3}z^2 = 1$ •
The z-intercepts are $z = \pm\sqrt{3}$.
(c) Set $y = 0$ and $z = 0$ in $\frac{1}{3}z^2 = x^2 + y^2 - 1$ • $x^2 - 1 = 0$ •
The x-intercepts are $x = \pm 1$.
Set $x = 0$ and $z = 0$ in $\frac{1}{3}z^2 = x^2 + y^2 - 1$ • $y^2 - 1 = 0$ •
The y-intercepts are $y = \pm 1$.
Set $x = 0$ and $y = 0$ in $\frac{1}{3}z^2 = x^2 + y^2 - 1$ • $\frac{1}{3}z^2 = -1$ • There are no z-intercepts.

Response 4   Differentiating with respect to y first yields $h_y = \dfrac{\partial h}{\partial y}(e^x \sin y \cos z) = e^x \cos y \cos z$

and $h_{yz} = \dfrac{\partial h}{\partial z}(e^x \cos y \cos z) = -e^x \cos y \sin z$. •

Differentiating with respect to z first gives $h_z = \dfrac{\partial h}{\partial z}(e^x \sin y \cos z) = -e^x \sin y \sin z$

and $h_{zy} = \dfrac{\partial h}{\partial y}(-e^x \sin y \sin z) = -e^x \cos y \sin z$.

Response 5 $\dfrac{\partial R}{\partial v}(5,10) = \left[\dfrac{\partial}{\partial v}[Q(x(u,v),y(u,v),z(u,v))]\right]_{x=5,y=10}$
$= Q_x(x(5,10),y(5,10),z(5,10))x_v(5,10)$
$+Q_y(x(5,10),y(5,10),z(5,10))y_v(5,10)$
$+Q_z(x(5,10),y(5,10),z(5,10))z_v(5,10)$

Response 6 The minimum directional derivative of $g(x,y,z) = x^3y^2z$ at $(-1,3,2)$ is the negative $-\sqrt{3141}$ of the maximum directional derivative and occurs in the opposite direction, which is the direction of the negative $\langle -54, 12, 9\rangle/\sqrt{3141}$ of the vector **u** from Example 9.

Responses 11.5

Response 1 Since $f(x,y)$ is the square of the distance from (x,y) to the origin, the minimum of f in the disk R is $f(0,0) = 0^2 + 0^2 = 0$ at $(0,0)$, and the maximum is $f(3,3) = 3^2 + 3^2 = 18$ at $(3,3)$.

Response 2 $f(x,y) = x^2 + y^2$ • $f_x(x,y) = 2x$, $f_y(x,y) = 2y$ • $f(0,0) = 0$, $f_x(0,0) = 0$, $f_y(0,0) = 0$ • Tangent plane: $z = 0$ (the xy-plane)

Response 3 (a) Because $e^{-x^2-y^2} = 1$ for $(x,y) = (0,0)$ and $e^{-x^2-y^2} < 1$ for $(x,y) \neq (0,0)$, the global maximum of $h(x,y) = 3e^{-x^2-y^2}$ is 3 at $(0,0)$.

(b) $h_x = \dfrac{\partial}{\partial x}(3e^{-x^2-y^2}) = 3e^{-x^2-y^2}\dfrac{\partial}{\partial x}(-x^2-y^2) = -6xe^{-x^2-y^2}$ •

$h_y = \dfrac{\partial}{\partial y}(3e^{-x^2-y^2}) = 3e^{-x^2-y^2}\dfrac{\partial}{\partial y}(-x^2-y^2) = -6ye^{-x^2-y^2}$ •

Since the exponential function is never zero, $h_x = 0$ and $h_y = 0$ only at $(0,0)$. • The global and local maximum is 3 at the one critical point $(0,0)$.

Response 4 (a) Because $k(x,y) = \sqrt{x^2+y^2}$ is zero at $(0,0)$ and positive elsewhere, its global minimum is 0 at the origin. This is also a local minimum because $k(x,y)$ is defined for all (x,y).

(b) $k(x,0) = \sqrt{x^2} = |x|$ does not have an x-derivative at $x = 0$ and $k(0,y) = \sqrt{y^2} = |y|$ does not have a y-derivative at $y = 0$ because their graphs consist of lines of different slopes that join at the origin (Figures R4a and R4b). • $(0,0)$ is a critical point because $k_x(0,0)$ and $k_y(0,0)$ do not exist. • The local minimum occurs at a critical point, as required by Theorem 1.

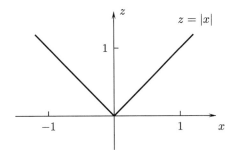
Figure R4a

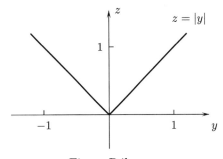
Figure R4b

Response 5 [Cost of the box] = [Cost of the bottom] + [Cost of the front and back] + [Cost of the two sides]
= [xy square feet][6 dollars per square foot]
+[$2xz$ square feet][2 dollars per square foot]
+[$2yz$ square feet][1 dollar per square foot] $= 6xy + 4xz + 2yz$ dollars

Response 6   The minimum cost is $C(1,2) = 6(1)(2) + \dfrac{24}{2} + \dfrac{12}{1} = 36$ dollars.

Response 7 $\dfrac{\partial N}{\partial x} = \dfrac{\partial}{\partial x}[x^2 e^{-x} + 4y - y^4 - z^2] = x^2 \dfrac{\partial}{\partial x}(e^{-x}) + e^{-x}\dfrac{\partial}{\partial x}(x^2)$
$= -x^2 e^{-x} + 2xe^{-x} = (2-x)xe^{-x}$ •
$\dfrac{\partial N}{\partial y} = \dfrac{\partial}{\partial y}[x^2 e^{-x} + 4y - y^4 - z^2] = 4 - 4y^3 = 4(1-y^3)$ •
$\dfrac{\partial N}{\partial z} = \dfrac{\partial}{\partial z}[x^2 e^{-x} + 4y - y^4 - z^2] = -2z$ •
All three derivatives are zero at $(0,1,0)$ and $(2,1,0)$, the two critical points. •
$N(0,1,0) = 0^2 e^0 + 4(1) - 1^2 - 0^2 = 3$ • $N(2,1,0) = 2^2 e^{-2} + 4(1) - 1^2 - 0^2 = 4e^{-2} + 3$
• Because we are told that N has a global maximum and it is defined for all (x,y,z), the global maximum is a local maximum and is $N(2,1,0) = 4e^{-2} + 3$.

Response 8 $P(x,y) = \tfrac{1}{2}A(x-x_0)^2 + B(x-x_0)(y-y_0) + \tfrac{1}{2}C(y-y_0)^2 + f(x_0, y_0)$ •
$P(x_0, y_0) = \tfrac{1}{2}A(0)^2 + B(0)(0) + \tfrac{1}{2}C(0)^2 + f(x_0, y_0) = f(x_0, y_0)$ •
$P_x(x,y) = A(x-x_0) + B(y-y_0)$ •
$P_x(x_0, y_0) = A(0) + B(0) = 0 = f_x(x_0, y_0)$ •
$P_y(x,y) = B(x-x_0) + C(y-y_0)$ •
$P_y(x_0, y_0) = B(0) + C(0) = 0 = f_y(0,0)$ •
$P_{xx}(x,y) = A, P_{xy}(x,y) = B, P_{yy}(x,y) = C$ •
$P_{xx}(x_0, y_0) = A = f_{xx}(x_0, y_0)$ •
$P_{xy}(x_0, y_0) = B = f_{xy}(x_0, y_0)$ •
$P_{yy}(x_0, y_0) = C = f_{yy}(x_0, y_0)$

Response 9   (a) $f = x^2 + y^2$ • $f_x = 2x, f_y = 2y$ • Critical point: $(0,0)$ •
$f_{xx} = 2, f_{xy} = 0, f_{yy} = 2$ • $A = 2, B = 0, C = 2$, and $AC - B^2 = 4$ •
$AC - B^2, A,$ and C are positive • Local minimum
(b) $g = -x^2 - y^2$ • $g_x = -2x, g_y = -2y$ • Critical point: $(0,0)$ •
$g_{xx} = -2, g_{xy} = 0, g_{yy} = -2$ • $A = -2, B = 0, C = -2$, and $AC - B^2 = 4$ •
$AC - B^2$ is positive and A and C are negative • Local maximum
(c) $h = x^2 - y^2$ • $h_x = 2x, h_y = -2y$ • Critical point: $(0,0)$ •
$h_{xx} = 2, h_{xy} = 0, h_{yy} = -2$ • $A = 2, B = 0, C = -2$, and $AC - B^2 = -4$ •
$AC - B^2$ is negative • Saddle point
(d) $k = xy$ • $k_x = y, k_y = x$ • $k_{xx} = 0, k_{xy} = 1, k_{yy} = 0$ • $A = 0, B = 1, C = 0$, and $AC - B^2 = -1$ •
$AC - B^2$ is negative • Saddle point

Response 10 $\nabla(2x + 2y) = \lambda \nabla(xy)$ • $\langle 2, 2 \rangle = \lambda \langle y, x \rangle$ • $2 = \lambda y, 2 = \lambda x$ •
One approach: $x = \dfrac{2}{\lambda}, y = \dfrac{2}{\lambda}$ • Since $xy = 9$, $\dfrac{4}{\lambda^2} = 9, \lambda = \tfrac{2}{3}$ • $x = 3, y = 3$
• Another approach: $x = y$ and $xy = 9$, so $x = 3, y = 3$

Response 11 $xy = 9$ • $y = 9/x$ • The perimeter $P = 2x + 2y$ as a function of x is
$P = 2x + 18x^{-1}$. • $\dfrac{dP}{dx} = 2 - 18x^{-2} = \dfrac{2x^2 - 18}{x^2}$ is zero at $x = 3$, negative for $0 < x < 3$, and positive for $x > 3$. The minimum perimeter is at $x = 3$, for which $y = 3$.

CHAPTER 11 ANSWERS

Tune-up exercises 11.1

T1. $f(x,y) = x^2 y^3$ • $f(2,3) = 2^2 3^3 = 108$ • $f(-1, 10) = (-1)^2 10^3 = 1000$

T2. (a) $g(x,y) = x^3 y$ • For $y = 10$, $g(x, 10) = 10x^3$ is an increasing function of x.
(b) For $x = -2$, $g(-2, y) = (-2)^3 y = -8y$ is a decreasing function of y.

T3. Because $(x-y)^2$ is zero for $x = y$ and positive for $x \neq y$, the global minimum of $h(x,y) = (x-y)^2 + 10$ is 10 and occurs all along the line $y = x$.

T4. Because $x^2 + y^2$ is zero at $(0,0)$ and positive for $(x,y) \neq (0,0)$, the global maximum of $k(x,y) = \dfrac{6}{x^2 + y^2 + 2}$ is 3 and occurs at the origin $(0,0)$.

T5. (a) The intersection of $z = x^2 + y^2 + 6$ with $x = c$ is a parabola $z = c^2 + y^2 + 6, x = c$ that opens upward. • The intersection of $z = x^2 + y^2 + 6$ with $y = c$ is a parabola $z = x^2 + c^2 + 6, y = c$ that opens upward. • The intersection of $z = x^2 + y^2 + 6$ with $z = c$ is the circle $x^2 + y^2 = c - 6, z = c$ if $c > 6$, is the point $(0, 0, 6)$ if $c = 6$ and is empty if $c < 6$.
(b) The lowest point on the surface is its z-intercept $z = 6$. • Figure T5

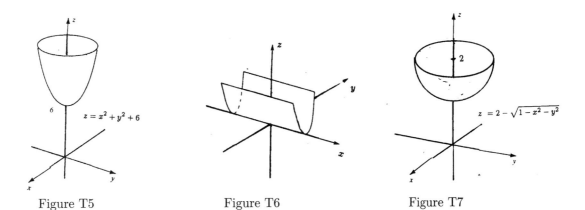

Figure T5 Figure T6 Figure T7

T6. (a) The intersection of $z = y^2$ with $x = c$ is a parabola $z = y^2, x = c$ that opens upward. • The intersection of $z = y^2$ with $y = c$ is a horizontal line $z = c^2, y = c$. • The intersection of $z = y^2$ with $z = c$ is the parallel lines $y = \pm\sqrt{c}, z = c$ if $c > 0$, is the line $y = 0, z = c$ if $c = 0$, and is empty if $c < 0$. (b) Figure T6

T7. (a) An equation $z = -\sqrt{1 - x^2 - y^2}$ for the lower half of the sphere $x^2 + y^2 + z^2 = 1$ is obtained by solving for $z = \pm\sqrt{1 - x^2 - y^2}$ and taking the minus sign. (b) The surface $z = 2 - \sqrt{1 - x^2 - y^2}$ is the surface $z = -\sqrt{1 - x^2 - y^2}$ of Figure 41 raised 2 units. • Figure T7

T8. The graph of $L(x,y) = -3$ is the horizontal plane $z = -3$, three units below the xy-plane. • Figure T8

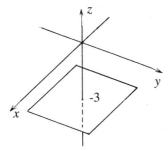

Figure T8

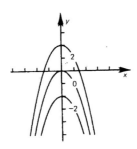

Figure T10

T9. (a) Since the numerator of $\dfrac{x}{\sqrt{y}}$ is defined for all (x, y) and the denominator for $y > 0$, the domain is the open upper half-plane $\{(x, y) : y > 0\}$. •
(b) $z(x, y)$ is positive for $x > 0, y > 0$, zero for $x = 0, y > 0$, and negative for $x < 0, y > 0$.

T10. $x^2 + y = c \iff y = c - x^2$ • Figure T10

T11. $y - \tfrac{1}{2}x = c \iff y = \tfrac{1}{2}x + c$ • Figure T11

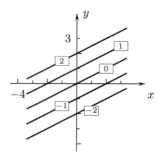

Figure T11

T12. The y-intercept of the innermost ellipse in Figure 42 is 1. • $P(0, 1) = 0^2 + 4(1^2) = 4$ • $P = 4$ on the innermost ellipse. • $P(0, 2) = 4(2^2) = 16$ • $P = 16$ on the middle ellipse. • $P(3, 0) = 4(3^2) = 36$ • $P = 36$ on the outer ellipse

T13. (a) As $(x, y) \to (3, 2)$: $\sin(x - y) \to \sin(3 - 2) = \sin(1)$ and $1 + x^2 + y^2 \to 1 + 3^2 + 2^2 = 14$, which is not zero. • $\dfrac{\sin(x - y)}{1 + x^2 + y^2} \to \tfrac{1}{14} \sin(1)$
(b) As $(x, y) \to (3, 0)$ with $y > 0$: $x^2 \to 3^2 = 9$ and $\cos(\sqrt{y}) \to \cos(0) = 1$, which is not zero. • $\dfrac{x^2}{\cos(\sqrt{y})} \to 9$

Problems 11.1

1. (a) $f = x^2 e^{-y} - y^3 e^x$ • $f(1, -1) = 1^2 e^{-(-1)} - (-1)^3 e^1 = 2e$
   (b) $g = \dfrac{x - y}{x + y}$ • $g(1, 3) = \dfrac{1 - 3}{1 + 3} = -\tfrac{1}{2}$

2. $M(x, y) = \dfrac{1 + \cos x}{2 + \sin y}$ • $M(1, 2) = \dfrac{1 + \cos(1)}{2 + \sin(2)} \doteq 0.529441$ • $M(2, 1) = \dfrac{1 + \cos(2)}{2 + \sin(1)} \doteq 0205476$ •
   $M(2, 2) = \dfrac{1 + \cos(2)}{2 + \sin(2)} \doteq 0.200685$ • The greatest is $M(1, 2)$ and the least is $M(2, 2)$.

3. (a) $\dfrac{1}{R} = \dfrac{r_2 + r_1}{r_1 r_2}$ • $R = \dfrac{r_1 r_2}{r_1 + r_2}$
   (b) $R(r_1, r_2)$ increases as r_1 or r_2 increases because $\dfrac{1}{R} = \dfrac{1}{r_1} + \dfrac{1}{r_2}$ decreases.

6. $\dfrac{x + 2y}{3x + y} = 1 \iff 3x + y = x + 2y$ and $y \neq -3x \iff y = 2x$ and $y \neq -3x$ •
   The level curve is the line $y = 2x$ with the origin removed.

7. $(3, 0)$ is on the outer square and $L(3, 0) = |3| + |0| = 3$. • $L = 3$ on the outer square. •
   $(2, 0)$ is on the middle square and $L(2, 0) = |2| + |0| = 2$ • $L = 2$ on the middle square •
   $(1, 0)$ is on the inner square and $L(1, 0) = |1| + |0| = 1$. • $L = 1$ on the inner square.

9. Because $F(x, y)$ is linear, the level lines in Figure 46 are at $F = 6, 4, 2, 0, -2, -4, -6$. • $A = -2, B = 1, C = 0$

11. (a) Figures A11a, A11b, and A11c (b) f and h are defined for all (x, y). •
    g is defined for $x^2 + y^2 \geq \tfrac{1}{2}$.

Answers 11.1

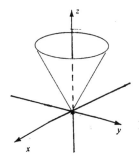
$z = f(x,y)$
Figure A11a

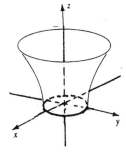
$z = g(x,y)$
Figure A11b

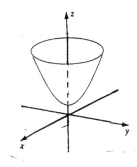
$z = h(x,y)$
Figure A11c

14a.   Figure A14

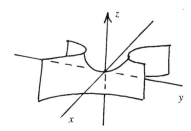

Figure A14

15.   (a) Figure A15   (b) $R = 0$ on the vertical lines $x = 2$ and $x = -2$. • $R = 1$ on the vertical lines $x = \sqrt{3}$ and $x = -\sqrt{3}$. • $R = 2$ on the y-axis.

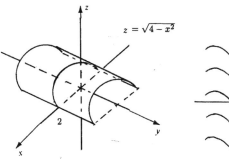

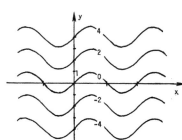

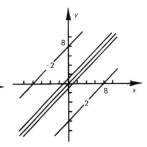

Figure A15   Figure A18   Figure A19

18.   Figure A18
19.   Figure A19
22.   (a) Figure A22a

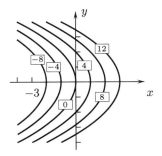

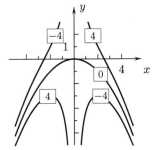

Figure A22a   Figure A30

26. (a) Figure 57 (b) Figure 55 (d) Figure 59

27. (a) Figure 60 (b) Figure 65 (d) Figure 64

30. $p = c \iff xy + \frac{1}{4}x^3 = c \iff y = \frac{c}{x} - \frac{1}{4}x^2$ • Figure A30

38. (a) $\ln(xy)$ is defined where $xy > 0$, which is for $x > 0, y > 0$ and for $x < 0, y < 0$.
(b) $\ln(xy) = \ln(c)$ on the level curve $xy = c$ for $c > 0$

41. (a) $-\frac{375}{22}$ (b) $-2\sin(-14) = 2\sin(14)$

42. (a) The limit is 0. (b) The limit does not exist. (c) The limit is 1.

47. [Global maximum] = 10 at $x = 0$

Tune-up exercises 11.2

T1. (a) $\frac{\partial}{\partial x}(xy^5 - 4y^2 + 6x^4y^7) = y^5 \frac{\partial}{\partial x}(x) - \frac{\partial}{\partial x}(4y^2) + 6y^7 \frac{\partial}{\partial x}(x^4) = y^5 + 24x^3y^7$
(b) $\frac{\partial}{\partial y}(xy^5 - 4y^2 + 6x^4y^7) = x \frac{\partial}{\partial y}(y^5) - \frac{\partial}{\partial y}(4y^2) + 6x^4 \frac{\partial}{\partial y}(y^7) = 5xy^4 - 8y + 42x^4y^6$

T2. $G_y(x,y) = \frac{\partial}{\partial y}[x^2 \sin(xy) + y - x] = x^2 \cos(xy)\frac{\partial}{\partial y}(xy) + 1 - 0 = x^3 \cos(xy) + 1$

T3. (a) One answer: $g_x(2,5) \approx \frac{g(2.5,5) - g(2,5)}{2.5 - 2} = \frac{223 - 212}{0.5} = 22$ •

Another answer: $g_x(2,5) \approx \frac{g(2,5) - g(1.5,5)}{2 - 1.5} = \frac{212 - 200}{0.5} = 24$ •

A third answer: $g_x(2,5) \approx \frac{g(2.5,5) - g(1.5,5)}{2.5 - 1.5} = \frac{223 - 200}{1} = 23$

(b) One answer: $g_y(2,5) \approx \frac{g(2,5.2) - g(2,5)}{5.2 - 5} = \frac{172 - 212}{0.2} = -200$ •

Another answer: $g_y(2,5) \approx \frac{g(2,5) - g(2,4.8)}{5 - 4.8} = \frac{212 - 253}{0.2} = -205$ •

A third answer: $g_y(2,5) \approx \frac{g(2,5.2) - g(2,4.8)}{5.2 - 4.8} = \frac{172 - 253}{0.4} = -202.5$

T4. From P to Q in Figure T4: $\Delta G = 40 - 30 = 10, \Delta x \approx 1.3$ • $\frac{\partial G}{\partial x}(3,3) \approx \frac{\Delta G}{\Delta x} \approx \frac{10}{1.3} \doteq 7.7$
From R to S: $\Delta G = 40 - 30 = 10, \Delta y \approx 1.5$ • $\frac{\partial G}{\partial y}(3,3) \approx \frac{\Delta G}{\Delta y} \approx \frac{\Delta G}{\Delta y} \approx \frac{10}{1.5} \doteq 6.7$

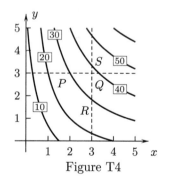
Figure T4

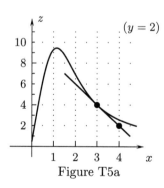
Figure T5a

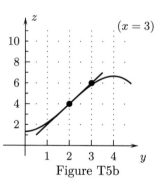
Figure T5b

T5. Draw an approximate tangent line at $x = 3$ to the graph of $z = P(x,2)$ as in Figure T5a. • From the left dot to the right dot: $\Delta x = 1, \Delta z \approx -2$. • $\frac{\partial P}{\partial x}(3,2) \approx \frac{\Delta z}{\Delta x} \approx \frac{-2}{1} = -2$ •
Draw an approximate tangent line at $y = 2$ to the graph of $z = P(3,y)$ as in Figure T5b. • From the left dot to the right dot: $\Delta y = 1, \Delta z \approx 2$. • $\frac{\partial P}{\partial y}(3,2) \approx \frac{\Delta z}{\Delta y} \approx \frac{2}{1} = 2$ •

Answers 11.2

T6. (a) $H(x,2) = \frac{1}{2}(2^2) - x^2 = 2 - x^2$ • $\left[2 - x^2\right]_{x=1} = 1$ •
$\frac{\partial H}{\partial x}(1,2) = \left[\frac{d}{dx}(2-x^2)\right]_{x=1} = \left[-2x\right]_{x=1} = -2$ • Tangent line: $z = 1 - 2(x-1)$ • Figure T6a
(b) $H(1,y) = \frac{1}{2}(y^2) - 1^2 = \frac{1}{2}y^2 - 1$ • $\left[\frac{1}{2}y^2 - 1\right]_{y=2} = 1$ •
$\frac{\partial H}{\partial y}(1,2) = \left[\frac{d}{dy}(\frac{1}{2}y^2 - 1)\right]_{y=2} = \left[y\right]_{y=2} = 2$ • Tangent line: $z = 1 + 2(y-2)$ • Figure T6b

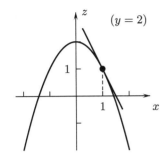

Figure T6a

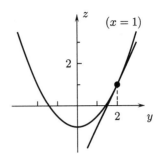

Figure T6b

T7. $f(x(t), y(t)) = x(t)\sin(x(t)y(t)) = t^5 \sin(t^5 t^{-3}) = t^5 \sin(t^2)$

T8. $\frac{\partial}{\partial x}[W(x^3 y^5)] = W'(x^3 y^5)\frac{\partial}{\partial x}(x^3 y^5) = 3x^2 y^5 W'(x^3 y^5)$
$\frac{\partial}{\partial y}[W(x^3 y^5)] = W'(x^3 y^5)\frac{\partial}{\partial y}(x^3 y^5) = 5x^3 y^4 W'(x^3 y^5)$

T9. $\left[\frac{d}{dt}[F(x(t), y(t))]\right]_{t=0} = \frac{\partial F}{\partial x}(x(0), y(0))\frac{dx}{dt}(0) + \frac{\partial F}{\partial y}(x(0), y(0))\frac{dy}{dt}(0)$
$= \frac{\partial F}{\partial x}(3,7)\frac{dx}{dt}(0) + \frac{\partial F}{\partial y}(3,7)\frac{dy}{dt}(0) = (8)(-4) + (6)(2) = -20$

T10. (a) $\frac{\partial f}{\partial u} = \frac{\partial f}{\partial x}\frac{\partial x}{\partial u} + \frac{\partial f}{\partial y}\frac{\partial y}{\partial u}$
$= \frac{\partial}{\partial x}(3x - 4y + 7)\frac{\partial}{\partial u}(2u - 5v) + \frac{\partial}{\partial y}(3x - 4y + 7)\frac{\partial}{\partial u}(3u + 8v) = (3)(2) + (-4)(3) = -6$
$\frac{\partial f}{\partial v} = \frac{\partial f}{\partial x}\frac{\partial x}{\partial v} + \frac{\partial f}{\partial y}\frac{\partial y}{\partial v}$
$= \frac{\partial}{\partial x}(3x - 4y + 7)\frac{\partial}{\partial v}(2u - 5v) + \frac{\partial}{\partial y}(3x - 4y + 7)\frac{\partial}{\partial v}(3u + 8v) = (3)(-5) + (-4)(8) = -47$
(b) $f = 3x - 4y + 7 = 3(2u - 5v) - 4(3u + 8v) + 7 = -6u - 47v + 7$ •
$\frac{\partial f}{\partial u} = -6$ • $\frac{\partial f}{\partial v} = -47$

T11. $\left[\frac{\partial U}{\partial t}(s,t)\right]_{s=2, t=6} = \left[\frac{\partial}{\partial t}[V(x(s,t), y(s,t))]\right]_{s=2, t=6}$
$= \left[\frac{\partial V}{\partial x}(x(s,t), y(s,t))\frac{\partial x}{\partial t}(s,t) + \frac{\partial V}{\partial y}(x(s,t), y(s,t))\frac{\partial y}{\partial t}(s,t)\right]_{s=2, t=6}$
$= \frac{\partial V}{\partial x}(x(2,6), y(2,6))\frac{\partial x}{\partial t}(2,6) + \frac{\partial V}{\partial y}(x(2,6), y(2,6))\frac{\partial y}{\partial t}(2,6)$
$= \frac{\partial V}{\partial x}(5,-7)\frac{\partial x}{\partial t}(2,6) + \frac{\partial V}{\partial y}(5,-7)\frac{\partial y}{\partial t}(2,6) = (11)(8) + (12)(10) = 208$

T12. $R_x = \frac{\partial}{\partial x}(x^{1/2} y^{1/3}) = \frac{1}{2}x^{-1/2} y^{1/3}$ • $R_y = \frac{\partial}{\partial y}(x^{1/2} y^{1/3}) = \frac{1}{3}x^{1/2} y^{-2/3}$ •
$R_{xx} = \frac{\partial}{\partial x}(\frac{1}{2}x^{-1/2} y^{1/3}) = -\frac{1}{4}x^{-3/2} y^{1/3}$ •
$R_{xy} = \frac{\partial}{\partial y}(\frac{1}{2}x^{-1/2} y^{1/3}) = \frac{1}{6}x^{-1/2} y^{-2/3}$ •
$R_{yx} = \frac{\partial}{\partial x}(\frac{1}{3}x^{1/2} y^{-2/3}) = \frac{1}{6}x^{-1/2} y^{-2/3}$ •
$R_{yy} = \frac{\partial}{\partial y}(\frac{1}{3}x^{1/2} y^{-2/3}) = -\frac{2}{9}x^{1/2} y^{-5/3}$ • (R_{xy} and R_{yx} should be and are equal.)

Problems 11.2

1. (a) $3x^2y^2 - 1$ (b) $3x^2e^{3y} + 2ye^{3x}$

2. (a) $F_x = 2xy^4 \cos(x^2y^4)$ • $F_y = 4x^2y^3 \cos(x^2y^4)$ (b) $G_x = \dfrac{-y}{1-xy}$ • $G_y = \dfrac{-x}{1-xy}$

3. $V_r = \frac{2}{3}\pi rh$ cubic meters per meter

6. (a) The volume increases when the temperature increases. (b) The solution with the smaller concentration has the greater volume. (c) $V_p(15,80)$ is the rate of change of the volume with respect to the percentage of acid in the solution and $\approx -1 \times 10^{-4}$ cubic feet per percent (d) $V_T \approx 1 \times 10^{-5}$ cubic feet per per degree

9. From the level curve $K = -18$ to the level curve $K = -10$ along $y = 2$: $\Delta K = -20 - (-18) = -2$ and $\Delta x \approx 2$ • $\dfrac{\partial K}{\partial x}(6,2) \approx \dfrac{\Delta K}{\delta x} \approx \dfrac{-2}{2} = -1$ •
From the level curve $K = -18$ to the level curve $K = -10$ along $x = 6$: $\Delta K = -20 - (-18) = -2$ and $\Delta y \approx 0.3$ • $\dfrac{\partial K}{\partial y}(6,2) \approx \dfrac{\Delta K}{\Delta y} \approx \dfrac{-2}{0.3} \doteq -6.7$

12. $\dfrac{dS}{dg}(6,14) \approx -1330$ pounds per square inch per gallon • $\dfrac{dS}{dt}(6,14) \approx 170$ pounds per square inch per day •

18. (a) $p(0,y) = \sin y$ • $p(0, \frac{3}{4}\pi) = \frac{1}{2}\sqrt{2}$ • $\dfrac{\partial p}{\partial y} = \cos y$ • $\dfrac{\partial p}{\partial y}(0, \frac{3}{4}\pi) = -\frac{1}{2}\sqrt{2}$ •
Tangent line: $z = \frac{1}{2}\sqrt{2} - \frac{1}{2}\sqrt{2}(y - \frac{3}{4}\pi)$ • Figure A18a
(b) $p(x, \frac{3}{4}\pi) = \frac{1}{2}\sqrt{2} - \frac{1}{9}x^3$ • $p(0, \frac{3}{4}\pi) = \frac{1}{2}\sqrt{2}$ • $\dfrac{\partial p}{\partial x} = -\frac{1}{3}x^2$ • $\dfrac{\partial p}{\partial x}(0, \frac{3}{4}\pi) = 0$ •
Tangent line: $z = \frac{1}{2}\sqrt{2}$ • Figure A18b

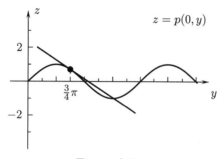

Figure A18a

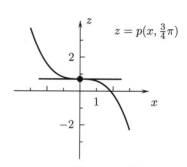

Figure A18b

19. (a) $z = x^2$ • Tangent line: $z = 4 + 4(x-2)$ • $\dfrac{\partial f}{\partial x}(2,-1) = 4$ • Figure A19a
(b) $z = 4e^{(1-y^2)/2}$ • Tangent line: $z = 4 + 4(y+1)$ • $\dfrac{\partial f}{\partial y}(2,-1) = 4$ • Figure A19b

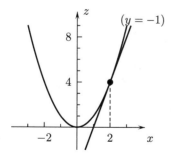

Figure A19a

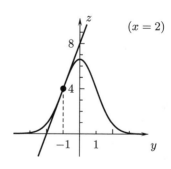

Figure A19b

Answers 11.2

22. (a) $W(96,73) \approx W(96,75) - 2\dfrac{\partial W}{\partial H}(96,75) = 17.5 - 2(0.025) = 17.45$ percent

(b) $W(98,75) \approx W(96,75) + 2\dfrac{\partial W}{\partial T}(96,75) = 17.5 + 2(-0.06) = 17.38$ percent

(c) $\dfrac{d}{dt}(W) = \dfrac{\partial W}{\partial T}\dfrac{dT}{dt} + \dfrac{\partial W}{\partial H}\dfrac{dH}{dt}$

$= \left[-0.06\,\dfrac{\text{percent}}{\text{degree}}\right]\left[3\,\dfrac{\text{degrees}}{\text{hour}}\right] + \left[0.025\,\dfrac{\text{percent}}{\text{percent}}\right]\left[0.4\,\dfrac{\text{percent}}{\text{hour}}\right] = -0.17$ percent per hour

25. (a) $F(t) = t^5 \sin(t^2)$ (b) $P(u,v) = ue^v \ln(u^3 v^3 e^v + 1)$

26. (a) $g(2) = f(2^3, 2^4) = f(8, 16) = 3$

(b) $\dfrac{dg}{dt}(t) = f_x(t^3, t^4)\dfrac{d}{dt}(t^3) + f_y(t^3, t^4)\dfrac{d}{dt}(t^4) = f_x(t^3, t^4)(3t^2) + f_y(t^3, t^4)(4t^3)$ •

$\dfrac{dg}{dt}(2) = f_x(8,16)(12) + f_y(8,16)(32) = 12(5) + 32(-7) = -164$

27. (a) $h(3) = 0$ (b) $\dfrac{dh}{dt}(3) = -44$

30. (a) $\dfrac{\partial T}{\partial x}(x,y) = \dfrac{dU}{dv}(v(x,y))\dfrac{\partial v}{\partial x}(x,y)$ •

$\dfrac{\partial T}{\partial x}(1,2) = \dfrac{dU}{dv}(v(1,2))\dfrac{\partial v}{\partial x}(1,2) = \dfrac{dU}{dv}(3)\dfrac{\partial v}{\partial x}(1,2) = (9)(5) = 45$

(b) $\dfrac{\partial T}{\partial y}(x,y) = \dfrac{dU}{dv}(v(x,y))\dfrac{\partial v}{\partial y}(x,y)$ •

$\dfrac{\partial T}{\partial y}(1,2) = \dfrac{dU}{dv}(v(1,2))\dfrac{\partial v}{\partial y}(1,2) = \dfrac{dU}{dv}(3)\dfrac{\partial v}{\partial y}(1,2) = (9)(7) = 63$

31. (a) $J_x(1,1) = \dfrac{25}{6}$ • $J_y(1,1) = \dfrac{8}{3}$

33. (a) $F(0,2) = f(0,0) = 4$ •

(b) $F_u(u,v) = f_x(v\sin u, u\sin v)\dfrac{\partial}{\partial u}(v\sin u) + f_y(v\sin u, u\sin v)\dfrac{\partial}{\partial u}(u\sin v)$

$= f_x(v\sin u, u\sin v)(v\cos u) + f_y(v\sin u, u\sin v)(\sin v)$ •

$F_u(0,2) = f_x(0,0)[2\cos(0)] + f_y(0,0)[\sin(2)] = 20 + 2\sin(2)$

(c) $F_v(u,v) = f_x(v\sin u, u\sin v)\dfrac{\partial}{\partial v}(v\sin u) + f_y(v\sin u, u\sin v)\dfrac{\partial}{\partial v}(u\sin v)$

$= f_x(v\sin u, u\sin v)(\sin u) + f_y(v\sin u, u\sin v)(u\cos v)$ •

$F_v(0,2) = f_x(0,0)[\sin(0)] + f_y(0,0)[0\cos(2)] = 0$

34. (a) $G(1,2) = 10$ (b) $G_u(1,2) = 77$ (c) $G_v(1,2) = 53$

35. (a) $\left[\dfrac{\partial}{\partial x}[P(q(x,y))]\right]_{x=3, y=7} = \left[P'(q(x,y))\dfrac{\partial q}{\partial x}(x,y)\right]_{x=3, y=7} = P'(q(3,7))\dfrac{\partial q}{\partial x}(3,7)$

$= P'(9)\dfrac{\partial q}{\partial x}(3,7) = (-11)(6) = -66$

(b) $\left[\dfrac{\partial}{\partial y}[P(q(x,y))]\right]_{x=3, y=7} = \left[P'(q(x,y))\dfrac{\partial q}{\partial y}(x,y)\right]_{x=3, y=7} = P'(q(3,7))\dfrac{\partial q}{\partial y}(3,7)$

$= P'(9)\dfrac{\partial q}{\partial y}(3,7) = (-11)(10) = -110$

36. (a) $P(3,2) = 8$ (b) $P_u(3,2) = 115$ (c) $P_v(3,2) = 94$

38. (a) $f = x^4 y^5$ • $f_x = 4x^3 y^5, f_y = 5x^4 y^4$ • $f_{xx} = 12x^2 y^5, f_{xy} = f_{yx} = 20x^3 y^4, f_{yy} = 20x^4 y^3$

(b) $k_{xx} = -4(2x-3y)^{-2}, k_{xy} = k_{yx} = 6(2x-3y)^{-2}, k_{yy} = -9(2x-3y)^{-2}$

(c) $M_x = 10(1+xy)^9 \dfrac{\partial}{\partial x}(1+xy) = 10y(1+xy)^9$

$M_y = 10(1+xy)^9 \dfrac{\partial}{\partial y}(1+xy) = 10x(1+xy)^9$ • $M_{xx} = 90y(1+xy)^8 \dfrac{\partial}{\partial x}(1+xy) = 90y^2(1+xy)^8$ •

Product Rule: $M_{xy} = \dfrac{\partial}{\partial y}[10y(1+xy)^9] = 10y\dfrac{\partial}{\partial y}[(1+xy)^9] + 10(1+xy)^9 \dfrac{\partial}{\partial y}(y)$

$= 90y(1+xy)^8 \dfrac{\partial}{\partial y}(1+xy) + 10(1+xy)^9 = 90xy(1+xy)^8 + 10(1+xy)^9$ •

$M_{yy} = 90x(1+xy)^8 \dfrac{\partial}{\partial y}(1+xy) = 90x^2(1+xy)^8$

39. (a) $24xy^3$

Tune-up exercises 11.3

T1. (a) $x = \frac{3}{5}s, y = \frac{4}{5}s$ • $F(s) = \frac{3}{5}s + \sin(\frac{4}{5}s) + 2$

(b) $F'(s) = \frac{3}{5} + \cos(\frac{4}{5}s)\frac{d}{ds}(\frac{4}{5}s) = \frac{3}{5} + \frac{4}{5}\cos(\frac{4}{5}s)$ • $D_\mathbf{u}f(0,0) = F'(0) = \frac{3}{5} + \frac{4}{5} = \frac{7}{5}$

(c) $F(0) = 2$ • Tangent line: $w = 2 + \frac{7}{5}s$ • Figure T1

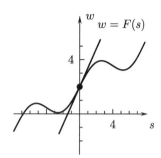

Figure T1

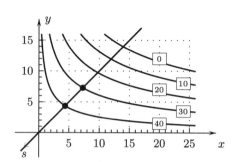

Figure T3

T2. $g = x^2y^3$ • $g_x = 2xy^3, g_y = 3x^2y^2$ • $g_x(1,-1) = 2(1)(-1)^3 = -2, g_y(1,-1) = 3(1)^2(-1)^2 = 3$ •
For $P = (1,-1)$ and $Q = (2,2)$: $\overrightarrow{PQ} = \langle 2-1, 2-(-1)\rangle = \langle 1, 3\rangle$ •
$|\overrightarrow{PQ}| = \sqrt{1^2 + 3^2} = \sqrt{10}$ • $\mathbf{u} = \frac{\overrightarrow{PQ}}{|\overrightarrow{PQ}|} = \frac{\langle 1, 3\rangle}{\sqrt{10}}$ • $u_1 = \frac{1}{\sqrt{10}}, u_2 = \frac{3}{\sqrt{10}}$ •
$D_\mathbf{u}g(1,-1) = g_x u_1 + g_y u_2 = (-2)\left(\frac{1}{\sqrt{10}}\right) + 3\left(\frac{3}{\sqrt{10}}\right) = \frac{7}{\sqrt{10}}$

T3. Draw an s axis through $(0,0)$ with $s = 0$ at $(5,5)$ as in Figure T3. • From the upper dot to the lower dot on the s-axis: $\Delta K = 40 - 30 = 10, \Delta s \approx 4$ • $D_\mathbf{u}K(5,5) \approx \frac{\Delta K}{\Delta s} \approx \frac{10}{4} = 2.5$

T4. $z = x^2 y^{-3}$ • $z_x = 2xy^{-3}, z_y = -3x^2y^{-4}$ •
$z(2,1) = (2)^2(1)^{-3} = 4$ • $z_x(2,1) = 2(2)(1)^{-3} = 4, z_y(2,1) = -3(2)^2(1)^{-4} = -12$ •
Tangent plane: $z = 4 + 4(x-2) - 12(y-1)$

T5. $P(0,0) = 15$ • Since we are told $P(x,y)$ is linear we can use any pair of values in any row and any pair of values in any column. •
$\frac{\partial P}{\partial x} = \frac{P(5,0) - P(0,0)}{5 - 0} = \frac{25 - 15}{5} = 2$ and
$\frac{\partial P}{\partial y} = \frac{P(0,5) - P(0,0)}{5 - 0} = \frac{-35 - 15}{5} = -10$ •
$P(x,y) = 2x - 10y + 15$

T6. $Q(x,y) = 600x - 200y + 400$ • $Q = c \iff 600x - 200y + 400 = c \iff y = 3x + 2 - \frac{1}{200}c$ •
Figure T6 uses $c = -500, 0, 500, 1000, 1500$.

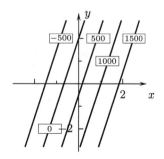

Figure T6

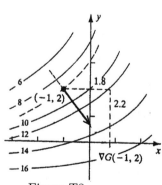

Figure T8

T7. $f = 2x - \ln y$ • $f_x = 2, f_y = -1/y$ • $\nabla f(0,3) = \langle f_x(0,3), f_y(0,3)\rangle = \langle 2, -\frac{1}{3}\rangle$

T8. One solution: Draw an approximate level curve through $(-1,2)$ and a normal line to it as in Figure T8. • The distance along the normal line between the level curves $G = 8$ and $G = 10$ above and below $(-1,2)$ is $\Delta s \approx 1$ and the corresponding change in G is $\Delta G = 10-8 = 2$. • The maximum directional derivative of G at $(-1,2)$ is $\approx \Delta G/\Delta s \approx 2$. • Draw the gradient vector of this length, perpendicular to the level curve, and pointing in the direction that G increases, as in Figure T8. • Measure its components: $\nabla G(-1,2) \approx \langle 1, -1.4\rangle$

T9. $f = x\sin y$ • $\nabla f = \langle \sin y, x\cos y\rangle$ • $\nabla f(5, \frac{1}{3}\pi) = \langle \frac{1}{2}\sqrt{3}, \frac{5}{2}\rangle$ •
The maximum directional derivative is $|\nabla f(5, \frac{1}{3}\pi)| = \sqrt{(\frac{1}{2}\sqrt{3})^2 + (\frac{5}{2})^2} = \sqrt{7}$ •
The unit vector in the corresponding direction is $\dfrac{\nabla f(5, \frac{1}{3}\pi)}{|\nabla f(5, \frac{1}{3}\pi)|} = \dfrac{\langle \sqrt{3}, 5\rangle}{2\sqrt{7}}$

T10. $g = x^2 - 3y^3$ • $\nabla g = \langle 2x, -9y^2\rangle$ • $\nabla g(3, 12) = \langle 6, -36\rangle$ •
The vectors $\pm\langle 36, 6\rangle = \pm 6\langle 6, 1\rangle$ are perpendicular to the gradient vector. •
Unit vectors: $\mathbf{u} = \dfrac{\pm\langle 6,1\rangle}{|\langle 6,1\rangle|} = \dfrac{\pm\langle 6,1\rangle}{\sqrt{37}}$

T11. $Z = x^2 y^3$ • $Z_x = 2xy^3, Z_y = 3x^2 y^2$ •
$Z_x(1,-1) = 2(1)(-1)^3 = -2, Z_y(1,-1) = 3(1)^2(-1)^2 = 3$ •
At $(1,-1)$: $dZ = -2\,dx + 3\,dy$ and $|dZ| \leq 2|dx| + 3|dy|$ •
For $|dx| \leq 3 \times 10^{-4}$, and $|dy| \leq 5 \times 10^{-4}$:
[Maximum error] $\approx$ [Maximum $|dZ|$] $\leq 2(3 \times 10^{-4}) + 3(5 \times 10^{-4}) \doteq 0.002$

Problems 11.3

1. (a) $f = x^2 - y^2$ • $(x_0, y_0) = (2,1)$ • $u_1 = \dfrac{1}{\sqrt{5}}, u_2 = -\dfrac{2}{\sqrt{5}}$ •

$x = 2 + \dfrac{s}{\sqrt{5}}, y = 1 - \dfrac{2s}{\sqrt{5}}$ • $F(s) = \left(2 + \dfrac{s}{\sqrt{5}}\right)^2 - \left(1 - \dfrac{2s}{\sqrt{5}}\right)^2$

$= (4 + \dfrac{4s}{\sqrt{5}} + \frac{1}{5}s^2) - (1 - \dfrac{4s}{\sqrt{5}} + \frac{4}{5}s^2) = 3 + \dfrac{8s}{\sqrt{5}} - \frac{3}{5}s^2$

(b) $D_{\mathbf{u}}f(2,1) = F'(0) = \left[\dfrac{8}{\sqrt{5}} - \frac{6}{5}s\right]_{s=0} = \dfrac{8}{\sqrt{5}}$ • $F(0) = f(2,1) = 3$

Tangent line: $w = 3 + \dfrac{8s}{\sqrt{5}}$ • Figure T1

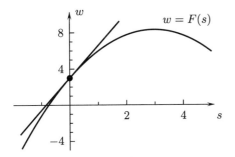
Figure A1

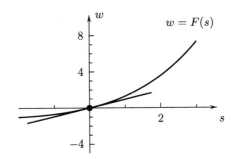
Figure A2

2. (a) $g = xe^y$ • $(x_0, y_0) = (0,0)$ • $u_1 = \dfrac{3}{\sqrt{10}}, u_2 = \dfrac{1}{\sqrt{10}}$ • $x = \dfrac{3s}{\sqrt{10}}, y = \dfrac{s}{\sqrt{10}}$

$F(s) = \dfrac{3s}{\sqrt{10}} e^{s/\sqrt{10}}$ • Figure A2

(b) $F'(s) = \dfrac{3s}{\sqrt{10}} \dfrac{d}{ds}\left(e^{s/\sqrt{10}}\right) + e^{s/\sqrt{10}}\dfrac{d}{ds}\left(\dfrac{3s}{\sqrt{10}}\right) = \left(\dfrac{3}{10}s + \dfrac{3}{\sqrt{10}}\right)e^{s/\sqrt{10}}$ •

$F(0) = 0, F'(0) = \dfrac{3}{\sqrt{10}}$ • Tangent line: $w = \dfrac{3s}{\sqrt{10}}$ • Figure A2

5. (a) One answer: Figure A5 (b) The distance from the point where the s-axis in Figure 33 crosses $D = 2000$ to where it crosses $D = 1500$ is $\Delta s \approx 2.5$. • The corresponding change in the depth is $\Delta D = -500$. • $D_{\mathbf{u}}(P) \approx \dfrac{\Delta D}{\Delta s} = \dfrac{-500}{2.5} \doteq -200$ feet per nautical mile

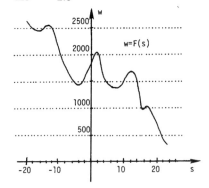

Figure A5

7. $f = x^2 y - xy^3$ • $\nabla f = \langle 2xy - y^3, x^2 - 3xy^2 \rangle$ • $\nabla f(3, -2) = \langle -4, -27 \rangle$ •
For $P = (3, -2)$ and $Q = (5, 6)$: $\overrightarrow{PQ} = \langle 5 - 3, 6 - (-2) \rangle = \langle 2, 8 \rangle = 2\langle 1, 4 \rangle$ • $|\langle 1, 4 \rangle| = \sqrt{17}$ •
$\mathbf{u} = \dfrac{\langle 1, 4 \rangle}{\sqrt{17}}$ • $D_{\mathbf{u}} f(3, -2) = \nabla f(3, -2) \cdot \mathbf{u} = \dfrac{\langle -4, -27 \rangle \cdot \langle 1, 4 \rangle}{\sqrt{17}} = \dfrac{-112}{\sqrt{17}}$

8. $\dfrac{2\pi - 3}{2\sqrt{10}}$

11. Along an s-axis through $(-1, 1)$ in the direction of the s-axis through the point $(0, -2)$ in Figure A11, the level curves $g = 30$ and $g = 20$ are approximately 2 units apart. • $\Delta g = -10, \Delta s \approx 2$ •
$D_{\mathbf{u}} g(-1, 1) \approx \dfrac{\Delta g}{\Delta s} \approx \dfrac{-10}{5} = -5$

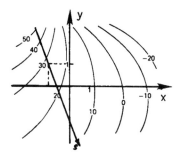

Figure A11

13. $L(x, y) = 6x - 3y + 7$

16. (a) $z = x^2 y^{-3}$ • $z_x = 2xy^{-3}, z_y = -3x^2 y^{-4}$ • $z(2, 1) = 4, z_x(2, 1) = 4, z_y(2, 1) = -12$ •
Tangent plane: $z = 4 + 4(x - 2) - 12(y - 1)$ (b) Tangent plane: $z = 3 + (x - 3) - 3y$

17. (a) From Problem 16a: $z_x(2, 1) = 4, z_y(2, 1) = -12$ •
Normal vector: $\mathbf{n} = \langle z_x(2, 1), z_y(2, 1), -1 \rangle = \langle 4, -12, -1 \rangle$ (b) $\mathbf{n} = \langle 1, -3, -1 \rangle$

18. (a) $z = e^{2x} \sin y$ • $z(2, \tfrac{1}{6}\pi) = \tfrac{1}{2} e^4$ • $z_x(2, \tfrac{1}{6}\pi) = e^4$ • $z_y(2, \tfrac{1}{6}\pi) = \tfrac{1}{2} e^4 \sqrt{3}$ •
Tangent plane: $z = \tfrac{1}{2} e^4 + e^4(x - 2) + \tfrac{1}{2} e^4 \sqrt{3}(y - \tfrac{1}{6}\pi)$
(b) Tangent plane: $z = \ln(5) + \tfrac{3}{25}(x - 3) + \tfrac{4}{25}(y - 4)$

19. (a) From Problem 18a: $z_x(2, \tfrac{1}{6}\pi) = e^4$ and $z_y(2, \tfrac{1}{6}\pi) = \tfrac{1}{2} e^4 \sqrt{3}$ •
Normal vector: $\langle e^4, \tfrac{1}{2} e^4 \sqrt{3}, -1 \rangle$ (b) Normal vector: $\langle \tfrac{3}{25}, \tfrac{4}{25}, -1 \rangle$

22. (a) Tangent plane: $z = 15 + 5(x - 3) + 3(x - 5)$
(b) $L(x, y) = 15 + 5(x - 3) + 3(x - 5)$ • [Error] $= |f(3.01, 5.01) - L(3.01, 5.01)| = 10^{-4}$
(c) [Error]$= |f(3.001, 5.001) - L(3.001, 5.001)| = 10^{-6}$

23. (a) $L(x, y) = -1 - \pi(x - 1)$ (b) [Error]$\doteq 4.5 \times 10^{-3}$

Answers 11.3

25. (a) $f = \ln(xy)$ • $\nabla f = \langle \frac{1}{x}, \frac{1}{y} \rangle$ • $\nabla f(5, 10) = \langle \frac{1}{5}, \frac{1}{10} \rangle$ (b) $\nabla g(-1, 1) = \langle 5, -20 \rangle$

26. $P_x(0, 0) = 4, P_y(0, 0) = 2$ • For $P = (0, 0)$ and $Q = (2, 2)$: $\overrightarrow{PQ} = \langle 2, 2 \rangle = 2\langle 1, 1 \rangle$ • $\mathbf{u} = \frac{\langle 1, 1 \rangle}{\sqrt{2}}$ • $D_{\mathbf{u}}P(0, 0) = 4\left(\frac{1}{\sqrt{2}}\right) + 2\left(\frac{1}{\sqrt{2}}\right) = 3\sqrt{2}$

27. $f = x^5 e^{4y}$ • $\nabla f = \langle 5x^4, 4e^{4y} \rangle$ • $\nabla f(1, 0) = \langle 5, 4 \rangle$ • [Minimum directional derivative] $= -|\langle 5, 4 \rangle| = -\sqrt{41}$

28. $\mathbf{u} = \pm \frac{\langle 5, -1 \rangle}{\sqrt{26}}$

29. $f(x, y) = y - \frac{1}{4}x^2$ equals 0 at $(0, 0)$, equals $-\frac{9}{4}$ at $(-1, -2)$, and equals $\frac{11}{4}$ at $(-1, 3)$. • The curves are $y = \frac{1}{4}x^2$, $y = \frac{1}{4}x^2 - \frac{9}{4}$, and $y = \frac{1}{4}x^2 + \frac{11}{4}$. • $\nabla f(x, y) = \langle -\frac{1}{2}x, 1 \rangle$ equals $\langle 0, 1 \rangle$ at $(0, 0)$, equals $\langle \frac{1}{2}, 1 \rangle$ at $(-1, -2)$, and equals $\langle \frac{1}{2}, 1 \rangle$ at $(-1, 3)$. • Figure A29

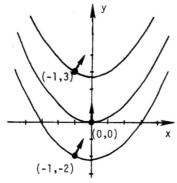

Figure A29

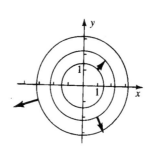

Figure A30

30. $\nabla g(1, 1) = \langle \frac{1}{2}, \frac{1}{2} \rangle, \nabla g(1, -2) = \langle \frac{1}{2}, -1 \rangle, \nabla g(-3, -1) = \langle -\frac{3}{2}, -\frac{1}{2} \rangle$ • Figure A30

34. Draw an s-axis perpendicular to the level curve through $(2, 1)$ as in Figure A34. • The distance along the s-axis from the level curve $f = 2$ to the level curve $f = 4$ is $\Delta s \approx \frac{3}{8}$. • $\Delta f = 2$ • [Maximum directional derivative] $= |\nabla f(2, 1)| \approx \frac{\Delta f}{\Delta s} \approx 5.3$ • The gradient vector is drawn in Figure A34 in the positive s-direction using the scale below the drawing to measure its length. • Measure the components: $\nabla f(2, 1) \approx \langle 1.5, 5 \rangle$

Figure A34

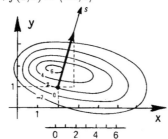

35. One answer: $\nabla g(2, 2) \approx \langle -9, -9 \rangle$

40. (a) $f = xy^3 + 6x^2y$ • $\nabla f = \langle y^3 + 12xy, 3xy^2 + 6x^2 \rangle$ • $\nabla f(1, -1) = \langle -13, 9 \rangle$ • Unit normal vectors: $\pm \frac{\langle -13, 9 \rangle}{5\sqrt{10}}$ (b) Unit normal vectors: $\pm \frac{\langle 1, -4 \rangle}{\sqrt{17}}$

41. (a) $\nabla F(5, -1) \approx \langle -3, \frac{1}{2} \rangle$ • $F_x(5, -1) \approx -3$ (b) $F_y(3, 3) \approx -\frac{3}{4}$ (c) $F_x(2, 1) \approx 0$ (e) $D_{\mathbf{u}}F(2, 1) \approx 0.7$

Tune-up exercises 11.4

T1. (a) $f = \sqrt{25 - x^2 - y^2 - z^2}$ is defined for $x^2 + y^2 + z^2 \leq 25$. •
Its domain is the closed ball of radius 5 with its center at the origin of xyz-space.
(b) $f = 4 \iff 25 - x^2 - y^2 - z^2 = 16 \iff x^2 + y^2 + z^2 = 9$ •
The level surface $f = 4$ is the sphere of radius 3 with its center at the origin.

T2. $R = (x + 2y + 3z)^{10}$ • $R_x = \dfrac{\partial}{\partial x}[(x+2y+3z)^{10}] = 10(x+2y+3z)^9 \dfrac{\partial}{\partial x}(x+2y+3z) = 10(x+2y+3z)^9$
$R_{xy} = \dfrac{\partial}{\partial y}[10(x+2y+3z)^9] = 90(x+2y+3z)^8 \dfrac{\partial}{\partial y}(x+2y+3z) = 180(x+2y+3z)^8$

T3. $\left[\dfrac{d}{dt}[G(x(t), y(t), z(t))]\right]_{t=10}$
$= \dfrac{\partial G}{\partial x}(x(10), y(10), z(10))\dfrac{dx}{dt}(10) + \dfrac{\partial G}{\partial y}(x(10), y(10), z(10))\dfrac{dy}{dt}(10) + \dfrac{\partial G}{\partial z}(x(10), y(10), z(10))\dfrac{dz}{dt}(10)$
$= \dfrac{\partial G}{\partial x}(1,2,3)\dfrac{dx}{dt}(10) + \dfrac{\partial G}{\partial y}(1,2,3)\dfrac{dy}{dt}(10) + \dfrac{\partial G}{\partial z}(1,2,3)\dfrac{dz}{dt}(10) = 50(8) + 60(7) + 70(6) = 1240$

T4. (a) $H = x^2 + y^3 + z^4$ • $\nabla H = \langle 2x, 3y^2, 4z^3 \rangle$ • $\nabla H(3,2,1) = \langle 6, 12, 4 \rangle$
(b) Set $P = (3,2,1)$ and $Q = (4,4,3)$. • $\overrightarrow{PQ} = \langle 4-3, 4-2, 3-1 \rangle = \langle 1, 2, 2 \rangle$ •
$|\langle 1,2,2\rangle| = \sqrt{1^2 + 2^2 + 2^2} = 3$ • $\mathbf{u} = \overrightarrow{PQ}/|\overrightarrow{PQ}| = \langle \tfrac{1}{3}, \tfrac{2}{3}, \tfrac{2}{3}\rangle$
$D_{\mathbf{u}}H(3,2,1) = \nabla H \cdot \mathbf{u} = \langle 6,12,4\rangle \cdot \langle \tfrac{1}{3}, \tfrac{2}{3}, \tfrac{2}{3}\rangle = 6(\tfrac{1}{3}) + 12(\tfrac{2}{3}) + 4(\tfrac{2}{3}) = \tfrac{38}{3}$
(c) [Maximum directional derivative] $= |\nabla H(3,2,1)| = |\langle 6,12,4\rangle| = \sqrt{6^2 + 12^2 + 4^2} = 14$ •
The unit vector in that direction: $\mathbf{u} = \dfrac{\langle 6,12,4\rangle}{|\langle 6,12,4\rangle|} = \langle \tfrac{3}{7}, \tfrac{6}{7}, \tfrac{2}{7}\rangle$

T5. From Exercise T4, $\nabla H(3,2,1) = \langle 6,12,4\rangle = 2\langle 3,6,2\rangle$ is normal to the level surface and to its tangent plane at $(3,2,1)$. • Use $\mathbf{n} = \langle 3,6,2\rangle$ • Tangent plane: $3(x-3) + 6(y-2) + 2(z-1) = 0$

T6. $g = (\sin x)(\cos y)(\ln z)e^{2w}$ • $g_x = (\cos x)(\cos y)(\ln z)e^{2w}$ • $g_{xy} = -(\cos x)(\sin y)(\ln z)e^{2w}$ •
$g_{xyz} = -(\cos x)(\sin y)(z^{-1})e^{2w}$ • $g_{xyzw} = -2(\cos x)(\sin y)(z^{-1})e^{2w}$

Problems 11.4

1. $R(10,8,36) = 5.85$ • Your boat would not qualify.

4. (a) $f = x^{1/2}y^{1/4} + y^{1/5}z^{1/6}$ • $f_x = \tfrac{1}{2}x^{-1/2}y^{1/4}$ •
$f_y = \tfrac{1}{4}x^{1/2}y^{-3/4} + \tfrac{1}{5}y^{-4/5}z^{1/6}$ • $f_z = \tfrac{1}{6}y^{1/5}z^{-5/6}$
(b) $g_x = \dfrac{1}{x + 3y - 4z}$ • $g_y = \dfrac{3}{x + 3y - 4z}$ • $g_z = \dfrac{-4}{x + 3y - 4z}$

5. (a) $f = x^2 y^3 (1 + xyz)^2$ • $f_x = x^2 y^3 \dfrac{\partial}{\partial x}[(1 + xyz)^2] + (1 + xyz)^2 \dfrac{\partial}{\partial x}(x^2 y^3)$
$= 2x^2 y^3 (1 + xyz)\dfrac{\partial}{\partial x}(xyz) + 2xy^3(1 + xyz)^2 = 2x^2 y^4 z(1 + xyz) + 2xy^3(1+xyz)^2$
$f_y = x^2 y^3 \dfrac{\partial}{\partial y}[(1 + xyz)^2] + (1 + xyz)^2 \dfrac{\partial}{\partial y}(x^2 y^3) = 2x^2 y^3 (1 + xyz)\dfrac{\partial}{\partial y}(xyz) + 3x^2 y^2(1 + xyz)^2$
$= 2x^3 y^3 z(1 + xyz) + 3x^2 y^2 (1 + xyz)^2$
$f_z = x^2 y^3 \dfrac{\partial}{\partial z}[(1 + xyz)^2] = 2x^2 y^3 (1 + xyz)\dfrac{\partial}{\partial z}(xyz) = 2x^3 y^4 (1 + xyz)$
(b) $g = xy \tan^{-1}(yz)$ • $g_x = y\tan^{-1}(yz)$ • $g_y = x\tan^{-1}(yz) + \dfrac{xyz}{1 + (yz)^2}$ • $g_z = \dfrac{xy^2}{1 + (yz)^2}$

6. $M_{xy} = -6xe^{-3y}\sin(4z)$ • $M_{yz} = -12x^2 e^{-3y}\cos(4z)$ • $M_{xz} = 8xe^{-3y}\cos(4z)$

9. (a) $\nabla f(x, y, z) = \langle \sin(yz^2), xz^2 \cos(yz^2), 2xyz \cos(yz^2)\rangle$ • $\nabla f(3,2,1) = \langle \sin(2), 3\cos(2), 12\cos(2)\rangle$
(b) $\mathbf{A} = \langle -3, -2, -1\rangle$ points from $(3,2,1)$ toward the origin. • $\mathbf{u} = \dfrac{\langle -3, -2, -1\rangle}{|\langle -3, -2, -1\rangle|} = \dfrac{\langle -3, -2, -1\rangle}{\sqrt{14}}$
• $D_{\mathbf{u}}f(3,2,1) = \dfrac{\langle \sin(2), 3\cos(2), 12\cos(2)\rangle \cdot \langle -3, -2, -1\rangle}{\sqrt{14}} = \dfrac{-3\sin(2) - 18\cos(2)}{\sqrt{14}}$

10. (a) $V = \pi r^2 H + \tfrac{1}{3}\pi r^2 h$

Answers 11.4

11. $\dfrac{\partial c}{\partial a} = \dfrac{a - b\cos\theta}{\sqrt{a^2 + b^2 - 2ab\cos\theta}}$ • $\dfrac{\partial c}{\partial b} = \dfrac{b - a\cos\theta}{\sqrt{a^2 + b^2 - 2ab\cos\theta}}$ • $\dfrac{\partial c}{\partial \theta} = \dfrac{ab\sin\theta}{\sqrt{a^2 + b^2 - 2ab\cos\theta}}$

13. $f = x^4 y^{-2} z^3$ • (a) $\nabla f = \langle 4x^3 y^{-2} z^3, -2x^4 y^{-3} z^3, 3x^4 y^{-2} z^2\rangle$ • $\nabla f(1,2,3) = \langle 27, -\frac{27}{4}, \frac{27}{4}\rangle$ •
$\mathbf{u} = \dfrac{\langle 3, -4, -12\rangle}{|\langle 3, -4, -12\rangle|} = \langle \frac{3}{13}, -\frac{4}{13}, -\frac{12}{13}\rangle$ • $D_{\mathbf{u}}f(1,2,3) = \langle 27, -\frac{27}{4}, \frac{27}{4}\rangle \cdot \langle \frac{3}{13}, -\frac{4}{13}, -\frac{12}{13}\rangle = \frac{27}{13}$
(b) $2\sqrt{3}$

14. (a) $2\sqrt{61}\, e^{24}$

15. (a) Set $f = 3xy^2 + 2yz^3 + xz$ • $\nabla f = \langle 3y^2 + z, 6xy + 2z^3, 6yz^2 + x\rangle$ •
$\nabla f(-1, 2, 3) = \langle 15, 42, 107\rangle$ •
Tangent plane: $15(x+1) + 42(y-2) + 107(z-3) = 0$
(b) Tangent plane: $(x - 5) + 25(z - \frac{1}{5}) = 0$

16. (a) $F(t) = 16t^{16}$

17. $P(2) = f(2,4,8) = 3$ • $P'(t) = f_x(t, t^2, t^3)\dfrac{d}{dt}(t) + f_y(t, t^2, t^3)\dfrac{d}{dt}(t^2) + f_z(t, t^2, t^3)\dfrac{d}{dt}(t^3)$
$= f_x(t, t^2, t^3) + 2t f_y(t, t^2, t^3) + 3t^2 f_z(t, t^2, t^3)$ •
$P'(2) = f_x(2,4,8) + 4 f_y(2,4,8) + 12 f_z(2,4,8) = 4 + 4(5) + 12(-6) = -48$

18. 250

19. $G_x(-5,0,5) = 40, G_y(-5,0,5) = 80, G_z(-5,0,5) = 100$

21. $B_x(3,2,1) = \frac{1}{4}\sqrt{3}\,\pi$

23. $G(1,2,3) = 35, G_x(1,2,3) = 18, G_y(1,2,3) = -120, G_z(1,2,3) = 30$

26. (a) Domain: All (x,y,z) with $(x,y) \neq 0$ (All of xyz-space except the z-axis)
sbeb $f = 1$ is the half-cone $z = \frac{1}{2}\sqrt{x^2 + y^2}$ • $f = -2$ is the half-cone $z = -\sqrt{x^2 + y^2}$ • Figure A26

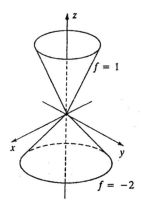

Figure A26

32. (a) $f = x^2 y^3 e^{4z}$ • $f_x = 2xy^3 e^{4z}, f_y = 3x^2 y^2 e^{4z}, f_z = 4x^2 y^3 e^{4z}$ •
$f_x(2,-1,0) = -4, f_y(2,-1,0) = 12, f_z(2,-1,0) = -16$ •
Total differential at $(2,-1,0)$: $df = -4\,dx + 12\,dy - 16\,dz$
(b) Total differential: $dg = 20 e^5 \,dx + 300 e^5 \,dy - 400 e^5 \,dz$

33. (a) $g_x = 2xe^{3y - 4z}, g_y = 3x^2 e^{3y - 4z}, g_z = -4x^2 e^{3y - 4z}$ •
Total differential: $dg = 2xe^{3y-4z}\,dx + 3x^2 e^{3y-4z}\,dy - 4x^2 e^{3y-4z}\,dz$
(b) $dQ = y(1 + e^{yz})\,dx + [x(1 + e^{yz}) + xyze^{yz}]\,dy + xy^2 e^{yz}\,dz$

35. [Maximum error] $\approx 50\sqrt{3}(0.1) + 5\sqrt{3}(\frac{1}{12}) + 500(\frac{1}{180}\pi) \doteq 18.1$ foot-pounds

36. (a) 1.8×10^{-4} inches (b) [Maximum error] $\approx 7.6 \times 10^{-7}$ inches

41. (a) $x = 2 + 3t, y = 1 + 10t, z = -1 - 14t$

42. $J_y(5, -4, 3, -2, 1) = 36{,}000$

Tune-up exercises 11.5

T1. $f = x^2 - xy + y$ • $\begin{cases} f_x = 2x - y \\ f_y = -x + 1 \end{cases}$ • Solve $\begin{cases} 2x - y = 0 \\ -x + 1 = 0. \end{cases}$ •
The second equation gives $x = 1$ and then the first gives $y = 2$. • Critical point: $(1, 2)$

T2. $g = e^{xy}$ • $g_x = e^{xy} \dfrac{\partial}{\partial x}(xy) = ye^{xy}$ • $g_y = e^{xy} \dfrac{\partial}{\partial y}(xy) = xe^{xy}$ • Solve $\begin{cases} ye^{xy} = 0 \\ xe^{xy} = 0. \end{cases}$ •
Since $e^{xy} > 0$, the one critical point is $(0, 0)$.

T3. $h = 2xy^3 - x^2 + 6y$ • $\begin{cases} h_x = 2y^3 - 2x = 2(y^3 - x) \\ h_y = 6xy^2 + 6 = 6(xy^2 + 1) \end{cases}$ • Solve $\begin{cases} y^3 - x = 0 \\ xy^2 + 1 = 0. \end{cases}$ •
The first equation gives $x = y^3$ and then the second gives $y^5 + 1 = 0$ • $y = -1$ •
$x = (-1)^3 = -1$ • The one critical point is $(-1, -1)$.

T4. $g = 4x - x^4 + 6y - y^2$ • $\begin{cases} g_x = 4 - 4x^3 = 4(1 - x^3) \\ g_y = 6 - 2y = 2(3 - y) \end{cases}$ • Solve $\begin{cases} 1 - x^3 = 0 \\ 3 - y = 0. \end{cases}$
The first equation gives $x = 1$ and the second gives $y = 3$. • Because we are told that g has a global maximum and g is defined for all (x, y), the global maximum is a local maximum and has to be at the critical point $(1, 3)$. • [Global maximum] $= g(1, 3) = 4(1) - (1)^4 + 6(3) - 3^2 = 12$

T5. $h = x^4 + 4x - xy$ • $\begin{cases} h_x = 4x^3 + 4 - y \\ h_y = -x \end{cases}$ • Solve $\begin{cases} 4x^3 + 4 - y = 0 \\ -x = 0. \end{cases}$ •
The second equation gives $x = 0$ and then the first gives $y = 4$. • Critical point: $(0, 4)$ •
$h_{xx} = 12x^2, h_{xy} = -1, h_{yy} = 0$ • $A = h_{xx}(0, 4) = 0, B = h_{xy}(0, 4) = -1, C = h_{yy}(0, 4) = 0$ •
$AC - B^2 = (0)(0) - (-1)^2 = -1$ is negative so there is a saddle point at $(0, 4)$.

T6. (a) $f = x^2 + y^2$ • Set $g = 3x + y$ • Constraint: $g = 20$ •
$\nabla f = \lambda \nabla g \iff \langle 2x, 2y \rangle = \lambda \langle 3, 1 \rangle \iff \begin{cases} 2x = 3\lambda \\ 2y = \lambda \end{cases}$ • $x = \tfrac{3}{2}\lambda, y = \tfrac{1}{2}\lambda$ •
$3x + y = 10 \implies 3(\tfrac{3}{2}\lambda) + (\tfrac{1}{2}\lambda) = 20 \implies 5\lambda = 20 \implies \lambda = 4$ • $x = 6, y = 2$ •
[Minimum] $= f(6, 2) = 6^2 + 2^2 = 40$
(b) Solve $3x + y = 20$ for $y = 20 - 3x$ • $f(x, y) = x^2 + (20 - 3x)^2$ •
Find the minimum of $F(x) = x^2 + (20 - 3x)^2$. •
$\dfrac{dF}{dx}(x) = 2x + 2(20 - 3x)\dfrac{d}{dx}(20 - 3x) = 2x - 6(20 - 3x) = 20x - 120 = 20(x - 6)$ •
$\dfrac{dF}{dx}(x)$ is negative for $x < 6$ and positive for $x > 6$, so $F(x)$ has a global minimum at $x = 6$. •
[Minimum] $= F(6) = 6^2 + [20 - 3(6)]^2 = 40$

Problems 11.5

1. $f = x^4 + 32x - y^3$ • $\begin{cases} f_x = 4x^3 + 32 = 4(x^3 + 8) \\ f_y = -3y^2 \end{cases}$ • Solve $\begin{cases} x^3 + 8 = 0 \\ y^2 = 0. \end{cases}$ •
The first equation gives $x = -2$ and the second gives $y = 0$. • Critical point: $(-2, 0)$

2. $g = x^2 + \sin y$ • $g_x = 2x, g_y = \cos y$ • Solve $\begin{cases} 2x = 0 \\ \cos y = 0. \end{cases}$ •
Critical points: $(0, (n + \tfrac{1}{2})\pi)$ for all integers $n = 0, \pm 1, \pm 2, \ldots$

3. $h = 3xy^2 + x^3 - 3x$ • $\begin{cases} h_x = 3y^2 + 3x^2 - 3 = 3(y^2 + x^2 - 1) \\ h_y = 6xy \end{cases}$ • Solve $\begin{cases} y^2 + x^2 - 1 = 0 \\ xy = 0. \end{cases}$ •
The second equation holds if $x = 0$ or $y = 0$.
For $x = 0$: The first equation is $y^2 - 1 = 0$ and has the solutions $y = \pm 1$ •
For $y = 0$: The first equation is $x^2 - 1 = 0$ and has the solutions $x = \pm 1$ •
Critical points: $(0, 1), (0, -1), (1, 0), (-1, 0)$

Answers 11.5

5. Critical points: $(2,0), (-2,0), (2,\sqrt{18}), (2,-\sqrt{18}), (-2,\sqrt{18}), (-2,-\sqrt{18})$

7. [Global maximum] $= e^{13}$

10a. Critical points: $(0, y)$ for all y (the y-axis)

12. $f(x,y) = [\text{Distance}]^2 = (x-6)^2 + (y-4)^2 + (\sqrt{x^2+y^2+3})^2 = (x-6)^2 + (y-4)^2 + x^2 + y^2 + 3$
$\begin{cases} f_x = 2(x-6) + 2x = 4(x-3) \\ f_y = 2(y-4) + 2y = 4(y-2). \end{cases}$ • Solve $\begin{cases} x - 3 = 0 \\ y - 2 = 0. \end{cases}$
Critical point: $x = 3, y = 2$ • The global minimum of $f(x,y)$ is at the critical point, where $z = \sqrt{3^2 + 2^2 + 3} = 4$. • Closest point: $(3, 2, 4)$

13a. $C(x,y) = 72x^{-1} + 108y^{-1} + 6xy$

14. $f = x^3 y - 12xy$ • $\begin{cases} f_x = 3x^2 y - 12y = 3y(x^2 - 4) \\ f_y = x^3 - 12x = x(x^2 - 12) \end{cases}$ • Solve $\begin{cases} y(x^2 - 4) = 0 \\ x(x^2 - 12) = 0. \end{cases}$ •
The first equation holds if $y = 0$ or $x = \pm 2$. •
For $y = 0$, the second equation holds if $x = 0$ or $x = \sqrt{12}$ or $x = -\sqrt{12}$. •
For $x = \pm 2$, the second equation is never valid. • Critical points: $(0,0), (\sqrt{12}, 0), (-\sqrt{12}, 0)$ •
$f_{xx} = 6xy, f_{xy} = 3x^2 - 12, f_{yy} = 0$

Critical point	$A = f_{xx}$	$B = f_{xy}$	$C = f_{yy}$	$AC - B^2$	Type
$(0,0)$	0	-12	0	-144	Saddle point
$(\sqrt{12}, 0)$	0	24	0	-576	Saddle point
$(-\sqrt{12}, 0)$	0	24	0	-576	Saddle point

15. The critical points are $(0, n\pi)$ with integers $n = 0, \pm 1, \pm 2, \ldots$ and are all saddle points

18a. $f = x^3 - y^3 - 3xy + \frac{1}{8}$ • $f_x = 3(x^2 - y), f_y = -3(y^2 + x)$ •
Solve $\begin{cases} x^2 - y = 0 \\ y^2 + x = 0. \end{cases}$ • $x = -x^4$ • $x(x^3 + 1) = 0$ • $x = 0$ or -1 and $y = x^2$ •
Critical points: $(0,0)$ and $(-1, 1)$ • $f_{xx} = 6x, f_{xy} = -3, f_{yy} = -6y$ • See the table • Figure 33

Critical point	$A = f_{xx}$	$B = f_{xy}$	$C = f_{yy}$	$AC - B^2$	Type
$(0,0)$	0	-3	0	-9	Saddle point
$(-1, 1)$	-6	-3	-6	27	Local maximum

18b. Saddle points at $(\frac{1}{2}\sqrt{6}, 0)$ and $(-\frac{1}{2}\sqrt{6}, 0)$ • Local maximum at $(0, 1)$ • Local minimum at $(0, -1)$
• Figure 35

18d $f = -2x^3 - 3y^4 + 6xy^2 + \frac{1}{16}$ • $f_x = -6(x^2 - y^2), f_y = -12y(y^2 - x)$ •
Solve $\begin{cases} x^2 - y^2 = 0 \\ y(y^2 - x) = 0. \end{cases}$ • The second equation gives $y = 0$ or $x = y^2$.
For $y = 0$ the first equation is $x = 0$. • Critical point: $(0, 0)$ •
For $x = y^2$ the first equation is $y^4 - y^2 = 0$ or $y^2(y^2 - 1) = 0$ and gives $y = 0, y = 1, y = -1$ with $x = y^2$. • Critical points: $(0,0), (1,1), (1,-1)$
$f_{xx} = -12x, f_{xy} = 12y, f_{yy} = -36y^2 + 12x$ • See the table below. • Figure 33

Critical point	$A = f_{xx}$	$B = f_{xy}$	$C = f_{yy}$	$AC - B^2$	Type
$(0,0)$	0	0	0	0	The test fails
$(1,1)$	-12	12	-24	144	Local maximum
$(1,-1)$	-12	-12	-24	144	Local maximum

19. **(a)** Figure 41 because of the saddle point at the origin and the local maximum at $(-1, 1)$
    (b) Figure 39 **(d)** Figure 42

21. $f = 2x + y$ and $g = x^2 + 2y^2$ • $\langle 2, 1 \rangle = \lambda \langle 2x, 4y \rangle$ • $x = \dfrac{1}{\lambda}, y = \dfrac{1}{4\lambda}$ •
    $\left(\dfrac{1}{\lambda}\right)^2 + 2\left(\dfrac{1}{4\lambda}\right)^2 = 18$ • $\dfrac{9}{8\lambda^2} = 18$ • $\lambda = \pm\dfrac{1}{4}$ •
    For $\lambda = \dfrac{1}{4}$: $x = 4, y = 1$ • For $\lambda = -\dfrac{1}{4}$: $x = -4, y = -1$ •
    [Maximum] $= f(4, 1) = 2(4) + 1 = 9$ at $(4, 1)$ •
    [Minimum] $= f(-4, -1) = 2(-4) - 1 = -9$ at $(-4, -1)$

22. [Maximum] $= f(\tfrac{1}{2}, 4) = \dfrac{29}{2}$ • [Minimum] $= f(-\tfrac{1}{2}, -4) = -\dfrac{37}{2}$

24. $f = x + 5z$ and $g = 2x^2 + 3y^2 + 5z^2$ • $\langle 1, 0, 5 \rangle = \lambda \langle 4x, 6y, 10z \rangle$ • $x = \dfrac{1}{4\lambda}, y = 0, z = \dfrac{1}{2\lambda}$ •
    $2\left(\dfrac{1}{4\lambda}\right)^2 + 3(0)^2 + 5\left(\dfrac{1}{2\lambda}\right)^2 = 198$ • $\lambda = \pm\dfrac{1}{12}$ •
    For $\lambda = \dfrac{1}{12}$: $x = 3, y = 0, z = 6$ • For $\lambda = -\dfrac{1}{12}$: $x = -3, y = 0, z = -6$ •
    [Maximum] $= f(3, 0, 6) = 33$ • [Minimum] $= f(-3, 0, -6) = -33$

26. $f = xy$ and $g = 4x^2 + 9y^2$ • $\langle y, x \rangle = \lambda \langle 8x, 18y \rangle$ • Solve $\begin{cases} y = 8\lambda x \\ x = 18\lambda y \end{cases}$ •
    If λ were 0, then x and y would be zero, but $(0, 0)$ is not on the ellipse $4x^2 + 9y^2 = 72$ •
    $\lambda \neq 0, x \neq 0, y \neq 0$ • Multiply the first equation by x and the second by y • $8\lambda x^2 = 18\lambda y^2$ •
    $4x^2 = 9y^2$ • $4x^2 + 9y^2 = 72 \implies 4x^2 = 36$ and $9y^2 = 36$ • $x = \pm 3, y = \pm 2$ •
    [Maximum] $= f(3, 2) = f(-3, -2) = 6$ • [Minimum] $= f(3, -2) = f(-3, 2) = -6$

27. [Maximum] $= f(2, \pm\sqrt{18}) = 26$ • [Minimum] $= f(-4, 0) = 8$

31. Since $f = x^2 y^2$ is zero for $x = 0$ and for $y = 0$ and positive elsewhere, its global minimum is 0. •
    For the maximum on the boundary: $g = x^2 + 4y^2$ • $\langle 2xy^2, 2x^2 y \rangle = \lambda \langle 2x, 8y \rangle$ •
    Solve $\begin{cases} 2xy^2 = 2\lambda x \\ 2x^2 y = 8\lambda y \end{cases}$ with $x \neq 0, y \neq 0$. • $y^2 = \lambda$ and $x^2 = 4\lambda$ • $x^2 + 4y^2 = 24 \implies \lambda = 3$
    • $x^2 = 12, y^2 = 3$ • [Maximum] $= 36$ at $(\sqrt{12}, \sqrt{3}), (-\sqrt{12}, \sqrt{3}), (\sqrt{12}, -\sqrt{3})$, and $(-\sqrt{12}, -\sqrt{3})$

32. [Maximum] $= f(2, 0) = 8$ • [Minimum] $= f(-1, 0) = -1$

34a. $f = 4x^2 y + 3xy^2 - 12xy$ • $f_x = y(8x + 3y - 12), f_y = 2x(2x + 3y - 6)$ •
    Solve $\begin{cases} y(8x + 3y - 12) = 0 \\ 2x(2x + 3y - 6) = 0. \end{cases}$ Case 1: $x = 0, y = 0$ • Critical point: $(0, 0)$ •
    Case 2: $y = 0$ and $2x + 3y = 6$ • Critical point: $(3, 0)$ •
    Case 3: $8x + 3y = 12$ and $x = 0$ • Critical point: $(0, 4)$ •
    Case 4: $8x + 3y = 12$ and $2x + 3y = 6$ • Critical point: $(1, \tfrac{4}{3})$ •
    $f_{xx} = 8y, f_{xy} = 8x + 6y - 12, f_{yy} = 6x$ • See the table below.

Critical point	$A = f_{xx}$	$B = f_{xy}$	$C = f_{yy}$	$AC - B^2$	Type
$(0, 0)$	0	-12	0	-144	Saddle point
$(3, 0)$	0	12	18	-144	Saddle point
$(0, 4)$	32	12	0	-144	Saddle point
$(1, \tfrac{4}{3})$	$\tfrac{32}{3}$	4	6	48	Local minimum

34b. Saddle point at $(0, 0)$ • Local maximum at $(1, 1)$

35a. The critical points form the lines $y = x + n\pi$ with integers n.

41. $f = 4x - \frac{1}{2}x^2 - \frac{1}{4}x^4 + 4y^2 + y - y^4$ • $f_x = 4 - x - x^3, f_y = 8y + 1 - 4y^3$ • $f_x = 0$ has the solution $x^* \approx 1.3787967$ • $f_y = 0$ has the solutions $y_1^* \approx -1.346997, y_1^* \approx -0.1260002$, and $y_3^* \approx 1.4729976$ • $f_{xx} = -1 - 3x^2, f_{xy} = 0, f_{yy} = 8 - 12y^2$ • See the table below.

Critical point	$A = f_{xx} \approx$	$B = f_{xy} \approx$	$C = f_{yy} \approx$	$AC - B^2 \approx$	Type
(x^*, y_1^*)	-6.70	0	-13.77	92.3	Local maximum
(x^*, y_2^*)	-6.70	0	7.81	-52.3	Saddle point
(x^*, y_3^*)	-6.70	0	-18.03	120.8	Local maximum

42. The Second-Derivative Test fails at the critical point $(0, 0)$. • There is a local minimum at the critical point $(x^*, y^*) \approx (2.148, 3.148)$.

45. Closest points: $(1, 1, 1), (1, -1, -1), (-1, 1, -1), (-1, -1, 1)$

CHAPTER 12
MULTIPLE INTEGRALS

In Chapters 4 and 6 we used definite integrals to find areas between curves, weights and centers of gravity of one-dimensional objects with variable density, and to find average values of functions in intervals. In the first two sections of this chapter we define DOUBLE and TRIPLE INTEGRALS (MULTIPLE INTEGRALS) and use them to solve similar problems with two and three variables. In Section 12.3 we show how double integrals can be evaluated using polar coordinates and triple integrals can be evaluated using CYLINDRICAL and SPHERICAL COORDINATES. Finally, In Section 12.4 we discuss other changes of variables in double and triple integrals.

Section 12.1

Double integrals

OVERVIEW: *In this section we first define* DOUBLE INTEGRALS *over bounded regions in xy-planes and see how they can be interpreted in terms of volumes. Then we use double integrals to calculate volumes of solids between graphs of functions, weights and centers of gravity of two-dimensional objects with variable density, quantities determined by other types of density, and average values of functions with two variables.*

Topics:

- *Double integrals*
- *Other volumes*
- *Density, weight, and centers of gravity*
- *Average values*

A volume and a price

We start with a Question and an Example that illustrate the basic idea of a double integral.

Question 1   Figure 1 shows a building that has the shape of three rectangular boxes. The largest is 80 feet high and its base is a 100-foot-wide square The smallest is 40 feet high and its base is a 50-foot wide square. The third is 60 feet high. Its base is a rectangle 50 feet wide and 100 feet long. What is the total volume of the building?

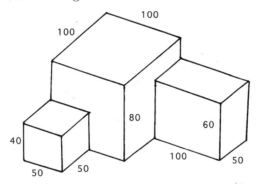

FIGURE 1

The roof of the building in Figure 1 is formed by the graph of the function $z = f(x, y)$ in Figure 2. The domain R of $f(x, y)$ in Figure 3 consists of the two squares R_1 and R_2 and the rectangle R_3; and $f(x, y)$ equals 80 on R_1, 40 on R_2, and 60 on R_3. We call $f(x, y)$ a STEP FUNCTION

♦ SUGGESTIONS TO INSTRUCTORS: The main topics in this section can be divided in two parts, which can be covered in two or more lectures.
Part 1 (Double integrals, volumes between graphs and the xy-plane, iterated integrals; Tune-up Exercises T1–T5, Problems 1–23, 37–51, 58) Have students begin work on Questions 1 through 7 before class. Problems 2, 3, 4, 8, 10, 12, 24, 46, 47a are good for in-class discussion.
Part 2 (Volumes between graphs, two-dimensional densities, centers of gravity, average values; Tune-up Exercises T6–T8, Problems 24–36, 52–57, 59–64) Have students begin Questions 8 through 11 before class. Problems 31, 35, 54 are good for in-class discussion.

because it is constant in the interior of each of these parts of its domain. We will see below that because the building lies between the graph of the posiitive function $f(x,y)$ and the xy-plane, its volume equals the DOUBLE INTEGRAL of $f(x,y)$ over its domain R.

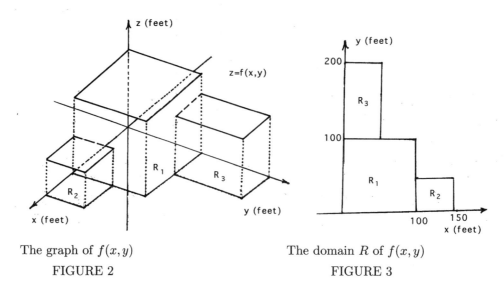

The graph of $f(x,y)$
FIGURE 2

The domain R of $f(x,y)$
FIGURE 3

In the following Example we use the PRICE DENSITY of a piece of property, measured in dollars per square meter, to calculate the value of the property.

Example 1   Imagine that Figure 4 is the map of a property in an xy-plane with distances measured in meters. The property consists of three parts labeled R_1, R_2, and R_3. Part R_1 extends for $0 \leq x \leq 200, 100 \leq y \leq 300$ and is worth \$4 per square meter; part R_2 extends for $200 \leq x \leq 400, 100 \leq y \leq 300$ and is worth \$3 per square meter, and part R_3 extends for $0 \leq x \leq 400, 300 \leq y \leq 500$ and is worth \$1 per square meter. What is the total value of the property?

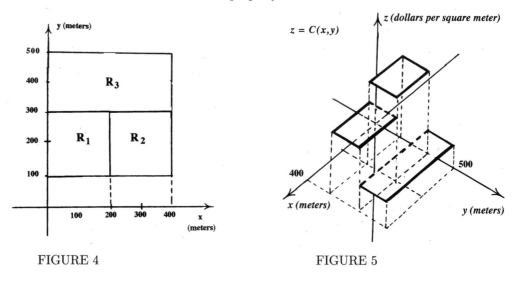

FIGURE 4

FIGURE 5

SOLUTION   Parts R_1 and R_2 of the property are 200-meter-wide squares with area $200(200) = 40{,}000$ square meters. Part R_3 is rectangle that is 400 meters wide and 200 meters deep and has area $400(200) = 80{,}000$ square meters. Because R_1, R_2, and R_3 are worth \$4, \$3, and \$1 per square meter, respectively, the total value of the property is

12.1 Double integrals

$$\left[40{,}000 \text{ square meters}\right]\left[\frac{4 \text{ dollars}}{\text{square meter}}\right]$$
$$+ \left[40{,}000 \text{ square meters}\right]\left[\frac{3 \text{ dollars}}{\text{square meter}}\right] \quad\quad (1)$$
$$+ \left[80{,}000 \text{ square meters}\right]\left[\frac{1 \text{ dollars}}{\text{square meter}}\right] = 360{,}000 \text{ dollars.} \ \square$$

The cost density of the property in Example 1 can given by the step function
$$C(x,y) = \begin{cases} 4 & \text{for} \quad 0 \leq x \leq 200, 100 \leq y \leq 300 \\ 3 & \text{for} \quad 200 \leq x \leq 400, 100 \leq y \leq 300 \\ 1 & \text{for} \quad 0 \leq x \leq 400, 300 \leq y \leq 500 \end{cases}$$

whose graph consists of the three of horizontal rectangles in Figure 5. The domain of $C(x,y)$ is the square R obtained by combining R_1, R_2, and R_3. We will see below that the cost of the property is equal to the double integral of the price density $C(x,y)$ over the square R.

Double integrals
Before we can define double integrals we need to describe the types of regions we will consider. We will say that a curve in an xy-plane is PIECEWISE SMOOTH if, like the curve in Figure 6, it consists of a finite number of graphs $y = f(x)$ or $x = g(y)$ of functions defined in finite closed intervals that have continuous first derivatives. Also, we will say that a region in an xy-plane is BOUNDED if it can be enclosed in a circle.

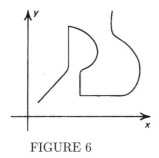

FIGURE 6

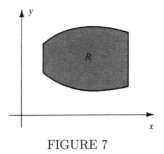

FIGURE 7

Consider a bounded region R with a piecewise smooth boundary, such as in Figure 7, and a function $f(x,y)$ that is defined in R except possibly at a finite number of points or on a finite number of piecewise smooth curves in R. To define the double integral $\iint_R f(x,y) \, dx\, dy$, we use piecewise smooth curves to form a PARTITION of R into subregions

$$R_1, R_2, \ldots, R_N. \quad\quad (2)$$

Figure 8 shows a partition consisting of four such subregions. We define the DIAMETER of each subregion to be the diameter of the smallest closed disk that contains it.

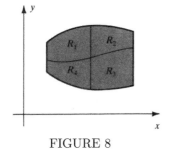

FIGURE 8

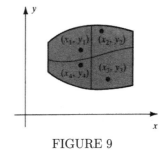

FIGURE 9

Next, we pick for each j a point (x_j, y_j) in the subregion R_j such that $f(x_j.y_j)$ is defined. Figure 9 shows such points for the partition in Figure 8. We construct the RIEMANN SUM

$$\sum_{j=1}^{N} f(x_j, y_j)[\text{Area of } R_j]. \tag{3}$$

The double integral is the limit of such Riemann sums:

Definition 1 *The double integral $\iint_R f(x,y)\,dx\,dy$ is the limit of the Riemann sums (3) as the number of subregions in the corresponding partitions tend to infinity and their diameters tend to zero, provided the limit exists and is finite.*[†]

It is shown in advanced calculus courses that integral in Definition 1 exists if if R is bounded and has a piecewise smooth boundary and $f(x,y)$ is PIECEWISE CONTINUOUS in R. The latter condition means that there is a (fixed) partition of R into subregions $R_1, R_2, \ldots, R_N$ such that in the interior of each R_j, the given function $f(x,y)$ equals another function $F_j(x,y)$ that is continuous in the closure of R_j. Figure 10 shows the graph of a piecewise continuous function.

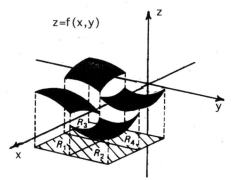

FIGURE 10

If $f(x,y)$ and $g(x,y)$ are piecewise continuous in the region R in Definition 1, then for any constants A and B, the Riemann sums for the integral of $Af(x,y) + Bg(x,y)$ over R are

$$\sum_{j=1}^{N} [Af(x_j, y_j) + Bg(x_j, y_j)][\text{Area of } R_j]$$

$$= A\sum_{j=1}^{N} f(x_j, y_j)[\text{Area of } R_j] + B\sum_{j=1}^{N} g(x_j, y_j)[\text{Area of } R_j].$$

The sums on the right of this equation are Riemann sums for the integrals of $f(x,y)$ and $g(x,y)$, so that when we let the number of subregions in the partitions tend to ∞ and their diameters to 0, we obtain the equation

$$\iint_R [Af(x,y) + Bg(x,y)]\,dx\,dy = A\iint_R f(x,y)\,dx\,dy + B\iint_R g(x,y)\,dx\,dy \tag{4}$$

which we will use when working with double integrals of linear combinations of functions.

[†]The $\epsilon\delta$-definition of this limit reads as follows: The limit of the Riemann sums is the number L if for every $\epsilon > 0$, there is a $\delta > 0$ such that $|[\text{Riemann sum}] - L| < \epsilon$ for all Riemann sums corresponding to partitions of R into subregions of diameter $< \delta$.

Double integrals and volumes

Consider the solid V in Figure 11, whose base in the xy-plane is the region R of Figure 7 and whose top is formed by the graph of a positive function $f(x, y)$. Figure 12 shows an approximation of V by four cylinder-shaped solids with vertical sides and horizontal tops and bottoms, whose bases are the subregions $R_1, R_2, R_3,$ and R_4 in Figure 8 and whose heights are the values of $f(x, y)$ at the four points in those subregions that are shown in Figure 9.

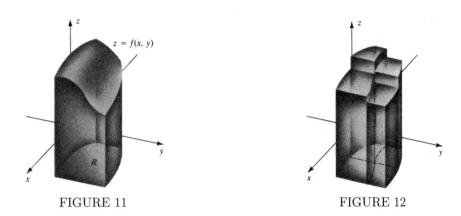

FIGURE 11   FIGURE 12

The volume of the jth cylinder is equal to the product $f(x_j, y_j)[\text{Area of } R_j]$ of its height and the area of its base for each j. Consequently, the total volume of the four cylinders is

$$\sum_{j=1}^{4} f(x_j, y_j)[\text{Area of } R_j].$$

This is a Riemann sum for the integral $\iint_R f(x, y)\, dx\, dy$.

The four cylinders in Figure 12 do not approximate the solid V in Figure 11 very well, but you can imagine that if we used a partition of R into a large number of subregions with very small diameters, the resulting collection of cylinders would approximate V very accurately. Because the volumes of such collections of cylinders are given by the corresponding Riemann sums (**3**), their limit, which is the integral of f over R, is defined to be the volume of V:

Definition 2 *Suppose that R is a bounded region with a piecewise-smooth boundary, that $f(x, y)$ is piecewise continuous and nonnegative in R, and that V is the solid below the graph of f and above R in the xy-plane. Then*

$$[\text{Volume of } V] = \iint_R f(x, y)\, dx\, dy. \tag{5}$$

Example 2   What is the value of $\iint_R \sqrt{9 - x^2 - y^2}\, dx\, dy$ where R is the disk $x^2 + y^2 \leq 9$?

SOLUTION   Because the hemisphere $z = \sqrt{9 - x^2 - y^2}$ is the top and the region R is the base of the half-ball of radius 3 in Figure 13, the integral is equal to the volume of the half-ball. We will see in Section 12.3 how this integral can be evaluated. Meanwhile, we can use the formula $\frac{4}{3}\pi r^3$ from geometry for the volume of a ball of radius r. By Definition 2, the integral equals the volume of a half-ball of radius 3, which is

$$\iint_R \sqrt{9 - x^2 - y^2}\, dx\, dy = [\text{Volume of the half-ball}] = \tfrac{1}{2}[\tfrac{4}{3}\pi(3^3)] = 18\pi. \ \square$$

278   Chapter 12: Multiple Integrals

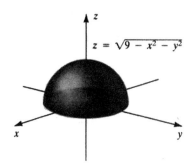

FIGURE 13

Question 2   Express the volume of the building in Figure 1 as an integral.

Question 3   What is the geometric significance of $\iint_R dx\,dy = \iint_R 1\,dx\,dy$ for a bounded region R with a piecewise-smooth boundary?

Recall that the integral $\int_a^b f(x)\,dx$ with $a < b$ of a function with one variable equals the area of the region below the graph of $f(x)$ and above the x-axis if $f(x) \geq 0$ in $[a,b]$ and equals the negative of the area of the region above the graph of $f(x)$ and below the x-axis if $f(x) \leq 0$ in $[a,b]$. If $f(x)$ has positive and negative values in $[a,b]$, then the integral equals the area below the graph and above the x-axis where $f(x) \geq 0$ minus the area above the graph and below the x-axis where $f(x) < 0$.

The situation is similar for the double integral of a function that has negative or positive and negative values. If $f(x_j, y_j)$ is negative, then the term $f(x_j, y_j)[\text{Area of } R_j]$ in a Riemann sum for $\iint_R f(x,y)\,dx\,dy$ is equal to the negative of the volume of the cylinder-shaped region beneath the xy-plane whose top is R_j and whose bottom contains the point at $x = x_j, y = y_j$ on the graph of f. Hence, the Riemann sum is equal to the total volumes of cylinders that approximate the solid between the graph of f and R where f is positive, minus the total volumes of cylinders that approximate the solid between the graph of f and R where f is negative. This leads to the following Rule.

Rule 1   If $f(x,y)$ is piecewise continuous in a bounded region R with a piecewise-smooth boundary, then $\iint_R f(x,y)\,dx\,dy$ equals the volume beneath the graph of f and above the xy-plane at those points (x,y) in R where $f(x,y)$ is positive, minus the volume above the graph of f and below the xy-plane at those points (x,y) in R where $f(x,y)$ is negative.

Question 4   What is the value of $\iint_R -5\,dx\,dy$, where R is the rectangle $0 \leq x \leq 3, 0 \leq y \leq 2$?

Double integrals as iterated integrals

Whereas some double integrals, such as in Example 2, can be evaluated by using formulas for volumes from geometry, most double integrals that we will encounter are evaluated by using integration techniques with one variable. This is done by expressing a double integral as an ITERATED INTEGRAL involving one integration with respect to x and another with respect to y.

Consider first the region R in Figure 14 that is bounded on the top by $y = g(x)$, on the bottom by $y = h(x)$ and that extends from $x = a$ on the left to $x = b$ on the right. Suppose that $f(x,y)$ is the positive continuous function whose graph is shown in Figure 14. Then, by Definition 1, the double integral $\iint_R f(x,y)\,dx\,dy$ is equal to the volume of the solid in Figure 14. We let $A(x)$ denote the area of the intersection of the solid with the plane perpendicular to the x-axis at x that is shown in

12.1 Double integrals

Figure 14. By the Method of Slicing from Section 6.1,

$$\iint_R f(x,y)\,dx\,dy = \begin{bmatrix} \text{The volume} \\ \text{of the solid} \end{bmatrix} = \int_{x=a}^{x=b} A(x)\,dx. \tag{6}$$

We have writen the limits of integration here as "$x=a$" and "$x=b$" to emphasize the fact that here the integration is with respect to x.

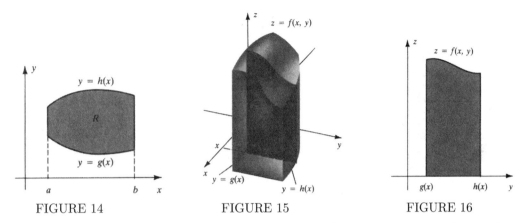

FIGURE 14   FIGURE 15   FIGURE 16

If we look straight down the x-axis, we obtain the view of the cross section in the yz-plane of Figure 16. As that view shows, the bottom of the cross section is at $z = 0$, the top is formed by the curve $z = f(x,y)$ and extends from $y = h(x)$ on the left to $y = g(x)$ on the right. (Here $z = f(x,y)$ is the graph of a function of y and $g(x)$ and $h(x)$ are constant because x is fixed.) Consequently, the area $A(x)$ of the cross section is

$$A(x) = \int_{y=h(x)}^{y=g(x)} f(x,y)\,dy. \tag{7}$$

Substituting **(7)** in **(6)** yields the following result:

Theorem 1 *Suppose that R is the region in an xy-plane bounded on the top by the graph $y = g(x)$, on the bottom by $y = h(x)$, and that extends from $x = a$ on the left to $x = b$ on the right (Figure 17). Also suppose that $g(x)$ and $h(x)$ are piecewise continuous in $[a,b]$. Then for any function $f(x,y)$ that is piecewise continuous in R*

$$\iint_R f(x,y)\,dx\,dy = \int_{x=a}^{x=b} \left\{ \int_{y=h(x)}^{y=g(x)} f(x,y)\,dy \right\} dx. \tag{8}$$

In practice we do not use the curly brackets in **(8)**. They are included here to emphasize that the inner integration with respect to y is carried out first with a fixed value of x.

Also, we do not need any three-dimensional sketches to determine how to apply Theorem 1. Instead, we draw the region of integration as in Figure 17 with a vertical line across it at a point x with $a < x < b$ with an upward-pointing arrow to indicate that in **(8)** we first integrate with respect to y for fixed x from the bottom $y = h(x)$ to the top $y = h(x)$ of the region. Then we put a horizontal arrow on the vertical line to indicate that we then integrate with respect to x from $x = a$ at the left to $x = b$ at the right. Notice that the vertical line from $y = g(x)$ to $y = h(x)$ sweeps out the region R as x runs from a to b.

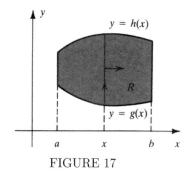
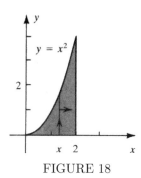

FIGURE 17   FIGURE 18

Example 3   Evaluate $\iint_R 2xy\,dx\,dy$, where R is the region bounded by the curve $y = x^2$ and the lines $y = 0$ and $x = 2$.

SOLUTION   The region of integration is shown in Figure 18 with arrows to indicate the integration procedure. We first integrate with respect to y for fixed x from $y = 0$ to $y = x^2$ and then with respect to x from $x = 0$ to $x = 2$:

$$\iint_R 2xy\,dx\,dy = \int_{x=0}^{x=2} \int_{y=0}^{y=x^2} 2xy\,dy\,dx. \tag{9}$$

In the inner integral x is constant, so we have

$$\int_{y=0}^{y=x^2} 2xy\,dy = \left[xy^2\right]_{y=0}^{y=x^2} = [x(x^2)^2] - [x(0)^2] = x^5.$$

We use this formula in (9) to obtain the final amswer:

$$\int_{x=0}^{x=2} \int_{y=0}^{y=x^2} 2xy\,dy\,dx = \int_{x=0}^{x=2} x^5\,dx = \left[\tfrac{1}{6}x^6\right]_{x=0}^{x=2} = \tfrac{32}{3}.\ \square$$

Question 5   What is the value of $\iint_R (2x + 3y^2)\,dx\,dy$ if R is the square $0 \leq x \leq 10, 0 \leq y \leq 10$?

Reasoning similar to that used to derive Theorem 1 yields the next result for evaluating double integrals by first integrating with respect to x.

Theorem 2   Suppose that R is the region in an xy-plane bounded on the left by the curve $x = M(y)$, on the right by $x = N(y)$ and that extends from $y = c$ on the bottom to $y = d$ on the top (Figure 19). Suppose also that $M(y)$ and $N(y)$ are piecewise continuous in $[c, d]$. Then for any function f that is piecewise continuous in R

$$\iint_R f(x,y)\,dx\,dy = \int_{y=c}^{y=d} \int_{x=M(y)}^{x=N(y)} f(x,y)\,dx\,dy. \tag{10}$$

12.1 Double integrals

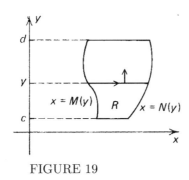
FIGURE 19

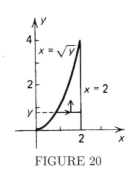
FIGURE 20

Example 4   Evaluate the integral of Example 3 by perfoming an x-integration first.

SOLUTION   Solving $y = x^2$ for x gives $x = \pm\sqrt{y}$. Therefore, the left side of the region in Figure 18 has the equation $x = \sqrt{y}$. The right side is $x = 2$ and the region extends from $y = 0$ to $y = 4$. As is indicated by the arrows in Figure 20, we integrate with respect to x for fixed y with $0 \le y \le 4$ from $x = \sqrt{y}$ at the left to $x = 2$ at the right and then integrate with respect to y from $y = 0$ at the bottom to $y = 4$ at the top:

$$\iint_R f(x,y)\,dx\,dy = \int_{y=0}^{y=4}\int_{x=\sqrt{y}}^{x=2} 2xy\,dx\,dy = \int_{y=0}^{y=4}\left[x^2 y\right]_{x=\sqrt{y}}^{x=2}\,dy$$

$$= \int_{y=0}^{y=4}[(2)^2 y - (\sqrt{y})^2 y]\,dy = \int_{y=0}^{y=4}(4y - y^2)\,dy$$

$$= \left[2y^2 - \tfrac{1}{3}y^3\right]_{y=0}^{y=4} = [2(4)^2 - \tfrac{1}{3}(4)^3] - [0] = 32 - \tfrac{64}{3} = \tfrac{32}{3}.$$

This is, of course, the value we obtained for the same integral in Example 3. □

Question 6   Evaluate the integral $\iint_R (2x + 3y^2)\,dx\,dy$ from Question 3, with R the square, $0 \le x \le 10, 0 \le y \le 10$ by carrying out the x-integration first.

Reversing the order of integration

In the next Example, we cannot perform the inner integration in the given iterated integral but we can evaluate the double integral by reversing the order of integration.

Example 5   Evaluate $\int_0^8 \int_{x^{1/3}}^2 \sin(y^4)\,dy\,dx$.

SOLUTION   In the symbol $\int_0^8 \int_{x^{1/3}}^2 \sin(y^4)\,dy\,dx$ the dy goes with the inner integral sign and the dx with the outer integral sign. Consequently, the given integral can be written

$$\int_{x=0}^{x=8} \int_{y=x^{1/3}}^{y=2} \sin(y^4)\,dy\,dx. \tag{11}$$

In (11) the inner integration is with respect to y from $y = x^{1/3}$ to $y = 2$ and the outer integration is with respect to x from $x = 0$ to $x = 8$. Because the curve $y = x^{1/3}$ intersects the line $y = 2$ at $x = 8$, the region of integration is bounded by the curve $y = x^{1/3}$, the line $y = 2$ and the y-axis, as shown in Figure 21.

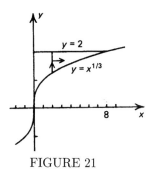
FIGURE 21

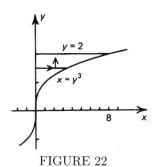
FIGURE 22

We cannot evaluate the inner integral $\int_{y=x^{1/3}}^{y=2} \sin(y^4)\, dy$ in **(11)**. (The substitution $u = y^4, du = 4y^3\, dy$ will not work, for instance, because there is no y^3-term in the integrand.) We try reversing the order of integration. This requires that we first express the sides of the region as graphs of equations where x is given in terms of y.

Solving $y = x^{1/3}$ for x gives $x = y^3$, so, as shown by the arrows in Figure 22, we can also evaluate the integral by integrating first with respect to x from $x = 0$ to $x = y^3$ and then with respect to y from $y = 0$ to $y = 2$, and the given integral equals

$$\int_{y=0}^{y=2} \int_{x=0}^{x=y^3} \sin(y^4)\, dx\, dy. \tag{12}$$

Because y is constant in the inner integral, we obtain

$$\int_{x=0}^{x=y^3} \sin(y^4)\, dx = \Big[x \sin(y^4)\Big]_{x=0}^{x=y^3} = y^3 \sin(y^4).$$

When this formula is substituted into **(12)**, we see that the original integral equals

$$\int_{y=0}^{y=2} \sin(y^4) y^3\, dy.$$

We use the substitution $u = y^4, du = 4y^3\, dy$ to show that the original integral equals

$$\tfrac{1}{4} \int_{y=0}^{y=2} \sin(y^4)(4y^3\, dy) = \int_{y=0}^{y=2} \sin u\, du = \Big[-\tfrac{1}{4} \cos u\Big]_{y=0}^{y=2} = \Big[-\tfrac{1}{4} \cos(y^4)\Big]_{y=0}^{y=2}$$
$$= [-\tfrac{1}{4}\cos(2^4)] - [-\tfrac{1}{4}\cos(0^2)] = \tfrac{1}{4} - \tfrac{1}{4}\cos(16). \ \square$$

Question 7 Reverse the order of integration in the integral $\int_0^2 \int_0^x g(x,y)\, dy\, dx.$

ΔxΔy-notation in Riemann sums

Occassionally it is convenient to view a double integral $\iint_R f(x,y)\, dx\, dy$ as a limit of Riemann sums formed from partitions into rectangles. If the region of integration R is a rectangle with horizontal and vertical sides, we set $S = R$. Otherwise, we choose such a rectangle $S : a \leq x \leq b, c \leq y \leq d$ that contains the region of integration R, as in Figure 23. We define, or redefine, $f(x,y)$ to be zero in the portion of S outside of R. so that

$$\iint_R f(x,y)\, dx\, dy = \iint_S f(x,y)\, dx\, dy.$$

12.1 Double integrals

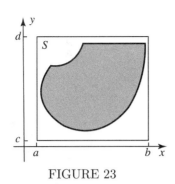

FIGURE 23

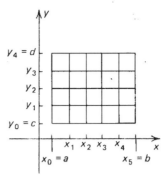

FIGURE 24

Next, we consider partitions

$$a = x_0 < x_1 < x_2 \cdots < x_N = b$$
$$c = y_0 < y_1 < y_2 \cdots < y_M = d \tag{13}$$

of the intervals $[a, b]$ and $[c, d]$ and let $\Delta x_j = x_j - x_{j-1}$ and $\Delta y_k = y_k - y_{k-1}$ be the widths of the subintervals. The vertical lines $x = x_j$ for $j = 0, 1, \ldots N$ and the horizontal lines $y = y_k$ for $k = 1, 2, \ldots, M$ partition S into rectangles (Figure 24). We pick a point (x_j^*, y_k^*) in the subregion $x_{j-1} \leq x \leq x_j, y_{k-1} \leq y \leq y_k$ for each j and k such that $f(x_j^*, y_j^*)$ is defined. Then the corresponding Riemann sum is

$$\sum_{j=1}^{N} \sum_{k=1}^{M} f(x_j^*, y_k^*) \Delta x_j \Delta y_k. \tag{14}$$

The integral is the limit of these Riemann sums as N and M tend to ∞ and the widths of the subintervals in the partitions **(13)** tend to zero. It is the terms $\Delta x_j \Delta y_k$ in the approximation **(14)** that leads to the terms $dx\,dy$ in the symbol for the double integral.

Other volumes

Recall that if a region in an xy-plane is bounded on the top by the curve $y = g(x)$ and on the bottom by $y = h(x)$ and it extends for $a \leq x \leq b$ where $h(x) \leq g(x)$ for $a \leq x \leq b$, then the area of the region is the integral $\int_a^b [g(x) - h(x)]\,dx$. We have the following analogous result concerning volumes of solids that are bounded by graphs of functions with two variables.

Rule 2 *Suppose that $g(x, y)$ and $h(x, y)$ are piecewise continuous in a bounded region R with a piecewise smooth boundary and that $g(x, y) \leq h(x, y)$ for all points (x, y) in R where $g(x, y)$ and $h(x, y)$ are defined. Let V be the solid consisting of the points above $z = g(x, y)$ and below $z = h(x, y)$ for (x, y) in R (Figure 25). Then*

$$[\text{Volume of } V] = \iint_R [h(x, y) - g(x, y)]\,dx\,dy \tag{15}$$

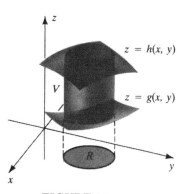

FIGURE 25

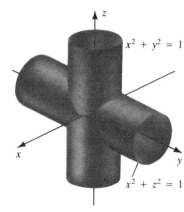

FIGURE 26

The region R in Rule 2 is the (ORTHOGONAL) PROJECTION of the solid V on the xy-plane. To apply the Rule, we determine the projection R and formulas for the functions $g(x, y)$ and $h(x, y)$ from the description of the solid and, if necessary, a rough sketch of the solid. Then we make a more careful drawing of R in an xy-plane to determine how to evaluate the integral **(15)** as an iterated integral.

Example 6   The solid V in Figure 26 is bounded by the circular cylinders $x^2 + y^2 = 1$ and $x^2 + z^2 = 1$. Find its volume.

SOLUTION   As can be seen from Figure 26, the projection of the solid on the xy-plane is determined by the vertical cylinder and is the disk $R : x^2 + y^2 \leq 1$ in Figure 27. Also, solving the equation $x^2 + z^2 = 1$ of the horizontal cylinder for z gives $z = \pm\sqrt{1 - x^2}$, so the top of V is formed by the surface $z = \sqrt{1 - x^2}$ in Figure 28 and its bottom is formed by $z = -\sqrt{1 - x^2}$ in Figure 29. Therefore,

$$[\text{Volume of } V] = \iint_R [\sqrt{1 - x^2} - (-\sqrt{1 - x^2})]\, dx\, dy$$
$$= \iint_R 2\sqrt{1 - x^2}\, dx\, dy. \tag{16}$$

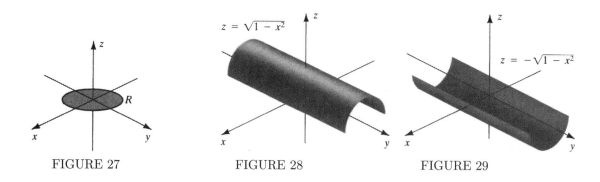

FIGURE 27   FIGURE 28   FIGURE 29

We solve the equation $x^2 + y^2 = 1$ for the boundary of the projection R for $y = \pm\sqrt{1 - x^2}$ to see that R extends from $y = -\sqrt{1 - x^2}$ to $y = \sqrt{1 - x^2}$ in the xy-plane (Figure 30), and the volume of V is

$$\int_{x=-1}^{x=1}\int_{y=-\sqrt{1-x^2}}^{y=\sqrt{1-x^2}} 2\sqrt{1-x^2}\,dy\,dx = \int_{x=-1}^{x=1}\left[2y\sqrt{1-x^2}\right]_{y=-\sqrt{1-x^2}}^{y=\sqrt{1-x^2}}dx$$

$$= \int_{x=-1}^{x=1}[2(\sqrt{1-x^2})^2 - 2(-\{\sqrt{1-x^2}\}^2)]\,dx = \int_{x=-1}^{x=1} 4(1-x^2)\,dx$$

$$= \left[4x - \tfrac{4}{3}x^3\right]_{x=-1}^{x=1} = [4(1) - \tfrac{4}{3}(1)^3] - [4(-1) - \tfrac{4}{3}(-1)^3] = \tfrac{16}{3}.\ \square$$

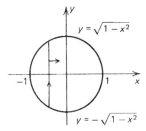

FIGURE 30

We can also find volumes of some solids by integrating over their projections on the xz- or yz-planes. If a solid consists of the points (x,y,z) with (x,z) in a region R of the xz-plane and with $g(x,z) \leq y \leq h(x,z)$ as in Figure 31, then its volume is

$$\iint_R [h(x,z) - g(x,z)]\,dx\,dz. \tag{17}$$

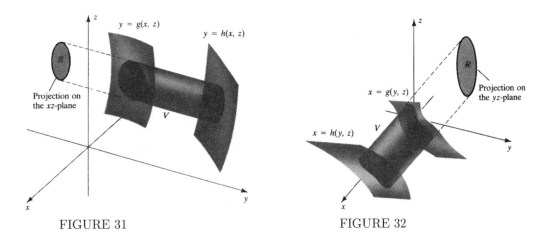

FIGURE 31   FIGURE 32

Similarly, if the solid consists of the points (x,y,z) with (y,z) in a region R of the yz-plane and with $g(y,z) \leq x \leq h(y,z)$ as in Figure 32, then its volume is

$$\iint_R [h(y,z) - g(y,z)]\,dy\,dz. \tag{18}$$

Two-dimensional densities

Suppose that a flat plate occupying the bounded region R with a piecewise smooth boundary in Figure 33 has density $\rho(x,y)$ pounds per square foot at (x,y). How much does the plate weigh? To answer this question we use piecewise-smooth curves to partition R into subregions R_j for $j = 1, 2, \ldots, N$, as in Figure 34.

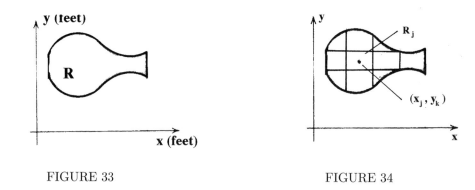

FIGURE 33   FIGURE 34

Then for each j we pick a point (x_j, y_j) in R_j where ρ is defined and approximate the given density by the step function whose value in the interior of R_j is $\rho(x_j, y_j)$. With this density, the weight of the part of the plate in R_j is $\rho(x_j, y_j)[\text{Area of } R_j]$, so the approximate weight of the entire actual plate is

$$\sum_{j=1}^{N} \rho(x_j, y_j)[\text{Area of } R_j]. \tag{19}$$

Because **(19)** is a Riemann sum for the integral of $\rho(x, y)$ over R and approaches it as the number of subregions in the partition tend to ∞ and their diameters tend to zero, we are led to the following Rule:

Rule 3 *If a flat plate occupies the bounded region R with a piecewise smooth boundary in an xy-plane and its density, measured in weight or mass per unit area, is $\rho(x,y)$ at (x,y), then the weight or mass of the plate is*

$$\iint_R \rho(x,y)\, dx\, dy. \tag{20}$$

Question 8   A plate that occupies the square $0 \leq x \leq 1, 0 \leq y \leq 1$ with distances measured in feet has density $72xy^2$ pounds per square foot at (x, y). How much does the plate weigh?

Rule 3 can also be with weight or mass replaced by other quantities and R the relevant region in an xy-plane, as in the next Question and Example.

Question 9   Express the total cost of the farm in Example 1 as an integral.

Example 7   Figure 35 shows level curves of the population density $p(x, y)$ (people per square mile) in a city. **(a)** Express the total population in the region $0 \leq x \leq 4, 0 \leq y \leq 4$ as an iterated integral. **(b)** Give the approximate value of the Riemann sum for the integral in part (a) corresponding to the partition of the region of integration into four equal squares and with the density evaluated at the centers of the squares.

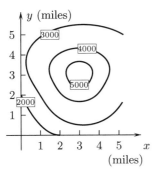
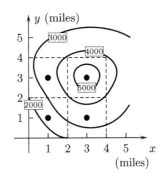

Level curves of $p(x,y)$
(people per square mile)

FIGURE 35

FIGURE 36

SOLUTION (a) Because $p(x,y)$ is the population density at (x,y), the population in the region $0 \leq x \leq 4, 0 \leq y \leq 4$ is $\int_{x=0}^{x=4} \int_{y=0}^{y=4} p(x,y)\, dy\, dx$.

(b) Figure 36 shows the partition of the region of integration into four equally shaped, square subregions and their midpoints $(1,1), (3,1), (1,3)$, and $(3,3)$. The area of each subregion is $2(2) = 4$ square miles, so the Riemann sum equals

$$p(1,1)(4) + p(1,3)(4) + p(3,1)(4) + p(3,3)(4).$$

From the level curves we obtain the approximate values $p(1,1) \approx 2400, p(3,1) \approx 3400, p(1,3) \approx 3500$, and $p(3,3) \approx 5200$. The Riemann sum is approximately equal to $2400(4) + 3400(4) + 3500(4) + 5200(4) = 58{,}000.$ □

Centers of gravity of plates

Recall from Section 4.6 that the center of gravity of a one-dimensional object, such as a rod, is the point at which it would balance and is such that the moment of the object about that point is zero. We used this fact to find the center of gravity from the rod's weight and its moment about the origin, calculated as integrals.

The CENTER OF GRAVITY of a two-dimensional object, such as a flat plate, is the point where it would balance on a pin (Figure 37). It can be found as the intersection of two nonparallel lines on each of which the plate would balance (Figure 38).

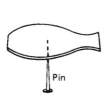

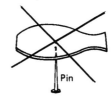

FIGURE 37

FIGURE 38

For a plate that occupies a bounded region R in an xy-plane we use horizontal and vertical lines. We begin with the following definition of the MOMENTS of a point mass about a vertical line $x = K$ and about a horizontal line $y = L$.

Definition 3 (a) *The* MOMENT *about the line* $x = K$ *of a point mass of weight* w *at* (x, y) *is the product* $w(x - K)$ *of its weight and its distance to the right of* $x = K$. (b) *The moment of the point mass about the line* $y = L$ *is the product* $w(y - L)$ *of its weight and its distance above* $y = L$.

The moment about $x = K$ is negative if the point is to the left of $x = K$ and the moment about $y = L$ is negative if the point is below $y = L$.

To extend Definition 3 to a plate with density $\rho(x, y)$, we use a partition of R into subregions to approximate the plate by point masses. We first approximate the weight of the portion of the plate in each R_j by $\rho(x_j, y_j)[\text{Area of } R_j]$ as we did in finding the integral that gives the area of the plate. Then we approximate the portion of the plate in R_j by a point mass of weight $\rho(x_j, y_j)[\text{Area of } R_j]$ at (x_j, y_j). The moment of the collection of point masses about $x = K$ is the sum

$$\sum_{j=1}^{N} (x_j - K)\rho(x_j, y_j)[\text{Area of } R_j] \tag{21}$$

of the moments of the N point masses. Similarly, the moment of the collection of N point masses about $y = L$ is

$$\sum_{j=1}^{N} (y_j - L)\rho(x_j, y_j)[\text{Area of } R_j]. \tag{22}$$

Because **(21)** is a Riemann sum for the double integral of $(x - K)\rho(x, y)$ over R and **(22)** is a Riemann sum for the double integral of $(y - L)\rho(x, y)$ over R, we are led to the next Definition.

Definition 4 *Suppose that R is a bounded region with a piecewise smooth boundary and that $\rho(x, y)$ is piecewise continuous in R. Then for a plate that occupies R and has density $\rho(x, y)$ at (x, y),*

$$[\text{Moment about } x = K] = \iint_R (x - K)\rho(x, y) \, dx \, dy \tag{23}$$

$$[\text{Moment about } y = L] = \iint_R (y - L)\rho(x, y) \, dx \, dy. \tag{24}$$

Question 10 The plate in Definition 4 would balance on the line $x = K$ if moment **(23)** is zero and would balance on the line $y = L$ if moment **(24)** is zero. Set the integrals on the right of these equations equal to zero and solve for K and L.

The coordinates of the center of gravity of the plate are the numbers $\bar{x} = K$ and $\bar{y} = L$ from Question 10:

Theorem 3 *If the conditions of Definition 5 are satisfied, then the center of gravity of the plate is $(\bar{x}, \bar{y})$, where*

$$\bar{x} = \frac{\iint_R x\, \rho(x, y) \, dx \, dy}{\iint_R \rho(x, y) \, dx \, dy} \quad \text{and} \quad \bar{y} = \frac{\iint_R y\, \rho(x, y) \, dx \, dy}{\iint_R \rho(x, y) \, dx \, dy}. \tag{25}$$

Notice that $\bar{x}$ in **(25)** is the moment of the plate about $x = 0$ (the y-axis) divided by the plate's weight, and that $\bar{y}$ is the moment of the plate about $y = 0$ (the x-axis) divided by the plate's weight.

12.1 Double integrals

Example 8 Find the center of gravity of the plate from Question 8 that occupies the square $R: 0 \leq x \leq 1, 0 \leq y \leq 1$ and has density $72xy^2$ pounds per square foot.

SOLUTION You found in answering Question 8 that the weight of the plate is

$$\iint_R xy^2 \, dx \, dy = 12 \text{ pounds} \tag{26}$$

Its moment about the y-axis is

$$\iint_R x(72xy^2) \, dx \, dy = \int_{x=0}^{x=1} \int_{y=0}^{y=1} 72x^2y^2 \, dy \, dx = \int_{x=0}^{x=1} \left[24x^2y^3\right]_{y=0}^{y=1} dx$$
$$= \int_{x=0}^{x=1} \left([24x^2] - [0]\right) = \left[8x^3\right]_{x=0}^{x=1} = 8 \text{ foot-pounds} \tag{27}$$

Similarly, the moment about the x-axis is

$$\iint_R y(72xy^2) \, dx \, dy = \int_{x=0}^{x=1} \int_{y=0}^{y=1} 72xy^3 \, dy \, dx = \int_{x=0}^{x=1} \left[18xy^4\right]_{y=0}^{y=1} dx$$
$$= \int_{x=0}^{x=1} \left([18x] - [0]\right) = \left[9x^2\right]_{x=0}^{x=1} = 9 \text{ foot-pounds.} \tag{28}$$

Dividing the moment (27) about $x = 0$ by the weight (26) gives the x-coordinate $\overline{x} = 8/12 = \frac{2}{3}$ of the center of gravity, and dividing the moment (28) about $y = 0$ by the weight (26) gives the y-coordinate $\overline{y} = 9/12 = \frac{3}{4}$. □

Question 11 Figure 39 shows the square R from Example 9 and its center of gravity. Why is the center of gravity to the right of and above the center of the square?

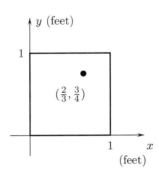

FIGURE 39

Average values of f(x,y)

Suppose that a step function $f(x,y)$ is constant in the interior of each of the subregions $R_1, R_2, \ldots, R_n$ of a partition of a bounded region R. We pick a point (x_j, y_j) in the interior of R_j for each j and define the AVERAGE VALUE of the function in R to be the sum from $j = 1$ to N of its value $f(x_j, y_j)$ in R_j multiplied by the fraction [Area of R_j]/[Area of R] of the area of R that is in R_j:

$$[\text{Average value of } f \text{ in } R] = \sum_{j=1}^{N} f(x_j, y_j) \frac{[\text{Area of } R_j]}{[\text{Area of } R]}$$
$$= \frac{1}{[\text{Area of } R]} \sum_{j=1}^{N} f(x_j, y_j)[\text{Area of } R_j]. \tag{29}$$

For a general piecewise-continuous function f, we define its average value in R to be the limit of the sums (29) as the number of subdivisions in the partitions tends to ∞ and their diameters tend to zero. In other words, we define the average value of f to equal its integral over R divided by the area of R:

Definition 5 If $f(x,y)$ is piecewise continuous in the bounded region R with piecewise-smooth boundary, then

$$[\text{The average value of } f \text{ in } R] = \frac{1}{[\text{Area of } R]} \iint_R f(x,y)\,dx\,dy. \tag{30}$$

Example 9 What is the average value of $f(x,y) = ye^{y^2}$ in the region $R: 0 \leq y \leq \sqrt{x}, 0 \leq x \leq 1$?

SOLUTION The region R is drawn in Figure 40. Its area is the integral of the function 1:

$$\begin{aligned}[\text{Area of } R] &= \iint_R 1\,dx\,dy = \int_{x=0}^{x=1} \int_{y=0}^{y=\sqrt{x}} 1\,dy\,dx \\ &= \int_{x=0}^{x=1} \left[y\right]_{y=0}^{y=\sqrt{x}} dx = \int_{x=0}^{x=1} x^{1/2}\,dx = \left[\tfrac{2}{3}x^{3/2}\right]_{x=0}^{x=1} = \tfrac{2}{3}.\end{aligned} \tag{31}$$

The integral of f over R is

$$\iint_R ye^{y^2}\,dx\,dy = \int_{x=0}^{x=1} \int_{y=0}^{y=\sqrt{x}} ye^{y^2}\,dy\,dx.$$

We use the substitution $u = y^2, du = 2y\,dy$ in the y-integral to obtain

$$\begin{aligned}\iint_R ye^{y^2}\,dx\,dy &= \tfrac{1}{2}\int_{x=0}^{x=1} \int_{y=0}^{y=\sqrt{x}} e^{y^2}(2y\,dy)\,dx = \tfrac{1}{2}\int_{x=0}^{x=1} \int_{y=0}^{y=\sqrt{x}} e^u\,du\,dx \\ &= \tfrac{1}{2}\int_{x=0}^{x=1} \left[e^u\right]_{y=0}^{y=\sqrt{x}} dx = \tfrac{1}{2}\int_{x=0}^{x=1} \left[e^{y^2}\right]_{y=0}^{y=\sqrt{x}} dx \\ &= \tfrac{1}{2}\int_{x=0}^{x=1} (e^x - 1)\,dx = \tfrac{1}{2}\left[e^x - x\right]_{x=0}^{x=1} \\ &= \tfrac{1}{2}[e-1] - [1-0] = \tfrac{1}{2}(e-2).\end{aligned} \tag{32}$$

With **(31)** and **(32)**, we see that the average value of f in R is

$$\frac{1}{[\text{Area of } R]} \iint_R ye^{y^2}\,dx\,dy = \frac{\tfrac{1}{2}(e-2)}{\tfrac{2}{3}} = \tfrac{3}{4}(e^2 - 1). \square$$

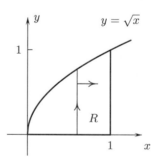

FIGURE 40

12.1 Double integrals

Principles and procedures

- The terms $dx\,dy$ in $\iint_R f(x,y)\,dx\,dy$ only indicate the variables and do not represent any order of integration. We could write $\iint_R f(x,y)\,dy\,dx$ for the same integral.

- To determine the iterated integral(s) to evaluate $\iint_R f(x,y)\,dx\,dy$, draw arrows as in Figures 17 through 22 to indicate the integration procedures. Also, use equations as limits of integration, such as $\int_{x=a}^{x=b}\int_{y=h(x)}^{y=g(x)}$, instead of just numbers and formulas. This will help you keep track of which variable is being used at each stage. Moreover, if you are given an iterated integral with just numbers or formulas as limits of integration, as in $\int_a^b \int_{h(x)}^{g(x)}$, convert the limits of integration to equations before you start working with the integral.

- In many contexts you can think of the double integral $\iint_R f(x,y)\,dx\,dy$ either as a difference of volumes, as in Rule 1, or as the total of a quantity in R where $f(x,y)$ is the density of the quantity, as in Rule 3 or Question 8.

- You usually can find a volume between graphs $z = h(x,y)$ and $z = g(x,y)$ using Rule 2 without drawing the solid. You can generally tell which function forms the top and which the bottom of the solid and determine the projection R on the xy-plane from the description of the solid or from the formulas used in the description.

- To help remember the formulas

$$\overline{x} = \frac{\iint_R x\,\rho(x,y)\,dx\,dy}{\iint_R \rho(x,y)\,dx\,dy} \quad \text{and} \quad \overline{y} = \frac{\iint_R y\,\rho(x,y)\,dx\,dy}{\iint_R \rho(x,y)\,dx\,dy}$$

for the center of gravity $(\overline{x}, \overline{y})$ of a plate, notice that the symbols in the denominator for the weight of the plate seem to cancel with all of the symbols in the numerators except the x and y.

- Notice that the formula $\dfrac{1}{[\text{Area of }R]}\iint_R f(x,y)\,dx\,dy$ for the average value of $f(x,y)$ in the region R is analogous to the formula $\dfrac{1}{b-a}\int_a^b f(x)\,dx$ for the average value of $f(x)$ in the interval $[a,b]$ from Section 4.6.

- The evaluation of an integral of a product $f(x)g(y)$ of a function of x and a function of y over a rectangle $R: a \leq x \leq b, c \leq y \leq d$ can be simplified by writing

$$\iint_R f(x)g(y)\,dx\,dy = \int_{x=a}^{x=b}\int_{y=c}^{y=d} f(x)g(y)\,dy\,dx = \int_{x=a}^{x=b} f(x)\left(\int_{y=c}^{y=d} g(y)\,dy\right)dx$$
$$= \left(\int_{x=a}^{x=b} f(x)\,dx\right)\left(\int_{y=c}^{y=d} g(y)\,dy\right). \tag{33}$$

Be careful, however, not to use this formula when it does not apply.

Tune-Up Exercises 12.1♦
A Answer provided. **O** Outline of solution provided.

T1.^O Find the value of $\iint_R f(x,y)\,dx\,dy$ where $f(x,y) = 10$ for $x \leq 1$, $f(x,y) = 20$ for $x > 1$, and $R = \{(x,y) : 0 \leq x \leq 3, 0 \leq y \leq 2\}$.
♦ Type 1, Double integral of a step function

T2.^O Calculate the Riemann sum for $\int_{x=0}^{x=1} \int_{y=0}^{y=1} g(x,y)\,dx\,dy$ corresponding to the partition of the region of integration into four equal squares and with the integrand evaluated at the centers of the squares, given that $g(0.25, 0.25) = 1.30, g(0.25, 0.75) = 2.80, g(0.75, 0.25) = 4.84$, and $g(0.75, 0.75) = 2.76$.
♦ Type 3, A Riemann sum

T3.^O Evaluate $\iint_R 10x^4 y\,dx\,dy$ where R is the triangle with vertices $(0,0), (0,2)$, and $(1,1)$.
♦ Type 1, value of a double integral

T4.^O Evaluate $\iint_R 2x\,dx\,dy$ where R is bounded by the curve $x = 1 - y^2$ and the y-axis.
♦ Type 1, value of a double integral

T5.^O Reverse the order of integration in $\int_0^4 \int_0^{2-x/2} f(x,y)\,dy\,dx$.
♦ Type 3, reversing the order of integration

T6.^O Find the volume of the solid bounded on the top by the hyperbolic paraboloid $z = xy + 2$ and on the bottom by the plane $z = -3$ for $0 \leq y \leq x^3, 0 \leq x \leq 1$.
♦ Type 1, a volume between graphs

T7.^O A plate in an xy-plane with distances measured in feet occupies the region bounded by the parabola $y = x - x^2$ and the x-axis and its density at (x,y) is $6x^2$ pounds per square foot. **(a)** How much does it weigh? **(b)** Where is its center of gravity?
♦ Type 3, weight and center of gravity

T8.^O What is the average value of $h = e^x \sin y$ in the rectangle $0 \leq x \leq 2, 0 \leq y \leq 1$?.
♦ Type 1, average value

Problems 12.1
A Answer provided. **O** Outline of solution provided.

1.^A What is the value of $\iint_R g(x,y)\,dx\,dy$ with R the rectangle $0 \leq x \leq 20, 0 \leq y \leq 20$ if g is the step function with the value 16 for $0 \leq x \leq 10, 0 \leq y \leq 10$, the value 17 for $0 \leq x \leq 10, 10 < y \leq 20$, the value 18 for $10 < x \leq 20, 0 \leq y \leq 10$, and the value 19 for $10 < x \leq 20, 10 < y \leq 20$?
♦ Type 1, double integral of a step function

2.^A Find the value of $\iint_R h(x,y)\,dx\,dy$ where R is the disk $x^2 + y^2 \leq 9$ and h is the step function with the value -7 for $x^2 + y^2 \leq 4$ and the value 8 for $x^2 + y^2 > 4$.
♦ Type 1, double integral of a step function

3. Evaluate $\iint_R k(x,y)\,dx\,dy$ where R is the rectangle $0 \leq x \leq 2, 0 \leq y \leq 1$ and k is the step function that equals 13 for $x < y$ and equals 25 for $x \geq y$.
♦ Type 1, double integral of a step function

4.º Estimate $\iint_R \sin(xy)\,dx\,dy$ where R is the square $0 \leq x \leq 1, 0 \leq y \leq 1$ by using the Riemann sum corresponding to a parition of R into four equal subsquares and with the function evaluated at their centers. Use three decimal places in the calculations.
♦ Type 3, a Riemann sum

♦The Tune-up Exercises and Problems are classified by type and content. The types are (1) basic, reactive; (2) basic reflective; (3) intermediate, reactive; (4) intermediate, reflective; (5) advanced, reactive; (6) advanced, reflective; and (7) advanced, theoretical.

12.1 Double integrals

5. Estimate $\int_0^{0.6} \int_0^{0.4} \ln(x + 2y + 1)\, dy\, dx$ by using a Riemann sum corresponding to a parition of R into six equal subsquares and with the function evaluated at their centers. Use four decimal places in the calculations.

♦ Type 3, a Riemann sum

6. Give the Riemann sum approximation of $\iint_R \sqrt{1 + x^2 + y^2}\, dy\, dx$ corresponding to the partition of $R = \{(x,y) : 0 \leq x \leq 3, 0 \leq y \leq 3\}$ into nine equal squares and with the integrand evaluated at their centers. Use two-decimal-place accuracy.

♦ Type 3, a Riemann sum

7.^A Give an approximate Riemann sum for $\iint_R M(x,y)\, dy\, dx$ with $R : 0 \leq x \leq 4, 0 \leq y \leq 4$ corresponding to the partition of R into four equal squares, where $M(x,y)$ is the function whose level curves are shown in Figure 41.

♦ Type 3, estimating a Riemann sum from level curves

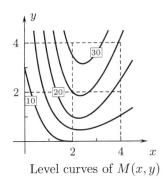
Level curves of $M(x,y)$

FIGURE 41

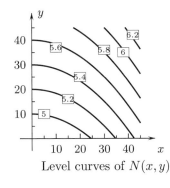
Level curves of $N(x,y)$

FIGURE 42

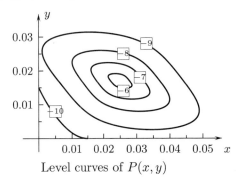
Level curves of $P(x,y)$

FIGURE 43

8. Is the integral $\iint_R N(x,y)\, dy\, dx$ with $R : 0 \leq x \leq 40, 0 \leq y \leq 40$ and the function $N(x,y)$ whose level curves are shown in Figure 42 closer to 200, 800, 2000, 8000, or 12,000?

♦ Type 3, estimating an integral sum from level curves

9. Which of the numbers $-5.50, -1.25, -0.550, -0.125, -0.0550$, and -0.0125 is closest to the integral $\int_0^{0.05} \int_0^{0.03} P(x,y)\, dy\, dx$ for the function $P(x,y)$ whose level curves are shown in Figure 43?

♦ Type 3, estimating an integral sum from level curves

Evaluate the integrals in Problems 10 through 23.

10.º $\iint_R 4x^3 y\, dx\, dy$ with R bounded by $y = x^2$ and $y = 2x$

♦ Type 1, evaluating an integral

11.^A $\iint_R 3x^2 y^2\, dx\, dy$ with R bounded by $y = x, y = 2x$, and $x = 1$

♦ Type 1, evaluating an integral

12. $\iint_R 2y\, dx\, dy$ with R bounded by $y = x^2$ and $y = 2x^2 - 1$

♦ Type 1, evaluating an integral

13. $\iint_R e^x \sin y\, dx\, dy$ with R the rectangle $-3 \leq x \leq 3, 0 \leq y \leq 5$

♦ Type 1, evaluating an integral

14.^A $\iint_R 3y^2 \sqrt{x}\, dx\, dy$ with R bounded by $y = x^2, y = -x^2$, and $x = 4$

♦ Type 1, evaluating an integral

15. $\iint_R 10x^3 y^4\, dx\, dy$ with R bounded by $y = x^3, y = 0$ and $x = 1$

♦ Type 1, evaluating an integral

16. $\iint_R y^2 \sin(x^2)\, dx\, dy$ with R bounded by $y = x^{1/3}, y = -x^{1/3}$ and $x = 8$

♦ Type 1, evaluating an integral

17.º $\iint_R y^2\, dx\, dy$ with R bounded by $x = y^2$ and $x = 2 - y^2$

♦ Type 1, evaluating an integral

18.^A $\iint_R e^{y^2}\,dx\,dy$ with R bounded by $x = -y, x = y$, and $y = 2$
♦ Type 1, evaluating an integral

19.º $\iint_R \sqrt{xy}\,dx\,dy$ with R the triangle with vertices $(0,0), (1,1)$, and $(4,1)$
♦ Type 2, evaluating an integral

20. $\iint_R 12x^y\,dx\,dy$ with R bounded by $x = -1 - y^2, x = 1 + y^2, y = -1$, and $y = 0$
♦ Type 1, evaluating an integral

21. $\iint_R (x + 2y)\,dx\,dy$ with R the portion of the disk $x^2 + y^2 \leq 1$ where $x \geq, y \geq 0$
♦ Type 2, evaluating an integral

22.^A $\iint_R (3y^2 + 4)\,dx\,dy$ with R bounded by $x = 1, x = 4, y = 1/x$, and $y = -1/x$
♦ Type 1, evaluating an integral

23. $\iint_R x\,dx\,dy$ with R bounded by $y = 2/x$ and $x + y = 3$
♦ Type 1, evaluating an integral

Find the volumes of the given solids in Problems 24 through 30.

24.º The solid between the paraboloid $z = x^2 + y^2$ and the plane $z = -1$ for (x,y) in the region bounded by $y = x^2$ and $y = 1$ in the xy-plane
♦ Type 1, volume of a solid between graphs

25.^A The solid between the surface $z = \sin x \sin y$ and the plane $z = -2$ for (x,y) in the square $0 \leq x \leq \pi, 0 \leq y \leq \pi$ in the xy-plane
♦ Type 1, volume of a solid between graphs

26. The solid between the surface $z = xe^y$ and the xy-plane for (x,y) in the the region bounded by $y = 0, y = x^2$ and $x = 1$ in the xy-plane
♦ Type 1, volume of a solid between graphs

27. The solid between the surfaces $z = e^x$ and $z = -e^y$ for (x,y) in the the rectangle $0 \leq x \leq 1, 0 \leq y \leq 1$ in the xy-plane
♦ Type 1, volume of a solid between graphs

28. The solid between the surfaces $z = e^x \sin(y^2) + 3$ and $z = e^x \sin(y^2)$ for (x,y) in the the disk $x^2 + y^2 \leq 16$ in the xy-plane
♦ Type 1, volume of a solid between graphs

29.º The tetrahedron bounded by the plane $3x + 2y + z = 6$ and the coordinate planes
♦ Type 2, volume of a solid between graphs

30. The tetrahedron bounded by the plane $z = 2x - y - 4$ and the coordinate planes
♦ Type 2, volume of a solid between graphs

31.º (a) Find the mass of a plate that occupies the quarter circle $x^2 + y^2 \leq 1, x \geq 0, y \geq 0$ with distances measured in centimeters and whose density at (x,y) is xy grams per square centimeter. (b) What is the center of gravity of the plane from part (a)?
♦ Type 3, weight and center of gravity

32.^A (a) Find the weight of a plate that occupies the rectangle $0 \leq x \leq 1, 0 \leq y \leq 2$ with distances measured in meters and whose density at (x,y) is $x + y^2$ newtons per square meter. (b) What is the center of gravity of the plane from part (a)?
♦ Type 3, weight and center of gravity

33. (a) How much does a plate weigh if it occupies the region bounded by $y = x^2$ and $y = x^3$ in an xy-plane with distances measured in feet and if the density of the plate at (x,y) is $\sqrt{xy}$ pounds per square foot? (b) What is the center of gravity of the plane from part (a)?
♦ Type 3, weight and center of gravity

34.º Find the average value of $f = x^2 y^2$ in the region bounded by $y = x^2$ and $y = x^3$ in an xy-plane
♦ Type 3, average value

35.^A (a) What the average value of $g = \sin x \sin y$ in the triangle bounded by $y = x, y = 0$, and $x = k$ in an xy-plane, where k is a positive constant? (b) What is the limit of the average value in part (a) as $k \to \infty$?
♦ Type 5, average value

36. What is the average value of $h = e^{x+2y}$ in the rectangle with vertices $(-2,-3), (3,-3), (3,4)$, and $(-2,4)$?
♦ Type 3, average value

12.1 Double integrals

37. Give the general Riemann sums for (a^A) $\iint_R e^x \sin(xy)\, dx\, dy$ and (b) $\iint_R \ln(x+y)\, dx\, dy$.
◆ Type 3, a Riemann sum

38.^{A} Evaluate $\iint_R (x+y)\, dx\, dy$, where R is bounded by the parabola $y = \frac{1}{2}x^2$ and the line $y = -x$.
◆ Type 2, a double integral

39.º Evaluate $\iint_R x^2\, dx\, dy$, where R is the triangle with vertices $(0,0), (1,3)$, and $(2,2)$. (Two interated integrals are required.)
◆ Type 4, a double integral

40.^{A} Find the value of $\iint_R 4xy\, dx\, dy$, where R is bounded by the hyperbola $y = 1/x$ and the lines $y = x, y = 0$, and $x = 2$. (Two interated integrals are required.)
◆ Type 4, a double integral

41. What is the value of the constant k such that $\iint_R ky\, dx\, dy = \frac{76}{15}$, where R is bounded by the curves $y = 2 - x^2$ and $y = |x|$? (Two interated integrals are required.)
◆ Type 5, a double integral

42. Express $\iint_R f(x,y)\, dx\, dy$ as a sum of iterated integrals where R is bounded by $y = 3/x, y = x/3$ and $y = 3x$
◆ Type 5, expressing a double integral as a sum of iterated integrals

43.º Show that $\iint_R xy^4\, dx\, dy$ with R the region bounded by $x = k^4 - y^4$ and $x = -k^4 + y^4$ has the same value for all positive constants k.
◆ Type 4, a double integral

44.^{A} What is the value of $\iint_R y^2 \cos x\, dx\, dy$ with R the region bounded by $x = y^3$ and the lines $y = 0$ and $x = \pi$.
◆ Type 4, a double integral

45. Find the value of $\iint_R y^2\, dx\, dy$ with R the region bounded by $x = y^2$ and $x = 2k^2 - y^2$, where k is a positive constant.
◆ Type 5, a double integral

46. Reverse the order of integration in (a^A) $\int_{-2}^{2}\int_{x^2}^{4} f(x,y)\, dy\, dx$, (b^A) $\int_{-1}^{1}\int_{-1}^{x^3} f(x,y)\, dy\, dx$, and (c) $\int_{0}^{1}\int_{0}^{2x} f(x,y)\, dy\, dx$.
◆ Type 5, reversing the order of integration

47. Reverse the order of integration in (a^A) $\int_{-2}^{1}\int_{y^3}^{1} f(x,y)\, dx\, dy$ and (b) $\int_{-1}^{3}\int_{-1}^{y} f(x,y)\, dx\, dy$.
◆ Type 5, reversing the order of integration

48. Use sums of iterated integrals as necessary to reverse the order of integration in
(a^A) $\int_{-1}^{2}\int_{x^2}^{x+2} f(x,y)\, dy\, dx$, (b^A) $\int_{0}^{1}\int_{y}^{2y} f(x,y)\, dx\, dy$, and (c) $\int_{0}^{2}\int_{x}^{4-x} f(x,y)\, dy\, dx$.
◆ Type 5, reversing the order of integration

49. Evaluate $\int_{0}^{4}\int_{y}^{4} e^{-x^2}\, dx\, dy$ by reversing the order of integration.
◆ Type 5, reversing the order of integration

50. Find the values of (a^A) $\iint_R y^{-2} e^{x/\sqrt{y}}\, dx\, dy$ where R is the rectangle $0 \leq x \leq 1, 1 \leq y \leq 2$ and
(b) $\iint_R x \sin(y^3)\, dx\, dy$ where R is the triangle with vertices $(0,0), (0,2)$, and $(1,2)$.
◆ Type 5, a double integral

51. Use a property of logarithms to evaluate $\iint_R \dfrac{x}{y}\, dx\, dy$ with R bounded by $x = 1, x = 3, y = x^3$, and $y = 4x^3$.

♦ Type 5, a double integral

52. Find the average value of $f = 15x^2y$ in the triangle R with vertices $(0,0), (1,1)$, and $(-1,1)$. Then draw R and the curve(s) in R on which f is equal to the average value.

♦ Type 4, average value and the curve where it is assumed

53.[A] Find the volume of the solid between the surface $y = x^2 + z^2$ and the plane $y = -3$ for (x, z) in the rectangle bounded by $x = 0, x = 4, z = 0$ and $z = 5$ in the xz-plane

♦ Type 3, volume of a solid

54. Find the volume of the solid between the surfaces $x = -z^2$ and $x = z^2 + 2$ for (y, z) in the rectangle bounded by $y = 0, y = 4, z = 0$ and $z = 1$ in the yz-plane

♦ Type 4, volume of a solid

55.[A] Find the volume of the tetrahedron bounded by the coordinate planes and the plane with x-intercept, y-intercept, and z-intercept all equal to 1

♦ Type 4, volume of a solid

56. What is the volume of the solid bounded by the xz-plane, the yz-plane, the plane $z = -2$, and the plane with x-intercept 2, y-intercept -2, and z-intercept 4?

♦ Type 4, volume of a solid

57. Find the volume of the solid bounded by the planes $x = 0, y = 0, 2x + 2y + z = 2$, and $4x + 4y - z = 4$.

♦ Type 4, volume of a solid

58. If you have access to a computer or a calculator to calculate sums automatically, find the approximate decimal values of the Riemann sums for the integral $\iint_R x^2 y^3\, dx\, dy$ for $R: 0 \le x \le 1, 0 \le y \le 1$ **(a**[O]**)** corresponding to a partition of R into 100 equal squares with the integrand evaluated at the upper right corner of each subregion, **(b)** corresponding to a partition into 100 equal squares with the integrand evaluated at the center of each subregion, **(c)** corresponding to a partition into 400 equal squares with the integrand evaluated at the upper right corner of each subregion, and **(d)** corresponding to a partition into 400 equal squares with the integrand evaluated at the center of each subregion. **(e) (a)** Find the exact value of the integral and calculate the errors made in approximating the integral by the Riemann sums in parts (a) through (d).

♦ Type 5, calculator/computer calculation of Riemann sums

59. **(a)** Find the average value of $f = 4xy$ in each of the subregions $R_1: 0 \le x \le \frac{1}{2}, 0 \le y \le \frac{1}{2}$; $R_2: \frac{1}{2} \le x \le 1, 0 \le y \le \frac{1}{2}; R_3: 0 \le x \le \frac{1}{2}, \frac{1}{2} \le y \le 1$ and $R_4: \frac{1}{2} \le x \le 1, \frac{1}{2} \le y \le 1$ of $R: 0 \le x \le 1, 0 \le y \le 1$. **(b)** Define a step function $S(x, y)$ that is constant in each of the subregions of part (a), that has three different values, and is such that $\iint_R S(x, y)\, dx\, dy = \iint_R f(x, y)\, dx\, dy$.

♦ Type 5, finding a step function with the same integral

60. (CENTROIDS) The centroid of a plane region is its center of gravity when it is given density $\rho = 1$. Find the centroids of **(a**[O]**)** the quarter circle lying between $y = \sqrt{1 - x^2}$ and the y-axis for $0 \le x \le 1$, **(b**[A]**)** the region bounded by $y = \sqrt{x}, y = 0$, and $x = 4$, **(c)** the region bounded by $y = x^3, y = -1$, and $x = 1$, and **(d)** the region bounded by $y = x^2$ and $y = \frac{1}{2}x^3$.

♦ Type 4, centroids

61. (PAPPUS' THEOREM FOR VOLUMES OF REVOLUTION) **(a)** Suppose that $h(x)$ and $g(x)$ are piecewise continuous and $0 \le h(x) \le g(x)$ for $a \le x \le b$ and R be the region R between the graphs for $a \le x \le b$. Prove that the volume generated by rotating R about the x-axis or about the y-axis equals the area of R multiplied by the distance traveled by its centoid. **(b)** Use Pappus' Theorem and the formula $\frac{4}{3}\pi r^3$ for the volume of a sphere of radius r to find the centroid of the quarter circle $0 \le y \le \sqrt{1 - x^2}, 0 \le x \le 1$ of Problem 60a. **(c)** Use Pappus' Theorem to find the volume of the TORUS that is generated when the disk $(x - b)^2 + y^2 \le a^2$ with $0 < a < b$ is rotated about the y axis.

♦ Type 7, proof of a therorem of Pappus

12.1 Double integrals

62. Figure 44 shows level curves of the elevation above sea level $h(x,y)$ (feet) on Mt. Shasta, California's second highest mountain. Here (x,y) are horizontal coordinates measured in feet with the scale on the right. Let R denote the region inside the 7500-foot level curve. **(a)** What does the integral
$$\iint_R [h(x,y) - 7500]\, dx\, dy$$
represent? **(b)** What is the approximate value of the integral in part (a)?

◆ Type 4, estimating a volume from level curves

FIGURE 44

63. Level curves of the intensity $p(x,y)$ of radio signals from a portion of the sky near the Crab Nebula in the constellation Taurus are shown in Figure 45. The radio signals have a wavelength of 21.3 centimeters and the intensity is measured in thousands of degrees Kelvin.[1] Estimate the average intensity in the square $R: 2 \le x \le 4, 2 \le y \le 4$ by using a Riemann sum with the partition of R into four equal subsquares.

◆ Type 4, estimating an average value from level curves

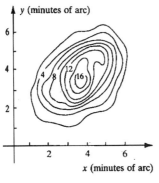

FIGURE 45

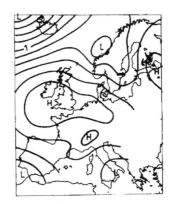

FIGURE 46

64. Figure 45 shows level curves of the barometric pressure in Europe on April 2, 1783 as compiled from records kept by organizations and individuals and in ship reports.[2] The level curve labeled "1" in the upper left corner represesents 1 atmosphere of pressure and the curves are at increments of 0.004 atmospheres. The three letters "L" indicate lows where the air pressure is a local minimum and the three letters "H" indicate highs where the air pressure is a local maximum. Based on this data, what was the approximate average air pressure **(a)** over England and **(b)** over Spain that day?

◆ Type 4, average values from a weather map

---

[1] Adapted from *The Radio Universe* by J. Hey, 3rd. edition, New York, NY: Pergamon Press, 1983.
[2] Data adapted from *The Weather of the 1780's over Europe* by J. Kington, Cambridge: Press Sindicate of the University of Cambridge, 1988, p. 94.

Section 12.2

Triple integrals

OVERVIEW: *In this section define* TRIPLE INTEGRALS *over bounded solids in xyz-space and use them in problems involving weights and centers of gravity of three-dimensional objects with variable density, amounts of other types given by densities, and average values of functions with three variables.*

Topics:
- *Triple integrals*
- *Density and weight*
- *Centers of gravity*
- *Average values*

Triple integrals

We begin with a calculation of weight from a three-dimensional density in a case that does not require an integral. The solid V in Figure 1 is defined by $0 \leq x \leq 3, 0 \leq y \leq 4, 0 \leq z \leq 2$ in xyz-space with distances measured in inches. V is divided into twelve one-inch-high, one-inch-wide, and two-inch-long subsolids, labeled $V_1, V_2, \ldots, V_{12}$, that are made of twelve different types of wood. The densities of the wood, measured in ounces per cubic inch, are given in the table below.[1]

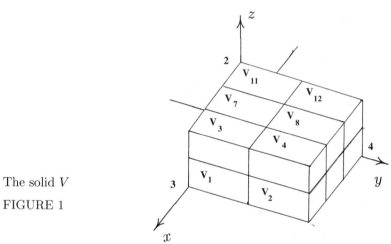

The solid V

FIGURE 1

TABLE 1. DENSITY (OUNCES PER CUBIC INCH) OF WOODS

Subsolid	Type of wood	Density	Piece	Type of wood	Density
V_1	Balsa	0.09	V_7	Cypress	0.26
V_2	Basswood	0.24	V_8	Ebony	0.59
V_3	Beech	0.43	V_9	Goncalvo Alves	0.55
V_4	Brazilwood	0.72	V_{10}	Larch	0.34
V_5	Cherry	0.43	V_{11}	Pecan	0.44
V_6	Chestnut	0.72	V_{12}	Willow	0.23

---

[1] Data adapted from *World Woods in Color* by W. Lincoln, New York, NY: Macmillan, 1986.

♦ SUGGESTIONS TO INSTRUCTORS: The main topics in this section can be covered in two or more class meetings. Have the students begin work on Questions 1 through 5 before class. Problems 7, 11, 15, 19, 24 are good for in-class discussion.

12.2 Triple integrals

Question 1 How much do the heaviest and the lightest subsolids of V weigh?

Example 1 What is the total weight of all twelve pieces of wood?

SOLUTION The volume of each piece is 2 cubic inches, so the total weight is

$$(0.09)(2) + (0.24)(2) + (0.43)(2) + (0.72)(2) + (0.43)(2)$$
$$+ (0.72)(2) + (0.26)(2) + (0.59)(2) + (0.55)(2) + (0.34)(2)$$
$$+ (0.44)(2) + (0.23)(2) = (5.04)(2) = 10.08 \text{ ounces. } \square$$

We can define a density $\rho(x, y, z)$ for the entire solid V of Figure 1 by giving it the values in Table 1 in the interiors of the twelve subsolids. Because $\rho(x, y, z)$ has only a finite number of values, we call it a STEP FUNCTION, by analogy with the definition of step-functions with one or two variables, even though in this case there is no step-like graph to visualize. We will see below that the total weight of the wood in Figure 1 is equal to the TRIPLE INTEGRAL of $\rho(x, y, z)$ over the solid V.

The definition of triple integrals is very similar to that of double integrals. We say that a solid V in xyz-space is BOUNDED if it can be enclosed in a ball. We say that a surface is PIECEWISE SMOOTH if it consists of a finite number of graphs $z = g(x, y), y = g(x, z)$, or $x = g(y, z)$, where in each case g has continuous first-order derivatives and is defined in a closed, bounded, plane region with a piecewise-smooth boundary.

Suppose that V is a bounded solid with a piecewise-smooth boundary. We use a finite number of piecewise smooth surfaces to divide V into a PARTITION of subsolids

$$V_1, V_2, V_3, \ldots, V_N. \tag{1}$$

The DIAMETER of a subsolid is the diameter of the smallest closed ball that contains it.

We say that $f(x, y, z)$ is PIECEWISE CONTINUOUS in V if it is defined in all of V, except possibly at a finite number of points or on a finite number of piecewise-smooth curves or surfaces, and if there is a fixed partition $V_1, V_2, V_3, \ldots, V_N$ of V such that in the interior of each V_j, $f(x, y, z) = F_j(x, y, z)$, where F_j is continuous in the closure of V_j.

Given any partition (1) of V, we pick a point (x_j, y_j, z_j) in V_j for $j = 1, 2, \ldots, N$ such that $f(x_j, y_j, z_j)$ is defined and form the RIEMANN SUM

$$\sum_{j=1}^{N} f(x_j, y_j, z_j) \, [\text{Volume of } V_j]. \tag{2}$$

The triple integral of f over V is a limit of such sums:

Definition 1 *If V is a bounded solid with a piecewise smooth boundary and $f(x, y, z)$ is piecewise continuous in V, then the* TRIPLE INTEGRAL

$$\iiint_V f(x, y, z) \, dx \, dy \, dz \tag{3}$$

is the limit of the Riemann sums (2) as the number of subsolids in the corresponding partitions (1) tends to infinity and their diameters tend to 0, provided the limit exists and is finite.[†]

Question 2 What is the geometric significance of $\iiint_V dx \, dy \, dz = \iiint_V 1 \, dx \, dy \, dz$?

[†]The limit of the Riemann sums is the number L if for each $\epsilon > 0$ there is a $\delta > 0$ such that $|[\text{Riemann sum}] - L| < \epsilon$ for all partititons into subsolids of diameters $< \delta$. The proof that the integral is defined under the conditions of Definition 1 is given in advanced-calculus courses.

Evaluating triple integrals by iteration

Suppose that V is the solid in Figure 2 that is bounded on the top by the surface $z = h(x,y)$ and on the bottom by $z = g(x,y)$ for (x,y) in the bounded region R with a piecewise-smooth boundary in the xy-plane. We pick constants a and b such that the horizontal plane $z = a$ is below V and the horizontal plane $z = b$ is above V and let W be the solid between these planes for (x,y) in R (Figure 3). Then we define, or redefine, f to be zero in the portion of W that is not in V, so that

$$\iiint_V f(x,y,z)\,dx\,dy\,dz = \iiint_W f(x,y,z)\,dx\,dy\,dz. \tag{4}$$

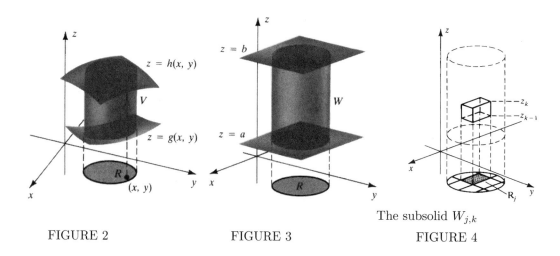

FIGURE 2   FIGURE 3   The subsolid $W_{j,k}$ FIGURE 4

Then we pick a partition

$$R_1, R_2, R_3, \ldots, R_{N-1}, R_N \tag{5}$$

of the region R in the xy-plane and a partition

$$a = z_0 < z_1 < z_2 < \cdots < z_{M-1} < z_M = b \tag{6}$$

of the interval $[a,b]$. For $j = 1, 2, \ldots, N$ and $k = 1, 2, \ldots, M$, we define $W_{j,k}$ to be the set of points (x,y,z) with (x,y) in R_j and $z_{k-1} \leq z \leq z_k$. Then the solids $W_{j,k}$ form a partition of W (Figure 4). We pick a point (x_j, y_j, z_k^*) in $W_{j,k}$ for each j and k such that $f(x_j, y_j, z_k^*)$ is defined. Then, because the volume of $W_{j,k}$ equals the area of R_j multiplied by the width $\Delta z_k = z_k - z_{k-1}$ of the subinteral, the corresponding Riemann sum for $\iiint_W f(x,y,z)\,dx\,dy\,dz$ is

$$\sum_{j=1}^{N}\sum_{k=1}^{M} f(x_j, y_j, z_k^*)\,[\text{Area of } R_j]\,\Delta z_k = \sum_{j=1}^{N}\left[\sum_{k=1}^{M} f(x_j, y_j, z_k^*)\,\Delta z_k\right][\text{Area of } R_j]. \tag{7}$$

If we let the number of subintervals in the partition of $[a,b]$ tend to ∞ while the widths of the subintervals tend to zero, then the inner sum on the right of (7) tends to $\int_{z=a}^{z=b} f(x,y,z)\,dz$, so that (7) tends to

$$\sum_{j=1}^{N}\left[\int_{z=a}^{z=b} f(x,y,z)\,dz\right][\text{Area of } R_j]. \tag{8}$$

12.2 Triple integrals

If we then let the number of subregions in the partition of R tend to ∞ while the diameters of the subregions tend to zero, then **(8)** approaches $\iint_R \left[\int_{z=a}^{z=b} f(x,y,z)\,dz \right] dx\,dy$, so that

$$\iiint_W f(x,y,z)\,dx\,dy\,dz = \iint_R \left[\int_{z=a}^{z=b} f(x,y,z)\,dz \right] dx\,dy.$$

Finally, because the triple integral over W equals the triple integral over V and $f(x,y,z) = 0$ for $z > h(x,y)$ and for $z < g(x,y)$, we obtain the following:

Theorem 1 *Suppose that R is a bounded region with a piecewise-smooth boundary in the xy-plane and that the solid V consists of the points (x,y,z) with (x,y) in R and $g(x,y) \leq z \leq h(x,y)$, where g and h are piecewise continuous in R. Then for any function $f(x,y,z)$ that is piecewise continuous in V*

$$\iiint_V f(x,y,z)\,dx\,dy\,dz = \iint_R \left\{ \int_{z=g(x,y)}^{z=h(x,y)} f(x,y,z)\,dz \right\} dx\,dy. \tag{9}$$

Often the curly brackets and the symbols "$z =$" in the limits of integration are omitted in **(9)**. Also, in applying **(9)**, we refer to R as the projection of V on the xy-plane.

Example 2 Evaluate $\displaystyle\iiint_V 2x\,dx\,dy\,dz$, where V is the tetrahedron bounded by the planes $3x + y + z = 3$ and the coordinate planes.

SOLUTION   Setting $x = 0$ and $y = 0$ in the equation $3x + y + z = 3$ for the plane shows that its z-intercept is 3; setting $x = 0$ and $z = 0$ shows that its y-intercept is 3; and setting $y = 0$ and $z = 0$ shows that its x-intercept is 1. Therefore, the tetrahedron is as shown in Figure 5. Its top is formed by the graph of $z = 3 - 3x - y$ and its projection on the xy-plane is its base, which is the triangle R shown in Figure 6.

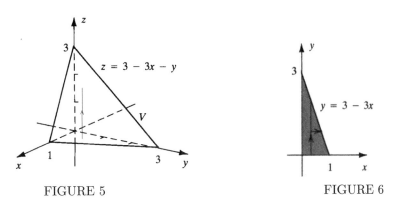

FIGURE 5   FIGURE 6

According to Theorem 1, we can integrate over V by integrating with respect to z from the bottom to the top of the tetrahedron for fixed (x,y) in R, as indicated by the vertical line with an arrow on it in Figure 5. Then we integrate the result with respect to x and y over R.

Setting $z = 0$ in the equation $z = 3 - 3x - y$ for the top of the tetrahedron shows that its intersection with the xy-plane is the line $3x + y = 3$ or $y = 3 - 3x$. This line forms the top of the region R, as is shown in Figure 6, and we can integrate over R by integrating with respect to y from $y = 0$ to $y = 3 - 3x$ for each x with $0 \leq x \leq 1$ and then integrating with respect to x from 0 to 1. Consequently,

$$\iiint_V 2x\,dx\,dy\,dz = \int_{x=0}^{x=1}\left\{\int_{y=0}^{y=3-3x}\left\{\int_{z=0}^{z=3-3x-y} 2x\,dz\right\}dy\right\}dx. \qquad (10)$$

Notice that the innermost limits of integration in **(10)** involve x and y; that the middle limits of integration involve only x; and the outermost limits of integration are constant.

In the innermost z-integral, x and y are constant and

$$\int_{z=0}^{z=3-3x-y} 2x\,dz = \Big[2xz\Big]_{z=0}^{z=3-3x-y} = 2x(3-3x-y)$$
$$= 6x - 6x^2 - 2xy.$$

Next, we use this result to carry out the middle y-integration with x constant:

$$\int_{y=0}^{y=3-3x}\int_{z=0}^{z=3-3x-y} 2x\,dz\,dy = \int_{y=0}^{y=3-3x}(6x - 6x^2 - 2xy)\,dy$$
$$= \Big[6xy - 6x^2y - xy^2\Big]_{y=0}^{y=3-3x}$$
$$= 6x(3-3x) - 6x^2(3-3x) - x(3-3x)^2$$
$$= 18x - 18x^2 - 18x^2 + 18x^3 - 9x + 18x^2 - 9x^3$$
$$= 9x - 18x^2 + 9x^3.$$

Finally, with **(10)**, we obtain

$$\iiint_V 2x\,dx\,dy\,dz = \int_{x=0}^{x=1}(9x - 18x^2 + 9x^3)\,dx$$
$$= \Big[\tfrac{9}{2}x^2 - 6x^3 + \tfrac{9}{4}x^4\Big]_0^1 = \tfrac{9}{2} - 6 + \tfrac{9}{4} = \tfrac{3}{4}. \quad\square$$

Question 3 What is the value of $\iiint_V (x+y+z)\,dx\,dy\,dz$ if V is the box defined by $0 \le x \le 3, 0 \le y \le 2, 0 \le z \le 1$?

Other orders of integration

Interchanging variables in Theorem 1 shows that triple integrals over solids of the type in Figure 7 may be expressed as

$$\iiint_V f(x,y,z)\,dx\,dy\,dz = \iint_R\left\{\int_{y=g(x,z)}^{y=h(x,z)} f(x,y,z)\,dy\right\}dx\,dz \qquad (11)$$

and triple integrals over solids of the type in Figure 8 may be expressed as

$$\iiint_V f(x,y,z)\,dx\,dy\,dz = \iint_R\left\{\int_{x=g(y,z)}^{x=g(y,z)} f(x,y,z)\,dx\right\}dy\,dz. \qquad (12)$$

The solid in Figure 7 is given by $g(x,z) \le y \le h(x,z)$ for (x,z) in R with R the projection of V on the xz-plane and the solid in Figure 8 by $g(y,z) \le x \le h(y,z)$ for (y,z) in R with R the projection of V on the yz-plane.

12.2 Triple integrals

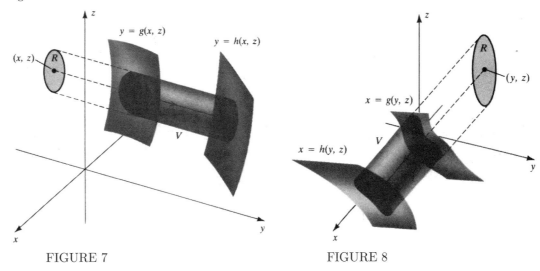

FIGURE 7　　　　　　　　　FIGURE 8

Density and weight

It can be shown that for a step function $f(x,y,z)$ that is constant in the interior of each subsolid of a partition of a bounded solid V, the integral $\iiint_V f(x,y,z)\,dx\,dy\,dz$ is equal to the sum of the values of f multiplied by the volumes of the corresponding subsolids. This shows, in particular, that the total weight of the wood occupying the solid V in Figure 1 is equal to $\iiint_V \rho(x,y,z)\,dx\,dy\,dz$, where $\rho(x,y,z)$ is the density of the wood at (x,y,z) in V.

For a density function $\rho(x,y,z)$ that is not a step function in a solid V, we use a partition $V_1, V_2, \ldots, V_N$ of V and a point (x_j, y_j, z_j) in V_j for each j to approximate ρ by the step function that equals $\rho(x_j, y_j, z_j)$ in the interior of each V_j. Then the weight of the entire solid is approximately equal to the Riemann sum $\sum_{j=1}^{N} \rho(x_j, y_j, z_j)\,[\text{Volume of } V_j]$, whose limit is the triple integral $\iiint_V \rho(x,y,z)\,dx\,dy\,dz$. This leads us to the next Definition.

Definition 2 *Suppose that V is a bounded solid with a piecewise smooth boundary that has density $\rho(x,y,z)$ (weight (or mass) per unit volume) at (x,y,z) and that ρ is piecewise continuous in V. Then*

$$[\text{Weight (or mass) of } V] = \iiint_V \rho(x,y,z)\,dx\,dy\,dz. \tag{13}$$

This Definition can also be used with other quantities in place of weight or mass.

Question 4　Express the weight of the solid V in Figure 1 as a definite integral.

Example 3　A solid V is given by $V = \{(x,y,z) : 0 \leq x \leq 1, 0 \leq y \leq x^2, 0 \leq z \leq 2\}$ in xyz-space with distances measured in centimeters, and its density at (x,y,z) is $\rho = xyz$ grams per cubic centimeter. Express its mass as an iterated integral.

SOLUTION　　The mass, measured in grams, is

$$\iiint_V xyz\,dx\,dy\,dz = \iint_R \int_{z=0}^{z=2} xyz\,dz\,dx\,dy$$
$$= \int_{x=0}^{x=1} \int_{y=0}^{y=x^2} \int_{z=0}^{z=2} xyz\,dz\,dy\,dx \tag{14}$$

where $R = \{(x,y) : 0 \leq x \leq 1, 0 \leq y \leq x^2\}$ is the region in Figure 9. □

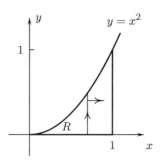

FIGURE 9

Question 5 Evaluate the iterated integral in **(14)** to find the mass of the solid in Example 3.

Moments and centers of gravity

If the weight or mass of a point mass at (x, y, z) in xyz-space is w, then its moment about the plane $x = 0$ is xw, its moment about the plane $y = 0$ is yw, and its moment about the plane $z = 0$ is zw. This principle leads to the following definition of moments of solids.

Definition 3 *Suppose that V is a bounded solid with a piecewise smooth boundary that has density $\rho(x, y, z)$ (weight or mass per unit volume) at (x, y, z) and that ρ is piecewise continuous in V. Then*

$$[\text{Moment of } V \text{ about } x = 0] = \iiint_V x\rho(x, y, z) \, dx \, dy \, dz \tag{15}$$

$$[\text{Moment of } V \text{ about } y = 0] = \iiint_V y\rho(x, y, z) \, dx \, dy \, dz \tag{16}$$

$$[\text{Moment of } V \text{ about } z = 0] = \iiint_V z\rho(x, y, z) \, dx \, dy \, dz. \tag{17}$$

The moments in Definition 3 are used to find the CENTER OF GRAVITY or CENTER OF MASS of the solid, just as moments of plates in an xy-plane are used to find their centers of gravity.

Rule 1 *If V satisfies the conditions of Definition 3, then its center of gravity $(\overline{x}, \overline{y}, \overline{z})$ is given by*

$$\overline{x} = \frac{[\text{Moment of } V \text{ about } x = 0]}{[\text{Weight or mass of } V]} = \frac{\iiint_V x\rho(x, y, z) \, dx \, dy \, dz}{\iiint_V \rho(x, y, z) \, dx \, dy \, dz} \tag{18}$$

$$\overline{y} = \frac{[\text{Moment of } V \text{ about } y = 0]}{[\text{Weight or mass of } V]} = \frac{\iiint_V y\rho(x, y, z) \, dx \, dy \, dz}{\iiint_V \rho(x, y, z) \, dx \, dy \, dz} \tag{19}$$

$$\overline{z} = \frac{[\text{Moment of } V \text{ about } z = 0]}{[\text{Weight or mass of } V]} = \frac{\iiint_V z\rho(x, y, z) \, dx \, dy \, dz}{\iiint_V \rho(x, y, z) \, dx \, dy \, dz} \tag{20}$$

12.2 Triple integrals

Example 4   What is the x-coordinate of the center of gravity of the solid V in Example 3 and Question 4?

SOLUTION   We first calculate the moment about $x = 0$, using the integration procedure in **(14)**:

$$[\text{Moment of } V \text{ about } x = 0] = \iiint_V x\rho \, dx \, dy \, dz = \iiint_V x^2 yz \, dx \, dy \, dz$$

$$= \int_{x=0}^{x=1} \int_{y=0}^{y=x^2} \int_{z=0}^{z=2} x^2 yz \, dz \, dy \, dx = \int_{x=0}^{x=1} \int_{y=0}^{y=x^2} \left[\tfrac{1}{2}x^2 y z^2\right]_{z=0}^{z=2} dy \, dx$$

$$= \int_{x=0}^{x=1} \int_{y=0}^{y=x^2} 2x^2 y \, dy \, dx = \int_{x=0}^{x=1} \left[x^2 y^2\right]_{y=0}^{y=x^2} dx$$

$$= \int_{x=0}^{x=1} x^6 \, dx = \left[\tfrac{1}{7} x^7\right]_{x=0}^{x=1} = \tfrac{1}{7}.$$

In Question 4, we saw that the mass of V is $\tfrac{1}{6}$. By **(18)**, we find the x-coordinate $\overline{x}$ of its center of gravity by dividing its moment about $x = 0$ by its mass. This gives $\overline{x} = \dfrac{1/7}{1/6} = \tfrac{6}{7}$. □

Average values

How do we find average values of functions with three variables? Suppose that $f(x, y, z)$ is piecewise continuous in a bounded solid V with a piecewise smooth boundary. For each partition $V_1, V_2, \ldots, V_N$ of V, we pick for each j a point (x_j, y_j, z_j) in V_j such that $f(x_j, y_j, z_j)$ is defined. Then we form

$$\sum_{j=1}^{N} f(x_j, y_j, z_j) \frac{[\text{Volume of } V_j]}{[\text{Volume of } V_j]} = \frac{1}{[\text{Volume of } V]} \sum_{j=1}^{N} f(x_j, y_j, z_j)[\text{Volume of } V_j]$$

which is the sum of the values of the function at the j points, each multiplied by the fraction of the volume of V that is in the corresponding subsolid. The average value of f in V is the integral with these Riemann sums.

Definition 4   *For a piecewise continuous function f in a bounded solid V with a piecewise smooth boundary,*

$$[\text{Average value of } f \text{ in } V] = \frac{1}{[\text{Volume of } V]} \iiint_V f(x, y, z) \, dx \, dy \, dz. \tag{21}$$

Question 6   What is the average density (ounces per cubic inch) of the stack of wood in Example 1?

Principles and procedures

- A definite integral $\int_a^b f(x) \, dx$ with one variable can be interpreted in terms of areas of regions between the graph of $f(x)$ and the x-axis, and a double integral $\iint_R f(x, y) \, dx \, dy$ can be interpreted in terms of volumes of solids between the graph of $f(x, y)$ and the xy-plane. There is no geometric interpretation of $\iiint_V f(x, y, z) \, dx \, dy \, dz$ in terms of the graph of $f(x, y, z)$ because it is a surface in four-dimensional space. Instead you can think of $\iiint_V f(x, y, z) \, dx \, dy \, dz$ either as the limit $\sum_{j=1}^{N} f(x_j, y_j, z_j) [\text{Volume of } V_j]$ of its Riemann sums, as the total weight of V if the values of f are nonnegative and its density at (x,y,z) is $f(x,y,z)$, or as the net total charge on V if $f(x,y)$ is its charge density (which can be posiitve, negative or zero) at (x,y,z).

- It is often difficult to draw a clear enough picture of the solid of integration V in a triple integral to help you decide how to express it as an iterated integral. In many cases, you can determine the projection R of the solid on the xy-plane and find formulas $z = g(x,y)$ and $z = h(x,y)$ for the surfaces that form its bottom and top by analyzing the description of the solid. Then
$$\iiint_V f\,dx\,dy\,dz = \iint_R \int_{z=g(x,y)}^{z=h(x,y)} f\,dz\,dx\,dy$$
and you can determine the appropriate x- and y-limits of integration from a sketch of R. In other cases you will need to use the projection of V on the xz- or yz-plane rather than on the xy-plane and use the integration procedures in **(11)** or **(12)**.

- In many problems in which the integrals can be evaluated by performing a z-integration first, the top and bottom of the solid V of integration are given as graphs $z = g(x,y)$ and $z = h(x,y)$ and the other conditions in the description of V determine its projection R on the xy-plane. In other cases, one edge or another of R is the intersection of the top and bottom, which has to be determined by solving the relevant equations. The solid $V = \{(x,y,z) : x^2 + y^2 \leq z \leq 1\}$, for example, consists of the points above the circular paraboloid $z = x^2 + y^2$ and beneath the plane $z = 1$, so its projection on the xy-plane is the disk $R = \{(x,y) : x^2 + y^2 \leq 1\}$.

- As an aid to remembering formula **(18)** in Rule 4 for the x-coordinate
$$\overline{x} = \frac{\iiint_V x\rho(x,y,z)\,dx\,dy\,dz}{\iiint_V \rho(x,y,z)\,dx\,dy\,dz}$$
of the center of gravity of a solid V with density ρ, notice that the symbols in the triple integral in the denominator seem to cancel with symbols in the numerator (they do not, of course), leaving only the symbol "x." Formulas **(19)** and **(20)** can be recalled similarly.

- An integral of a product $f(x)g(y)h(z)$ of a function of x, a function of y, and a function of z over a rectangular box $V : \alpha \leq x \leq \beta, \gamma \leq y \leq \delta, \kappa \leq z \leq \lambda$ can be evaluated by writing
$$\iiint_V f(x)g(y)h(z)\,dx\,dy\,dz = \int_{x=\alpha}^{x=\beta} \int_{y=\gamma}^{y=\delta} \int_{z=\kappa}^{z=\lambda} f(x)g(y)h(z)\,dz\,dy\,dx$$
$$= \int_{x=\alpha}^{x=\beta} f(x) \int_{y=\gamma}^{y=\delta} g(y) \int_{z=\kappa}^{z=\lambda} h(z)\,dz\,dy\,dx$$
$$= \left(\int_{x=\alpha}^{x=\beta} f(x)\,dx\right)\left(\int_{y=\gamma}^{y=\delta} g(y)\,dy\right)\left(\int_{z=\kappa}^{z=\lambda} h(z)\,dz\right).$$

Be careful, however, not to use this formula with sums of functions or in any other cases where it does not apply.

- To help remember formula **(21)** for average value, notice that if f is constant, its average value is equal to that constant.

- The definition of average value in Definition 4 provides another way to interpret
$$\iiint_V f(x,y,z)\,dx\,dy\,dz:$$ It equals the average value of f on V, multiplied by the volume of V.

12.2 Triple integrals

Tune-Up Exercises 12.2♦
AAnswer provided. **O**Outline of solution provided.

T1.^O Express $\iiint_V f(x,y,z)\,dx\,dy\,dz$ as an iterated integral, where
$V = \{(x,y,z) : 0 \le x \le 1, -1 \le y \le 1, 0 \le z \le 1\}$.
♦ Type 1, expressing a triple integral as an iterated integral

T2.^O Express $\iiint_V g(x,y,z)\,dx\,dy\,dz$ as an iterated integral, where V is the solid bounded by $z = 0, z = 1 - y^2, x = -2$, and $x = 0$.
♦ Type 1, expressing a triple integral as an iterated integral

T3.^O What is the value of $\iiint_V x^2 y^3 z^4 \,dx\,dy\,dz$ if V is the cube bounded by the planes $x = 1, y = 1$, and $z = 1$ and the coordinate planes?
♦ Type 1, Evaluating a triple integral

T4.^O Evaluate $\iiint_V xy \,dx\,dy\,dz$ where $V = \{(x,y,z) : 0 \le x \le 1, 0 \le y \le 1, -1 \le z \le x^2 + y^2\}$.
♦ Type 3, Evaluating a triple integral

T5.^O A rock occupying the solid V in xyz-space with distances measured in meters has density $\sqrt{1+xyz}$ kilograms per cubic meter at (x,y,z). Express the mass of the cylinder and the coordinates $(\overline{x},\overline{y},\overline{z})$ of its center of gravity in terms of triple integrals.
♦ Type 3, Coordinates of a center of gravity as ratios of integrals

T6.^O **(a)** What is the value of $\iiint_V F\,dx\,dy\,dz$ if F is the constant function $F(x,y,z) = 7$ and V is a bounded solid with piecewise smooth boundary whose volume is 10? **(b)** What is the average value of F in V for the function and solid of part (a)?
♦ Type 2, the average value of a constant

Problems 12.2
AAnswer provided. **O**Outline of solution provided.

Evaluate the integrals in Problems 1 through 10.

1.^O $\iiint_V (x^2 + y^2 + z^2)\,dx\,dy\,dz$ with $V = \{(x,y,x) : -1 \le x \le 2, 0 \le y \le 1, 0 \le z \le 2\}$
♦ Type 1, Evaluating a triple integral

2.^O $\iiint_V x^2 y e^{xyz}\,dx\,dy\,dz$ with V the box bounded by $x = 0, x = 1, y = 0, y = 4, z = 0$ and $z = 3$
♦ Type 1, Evaluating a triple integral

3.^A $\iiint_V xy^2 e^{-z}\,dx\,dy\,dz$ with $V = \{(x,y,x) : 0 \le x \le 2, 0 \le y \le 3, 0 \le z \le 4\}$
♦ Type 1, Evaluating a triple integral

4.º $\iiint_V z\,dx\,dy\,dz$ with V bounded by the cone $z = \sqrt{x^2 + y^2}$ and the planes $z = 0, x = \pm 1, y = \pm 1$
♦ Type 3, Evaluating a triple integral

5.^A $\iiint_V \sin y\, e^{3z}\,dx\,dy\,dz$ with V bounded by the planes $z = 0, z = 6, y = x, x = 0, x = 4$
♦ Type 1, Evaluating a triple integral

6. $\iiint_V 2z\,dx\,dy\,dz$ with V bounded by $z = -1, z = x^2 y^2, x = 0, x = 1, y = 0, y = 1$
♦ Type 3, Evaluating a triple integral

7.º $\iiint_V e^{2x+3y-4z}\,dx\,dy\,dz$ with V the box defined by $-4 \le x \le 4, -3 \le y \le 3, -2 \le z \le 2$
♦ Type 3, Evaluating a triple integral

♦The Tune-up Exercises and Problems are classified by type and content. The types are (1) basic, reactive; (2) basic reflective; (3) intermediate, reactive; (4) intermediate, reflective; (5) advanced, reactive; (6) advanced, reflective; and (7) advanced, theoretical.

8.^A $\iiint_V xz^2 \sin y \, dx \, dy \, dz$ with V the box bounded by $x = 0, x = 3, y = 1, y = 5, z = -3, z = 0$

♦ Type 1, Evaluating a triple integral

9. $\iiint_V (x^2 + 1)yz \sin y \, dx \, dy \, dz$ with $V = \{(x, y, z) : 0 \le x \le 5, 0 \le y \le 10, 0 \le z \le 20\}$

♦ Type 3, Evaluating a triple integral

10.^A $\iiint_V 2z \, dx \, dy \, dz$ with $V = \{(x, y, z) : 0 \le x \le 1, x^2 \le y \le 1, -2 \le z \le xy\}$

♦ Type 3, Evaluating a triple integral

11.º A block occupies the region bounded by $x = -2, x = 2, y = -2, y = 2, z = 1$, and $z = 2$ in xyz-space with distances measured in meters and its density at (x, y, z) is $x^2 e^y \sin z$ kilograms per cubic meter. What is its mass?

♦ Type 3, finding mass from density

12.^A A solid V bounded by $z = x^2 + 1, z = -y^2 - 1, x = 0, x = 2, y = -1$, and $y = 1$ in xyz-space with distances measured in feet contains electrical charges with density $8xy^2$ coulombs per cubic foot at (x, y, z). What is the overall net charge in V?

♦ Type 3, finding net total charge from charge density

13. The box $V = \{(x, y, z) : 0 \le x \le 2, 0 \le y \le 4, 0 \le z \le 6\}$ in xyz-space with distances measured in centimeters has density $x + y + z$ dynes per cubic centimeter at (x, y, z). (**a^O**) What is its weight? (**b^O**) What is the x-coordinate of its center of gravity? (**c^A**) What is the y-coordinate of its center of gravity? (**d^A**) What is the z-coordinate of its center of gravity?

♦ Type 3, mass and center of gravity

14. Find (**a**) the weight, (**b**) the x-coordinate of the center of gravity, (**c**) the y-coordinate of the center of gravity, and (**d**) the z-coordinate of the center of gravity of the box V bounded by $x = 1$, $x = 2, y = 1, y = 2, z = 1$ and $z = 2$ that has density $\dfrac{1}{xyz}$ pounds per cubic foot at (x, y, z). The coordinates are measured in feet.

♦ Type 3, center of gravity

15.º What is the average value of $f = xy^3 z^7$ in the the box V bounded by $x = 0, x = 2, y = 0, y = 2, z = 0$ and $z = 2$?

♦ Type 3, average value

16. Find the average value of $g = xyz$ in $V = \{(x, y, z) : 0 \le x \le 1, 0 \le y \le x^3, 0 \le z \le 8\}$.

♦ Type 3, average value

17. What is the center of gravity of the solid V bounded by $x = 0, x = 2, y = 0, y = 3, z = 0$, and $z = 4$ in xyz-space with distances measured in feet if its density at (x, y, z) is $x + y$ pounds per cubic foot?

♦ Type 4, center of gravity

18. Find the center of gravity of the solid V bounded by $x = 0, x = 1, y = 0, y = 1, z = 0$, and $z = x^2 + y^2$ in xyz-space with distances measured in inches and density xy ounces per cubic inch at (x, y, z)

♦ Type 4, center of gravity

Describe the solids of integration that lead to the iterated integrals in Problems 19 through 24.

19.º $\displaystyle\int_{x=0}^{x=2} \int_{y=0}^{y=\sqrt{4-x^2}} \int_{z=0}^{z=\sqrt{4-x^2-y^2}} f(x, y, z) \, dz \, dy \, dx$

♦ Type 4, describe a solid of integration

20.^A $\displaystyle\int_{-1}^{1} \int_{x^2}^{1} \int_{0}^{1-y} f(x, y, z) \, dz \, dy \, dx$

♦ Type 4, describe a solid of integration

21.^A $\displaystyle\int_{y=0}^{y=1} \int_{z=0}^{z=2-2y} \int_{x=0}^{x=4-4y-2z} f(x, y, z) \, dx \, dz \, dy$

♦ Type 4, describe a solid of integration

22. $\displaystyle\int_{z=-2}^{z=0} \int_{x=0}^{x=z+2} \int_{y=0}^{y=2-x+z} f(x, y, z) \, dy \, dx \, dz$

♦ Type 4, describe a solid of integration

12.2 Triple integrals

23.^A $\int_{x=-1}^{x=1} \int_{z=-\sqrt{1-x^2}}^{z=\sqrt{1-x^2}} \int_{y=0}^{y=4} f(x,y,z)\, dy\, dz\, dx$

♦ Type 4, describe a solid of integration

24. $\int_{x=-1}^{x=1} \int_{y=-\sqrt{1-x^2}}^{y=\sqrt{1-x^2}} \int_{z=-5}^{z=x^2-y^2} f(x,y,z)\, dz\, dy\, dx$

♦ Type 4, describe a solid of integration

Find the values of the integrals in Problems 25 through 31.

25.º $\iiint_V x^2 y^2 \cos z\, dx\, dy\, dz$ with V bounded by $x=0, x=1, y=1, z=-1$, and $z=y^3$

♦ Type 4, evaluate an integral

26.^A $\iiint_V z\, dx\, dy\, dz$ with V bounded by $x=1, y=1, z=xy$, and $z=2$

♦ Type 4, evaluate an integral

27. $\iiint_V x\, dx\, dy\, dz$ with V bounded by $x=1, z=0$, and $z=x-y^2$

♦ Type 4, evaluate an integral

28.^A $\iiint_V 15x^2 z^2\, dx\, dy\, dz$ with V bounded by $x^2+y^2=1$, and $x^2+z^2=1$

♦ Type 4, evaluate an integral

29. $\iiint_V 180 x^4 y^2 z^{-1}\, dx\, dy\, dz$ with V bounded by $y=1-x^2, y=x^2-1, z=1$, and $z=10$

♦ Type 4, evaluate an integral

30. $\iiint_V e^{x^4}\, dx\, dy\, dz$ with V bounded by $y=0, y=x, x=1, z=0$, and $z=y^2$

♦ Type 4, evaluate an integral

31.^A $\iiint_V x^2 y^2 z\, dx\, dy\, dz$ with $V = \{(x,y,z) : 0 \le x \le 1, 0 \le z \le x^2 - y^2\}$

♦ Type 4, evaluate an integral

32. Find the value of k such that $\iiint_V k(yz)^{1/2}\, dx\, dy\, dz = \frac{40}{7}$ where V is bounded by $z=0, z=y$, $y=x^2$, and $y=1$.

♦ Type 4, determine a parameter in an integral

33. Show that $\iiint_V \sin x \sin y \sin z\, dx\, dy\, dz$ with V bounded by $z=0, z=y, x=0, x=\frac{1}{2}\pi$, and $y=\pi$ equals an even integer.

♦ Type 4, evaluate an integral

34.^A Find k such that $\iiint_V x^2 z\, dx\, dy\, dz = \frac{1}{60}$ where V bounded by $z=0, z=x, x=0, y=0$, and $x+y=k$.

♦ Type 4, determine a parameter in an integral

35. Show that $\iiint_V y \cos z\, dx\, dy\, dz$ equals 0 if V is bounded by $x=0, x=5, z=y^2$, and $z=4$.

♦ Type 4, evaluate an integral

36. Show that $\iiint_V x\, dx\, dy\, dz = \frac{1016}{1155}$ where V is bounded by $z=-x^2-y^2, z=x^2+y^2, y=-x, y=x^3$, and $x=1$.

♦ Type 4, evaluate an integral

37. The tetrahedron with vertices $(0,0,0), (1,0,0), (0,1,0)$, and $(0,0,1)$ in xyz-space with distances measured in meters has density x^2 newtons per cubic meter at (x,y,z). Find its center of gravity. (Use symmetry.)

♦ Type 4, center of gravity

38.^A What is the center of gravity of the cylinder $V = \{(x,y,z) : x^2 + y^2 \leq 4, 0 \leq z \leq 1\}$ in xyz-space with distances measured in feet if its density at (x, y, z) is $\cos z$ pounds per square foot?

♦ Type 4, center of gravity

39. A solid occupies the region bounded by $z = y^2, x = 0, x = 2$, and $z = 1$ in xyz-space with distances measured in meters. Its mass at (x, y, z) is $z+2$ kilograms per square meter. Find its center of gravity.

♦ Type 4, center of gravity

In Problems 40 through 42 switch the order of integration to one where the z-integration is performed last.

40.^A $\int_0^4 \int_0^{4-x} \int_0^{4-x-y} f(x,y,z) \, dz \, dy \, dx$

♦ Type 5, switching orders of integration

41. $\int_{-2}^2 \int_{x^3}^8 \int_0^3 f(x,y,z) \, dz \, dy \, dx$

♦ Type 5, switching orders of integration

42. $\int_0^5 \int_x^5 \int_0^{5-y} f(x,y,z) \, dz \, dy \, dx$

♦ Type 5, switching orders of integration

43. The CENTROID of a solid is its center of gravity when it is given density 1. Find the centroid of $V = \{(x, y, z) : 0 \leq x \leq 1, 0 \leq y \leq 1, 0 \leq z \leq \sin y\}$

♦ Type 3, a centroid

44. The base of a solid V is the square $R : 0 \leq x \leq 2, 0 \leq y \leq 2$ in the xy-plane, the sides of V are vertical, and its top is formed by the graph $z = F(x,y)$ where $F(x,y) \approx 5$ for $0 \leq x \leq 1, 0 \leq y \leq 1$, $F(x,y) \approx 6$ for $1 < x \leq 2, 0 \leq y \leq 1$, $F(x,y) \approx 7$ for $0 \leq x \leq 1, 1 < y \leq 2$, and $F(x,y) \approx 8$ for $1 < x \leq 2, 1 < y \leq 2$. What is the approximate value of $\iiint_V 2z \, dx \, dy \, dz$?

♦ Type 4, estimating an integral

45. The function $G(x, y, z)$ equals $4xy$ for $0 \leq x \leq 1$ and all y and z, equals $9x^2z^2$ for $1 < x \leq 2$ and all y and z, and equals $16y^3z^3$ for $2 < x \leq 3$ and all y and z. What is the value of $\iiint_V G(x,y,z) \, dx \, dy \, dz$ if V is the cube $0 \leq x \leq 3, 0 \leq y \leq 3, 0 \leq z \leq 3$?

♦ Type 4, estimating an integral

46. The function $T(x, y, z)$ equals 1 for $0 \leq \sqrt{x^2+y^2+z^2} \leq 1$, equals 10 for $1 < \sqrt{x^2+y^2+z^2} \leq 2$, and equals 100 for $2 < \sqrt{x^2+y^2+z^2} \leq 3$. What is the value of $\iiint_V T(x,y,z) \, dx \, dy \, dz$ where V is the ball $\{(x,y,z) : \sqrt{x^2+y^2+z^2} < 3\}$?

♦ Type 4, evaluating an integral of as step function

Section 12.3

Integrals in polar, cylindrical, and spherical coordinates

OVERVIEW: *In this section we show how double integrals can be evaluated using polar coordinates and how triple integrals can be evaluated with* CYLINDRICAL *and* SPHERICAL *coordinates. These special coordinates are often used in integrals where the integrand or equations for the boundary of the region or solid of integration are given in terms of these particular variables. Polar coordinates $[r, \theta]$ were discussed in Section 9.3. Cylindrical coordinates $[r, \theta, z]$ use polar coordinates in place of the x and y in rectangular coordinates. Spherical coordinates $[\rho, \theta, \phi]$ use the distance ρ from a point to the origin in xyz-space and angles θ and ϕ that are similar to those used to designate longitude and latitude on the earth. At the end of the section we show how spherical coordinates can be used to find great-circle distances between points on the globe from their latitudes and longitudes.*

Topics:

- *Double integrals in polar coordinates*
- *Triple integrals in cylindrical coordinates*
- *Triple integrals in spherical coordinates*
- *Spherical coordinates and geography*

Double integrals in polar coordinates

We begin this section by showing how double integrals $\iint_R f(x, y)\ dx\, dy$ can be evaluated using polar coordinates. We will use polar coordinates with $r \geq 0$. Then the polar coordinates of any point other than the origin are $[r, \theta]$, where r is the distance from P to the origin and θ is an angle from the positive x-axis to the line segment from the origin to the point (Figure 1), and the polar coordinates of the origin are $[0, \theta]$ for any angle θ.

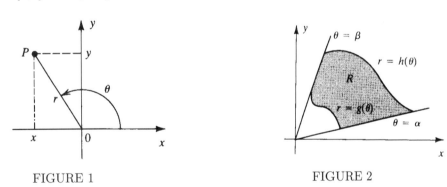

FIGURE 1   FIGURE 2

Suppose first that the region of integration R, like the region in Figure 2, is bounded by the rays $\theta = \alpha$ and $\theta = \beta$ through the origin, where $\alpha < \beta$ and $\beta - \alpha \leq 2\pi$, and by the curves with polar equations $r = g(\theta)$ and $r = h(\theta)$ such that $0 \leq g(\theta) \leq h(\theta)$ for $\alpha \leq \theta \leq \beta$ and $g(\theta)$ and $h(\theta)$ are piecewise smooth in $[\alpha, \beta]$. We will derive the basic integration formula for such a region by using rays $\theta = c$ and circles $r = c$ to partition the region of integration. ITo carry out the calculations we will need to to find areas of regions bounded by such rays and circles. We begin by deriving the formula

$$[\text{Area}] = 2\pi r^* \Delta r \qquad (1)$$

♦ SUGGESTIONS TO INSTRUCTORS: This section can be divided into two parts to be covered in two or more class meetings.
Part 1 (Integrals in polar and cylindrical coordinates; Tune-up Exercises T1, T2, T4a, T5–T6, T8a, T9; Problems 1–12, 14, 15, 17–27, 29–32, 34–35, 37, 39–43, 49) Have students begin work on Questions 1 through 4 before class. Tune-up Exercises T1, T2, T6 , T9 and Problems 2, 6, 7a, 15 29 are good for in-class discussions.
Part 2 (Spherical coordinates and geography; Tune-up Exercises T3, T4b, T7, T8b, T10; Problems 13, 14, 16, 28, 33, 36, 38, 44-48, 50, 51) Have students begin work on Questions 5 through 9 before class. Tune-up Exercises T8, T10 and Problems 16, 43, 47, 48 are good for in-class discussions

for the area of the ring-shaped region between concentric circles in Figure 3, where r^* is the average of the radii of the inner and outer circle and Δr is the thickness of the ring—the outer radius minus the inner radius. Notice that **(1)** equals the circumference $2\pi r^*$ of the circle of radius r^* at the center of the ring, multiplied by the thickness Δr of the ring.

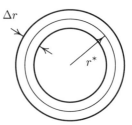

FIGURE 3

Question 1   Verify formula **(1)** **(a)** for the ring with inner radius 4 and outer radius 6 in Figure 4 and **(b)** for the ring with inner radius 8 and outer radius 10 in Figure 5.

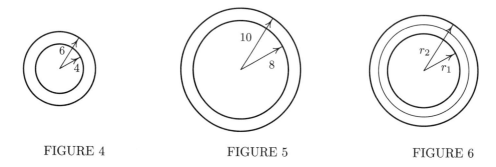

FIGURE 4   FIGURE 5   FIGURE 6

To verify formula **(1)** in general, we suppose that the inner radius in Figure 3 is r_1 and the outer radius is r_2, as in Figure 6. Then the area of the outer circle is $\pi(r_2)^2$, the area of the inner circle is $\pi(r_1)^2$, so the area of the ring is

$$\pi(r_2)^2 - \pi(r_1)^2 = \pi[(r_2)^2 - (r_1)^2] = \pi(r_2 + r_1)(r_2 - r_1)$$
$$= 2\pi\left[\tfrac{1}{2}(r_2 + r_1)\right](r_2 - r_1).$$

This equals $2\pi r^*\Delta r$ because $r^* = \tfrac{1}{2}(r_2 + r_1)$, and $\Delta r = r_2 - r_1$, so **(1)** is established.

Now that we have verified **(1)**, we can use it to find a formula for the area of the region in Figure 7 consisting of the portion of the ring in Figure 3 that is inside an angle of $\Delta\theta$ radians ($0 < \theta \leq 2\pi$) with its vertex at the center of the circle. Because a full revolution is 2π radians, the region in Figure 7 is the fraction $\Delta\theta/(2\pi)$ of the ring, and

$$[\text{Area of the region in Figure 7}] = \left(\frac{\Delta\theta}{2\pi}\right)(2\pi r^*\Delta r) = r^*\Delta r\Delta\theta. \tag{2}$$

12.3 Integrals in polar, cylindrical, and spherical coordinates,

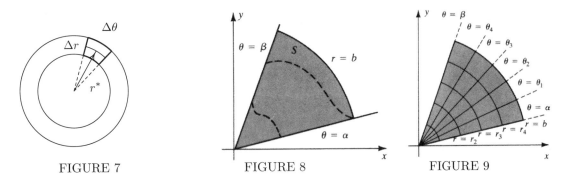

FIGURE 7   FIGURE 8   FIGURE 9

Now we can see how to express the integral $\iint_R f(x,y)\,dx\,dy$ in polar coordinates, where R is the region in Figure 2. We choose a number b that is greater than the values of $h(\theta)$ for $\alpha \leq \theta \leq \beta$, and let S be the region bounded by $\theta = \alpha, \theta = \beta$, and $r = b$ that contains R (Figure 8). We define, or redefine, $f(x,y)$ so that it equals zero on the part of S that is not in R and

$$\iint_R f(x,y)\,dx\,dy = \iint_S f(x,y)\,dx\,dy. \tag{3}$$

We pick a positive integer N and let

$$\alpha = \theta_0 < \theta_1 < \theta_2 < \cdots < \theta_N = \beta \tag{4}$$

be the partition of the θ-interval $[\alpha, \beta]$ into equal subintervals of length $\Delta\theta = (\beta - \alpha)/N$. We also pick another positive integer M and let

$$0 = r_0 < r_1 < r_2 < \cdots < r_M = b \tag{5}$$

be the partition of the r-interval $[0, b]$ into equal subintervals of length $\Delta r = b/M$. Then the rays $\theta = \theta_j$ for $j = 1, 2, \ldots, N-1$ and the circles $r = r_k$ for $k = 1, 2, \ldots, M-1$ divide S into subregions, as in Figure 9, where $N = 5$ and $M = 5$. We let $S_{j,k}$ denote the subregion defined by

$$S_{j,k} = \{[r, \theta] : \theta_{j-1} \leq \theta \leq \theta_j, r_{k-1} \leq r \leq r_k\}$$

as shown in Figure 10. The subregions for $j = 1, 2, \ldots, N, k = 1, 2, \ldots, M$ form a partition of the region S which we can use to construct Riemann sums for double integrals over S.

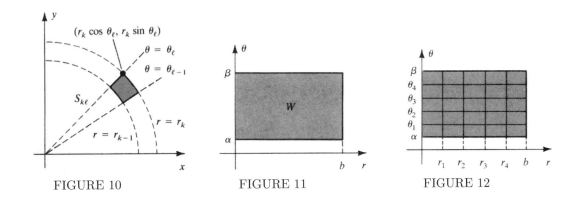

FIGURE 10   FIGURE 11   FIGURE 12

We let r_k^* denote the midpoint between r_{k-1} and r_k. Then the point with polar coordinates $[r_k^*, \theta_j]$ in $S_{j,k}$ (Figure 10) has rectangular coordinates

$$x = r_k^* \cos \theta_j, y = r_k^* \sin \theta_j$$

and with these points, we obtain

$$\sum_{j=1}^{N} \sum_{k=1}^{M} f(r_k^* \cos \theta_j, r_k^* \sin \theta_j) [\text{Area of } S_{j,k}] \tag{6}$$

which is a Riemann sum for $\iint_S f(x,y) \, dx \, dy$. Formula (2) shows that the area of $S_{j,k}$ is $r_k^* \Delta r \Delta \theta$, so that (6) equals

$$\sum_{j=1}^{N} \sum_{k=1}^{M} f(r_k^* \cos \theta_j, r_k^* \sin \theta_j) \, r_k^* \, \Delta r \, \Delta \theta \tag{7}$$

which is a Riemann sum for the double integral $\iint_W f(r \cos \theta, r \sin \theta) r \, dr \, d\theta$ over the rectangle $W = \{(r, \theta) : 0 \leq r \leq b, \alpha \leq \theta \leq \beta\}$ in the $r\theta$-plane of Figure 11, corresponding to the partition into rectangles determined by the partitions (4) and (5), as shown in Figure 12 with $N = 5$ and $M = 5$.

If we let N and M tend to ∞, then the widths of the subintervals tend to zero and the sum (6) tends, on the one hand, to the integral of $f(x,y)$ with respect to x and y over S in Figure 7, and, on the other hand to the integral of $f(r \cos \theta, r \sin \theta) \, r$ with respect to r and θ over the rectangle in Figure 12. Therefore

$$\iint_S f(x,y) \, dx \, dy = \iint_W f(r \cos \theta, r \sin \theta) \, r \, dr \, d\theta$$
$$= \int_{\theta=\alpha}^{\theta=\beta} \int_{r=0}^{r=b} f(r \cos \theta, r \sin \theta) \, r \, dr \, d\theta.$$

Finally, because the integral over S equal the integral over the original region R and $f = 0$ for $0 \leq r < g(\theta)$ and for $r > h(\theta)$, we have established the following result.

Theorem 1 *Suppose that R is bounded by the curves $r = g(\theta)$ and $r = h(\theta)$ for $\alpha \leq \theta \leq \beta$ where $0 < \beta - \alpha \leq 2\pi$ and $g(\theta)$ and $h(\theta)$ satisfy $0 \leq g(\theta) \leq h(\theta)$ and are piecewise continuous in the closed interval $[\alpha, \theta]$. Then for any piecewise continuous function $f(x,y)$ in R*

$$\iint_R f(x,y) \, dx \, dy = \int_{\theta=\alpha}^{\theta=\beta} \int_{r=g(\theta)}^{r=h(\theta)} f(r \cos \theta, r \sin \theta) \, r \, dr \, d\theta. \tag{8}$$

The integration procedure in (8) is represented by the lines and arrows in Figure 13. The radial line with an arrow on it from the inside to the outside of the region in the drawing represents the integration with respect to r from $r = g(\theta)$ to $r = h(\theta)$ for a fixed θ between α and β, and the curved arrow represents the integration with respect to θ from α to β. Notice that the radial line from the inner to the outer curve in Figure 13 sweeps out the region R as θ goes from α to β.

In applying (8) to change variables from x and y to r and θ we use the formulas

$$x = r \cos \theta, y = r \sin \theta, dx \, dy = r \, dr \, d\theta. \tag{9}$$

12.3 Integrals in polar, cylindrical, and spherical coordinates,

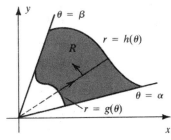

FIGURE 13

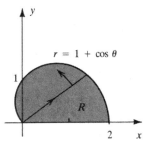

FIGURE 14

Example 1 Use polar coordinates to evaluate $\iint_R y \, dx \, dy$, where R is the region bounded by the x-axis and the upper half of the cardioid $r = 1 + \cos\theta$ in Figure 14.

SOLUTION We make the substitutions $y = r\sin\theta$ and $dx\,dy = r\,dr\,d\theta$ in the integral and integrate with respect to r from 0 to $1 + \cos\theta$ and then with respect to θ from 0 to π, as indicated by the lines and arrows in Figure 14. This gives

$$\iint_R y\,dx\,dy = \int_{\theta=0}^{\theta=\pi}\int_{r=0}^{r=1+\cos\theta}(r\sin\theta)r\,dr\,d\theta = \int_{\theta=0}^{\theta=\pi}\int_{r=0}^{r=1+\cos\theta} r^2 \sin\theta\,dr\,d\theta$$

$$= \int_{\theta=0}^{\theta=\pi}\left[\tfrac{1}{3}r^3 \sin\theta\right]_{r=0}^{r=1+\cos\theta} d\theta = \int_{\theta=0}^{\theta=\pi} \tfrac{1}{3}(1+\cos\theta)^3 \sin\theta\,d\theta.$$

To evaluate the last integral, we use the substitutions $u = 1 + \cos\theta$, $du = -\sin\theta\,d\theta$ to see that the original integral equals

$$-\tfrac{1}{3}\int_{\theta=0}^{\theta=\pi}(1+\cos\theta)^3(-\sin\theta\,d\theta) = -\tfrac{1}{3}\int_{\theta=0}^{\theta=\pi} u^3\,du$$

$$= -\tfrac{1}{3}\left[\tfrac{1}{4}u^4\right]_{\theta=0}^{\theta=\pi} = -\tfrac{1}{3}\left[\tfrac{1}{4}(1+\cos\theta)^4\right]_{\theta=0}^{\theta=\pi} = -\tfrac{1}{12}(0^4 - 2^4) = \tfrac{4}{3}. \quad \square$$

Question 2 Because the disk $R = \{(x,y) : x^2 + y^2 \leq 1\}$ in Figure 15 has radius 1 its area is $\pi(1)^2 = \pi$ and the average value of $f = \sqrt{x^2 + y^2 + 1}$ in R is

$$\frac{1}{\pi}\iint_R \sqrt{x^2 + y^2 + 1}\,dx\,dy. \tag{10}$$

Use polar coordinates to evaluate this integral.

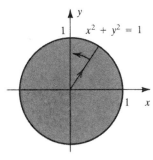

FIGURE 15

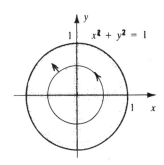

FIGURE 16

With the integration procedure indicated in Figure 15, we integrate with respect to r along a ray with θ equal to a constant from 0 to 1 and then integrate with respect to θ from 0 to 2π. We can also evaluate some integrals with the procedure shown in Figure 16, where we integrate first with respect to θ on a circle with r equal to a constant and then with respect to r.

Cylindrical coordinates

CYLINDRICAL COORDINATES $[r, \theta, z]$ for xyz-space are obtained by replacing x and y with polar coordinates in the xy-plane with $r \geq 0$. To use these new coordinates, we need to know the COORDINATE SURFACES where one of the variables is constant. The coordinate surface $r = c$ where r is constant is the z-axis if $c = 0$ and the cylinder of radius c with the z-axis at its center if $c > 0$ (Figure 17); the coordinate surface $\theta = c$ is a vertical half plane with its edge on the z-axis (Figure 18); and the coordinate surface $z = c$ is a horizontal plane (Figure 19).

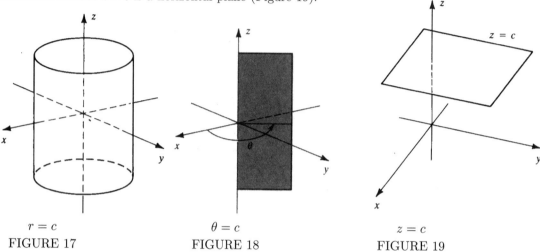

$r = c$
FIGURE 17

$\theta = c$
FIGURE 18

$z = c$
FIGURE 19

Example 2   Evaluate $\iiint_V x^2 z \, dx \, dy \, dz$ where V is the cylinder bounded by $x^2 + y^2 = 4$ and the planes $z = 2$ and $z = 6$.

SOLUTION   The cylinder is shown in Figure 20. Its projection on the xy-plane is the disk $R = \{(x, y) : x^2 + y^2 \leq 4\}$ in Figure 21, so that

$$\iiint_V x^2 z \, dx \, dy \, dz = \iint_R \int_{z=2}^{z=6} x^2 z \, dz \, dx \, dy.$$

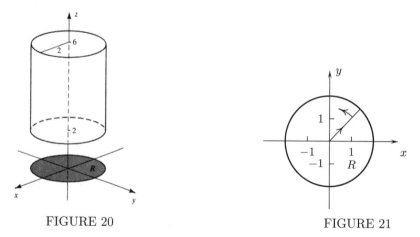

FIGURE 20

FIGURE 21

To use cylindrical coordinates, we switch the integral over R to polar coordinates with the substitutions $x = r \cos\theta$ and $dx \, dy = r \, dr \, d\theta$:

12.3 Integrals in polar, cylindrical, and spherical coordinates, 317

$$\iiint_V x^2 z\, dx\, dy\, dz = \int_{\theta=0}^{\theta=2\pi} \int_{r=0}^{r=2} \int_{z=2}^{z=6} (r\cos\theta)^2 z\, dz\, r\, dr\, d\theta \quad (11)$$
$$= \int_{\theta=0}^{\theta=2\pi} \int_{r=0}^{r=2} \int_{z=2}^{z=6} r^3 (\cos^2\theta)\, z\, dz\, dr\, d\theta.$$

Because the limits of integration on the right of **(11)** are constant and the integrand is the product of functions of $r, \theta,$ and z, we can write

$$\int_{\theta=0}^{\theta=2\pi} \int_{r=0}^{r=2} \int_{z=2}^{z=6} r^3 (\cos^2\theta)\, z\, dz\, dr\, d\theta$$
$$= \int_{\theta=0}^{\theta=2\pi} \cos^2\theta \int_{r=0}^{r=2} r^3 \int_{z=2}^{z=6} z\, dz\, dr\, d\theta \quad (12)$$
$$= \left\{\int_{\theta=0}^{\theta=2\pi} \cos^2\theta\, d\theta\right\} \left\{\int_{r=0}^{r=2} r^3\, dr\right\} \left\{\int_{z=2}^{z=6} z\, dz\right\}.$$

The θ-integration can be carried out with the trigonometric identity $\cos^2\theta = \frac{1}{2}[1 + \cos(2\theta)]$ and the substitutions $u = 2\theta, du = 2d\theta$:

$$\int_{\theta=0}^{\theta=2\pi} \cos^2\theta\, d\theta = \frac{1}{2}\int_{\theta=0}^{\theta=2\pi}[1+\cos(2\theta)]\, d\theta = \frac{1}{4}\int_{\theta=0}^{\theta=2\pi}[1+\cos(2\theta)]\,(2\,d\theta)$$
$$= \frac{1}{4}\int_{\theta=0}^{\theta=2\pi}(1+\cos u)\, du = \frac{1}{4}\Big[u + \sin u\Big]_{\theta=0}^{\theta=2\pi}$$
$$= \frac{1}{4}\Big[2\theta + \sin(2\theta)\Big]_{\theta=0}^{\theta=2\pi} = \frac{1}{4}[4\pi + \sin(4\pi)] - \frac{1}{4}[\sin(0)] = \pi.$$

The r- and z-integrals are

$$\int_{r=0}^{r=2} r^3\, dr = \Big[\tfrac{1}{4}r^4\Big]_{r=0}^{r=2} = 4$$
$$\int_{z=2}^{z=6} z\, dz = \Big[\tfrac{1}{2}z^2\Big]_{z=2}^{z=6} = \tfrac{1}{2}(36 - 4) = 16.$$

Substituting the last three values in **(12)** shows that the original integral equals $\pi(4)(16) = 64\pi$. □

The six orders of integration

With double integrals there are two variables and two possible orders of integration in iterated integrals, as is illustrated in Figures 15 and 16 for polar coordinates. With the three variables in triple integrals there are six possible orders of iterated integrations, one or more of which might be more convenient to use in a particular integral.

In the integral of Example 2, where the solid is bounded by coordinate surfaces and the limits of integration are all constant, any of the six orders is as good as the others. The order of integration in **(11)** is indicated in Figure 22. The innermost integral with respect to z is represented by the vertical line with an arrow on it. Imagine that it is drawn as z goes from 2 at the bottom of the cylinder to 6 at the top. The middle r-integration is represented by the horizontal straight arrow. Imagine that the vertical line sweeps out the shaded rectangle in Figure 22 as r goes from 0 at the center to 2 at the outside. Finally, the outermost θ-integration is represented by the curved horizontal arrow. Visualize how the shaded rectangle sweeps out the cylinder as θ goes from 0 to 2π.

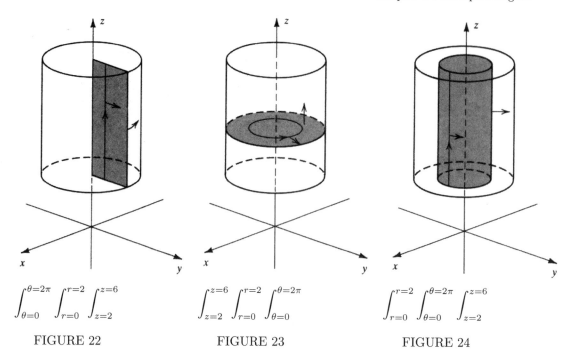

FIGURE 22 $\qquad$ FIGURE 23 $\qquad$ FIGURE 24

$$\int_{\theta=0}^{\theta=2\pi}\int_{r=0}^{r=2}\int_{z=2}^{z=6} \qquad \int_{z=2}^{z=6}\int_{r=0}^{r=2}\int_{\theta=0}^{\theta=2\pi} \qquad \int_{r=0}^{r=2}\int_{\theta=0}^{\theta=2\pi}\int_{z=2}^{z=6}$$

The integration order $\int_{z=2}^{z=6}\int_{r=0}^{r=2}\int_{\theta=0}^{\theta=2\pi}$ is represented in Figure 23. The horizontal circle with an arrow on it represents the inner θ-integration; the circle is drawn as θ goes from 0 to 2π. The horizontal arrow reprsents the middle r-integration; the circle sweeps out the disk as r goes from 0 to 2. The vertical arrow represents the z-integration; the disk sweeps out the cylinder as z goes from 0 to 6.

Similarly, Figure 24 shows the integration order $\int_{r=0}^{r=2}\int_{\theta=0}^{\theta=2\pi}\int_{z=2}^{z=6}$. The vertical line is drawn as z goes from 2 to 6; the shaded cylindrical surface is swept out by the vertical line as θ goes from 0 to 2π; and the cylindrical surface sweeps out the cylinder of integration as r goes from 0 to 2.

Question 3 What orders of integration are represented (a) in Figure 25, (b) in Figure 26, and (c) in Figure 27?

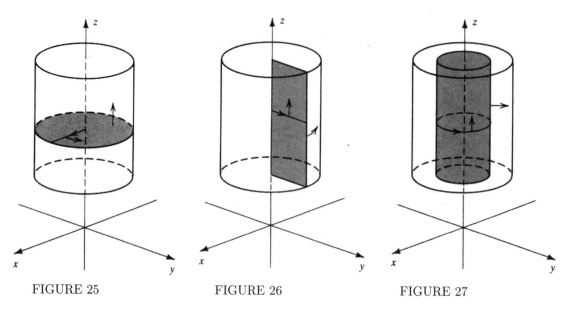

FIGURE 25 $\qquad$ FIGURE 26 $\qquad$ FIGURE 27

12.3 Integrals in polar, cylindrical, and spherical coordinates,

Example 3 Express $\iiint_V z\sqrt{x^2+y^2}\,dx\,dy\,dz$ as an iterated integral in cylindrical coordinates, where the solid of integration is the half ball $V = \{(x,y,z) : x^2+y^2+z^2 \leq 4, z \geq 0\}$.

SOLUTION It is natural to try cylindrical coordinates in the given integral because then the integrand $z\sqrt{x^2+y^2}$ becomes zr, which does not involve any square roots.

The solid V of integration is shown in Figure 28. It is bounded on the bottom by the coordinate surface $z=0$ and on the top by the hemisphere $\{(x,y,z) : x^2+y^2+z^2=4, z\geq 0\}$ of radius 2. This surface is the graph of the function $z=\sqrt{4-x^2-y^2}$, which is given in cylindrical coordinates by $z=\sqrt{4-r^2}$.

Figure 28 also shows a suitable order of integration. We integrate with respect to z from $z=0$ at the bottom of V to $z=\sqrt{4-r^2}$ at the top, as indicated by the vertical line. Next, we integrate with respect to θ from 0 to 2π since as θ goes from 0 to 2π the vertical line in Figure 28 sweeps out the shaded cylinder. Then, because the shaded cylinder sweeps out the half ball as r goes from 0 to 2, we integrate with respect to r from 0 to 2. We also make the substitution $dx\,dy = r\,dr\,d\theta$ and obtain

$$\iiint_V z\sqrt{x^2+y^2}\,dx\,dy\,dz = \int_{r=0}^{r=2}\int_{\theta=0}^{\theta=2\pi}\int_{z=0}^{z=\sqrt{4-r^2}} (zr)\,r\,dz\,d\theta\,dr \qquad (13)$$
$$= \int_{r=0}^{r=2}\int_{\theta=0}^{\theta=2\pi}\int_{z=0}^{z=\sqrt{4-r^2}} r^2 z\,dz\,d\theta\,dr. \quad \Box$$

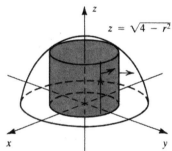

FIGURE 28

Question 4 Evaluate the iterated integral (13).

Spherical coordinates

SPHERICAL COORDINATES $[\rho, \theta, \phi]$ are often used in problems that involve the distance $\rho = \sqrt{x^2+y^2+z^2}$ from a point $P = (x,y,z)$ to the origin or that have other spherical symmetry. For a point P other than the origin, the angle θ is the same as in cylindrical coordinates, and ϕ is the angle with $0 \leq \phi \leq \pi$ from the positive z-axis to the line segment from P to the origin (Figure 29). The origin has spherical coordinates $[0, \theta, \phi]$ with arbitrary θ and arbitrary ϕ such that $0 \leq \phi \leq \pi$.

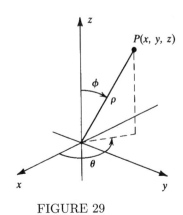

FIGURE 29

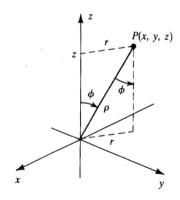

FIGURE 30

To find the xyz-coordinates of a point P with spherical coordinates $[\rho, \theta, \phi]$ with $\rho > 0$, we consider the two congruent right triangles in Figure 30 in the vertical plane containing the line segment from the origin to P. The length of the common hypotenuse of the two triangles is ρ and the length of the side opposite the angle ϕ is the distance r from P to the z-axis. Because $\sin\phi = r/\rho$ and $\cos\phi = z/\rho$, we have

$$r = \rho \sin\phi, z = \rho \cos\phi. \tag{14}$$

Then, because $x = r\cos\theta$ and $y = r\sin\theta$, we can write

$$x = \rho\cos\theta\sin\phi, y = \rho\sin\theta\sin\phi, z = \rho\cos\phi. \tag{15}$$

You can use a drawing such as in Figure 30 to recall formulas (14) and then (15) whenever you need them.

Example 4 (a) Plot the point P with spherical coordinates $[4, \frac{3}{4}\pi, \frac{1}{3}\pi]$. (You do not need to use scales on the coordinate axes or measure any angles or distances exactly.) (b) Find the rectangular coordinates of the point.

SOLUTION (a) To plot the point, we first draw the angle $\theta = \frac{3}{4}\pi$ in the xy-plane, as in Figure 31 and the angle $\phi = \frac{1}{3}\pi$ down from the positive z-axis, as in Figure 31. Then we draw a vertical line from the terminal side of the angle θ in the xy-plane up the terminal side of the angle ϕ and label the intersection P. Finally, we put the label "4" on the line segment from P to the origin.

(b) By formulas (15),

$$x = 4\cos(\tfrac{3}{4}\pi)\sin(\tfrac{1}{3}\pi) = 4(-\tfrac{1}{2}\sqrt{2})(\tfrac{1}{2}\sqrt{3}) = -\sqrt{6}$$
$$y = 4\sin(\tfrac{3}{4}\pi)\sin(\tfrac{1}{3}\pi) = 4(\tfrac{1}{2}\sqrt{2})(\tfrac{1}{2}\sqrt{3}) = \sqrt{6}$$
$$z = 4\cos(\tfrac{1}{3}\pi) = 2.$$

The point P has rectangular coordinates $(-\sqrt{6}, \sqrt{6}, 2)$. □

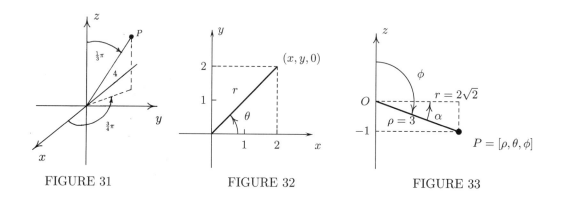

FIGURE 31   FIGURE 32   FIGURE 33

12.3 Integrals in polar, cylindrical, and spherical coordinates,

Example 5 Find spherical coordinates of the point P with rectangular coordinates $(2, 2, -1)$.

SOLUTION Because the x- and y-coordinates of P are both 2 (Figure 33), the point $(2, 2, 0)$ is on the line $y = x$ in the xy-plane and we can use $\theta = \frac{1}{4}\pi$.

The distance ρ from $(2, 2, -1)$ to the origin in space is $\sqrt{x^2 + y^2 + z^2} = \sqrt{2^2 + 2^2 + (-1)^2} = 3$. To find ϕ we calculate the distance $r = \sqrt{x^2 + y^2} = \sqrt{2^2 + 2^2} = 2\sqrt{2}$ from $(2, 2, 0)$ to the origin in the xy-plane and draw the vertical plane through the point P and the z-axis, as in Figure 33. The angle ϕ is the positive angle from the positive z-axis to the line segment OP from the origin to P. This line segment, the z-axis, and the horizontal line from P to the z-axis form a right triangle whose hypotenuse is of length $\rho = 3$, whose base is $r = 2\sqrt{2}$ units wide, and whose height is $|z| = 1$. Consequently, the angle α in Figure 33 equals $\sin^{-1}(\frac{1}{3})$ and $\phi = \alpha + \frac{1}{2}\pi = \sin^{-1}(\frac{1}{3}) + \frac{1}{2}\pi$. The spherical coordinates of P are $[3, \frac{1}{4}\pi, \sin^{-1}(\frac{1}{3}) + \frac{1}{2}\pi] \doteq [3, 0.7854, 1.9106]$. □

Notice that because θ is determined by x and y and ϕ is determined by ρ and r, we did not need to use a three-dimensional drawing to work Example 5.

Coordinate surfaces

The surface $\rho = c$ is the sphere of radius c with its center at the origin for positive c (Figure 34). It consists of a single point (the origin) if $c = 0$ and is empty if $c < 0$. The coordinate surface $\theta = c$, as with cylindrical coordinates, is a half plane with its edge on the z-axis (Figure 35).

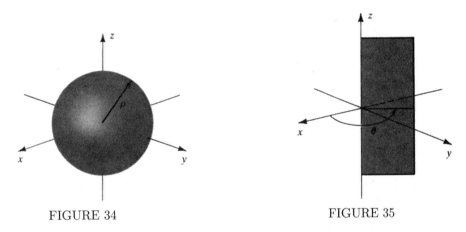

FIGURE 34   FIGURE 35

The surface $\phi = c$ is a half cone with vertex at the origin that opens upward for $0 < c < \frac{1}{2}\pi$ (Figure 36) and a half cone with vertex at the origin that opens downward for $\frac{1}{2}\pi < c < \pi$ (Figure 37).

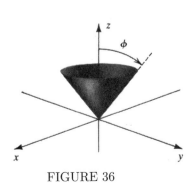
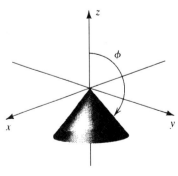

FIGURE 36   FIGURE 37

322   Chapter 12: Multiple Integrals

Question 5   What are (a) the coordinate surface $\phi = 0$, (b) the coordinate surface $\phi = \frac{1}{2}\pi$, and (c) the coordinate surface $\phi = \pi$ in spherical coordinates?

Integrals in spherical coordinates

We will now see how triple integrals can be expressed in spherical coordinates. We consider first a ball $V = \{(x,y,z) : x^2+y^2+z^2 \leq a^2\}$ with a positive radius a centered at the origin. It can be described in spherical coordinates by the inequalities $0 \leq \rho \leq a, 0 \leq \theta \leq 2\pi, 0 \leq \phi \leq \pi$. Accordingly, we partition it into subsolids by partitioning the ρ-interval $[0,a]$, the θ-interval $[0,2\pi]$, and the ϕ-interval $[0,\pi]$. We pick positive integers $J, K,$ and N and set $0 = \rho_0 < \rho_1 < \cdots < \rho_J = c, 0 = \theta_0 < \theta_1 < \cdots \theta_K = 2\pi$, and $0 = \phi_0 < \phi_1 < \cdots < \phi_N = \pi$, where the first partition is into equal subintervals of length $\Delta\rho = c/J$, the second partition is into equal subintervals of length $\Delta\theta = 2\pi/K$, and the third partition is into equal subintervals of length $\Delta\phi = \pi/N$. The corresponding coordinate surfaces partition the ball V into subsolids of the type in Figure 38, which shows the subsolid

$$V_{j,k,n} = \{[\rho, \theta, \phi] : \rho_{j-1} \leq \rho \leq \rho_j, \theta_{k-1} \leq \theta \leq \theta_k, \phi_{n-1} \leq \phi \leq \phi_n\}.$$

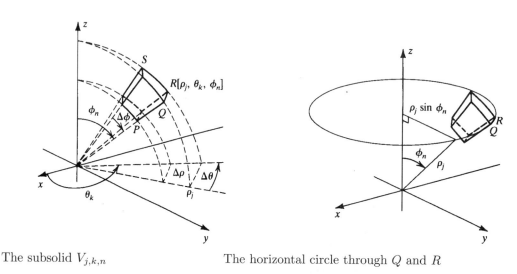

The subsolid $V_{j,k,n}$ The horizontal circle through Q and R

FIGURE 38   FIGURE 39

Notice that the subsolid in Figure 38 has four straight edges on rays from the origin and eight edges that are arcs of circles. The straight edges have the same length, which is the distance $\overline{PQ}$ from the point P to the point Q in the drawing. Since PQ extends from $\rho = \rho_{j-1}$ to $\rho = \rho_j$ on a ray from the origin,

$$\overline{PQ} = \Delta\rho. \tag{16}$$

The circular arc RS is subtended by the angle $\Delta\phi$ on a circle of radius ρ_j, so that

$$\widehat{RS} = \rho_j \Delta\phi. \tag{17}$$

The circular arc QR is on the horizontal circle shown in Figure 39. The radius of the circle is the side opposite the angle ϕ_n in the right triangle with hypoteuse ρ_j in the drawing and, therefore, equals $\rho_j \sin(\phi_n)$.

Question 6   Derive the following formula for the length $\widehat{QR}$ of the circular arc in Figures 38 and 39:

$$\widehat{QR} = \rho_j \sin(\phi_n)\, \Delta\theta. \tag{18}$$

12.3 Integrals in polar, cylindrical, and spherical coordinates,

If the subsolid in Figure 38 were a rectangular box, its volume would be the product of the lengths of its perpendicular sides. It is not a box, but the product of the lengths **(16)**, **(17)**, and **(18)** does give its approximate volume:

$$[\text{Volume of } V_{j,k,n}] \approx \overline{PQ}\, \widehat{RS}\, \widehat{QR} = (\Delta\rho)(\rho_j \Delta\phi)(\rho_j \sin(\phi_n)\, \Delta\theta) \qquad (19)$$
$$= (\rho_j)^2 \sin(\phi_n)\, \Delta\rho\, \Delta\theta\, \Delta\phi.$$

Now suppose that $f(x,y,z)$ is piecewise continuous in the ball V. Then the point with rectangular coordinates

$$P_{j,k,n} = \big(\rho_j \sin(\phi_n)\cos(\theta_k),\, \rho_j \sin(\phi_n)\sin(\theta_k),\, \rho\cos(\phi_n)\big)$$

is in the subsolid $V_{j,k,n}$ for each $j,k,$ and n. Consequently,

$$\sum_{j=1}^{J}\sum_{k=1}^{K}\sum_{n=1}^{N} f(P_{j,k,n})\,[\text{Volume of } V_{j,k,n}] \qquad (20)$$

is a Riemann sum for $\iiint_V f(x,y,z)\,dx\,dy\,dz$. Because of **(19)**, this Riemann sum is approximately equal to

$$\sum_{j=1}^{J}\sum_{k=1}^{K}\sum_{n=1}^{N} f(P_{j,k,n})(\rho_j)^2 \sin(\phi_n)\, \Delta\rho\, \Delta\theta\, \Delta\phi. \qquad (21)$$

The sum **(21)**, on the other hand, is a Riemann sum for the triple integral

$$\iiint_W f(P)\rho^2 \sin\phi\, d\rho\, d\theta\, d\phi$$

with $P = (\rho\sin\phi\cos\theta, \rho\sin\phi\sin\theta, \rho\cos\phi)$, integrated over the rectangle $W = \{(\rho,\theta,\phi) : 0 \leq \rho \leq a, 0 \leq \theta \leq 2\pi, 0 \leq \phi \leq \pi\}$ in $\rho\theta\phi$-space. It can be shown that the Riemann sum **(20)** for the original integral also tends to this integral as the number of subintervals in the partitions tend to ∞ and their widths tend to zero to establish the next result.

Theorem 2 If $f(x,y,z)$ is piecewise continuous in the sphere $V : x^2 + y^2 + z^2 \leq a^2$ with $a > 0$, then

$$\iiint_V f(x,y,z)\,dx\,dy\,dz = \int_{\rho=0}^{\rho=a}\int_{\theta=0}^{\theta=2\pi}\int_{\phi=0}^{\phi=\pi} f\, \rho^2 \sin\phi\, d\phi\, d\theta\, d\rho \qquad (22)$$

with f evaluated at the point $(\rho\sin\phi\cos\theta, \rho\sin\phi\sin\theta, \rho\cos\phi)$ with spherical coordinates $[\rho,\theta,\phi]$.

Because the limits of integration in **(22)** are constant, the integrations may be carried out in any of six possible orders. The order in **(22)** is indicated in Figure 40. The vertical half-circle with an arrow on it respesents the inner ϕ-integral. The horizontal curved arrow represents the middle θ-integration; the half-circle sweeps out the inner sphere as θ goes from 0 to 2π. The straight arrow respresents the outer ρ-integral; the inner sphere sweeps out the ball V as ρ goes from 0 to a.

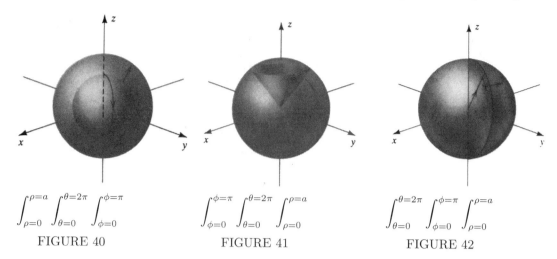

$$\int_{\rho=0}^{\rho=a}\int_{\theta=0}^{\theta=2\pi}\int_{\phi=0}^{\phi=\pi} \qquad \int_{\phi=0}^{\phi=\pi}\int_{\theta=0}^{\theta=2\pi}\int_{\rho=0}^{\rho=a} \qquad \int_{\theta=0}^{\theta=2\pi}\int_{\phi=0}^{\phi=\pi}\int_{\rho=0}^{\rho=a}$$

FIGURE 40 FIGURE 41 FIGURE 42

Question 7 (a) Explain how Figure 41 shows the order of integration given in the accompanying label. (b) Explain how Figure 42 shows the order of integration given in the label accompanying it.

Question 8 Give the orders of integration that are indicated (a) in Figure 43, (b) in Figure 44, and (c) in Figure 45.

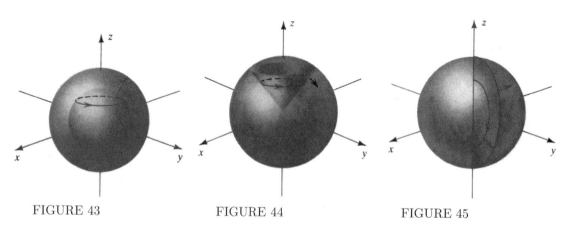

FIGURE 43 FIGURE 44 FIGURE 45

Example 6 Use spherical coordinates to evaluate $\iiint_V z\sqrt{x^2+y^2+z^2}\,dx\,dy\,dz$, where $V=\{(x,y,z): \sqrt{x^2+y^2} \leq z \leq \sqrt{1-x^2-y^2}\}$.

SOLUTION The definition of V shows that it is bounded on the bottom by the cone $z=\sqrt{x^2+y^2}$ and on the top by the hemisphere $z=\sqrt{1-x^2-y^2}$ (Figure 46). The hemisphere is the upper half of the sphere with the equation $\rho=1$ in spherical coordinates and the cone has the equation $\phi=\frac{1}{4}\pi$. As shown in Figure 47, we can integrate over V by integrating first with respect to ϕ from 0 to $\frac{1}{4}\pi$, then with respect to ρ from 0 to 1, and finally with respect to θ from 0 to 2π. We make the substitutions $z=\rho\cos\phi, \rho=\sqrt{x^2+y^2+z^2}$, and $dx\,dy\,dz=\rho^2\sin\phi\,d\rho\,d\theta\,d\phi$ to see that the given integral equals

$$\int_{\theta=0}^{\theta=2\pi}\int_{\rho=0}^{\rho=1}\int_{\phi=0}^{\phi=\pi/4}(\rho\cos\phi)(\rho)(\rho^2\sin\phi\,d\phi\,d\rho\,d\theta)$$
$$=\left(\int_0^{2\pi} d\theta\right)\left(\int_0^1 \rho^4\,d\rho\right)\left(\int_0^{\pi/4}\cos\phi\sin\phi\,d\phi\right). \qquad (23)$$

12.3 Integrals in polar, cylindrical, and spherical coordinates,

The first integral in (23) equals 2π, and the second equals $\frac{1}{5}$. The substitution $u = \sin\phi, du = \cos\phi\, d\phi$ can be used to show that the third integral equals $\left[\frac{1}{2}\sin^2\phi\right]_0^{\pi/4} = \frac{1}{4}$. Consequently, the integral (23) equals $(2\pi)(\frac{1}{5})(\frac{1}{4}) = \frac{1}{10}\pi$. □

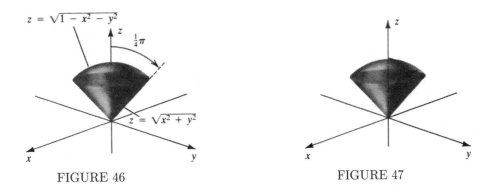

FIGURE 46   FIGURE 47

Spherical coordinates and geography

The angles θ and ϕ in spherical coordinates are closely related to LONGITUDE and LATITUDE as used to locate points on the surface of the earth (Figure 48). The great circles (circles with their centers at the center of the earth) that pass through the north and south pole are called MERIDIANS. They are identified by their angles, measured in degrees east or west from the PRIME MERIDIAN, which passes through a point in Greenwich, England. The PARALLELS are horizontal circles, indicated by their angles, measured in degrees, north or south of the equator. The longitude of a point is the longitude of the meridian passing through it and its latitude is the latitude of the parallel through it.

To relate longitude and latitude to spherical coordinates, we introduce xyz-coordinates with the origin at the center of the earth, with the positive x-axis pointing through the point where the prime meridian intersects the equation. Then we can convert longitude to the angle θ by converting degrees to radians and having longitudes west of the prime meridian be negative. Similarly, we can convert a latitude to ϕ by switching from degrees to radians and measuring the angles from the north pole instead of the equator.

FIGURE 48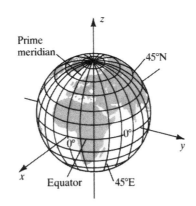

Example 7   Suppose that the scales on the coordinate axes in Figure 48 are given in miles. What are the rectangular coordinates of the town of Esquina, Argentina, which has a longitude of 60° W and a latitude of 30° S? Use 4000 miles as the radius of the earth.

SOLUTION   The town has spherical coordinate $\rho = 4000$ equal to the radius of the earth. Because 60° W is $\frac{1}{3}\pi$ radians west of the prime meridian, we can use $\theta = -\frac{1}{3}\pi$. Because 30° S is $\frac{1}{6}\pi$ radians south of the equator, $\phi = \frac{1}{2}\pi + \frac{1}{6}\pi = \frac{2}{3}\pi$.

By formulas **(15)**, Esquina has rectangular coordinates

$$x = 4000\sin(\tfrac{2}{3}\pi)\cos(-\tfrac{1}{3}\pi) = 4000(\tfrac{1}{2}\sqrt{3})(\tfrac{1}{2}) = 1000\sqrt{3} \text{ miles}$$
$$y = 4000\sin(\tfrac{2}{3}\pi)\sin(-\tfrac{1}{3}\pi) = 4000(\tfrac{1}{2}\sqrt{3})(-\tfrac{1}{2}\sqrt{3}) = -3000 \text{ miles}$$
$$z = 4000\cos(\tfrac{2}{3}\pi) = 4000(-\tfrac{1}{2}) = -2000 \text{ miles}. \ \square$$

Question 9 Find the rectangular coordinates, with the coordinate axes in Figure 46 and distances measured in miles, of **(a)** St. Petersburg, Russia (30° E, 60° N) and **(b)** Wassau, Wisconsin (90° W, 45° N).

The shortest path along the surface of the earth between two points P and Q is along an arc of the great circle formed by the intersection of the surface of the earth with the plane through the two points and the center of the earth. Great-circle distances can be found by using spherical coordinates and the dot product of vectors.

Example 8 What is the great-circle distance between the cities of St. Petersburg, Russia and Wassau, Wisconsin from Question 9?

SOLUTION We use coordinate axes as in Figure 46 and let **A** and **B** be the unit vectors pointing from the center of the earth toward St. Petersburg and Wassau, respectively. You saw in Question 9 that we can use $\theta = \tfrac{1}{6}\pi$ and $\phi = \tfrac{1}{6}\pi$ for St. Petersburg. Therefore,

$$\mathbf{A} = \langle \sin(\tfrac{1}{6}\pi)\cos(\tfrac{1}{6}\pi), \sin(\tfrac{1}{6}\pi)\sin(\tfrac{1}{6}\pi), \cos(\tfrac{1}{6}\pi) \rangle$$
$$= \langle (\tfrac{1}{2})(\tfrac{1}{2}\sqrt{3}), (\tfrac{1}{2})(\tfrac{1}{2}), \tfrac{1}{2}\sqrt{3} \rangle = \langle \tfrac{1}{4}\sqrt{3}, \tfrac{1}{4}, \tfrac{1}{2}\sqrt{3} \rangle. \quad (24)$$

Similarly, we saw that we can use $\theta = -\tfrac{1}{2}\pi$ and $\phi = \tfrac{1}{4}\pi$ for Wassau, so that

$$\mathbf{B} = \langle \sin(\tfrac{1}{4}\pi)\cos(-\tfrac{1}{2}\pi), \sin(\tfrac{1}{4}\pi)\sin(-\tfrac{1}{2}\pi), \cos(\tfrac{1}{4}\pi) \rangle$$
$$= \langle 0, -\tfrac{1}{2}\sqrt{2}, \tfrac{1}{2}\sqrt{2} \rangle. \quad (25)$$

Since **A** and **B** are unit vectors, their dot product equals the cosine of the angle ψ between them. Hence, by **(24)** and **(25)**

$$\cos\psi = \langle \tfrac{1}{4}\sqrt{3}, \tfrac{1}{4}, \tfrac{1}{2}\sqrt{3} \rangle \cdot \langle 0, -\tfrac{1}{2}\sqrt{2}, \tfrac{1}{2}\sqrt{2} \rangle$$
$$= (\tfrac{1}{4}\sqrt{3})(0) + \tfrac{1}{4}(-\tfrac{1}{2}\sqrt{2}) + (\tfrac{1}{2}\sqrt{3})(\tfrac{1}{2}\sqrt{2}) = -\tfrac{1}{8}\sqrt{2} + \tfrac{1}{4}\sqrt{6}.$$

The exact angle ψ is the inverse cosine of this amount: $\cos^{-1}(-\tfrac{1}{8}\sqrt{2} + \tfrac{1}{4}\sqrt{6}) \doteq 1.120$ radians. Since the great-circle distances is subtended by this angle on a circle of radius 4000 miles, it is approximately $4000(1.120) = 4480$ miles $\square$

Principles and procedures

- Polar coordinates are often used in double integrals when the region of integration is bounded by curves with convenient polar equations or if the integrand involves the expression $\sqrt{x^2 + y^2} = r$.
- Cylindrical coordinates are often used in triple integrals when the projection on the xy-plane of the solid of integration is bounded by curves with convenient polar equations or if the integrand involves $\sqrt{x^2 + y^2} = r$.
- Spherical coordinates are often used in triple integrals when the solid of integration is bounded by spheres ($\rho = c$), half-cones ($\phi = c$ with $0 < c < \tfrac{1}{2}\pi$ or $\tfrac{1}{2}\pi < c < \pi$), the xy-plane ($\phi = \tfrac{1}{2}\pi$), or vertical half-planes with their edges on the z-axis ($\theta = c$).
- To remember the the expression $r\,d\theta\,dr$ that replaces $dx\,dy$ when double integrals are converted from rectangular to polar coordinates, picture a typical subregion of a partition in polar coordinates. The typical subregion is cut by an angle $\Delta\theta$ from a ring with approximate radius r and small thickness Δr (Figure 49), so that its two sides are of length Δr, its inner and outer arcs are approximately $r\Delta\theta$ long, and its area is approximately $r\Delta\theta\,\Delta r$.

12.3 Integrals in polar, cylindrical, and spherical coordinates, 327

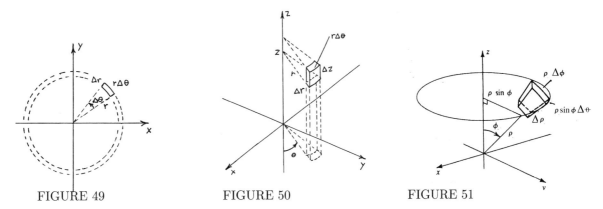

FIGURE 49   FIGURE 50   FIGURE 51

- You can remember the the expression $r\,d\theta\,dr\,dz$ that replaces $dx\,dy\,dz$ when triple integrals are converted from rectangular to cylindrical coordinates, either by referring back to the case of polar coordinates or by picturing a typical subregion of a partition in cylindrical coordinates. The typical subsolid is shown in Figure 50. If Δr is small, then its two curved sides are approximately rectangles of width $r\Delta\theta$ and height Δz and it is Δr units thick, so its volume is approximately $r\Delta\theta\,\Delta r\,\Delta z$.

- A drawing as in Figure 51 of a typical subregion of a partition in spherical coordinates can remind you of the expression $\rho^2\sin\phi\,d\rho\,d\theta\,d\phi$ that replaces $dx\,dy\,dz$ when triple integrals are converted from rectangular to spherical coordinates: If $\Delta\rho$ and $\Delta\phi$ are small, then the subsolid is close to being a rectangular box whose dimensions are $\Delta\rho$, $\rho\Delta\phi$, and $\rho\sin\phi\Delta\theta$, so the subsolid's volume is approximately $\rho^2\sin\phi\Delta r\,\Delta\theta\,\Delta\phi$.

- Some of the problems in this section use the polar equation $r = 2a\cos\theta$ of the circle of radius $|a|$ that passes through $(2a, 0)$ and the origin or the polar equation $r = 2a\sin\theta$ of the circle of radius $|a|$ that passes through $(0, 2a)$ and the origin.

- Many integrals in polar, cylindrical, and spherical coordinates involve integrals such as

    (i) $\int \sin^3\theta\,d\theta$ and $\int \sin^3\theta\cos^2\theta\,d\theta$ or (ii) $\int \sin^2\theta\,d\theta$ and $\int \sin^2\theta\cos^2\theta\,d\theta$. Recall from Section 6.3 that the integrals in (i) can be evaluated by using the Pythagorean identity $\sin^2\theta = 1 - \cos^2\theta$ to express all but one $\sin\theta$ in terms of $\cos\theta$ and then using the substitution $u = \cos\theta$. A similar technique can be used with any product of a positive, odd power of $\sin\theta$ or $\cos\theta$ with a power of the other function. Integrals of type (ii), which are products of positive, even powers of $\sin\theta$ and $\cos\theta$, can be integrated with one or more applications of the double-angle identities $\cos^2\theta = \frac{1}{2}[1 + \cos(2\theta)]$ and $\sin^2\theta = \frac{1}{2}[1 - \cos(2\theta)]$.

Tune-Up Exercises 12.3♦

A Answer provided. **O** Outline of solution provided.

T1.O Use an integral in polar coordinates to find the area inside the circle $r = 6\sin\theta$.

♦ Type 1, an integral in polar coordinates

T2.O Give the rectangular coordinates of the point with cylindrical coordinates $r = 4, \theta = \frac{7}{6}\pi, z = 10$..

♦ Type 1, converting cylindrical coordinates

T3.O Plot the point P with spherical coordinates $\rho = 5, \theta = \frac{1}{4}\pi, \phi = \frac{2}{3}\pi$. Then give its rectangular coordinates.

♦ Type 1, converting spherical coordinates

T4.O Give **(a)** cylindrical and **(b)** spherical coordinates of the point P with rectangular coordinates $(-2, 2\sqrt{3}, 4)$.

♦ Type 1, converting to cylindrical and spherical coordinates

---

♦The Tune-up Exercises and Problems are classified by type and content. The types are (1) basic, reactive; (2) basic reflective; (3) intermediate, reactive; (4) intermediate, reflective; (5) advanced, reactive; (6) advanced, reflective; and (7) advanced, theoretical.

T5.[o] Give equations in polar coordinates for the lines and curve **(a)** $x = 2$, **(b)** $y = -3$, **(c)** $x^2+y^2 = 19$, and **(d)** $y = x$ in an xy-plane.

♦ Type 1, giving equations in polar coordinates

T6.[o] Use cylindrical coordinates to describe the following surfaces and solids: **(a)** $z = x^2 + y^2$ **(b)** $x^2 + y^2 + z^2 = 4$ **(c)** $x = 2$ **(d)** $0 \leq z \leq x^2 + y^2 + 1$, **(e)** $y \leq 0$

♦ Type 1, describing surfaces given in cylindirical coordinates

T7.[o] Use spherical coordinates to describe the following surfaces and solids: **(a)** $x^2 + y^2 + z^2 \leq 9$ **(b)** $0 \leq z \leq \sqrt{x^2 + y^2}$ **(c)** $x \leq 0$ **(d)** $z < 0$.

♦ Type 1, describing surfaces given in spherical coordinates

T8[o] Calculate the volume of a sphere of radius R **(a)** by using cylindrical coordinates and **(b)** by using spherical coordinates.

♦ Type 1, a voluime by cylindrical and spherical cordinates

T9[o] Use cylindrical coordinates to evaluate $\iiint_R 4z^3 \, dx \, dy \, dz$, where V is the solid bounded by the cone $z = \sqrt{x^2 + y^2}$ and the plane $z = 1$.

♦ Type 2, evaluating an integral with cylindrical coordinates

T10[o] Use spherical coordinates to find the value of $\iiint_V z^2 \, dx \, dy \, dz$ for V the quarter-sphere $V = \{(x,y,z) : x^2 + y^2 + z^2 \leq 1, y \geq 0, z \geq 0\}$.

♦ Type 2, evaluating an integral with spherical coordinates

Problems 12.3

[A] Answer provided. [o] Outline of solution provided.

1.[o] Find the area of the region bounded by the limaçon with the polar equation $r = 5 + 4\sin\theta$.

♦ Type 1, evaluating an integral with polar coordinates

2.[o] What is the area inside one leaf of the three-leaved rose $r = 2\cos(3\theta)$?

♦ Type 2, evaluating an integral with polar coordinates

3.[o] Use polar coordinates to evaluate $\iint_V xy \, dx \, dy$, where R is bounded by the semicircle $y = \sqrt{x - x^2}$ and the x-axis.

♦ Type 2, evaluating an integral with polar coordinates

4. Find the area of the region bounded by the rays $\theta = 0$ and $\theta = \tfrac{1}{2}\pi$ and by the portion of the hyperbolic spiral $r = e^\theta$ for $0 \leq \theta \leq \tfrac{1}{2}\pi$.

♦ Type 2, evaluating an integral with polar coordinates

5.[A] What is the area of the region inside the limaçon $r = 2 - \sin\theta$?

♦ Type 2, evaluating an integral with polar coordinates

6.[o] Evaluate $\iint_R (x^2 + y^2)^{-2} \, dx \, dy$ with $R = \{(x,y) : 2 \leq x^2 + y^2 \leq 4\}$.

♦ Type 2, evaluating an integral with polar coordinates

7. Evaluate **(a**[o]**)** $\int_{x=-4}^{x=4} \int_{y=-\sqrt{16-x^2}}^{y=\sqrt{16-x^2}} e^{-x^2-y^2} \, dy \, dx$ and

(b) $\int_{x=0}^{x=3} \int_{y=x}^{y=\sqrt{18-x^2}} \sin(x^2 + y^2 + 1) \, dy \, dx$ by switching to polar coordinates.

♦ Type 2, evaluating an integral with polar coordinates

8. Calculate the volume of the ellipsoid $V = \{(x,y,z) : x^2 + y^2 + 4z^2 \leq 4\}$.

♦ Type 2, evaluating an integral with polar coordinates

9. Find the value of $\iint_R (x^2 + y^2)^{1/2} \, dx \, dy$ with R the circle $R = \{(x, .y) : (x - 1)^2 + y^2 \leq 1$.

♦ Type 2, evaluating an integral with polar coordinates

10. A solid is defined by $V = \{(x,y,z) : 0 \leq z \leq xy, x \geq 0, y \geq 0, x^2 + y^2 \leq 9\}$. Express its volume **(a)** as an iterated integral in rectangular coordinates and **(b)** as an iterated integral in polar coordinates. **(c)** Evaluate both integrals.

♦ Type 2, evaluating an integral with polar coordinates

12.3 Integrals in polar, cylindrical, and spherical coordinates,

11.º A flat disk occupies the disk $x^2 - 2x + y^2 \leq 0$ in an xy-plane with distances measure in inches, and its density at (x,y) is $5\sqrt{x^2+y^2}$ ounces per square inch. Where is its center of gravity?
♦ Type 3, evaluating an integral with polar coordinates

12. Give rectangular coordinates of the points with cylindrical coordinates (a^O) $[3, \frac{5}{6}\pi, 4]$, (b^A) $[1, \frac{3}{2}\pi, -2]$, (c) $[2, \frac{5}{4}\pi, 0]$, and (d) $[4, \pi, e]$
♦ Type 1, converting cylindrical coordinates

13. Give rectangular coordinates of the points with spherical coordinates (a^O) $[4, \frac{3}{4}\pi, \frac{1}{3}\pi]$, (b^A) $[0, 5, \pi]$, (c) $[4, \pi, \frac{5}{6}\pi]$, (d) $[5, 1, \pi]$, (e) $[4, \frac{7}{4}\pi, \frac{1}{2}\pi]$, (f) $[\sqrt{2}, \pi, 0]$, and (g) $[10, 0, \frac{1}{2}\pi]$
♦ Type 1, converting cylindrical coordinates

14. The following surfaces are given in either cylindrical or spherical coordinates. Describe the surfaces and give them equations in rectangular coordinates: (a^A) $r=2$, (b^A) $\theta = \frac{1}{4}\pi$, (c^A) $\theta = \frac{2}{3}\pi$, (d) $\theta = \frac{3}{2}\pi$, (e^A) $\rho = 2$, (f^A) $\phi = \frac{1}{4}\pi$, (g^A) $\phi = \frac{2}{3}\pi$, (h) $\phi = \frac{1}{2}\pi$, and (i) $\phi = \pi$.
♦ Type 4, describing surfaces given in cylindrical or spherical coordinates

15. Give equations in cylindrical coordinates for the following curves and surfaces: (a^A) $z = -\sqrt{x^2+y^2}$, (b^A) $y = x, x \geq 0$, (c^A) $z = -\sqrt{4-x^2-y^2}$, (d) $z = 5$, (e^A) $x^2 - 2x + y^2 = 0$, (f) $x^2 + y^2 = 9$, (g^A) $y = 3$, (h) $x = 0, y = 0, z \leq 0$, and (i) $z = \sqrt{3x^2 + 3y^2}$
♦ Type 3, giving equations for surfaces in cylindrical coordinates

16. Give equations in spherical coordinates for the lines and surfaces in Problem 15.
♦ Type 3, giving equations for surfaces in spherical coordinates

17.º Evaluate $\iiint_V z\sqrt{x^2+y^2+z^2}\, dx\, dy\, dz$ with $V = \{(x,y,z) : x^2 + y^2 \leq 16, 0 \leq z \leq 3\}$.
♦ Type 3, evaluating an integral with cylindrical coordinates

18.º Find the value of $\iiint_V \cos[(x^2+y^2+z^2)^{3/2}]\, dx\, dy\, dz$ with $V = \{(x,y,z) : 1 \leq x^2 + y^2 + z^2 \leq 36\}$.
♦ Type 3, evaluating an integral with cylindrical coordinates

19.^{A} A solid occupies the half-cylinder $V = \{(x,y,z) : x^2 + y^2 \leq 25, x \geq 0, 0 \leq z \leq 1\}$ in xyz-space with distances measured in meters. Its density at (x,y,z) is z^2 kilograms per square meter. What is its mass?
♦ Type 3, evaluating an integral with cylindrical coordinates

20.^{A} Find the value of the definite integral of $f(x,y,z) = \sqrt{z}$ over the hemisphere bounded by $z = \sqrt{4-x^2-y^2}$ and by the xy-plane.
♦ Type 3, evaluating an integral with cylindrical coordinates

21. What is the volume of the solid bounded by $z = x^2+y^2, z = 0$, and $x^2 + (y-1)^2 = 1$?
♦ Type 3, evaluating an integral with cylindrical coordinates

22. What is the integral of $(x^2+y^2)^{-1/2}$ over the solid that is bounded by $z = \sqrt{x^2+y^2}, x^2+y^2 = 4$, and $z = b$, where b is a constant > 2?
♦ Type 3, evaluating an integral with cylindrical coordinates

23. Evaluate $\iiint_V z\sqrt{x^2+y^2}\, dx\, dy\, dz$, where $V = \{(x,y,z) : x^2 + y^2 \leq z \leq 4\}$.
♦ Type 3, evaluating an integral with cylindrical coordinates

24.^{A} Express $\int_{-1}^{1}\int_{1}^{\sqrt{2-x^2}} \frac{x+1}{y+1}\, dy\, dx$ as an iterated integral in polar coordinates.
♦ Type 3, expressing an integral in polar coordinates

25.^{A} Find the value of $\int_{3\pi/4}^{4\pi/3}\int_{0}^{-5\sec\theta} r^3 \sin^2\theta\, dr\, d\theta$ by first switching it from polar to rectangular coordinates.
♦ Type 3, evaluating an integral by converting it from polar coordinates

26. Find the average value of $(x^2+y^2)^{-1/2}$ in $R = \{(x,y) : 0 \leq y \leq \sqrt{4x-x^2}, 2 \leq x \leq 4\}$.
♦ Type 3, finding an average value with cylindrical coordinates

27. Give cylindrical coordinates of the points with rectangular coordinates (a^O) $(2,-2,3)$, (b^A) $(-\sqrt{3},-1,-5)$, (c) $(0,-7,8)$, and (d) $(-3,-3,-3)$
♦ Type 3, converting rectangular to cylindrical coordinates

28. Give spherical coordinates of the points with rectangular coordinates (**a**^{O}) $(-3,-3,0)$, (**b**^{O}) $(\sqrt{2}, \sqrt{2}, -2)$, (**c**^{A}) $(-2, -2\sqrt{3}, 0)$, (**d**^{A}) $(0, 0, -100)$, (**e**) $(-5, 0, 5)$, (**f**^{A}) $(-\sqrt{3}, 1, -2\sqrt{3})$, (**g**) $(-1, -1, -\sqrt{2})$, (**h**) $(\sqrt{1.5}, \sqrt{1.5}, 1)$, (**i**) $(0, 0, 4)$, and (**j**) $(0, -5, 0)$.

♦ Type 3, converting rectangular to spherical coordinates

29.^{A} Find the volume of the solid that is inside the sphere $x^2 + y^2 + z^2 = 4$ and inside the cylinder $(x-1)^2 + y^2 = 1$.

♦ Type 3, evaluating an integral with cylindrical coordinates

30. What is the volume of the solid that is bounded by the circular paraboloid $z = x^2 + y^2$ and the plane $z = 2y$?

♦ Type 3, evaluating an integral with cylindrical coordinates

31. The integrals (**a**^{A}) $\int_0^{\pi/4} \int_{\sec\theta}^{2\cos\theta} \dfrac{r^2}{1 + r\sin\theta}\, dr\, d\theta$ and (**b**) $\int_{\pi/4}^{3\pi/4} \int_0^{4\csc\theta} r^5 \sin^2\theta\, dr\, d\theta$ are given in polar coordinates. Rewrite them as iterated integrals in rectangular coordinates.

♦ Type 5, converting integrals from polar coordinates

32. Find the value of the integral of $f(x, y, z) = z$ over the solid bounded by the half-cone $z = \sqrt{2x^2 + 2y^2}$ and the half-hyperbola $z = \sqrt{x^2 + y^2 + 1}$.

♦ Type 3, evaluating an integral with cylindrical coordinates

33. Evaluate $\iiint_V z\, dx\, dy\, dz$ with $V = \{(x, y, z) : 3 \le z \le \sqrt{25 - x^2 - y^2}\}$.

♦ Type 3, evaluating an integral with spherical coordinates

34. Calculate the integral of e^{-z} over the sphere with radius 6 centered at the origin.

♦ Type 3, evaluating an integral with cylindrical coordinates

35. What is the value of $\iiint_V z^2\, dx\, dy\, dz$ where V is the hemisphere bounded by $z = \sqrt{1 - x^2 - y^2}$, and $z = 0$?

♦ Type 3, evaluating an integral with cylindrical coordinates

36. What is the volume of the solid bounded by $z = x^2 + y^2$ and $x^2 + y^2 + z^2 = R^2$?

♦ Type 3, evaluating an integral with spherical coordinates

37.^{A} A solid is bounded by $z^2 = x^2 + y^2 + 1$ and $x^2 + y^2 = 1$. What is its volume?

♦ Type 3, evaluating an integral with cylindrical coordinates

38. A solid that occupies $\{(x, y, z) : \sqrt{x^2 + y^2} \le z \le 1\}$ in xyz-space with distances measured in feet has density $z\sqrt{x^2 + y^2 + z^2}$ pounds per square foot at (x, y, z). How much does it weigh?

♦ Type 3, evaluating an integral with spherical coordinates

39. A ball with its center at the origin in xyz-space with distances measured in meters has density $x^2 + y^2$ kilograms per cubic meter. Its mass is 2500π kilograms. What is its radius?

♦ Type 3, evaluating an integral with cylindrical coordinates

40. The solid bounded by the cylinder $x^2 + y^2 = 25$ and by the planes $z = 0$ and $x + y + z = k$ with $k > 0$ has volume 200π. What is the constant k?

♦ Type 3, evaluating an integral with cylindrical coordinates

41.^{A} What is the volume of the solid bounded by $z = e^{x^2 + y^2}$, $z = 0$, and $x^2 + y^2 = a^2$ with $a > 0$?

♦ Type 3, evaluating an integral with cylindrical coordinates

42. (**a**^{A}) $\int_0^{\pi/2} \int_0^2 \int_{-r^2}^4 \dfrac{zr^3}{r\sin\theta + 4}\, dz\, dr\, d\theta$ and (**b**) $\int_0^{2\pi} \int_0^3 \int_{-r}^{\sqrt{9-r^2}} e^z r^6 \cos^4\theta\, dz\, dr\, d\theta$ are given in cylindrical coordinates. Express them as iterated integrals in rectangular coordinates. Do not evaluate them.

♦ Type 3, converting integrals from cylindrical to rectangular coordinates

43. (**a**^{A}) $\int_0^\pi \int_{3\pi/4}^\pi \int_0^1 \rho^5 \cos\theta \sin^2\phi\, d\rho\, d\phi\, d\theta$ and (**b**) $\int_{\pi/2}^{3\pi/2} \int_{\pi/2}^\pi \int_0^5 \dfrac{\rho^4 \sin^2\theta}{\rho^2 + 1}\, d\rho\, d\phi\, d\theta$ are given in spherical coordinates. Express them as iterated integrals in rectangular coordinates. Do not evaluate them.

♦ Type 3, converting integrals from cylindrical to rectangular coordinates

12.3 Integrals in polar, cylindrical, and spherical coordinates,

44. The integral $\int_0^\pi \int_0^5 \int_0^{\sqrt{25-z^2}} r^2 \sin\theta \, dz \, dr \, d\theta$ is in cylindrical coordinates. Express it as iterated integrals in spherical coordinates. Do not evaluate them.

♦ Type 3, converting an integral from cylindrical to spherical coordinates

45. Evaluate **(a)** $\iiint_V z\sqrt{x^2+y^2} \, dx \, dy \, dz$ with $V = \{(x,y,z) : (x-1)^2 + y^2 \leq 1, 0 \leq z \leq 4\}$ and **(b)** $\int_{-2}^{2} \int_{-\sqrt{4-x^2}}^{\sqrt{4-x^2}} \int_0^{\sqrt{4-x^2-y^2}} z(x^2+y^2+z^2)^{3/2} \, dz \, dy \, dx$.

♦ Type 3, evaluating an integral with cylindrical or spherical coordinates

46.^A The eighth-sphere $V = \{(x,y,z) : x^2+y^2+z^2 \leq 4, x \geq 0, y \leq 0, z \leq 0\}$ has density xyz at (x,y,z). What is its center of gravity?

♦ Type 4, finding a center of gravity with cylindrical or spherical coordinates

47. The table below gives longitudes and latitudes of ten cities. Use these numbers to find the rectangular coordinates of **(a^O)** Tambu, Indonesia, **(b^A)** Davyhurst, Australia, and **(c)** Chaingchunmiao, China. Use xyz-axes with origin at the center of the earth, the positive x-axis passing through the prime meridian, the positive z-axis going through the North Pole, and with distances measured in miles. Use 4000 miles as the radius of the earth.

♦ Type 3, finding rectangular coordinates from longitude and latitude

Bennett, British Columbia	135° W, 60° N	St. Petersburg, Russia	30° E, 60° N
Calais, France	0° E, 45° N	Tambu, Indonesia	120° E, 0° S
Chaingchunmiao, China	90° E, 45° N	Timimoun, Algeria	0° W, 30° N
Davyhurst, Australia	120° E, 30° S	Wausau, Wisconsin	90° W, 45° N
Esquina, Argentina	60° W, 30° S	Yartsevo, Russia	90° W, 60° N

48. Use the longitudes and latitudes in the table above to find the approximate great-circle distances **(a^O)** between Timimoun, Algeria and Bennett, British Columbia, **(b^A)** between Chiangchunmiao, China and Esquina, Argentina, **(c)** between Davyhurst, Australia and Calais, France, and **(d)** Yartsevo, Russia and Tambu, Indonesia. Use 4000 miles as the radius of the earth.

♦ Type 3, finding great-circle distances

49. (CAS) Evaluate $\iint_R (x^2+y^2) \, dx \, dy$ where R is the region inside $x^2 - 4x + y^2 = 1$ and outside $x^2 - 2x + y^2 = 0$

♦ Type 3, Evaluating an integral with cylindrical coordinates

50. Evaluate $\iiint_V \sqrt{x^2+y^2+z^2} \, dx \, dy \, dz$, where V is the region inside $x^2+y^2+z^2 = 4$ and above $z = -\sqrt{3x^2+3y^2}$

♦ Type 3, Evaluating an integral with spherical coordinates

51. The solid V is defined by $V = \{(x,y,z) : 0 \leq z \leq 4x^2 + 4y^2, x \leq y \leq \sqrt{1-x^2}, 0 \leq x \leq 2^{-1/2}\}$. Find the value of $\iiint_V \sin z \, dx \, dy \, dz$.

♦ Type 3, Evaluating an integral with cylindrical or spherical coordinates

Section 12.4

Other changes of variables: Jacobians

OVERVIEW: *In Section 12.3 we saw how double integrals in rectangular coordinates could be changed to polar coordinates and how triple integrals in rectangular coordinates could be changed to cylindrical and spherical coordinates. Here we will study other changes of variables in multiple integrals, with an emphasis on* AFFINE TRANSFORMATIONS *of variables.*

Topics:

- **Affine transformations in the plane**
- **Changing variables in double integrals: Jacobians**
- **Changing variables in triple integrals**

Affine transformations in the plane

An AFFINE TRANSFORMATION or AFFINE MAPPING T from a uv-coordinate plane to an xy-coordinate plane is a pair of linear functions

$$T : \begin{cases} x(u,v) = x_0 + a_1 u + b_1 v \\ y(u,v) = y_0 + a_2 u + b_2 v \end{cases} \tag{1}$$

that give xy-coordinates as functions of the uv-coordinates. Here x_0, y_0, a_1, a_2, b_1, and b_2 are constants. For any point (u,v) in the uv-plane, we call the point $(x,y) = \big(x(u,v), y(u,v)\big)$ given by **(1)** the IMAGE of (u,v) under the transformation T and we write $T : (u,v) \to \big(x(u,v), y(u,v)\big)$ (Figure 1).

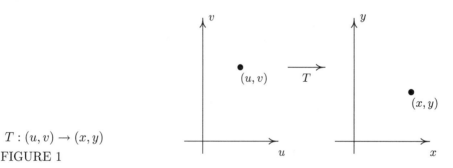

$T : (u,v) \to (x,y)$
FIGURE 1

When we refer to the IMAGE of a set of points in the uv-plane, we mean the set of the images of those points. Because the two functions in **(1)** are linear, the images of parallel lines in the uv-plane under the transformation T are parallel lines in the xy-plane.

Example 1 Find the image of the square $\tilde{R} = \{(u,v) : 0 \le u \le 1, 0 \le v \le 1\}$ in Figure 2 under the affine transformation

$$T_1 : \begin{cases} x(u,v) = 7u + 3v \\ y(u,v) = u + 4v. \end{cases} \tag{2}$$

SOLUTION Because T_1 maps parallel lines into parallel lines, it maps the square in Figure 2 into a parallelogram which we can draw by finding the images of the three adjacent corners $\tilde{O} = (0,0), \tilde{A} = (1,0)$, and $\tilde{B} = (0,1)$ of the square. By formulas **(2)**,

---

♦ SUGGESTIONS TO INSTRUCTORS: The material in this section can be divided into two parts, which can be covered in two or more class meetings.
Part 1 (Affine transformations in the xy-plane and in double integrals; Tune-up Exercises T1–T3, Problems 1–10, 12, 13, 16, 19, 20, 21): Have students begin work on Questions 1–4 before class. Problems 1, 2, 4, 6, 7, 12 are good for in-class discussions.
Part 2 (Nonaffine transformations in double integrals and affine transformations in triple integrals; Tune-up Exercises T4, T5; Problems 11, 14, 15, 17, 18, 22–28): Have students begin work on Question 5 before class. Problems 15, 22, 23 are good for in-class discussions.

12.4 Other changes of variables: Jacobians

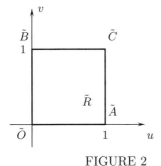

FIGURE 2

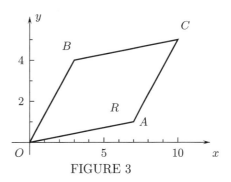

FIGURE 3

By formulas **(2)**,

$$T_1 : (0,0) \to \big(x(0,0), y(0,0)\big) = (7(0) + 3(0), 1(0) + 4(0)) = (0,0)$$
$$T_1 : (1,0) \to \big(x(1,0), y(1,0)\big) = (7(1) + 3(0), 1(1) + 4(0)) = (7,1)$$
$$T_1 : (0,1) \to \big(x(0,1), y(0,1)\big) = (7(0) + 3(1), 1(0) + 4(1)) = (3,4).$$

The images of $\tilde{O} = (0,0), \tilde{A} = (1,0)$, and $\tilde{B} = (0,1)$ are $O = (0,0), A = (7,1)$, and $B = (3,4)$, and the image of $\tilde{R}$ in Figure 2 is the parallelogram R in Figure 3. □

Question 1 (a) Use formulas **(2)** to find the image C under the trasformation **(2)** of the fourth corner $\tilde{C} = (1,1)$ of the square in Figure 2. (b) Use vectors to show that the quadrilateral R in Figure 3, whose vertices were found in Example 1 and part (a) of this Question, is a parallelogram.

How areas change under affine transformations

In order to see how to make affine changes of variables in double integrals, we need to know how affine transformations affect areas. We first use equations **(1)** to give a formula for the position vector $\mathbf{R}(u,v)$ of the image of the point (u,v) under the transformation T. We obtain

$$\begin{aligned}
\mathbf{R}(u,v) &= \big(x(u,v), y(u,v)\big) = \langle x_0 + a_1 u + b_1 v, y_0 + a_2 u + b_2 v \rangle \\
&= \langle x_0, y_0 \rangle + \langle a_1 u, a_2 u \rangle + \langle b_1 v, b_2 v \rangle \\
&= \langle x_0, y_0 \rangle + u \langle a_1, a_2 \rangle + v \langle b_1, b_2 \rangle \\
&= \mathbf{R}_0 + u\mathbf{A} + v\mathbf{B}.
\end{aligned} \qquad (3)$$

Here $\mathbf{R}_0 = \langle x_0, y_0 \rangle$ is the position vector of the point (x_0, y_0) and $\mathbf{A} = \langle a_1, a_2 \rangle$ and $\mathbf{B} \langle b_1, b_2 \rangle$ are constant vectors.

Question 2 Consider the rectangle $\tilde{R} : 0 \leq x \leq h, 0 \leq y \leq k$ in Figure 4, where h and k are arbitrary positive numbers. Find the position vectors of the image under T of (a) its lower left corner $(0,0)$, (b) its lower right corner $(h,0)$, and (c) its upper right corner $(0,k)$.

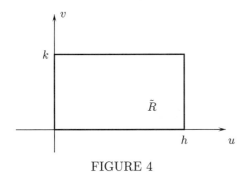

FIGURE 4

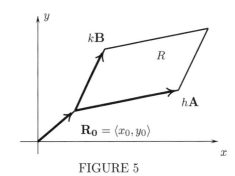

FIGURE 5

The results of Question 2 show that the image of the rectangle $\tilde{R}$ under the transformation T is the parallelogram R in Figure 5 with one vertex at the point (x_0, y_0) and with two sides formed by the vectors $h\mathbf{A}$ and $k\mathbf{B}$. We need a formula for the area of this parallogram.

We saw in Section 10.3 that if the sides of a parallelogram are formed by vectors $\mathbf{A}$ and $\mathbf{B}$ in xyz-space, then the area of the parallelogram is the absolute value $|\mathbf{A} \times \mathbf{B}|$ of the cross product of the vectors. To apply this result to Figure 5, we convert $\mathbf{A} = \langle a_1, a_2 \rangle$ and $\mathbf{B} = \langle b_1, b_2 \rangle$ to vectors in space by giving them zero z-components. Then $h\mathbf{A} = \langle ha_1, ha_2, 0 \rangle$ and $k\mathbf{B} = \langle kb_1, kb_2, 0 \rangle$, and we obtain

$$(h\mathbf{A}) \times (k\mathbf{B}) = \begin{vmatrix} \mathbf{i} & \mathbf{j} & \mathbf{k} \\ ha_1 & ha_2 & 0 \\ kb_1 & kb_2 & 0 \end{vmatrix} = \mathbf{i} \begin{vmatrix} ha_2 & 0 \\ kb_2 & 0 \end{vmatrix} - \mathbf{j} \begin{vmatrix} ha_1 & 0 \\ kb_1 & 0 \end{vmatrix} + \mathbf{k} \begin{vmatrix} ha_1 & ha_2 \\ kb_1 & kb_2 \end{vmatrix} \quad (4)$$

$$= (0)\mathbf{i} - (0)\mathbf{j} + hk[a_1 b_2 - a_2 b_1]\mathbf{k} = hk \begin{vmatrix} a_1 & b_1 \\ a_2 & b_2 \end{vmatrix} \mathbf{k}.$$

The determinant $\begin{vmatrix} a_1 & b_1 \\ a_2 & b_2 \end{vmatrix}$ on the right of **(4)** is called the JACOBIAN of the affine transformation

$$T : \begin{cases} x(u,v) = x_0 + a_1 u + b_1 v \\ y(u,v) = y_0 + a_2 u + b_2 v. \end{cases} \quad (1)$$

The Jacobian is denoted by the symbol $\dfrac{\partial(x,y)}{\partial(u,v)}$ because it can be determined from formulas **(1)** by taking partial derivatives:

$$\frac{\partial(x,y)}{\partial(u,v)} = \begin{vmatrix} \dfrac{\partial x}{\partial u} & \dfrac{\partial x}{\partial v} \\ \dfrac{\partial y}{\partial u} & \dfrac{\partial y}{\partial v} \end{vmatrix} = \begin{vmatrix} \dfrac{\partial}{\partial u}(x_0 + a_1 u + b_1 v) & \dfrac{\partial}{\partial v}(x_0 + a_1 u + b_1 v) \\ \dfrac{\partial}{\partial u}(y_0 + a_2 u + b_2 v) & \dfrac{\partial}{\partial v}(y_0 + a_2 u + b_2 v) \end{vmatrix} = \begin{vmatrix} a_1 & b_1 \\ a_2 & b_2 \end{vmatrix}.$$

With this notation for the Jacobian, **(4)** gives $(h\mathbf{A}) \times (k\mathbf{B}) = (hk)\dfrac{\partial(x,y)}{\partial(u,v)} \mathbf{k}$. Then, since $\mathbf{k}$ is a unit vector and the area of the rectangle $\tilde{R}$ in Figure 4 is hk, we obtain

$$[\text{Area of } R] = |(h\mathbf{A}) \times (k\mathbf{B})| = \left| \frac{\partial(x,y)}{\partial(u,v)} \right| (hk)$$

$$= \left| \frac{\partial(x,y)}{\partial(u,v)} \right| [\text{Area of } \tilde{R}].$$

This formula can be applied to any rectangle $\tilde{R}$ and then can be used to derive the following result by approximating other regions $\tilde{R}$ by collections of rectangles.

Theorem 1 *If $\tilde{R}$ is a bounded region with a piecewise-smooth boundary in the uv-plane and R is its image in the xy-plane under the affine transformation*

$$T : \begin{cases} x(u,v) = x_0 + a_1 u + b_1 v \\ y(u,v) = y_0 + a_2 u + b_2 v \end{cases}$$

Then

$$[\text{Area of } R] = \left| \frac{\partial(x,y)}{\partial(u,v)} \right| [\text{Area of } \tilde{R}]. \quad (5)$$

where $\dfrac{\partial(x,y)}{\partial(u,v)} = \begin{vmatrix} a_1 & b_1 \\ a_2 & b_2 \end{vmatrix}$ is the Jacobian of the transformation.

12.4 Other changes of variables: Jacobians

Example 2 What is the area of the parallelogram R in Figure 3?

SOLUTION The parallelogram R in Figure 3 is the image of the square $\tilde{R}: 0 \leq x \leq 1, 0 \leq y \leq 1$ in Figure 2 under the affine transformation (2), whose Jacobian is

$$\frac{\partial(x,y)}{\partial(u,v)} = \begin{vmatrix} \frac{\partial x}{\partial u} & \frac{\partial x}{\partial v} \\ \frac{\partial y}{\partial u} & \frac{\partial y}{\partial v} \end{vmatrix} = \begin{vmatrix} \frac{\partial}{\partial u}(7u+3v) & \frac{\partial}{\partial v}(7u+3v) \\ \frac{\partial}{\partial u}(u+4v) & \frac{\partial}{\partial v}(u+4v) \end{vmatrix} = \begin{vmatrix} 7 & 3 \\ 1 & 4 \end{vmatrix} = 7(4) - 3(1) = 25.$$

Since the square $\tilde{R}$ has area 1, the area of its image R is $|25|(1) = 25$, by Rule 1. $\square$

If the Jacobian of an affine transformation (1) is zero, as in the next Example, then it maps the entire uv-plane into a line in the xy-plane and all rectangles in the uv-plane are mapped into line segments with zero area.

Example 3 (a) What is the Jacobian of the affine transformation $T: \begin{cases} x = u + v \\ y = 2u + 2v \end{cases}$?

(b) What is the image of the uv-plane under this transformation?

SOLUTION (a) The Jacobian of T is

$$\frac{\partial(x,y)}{\partial(u,v)} = \begin{vmatrix} \frac{\partial}{\partial u}(u+v) & \frac{\partial}{\partial v}(u+v) \\ \frac{\partial}{\partial u}(2u+2v) & \frac{\partial}{\partial v}(2u+2v) \end{vmatrix} = \begin{vmatrix} 1 & 1 \\ 2 & 2 \end{vmatrix} = 1(2) - 2(1) = 0.$$

(b) The equations for T imply that $y = 2x$ for all (u,v), so the image of the uv-plane is the line $y = 2x$ in the xy-plane. $\square$

Inverse transformations

If the affine transformation (1) has a nonzero determinant, then its INVERSE T^{-1} maps the xy-plane to the uv-plane. To find formulas for it, we first rewrite (1) in the form

$$\begin{cases} a_1 u + b_1 v = x - x_0 \\ a_2 u + b_2 v = y - y_0. \end{cases} \tag{6}$$

These can be solved by using CRAMER'S RULE, which gives u and v as the ratios of determinants:

$$u = \frac{\begin{vmatrix} x - x_0 & b_1 \\ y - y_0 & b_2 \end{vmatrix}}{\begin{vmatrix} a_1 & b_1 \\ a_2 & b_2 \end{vmatrix}}, \quad v = \frac{\begin{vmatrix} a_1 & x - x_0 \\ a_2 & y - y_0 \end{vmatrix}}{\begin{vmatrix} a_1 & b_1 \\ a_2 & b_2 \end{vmatrix}}. \tag{7}$$

Example 4 Find the inverse of the transformation T_1, given by (2).

SOLUTION The inverse transformation is obtained by solving equations (2) to obtain formulas for u and v in terms of x and y. By Cramer's Rule, T_1^{-1} is given by

$$u = \frac{\begin{vmatrix} x & 3 \\ y & 4 \end{vmatrix}}{\begin{vmatrix} 7 & 3 \\ 1 & 4 \end{vmatrix}} = \frac{4x - 3y}{7(4) - 3(1)} = \tfrac{4}{25}x - \tfrac{3}{25}y$$

$$v = \frac{\begin{vmatrix} 7 & x \\ 1 & y \end{vmatrix}}{\begin{vmatrix} 7 & 3 \\ 1 & 4 \end{vmatrix}} = \frac{-x + 7y}{7(4) - 3(1)} = -\tfrac{1}{25}x + \tfrac{7}{25}y. \square \tag{8}$$

Question 3 Verify directly that T^{-1}, given by **(8)**, transforms the parallelogram R in Figure 3 into the square $\tilde{R}$ in Figure 2.

Question 4 (a) What is the Jacobian of the inverse transformation **(8)**? (b) How are the Jacobians of T_1 and T_1^{-1} related? (c) How does this fact relate to Theorem 1?

The relationship in Question 4 between the Jacobian of an affine transformation and the Jacobian of its inverse holds in general. Let D denote the nonzero Jacobian $\dfrac{\partial(x,y)}{\partial(u,v)}$ of the affine transformation **(1)**. Then equations **(7)** imply that the Jacobian of the inverse transformation is

$$\frac{\partial(u,v)}{\partial(x,y)} = \begin{vmatrix} \dfrac{b_2}{D} & -\dfrac{b_1}{D} \\ -\dfrac{a_2}{D} & \dfrac{a_1}{D} \end{vmatrix} = \frac{a_1 b_2 - b_1 a_2}{D^2} = \frac{1}{D} = \frac{1}{\dfrac{\partial(x,y)}{\partial(u,v)}}.$$

Thus, the Jacobian of the inverse transformation is the reciprocal of the Jacobian of the original transformation in all cases of affine transformations with nonzero Jacobians.

Affine changes of variables in double integrals

Given an integral $\iint_R f(x,y)\,dx\,dy$ of a piecewise-continuous function $f(x,y)$ over the bounded region R with a piecewise-smooth boundary in Figure 7, we suppose that R is the image of the region $\tilde{R}$ in the uv-plane of Figure 6 under an affine transformation $T : x = x(u,v), y = y(u,v)$ with nonzero Jacobian. We would like to change variables in the integral to u and v to obtain an integral over $\tilde{R}$.

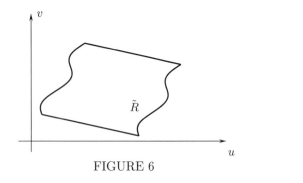
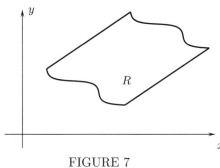

FIGURE 6 FIGURE 7

We form a partition $\tilde{R}$ into N subregions $\tilde{R}_1, \tilde{R}_2, \ldots, \tilde{R}_N$ as shown in Figure 8 with $N = 9$. The images $R_1, R_2, \ldots, R_N$ of the subregions under T form a partition of R (Figure 9), and by Theorem 1, we have for each j

$$[\text{Area of } R_j] = \left|\frac{\partial(x,y)}{\partial(u,v)}\right| [\text{Area of } \tilde{R}_j].$$

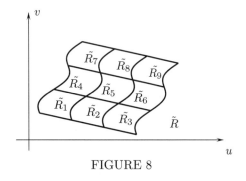
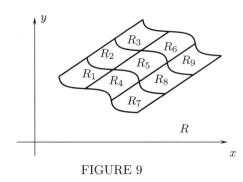

FIGURE 8 FIGURE 9

12.4 Other changes of variables: Jacobians

We pick a point (u_j, v_j) in $\tilde{R}_j$ for each j and let $(x_j, y_j) = \big(x(u_j, v_j), y(u_j, v_j)\big)$ be its image in R_j under T. Then the corresponding Riemann sum for $\iint_R f(x,y)\,dx\,dy$ is

$$\sum_{j=1}^N f(x_j, y_j)\,[\text{Area of } R_j] = \sum_{j=1}^N f\big(x(u_j, v_j), y(u_j, v_j)\big) \left|\frac{\partial(x,y)}{\partial(u,v)}\right| [\text{Area of } \tilde{R}_j]. \tag{9}$$

The sum on the right of (9) is a Riemann sum for $\iint_{\tilde{R}} f\big(x(u,v), y(u,v)\big) \left|\dfrac{\partial(x,y)}{\partial(u,v)}\right|\,du\,dv$. Each Riemann sum tends to the corresponding integral as the number of subregions in the partitions tends to ∞ and their diameters tend to 0, so we have the following result.

Theorem 2 *Suppose that R is a bounded region with piecewise-smooth bounary in the xy-plane that is the image of the region $\tilde{R}$ under an affine transformation $x = x(u,v), y = y(u,v)$ with nonzero Jacobian. Then for any $f(x,y)$ that is piecewise-continuous in R*

$$\iint_R f(x,y)\,dx\,dy = \iint_{\tilde{R}} f\big(x(u,v), y(u,v)\big) \left|\frac{\partial(x,y)}{\partial(u,v)}\right|\,du\,dv. \tag{10}$$

Formula (10) is expressed in part by the symbolic equation

$$dx\,dy = \left|\frac{\partial(x,y)}{\partial(u,v)}\right|\,du\,dv. \tag{11}$$

To remember this formula, notice that the u and v in the denominator on the right seem to cancel with the u and v in the numerator.

Example 5 Evaluate $\iint_R (x - 7y)\,dx\,dy$ with R the parallelogram in Figure 3 by making the affine change of variables $T_1: x = 7u + 3v, y = u + 4v$, given by (2).

SOLUTION We saw in the solution of Example 2 that the Jacobian of the transformation T_1 is 25, and that R is the image under the transformation of the rectangle $\tilde{R} = \{(u,v): 0 \leq u \leq 1, 0 \leq v \leq 1\}$ in Figure 2. Therefore,

$$\iint_R (x - 7y)\,dx\,dy = \iint_{\tilde{R}} [(7u + 3v) - 7(u + 4v)] \left|\frac{\partial(x,y)}{\partial(u,v)}\right|\,du\,dv$$

$$= \int_{u=0}^{u=1} \int_{v=0}^{v=1} (-25v)(25)\,dv\,du$$

$$= -625 \left(\int_{u=0}^{u=1} du\right)\left(\int_{v=0}^{v=1} v\,dv\right)$$

$$= -625 \Big[u\Big]_{u=0}^{u=1} \Big[\tfrac{1}{2}v^2\Big]_{v=0}^{v=1}\,du = -625(1)(\tfrac{1}{2}) = -\tfrac{625}{2}.\ \square$$

Example 6 Evaluate $\iint_R e^{3x}\,dx\,dy$ with R the parallellogram bounded by the lines $y = 2x, y = 2x - 6, x = 2y$, and $x = 2y - 2$.

SOLUTION The parallelogram R is shown in Figure 11. We rewrite the equations for its sides in the form $2x - y = 0, 2x - y = 6, x - 2y = 0$, and $x - 2y = -2$. Then we use the two formulas on the left of these equations to define the affine transformation

$$\begin{cases} u = 2x - y \\ v = x - 2y. \end{cases} \tag{12}$$

With this definition, the sides of R are the level curves $u = 0, u = 6, v = 0$, and $v = -2$, and the image of R under this transformation is the rectangle $\tilde{R} = \{(u,v) : 0 \leq u \leq 6, -2 \leq v \leq 0\}$ in Figure 10.

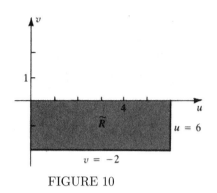
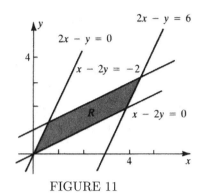

FIGURE 10   FIGURE 11

To change variables in the integral, we need the inverse of the transformation (**12**). Cramer's Rule gives

$$x = \frac{\begin{vmatrix} u & -1 \\ v & -2 \end{vmatrix}}{\begin{vmatrix} 2 & -1 \\ 1 & -2 \end{vmatrix}} = \frac{-2u + v}{2(-2) - (-1)(1)} = \tfrac{2}{3}u - \tfrac{1}{3}v$$

$$y = \frac{\begin{vmatrix} 2 & u \\ 1 & v \end{vmatrix}}{\begin{vmatrix} 2 & -1 \\ 1 & -2 \end{vmatrix}} = \frac{-u + 2v}{2(-2) - (-1)(1)} = \tfrac{1}{3}u - \tfrac{2}{3}v.$$

These equations show that $e^{3x} = e^{2u-v}$ and

$$\frac{\partial(x,y)}{\partial(u,v)} = \begin{vmatrix} x_u & x_v \\ y_u & y_v \end{vmatrix} = \begin{vmatrix} \tfrac{2}{3} & -\tfrac{1}{3} \\ \tfrac{1}{3} & -\tfrac{2}{3} \end{vmatrix} = \tfrac{2}{3}(-\tfrac{2}{3}) - \tfrac{1}{3}(-\tfrac{1}{3}) = -\tfrac{1}{3}.$$

(We could also find this Jacobean as the reciprocal of the Jacobian $\dfrac{\partial(u,v)}{\partial(x,y)}$, which we could find from (**12**).) By Theorem 2,

$$\iint_R e^{3x}\, dx\, dy = \iint_{\tilde{R}} e^{2u-v} \left| \frac{\partial(x,y)}{\partial(u,v)} \right| du\, dv = \tfrac{1}{3} \int_{u=0}^{u=6} \int_{v=-2}^{v=0} e^{2u} e^{-v}\, dv\, du$$

$$= \tfrac{1}{3} \left(\int_{u=0}^{u=6} e^{2u}\, du \right) \left(\int_{v=-2}^{v=0} e^{-v}\, dv \right)$$

$$= \tfrac{1}{3} \left[\tfrac{1}{2} e^{2u} \right]_{u=0}^{u=6} \left[-e^{-v} \right]_{v=-2}^{v=0} = \tfrac{1}{6}(e^{12} - 1)(e^2 - 1). \;\square$$

12.4 Other changes of variables: Jacobians

General transformations

Now suppose we want to change variables in a double integral $\iint_R f(x,y)\,dx\,dy$ by using a general transformation

$$T: x = x(u,v), y = y(u,v) \tag{13}$$

that maps a region $\tilde{R}$ in the uv-plane to the region of integration R for the given integral (Figure 12).

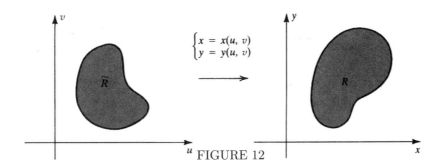

FIGURE 12

We need to be sure that the region $\tilde{R}$ maps onto all of R without any part of R being covered more than once. We do this by assuming that T is ONE-TO-ONE and ONTO, according to the following definition.

Definition 1 *A mapping from one set of points $\tilde{R}$ to another set of points R is ONTO if every point in R is the image of at least one point in $\tilde{R}$. The mapping is ONE-TO-ONE if no two points in $\tilde{R}$ are mapped into the same point in R.*

We can see how to change variables in an integral $\iint_R f(x,y)\,dx\,dy$ by modifying the reasoning we used to derive Theorem 2 for the affine case. We form a partition $R_1, R_2, \ldots, R_N$ of R. Then the subregions $\tilde{R}_1, \tilde{R}_2, \ldots, \tilde{R}_N$ that are mapped by the transformation into $R_1, R_2, \ldots, R_N$ form a partition of $\tilde{R}$. We pick a point (u_j, v_j) in $\tilde{R}_j$ and set $(x_j, y_j) = \bigl(x(u_j, v_j), y(u_j, v_j)\bigr)$ for each j.

We assume that the functions in (13) have continuous first derivatives in an open set containing $\tilde{R}$. Then for each j, the tangent-plane approximations of $x(u,v)$ and $y(u,v)$ near (u_j, v_j) are

$$x(u,v) \approx x(u_j, v_j) + \frac{\partial x}{\partial u}(u_j, v_j)(u - u_j) + \frac{\partial x}{\partial v}(u_j, v_j)(v - v_j)$$

$$y(u,v) \approx y(u_j, v_j) + \frac{\partial y}{\partial u}(u_j, v_j)(u - u_j) + \frac{\partial y}{\partial y}(u_j, v_j)(v - v_j).$$

These form an affine approximation T_A of T for (u,v) near (u_j, v_j), given by

$$T_A : \begin{cases} x_A = \dfrac{\partial x}{\partial u}(u_j, v_j) u + \dfrac{\partial y}{\partial v}(u_j, v_j) v + \text{[a constant]} \\ y_A = \dfrac{\partial y}{\partial u}(u_j, v_j) u + \dfrac{\partial x}{\partial v}(u_j, v_j) v + \text{[a constant]}. \end{cases}$$

The Jacobian of this affine transformation is the Jacobian

$$\frac{\partial(x,y)}{\partial(u,v)} = \begin{vmatrix} \dfrac{\partial x}{\partial u}(u,v) & \dfrac{\partial x}{\partial v}(u,v) \\ \dfrac{\partial y}{\partial u}(u,v) & \dfrac{\partial y}{\partial v}(u,v) \end{vmatrix} \tag{14}$$

of the original transformation (13) at (u_j, v_j). Moreover, if the number of subregions in the partitions is large and their diameters are small, then the area of each R_j is very close to the area of the image of $\tilde{R}_j$ under the affine approximation (14), and for each j,

$$[\text{Area of } R_j] \approx \left|\frac{\partial(x,y)}{\partial(u,v)}(u_j, v_j)\right| [\text{Area of } \tilde{R}_j].$$

Consequently,

$$\sum_{j=1}^{N} f(x_j, y_j)[\text{Area of } R_j] \approx \sum_{j=1}^{N} f(x(x(u_j, y_l), y(u_j, y_j))\left|\frac{\partial(x,y)}{\partial(u,v)}(u_j, v_j)\right| [\text{Area of } \tilde{R}_j].$$

As the number of subregions in the partitions tends to ∞ and their diameters tend to zero, these approximations become increasingly accurate. Also, the Riemann sum on the left tends to $\iint_R f(x,y)\,dx\,dy$ and the Riemann sum on the right tends to $\iint_{\tilde{R}} f(x(u,v), y(u,v))\left|\frac{\partial(x,y)}{\partial(u,v)}\right|\,du\,dv$, so we are led to the following result.†

Theorem 3 *Suppose that $x(u,v)$ and $y(u,v)$ have continuous first-order derivatives in a closed, bounded region $\tilde{R}$ with a piecewise-smooth boundary, that the Jacobian (14) is not zero in $\tilde{R}$, and that the mapping $T : x = x(u,v), y = y(u,v)$ of $\tilde{R}$ to R is one-to-one and onto. If $f(x,y)$ is piecewise continuous in R, then*

$$\iint_R f(x,y)\,dx\,dy = \iint_{\tilde{R}} f(x(u,v), y(u,v))\left|\frac{\partial(x,y)}{\partial(u,v)}\right|\,du\,dv. \tag{15}$$

This theorem can also be applied in may cases where the Jacobian is zero or the mapping is not one-to-one at isolated points or on curves in $\tilde{R}$.

Formula (15) is the same as formula (10) for the case of an affine transformation. The difference is that the Jacobian for most nonaffine transformations is not a constant, but is a variable function of (u,v). The equation

$$\frac{\partial(u,v)}{\partial(x,y)} = \frac{1}{\frac{\partial(x,y)}{\partial(u,v)}} \tag{16}$$

relating the Jacobian of $x = x(u,v), y = y(u,v)$ to that of its inverse $u = u(x,y), v = v(x,y)$ also holds for nonaffine transformations, provided that the inverse transformation is defined and the Jacobians are not zero.

Example 7 Use the transformation

$$u = xy, \ v = x^2 - y^2 \tag{17}$$

to evaluate $\iint_R (x^2 + y^2)\cos(xy)\,dx\,dy$, where R is the region to the right of the y-axis that is bounded by the hyperbolas $xy = 3, xy = -3, x^2 - y^2 = 1$, and $x^2 - y^2 = 9$.

---

†This reasoning can be made more complete by using Green's Theorem, which we will study in Section 13.2, to show that for each j there is a point (u_j, v_j) in $\tilde{R}_j$ such that $[\text{Area of } R_j] = \left|\frac{\partial(x,y)}{\partial(u,v)}(u_j, v_j)\right| [\text{Area of } \tilde{R}_j]$.

12.4 Other changes of variables: Jacobians

SOLUTION   Notice that the transformation **(17)** gives u and v in terms of x and y, whereas in earlier Examples we began with formulas for x and y in terms of u and v. The region R is shown in Figure 13. Because its top and bottom are formed by the level curves $u = 3$ and $u = -3$ and its left and right sides are formed by the level curves $v = 1$ and $v = 9$, its image under **(17)** is the rectangle $\tilde{R} = \{(u,v) : -3 \le u \le 3, 1 \le v \le 9\}$ in Figure 14.

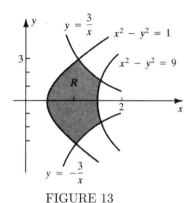

FIGURE 13   FIGURE 14

The Jacobian of **(17)** is

$$\frac{\partial(u,v)}{\partial(x,y)} = \begin{vmatrix} u_x & u_y \\ v_x & v_y \end{vmatrix} = \begin{vmatrix} y & x \\ 2x & -2y \end{vmatrix} = -2(x^2 + y^2)$$

and by **(16)**, the Jacobean of the inverse is the reciprocal

$$\frac{\partial(x,y)}{\partial(u,v)} = \frac{1}{-2(x^2+y^2)}.$$

Consequently, we need to make the substitution

$$dx\,dy = \left|\frac{\partial(x,y)}{\partial(u,v)}\right| du\,dv = \frac{1}{2(x^2+y^2)}\,du\,dv$$

in the intergal. The integrand in the given integral is such that we can transform it into an integral with respect to u and v without solving **(17)**. We obtain

$$\iint_R (x^2+y^2)\cos(xy)dx\,dy = \iint_{\tilde{R}} (x^2+y^2)\cos(xy) \left|\frac{\partial(x,y)}{\partial(u,v)}\right| du\,dv$$

$$= \tfrac{1}{2}\iint_{\tilde{R}} \cos u\,du\,dv = \tfrac{1}{2}\int_{u=-3}^{u=3}\int_{v=1}^{v=9} \cos u\,dv\,du$$

$$= \tfrac{1}{2}\int_{u=-3}^{u=3} 8\cos u\,du = \Big[4\sin u\Big]_{u=-3}^{u=3} = 4[\sin(3) - \sin(-3)] = 8\sin(3). \;\square$$

Changing variables in triple integrals

The procedure for making a change of variables

$$x = x(u,v,w), y = y(u,v,w), z = z(u,v,w) \tag{18}$$

in a triple integral is similar to that for double integrals. We use the following result, which we state without proof.

Theorem 4 *Suppose that $\tilde{V}$ is a closed bounded solid with a piecewise-smooth boundary in uvw-space, that $x(u,v,w), y(u,v,w)$, and $z(u,v,w)$ have continuous first-order derivatives in $\tilde{V}$, that the mapping* **(18)** *is one-to-one onto $\tilde{V}$, and that the Jacobian*

$$\frac{\partial(x,y,z)}{\partial(u,v,w)} = \begin{vmatrix} x_u & x_v & x_w \\ y_u & y_v & y_w \\ z_u & z_v & z_w \end{vmatrix} \tag{19}$$

is nonzero in $\tilde{V}$. If $f(x,y,z)$ is piecewise-continuous in the image V of $\tilde{V}$ under **(18)**, *then*

$$\iiint_V f(x,y,z)\,dx\,dy\,dz = \iiint_{\tilde{V}} f(x,y,z) \left| \frac{\partial(x,y,z)}{\partial(u,v,w)} \right| du\,dv\,dw \tag{20}$$

with $f(x,y,z) = f\bigl(x(u,v,w), y(x,v,w), z(u,v,w)\bigr)$.

This result can also be applied in many cases where the Jacobian is zero or the mapping is not one-to-one at a finite number of points or on a finite number of curves or surfaces in $\tilde{V}$. Equation **(20)** is expressed partially by the symbolic equation

$$dx\,dy\,dz = \left| \frac{\partial(x,y,z)}{\partial(u,v,w)} \right| du\,dv\,dw. \tag{21}$$

Question 5 Use **(21)** to express $dx\,dy\,dz$ in terms of $r, dr, d\theta$, and dz in the case of cylindrical coordinates.

A somewhat lengthy calculation similar to that in the Response to Question 5 shows that for spherical coordinates $x = \rho\sin\phi\cos\theta, y = \rho\sin\phi\sin\theta, z = \rho\cos\phi$,

$$\frac{\partial(x,y,z)}{\partial(\rho,\theta,\phi)} = -\rho^2 \sin\phi$$

so that, because $\sin\phi \geq 0$, $dx\,dy\,dz = \rho^2 \sin\phi\,d\rho\,d\theta\,d\phi$.[†]

Under the conditions of Theorem 4, the transformation **(18)** has an inverse

$$u = u(x,y,z), v = v(x,y,z), w = w(x,y,z)$$

whose Jacobian is the reciprocal of the Jacobian $\dfrac{\partial(x,y,z)}{\partial(u,v,w)}$.

We will apply Theorem 4 mostly to affine transformations

$$\begin{cases} x = a_1 u + b_1 v + c_1 w + d_1 \\ y = a_2 u + b_2 v + c_2 w + d_2 \\ z = a_3 u + b_3 v + c_3 w + d_3. \end{cases}$$

[†]See Problem 27.

12.4 Other changes of variables: Jacobians

Question 6 Find **(a)** the Jacobian of the affine transformation

$$\begin{cases} u = x + y + z \\ v = x + 2y \\ w = y - 3z \end{cases} \tag{22}$$

and **(b)** the Jacobian of the inverse transformation.

Example 8 Use the transformation (22) to find the value of $\iiint_V \sqrt{x+y+z}\, dx\, dy\, dz$ with V the parallelopiped $\{(x,y,z) : 0 \leq x+y+z \leq 9, 1 \leq x+2y \leq 4, 2 \leq y - 3z \leq 6\}$.

SOLUTION The change of variables (22) has been chosen so that the image of V is the box $\tilde{V} = \{(u,v,w) : 0 \leq u \leq 9, 1 \leq v \leq 4, 2 \leq w \leq 6\}$ in uvw-space. Also because the integrand $\sqrt{x+y+z}$ equals $\sqrt{u}$, we do not need to solve (22) for x, y, and z. You saw in Question 6 that the Jacobian $\dfrac{\partial(x,y,z)}{\partial(u,v,w)}$ is $-\tfrac{1}{2}$. Therefore,

$$\iiint_V \sqrt{x+y+z}\, dx\, dy\, dz = \iiint_{\tilde{V}} \sqrt{u}\, \left|\frac{\partial(x,y,z)}{\partial(u,v,w)}\right|\, du\, dv\, dw$$

$$= \int_{u=0}^{u=9} \int_{v=1}^{v=4} \int_{w=2}^{w=6} \tfrac{1}{2}\sqrt{u}\, dw\, dv\, du = 6\int_{u=0}^{u=9} u^{1/2}\, du = 4\Big[u^{3/2}\Big]_0^9 = 108. \ \square$$

Principles and procedures

- Affine transformations map parallel lines into parallel lines and have constant Jacobians because they are given by linear functions which have constant partial derivatives.

- Cramer's Rule with two variables states that if the determinant $\begin{vmatrix} a_1 & b_1 \\ a_2 & b_2 \end{vmatrix}$ is not zero, then the unique solution of the system of equations

$$\begin{cases} a_1 x + b_1 y = c_1 \\ a_2 x + b_2 y = c_2 \end{cases} \tag{27}$$

is given by

$$x = \frac{\begin{vmatrix} c_1 & b_1 \\ c_2 & b_2 \end{vmatrix}}{\begin{vmatrix} a_1 & b_1 \\ a_2 & b_2 \end{vmatrix}}, \quad y = \frac{\begin{vmatrix} b_1 & c_1 \\ b_1 & c_2 \end{vmatrix}}{\begin{vmatrix} a_1 & b_1 \\ a_2 & b_2 \end{vmatrix}}$$

where the determinant in the numerator in the formula for x is obtained by replacing the first column of the denominator by the numbers on the right of (27) and the determinant in the numerator in the formula for y is obtained by replacing the second column of the determinant in the denominator by these numbers.

- The Jacobian is needed in changing variables in double or triple integrals because the subregions or subsolids in a partition for the original integral are transformed into subregions or subsolids with different areas or volumes with the new coordinates.

- The absolute values of Jacobians are used in changing variables because Jacobians can be negative. (See Problem 28.)

- The symbols for Jacobians make formulas such as $dx\,dy = \left|\dfrac{\partial(x,y)}{\partial(u,v)}\right| du\,dv$, and $\dfrac{\partial(u,v)}{\partial(x,y)} = \left[\dfrac{\partial(u,v)}{\partial(x,y)}\right]^{-1}$ easier to remember because they look somewhat like relations among fractions.

Tune-Up Exercises 12.4♦

A Answer provided. **O** Outline of solution provided.

T1.$^\text{O}$ Draw in an xy-plane the image of the four squares in Figure 15 under the affine transformation $x = 1 + 2u + v, y = 2 - u + 2v$.

♦ Type 1, drawing an image under an affine transformation

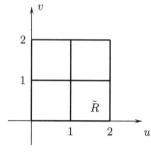

FIGURE 15

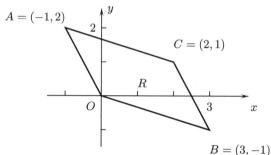

FIGURE 16

T2.$^\text{O}$ (a) Find an affine transformation $T: x = a_1 u + b_1 v + c_1, y = a_2 u + b_2 v + c_2$ that maps the square with corners $\tilde{O} = (0,0), \tilde{A} = (1,0), \tilde{B} = (0,1)$, and $\tilde{C} = (1,1)$ in a uv-plane into the parallelogram in Figure 16. (b) What is the Jacobian of the transformation of part (a)? (c) Find the inverse of the transformation of part (a). (d) What is the Jacobian of the inverse transformation?

♦ Type 1, finding an affine transformation

T3.$^\text{O}$ Figure 17 shows a region $\tilde{R}$ in a uv-plane and Figure 18 shows its image R under the affine transformation $x = u + v, y = u - v + 2$. Use this transformation to convert $\displaystyle\iint_R \dfrac{x-y}{x+y}\,dx\,dy$ into an integral with respect to u and v over $\tilde{R}$.

♦ Type 1, transforming an integral under an affine transformation

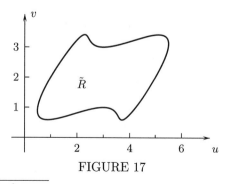

FIGURE 17

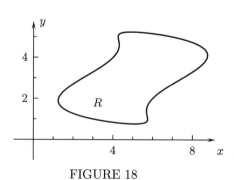

FIGURE 18

♦The Tune-up Exercises and Problems are classified by type and content. The types are (1) basic, reactive; (2) basic reflective; (3) intermediate, reactive; (4) intermediate, reflective; (5) advanced, reactive; (6) advanced, reflective; and (7) advanced, theoretical.

12.4 Other changes of variables: Jacobians

T4.^{O} Figure 19 shows a region $\tilde{R}$ in a uv-plane and Figure 20 shows its image R under the transformation $x = 3/u, y = 4v^{1/3} - 3$. Use this transformation to convert $\iint_R x(y+3)\,dx\,dy$ into an integral with respect to u and v over $\tilde{R}$.

◆ Type 2, transforming an integral under an nonaffine transformation

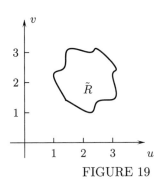
FIGURE 19

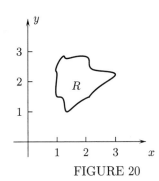
FIGURE 20

T5^O (a) What is the Jacobian of the affine transformation $x = u + 2v + 3w, y = 4u - 5v, z = 6v + 2w$?
(b) Suppose that a solid $\tilde{V}$ in uvw-space has volume 100 cubic meters. What is the volume of its inverse under the transformation of part (a)?

◆ Type 2, Jacobian of an affine transformation in space; finding the volume of an image

Problems 12.4
^{A}Answer provided. ^{O}Outline of solution provided.

1.^{A} Draw in an xy-plane the image under the affine transformation $x = 2u + 5v, y = -2u + 2v$ of the pentagon in Figure 21.

◆ Type 1, drawing an image under an affine transformation

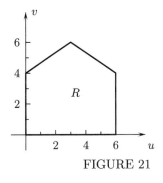
FIGURE 21

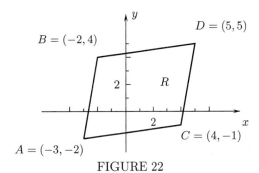
FIGURE 22

2.^{A} Find an affine transformation $T: x = a_1 u + b_1 v + c_1, y = a_2 u + b_2 v + c_2$ that transforms the square with corners $\tilde{A} = (0,0), \tilde{B} = (1,0), \tilde{C} = (0,1)$, and $\tilde{D} = (1,1)$ in a uv-plane into the parallelogram in Figure 22.

◆ Type 1, finding an affine transformation

3. Find the inverse of the affine transformations (a^A) $x = \frac{1}{4}u + \frac{1}{8}v, y = -\frac{1}{4}u + \frac{3}{8}v$ and
(b) $x = 10u - 20v, y = 5u + 5v$.

◆ Type 1, finding the inverse of an affine transformation

4.º Solve $x = u - 2v, y = 3u + v$ for u and v and calculate $\dfrac{\partial(x,y)}{\partial(u,v)}$ and $\dfrac{\partial(u,v)}{\partial(x,y)}$ with these variables.

◆ Type 1, finding the inverse of an affine transformation and Jacobians

5. Find the inverse transformations (a^A) of $u = 2x - y, v = x - 2y$, (b) of $u = \frac{1}{4}x + \frac{1}{4}y, v = -\frac{1}{4}x + \frac{3}{4}y$, and (c) of $u = x + 4y, v = x - 4y$.

◆ Type 1, finding the inverse of an affine transformation

6. The ellipse R in Figure 24 is the image of the circle $\tilde{R} : x^2 + y^2 \leq 4$ in Figure 23 under the affine transformation $x = 2u + v + 6, y = -u + v + 4$. What is the area of the ellipse?

♦ Type 2, finding the area of an image under an affine transformation

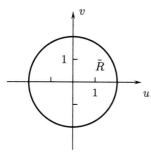

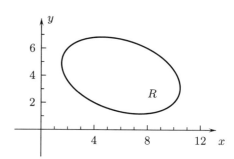

FIGURE 23   FIGURE 24

7.º Use $u = x + y, v = x - y$ to evaluate $\iint_R \dfrac{x+y}{x-y+1}\, dx\, dy$ with R the square bounded by the lines $y = x, y = -x, y = x - 4$, and $y = -x + 4$.

♦ Type 1, evaluating an integral with an affine transformation

8.ᴬ Use $u = x + y, v = x - 2y$ to evaluate $\iint_R (x+y)^3 \sin(x - 2y)\, dx\, dy$, where R is the parallelogram bounded by the lines $y = -x, y = -x + 4, y = \tfrac{1}{2}x$, and $y = \tfrac{1}{2}x + 3$.

♦ Type 2, evaluating an integral with an affine transformation

9. Evaluate $\iint_R (x^2 - y^2)^{10}\, dx\, dy$ with R the square bounded by the lines $y = x + 1, y = x - 1$, $y = -x + 1$, and $y = -x - 1$ by using the transformation $u = x + y, v = x - y$.

♦ Type 1, evaluating an integral with an affine transformation

10. Evaluate $\iint_R e^x\, dx\, dy$ with R the square bounded by $y = 3x + 1, y = 3x - 3, y = -x + 1$, and $y = -x + 5$ by using the transformation $u = 3x - y, v = x + y$.

♦ Type 1, evaluating an integral with an affine transformation

11. Give formulas in terms of x and y for $\dfrac{\partial(u,v)}{\partial(x,y)}$ and $\dfrac{\partial(x,y)}{\partial(u,v)}$ (**a**ᴬ) where $u = x^2 - y^2, v = xy$, (**b**) where $u = x^{-1}y, v = x^2 y$, and (**c**) where $u = e^x \sin y, v = e^x \cos y$.

♦ Type 3, Jacobians of nonaffine transformations and their inverses

Evaluate the integrals in Problems 12 through 21 by making suitable transformations of the variables.

12.º $\iint_R \cos(5x)\, dx\, dy$ with R bounded by $y + 4x = 0, y + 4x = -2, y - x = 0$, and $y - x = 1$

♦ Type 3, evaluating an integral with an affine transformation

13.ᴬ $\iint_R e^{4x}\, dx\, dy$ with R bounded by $y - 2x = 2, y - 2x = -2, y + 2x = 0$, and $y + 2x = 6$

♦ Type 3, evaluating an integral with an affine transformation

14. $\iint_R \dfrac{x + 2y^2}{xy}\, dx\, dy$ with R bounded by $x - y^2 = 0, x - y^2 = 2, xy = 1$, and $xy = 3$

♦ Type 3, evaluating an integral with a nonaffine transformation

15.ᴬ $\iint_R y e^{x^2 y}\, dx\, dy$ with R bounded by $x^2 y = 1, x^2 y = 4, y/x = \tfrac{1}{3}$, and $y/x = 3$

♦ Type 3, evaluating an integral with a nonaffine transformation

12.4 Other changes of variables: Jacobians

16. $\iint_R \sin\left((x+y)^2\right) \, dx\, dy$ where R is the triangle with vertices $(0,0), (1,0)$, and $(0,1)$ (Use $u = x+y, v = -x-y$.)

♦ Type 3, evaluating an integral with an affine transformation

17.^{A} $\iint_R e^{xy} \, dx\, dy$ for R bounded by $xy = 1, xy = 4, y = 1$, and $y = 3$

♦ Type 3, evaluating an integral with a nonaffine transformation

18. $\iint_R (3x^2 + 1)\sqrt{x+y} \, dx\, dy$ for R bounded by $y = x^3, y = x^3 + 2, y = -x$, and $y = -x+2$

♦ Type 3, evaluating an integral with a nonaffine transformation

19. $\iint_R \dfrac{1}{1+(x+y)^2} \, dx\, dy$, where R is bounded by $y = x-2, y = -x$, and $x = 0$

♦ Type 3, evaluating an integral with an affine transformation

20. $\iint_R (x-y)^3 \, dx\, dy$, where R is bounded by $y = x, y = 3x$, and $x = \frac{3}{2}$

♦ Type 3, evaluating an integral with an affine transformation

21.^{A} $\iint_R \sin\left(\frac{1}{4}x^2 + \frac{1}{9}y^2\right) \, dx\, dy$, with R the ellipse $\frac{1}{4}x^2 + \frac{1}{9}y^2 \leq 1$ (Use $x = 2u, y = 3v$ and polar coordinates in the uv-plane.)

♦ Type 3, evaluating an integral with an affine transformation and polar coordinates

22.º Evaluate $\iiint_V \dfrac{2y}{x+2y+4z} \, dx\, dy\, dz$, with V the parallelopiped $V = \{(x,y,z) : 1 \leq x+2y+4z \leq 4, 0 \leq y-2z \leq 1, 0 \leq y+2z \leq 2\}$ by making an affine change of variables.

♦ Type 4, evaluating a triple integral with an affine transformation

In Problems 23 through 25 find the values of the integrals by making affine changes of variables to obtain integrals over boxes with sides parallel to the coordinate planes.

23.º $\iiint_V \dfrac{x+y-z}{1+(y+2z)^2} \, dx\, dy\, dz$, with $V = \{(x,y,z) : 0 \leq x+y-z \leq 2, 0 \leq x-y+z \leq 3, 0 \leq y+2z \leq 4\}$

♦ Type 4, evaluating a triple integral with an affine transformation

24.^{A} $\iiint_V (x^2 - y^2) \, dx\, dy\, dz$, with $V = \{(x,y,z) : 1 \leq x+y \leq 2, 3 \leq x-y \leq 4, 4 \leq x+y+z \leq 5\}$

♦ Type 4, evaluating a triple integral with an affine transformation

25. $\iiint_V \dfrac{x-y-2z}{3x+2y} \, dx\, dy\, dz$, with $V = \{(x,y,z) : 1 \leq 3x+2y \leq 4, 0 \leq x-y-2z \leq 1, 0 \leq z \leq 5\}$

♦ Type 4, evaluating a triple integral with an affine transformation

26. Evaluate $\iiint_V \sqrt{4x^2 + 9y^2 + 16z^2} \, dx\, dy\, dz$, with $V = \{(x,y,z) : 1 \leq 4x^2 + 9y^2 + 16z^2 \leq 54\}$ by using the change of variables $u = 2x, v = 3y, w = 4z$.

♦ Type 4, evaluating a triple integral with an affine transformation

27. Show that with spherical coordinates $\left|\dfrac{\partial(x,y,z)}{\partial(\rho,\theta,\phi)}\right| = \rho^2 \sin\phi$,

♦ Type 4, finding the Jacobian for spherical coordinates

28. What is the significance of the sign of the Jacobian in the case of two variables? (Compare the boundaries of the images of the triangle with vertices $(0,0), (1,0)$, and $(0,1)$ under $x = au + bv$, $y = cu + dv$ and under $x = au + bv, y = -cu - dv$ for various choices of a, b, c and d.)

♦ Type 7, explaining the sign of the Jacobian

RESPONSES TO QUESTIONS IN CHAPTER 12

Responses 12.1

Response 1   [Volume] $= (80 \times 100 \times 100) + (60 \times 100 \times 50) + (40 \times 50 \times 50)$
$= 800{,}000 + 300{,}000 + 100{,}000 = 1{,}200{,}000$ cubic meters

Response 2   Because the base of the building is the region R in the xy-plane of Figure 3 and its top is formed by the graph of the positive function $f(x,y)$ of Figure 2, the volume of the building is $\iint_R f(x,y)\,dx\,dy$.

Response 3   The Riemann sum for $\iint_R 1\,dx\,dy$ corresponding to any partition $R_1, R_2, \ldots R_N$ of R is $\sum_{j=1}^{N} [\text{Area of } R_j] = [\text{Area of } R]$ • Therefore, the integral, which is the limit of the Riemann sums, is also equal to the area of R.

Response 4   One answer: Because the integrand -5 is negative, the integral equals the negative of the volume of the rectangular box $0 \le x \le 3, 0 \le y \le 2, -5 \le z \le 0$. •
The box has volume $3 \times 2 \times 5 = 30$, so $\iint_R -5\,dx\,dy = -30$
Another answer: By formula (4) and the response to Question 3,
$\iint_R -5\,dx\,dy = -5\iint_R dx\,dy = -5[\text{Area of } R] = -5(6) = -30$

Response 5   The region of integration and integration procedure are shown in Figure R5. •
Integrate with respect to y from $y = 0$ to $y = 10$ with x constant and then with respect to x from 0 to 10. •
$$\iint_R (2x + 3y^2)\,dx\,dy = \int_{x=0}^{x=10} \int_{y=0}^{y=10} (2x + 3y^2)\,dy\,dx$$
$$= \int_{x=0}^{x=10} \left[2xy + y^3\right]_{y=0}^{y=10} dx = \int_{x=0}^{x=10} (20x + 1000)\,dx$$
$$= \left[10x^2 + 1000x\right]_0^{10} = 11{,}000$$

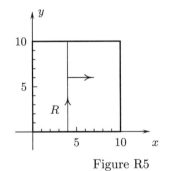

Figure R5

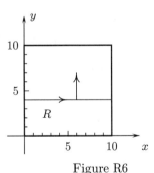

Figure R6

Response 6   The region of integration and integration procedure are shown in Figure R6 •
$$\iint_R (2x + 3y^2)\,dx\,dy = \int_{y=0}^{y=10} \int_{x=0}^{x=10} (2x + 3y^2)\,dx\,dy$$
$$= \int_{y=0}^{y=10} \left[x^2 + 3xy^2\right]_{x=0}^{x=10} dy = \int_{y=0}^{y=10} (30y^2 + 100)\,dy$$
$$= \left[10y^3 + 100y\right]_0^{10} = 11{,}000$$

Response 7 Writing the integral as $\int_{x=0}^{x=2}\int_{y=0}^{y=x} g(x,y)\,dy\,dx$ shows that the region of integration is the triangle in Figure R7a. • Figure R7b shows the integration procedure where the x-integration is done first. •
$$\int_{y=0}^{y=2}\int_{x=y}^{x=2} g(x,y)\,dx\,dy = \int_0^2\int_y^2 g(x,y)\,dx\,dy$$

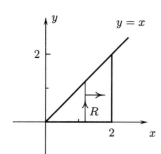
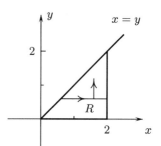

Figure R7a Figure R7b

Response 8 $[\text{Weight}] = \int_{x=0}^{x=1}\int_{y=0}^{y=1} 72xy^2 \left[\dfrac{\text{pounds}}{\text{square foot}}\right] dy\,dx\ [\text{square feet}]$

$= \int_{x=0}^{x=1}\int_{y=0}^{y=1} 72xy^2\,dy\,dx\ \text{pounds} = \int_{x=0}^{x=1} \Big[24xy^3\Big]_{y=0}^{y=1} dx = \int_{x=0}^{x=1} 24x\,dx$

$= \Big[12x^2\Big]_{x=0}^{x=1} = 12\ \text{pounds}$

Response 9 Because the property occupies the square R formed by the subregions R_1, R_2, and R_3 in Figure 4 and its cost density (dollars per square meter) is the function $C(x,y)$ of Figure 5, the price of the property is $\iint_R C(x,y)\,dx\,dy$ dollars.

Response 10 $[\text{Moment about } x = K] = 0 \implies \iint_R x\rho(x,y)\,dx\,dy = K\iint_R \rho(x,y)\,dx\,dy$

$\implies K = \dfrac{\iint_R x\rho(x,y)\,dx\,dy}{\iint_R \rho(x,y)\,dx\,dy}$

$[\text{Moment about } y = L] = 0 \implies \iint_R y\rho(x,y)\,dx\,dy = L\iint_R \rho(x,y)\,dx\,dy$

$\implies L = \dfrac{\iint_R y\rho(x,y)\,dx\,dy}{\iint_R \rho(x,y)\,dx\,dy}$

Response 11 One answer: The center of gravity is to the right of the center $(\tfrac{1}{2},\tfrac{1}{2})$ of the square because for each y the density xy^2 is greater for $\tfrac{1}{2} < x < 1$ than for $0 < x < \tfrac{1}{2}$.
•
The center of gravity is above the center $(\tfrac{1}{2},\tfrac{1}{2})$ of the square because for each x the density xy^2 is greater for $\tfrac{1}{2} < y < 1$ than for $0 < y < \tfrac{1}{2}$.

Responses 12.2

Response 1 The heaviest subsolids are the piece of brazilwood V_4 and the piece of chestnut V_6, whose volumes are $1 \times 1 \times 2 = 2$ cubic inches and whose densities are 0.72 ounces per cubic inch. They each weigh $\left[0.72 \, \dfrac{\text{ounces}}{\text{cubic inch}}\right] [2 \text{ cubic inches}] = 1.44$ ounces.
• The lightest subsolid is the piece of balsa V_1, which has the same volume and density 0.09 ounces per cubic inch. It weighs $\left[0.09 \, \dfrac{\text{ounces}}{\text{cubic inch}}\right] [2 \text{ cubic inches}] = 0.18$ ounces.

Response 2 For any partition of V, the Riemann sum $\sum_{j=1}^{N} 1 \, [\text{Volume of } V_j]$ for the integral
$$\iiint_V 1 \, dx \, dy \, dz$$
is the sum of the volumes of the subsolids and equals the volume of V, so the limit of the Riemann sums, which is the integral, also equals the volume of V.

Response 3
$$\iiint_V (x+y+z) \, dx \, dy \, dz = \int_{x=0}^{x=3} \int_{y=0}^{y=2} \int_{z=0}^{z=1} (x+y+z) \, dz \, dy \, dx$$
$$= \int_{x=0}^{x=3} \int_{y=0}^{y=2} \left[xz + yz + \tfrac{1}{2}z^2\right]_{z=0}^{z=1} dy \, dx = \int_{x=0}^{x=3} \int_{y=0}^{y=2} (x+y+\tfrac{1}{2}) \, dy \, dx$$
$$= \int_{x=0}^{x=3} \left[xy + \tfrac{1}{2}y^2 + \tfrac{1}{2}y\right]_{y=0}^{y=2} dx = \int_{x=0}^{x=3} (2x+3) \, dx$$
$$= \left[x^2 + 3x\right]_{x=0}^{x=3} = 18$$

Response 4 The weight of V is $\iiint_V \rho(x,y,z) \, dx \, dy \, dz$, where $\rho(x,y,z)$ is is defined in the interior of each of the subsolids by the values in Table 1.

Response 5
$$\int_{x=0}^{x=1} \int_{y=0}^{y=x^2} \int_{z=0}^{z=2} xyz \, dz \, dy \, dx = \int_{x=0}^{x=1} \int_{y=0}^{y=x^2} \left[\tfrac{1}{2}xyz^2\right]_{z=0}^{z=2} dy \, dx$$
$$= \int_{x=0}^{x=1} \int_{y=0}^{y=x^2} 2xy \, dy \, dx = \int_{x=0}^{x=1} \left[xy^2\right]_{y=0}^{y=x^2} dx$$
$$= \int_{x=0}^{x=1} x^5 \, dx = \left[\tfrac{1}{6}x^6\right]_{x=0}^{x=1} = \tfrac{1}{6}$$

Response 6 From Example 1, the wood weighs 10.08 ounces. • The total volume of the twelve pieces is $2 \times 12 = 24$ cubic inches. • $[\text{Average density}] = \dfrac{10.08 \text{ ounces}}{24 \text{ cubic inches}}$
$= 0.42$ ounces per cubic inch • (This is the result that would be obtained from Definition 4 because the integral of the density equals the total weight of the wood.)

Responses 12.3

Response 1 (a) The outer circle of radius 6 in Figure 4 has area $\pi(6)^2 = 36\pi$ and the inner circle of radius 4 has area $\pi(4)^2 = 16\pi$. • The area of the ring between the two circles is $36\pi - 16\pi = 20\pi$. • The average r^* of the two radii is $\frac{1}{2}(6+4) = 5$ and its thickness is $\Delta r = 6 - 4 = 2$. • Formula **(1)** gives the the area as $2\pi r^* \Delta r = 2\pi(5)(2) = 20\pi$, which is the same value.
(b) The outer circle in Figure 5 has area $\pi(10)^2 = 100\pi$ and the inner circle has area $\pi(8)^2 = 64\pi$. • The area of the ring is $100\pi - 64\pi = 36\pi$. •
$r^* = \frac{1}{2}(10+8) = 9$ and $\Delta r = 10 - 8 = 2$ • Formula **(1)** gives
$2\pi r^* \Delta r = 2\pi(9)(2) = 36\pi$, which is the same value.

Response 2 Use the substitutions $\sqrt{x^2 + y^2 + 1} = \sqrt{r^2 + 1}$ and $dx\,dy = r\,dr\,d\theta$ and integrate with respect to r from 0 to 1 and then with respect to θ from 0 to 2π, as indicated in Figure 15. •
$$\frac{1}{\pi}\iint_R \sqrt{x^2 + y^2 + 1}\,dx\,dy = \frac{1}{\pi}\int_{\theta=0}^{\theta=2\pi}\int_{r=0}^{r=1}\sqrt{r^2+1}\,r\,dr\,d\theta$$
$$= \left(\frac{1}{\pi}\int_{\theta=0}^{\theta=2\pi}d\theta\right)\left(\int_{r=0}^{r=1}\sqrt{r^2+1}\,r\,dr\right) = \int_{r=0}^{r=1}(r^2+1)^{1/2}(2r\,dr)$$
$$= \int_{r=0}^{r=1}u^{1/2}\,du \text{ with } u = r^2 + 1, du = 2r\,dr \bullet \text{ The average value is}$$
$$\left[\tfrac{2}{3}u^{3/2}\right]_{r=0}^{r=1} = \left[\tfrac{2}{3}(r^2+1)^{3/2}\right]_{r=0}^{r=1} = \tfrac{2}{3}(2^{3/2} - 1).$$

Response 3 (a) The arrow on the radius in Figure 25 represents the innermost integral with respect to r; the radius is genrated as r goes from 0 to 2. • The curved arrow represents the middle integral with respect to θ; the radius generates the disk as θ goes from 0 to 2π. • The vertical arrow represents the outermost integral with respect to z; the disk generates the cylinder as z goes from 2 to 6. •
$$\int_{z=2}^{z=6}\int_{\theta=0}^{\theta=2\pi}\int_{r=0}^{r=2}$$
(b) The arrow on the horizontal line in Figure 26 represents the innermost integral with respect to r; the line is generated as r goes from 0 to 2. • The vertical arrow represents the middle integral with respect to z; the horizontal line generates the rectangle as z goes from 0 to 6. • The curved arrow represents the outermost integral with respect to θ; the rectangle generates the cylinder as θ goes from 0 to 2π. • $\int_{\theta=0}^{\theta=2\pi}\int_{z=2}^{z=6}\int_{r=0}^{r=2}$
(c) The arrow on the horizontal circle in Figure 27 represents the innermost integral with respect to θ; the circle is generated as θ goes from 0 to 2π. • The vertical arrow represents the middle integral with respect to z; the circle generates the cylindrical surface as z goes from 2 to 6. • The horizontal arrow represents the outermost integral with respect to r; the cylindrical surface generates the cylinder as r goes from 0 to 2. • $\int_{r=0}^{r=2}\int_{z=2}^{z=6}\int_{\theta=0}^{\theta=2\pi}$

Response 4
$$\int_{r=0}^{r=2}\int_{\theta=0}^{\theta=2\pi}\int_{z=0}^{z=\sqrt{4-r^2}} r^2 z\,dz\,d\theta\,dr = \int_{r=0}^{r=2}\int_{\theta=0}^{\theta=2\pi}\left[\tfrac{1}{2}r^2 z^2\right]_{z=0}^{z=\sqrt{4-r^2}}d\theta\,dr$$
$$= \int_{r=0}^{r=2}\int_{\theta=0}^{\theta=2\pi}\tfrac{1}{2}r^2(4-r^2)\,d\theta\,dr = \int_{r=0}^{r=2}\int_{\theta=0}^{\theta=2\pi}(2r^2 - \tfrac{1}{2}r^4)\,d\theta\,dr$$
$$= \int_{r=0}^{r=2}\left[(2r^2 - \tfrac{1}{2}r^4)\theta\right]_{\theta=0}^{\theta=2\pi}dr = 2\pi\int_{r=0}^{r=2}(2r^2 - \tfrac{1}{2}r^4)\,dr$$
$$= 2\pi\left[\tfrac{2}{3}r^3 - \tfrac{1}{10}r^5\right]_{r=0}^{r=2} = \tfrac{64}{15}\pi$$

Response 5 (a) The surface $\phi = 0$ is the positive z-axis.
(b) The surface $\phi = \frac{1}{2}\pi$ is the xy-plane.
(c) The surface $\phi = \pi$ is the negative z-axis.

Response 6 The arc QR is subtended by the angle $\Delta\theta$ on a circle of radius $\rho_j \sin(\phi_n)$, so its (curved) length is $\widehat{QR} = \rho_j \sin(\phi_n)\,\Delta\theta$.

Response 7 (a) The arrow on the radius in Figure 41 indicates the first integration with respect to ρ; the radius is generated as ρ goes from 0 to a. • The curved arrow on the cone represents the middle integral with respect to θ; the radius generates the cone as θ goes from 0 to 2π. • The outer curved arrow represents the outer integral with respect to ϕ; the cone generates the ball as ϕ goes from 0 to π.
(b) The arrow on the radius indicates the first integration with respect to ρ; the radius is generated as ρ goes from 0 to a. • The curved arrow on the half-disk represents the middle integral with respect to θ; the radius generates the half-disk as ϕ goes from 0 to π. • The outer curved arrow represents the outer integral with respect to θ; the half-disk generates the ball as θ goes from 0 to 2π.

Response 8 (a) $\displaystyle\int_{\rho=0}^{\rho=a}\int_{\phi=0}^{\phi=\pi}\int_{\theta=0}^{\theta=2\pi}$ (b) $\displaystyle\int_{\phi=0}^{\phi=\pi}\int_{\rho=0}^{\rho=a}\int_{\theta=0}^{\theta=2\pi}$ (c) $\displaystyle\int_{\theta=0}^{\theta=2\pi}\int_{\rho=0}^{\rho=a}\int_{\phi=0}^{\phi=\pi}$

Response 9 (a) St. Petersburg has spherical coordinates $\rho = 4000$ because the radius of the earth is 4000 miles, $\theta = \frac{1}{6}\pi$ because its longitude is $30°\text{E}$, and $\phi = \frac{1}{6}\pi$ because its latitude is $60°\text{N}$. • Its rectangular coordinates are $x = 4000\sin(\frac{1}{6}\pi)\cos(\frac{1}{6}\pi) = 1000\sqrt{3}$, $y = 4000\sin(\frac{1}{6}\pi)\sin(\frac{1}{6}\pi) = 1000$, and $z = 4000\cos(\frac{1}{6}\pi) = 2000\sqrt{3}$.
(b) Wassau has spherical coordinates $\rho = 4000$, $\theta = -\frac{1}{2}\pi$, and $\phi = \frac{1}{4}\pi$ because its latitude is $45°\text{N}$. • Its rectangular coordinates are $x = 4000\sin(\frac{1}{4}\pi)\cos(-\frac{1}{2}\pi) = 0$, $y = 4000\sin(\frac{1}{4}\pi)\sin(-\frac{1}{2}\pi) = -2000\sqrt{2}$, and $z = 4000\cos(\frac{1}{4}\pi) = 2000\sqrt{2}$.

Responses 12.4

Response 1 (a) $T_1 : C = (1,1) \to \big(x(1,1), y(1,1)\big) = \big(7(1) + 3(1), 1 + 4(1)\big) = (10, 5) = C$
(b) $A = (7, 1)$ and $B = (3, 4)$ from Example 1 •
$\overrightarrow{OA} = \langle 7-0, 1-0\rangle = \langle 7, 1\rangle$ and $\overrightarrow{BC} = \langle 10-3, 5-4\rangle = \langle 7, 1\rangle$ •
The region R in Figure 3 is a parallelogram because $\overrightarrow{OA}$ and $\overrightarrow{BC}$ are equal.

Response 2 Use (3). (a) The position vector of the image of $(0,0)$ is $\mathbf{R}(0,0) = \mathbf{R_0}$.
(b) The position vector of the image of $(h,0)$ is $\mathbf{R}(h,0) = \mathbf{R_0} + h\mathbf{A}$.
(c) The position vector of the image of $(0,k)$ is $\mathbf{R}(0,k) = \mathbf{R_0} + k\mathbf{B}$.

Response 3 $T_1^{-1} : O = (0,0) \to \big(\frac{4}{25}(0) - \frac{3}{25}(0), -\frac{1}{25}(0) + \frac{7}{25}(0)\big) = (0, 0) = \tilde{O}$ •
$T_1^{-1} : A = (7,1) \to \big(\frac{4}{25}(7) - \frac{3}{25}(1), -\frac{1}{25}(7) + \frac{7}{25}(1)\big) = (1, 0) = \tilde{A}$ •
$T_1^{-1} : B = (3,4) \to \big(\frac{4}{25}(3) - \frac{3}{25}(4), -\frac{1}{25}(3) + \frac{7}{25}(4)\big) = (0, 1) = \tilde{B}$ •
$T_1^{-1} : C = (10,5) \to \big(\frac{4}{25}(10) - \frac{3}{25}(5), -\frac{1}{25}(10) + \frac{7}{25}(5)\big) = (1, 1) = \tilde{C}$

Response 4 (a) The Jacobian of T_1^{-1} is $\dfrac{\partial(u,v)}{\partial(x,y)} = \begin{vmatrix} u_x & u_y \\ v_x & v_y \end{vmatrix}$
$= \begin{vmatrix} \frac{4}{25} & -\frac{3}{25} \\ -\frac{1}{25} & \frac{7}{25} \end{vmatrix} = \frac{4}{25}(\frac{7}{25}) - (-\frac{1}{25})(-\frac{3}{25}) = \dfrac{28 - 3}{25^2} = \dfrac{1}{25}$
(b) Since the Jacobian of T_1 is 25, the Jacobians of T_1 and T_1^{-1} are reciprocals of each other.
(c) According to Theorem 1, T_1 magnifies areas by the factor 25 and T_1^{-1} reduces areas by the factor $\frac{1}{25}$, which is consistent.

Response 5 (a) With cylindrical coordinates $x = r\cos\theta, y = r\sin\theta, z = z$:

$$\frac{\partial(x,y,z)}{\partial(r,\theta,z)} = \begin{vmatrix} \frac{\partial}{\partial r}(r\cos\theta) & \frac{\partial}{\partial \theta}(r\cos\theta) & \frac{\partial}{\partial z}(r\cos\theta) \\ \frac{\partial}{\partial r}(r\sin\theta) & \frac{\partial}{\partial \theta}(r\sin\theta) & \frac{\partial}{\partial z}(r\sin\theta) \\ \frac{\partial}{\partial r}(z) & \frac{\partial}{\partial \theta}(z) & \frac{\partial}{\partial z}(z) \end{vmatrix} = \begin{vmatrix} \cos\theta & -r\sin\theta & 0 \\ \sin\theta & r\cos\theta & 0 \\ 0 & 0 & 1 \end{vmatrix}$$

$$= \begin{vmatrix} \cos\theta & -r\sin\theta \\ \sin\theta & r\cos\theta \end{vmatrix} = r(\cos^2\theta + \sin^2\theta) = r \bullet \quad dx\,dy\,dz = r\,dr\,d\theta\,dz$$

Response 6
$$\frac{\partial(u,v,w)}{\partial(x,y,z)} = \begin{vmatrix} u_x & u_y & u_z \\ v_x & v_y & v_z \\ w_x & w_y & w_z \end{vmatrix} = \begin{vmatrix} 1 & 1 & 1 \\ 1 & 2 & 0 \\ 0 & 1 & -3 \end{vmatrix}$$

$$= (1)\begin{vmatrix} 2 & 0 \\ 1 & -3 \end{vmatrix} - (1)\begin{vmatrix} 1 & 0 \\ 0 & -3 \end{vmatrix} + (1)\begin{vmatrix} 1 & 2 \\ 0 & 1 \end{vmatrix} = -6 + 3 + 1 = -2 \bullet$$

$$\frac{\partial(x,y,z)}{\partial(u,v,w)} = \left[\frac{\partial(u,v,w)}{\partial(x,y,z)}\right]^{-1} = -\tfrac{1}{2}$$

CHAPTER 12 ANSWERS

Tune-Up Exercises 12.1

T1. Set $R_1 = \{(x,y) : 0 \leq x \leq 1, 0 \leq y \leq 2\}$ and $R_2 = \{(x,y) : 1 < x \leq 3, 0 \leq y \leq 2\}$. (Figure T1) •
[Area of R_1] = 2, [Area of R_2] = 4. •
$$\iint_R f(x,y)\,dx\,dy = \iint_{R_1} f(x,y)\,dx\,dy + \iint_{R_2} f(x,y)\,dx\,dy = 10(2) + 20(4) = 100$$

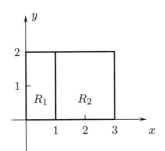
Figure T1

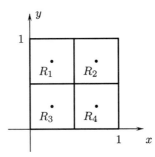
Figure T2

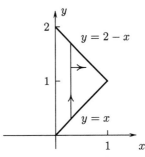
Figure T3

T2. Figure T2 • $\displaystyle\int_{x=0}^{x=1}\int_{y=0}^{y=1} g(x,y)\,dx\,dy \approx 2.80(\tfrac{1}{4}) + 4.84(\tfrac{1}{4}) + 1.30(\tfrac{1}{4}) + 2.76(\tfrac{1}{4}) = 2.925$

T3. Figure T3 • $\displaystyle\iint_R 10x^4 y\,dx\,dy = \int_{x=0}^{x=1}\int_{y=x}^{y=2-x} 10x^4 y\,dy\,dx = \int_{x=0}^{x=1}\left[5x^4 y^2\right]_{y=x}^{y=2-x} dx$
$= \displaystyle\int_{x=0}^{x=1} 5x^4[(2-x)^2 - x^2]\,dx = \int_{x=0}^{x=1}(20x^4 - 20x^5)\,dx = \left[4x^5 - \tfrac{20}{6}x^6\right]_0^1 = 4 - \tfrac{10}{3} = \tfrac{2}{3}$

T4. Figure T4 • $\displaystyle\iint_R 2x\,dx\,dy = \int_{y=-1}^{y=1}\left[x^2\right]_{x=0}^{x=1-y^2} dx = \int_{y=-1}^{y=1}(1-y^2)^2\,dx = 2\int_{y=0}^{y=1}(1-2y^2+y^4)\,dy$
$= 2\left[y - \tfrac{2}{3}y^3 + \tfrac{1}{5}y^5\right]_0^1 = 2(1 - \tfrac{2}{3} + \tfrac{1}{5}) = \tfrac{16}{15}$

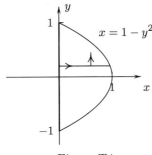
Figure T4

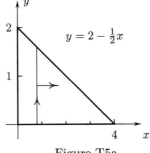
Figure T5a

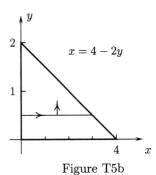
Figure T5b

T5. R is the triangle in Figure T5a. • Solve $y = 2 - \tfrac{1}{2}x$ for $x = 4 - 2y$: R is also given by $0 \leq y \leq 2$, $0 \leq x \leq 4 - 2y$ (Figure T5b) • $\displaystyle\int_{y=0}^{y=2}\int_{x=0}^{x=4-2y} f(x,y)\,dx\,dy$ or $\displaystyle\int_0^2 \int_0^{4-2y} f(x,y)\,dx\,dy$

T6. The projection of the solid on the xy-plane is the region R in Figure T6 •

$$[\text{Volume}] = \iint_R [(xy+2)-(-3)]\,dx\,dy = \int_{x=0}^{x=1}\int_{y=0}^{y=x^3}(xy+5)\,dx\,dy = \int_{x=0}^{x=1}\left[\tfrac{1}{2}xy^2+5y\right]_{y=0}^{y=x^3}dx$$

$$= \int_{x=0}^{x=1}[\tfrac{1}{2}x(x^3)^2+5x^3]\,dx = \int_{x=0}^{x=1}(\tfrac{1}{2}x^7+5x^3)\,dx = \left[\tfrac{1}{16}x^8+\tfrac{5}{4}x^4\right]_0^1 = \tfrac{21}{16}$$

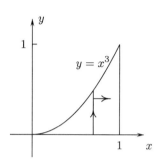

Figure T6

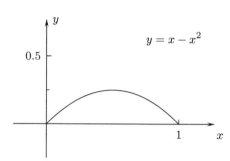

Figure T7

T7. (a) $[\text{Volume}] = \iint_R 6x^2\,dx\,dy = \int_{x=0}^{x=1}\int_{y=0}^{y=x-x^2}6x^2\,dy\,dx = \int_{x=0}^{x=1}(6x^3-6x^4)\,dx = \tfrac{3}{10}$ pounds

(b) $[\text{Moment about the }y\text{-axis}] = \iint_R x(6x^2)\,dx\,dy = \int_{x=0}^{x=1}\int_{y=0}^{y=x-x^2}6x^3\,dy\,dx$

$= \int_{x=0}^{x=1}(6x^4-6x^5)\,dx = \tfrac{1}{5}$ foot-pounds • $[\text{Moment about the }x\text{-axis}] = \iint_R y(6x^2)\,dx\,dy$

$= \int_{x=0}^{x=1}\int_{y=0}^{y=x-x^2}6x^2 y\,dy\,dx = \int_{x=0}^{x=1}3x^2(x-x^2)^2\,dx = \int_{x=0}^{x=1}(3x^4-6x^5+3x^6)\,dx$

$= \tfrac{1}{35}$ foot-pounds • $\bar{x} = \dfrac{1/5}{3/10} = \tfrac{2}{3},\ \bar{y} = \dfrac{1/35}{3/10} = \tfrac{2}{21}$

T8. $[\text{Area of }R] = 2$ • $[\text{Average value of }h\text{ over }R] = \tfrac{1}{2}\iint_R e^x\sin y\,dx\,dy = \tfrac{1}{2}\int_{x=0}^{x=2}\int_{y=0}^{y=1}e^x\sin y\,dy\,dx$

$= \tfrac{1}{2}\int_{x=0}^{x=2}e^x(1-\cos(1))\,dx = \tfrac{1}{2}(1-\cos(1))\left[e^x\right]_0^2 = \tfrac{1}{2}(1-\cos(1))(e^2-1)$

Problems 12.1

1. $\iint_R g(x,y)\,dx\,dy = 7000$

2. $\iint_R h(x,y)\,dx\,dy = 12\pi$

4. $\sin[(0.25)(0.25)] = \sin(0.065) \doteq 0.062,\ \sin[(0.25)(0.75)] = \sin[(0.75)(0.25)] \doteq 0.186,$
   $\sin[(0.75)(0.75)] \doteq 0.533$ • $\iint_R \sin(xy)\,dx\,dy \approx (0.062+0.186+0.186+0.533)(\tfrac{1}{4}) \doteq 0.242$

7. One possible answer: $\iint_R M(x,y)\,dx\,dy \approx 308$

10. $\iint_R 4x^3 y\,dx\,dy = \int_{x=0}^{x=2}\int_{y=x^2}^{y=2x}4x^3 y\,dy\,dx = \int_{x=0}^{x=2}\left[2x^3 y^2\right]_{y=x^2}^{y=2x}dx = \int_{x=0}^{x=2}(8x^5-2x^7)\,dx$

$= \left[\tfrac{4}{3}x^6-\tfrac{1}{4}x^8\right]_0^2 = \tfrac{64}{3}$

Answers 12.1

11. $\frac{7}{6}$

14. $\frac{1}{15}2^{17}$

17. $\iint_R y^2 \, dx \, dy = \int_{y=-1}^{y=1} \int_{x=y^2}^{x=2-y^2} y^2 \, dx \, dy = \int_{y=-1}^{y=1} \left[y^2 x \right]_{x=y^2}^{x=2-y^2} dy = \int_{y=-1}^{y=1} (2y^2 - 2y^4) \, dy$

$= \left[\frac{2}{3}y^3 - \frac{2}{5}y^5 \right]_{-1}^{1} = \frac{8}{15}$

18. $e^4 - 1$

19. $\iint_R \sqrt{xy} \, dx \, dy = \int_{y=0}^{y=1} \int_{x=y}^{x=4y} x^{1/2} y^{1/2} \, dx \, dy = \int_{y=0}^{y=1} \left[\frac{2}{3} x^{3/2} y^{1/2} \right]_{x=y}^{x=4y} dy$

$= \int_{y=0}^{y=1} \frac{2}{3} y^{1/2} (8y^{3/2} - y^{3/2}) \, dy = \left[\frac{14}{9} y^3 \right]_0^1 = \frac{14}{9}$

22. $8 \ln(4) + \frac{15}{16}$

24. $\iint_R [x^2 + y^2 - (-1)] \, dx \, dy = \int_{x=-1}^{x=1} \int_{y=x^2}^{y=1} (x^2 + y^2 + 1) \, dy \, dx = \int_{x=-1}^{x=1} \left[x^2 y + \frac{1}{3}y^3 + y \right]_{y=x^2}^{y=1} dx$

$= \int_{x=-1}^{x=1} (\frac{4}{3} - x^4 - \frac{1}{3}x^6) \, dx = \left[\frac{4}{3}x - \frac{1}{5}x^5 - \frac{1}{21}x^7 \right]_{-1}^{1} = \frac{76}{35}$

25. $4 + 2\pi^2$

29. $\iint_R (6 - 3x - 2y) \, dx \, dy = \int_{x=0}^{x=2} \int_{y=0}^{y=3-3x/2} (6 - 3x - 2y) \, dy \, dx$

$= \int_{x=0}^{x=2} \left[6y - 3xy - y^2 \right]_{y=0}^{y=3-3x/2} dx = 9 \int_{x=0}^{x=2} (1 - \frac{x}{2})^2 \, dx = \left[-6(1 - \frac{x}{2})^3 \right]_0^2 = 6$

31. (a) $\iint_R xy \, dx \, dy = \int_{x=0}^{x=1} \int_{y=0}^{y=\sqrt{1-x^2}} xy \, dy \, dx = \int_{x=0}^{x=2} \left[\frac{1}{2} y^2 x \right]_{y=0}^{y=\sqrt{1-x^2}} dx$

$= \int_{x=0}^{x=1} \frac{1}{2}(x - x^3) \, dx = \frac{1}{8}$ grams

(b) $\bar{x} = 8 \int_{x=0}^{x=1} \int_{y=0}^{y=\sqrt{1-x^2}} x^2 y \, dy \, dx$

$= \int_{x=0}^{x=2} \left[\frac{1}{2} y^2 x^2 \right]_{y=0}^{y=\sqrt{1-x^2}} dx = \int_{x=0}^{x=1} \frac{1}{2}(x^2 - x^4) \, dx = \frac{8}{15}$ • $\bar{x} = \bar{y}$ by symmetry • $(\frac{8}{15}, \frac{8}{15})$

32. (a) $\frac{8}{3}$ newtons (b) Center of gravity: $(\frac{5}{4}, \frac{9}{16})$

34. $[\text{Area}] = \int_0^1 (x^2 - x^3) \, dx = \frac{1}{12}$ • [Average value of f over R] $= 12 \iint_R x^2 y^2 \, dx \, dy$

$= 12 \int_{x=0}^{x=1} \int_{y=x^3}^{y=x^2} x^2 y^2 \, dy \, dx = 4 \int_{x=0}^{x=1} \left[y^3 x^2 \right]_{y=x^3}^{y=x^2} dx = 4 \int_{x=0}^{x=1} (x^8 - x^{11}) \, dx = \frac{1}{9}$

35. (a) The triangle R has area $\frac{1}{2}k^2$ • [Average value of g in R] $= \frac{2}{k^2} \iint_R \sin x \sin y \, dx \, dy$

$= \frac{2}{k^2} \int_{x=0}^{x=k} \int_{y=0}^{y=x} \sin x \sin y \, dy \, dx = \frac{2}{k^2} \int_{x=0}^{x=k} \left[-\sin x \cos y \right]_{y=0}^{y=x} dx$

$= \frac{2}{k^2} \int_{x=0}^{x=k} (\sin x - \sin x \cos x) \, dx = \frac{2}{k^2} \left[-\cos x + \frac{1}{2} \cos^2 x \right]_{x=0}^{x=k} = \frac{\cos^2(k) - 2\cos(k) + 1}{k^2}$

(b) [Average value] $\to 0$ as $k \to \infty$

37. (a) $\sum_{j=1}^{N} e^{x_j} \sin(x_j y_j) [\text{Area of } R_j]$ with $R_1, R-2, \ldots, R_N$ a partition of R

38. $\iint_R (x + y) \, dx \, dy = -\frac{2}{15}$

39. $\iint_R x^2 \, dx \, dy = \int_{x=0}^{x=1} \int_{y=x}^{y=3x} x^2 \, dy \, dx + \int_{x=1}^{x=2} \int_{y=x}^{y=4-x} x^2 \, dy \, dx$

$= \int_{x=0}^{x=1} 2x^3 \, dx + \int_{x=1}^{x=2} (4x^2 - 2x^3) \, dx = \frac{7}{3}$

40. $\iint_R 4xy \, dx \, dy = \frac{1}{2} + 2\ln(2)$

43. $\iint_R xy^4 \, dx \, dy = \int_{y=-\sqrt{k}}^{y=\sqrt{k}} \int_{x=-k+y^2}^{x=k+y^2} xy^4 \, dx \, dy$

$= \int_{y=-\sqrt{k}}^{y=\sqrt{k}} \left[x^2 y^4\right]_{x=-k+y^2}^{x=k-y^2} dy = \int_{y=-\sqrt{k}}^{y=\sqrt{k}} 0 \, dy = 0$

44. $-\frac{2}{3}$

46. (a) $\int_{y=0}^{y=4} \int_{x=-\sqrt{y}}^{x=\sqrt{y}} f(x,y) \, dx \, dy$ (b) $\int_{y=-1}^{y=1} \int_{x=y^{1/3}}^{x=1} f(x,y) \, dx \, dy$

47. (a) $\int_{x=-8}^{x=1} \int_{-2}^{y=x^{1/3}} f(x,y) \, dy \, dx$

48. (a) $\int_{y=0}^{y=1} \int_{x=-y^{1/2}}^{x=y^{1/2}} f(x,y) \, dx \, dy + \int_{y=1}^{y=4} \int_{x=y-2}^{x=y^{1/2}} f(x,y) \, dx \, dy$

(b) $\int_{x=0}^{x=1} \int_{y=x/2}^{y=x} f(x,y) \, dy \, dx + \int_{x=1}^{x=2} \int_{y=x/2}^{y=1} f(x,y) \, dy \, dx$

50. (a) $\sqrt{2} - 2 + 2(e - e^{1/\sqrt{2}})$

53. $\frac{1000}{3}$

55. $\frac{1}{6}$

58a. Set $x_j = j/10, y_k = k/10$ for $j, k = 1, 2, \ldots, 10$ •

[Riemann sum] $= \frac{1}{100} \sum_{j=1}^{100} \sum_{k=1}^{100} (\frac{1}{10}j)^2(\frac{1}{10}k)^3 = \frac{9317}{8000} \doteq 0.116463$

60a (a) [Area] $= \int_{x=0}^{x=1} \int_{y=0}^{y=\sqrt{1-x^2}} dy \, dx = \frac{1}{4}\pi$ • $\bar{x} = \frac{4}{\pi} \int_{x=0}^{x=1} \int_{y=0}^{y=\sqrt{1-x^2}} x \, dy \, dx = \frac{4}{3\pi}$ •

$\bar{y} = \bar{x}$ by symmetry • Centroid: $\left(\frac{4}{3\pi}, \frac{4}{3\pi}\right)$ (b) Centroid: $(\frac{12}{5}, \frac{3}{4})$

Tune-Up Exercises 12.2

T1. $\iiint_V f(x,y,z) \, dx \, dy \, dz = \iint_R \int_{z=0}^{z=1} f(x,y,z) \, dz \, dy \, dx$ with R the rectangle

$0 \le x \le 1, -1 \le y \le 1$ • Figure T1 • $\int_{x=0}^{x=1} \int_{y=-1}^{y=1} \int_{z=0}^{z=1} f(x,y,z) \, dz \, dy \, dx$

T2. The top of V is formed by $z = 1 - y^2$ and the bottom by $z = 0$. • The conditions $x = -2$ and $x = 0$ give only two sides of the projection R of V on the xy-plane. • The top and bottom interseect in the lines $y = \pm 1$ in the xy-plane, so R is the square $-2 \le x \le 0, -1 \le y \le 1$ in Figure T2. •

$\iiint_V g(x,y,z) \, dx \, dy \, dz = \iint_R \int_{z=0}^{z=1-y^2} g(x,y,z) \, dz \, dy \, dx$

$= \int_{x=-2}^{x=0} \int_{y=-1}^{y=1} \int_{z=0}^{z=1-y^2} g(x,y,z) \, dz \, dy \, dx$

Answers 12.2

Figure T1   Figure T2

T3. $\iiint_V x^2 y^3 z^4 \, dx \, dy \, dz = \int_{x=0}^{x=1} \int_{y=0}^{y=1} \int_{z=0}^{z=1} x^2 y^3 z^4 \, dz \, dy \, dx$

$= \left(\int_0^1 x^2 \, dx\right)\left(\int_0^1 y^3 \, dy\right)\left(\int_0^1 z^4 \, dz\right) = (\frac{1}{3})(\frac{1}{4})(\frac{1}{5}) = \frac{1}{60}$

T4. Because $x^2 + y^2 \geq 0$, the top of V is formed by $z = x^2 + y^2$ and the bottom by $z = -1$. The projection of V on the xy-plane is the square $R: 0 \leq x \leq 1,\ 0 \leq y \leq 1$. •

$\iiint_V xy \, dx \, dy \, dz = \int_{x=0}^{x=1} \int_{y=0}^{y=1} \int_{z=-1}^{z=x^2+y^2} xy \, dz \, dy \, dx = \int_{x=0}^{x=1} \int_{y=0}^{y=1} \Big[xyz\Big]_{z=-1}^{z=x^2+y^2} dy \, dx$

$= \int_{x=0}^{x=1} \int_{y=0}^{y=1} [xy(x^2+y^2) + xy] \, dy \, dx = \int_{x=0}^{x=1} \int_{y=0}^{y=1} (x^3 y + xy^3 + xy) \, dy \, dx$

$= \int_{x=0}^{x=1} \left[\frac{1}{2}x^3 y^2 + \frac{1}{4}xy^4 + \frac{1}{2}xy^2\right]_{y=0}^{y=1} dx = \int_{x=0}^{x=1} (\frac{1}{2}x^3 + \frac{3}{4}x) \, dx = \left[\frac{1}{8}x^4 + \frac{3}{8}x^2\right]_{x=0}^{x=1} = \frac{1}{2}$

T5. $[\text{Mass}] = \iiint_V \sqrt{1 + xyz} \, dx \, dy \, dz$ kilograms • $\bar{x} = \dfrac{\iiint_V x\sqrt{1+xyz}\, dx\,dy\,dz}{\iiint_V \sqrt{1+xyz}\, dx\,dy\,dz}$ meters •

$\bar{y} = \dfrac{\iiint_V y\sqrt{1+xyz}\, dx\,dy\,dz}{\iiint_V \sqrt{1+xyz}\, dx\,dy\,dz}$ meters • $\bar{z} = \dfrac{\iiint_V z\sqrt{1+xyz}\, dx\,dy\,dz}{\iiint_V \sqrt{1+xyz}\, dx\,dy\,dz}$ meters

T6. (a) $\iiint_V 7 \, dx\, dy\, dz = 7 \iiint_V dx\, dy\, dz = 7[\text{Volume of } V] = 70$

(b) $[\text{Average value of 7 in } V] = \dfrac{1}{[\text{Volume of } V]} \iiint_V 7 \, dx\, dy\, dz = 7$

Problems 12.2

1. $\iiint_V (x^2 + y^2 + z^2) \, dx\, dy\, dz = \int_{x=-1}^{x=2} \int_{y=0}^{y=1} \int_{z=0}^{z=2} (x^2 + y^2 + z^2) \, dz\, dy\, dx$

$= \int_{x=-1}^{x=2} \int_{y=0}^{y=1} (2x^2 + 2y^2 + \frac{8}{3}) \, dy \, dx = \int_{x=-1}^{x=2} (2x^2 + \frac{10}{3}) \, dx = \left[\frac{2}{3}x^3 + \frac{10}{3}x\right]_{-1}^{2}$

$= \frac{16}{3} + \frac{20}{3} - (-\frac{2}{3} - \frac{10}{3}) = \frac{48}{3} = 16$

2. $\iiint_V (x^2 y e^{xyz}) \, dx \, dy \, dz = \int_{x=0}^{x=1} \int_{y=0}^{y=4} \int_{z=0}^{z=3} (x^2 y e^{xyz}) \, dz \, dy \, dx = \int_{x=0}^{x=1} \int_{y=0}^{y=4} (xe^{3xy} - x) \, dy \, dx$

$= \int_{x=0}^{x=1} \left[\tfrac{1}{3} e^{3xy} - xy \right]_0^4 dx = \int_{x=0}^{x=1} (\tfrac{1}{3} e^{12x} - 4x - \tfrac{1}{3}) \, dx = \left[\tfrac{1}{36} e^{12x} - 2x^2 - \tfrac{1}{3} x \right]_0^1 = \tfrac{1}{36}(e^{12} - 85)$

3. $18(1 - e^{-4})$

4. $\iiint_V z \, dx \, dy \, dz = \int_{x=-1}^{x=1} \int_{y=-1}^{y=1} \int_{z=0}^{z=\sqrt{x^2+y^2}} z \, dz \, dy \, dx = \int_{x=-1}^{x=1} \int_{y=-1}^{y=1} \left[\tfrac{1}{2} z^2 \right]_{z=0}^{z=\sqrt{x^2+y^2}} dy \, dx$

$= \tfrac{1}{2} \int_{x=-1}^{x=1} \int_{y=-1}^{y=1} (x^2 + y^2) \, dy \, dx = 2 \int_{x=0}^{x=1} \left[x^2 y + \tfrac{1}{3} y^3 \right]_{y=0}^{y=1} dx$

$= \int_{x=0}^{x=1} 2x^2 + \tfrac{2}{3} \, dx = \left[\tfrac{2}{3} x^3 + \tfrac{2}{3} x \right]_{x=0}^{x=1} = \tfrac{4}{3}$

5. $\tfrac{1}{3}(e^{18} - 1)[4 - \sin(4)]$

7. $\int_{z=-2}^{z=2} \int_{y=-3}^{y=3} \int_{x=-4}^{x=4} e^{2x+3y-4z} \, dx \, dy \, dz = \left(\int_{-2}^{2} e^{-4z} \, dz \right) \left(\int_{-3}^{3} e^{3y} \, dy \right) \left(\int_{-4}^{4} e^{2x} \, dx \right)$

$= [-\tfrac{1}{4}(e^{-8} - e^8)][\tfrac{1}{3}(e^9 - e^{-9})][\tfrac{1}{2}(e^8 - e^{-8})] = \tfrac{1}{24}(e^8 - e^{-8})^2(e^9 - e^{-9})$

8. $\tfrac{81}{2}[\cos(1) - \cos(5)]$

10. $-\tfrac{70}{27}$

11. $\int_{z=1}^{z=2} \int_{y=-2}^{y=2} \int_{x=-2}^{x=2} x^2 e^y \sin z \, dx \, dy \, dz = \left(\int_1^2 \sin z \, dz \right) \left(\int_{-2}^{2} e^y \, dy \right) \left(\int_{-2}^{2} x^2 \, dx \right)$

$= [\cos(1) - \cos(2)][(e^2 - e^{-2})][\tfrac{1}{3}(2^3 - (-2)^3)] = \tfrac{16}{3}[\cos(1) - \cos(2)](e^2 - e^{-2})$

12. $\tfrac{736}{15}$ coulombs

13. **(a)** $[\text{Weight}] = \int_{x=0}^{x=2} \int_{y=0}^{y=4} \int_{z=0}^{z=6} (x + y + z) \, dz \, dy \, dx = \int_{x=0}^{x=2} \int_{y=0}^{y=4} \left[xz + yz + \tfrac{1}{2} z^2 \right]_{z=0}^{z=6} dy \, dx$

$= \int_{x=0}^{x=2} \int_{y=0}^{y=4} (6x + 6y + 18) \, dy \, dx = \int_{x=0}^{x=2} \left[6xy + 3y^2 + 18y \right]_{y=0}^{y=4} dx = \int_{x=0}^{x=2} (24x + 48 + 72) \, dx$

$= \left[12x^2 + 120x \right]_{x=0}^{x=2} = 288$ dynes

(b) $[\text{Moment about } x = 0] = \int_{x=0}^{x=2} \int_{y=0}^{y=4} \int_{z=0}^{z=6} (x^2 + xy + xz) \, dz \, dy \, dx$

$= \int_{x=0}^{x=2} \int_{y=0}^{y=4} \left[x^2 z + xyz + \tfrac{1}{2} xz^2 \right]_{z=0}^{z=6} dy \, dx = \int_{x=0}^{x=2} \int_{y=0}^{y=4} (6x^2 + 6xy + 18x) \, dy \, dx$

$= \int_{x=0}^{x=2} \left[6x^2 y + 3xy^2 + 18xy \right]_{y=0}^{y=4} dx = \int_{x=0}^{x=2} (24x^2 + 120x) \, dx = \left[8x^3 + 60x^2 \right]_{x=0}^{x=2} = 304$ •

$\bar{x} = \tfrac{304}{288} = \tfrac{19}{18}$ **(c)** $\bar{y} = \tfrac{20}{9}$ **(d)** $\bar{z} = \tfrac{7}{2}$ centimeters

15. $[\text{Volume of } V] = 8$ cubic feet • $\iiint_V xy^3 z^7 \, dx \, dy \, dz = \int_{x=0}^{x=2} \int_{y=0}^{y=2} \int_{z=0}^{z=2} xy^3 z^7 \, dz \, dy \, dx$

$= \left(\int_{z=0}^{z=2} z^7 \, dz \right) \left(\int_{y=0}^{y=2} y^3 \, dy \right) \left(\int_{x=0}^{x=2} x \, dx \right) = \left[\tfrac{1}{2} x^2 \right]_0^2 \left[\tfrac{1}{4} y^4 \right]_0^2 \left[\tfrac{1}{8} z^8 \right]_0^2 = (2)(4)(32) = 256$ •

[Average value of $xy^3 z^7$ in V] $= \tfrac{256}{8} = 32$

19. Because the integration with respect to z is from the xy-plane ($z = 0$) to $z = \sqrt{4 - x^2 - y^2}$ the solid is part of the upper hemisphere $x^2 + y^2 + z^2 = 4$. • Because the integration with respect to y is from the x-axis ($y = 0$) to $z = \sqrt{4 - y^2}$ and with respect to x from 0 to 2, the solid is the eighth-sphere $\{(x, y, z) : 0 \leq z \leq \sqrt{4 - x^2 - y^2}, x \geq 0, y \geq 0\}$.

Answers 12.3

20. The cylinder with base formed by the region between $y = x^2$ and $y = 1$ in the xy-plane, with vertical sides, and with top formed by the plane $z = 1 - y$

21. The tetrahedron bounded by the plane $x = 4 - 4y - 2z$ and the coordinate planes.

23. The circular cylinder of radius 1 with the y-axis as axis and extending from $y = 0$ to $y = 4$.

25. $z = y^3$ intersects $z = -1$ in the line $y = -1$, $z = -1$ • The top of V is formed by $z = y^3$ and the bottom by $z = -1$ and its projection on the xy-plane is the rectangle, $0 \le x \le 1$, $-1 \le y \le 1$ •

$$\iiint_V x^2 y^2 \cos z \, dx \, dy \, dz = \int_{x=0}^{x=1} \int_{y=-1}^{y=1} \int_{z=-1}^{z=y^3} x^2 y^2 \cos z \, dz \, dy \, dx$$

$$= \int_{x=0}^{x=1} \int_{y=-1}^{y=1} [x^2 y^2 \sin(y^3) - x^2 y^2 \sin(-1)] \, dy \, dx$$

$$= \int_{x=0}^{x=1} \left[-\tfrac{1}{3} x^2 \cos(y^3) - \tfrac{1}{3} x^2 y^3 \sin(-1) \right]_{y=-1}^{y=1} dx = \int_{x=0}^{x=1} \tfrac{2}{3} x^2 \sin(1) \, dx = \tfrac{2}{9} \sin(1)$$

26. $\tfrac{8}{3} \ln(2) - \tfrac{29}{18}$

28. $\tfrac{64}{21}$

31. $\tfrac{4}{525}$

34. $k = 1$

38. $(0, 0, \dfrac{\sin(1) + \cos(1) - 1}{\sin(1)})$

40. $\displaystyle\int_{z=0}^{z=4} \int_{x=0}^{x=4-z} \int_{y=0}^{y=4-x-z} f(x,y,z) \, dy \, dx \, dz$ or $\displaystyle\int_{z=0}^{z=4} \int_{y=0}^{y=4-z} \int_{x=0}^{x=4-y-z} f(x,y,z) \, dx \, dy \, dz$

Tune-Up Exercises 12.3

T1. Figure T1 • $\displaystyle\int_{\theta=0}^{\theta=\pi} \int_{r=0}^{r=6\cos\theta} r \, dr \, d\theta = \int_{\theta=0}^{\theta=\pi} \left[\tfrac{1}{2} r^2\right]_{r=0}^{r=6\cos\theta} d\theta = \tfrac{1}{2} \int_0^\pi (6\sin\theta)^2 \, d\theta$

$= \tfrac{36}{4} \displaystyle\int_0^\pi [1 - \cos(2\theta)] \, d\theta = 9 \left[\theta - \tfrac{1}{2}\sin(2\theta)\right]_0^\pi = 9\pi$

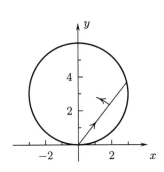

Figure T1

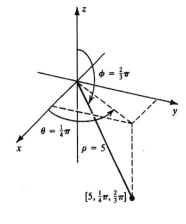

Figure T3

T2. $x = r\cos\theta = 4\cos(\tfrac{7}{6}\pi) = -2\sqrt{3}, y = r\sin\theta = 4\sin(\tfrac{7}{6}\pi) = -2$ • $(-2\sqrt{3}, -2, 10)$

T3. The ray from the origin to the point makes an angle of $\phi = \frac{2}{3}\pi$ with the positive z-axis and its projection on the xy-plane makes an angle of $\theta = \frac{1}{4}\pi$ with the positive x-axis (Figure T3). • $x = \rho\sin\phi\cos\theta = 5\sin(\frac{2}{3}\pi)\cos(\frac{1}{4}\pi) = 5(\frac{1}{2}\sqrt{3})(\frac{1}{2}\sqrt{2}) = \frac{5}{4}\sqrt{6}$ • $y = \rho\sin\phi\sin\theta = 5\sin(\frac{2}{3}\pi)\sin(\frac{1}{4}\pi) = 5(\frac{1}{2}\sqrt{3})(\frac{1}{2}\sqrt{2}) = \frac{5}{4}\sqrt{6}$ • $z = \rho\cos\phi = 5\cos(\frac{2}{3}\pi) = 5(-\frac{1}{2}) = -\frac{5}{2}$ • $(\frac{5}{4}\sqrt{6}, \frac{5}{4}\sqrt{6}, -\frac{5}{2})$

T4. (a) The projection of $P(-2, 2\sqrt{3}, 4)$ on the xy-plane is the point $Q(-2, 2\sqrt{3}, 0)$. (Figure T4a) • The polar coordinates of Q are $r = 4$ and $\theta = \frac{2}{3}\pi$ (Figure T4b). • The cylindrical coordinates of P are $r = 4$, $\theta = \frac{2}{3}\pi$, $z = 4$.

(b) The sketch of the vertical plane containing P and the z-axis in Figure Tc shows that the spherical coordinates are $\rho = 4\sqrt{2}$, $\theta = \frac{2}{3}\pi$, $\phi = \frac{1}{4}\pi$.

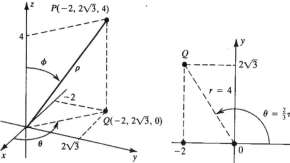
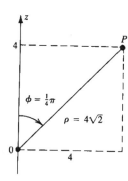

Figure T4a   Figure T4b   Figure T4c

T5. (a) $2 = r\cos\theta$ • $r = 2\sec\theta$ (b) $-3 = r\sin\theta$ • $r = -3\csc\theta$ (c) $r^2 = 19$ • $r = \sqrt{19}$ (d) $r\cos\theta = r\sin\theta$ • $r = 0$ or $\cos\theta = \sin\theta$ • $r = 0$ or $\tan\theta = 1$ • $\theta = \frac{1}{4}\pi$ or $\theta = \frac{5}{4}\pi$ (since we use only $r \geq 0$)

T6. (a) $z = r^2$ (b) $r^2 + z^2 = 4$ or $z = \pm\sqrt{4 - r^2}$ (c) $r\cos\theta = 2$ • $r = 2\sec\theta$ (d) $0 \leq z \leq r^2 + 1$
(e) One answer: $\pi \leq \theta \leq 2\pi$ since the half-rays $\theta = \pi$ and $\theta = 2\pi$ combine to form the x-axis

T7. (a) $0 \leq \rho \leq 3$ since $x^2 + y^2 + z^2 = 9$ is the sphere $\rho = 3$ (b) $\frac{1}{4}\pi \leq \phi \leq \pi$ since $z = \sqrt{x^2 + y^2}$ is the half-cone $\phi = \frac{1}{4}\pi$ (c) One answer: $\frac{1}{2}\pi \leq \theta \leq \frac{3}{2}\pi$ (d) $\frac{1}{2}\pi < \phi \leq \pi$ with $\rho > 0$

T8. (a) $\int_0^{2\pi}\int_0^R \int_{-\sqrt{R^2-r^2}}^{\sqrt{R^2-r^2}} r\,dz\,dr\,d\theta = \int_0^{2\pi}\int_0^R 2r(R^2-r^2)^{1/2}\,dr\,d\theta = \int_0^{2\pi}\left[-\frac{2}{3}(R^2-r^2)^{3/2}\right]_0^R d\theta$
$= \int_0^{2\pi} \frac{2}{3}R^3\,d\theta = \frac{4}{3}\pi R^3$ (b) $\int_0^{2\pi}\int_0^\pi \int_0^R \rho^2\sin\phi\,d\rho\,d\phi\,d\theta = \frac{1}{3}\int_0^{2\pi}\int_0^\pi R^3\sin\phi\,d\phi\,d\theta = \frac{2}{3}\int_0^{2\pi} R^3\,d\theta = \frac{4}{3}\pi R^3$

T9. The cone $z = \sqrt{x^2+y^2}$ has the equation $z = r$ in cylindrical coordinates. • The projection of the solid on the xy-plane is the disk $0 \leq r \leq 1$ (Figure T9) • $dx\,dy\,dz = r\,dr\,d\theta\,dz$ •
$\iiint_R 4z^3\,dx\,dy\,dz = \int_{\theta=0}^{\theta=2\pi}\int_{r=0}^{r=1}\int_{z=r}^{z=1} 4z^3 r\,dz\,dr\,d\theta = \int_{\theta=0}^{\theta=2\pi}\int_{r=0}^{r=1}\left[z^4 r\right]_{z=r}^{z=1} dr\,d\theta$
$= \int_{\theta=0}^{\theta=2\pi}\int_{r=0}^{r=1}(r - r^5)\,dr\,d\theta = \int_{\theta=0}^{\theta=2\pi}\left[\frac{1}{2}r^2 - \frac{1}{6}r^6\right]_0^1 d\theta = \int_{\theta=0}^{\theta=2\pi}(\frac{1}{2} - \frac{1}{6})\,d\theta = \frac{2}{3}\pi$

T10. $\int_{\phi=0}^{\phi=\pi/2}\int_{\theta=0}^{\theta=\pi}\int_{\rho=0}^{\rho=1} \rho^2\cos^2\phi\rho^2\sin\phi\,d\rho\,d\theta\,d\phi = \int_{\phi=0}^{\phi=\pi/2}\int_{\theta=0}^{\theta=\pi} \frac{1}{5}\cos^2\phi\sin\phi\,d\theta\,d\phi$
$= \frac{1}{5}\pi\int_{\phi=0}^{\phi=\pi/2} \cos^2\phi\sin\phi\,d\phi = -\frac{1}{15}\pi\left[\cos^3\phi\right]_0^{\pi/2} = \frac{1}{15}\pi$

Answers 12.3

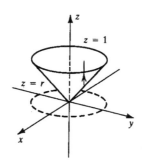

Figure T9

Problems 12.3

1. Figure A1 • [Area] = $\int_{\theta=0}^{\theta=2\pi} \int_{r=0}^{r=4+4\sin\theta} r\, dr\, d\theta = \frac{1}{2}\int_0^{2\pi} (5+4\sin\theta)^2\, d\theta$

$= \frac{1}{2}\int_0^{2\pi} [25 + 40\sin\theta + 8 - 8\cos(2\theta)]\, d\theta = \frac{1}{2}\Big[33\theta - 40\cos\theta - 4\sin(2\theta)\Big]_0^{2\pi} = 33\pi$

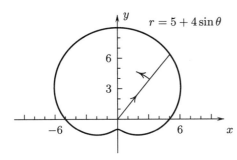
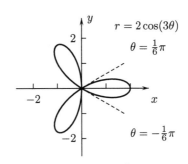

Figure A1   Figure A2

2. Figure A2 • $r = 2\cos(3\theta)$ is zero at $\theta = \pm\frac{1}{6}\pi$. • [Area] $= \frac{1}{2}\int_{-\frac{\pi}{6}}^{\frac{\pi}{6}} (2\cos 3\theta)^2\, d\theta = 4\int_0^{\frac{\pi}{6}} \cos^2 3\theta\, d\theta$

$= 2\int_0^{\frac{\pi}{6}} (1 + \cos 6\theta)\, d\theta = 2\Big[\theta + \frac{1}{6}\sin(6\theta)\Big]_0^{\frac{\pi}{6}} = \frac{1}{3}\pi$

3. $y = \sqrt{x - x^2} \iff y^2 = x - x^2$ with $y \geq 0$ • Complete the square: $x^2 - x + \frac{1}{4} + y^2 = \frac{1}{4}$ with $y \geq 0$ • $(x - \frac{1}{2})^2 + y^2 = \frac{1}{4}$ • The semicircle is $r = \cos\theta$ for $0 \leq \theta \leq \frac{1}{2}\pi$ • Figure A3

$\int_{\theta=0}^{\theta=\pi/2} \int_{r=0}^{r=\cos\theta} (r\cos\theta)(r\sin\theta)r\, dr\, d\theta = \int_{\theta=0}^{\theta=\pi/2} \int_{r=0}^{r=\cos\theta} r^3 \cos\theta \sin\theta\, dr\, d\theta$

$= \int_{\theta=0}^{\theta=\pi/2} \Big[\frac{1}{4}r^4 \cos\theta \sin\theta\Big]_{r=0}^{r=\cos\theta} d\theta = \frac{1}{4}\int_0^{\pi/2} \cos^5\theta \sin\theta\, d\theta = \frac{1}{4}\Big[-\frac{1}{6}\cos^6\theta\Big]_0^{\pi/2} = \frac{1}{24}$

Figure A3

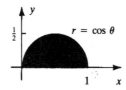

5. $\frac{9}{2}\pi$

6. $\displaystyle\int_{\theta=0}^{\theta=2\pi}\int_{r=\sqrt{2}}^{r=2} r^{-4}\, r\, dr\, d\theta = \left(\int_{\theta=0}^{\theta=2\pi} d\theta\right)\left(\int_{r=\sqrt{2}}^{r=2} r^{-3}\, dr\right) = 2\pi\left[-\tfrac{1}{2}r^{-2}\right]_{r=\sqrt{2}}^{r=2} = \tfrac{1}{4}\pi$

7. (a) $\displaystyle\int_{x=-4}^{x=4}\int_{-\sqrt{16-x^2}}^{\sqrt{16-x^2}} e^{-x^2-y^2}\, dy\, dx = \int_{\theta=0}^{\theta=2\pi}\int_{r=0}^{r=4} e^{-r^2} r\, dr\, d\theta = \int_{\theta=0}^{\theta=2\pi}\left[-\tfrac{1}{2}e^{-r^2}\right]_{r=0}^{r=4} d\theta$

$= \tfrac{1}{2}\displaystyle\int_{\theta=0}^{\theta=2\pi}(1-e^{-16})\, d\theta = \pi(1-e^{-16})$

11. $\bar{y}=0$ by symmetry • [Weight] $= \displaystyle\int_{\theta=-\pi/2}^{\theta=\pi/2}\int_{r=0}^{r=2\cos\theta} 5r^2\, dr\, d\theta = \int_{\theta=-\pi/2}^{\theta=\pi/2}\left[\tfrac{5}{3}r^3\right]_0^{2\cos\theta} d\theta$

$= \displaystyle\int_{\theta=-\pi/2}^{\theta=\pi/2} \tfrac{40}{3}\cos^3\theta\, d\theta = \tfrac{160}{9}$ • $\bar{x} = \tfrac{9}{160}\displaystyle\int_{\theta=-\pi/2}^{\theta=\pi/2}\int_{r=0}^{r=2\cos\theta} 5r^3\cos\theta\, dr\, d\theta$

$= \tfrac{9}{160}\displaystyle\int_{\theta=-\pi/2}^{\theta=\pi/2}\left[\tfrac{5}{4}r^4\cos\theta\right]_0^{2\cos\theta} d\theta = \tfrac{9}{8}\int_{\theta=-\pi/2}^{\theta=\pi/2}\cos^5\theta\, d\theta = \tfrac{6}{5}$ • $(\tfrac{6}{5},0)$

12. (a) $x=3\cos(\tfrac{5}{6}\pi)=-\tfrac{3}{2}\sqrt{3}$ • $y=3\sin(\tfrac{5}{6}\pi)=\tfrac{3}{2}$ • $(-\tfrac{3}{2}\sqrt{3},\tfrac{3}{2},4)$ **(b)** $(0,-1,-2)$

13. (a) $x=4\sin(\tfrac{1}{3}\pi)\cos(\tfrac{3}{4}\pi)=4(\tfrac{1}{2}\sqrt{3})(-\tfrac{1}{2}\sqrt{2})=-\sqrt{6}$ • $y=4\sin(\tfrac{3}{4}\pi)\sin(\tfrac{1}{3}\pi)=4(\tfrac{1}{2}\sqrt{3})(\tfrac{1}{2}\sqrt{2})$
$=\sqrt{6}$ • $z=4\cos(\tfrac{1}{3}\pi)=4(\tfrac{1}{2})=2$ • $(-\sqrt{6},\sqrt{6},2)$ **(b)** $(0,0,0)$

14. (a) Cylinder: $x^2+y^2=4$ **(b)** Half-plane: $y=x$, $x\geq 0$ **(c)** Half-plane: $y=-\sqrt{3}x$, $x\leq 0$
(e) Sphere: $x^2+y^2+z^2=4$ **(f)** Half-cone: $z=\sqrt{x^2+y^2}$ **(g)** Half-cone: $z=\dfrac{\sqrt{x^2+y^2}}{\sqrt{3}}$

15. (a) $z=-r$ **(b)** $\theta=\tfrac{1}{4}\pi$ **(c)** $z=-\sqrt{4-r^2}$ **(e)** $r=2\cos\theta$ **(g)** $r=3\csc\theta$

17. $\displaystyle\iiint_V z\sqrt{x^2+y^2+z^2}\, dx\, dy\, dz = \int_{\theta=0}^{\theta=2\pi}\int_{r=0}^{r=4}\int_{z=0}^{z=3} z\sqrt{r^2+z^2}\, r\, dz\, dr\, d\theta$

$= \displaystyle\int_{\theta=0}^{\theta=2\pi}\int_{r=0}^{r=4} r\left[\tfrac{1}{3}(r^2+z^2)^{3/2}\right]_{z=0}^{z=3} dr\, d\theta = 2\pi\int_{r=0}^{r=4}\left[\tfrac{1}{3}r(r^2+9)^{3/2}-\tfrac{1}{3}r^4\right] dr$

$= 2\pi\left[\tfrac{1}{15}(r^2+9)^{5/2}-\tfrac{1}{15}r^5\right]_0^4 = \tfrac{3716}{15}\pi$

18. $\displaystyle\iiint_V \cos[(x^2+y^2+z^2)^{3/2}]\, dx\, dy\, dz = \int_{\theta=0}^{\theta=2\pi}\int_{\phi=0}^{\phi=\pi}\int_{\rho=1}^{\rho=6}\cos(\rho^3)\rho^2\sin\phi\, d\rho\, d\phi\, d\theta$

$= \left(\displaystyle\int_{\theta=0}^{\theta=2\pi} d\theta\right)\left(\int_{\phi=0}^{\phi=\pi}\sin\phi\, d\phi\right)\left(\int_{\rho=0}^{\rho=6}\cos(\rho^3)\rho^2\, d\rho\right) = 2\pi\left[-\cos\phi\right]_{\phi=0}^{\phi=\pi}\left[\tfrac{1}{3}\sin(\rho^3)\right]_{\rho=1}^{\rho=6}$

$= \tfrac{4}{3}\pi[\sin(6^3)-\sin(1)]$

19. $\tfrac{25}{6}\pi$

20. $\tfrac{64}{21}\pi\sqrt{2}$

24. $\displaystyle\int_{\theta=\pi/4}^{\theta=3\pi/4}\int_{r=1/\sin\theta}^{r=\sqrt{2}}\left(\dfrac{r\cos\theta+1}{r\sin\theta+1}\right) r\, dr\, d\theta$

25. $\tfrac{625}{12}(3\sqrt{3}+1)$

27. (a) $r=\sqrt{(2)^2+(-2)^2}=2\sqrt{2}$ • $\theta=\tfrac{7}{4}\pi$ • $[2\sqrt{2},\tfrac{7}{4}\pi,3]$ **(b)** $[2,\tfrac{7}{6}\pi,-5]$

28. (a) $\rho=\sqrt{(-3)^2+(-3)^2+(0)^2}=3\sqrt{2}$ • $[3\sqrt{2},\tfrac{5}{4}\pi,\tfrac{1}{2}\pi]$ **(b)** $\rho=\sqrt{(\sqrt{2})^2+(\sqrt{2})^2+(-2)^2}$
$=2\sqrt{2}$ • $[2\sqrt{2},\tfrac{1}{4}\pi,\tfrac{3}{4}\pi]$ **(c)** $[4,\tfrac{4}{3}\pi,\tfrac{1}{2}\pi]$ **(d)** $[100,\theta,\pi]$ with any θ **(f)** $[4,\tfrac{5}{6}\pi,\tfrac{5}{6}\pi]$

29. $\tfrac{16}{3}\pi - \tfrac{64}{9}$

31. (a) $\displaystyle\int_{y=0}^{y=1}\int_{x=1}^{x=1+\sqrt{1-y^2}}\dfrac{\sqrt{x^2+y^2}}{1+y}\, dx\, dy$ or $\displaystyle\int_{x=1}^{x=2}\int_{y=0}^{y=\sqrt{1-(x-1)^2}}\dfrac{\sqrt{x^2+y^2}}{1+y}\, dy\, dx$

Answers 12.4

37. $\frac{4}{3}\pi(2^{2/3} - 1)$

41. $\pi(e^{a^2} - 1)$

42. (a) $\displaystyle\int_{x=0}^{x=2}\int_{y=0}^{y=\sqrt{4-x^2}}\int_{z=-(x^2+y^2)}^{z=4}\frac{z(x^2+y^2)}{y+4}\,dz\,dy\,dx$

43. (a) $\displaystyle\int_{x=-1}^{x=1}\int_{y=0}^{y=\sqrt{1-x^2}}\int_{z=-\sqrt{1-x^2-y^2}}^{z=-\sqrt{x^2+y^2}}x(x^2+y^2+z^2)\,dz\,dy\,dx$

46. $\left(\frac{32}{35}, -\frac{32}{35}, -\frac{32}{35}\right)$

47. (a) $x = 4000\sin(\frac{1}{2}\pi)\cos(\frac{2}{3}\pi)$, $y = 4000\sin(\frac{1}{2}\pi)\sin(\frac{2}{3}\pi)$, $z = 4000\cos(\frac{1}{2}\pi)$ • $(-2000, 2000\sqrt{3}, 0)$
(b) $(-1000\sqrt{3}, 3000, -2000)$

48. (a) Timimoun has spherical coordinates $[4000, 0, \frac{1}{3}\pi]$ and Bennet has spherical coordinates $[4000, -\frac{3}{4}\pi, \frac{1}{6}\pi]$ • The unit vector $\vec{A} = \langle\sin(\frac{1}{3}\pi)\cos(0), \sin(\frac{1}{3}\pi)\sin(0), \cos(\frac{1}{3}\pi)\rangle = \langle\frac{1}{2}\sqrt{3}, 0, \frac{1}{2}\rangle$ points from the center of the earth toward Timimoun and the unit vector
$\vec{B} = \langle\sin(\frac{1}{6}\pi)\cos(-\frac{3}{4}\pi), \sin(\frac{1}{6}\pi)\sin(-\frac{3}{4}\pi), \cos(\frac{1}{6}\pi)\rangle = \left\langle-\frac{1}{2\sqrt{2}}, -\frac{1}{2\sqrt{2}}, \frac{\sqrt{3}}{2}\right\rangle$ points toward Bennet.
• The smallest positive angle α between $\vec{A}$ and $\vec{B}$ satisfies $\cos\alpha = \vec{A}\cdot\vec{B} = \frac{1}{4}\sqrt{3} - \frac{1}{8}\sqrt{6}$ •
[Distance] $= 4000\cos^{-1}(\frac{1}{4}\sqrt{3} - \frac{1}{8}\sqrt{6}) \doteq 5775$ miles
(b) $\approx 10,620$ miles

Tune-Up Exercises 12.4

T1. Figure T1

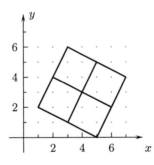

Figure T1

T2. (a) $T : (0,0) \to (0,0) \implies c_1 = 0, c_2 = 0$ • $T : (1,0) \to (-1,2) \implies a_1 = -1, a_2 = 2$ •
$T : (0,1) \to (3,-1) \implies b_1 = 3, b_2 = -1$ • $x = -u + 3v$, $y = 2u - v$

(b) $\dfrac{\partial(x,y)}{\partial(u,v)} = \begin{vmatrix} x_u & x_v \\ y_u & y_v \end{vmatrix} = \begin{vmatrix} -1 & 3 \\ 2 & -1 \end{vmatrix} = (-1)(-1) - 2(3) = -5$

(c) $u = \dfrac{\begin{vmatrix} x & 3 \\ y & -1 \end{vmatrix}}{-5} = \frac{1}{5}x + \frac{3}{5}y$ • $v = \dfrac{\begin{vmatrix} -1 & x \\ 2 & y \end{vmatrix}}{-5} = \frac{2}{5}x + \frac{1}{5}y$

(d) $\dfrac{\partial(u,v)}{\partial(x,y)} = \begin{vmatrix} u_x & u_y \\ v_x & v_y \end{vmatrix} = \begin{vmatrix} \frac{1}{5} & \frac{3}{5} \\ \frac{2}{5} & \frac{1}{5} \end{vmatrix} = (\frac{1}{5})(\frac{1}{5}) - (\frac{2}{5})(\frac{3}{5}) = -\frac{1}{5}$

T3. $\dfrac{\partial(x,y)}{\partial(u,v)} = \begin{vmatrix} x_u & x_v \\ y_u & y_v \end{vmatrix} = \begin{vmatrix} 1 & 1 \\ 1 & -1 \end{vmatrix} = -2$ • $\displaystyle\iint_R \frac{x-y}{x+y}\,dx\,dy$
$= \displaystyle\iint_{\tilde{R}} \frac{(u+v)-(u-v+2)}{(u+v)+(u-v+2)}\left|\frac{\partial(x,y)}{\partial(u,v)}\right|\,du\,dv = \iint_{\tilde{R}} \frac{2v-2}{2u+2}(2)\,du\,dv$

T4. $\dfrac{\partial(x,y)}{\partial(u,v)} = \begin{vmatrix} -3u^{-2} & 0 \\ 0 & \frac{4}{3}v^{-2/3} \end{vmatrix} = -4u^{-2}v^{-2/3}$ • $x(y+3) = 3u^{-1}(4v^{1/3})$ •

$\displaystyle\iint_R x(y+3)\,dx\,dy = \iint_{\tilde R}(3u^{-1})(4v^{1/3})(4u^{-2}v^{-2/3})\,du\,dv = 48\iint_{\tilde R} u^{-3}v^{-1/3}\,du\,dv$

T5. (a) $\dfrac{\partial(x,y,z)}{\partial(u,v,w)} = \begin{vmatrix} x_u & x_v & x_w \\ y_u & y_v & y_w \\ z_u & z_v & z_w \end{vmatrix} = \begin{vmatrix} 1 & 2 & 3 \\ 4 & -5 & 0 \\ 0 & 6 & 2 \end{vmatrix} = (1)\begin{vmatrix} -5 & 0 \\ 6 & 2 \end{vmatrix} - (2)\begin{vmatrix} 4 & 0 \\ 0 & 2 \end{vmatrix} + (3)\begin{vmatrix} 4 & -5 \\ 0 & 6 \end{vmatrix}$

$= -10 - 2(8) + 3(24) = 46$

(b) [Volume V] $= \displaystyle\iiint_V dx\,dy\,dz = \iiint_{\tilde V}\left|\dfrac{\partial(x,y,z)}{\partial(u,v,w)}\right|\,du\,dv\,dw$

$= 46[\text{Volume of } \tilde V] = 4600$ cubic meters

Problems 12.4

1. Figure A1

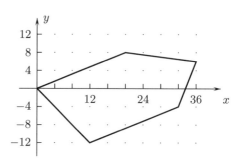

Figure A1

2. $x = u + 7v - 3,\ y = 6u + v - 2$

3a. $u = 3x - y,\ v = 2x + 2y$

4. $\dfrac{\partial(x,y)}{\partial(u,v)} = \begin{vmatrix} 1 & -2 \\ 3 & 1 \end{vmatrix} = 7$ • $u = \frac{1}{7}\begin{vmatrix} x & -2 \\ y & 1 \end{vmatrix} = \frac{1}{7}x + \frac{2}{7}y$ • $v = \frac{1}{7}\begin{vmatrix} 1 & x \\ 3 & y \end{vmatrix} = -\frac{3}{7}x + \frac{1}{7}y$ •

• $\dfrac{\partial(u,v)}{\partial(x,y)} = \left[\dfrac{\partial(x,y)}{\partial(u,v)}\right]^{-1} = \frac{1}{7}$ or $\dfrac{\partial(u,v)}{\partial(x,y)} = \begin{vmatrix} \frac{1}{7} & \frac{2}{7} \\ -\frac{3}{7} & \frac{1}{7} \end{vmatrix} = \frac{1}{7}$

5a. $x = \frac{2}{3}u - \frac{1}{3}v,\ y = \frac{1}{3}u - \frac{2}{3}v$

7. $\dfrac{\partial(u,v)}{\partial(x,y)} = \begin{vmatrix} 1 & 1 \\ 1 & -1 \end{vmatrix} = -2$ • $\dfrac{\partial(x,y)}{\partial(u,v)} = \left[\dfrac{\partial(u,v)}{\partial(x,y)}\right]^{-1} = \begin{vmatrix} 1 & 1 \\ 1 & -1 \end{vmatrix}^{-1} = (-2)^{-1} = -\frac{1}{2}$ •

$\displaystyle\iint_R \frac{x+y}{1+x-y}\,dx\,dy = \frac{1}{2}\iint_{\tilde R}\frac{u}{1+v}\,du\,dv = \frac{1}{2}\left(\int_0^4 u\,du\right)\left(\int_0^4 \frac{1}{1+v}\,dv\right)$

$= \frac{1}{2}\left[\frac{1}{2}u^2\right]_0^4 \left[\ln(1+v)\right]_0^4 = \frac{1}{2}(\frac{1}{2}4^2)\ln(5) = 4\ln(5)$

8. $\frac{64}{3}[\cos(6) - 1]$

11a. $\dfrac{\partial(u,v)}{\partial(x,y)} = 2(x^2+y^2)$ • $\dfrac{\partial(x,y)}{\partial(u,v)} = \dfrac{1}{2(x^2+y^2)}$

Answers 12.4

12. Let $u = x - y$, $v = 4x + y$ • R is the image of $\tilde{R}: -1 \leq u \leq 0, -2 \leq v \leq 0$. • $5x = u + v$ •
$\dfrac{\partial(u,v)}{\partial(x,y)} = \begin{vmatrix} 1 & -1 \\ 4 & 1 \end{vmatrix} = 5$ • $\dfrac{\partial(x,y)}{\partial(u,v)} = \frac{1}{5}$ • $\displaystyle\iint_R \cos(5x)\, dx\, dy = \iint_{\tilde{R}} \cos(u+v)(\tfrac{1}{5})\, du\, dv$

$= \dfrac{1}{5}\displaystyle\int_{v=-2}^{v=0}\int_{u=-1}^{u=0} \cos(u+v)\, du\, dv = \dfrac{1}{5}\int_{v=-2}^{v=0}\Big[\sin(u+v)\Big]_{u=-1}^{u=0} dv = \dfrac{1}{5}\int_{v=-2}^{v=0}[\sin v - \sin(v-1)]\, dv$

$= \dfrac{1}{5}\Big[-\cos(v) + \cos(v-1)\Big]_{-2}^{0} = \tfrac{1}{5}[-1+\cos(-2)+\cos(-1)-\cos(-3)] = \tfrac{1}{5}[-1+\cos(2)+\cos(1)-\cos(3)]$

13. $\frac{1}{4}[e^2 - e^{-2}][e^6 - 1]$

15. $\frac{8}{9}(e^4 - e)$

17. $(e^4 - e)\ln(3)$

21. $6\pi[1 - \cos(1)]$

22. Let $u = x + 2y + 4z$, $v = y - 2z$, $w = y + 2z$. • V is the image of $\tilde{V}: 1 \leq u \leq 4, 0 \leq v \leq 1, 0 \leq w \leq 2$.

• $\dfrac{\partial(u,v,w)}{\partial(x,y,z)} = \begin{vmatrix} 1 & 2 & 4 \\ 0 & 1 & -2 \\ 0 & 1 & 2 \end{vmatrix} = 1\begin{vmatrix} 1 & -2 \\ 1 & 2 \end{vmatrix} - 2\begin{vmatrix} 0 & -2 \\ 0 & 2 \end{vmatrix} + 4\begin{vmatrix} 0 & 1 \\ 0 & 1 \end{vmatrix} = 1(4) - 2(0) + 4(0) = 4$ •

$\dfrac{\partial(x,y,z)}{\partial(u,v,w)} = \left[\dfrac{\partial(u,v,w)}{\partial(x,y,z)}\right]^{-1} = \tfrac{1}{4}$ • $2y = v + w$ •

$\displaystyle\iiint_V \dfrac{2y}{x+2y+4z}\, dx\, dy\, dz = \iiint_{\tilde{V}} \dfrac{v+w}{u}\left|\dfrac{\partial(x,y,z)}{\partial(u,v,w)}\right| du\, dv\, dw$

$= \dfrac{1}{4}\displaystyle\int_{u=1}^{u=4}\int_{v=0}^{v=1}\int_{w=0}^{w=2} (v+w)u^{-1}\, dw\, dv\, du = \dfrac{1}{4}\int_{u=1}^{u=4} u^{-1}\, du \int_{v=0}^{v=1}\int_{w=0}^{w=2} (v+w)\, dw\, dv$

$= \tfrac{1}{4}\ln(4)\displaystyle\int_{v=0}^{v=1}\Big[vw + \tfrac{1}{2}w^2\Big]_{w=0}^{w=2} dv = \tfrac{1}{4}\ln(4)\int_0^1 (2v+2)\, dv = \tfrac{3}{4}\ln(4)$

23. Set $u = x + y - z$, $v = x - y + z$, $w = y + 2z$ • V is the image of $\tilde{V}: 0 \leq u \leq 2, 0 \leq v \leq 3, 0 \leq w \leq 4$. •

$\dfrac{\partial(u,v,w)}{\partial(x,y,z)} = \begin{vmatrix} 1 & 1 & -1 \\ 1 & -1 & 1 \\ 0 & 1 & 2 \end{vmatrix} = 1\begin{vmatrix} -1 & 1 \\ 1 & 2 \end{vmatrix} - 1\begin{vmatrix} 1 & 1 \\ 0 & 2 \end{vmatrix} - 1\begin{vmatrix} 1 & -1 \\ 0 & 1 \end{vmatrix} = 1(-3) - 1(2) - 1(1) = -6$ •

$\dfrac{\partial(x,y,z)}{\partial(u,v,w)} = \left[\dfrac{\partial(u,v,w)}{\partial(x,y,z)}\right]^{-1} = -\tfrac{1}{6}$ • $\displaystyle\iiint_V \dfrac{x+y-z}{1+(y+2z)^2}\, dx\, dy\, dz$

$= \displaystyle\iiint_{\tilde{V}} \dfrac{u}{1+w^2}\left|\dfrac{\partial(x,y,z)}{\partial(u,v,w)}\right| du\, dv\, dw = \dfrac{1}{6}\int_{u=0}^{u=2}\int_{v=0}^{v=3}\int_{w=0}^{w=4} \dfrac{u}{1+w^2}\, dw\, dv\, du$

$= \dfrac{1}{6}\displaystyle\int_{u=0}^{u=2} u\, du \int_{v=0}^{v=3} dv \int_{w=0}^{w=4} \dfrac{1}{1+w^2}\, dw\, dv = \tfrac{1}{6}\Big[\tfrac{1}{2}u^2\Big]_{u=0}^{u=2}\Big[v\Big]_{v=0}^{v=3}\Big[\tan^{-1} w\Big]_{w=0}^{w=4} = \tan^{-1}(4)$

24. $\dfrac{21}{8}$

CHAPTER 13
VECTOR ANALYSIS

In Chapter 10 we used vectors to study the motion of single objects. In this chapter we study vector-valued functions called VECTOR FIELDS, including VELOCITY FIELDS, which are used to study the flow of fluids, and FORCE FIELDS, which are used in the study of gravitational, electromagnetic, and other forces. We discuss vector fields and LINE INTEGRALS in Section 13.1. Section 13.2 deals with theorems that relate double integrals of derivatives of functions in the xy-plane to line integrals of the functions and that are used to determine whether line integrals in the plane are PATH INDEPENDENT. In Section 13.3 we define SURFACE INTEGRALS and use them in theorems that relate triple integrals of derivatives to surface integrals of the functions, that relate surface integrals of derivatives of functions to line integrals of the functions, and that are used to study path independence in space.

Section 13.1

Vector fields and line integrals

OVERVIEW: *In this section we discuss* VECTOR FIELDS *in the xy-plane and in xyz-space and* STREAMLINES *of velocity fields. Next, we define the* FLUX *of a constant velocity field across a line segment in the xy-plane and the* WORK *done by a constant force field on an object that traverses a line segment in the xy-plane or in xyz-space. Then we discuss* LINE INTEGRALS *and use them to calculate flux and work for nonconstant vector fields and curved paths.*

Topics:

- **Vector fields**
- **Streamlines of velocity fields**
- **Flux and work for constant fields and line segments**
- **Line integrals**
- **Flux and work for general fields and paths**

Vector fields

A two-dimensional VECTOR FIELD is a rule that assigns a two-dimensional vector $\mathbf{A}(x,y) = p(x,y)\mathbf{i} + q(x,y)\mathbf{j} = \langle p(x,y), q(x,y) \rangle$ with its base at each point (x,y) in a portion or all of the xy-plane. A three-dimensional vector field assigns a three-dimensional vector $\mathbf{A}(x,y,z) = p(x,y,z)\mathbf{i} + q(x,y,z)\mathbf{j} + r(x,y,z)\mathbf{k} = \langle p(x,y,z), q(x,y,z), r(x,y,z) \rangle$ with its base at the points (x,y,z) in a portion or all of xyz-space.

A two-dimensional vector field is sketched by drawing the vectors at an array of points, with the lengths of the arrows measured by a scale that is usually chosen to make the drawing as clear as possible.

Example 1 Figure 1 shows the vector field $\mathbf{A}(x,y) = \frac{1}{2}x\mathbf{i} + \frac{1}{2}y\mathbf{j} = \langle \frac{1}{2}x, \frac{1}{2}y \rangle$ with the lengths of the arrows measured by the scales on the coordinate axes. **(a)** What are the values of $\mathbf{A}(x,y)$ at $(0,1), (1,1)$ and $(2,1)$? **(b)** Use the formula for the vector field to explain the lengths and orientations of the vectors in Figure 1.

SOLUTION **(a)** We calculate $\mathbf{A}(0,1) = \langle \frac{1}{2}(0), \frac{1}{2}(1) \rangle = \langle 0, \frac{1}{2} \rangle$, $\mathbf{A}(1,1) = \langle \frac{1}{2}(1), \frac{1}{2}(1) \rangle = \langle \frac{1}{2}, \frac{1}{2} \rangle$, and $\mathbf{A}(2,1) = \langle \frac{1}{2}(2), \frac{1}{2}(1) \rangle = \langle 1, \frac{1}{2} \rangle$. Notice the arrows at $(0,1), (1,1)$, and $(2,1)$ that represente these vectors in Figure 1.

♦ SUGGESTIONS TO INSTRUCTORS: The main topics of this section can be divided into two parts to be covered in several class meetings.
Part 1 (Vector fields, streamlines, flux of a constant velocity field, work by a constant force field, line integrals with respect to dx, dy, dz and ds; Tune-up Exercises T1–T4; Problems 1–18, 20–27) Have students begin work on Questions 1 through 7 before class. Tune-up Exercises T1, T2, T3 and Problems 1, 2, 3, 5, 8, 11, 16 are good for in-class discussion.
Part 2 (Interpreting line integrals with respect to arclength, unit tangent and normal vectors, relating $dx, dy,$ and dz to ds, work and flux for nonconstant fields and curves, average values; Tune-up Problems T5–T7; Problems 19, 28–43) Have students begin work on Questions 8 and 9 and Tune-up Exercises T3 and T6 before class. Problems 19, 28, 33, 37, 39 are good for in-class discussion.

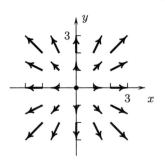

FIGURE 1

The vector field $\langle \frac{1}{2}x, \frac{1}{2}y \rangle$

(b) For each (x,y), $\mathbf{A}(x,y) = \langle \frac{1}{2}x, \frac{1}{2}y \rangle = \frac{1}{2}\langle x, y \rangle$ is half the position vector $\langle x, y \rangle$ of the point, so for $(x,y) \neq (0,0)$, the vector $\mathbf{A}(x,y)$ at (x,y) points away from the origin and its length is half the distance from the origin to the point. $\mathbf{A}(0,0)$ is the zero vector, shown by a dot in Figure 1. □.

Question 1   Sketch the constant vector field $\mathbf{B} = \langle -1, -\frac{1}{2} \rangle$.

The velocity field of a fluid

Figure 3 shows some of the vectors in the constant velocity field $\mathbf{v} = \langle \frac{3}{4}, 1 \rangle$ (feet per second) of a flowing horizontal layer of water. Because the velocity field is constant, all particles of the water have the same velocity at all times. With this, as with all two-dimensional models of fluid flow, we assume that all of the particles of water on each vertical line segment move together so that we can study the flow by studying the motion of the particles on the surface.

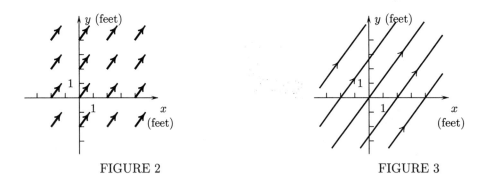

FIGURE 2   FIGURE 3

The STREAMLINES of a fluid flow are the paths $C : x = x(t), y = y(t), -\infty < t < \infty$ taken by the particles of water in the fluid. Figure 3 shows five streamlines for the velocity field in Figure 2. They are parallel lines because the velocity field is constant. In fact, the velocity vector $\mathbf{v}(t) = \langle \frac{dx}{dt}(t), \frac{dy}{dt}(t) \rangle$ along each streamline equals $\mathbf{v} = \langle \frac{3}{4}, 1 \rangle$ so that

$$\frac{dx}{dt}(t) = \tfrac{3}{4} \text{ and } \frac{dy}{dt}(t) = 1. \tag{1}$$

Taking antiderivatives of these equations yields parametric equations

$$C : x = \tfrac{3}{4}t + c_1, \ y = t + c_2 \tag{2}$$

for the streamlines with the time as parameter, where the constants c_1 and c_2 can be determined from the xy-coordinates of any point on the streamline at any particular time.

We can find equations in x and y for the streamlines (2) by eliminating t from the parametric equations.

Question 2   Give an equation in x and y, with parameters c_1 and c_2, for the streamlines in Figure 3.

13.1 Vector fields and line integrals

Example 2 (a) Suppose that the vector field $\mathbf{A} = \langle \frac{1}{2}x, \frac{1}{2}y \rangle$ of Figure 1 is the velocity field of a fluid flow. (a) Find parametric equations for the path of a fluid particle (a streamline) with time t as parameter under the assumption that the particle is at (x_0, y_0) at $t = 0$ (seconds). (b) What is the streamline that contains the origin? (c) Give equations in terms of x and y for the streamlines that do not pass through the origin and describe the motion on those streeamlines.

SOLUTION (a) Because the velocity field is $\langle \frac{1}{2}x, \frac{1}{2}y \rangle$, the functions $C: x = x(t), y = y(t)$ that give a particle's coordinates as functions of the time t satisfy

$$\frac{dx}{dt}(t) = \tfrac{1}{2}x(t) \text{ and } \frac{dy}{dt}(t) = \tfrac{1}{2}y(t).$$

The solutions of these differential equations (by Theorem 4 of Section 3.4) are $x(t) = c_1 e^{t/2}$ and $y(t) = c_2 e^{t/2}$ with constants c_1 and c_2. Setting $x = x_0$ and $y = y_0$ at $t = 0$ gives $c_1 = x_0$ and $c_2 = y_0$, so that

$$x(t) = x_0 e^{t/2} \text{ and } y(t) = y_0 e^{t/2}. \tag{3}$$

(b) If a particle is ever at the origin, then $x_0 = 0$ and $y_0 = 0$, the functions (3) are zero for all t, and the particle stays at the origin. Hence, the streamline through the origin consists of the origin alone.

(c) To deal with the other cases, notice that the function $e^{t/2}$ in (3) is positive for all t, tends to ∞ as $t \to \infty$, and tends to 0 as $t \to -\infty$.

If $x_0 = 0$ and $y_0 > 0$, then $x(t) = 0$ for all t, $y(t) \to \infty$ as $t \to \infty$, and $y(t) \to 0$ as $t \to -\infty$, so the streamline is the positive y-axis, oriented upward. If $x_0 = 0$ and $y_0 < 0$, then $x(t) = 0$, for all t, $y(t) \to -\infty$ as $t \to \infty$, and $y(t) \to 0$ as $t \to -\infty$ and the streamline is the negative y-axis, oriented downward. (See Figure 4.)

If $x_0 > 0$ then $x(t)$ tends to ∞ as $t \to \infty$ and tends to 0 as $t \to -\infty$, and if $x_0 < 0$ then $x(t)$ tends to $-\infty$ as $t \to \infty$ and tends to 0 as $t \to -\infty$ In both cases equations (3) give $e^{t/2} = x(t)/x_0$ and then $y(t) = (y_0/x_0)x$, so the streamline is the ray of slope y_0/x_0 oriented away from the origin, as shown in Figure 4. □

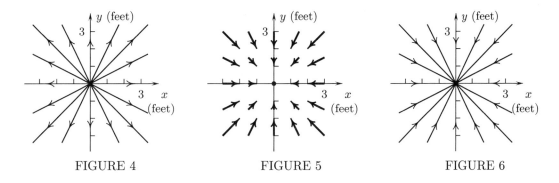

FIGURE 4   FIGURE 5   FIGURE 6

Question 3 Figure 5 shows the velocity field $\mathbf{v} = \langle -\tfrac{1}{3}x, -\tfrac{1}{3}y \rangle$. (a) Find parametric equations of the streamlines with the time t as parameter in terms of the position (x_0, y_0) of a particle of fluid at time $t = 0$. (b) Explain why the streamlines are the rays in Figure 6.

In most cases with nonconstant velocity fields, the streamlines are not straight lines. Figure 7, for example, shows the velocity field $\mathbf{v} = \langle \frac{1}{2}x, -\frac{1}{2}y \rangle$, and Figure 8 shows some of its streamlines. Calculations similar to those in the solution of Example 2 and the response to Question 3 show that the streamlines are the half-lines $y = x$ and $y = -x$ with positive or negative x and the half-hyperbolas $y = c/x$ with nonzero constants c and positive or negative x (see Problem 4).

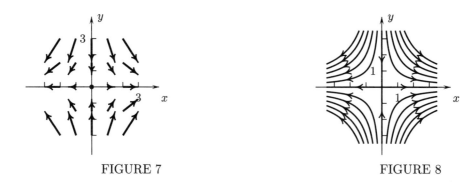

FIGURE 7 FIGURE 8

A rotational flow

Now we will consider a case where we derive a velocity field from its streamlines. Figure 9 shows streamlines of a body of water that is rotating about the origin at a constant rate. We suppose that each particle of water rotates once counterclockwise around the origin every 2π seconds. Then a point, as in Figure 10, that moves on the circle of radius r and is on the positive x-axis at $t = 0$ is at the point $(x, y) = (r\cos t, r\sin t)$ at time t. The point in Figure 10 has position vector $\mathbf{R}(t) = \langle r\cos t, r\sin t \rangle$ at time t, so its velocity vector at that time is

$$\mathbf{v}(t) = \frac{d\mathbf{R}}{dt}(t) = \frac{d}{dt}\langle r\cos t, r\sin t \rangle = \langle -r\sin t, r\cos t \rangle. \tag{4}$$

The velocity field of the entire body of water is shown in Figure 11, where the scale for measuring the arrows is half that on the coordinate axes. We can obtain a formula for this velocity field by expressing the velocity (4) of the point at (x, y) in terms of x and y instead of t.

Question 4   Give a formula in terms of x and y for $\mathbf{v}$ in (4).

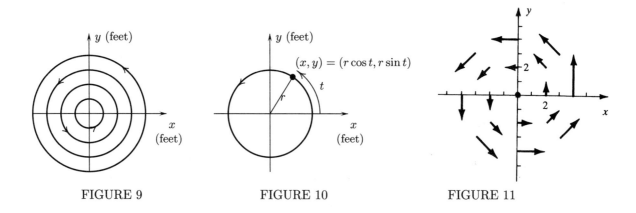

FIGURE 9 FIGURE 10 FIGURE 11

A gravitational force field

It is convenient in studying the gravitational attraction of the sun to imagine that the sun creates a GRAVITATIONAL FORCE FIELD, which in turn exerts force on the planets and other objects. The gravitational force field of the sun is a vector field $\mathbf{F}(x, y, z)$ which at each point (x, y, z) other than the origin equals the force of the sun's gravity on an object at that point whose mass is 1, relative to whatever units are being used.

According to NEWTON'S LAW OF GRAVITY, $\mathbf{F}(x, y, z)$ is directed from the point (x, y, z) toward the sun and its magnitude is $\dfrac{MG}{\rho^2}$, where M is the mass of the sun, ρ is the distance from (x, y, z) to the origin, and G is a universal constant whose value depends on the units used for mass, force, and distance. Figure 12 shows some of those vectors in a plane through the sun.

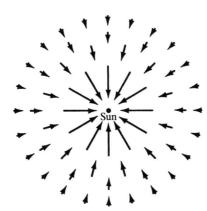

The gravitational field $\mathbf{F}(x, y, z)$ of the sun

FIGURE 12

Example 3   Give a formula for the gravitiational field $\mathbf{F}(x, y, z)$ of the sun in xyz-space with the sun at the origin.

SOLUTION   Because the distance from (x, y, z) to the origin is $\rho = \sqrt{x^2 + y^2 + z^2}$, the magnitude of $\mathbf{F}(x, y, z)$ is $\dfrac{MG}{\rho^2} = \dfrac{MG}{x^2 + y^2 + z^2}$. Then, since $\mathbf{F}(x, y, z)$ is directed toward the origin, $\mathbf{F}(x, y, z) = \left[\dfrac{MG}{x^2 + y^2 + z^2}\right] \mathbf{u}$ with $\mathbf{u}$ the unit vector pointing from (x, y, z) toward the origin. The vector $\langle -x, -y, -z\rangle$ has the direction from (x, y, z) to the origin, so $\mathbf{u} = \dfrac{\langle -x, -y, -z\rangle}{\sqrt{x^2 + y^2 + z^2}}$ and

$$\mathbf{F}(x, y, z) = \dfrac{MG}{x^2 + y^2 + z^2} \dfrac{\langle -x, -y, -z\rangle}{\sqrt{x^2 + y^2 + z^2}} = \dfrac{MG\langle -x, -y, -z\rangle}{(x^2 + y^2 + z^2)^{3/2}}. \quad \square \qquad (5)$$

By Newton's law, the force on an object is equal to the object's mass multiplied by its acceleration vector, which is the second derivative $\dfrac{d^2\mathbf{R}}{dt^2}$ of its position vector $\mathbf{R} = \langle x, y, z\rangle$. Consequently, formulas such as (5) for force fields lead to systems of second-order differential equations for $x(t), y(t)$, and $z(t)$ which—unlike the first-order equations arising from velocity fields that we study in this section—are very difficult to analyze. (See, for example, the discussion of Newton's Law of Gravity and Kepler's Laws in Section 10.5.)

Flux of a constant velocity field

Figure 13 shows a constant velocity field **v** in an xy-plane of a fluid that flows across a line segment PQ. The magnitude of the velocity field is 50 meters per minute, the line segment is 20 meters long, and the vectors **v** make an angle of $\frac{1}{6}\pi$ with normal lines to PQ. Because the velocity field is constant, a particle of water that is at the base of a vector **v** at one moment is at the tip of that vector one minute later. Consequently, in one minute enough water crosses the line segment to fill the parallelogram in Figure 13.

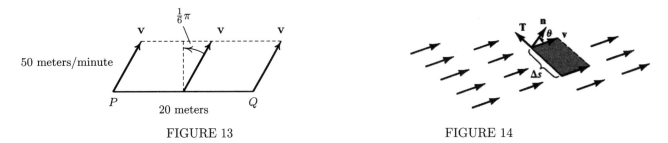

FIGURE 13   FIGURE 14

Question 5 (a) How much water crosses the line segment PQ in Figure 13 every minute?
(b) What is the instantaneous rate of flow of water across the line segment?

The rate of flow of fluid across the line segment PQ in Figure 13 is called the FLUX of the velocity field across the line segment. The sort of calculation in Question 5 can be used to find the flux of any constant velocity field across any line segment. Consider the velocity field **v** and line segment of length Δs in Figure 14. We let **n** denote the unit normal vector to the line segment that points in the direction of the flow and let θ be the angle between **v** and **n**. Then since **n** is a unit vector, the height of the parallelogram is $|\mathbf{v}|\cos\theta = \mathbf{v}\cdot\mathbf{n}$. Its width is Δs, so its area is $\mathbf{v}\cdot\mathbf{n}\Delta s$ and

$$[\text{Flux of } \mathbf{v} \text{ across the line segment}] = \mathbf{v}\cdot\mathbf{n}\,\Delta s. \tag{6}$$

We will use this formula later in this section when we consider the flux of general velocity fields across curves.

Work by a constant force field

In Section 6.7 we defined the work done by a constant force on an object that moves from one point or another on an s-axis that is parallel to the direction of the force as follows: If the force in the positive s-direction is F, then the work done by the force on the object as it goes from $s = a$ to $s = b$ is $F(b-a)$.

For a constant force that is not parallel to the motion of the object we use the component of the force in the direction of the motion. Suppose that as an object goes along the line segment from the point P at the left to the point Q at the right of the line segment in Figure 15, it is subject to the constant vector force **F** that makes an angle θ with the direction of motion. Denote the distance between P and Q by Δs and let **T** be the unit tangent vector in the direction of the motion. Then the component of **F** in the direction of **T** is $|\mathbf{F}|\cos\theta = \mathbf{F}\cdot\mathbf{T}$ and we define

$$\begin{bmatrix} \text{Work done by } \mathbf{F} \text{ as the} \\ \text{object goes from } P \text{ to } Q \end{bmatrix} = \mathbf{F}\cdot\mathbf{T}\,\Delta s \tag{7}$$
$$= \mathbf{F}\cdot\overrightarrow{PQ}.$$

The second formula in **(7)** follows from the first because $\mathbf{T}\,\Delta s = \overrightarrow{PQ}$.

FIGURE 15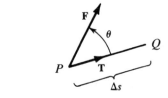

Question 6 How much work is done by the constant force $\mathbf{F} = \langle 4, 7 \rangle$ (dynes) on an object as it traverses the line segment from $P = (5, 10)$ to $Q = (20, 30)$ in an xy-plane with distances measured in centimeters?

More on parametrizations of curves

In previous chapters we oriented curves in the directions in which their parameters increased. It will be convenient in this chapter to also allow orientations in the direction in which parameters decrease by writing $C : x = x(t), y = y(t), a \xrightarrow[t]{} b$ for a curve in an xy-plane and $C : x = x(t), y = y(t), z = z(t), a \xrightarrow[t]{} b$ for a curve in space. In both cases the curve is considered to be oriented from the point at $t = a$ to the point at $t = b$. This is in the direction of increasing t if $a < b$ and in the direction of decreasing t if $a > b$. This convention gives us more flexibility in defining parametrizations of curves and enables us to reverse a curve's orientation by interchanging the numbers a and b.

We can express parametrizations of curves in vector form by writing $C : \mathbf{R} = \mathbf{R}(t), a \xrightarrow[t]{} b$ where $\mathbf{R}(t) = \langle x(t), y(t) \rangle$ or $\mathbf{R}(t) = \langle x(t), y(t), z(t) \rangle$ is the position vector of a point on the curve at the value t of its parameter. We say that this curve is CONTINUOUS if $\mathbf{R}(t)$ is continuous in the closed interval with endpoints a and b and that it is SMOOTH if it is continuous and if its velocity vector $\mathbf{R}'(t)$ is continuous and nonzero in the interior of the interval and has nonzero limits as t tends to a and b from the interior of the interval. A curve is PIECEWISE SMOOTH if it consists of a finite number of smooth curves.

Example 4 Sketch the curve $C : x = 2\cos t, y = \sin t, \frac{3}{4}\pi \xrightarrow[t]{} -\frac{1}{4}\pi$, showing its orientation.

SOLUTION The curve is the half-ellipse shown in Figure 16, oriented clockwise. □

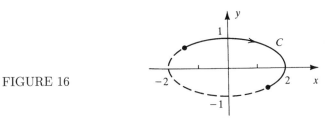

FIGURE 16

If a curve is a portion of the graph $y = f(x)$ of a function, as in the next Question, we can use x as the parameter.

Question 7 Give a parametric equation of the portion of $y = x^3$ for $-1 \le x \le 1$, oriented from right to left, with x as parameter. Then draw the curve.

Line integrals

To define the flux of a velocity field across a curve in the xy-plane or the work done by a force field on an object as it traverses a curve in the xy-plane or xyz-space, we will need LINE INTEGRALS. For functions $f(x, y)$ of two variables and oriented curves in an xy-plane, there are line integrals with respect to x, y, and arclength s, which are denoted $\int_C f\, dx, \int_C f\, dy$, and $\int_C f\, ds$, respectively. For functions $f(x, y, z)$ of three variables and curves in space there are also line integrals $\int_C f\, dz$ with respect to z.

Each of these is defined as a limit of approximating sums. We pick points $P_0, P_1, P_2, \ldots, P_N$ on the curve C with P_0 the beginning, P_N the end, and the other points numbered in order following the orientation of the curve (Figure 17). The points form a PARTITION of the curve C into N subcurves. We denote the coordinates of P_j by (x_j, y_j) or (x_j, y_j, z_j) and write

$$\Delta x_j = x_j - x_{j-1}, \Delta y_j = y_j - y_{j-1}, \Delta z_j = z_j - z_{j-1} \tag{8}$$

for the changes in x, y, and, in the three-variable case, of z across the subcurves. Also, we let Δs_j

denote the (positive) distance between the points P_{j-1} and P_j, so that in the two-variable case

$$\Delta s_j = \sqrt{(x_j - x_{j-1})^2 + (y_j - y_{j-1})^2} \tag{9}$$

and in the case of three variables

$$\Delta s_j = \sqrt{(x_j - x_{j-1})^2 + (y_j - y_{j-1})^2 + (z_j - z_{j-1})^2}. \tag{10}$$

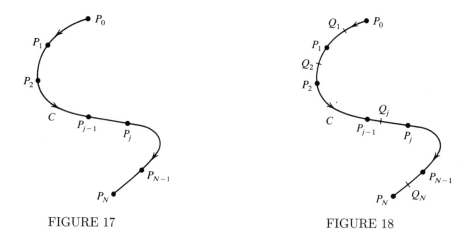

FIGURE 17 · · · FIGURE 18

We say that $f = f(x,y)$ or $f = f(x,y,z)$ is PIECEWISE CONTINUOUS on the curve C if it is equal on C to a function that is piecewise condinuous in an open set containing C. If f is such a function, we pick, for each j, a point Q_j in the jth subcurve such that $f(Q_j)$ is defined (Figure 18). Then we make the following definition.

Definition 1 *For a piecewise-smooth curve C in an xy-plane or in xyz-space and a function f that is piecewise continuous on C,*

$$\int_C f\,dx = \lim \sum_{j=1}^N f(Q_j)\,\Delta x_j \tag{11}$$

$$\int_C f\,dy = \lim \sum_{j=1}^N f(Q_j)\,\Delta y_j \tag{12}$$

$$\int_C f\,ds = \lim \sum_{j=1}^N f(Q_j)\,\Delta s_j \tag{13}$$

and for a curve C in space and a function of three variables,

$$\int_C f\,dz = \lim \sum_{j=1}^N f(Q_j)\,\Delta z_j. \tag{14}$$

In this definition the limit is taken as the number of points P_j in the partitions tends to infinity and the distances Δs_j betweem them tend to 0.† The limits exist and are finite under the conditions stated in the definition.

---

†The $\epsilon\delta$-definition of these limits reads as follows: The limit of an approximating sum is the number L if for every $\epsilon > 0$ there is a $\delta > 0$ such that $|[\text{Approximating sum}] - L| < \epsilon$ for all partitions of the curve with $\Delta s_j < \delta$ for all j.

13.1 Vector fields and line integrals

It follows from Definition 1 that any line integral of a linear combination $Af + Bg$ of functions f and g with constants A and B equals the same linear combination of the line integrals of the functions, as in the equation

$$\int_C (Af + Bg)\, dx = A \int_C f\, dx + B \int_C g\, dx$$

for integrals with respect to x.

Example 5 What is the value of $\int_C dx$ for the curve $C : \mathbf{R} = \mathbf{R}(t), a \xrightarrow{t} b$ for $\mathbf{R}(t) = \langle x(t), y(t)\rangle$ or $\mathbf{R}(t) = \langle x(t), y(t), z(t)\rangle$?

SOLUTION   For each partition, the approximating sum $\sum_{j=1}^{N} \Delta x_j$ for $\int_C dx$ equals the sum of the changes in x from P_{j-1} to P_j for $j = 1, \ldots N$ and, consequently, equals the value $x(b)$ of x at the end of the curve, minus the value $x(a)$ of x at the beginning of the curve. Therefore, if $\mathbf{R} = \langle x(t), y(t)\rangle$ for a plane curve or $\mathbf{R}(t) = \langle x(t), y(t), z(t)\rangle$ for a curve in space, then

$$\int_C dx = \lim [x(b) - x(a)] = x(b) - x(a). \square \tag{15}$$

Question 8 Derive the formulas

$$\int_C dy = y(b) - y(a) \tag{16}$$

$$\int_C dz = z(b) - z(a) \tag{17}$$

where $C : \mathbf{R} = \mathbf{R}(t), a \xrightarrow{t} b$ is a curve in the xy-plane or in xyz-space for **(16)** and is a curve in space for **(17)**.

The approximating sum in **(13)** for $f = 1$ is the length of a polygonal line that approximates the curve, whose length tends to the length of the curve, as defined in Section 4.6. Consequently,

$$\int_C ds = [\text{Length of } C]. \tag{18}$$

The sum and difference of two curves

The SUM $C_1 + C_2$ of two oriented curves is formed by following C_1 with C_2 (Figure 19). It is a consequence of Definition 1 that any line integral with respect to x, y, z or s over $C_1 + C_2$ is the sum of the line integrals over C_1 and C_2, as expressed in the equation

$$\int_{C_1 + C_2} f\, dx = \int_{C_1} f\, dx + \int_{C_2} f\, dx$$

for line integrals with respect to x.

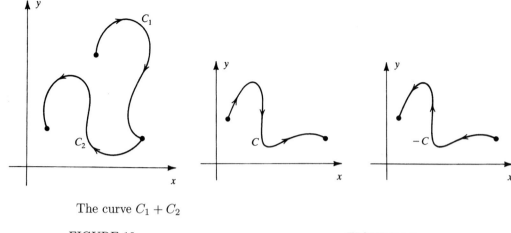

The curve $C_1 + C_2$

FIGURE 19   FIGURE 20

The NEGATIVE $-C$ of a curve C is obtained by reversing its orientation (Figure 20). When the orientation is reversed, the terms $\Delta x_j, \Delta y_j$, and Δz_j in an approximating sum are replaced by their negatives. Consequently, the line integrals with respect to x, y, and z over $-C$ are the negatives of the integrals over C:

$$\int_{-C} f\, dx = -\int_C f\, dx,\quad \int_{-C} f\, dy = -\int_C f\, dy,\quad \int_{-C} f\, dz = -\int_C f\, dz. \qquad (19)$$

In contrast, the distances Δs_j are not changed when the orientation of the curve is reversed, so the integrals with respect to s over C and $-C$ are equal:[†]

$$\int_{-C} f\, ds = \int_C f\, ds. \qquad (20)$$

The DIFFERENCE $C_1 - C_2$ of two curves is defined to be the sum of C_1 and $-C_2$.

Example 5   Sketch $C = C_1 - C_2$ where $C_1 : y = x, 0 \xrightarrow[x]{} 1$ and $C_2 : x = 2 - y^2, -2 \xrightarrow[y]{} 1$.

SOLUTION C_1 is the line segment between $(0,0)$ and $(1,1)$, oriented from left to right, and C_2 is the portion of the parabola $x = 2 - y^2$ for $-2 \le y \le 1$, oriented upward. Because $-C_2$ is oriented downward, $C_1 - C_2$ is the curve in Figure 21. □

The curve $C_1 - C_2$

FIGURE 21

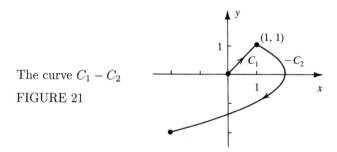

[†]We define line integrals with respect to s to be independent of the orientation of the curve, so that the relations $\mathbf{T}\, ds = \langle dx, dy \rangle$ for a curve in the plane and $\mathbf{T}\, ds = \langle dx, dy, dz \rangle$ for a curve in space, which we establish below, are valid with $\mathbf{T}$ the unit tangent vector in the direction of the orientation of the curve.

13.1 Vector fields and line integrals

Evaluating line integrals

Suppose that the points $P_j = (x_j, y_j)$ for $j = 0, 1, \ldots, N$ form a partition of a smooth curve $C : x = x(t), y = y(t), a \xrightarrow{t} b$ with $a < b$ in the xy-plane and $x_j = x(t_j)$ for each j. Set $\Delta t_j = t_j - t_{j-1}$. By the Mean-Value Theorem of Section 3.1, there are number c_j with $t_{j-1} < c_j < t_j$ such that

$$\Delta x_j = x_j - x_{j-1} = x(t_j) - x(t_{j-1}) = \frac{dx}{dt}(c_j) \Delta t_j. \tag{21}$$

If $f(x, y)$ is piecewise continuous on C, we pick numbers d_j with $t_{j-1} \le d_j \le c_j$ such that f is defined at $Q_j = (x(d_j), y(d_j))$. Then we substitute formulas (21) in the approximating sum (11) for $\int_C f(x, y)\, dx$ and obtain

$$\sum_{j=1}^{N} f(x(d_j), y(d_j)) \, \Delta x_j = \sum_{j=1}^{N} f(x(d_j), y(d_j)) \frac{dx}{dt}(c_j) \Delta t_j. \tag{22}$$

The sum on the left of (22) is an approximating sum for $\int_C f\, dx$, the sum on the right is an approximating sum for $\int_a^b f(x(t), y(t)) \frac{dx}{dt}(x)\, dt$, and the sums tend to the respective integrals as the numbers of points in the partitions tend to infinity and the numbers Δt_j tend to zero. Therefore the two integrals are equal. This establishes formula (23) in the following Theorem. Similar lines of reasoning can be used to establish the other statements in it.

Theorem 1 (a) Suppose that $C : x = x(t), y = y(t), a \xrightarrow{t} b$ is a piecewise-smooth curve in the xy-plane and that $f(x, y)$ is piecewise continuous on C. Then

$$\int_C f\, dx = \int_a^b f \frac{dx}{dt}\, dt \tag{23}$$

$$\int_C f\, dy = \int_a^b f \frac{dy}{dt}\, dt \tag{24}$$

$$\int_C f\, ds = \pm \int_a^b f \sqrt{\left[\frac{dx}{dt}\right]^2 + \left[\frac{dy}{dt}\right]^2}\, dt \tag{25}$$

where f denotes $f(x, y)$ in the integrals on the left and $f(x(t), y(t))$ in the integrals on the right, and the plus sign is used in (25) if $a < b$ and the minus sign if $b < a$.

(b) Suppose that $C : x = x(t), y = y(t), z = z(t), a \xrightarrow{t} b$ is a piecewise-smooth curve in xyz-space and that $f(x, y, z)$ is piecewise continuous on C. Then equations (23) and (24) hold with f denoting $f(x, y, z)$ in the integrals on the left and $f(x(t), y(t), z(t))$ in the integrals on the right. Moreover

$$\int_C f\, dz = \int_a^b f \frac{dz}{dt}\, dt \tag{26}$$

$$\int_C f\, ds = \pm \int_a^b f \sqrt{\left[\frac{dx}{dt}\right]^2 + \left[\frac{dy}{dt}\right]^2 + \left[\frac{dz}{dt}\right]^2}\, dt \tag{27}$$

with the plus sign in (27) if $a < b$ and the minus sign if $b < a$.

The formulas in Theorem 1 can be recalled with the symbolic equations

$$dx = \frac{dx}{dt}\,dt,\ dy = \frac{dy}{dt}\,dt,\ dz = \frac{dz}{dt}\,dt \tag{28}$$

$$ds = \pm\sqrt{\left[\frac{dx}{dt}\right]^2 + \left[\frac{dy}{dt}\right]^2}\,dt \tag{29}$$

$$ds = \pm\sqrt{\left[\frac{dx}{dt}\right]^2 + \left[\frac{dy}{dt}\right]^2 + \left[\frac{dz}{dt}\right]^2}\,dt \tag{30}$$

where **(29)** is for curves in the xy-plane and **(30)** is for curves in xyz-space. The plus signs are used in the last two formulas if $a < b$ and the minus signs if $b < a$.

Example 6 Evaluate $\int_C x^2 y\,dx$, where C is the curve $x = t^3, y = t^2 - 2t, 0 \xrightarrow{t} 1$.

SOLUTION Since $x = t^3$, $dx = \dfrac{dx}{dt}\,dt = \dfrac{d}{dt}(t^3)\,dt = 3t^2\,dt$. With these formulas and $y = t^2 - 2t$, **(23)** yields

$$\int_C x^2 y\,dx = \int_0^1 (t^3)^2(t^2 - 2t)(3t^2\,dt) = \int_0^1 (3t^{10} - 6t^9)\,dt$$

$$= \left[\tfrac{3}{11}t^{11} - \tfrac{3}{5}t^{10}\right]_0^1 = \tfrac{3}{11} - \tfrac{3}{5} = -\tfrac{18}{55}.\ \square$$

Question 9 What is the value of $\int_C (x+y)\,dy$ for $C: y = x^2, 1 \xrightarrow{x} 0$?

Example 7 Draw $C: x = 2\cos t, y = 2\sin t, \tfrac{1}{2}\pi \xrightarrow{t} -\tfrac{1}{2}\pi$. and evaluate $\int_C xy^2\,ds$.

SOLUTION The curve C is the semicircle in Figure 22, oriented clockwise.
Here $\dfrac{dx}{dt} = \dfrac{d}{dt}(2\cos t) = -2\sin t$ and $\dfrac{dy}{dt} = \dfrac{d}{dt}(2\sin t) = 2\cos t$. Also, because the curve is oriented with decreasing t, we need to use the minus sign in **(25)**. We obtain

$$\int_C xy^2\,ds = -\int_{\pi/2}^{-\pi/2} (2\cos t)(2\sin t)^2 \sqrt{(-2\sin t)^2 + (2\cos t)^2}\,dt$$

$$= -16\int_{\pi/2}^{-\pi/2} \sin^2 t \cos t\,dt = -16\int_{t=\pi/2}^{t=-\pi/2} u^2\,du$$

$$= -16\left[\tfrac{1}{3}u^3\right]_{t=\pi/2}^{t=-\pi/2} = -16\left[\tfrac{1}{3}\sin^3 t\right]_{\pi/2}^{-\pi/2} = -16(-\tfrac{1}{3} - \tfrac{1}{3}) = \tfrac{32}{3}.$$

Here we used the substitution $u = \sin t, du = \cos t\,dt.\ \square$

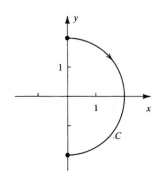

FIGURE 22

13.1 Vector fields and line integrals

Interpreting line integrals with respect to arclength

A line integral $\int_C f\, ds$ with respect to arclength in the xy-plane can be interpreted as a difference of areas, just as a definite integral $\int_a^b F(x)\, dx$ with $a < b$ is a difference of areas. If $f(x, y)$ is nonnegative on the curve, then the integral equals the area of the vertical curtain-like surface between the curve C and the graph of $f(x, y)$ (Figure 23). This is the case because for each partition, the approximating sum $\sum_{j=1}^{N} f(Q_j)\, \Delta s_j$ on the right of **(13)** equals the area of an approximation of this surface by vertical rectangles of widths Δs_j and heights $f(Q_j)$ (Figure 24).

If f is negative on C then $\int_C f\, ds$ is the negative of the area of such a curtain-like surface below the xy-plane. If f has positive and negative values on C, then the integral equals the area of the curtain-like surface extending from C in the xy-plane up to the graph of f, where f is positive, minus the area of the curtain-like surface extending from C in the xy-plane down the graph of f, where f is negative.

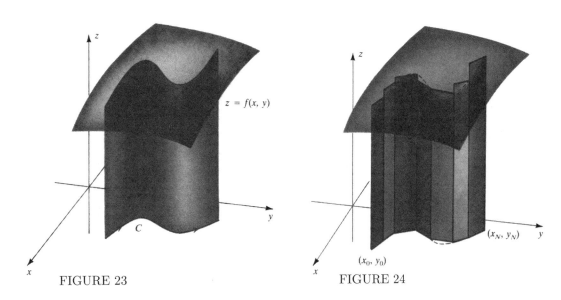

FIGURE 23   FIGURE 24

The integral $\int_C f\, ds$ can also be interpreted in terms of weight or electric charge, and these interpretations apply to curves in space as well as to plane curves. If f is nonnegative on C we can think of it as the density of the curve, measured in weight per unit length. Then $\int_C f\, ds$ is the total weight of the curve. If f has negative values on C, we can think of it as an electric-charge density. Then the integral is the total, net charge on the curve—the total positive charge, minus the total negative charge.

Relating dx, dy and dz to ds

Suppose that $C : x = x(t), y = y(t), a \xrightarrow{t} b$ is a smooth curve in the xy-plane. Then for t between a and b, $\langle x'(t), y'(t) \rangle$ is a nonzero tangent vector to C at $(x(t), y(t))$ that points in the direction of increasing t and, consequently,

$$\mathbf{T} = \langle u_1, u_2 \rangle = \frac{\pm \langle x'(t), y'(t) \rangle}{\sqrt{[x'(t)]^2 + [y'(t)]^2}} \tag{31}$$

is a unit vector in the direction of the orientation of the curve, with the plus sign if $a < b$ and the minus sign if $b < a$.

On the other hand, formula **(29)** reads

$$ds = \pm\sqrt{[x'(t)]^2 + [y'(t)]^2}\, dt \tag{32}$$

again with the plus sign if $a < b$ and a minus sign if $b < a$. Combining **(31)** and **(32)** shows that

$$\mathbf{T}\, ds = \langle x'(t), y'(t)\rangle\, dt = \langle dx, dy\rangle.$$

This gives the first part of the following rule. A similar calculation for a curve C in space gives the second part.

Rule 1 (a) *For a smooth curve in the xy-plane*

$$\langle dx, dy\rangle = \mathbf{T}\, ds \tag{33a}$$

or equivalently

$$dx = u_1\, ds, \quad dy = u_2\, ds \tag{33b}$$

where $\mathbf{T} = \langle u_1, u_2\rangle$ is the unit tangent vector in the direction of the orientation of the curve.
(b) *For a smooth curve in xyz-space*

$$\langle dx, dy, dz\rangle = \mathbf{T}\, ds \tag{34a}$$

or equivalently

$$dx = u_1\, ds, \quad dy = u_2\, ds, \quad dz = u_3\, ds \tag{34b}$$

where $\mathbf{T} = \langle u_1, u_2, u_3\rangle$ is the unit tangent vector in the direction of the orientation of the curve.

Example 8   A function $f(x,y)$ has negative values on the curve C in Figure 25. Is $\displaystyle\int_C f\, dx$ positive or negative?

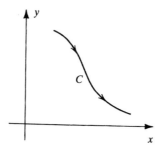

FIGURE 25

SOLUTION   With the first equation in **(33)**, we have $\displaystyle\int_C f\, dx = \int_C f u_1\, ds$ with u_1 the x-component of the unit tangent vector to the curve. Since u_1 is positive all along C and f is negative there, the integral is negative. □

Question 10   Are (a) $\displaystyle\int_C f\, dy$ and (b) $\displaystyle\int_C f\, ds$ positive or negative if C is the curve in Figure 17 and f is negative on C?

We might think of dx and ds as being "positive" and dy as being "negative" on the curve in Figure 25.

Unit normal vectors to oriented plane curves

We define the UNIT NORMAL VECTOR $\mathbf{n} = \mathbf{n}(x,y)$ at a point (x,y) on an oriented curve in the xy-plane to be the unit vector perpendicular to the curve at (x,y) that points to the right of the unit tangent vector $\mathbf{T}$ in the direction of the curve's orientation (Figure 26).

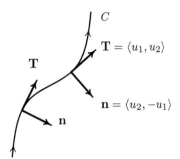

FIGURE 26

If, as is our convention, we write $\mathbf{T} = \langle u_1, u_2 \rangle$, then
$$\mathbf{n} = \langle u_2, -u_1 \rangle. \tag{35}$$

To recall this definition, notice that $\mathbf{n}$ is a unit vector with $\mathbf{T} \cdot \mathbf{n} = \langle u_1, u_2 \rangle \cdot \langle u_2, -u_1 \rangle = u_1 u_2 - u_2 u_1 = 0$ and that if $\mathbf{T}$ has positive components, as in Figure 26, then $\mathbf{n}$ has a positive x-component and a negative y-component.

Formula (35) with Rule 1 yields the next Rule for relating integrals of normal components of vectors with respect to arclength in the xy-plane to line integrals with respect to x and y.

Rule 2 *For a curve in the xy-plane with the unit tangent vectors $\mathbf{T} = \langle u_1, u_2 \rangle$ and unit normal vectors $\mathbf{n} = \langle u_2, -u_1 \rangle$*
$$\mathbf{n}\, ds = \langle u_2\, ds, -u_1\, ds \rangle = \langle dy, -dx \rangle. \tag{36}$$

Flux of a vector field across a plane curve

The FLUX of the velocity field $\mathbf{v}$ of a fluid across a smooth, oriented, plane curve C is defined to be the rate of flow of the fluid across C in the direction of the unit normal vectors $\mathbf{n}$. To find a formula for the flux, we consider a partition $P_0, P_1, P_2, \ldots, P_N$ of C and, as in the definition of line integrals with respect to arclength, let Δs_j denote the distance between the points P_{j-1} and P_j in the partition. By formula (6), the quantity
$$\mathbf{v}(P_j) \cdot \mathbf{n}(P_j)\, \Delta s_j \tag{37}$$
is the approximate flux of the fluid in the direction of $\mathbf{n}$ across the line segment joining P_{j-1} and P_j. The sum of the quantities (37) is the approximate flux of the fluid across the entire curve and its limit is the integral of $\mathbf{v} \cdot \mathbf{n}$ with respect to arclength. We define this integral to be the exact flux:

Definition 2 *If $\mathbf{v}(x,y) = \langle p(x,y), q(x,y) \rangle$ is a piecewise-continuous vector field on the oriented, piecewise-smooth curve with unit normal vectors $\mathbf{n}$, then*
$$\begin{bmatrix} \text{Flux of } \mathbf{v} \text{ across } C \\ \text{in the direction of } \mathbf{n} \end{bmatrix} = \int_C \mathbf{v} \cdot \mathbf{n}\, ds$$
$$= \int_C \langle p, q \rangle \cdot \langle dy, -dx \rangle \tag{38}$$
$$= \int_C (p\, dy - q\, dx).$$

The second formula on the right of (38) follows from the equation $\mathbf{n}\, ds = \langle dy, -dx \rangle$ in Rule 2.

Example 9   A fluid in the xy-plane with distances measured in feet has the velocity field $\mathbf{v} = x^3\mathbf{i} + \mathbf{j}$ (feet per minute) at (x, y). What is the rate of flow of the fluid from left to right across the portion of the parabola $x = y^2$ for $-1 \leq y \leq 1$?

SOLUTION   To have the unit normal vector $\mathbf{n}$ point to the right, we orient the parabola from the bottom to the top (Figure 27). We use y as parameter and write $C : x = y^2, -1 \underset{y}{\longrightarrow} 1$, for which $dx = 2y\,dy$. By the third of formulas **(38)** with $p = x^3$ and $q = 1$, the rate of flow is

$$[\text{Flux}] = \int_C (x^3\,dy - dx) = \int_{-1}^{1} [(y^2)^3 - 2y]\,dy = \int_{-1}^{1} (y^6 - 2y)\,dy$$

$$= \left[\tfrac{1}{7}y^7 - y^2\right]_{-1}^{1} = [\tfrac{1}{7} - 1] - [-\tfrac{1}{7} - 1] = \tfrac{2}{7} \text{ square feet per minute.} \quad \square$$

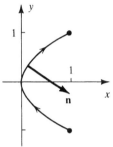

FIGURE 27

Work on an object that traverses a curve

Suppose that a piecewise-continuous force field $\mathbf{F} = \mathbf{F}(x, y)$ or $\mathbf{F} = \mathbf{F}(x, y, x)$ is defined on a smooth curve C in the xy-plane or in xyz-space. We want to define the work done by the force field on an object that traverses the curve. We let $P_0, P_1, P_2, \ldots, P_N$ be a partition of C and, as usual, let Δs_j be the distance between P_{j-1} and P_j. If the vector field had the constant value $\mathbf{F}(P_j)$ on the line segment from P_{j-1} to P_j, then, by **(7)**, the work done by the force on an object traversing the line segment would equal $\mathbf{F}(P_j)\cdot \mathbf{T}\,\Delta s_j$ with $\mathbf{T}$ the unit tangent vector to the line segment. If the numbers Δs_j are small, then this tangent vector is close to the unit tangent vector $\mathbf{T}(P_j)$ to C at P_j, and the sum

$$\sum_{j=1}^{N} \mathbf{F}(P_j) \cdot \mathbf{T}(P_j)\,\Delta s_j$$

approximates what we will define to be the work done by the force. The corresponding line integral gives the exact work:

Definition 3   If $\mathbf{F}$ is a piecewise-continuous force field defined on an oriented, piecewise-smooth curve C in the xy-plane or xyz-space, then

$$\begin{bmatrix} \text{The work done by } \mathbf{F} \text{ on} \\ \text{an object that traverses } C \end{bmatrix} = \int_C \mathbf{F}\cdot \mathbf{T}\,ds \tag{39}$$

$$= \begin{cases} \int_C (p\,dx + q\,dy) \\ \int_C (p\,dx + q\,dy + r\,dz) \end{cases} \tag{40}$$

where the first formula on the right of **(40)** applies to a force field $\mathbf{F} = \langle p(x, y), q(x, y)\rangle$ in the xy-plane and the second formula to a force field $\mathbf{F} = \langle p(x, y, z), q(x, y, z), r(x, y, z)\rangle$ in xyz-space.

Formulas **(40)** follow from **(39)** by Rule 1, which states that $\mathbf{T}\,ds = \langle dx, dy\rangle$ in the xy-plane and $\mathbf{T}\,ds = \langle dx, dy, dz\rangle$ in xyz-space.

13.1 Vector fields and line integrals

Example 10 How much work is done by the force field $\mathbf{F} = \langle xy, -yz, x \rangle$ (pounds) on an object that traverses the curve $C : x = t^4, y = -t^2, z = t^3, 0 \xrightarrow{t} 1$? Distances are measured in feet.

SOLUTION By the second formula in **(40)** with $p = xy, q = -yz$, and $r = x$, the work is

$$\int_C \mathbf{F} \cdot \mathbf{T} \, ds = \int_C (p \, dx + q \, dy + r \, dz) = \int_C (xy \, dx - yz \, dy + x \, dz).$$

We make the substitutions $x = t^4, y = -t^2, z = t^3, dx = 4t^3 \, dt, dy = -2t \, dt$, and $dz = 3t^2 \, dz$ to obtain

$$\int_0^1 [(t^4)(-t^2)(4t^3 \, dt) - (-t^2)(t^3)(-2t \, dt) + (t^4)(3t^2 \, dt)]$$

$$= \int_0^1 (-4t^9 + t^6) \, dt = \left[-\tfrac{2}{5} t^{10} + \tfrac{1}{7} t^7 \right]_0^1$$

$$= \left[-\tfrac{2}{5} + \tfrac{1}{7} \right] - [0] = -\tfrac{9}{35} \text{ foot-pounds.} \ \square$$

Example 11 Figure 28 shows a force field $\mathbf{F}(x, y)$ (newtons) along a curve C in an xy-plane. The lengths of the arrows are measured by using the scales on the coordinate axes. What is the approximate work done by the force field on an object that traverses C?

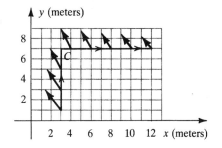

FIGURE 28

SOLUTION Along the vertical portion C_1 of C, $\mathbf{F} \cdot \mathbf{T}$ is approximately 2 newtons and along the horizontal portion C_2, $\mathbf{F} \cdot \mathbf{T}$ is approximately -1 newton. Therefore, the work is

$$\int_C \mathbf{F} \cdot \mathbf{T} \, ds = \int_{C_1} \mathbf{F} \cdot \mathbf{T} \, ds + \int_{C_2} \mathbf{F} \cdot \mathbf{T} \, ds$$

$$\approx \int_{C_1} 2 \, ds + \int_{C_2} (-1) \, ds$$

$$= 2[\text{Length of } C_1] - [\text{Length of } C_2].$$

C_1 is approximately 6 meters long and C_2 is approximately 9 meters long, so the work is approximately $2(6) - 9 = 3$ newton-meters (joules). $\square$

Average values

In Section 4.6 we defined the average value of $F(x)$ in an interval $[a, b]$ to be the integral of F over the interval divided by the length of the interval: $\dfrac{1}{b-a}\displaystyle\int_a^b F(x)\,dx$. The definition of average value on a curve is similar. It uses line integrals with respect to arclength:

Definition 4 The AVERAGE VALUE of $f = f(x, y)$ on a curve C in the xy-plane or of $f = f(x, y, z)$ on a curve C in xyz-space is

$$\frac{1}{[\text{Length of } C]} \int_C f\, ds. \tag{41}$$

Example 12   Express the average value of $x^2 y^2$ on the portion of the curve $y = x^3$ for $0 \leq x \leq 2$ as a quotient of integrals.

SOLUTION   We describe the curve with x as parameter by writing $C : x = x, y = x^3, 0 \xrightarrow{x} 2$. Then $\dfrac{dx}{dx} = 1$ and $\dfrac{dy}{dx} = 3x^2$, so the length of C is

$$\int_C ds = \int_0^2 \sqrt{\left[\frac{dx}{dx}\right]^2 + \left[\frac{dy}{dx}\right]^2}\, dx = \int_0^2 \sqrt{1 + 9x^4}\, dx.$$

Similarly, the integral of $x^2 y^2$ with respect to s equals

$$\int_C x^2 y^2\, ds = \int_0^2 x^2 (x^3)^2 \sqrt{\left[\frac{dx}{dx}\right]^2 + \left[\frac{dy}{dx}\right]^2}\, dx = \int_0^2 x^8 \sqrt{1 + 9x^4}\, dx.$$

The average value is

$$\frac{\displaystyle\int_0^2 x^8 \sqrt{1 + 9x^4}\, dx}{\displaystyle\int_0^2 \sqrt{1 + 9x^4}\, dx}. \quad \square$$

Question 11   Use a calculator or computer algorithm to find the approximate decimal value of the average value in Example 9.

Principles and procedures

- The velocity and force fields discussed in this chapter are determined completely by the coordinates of the points where they are positioned and does not vary with time or depend on any other factors.

- The notation $a \xrightarrow{t} b$ is used to describe the orientation of curves, rather than $a \leq t \leq b$ so that b can be less than a and the negative of a curve can be obtained by switching a and b and writing $b \xrightarrow{t} a$.

- An integral $\displaystyle\int_C f\, ds$ with respect to arclength in the xy-plane can be interpreted as the area of the curtain-like surface between the curve and the graph $z = f(x, y)$ where f is positive minus the area of the curtain-like surface between the curve and the graph $z = f(x, y)$ where f is negative.

- An integral $\displaystyle\int_C f\, ds$ with respect to arclength in the xy-plane or in xyz-space can be interpreted for nonnegative f as the weight of the curve if f is its density and for general f as the total, net charge on the curve if f is its charge density.

13.1 Vector fields and line integrals

- One way to understand an integral $\int_C f\, dx$ is to think of it as the limit of approximating sums $\sum_{j=1}^N f(Q_j)\Delta x_j$ of the product of the value $f(Q_j)$ at a point in the subcurve between points P_{j-1} and P_j, multiplied by the change in x from P_{j-1} to P_j. Another approach is to interpret $\int_C f\, dx$ as the integral $\int_C f\, u_1\, ds$ with respect to arclength of the function f multiplied by the x-component u_1 of the unit tangent vector to the curve, and then interpret the integral with respect to arclength. Similar reasoning can be used for integrals with respect to y and z.

- Definition 4 of average value provides another interpretation of integrals with respect to arclength because the average value equals the integral $\int_C f\, ds$, divided by the length of the curve, so that

$$\int_C f\, ds = [\text{Average value of } f \text{ on } C][\text{Length of } C].$$

- The symbolic equations $\mathbf{T}\, ds = \langle dx, dy\rangle$ and $\mathbf{n}\, ds = \langle dy, -dx\rangle$ for an oriented, smooth plane curve can be recalled with a sketch as in Figure 29, where the curve is oriented toward the upper right and $\mathbf{T}\, ds$ and $\mathbf{n}\, ds$ are visualized as tangent and normal vectors, respectively. In this picture dx and dy are positive because $\mathbf{T}\, ds$ points up to the right. This causes $\mathbf{n}\, ds = \langle dy, -dx\rangle$ to have a positive x-component and a negative y-component because it points down to the right.

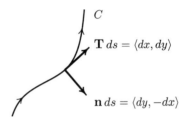

FIGURE 29

- The flux $\int_C \mathbf{v}\cdot\mathbf{n}\, ds$ of a vector field across a curve C in the xy-plane is the integral with respect to arclength of the normal component $\mathbf{v}\cdot\mathbf{n}$ of $\mathbf{v}$ in the direction of the unit normal vector $\mathbf{n}$.

- The work $\int_C \mathbf{F}\cdot\mathbf{T}\, ds$ done by a force field on an object that traverses a curve C in the xy-plane or in xyz-space is the integral with respect to arclength of the tangential component $\mathbf{F}\cdot\mathbf{T}$ of $\mathbf{v}$ in the direction of the unit tangent vector $\mathbf{T}$.

- We will discuss the flux of vector fields across surfaces in xyz-space in Section 13.3.

Tune-Up Exercises 13.1♦

A *Answer provided.* **O** *Outline of solution provided.*

T1.^O Match the vector fields **(a)** $\langle -\frac{1}{2}, \frac{1}{2}|x|\rangle$, **(b)** $\langle -\frac{1}{2}, 1\rangle$, and **(c)** $\langle \frac{1}{2}x, \frac{1}{2}x\rangle$ to their sketches in Figures 30 through 32. Explain how each formula determines the pattern of the arrows in the drawings.

♦ Type 1, matching vector fields to their formulas

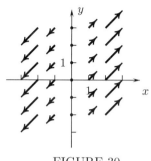
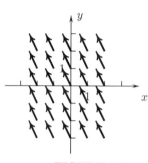
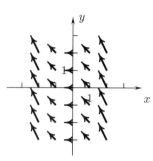

FIGURE 30 FIGURE 31 FIGURE 32

T2.^O What are the values of **(a)** $\int_C 5\,dx$, **(b)** $\int_C 10\,dy$, and **(c)** $\int_C 15\,ds$ for the curve C in Figure 33?

♦ Type 1, evaluating line integrals of constants

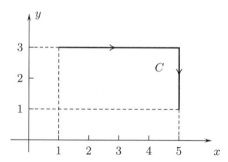

FIGURE 33

T3.^O What are the values of **(a)** $\int_C 5\,dx$, **(b)** $\int_C 10\,dy$, **(c)** $\int_C 15\,dz$, and **(d)** $\int_C 20\,ds$ if C is a curve in space that runs from $(1, 2, 3)$ to $(6, -3, 0)$ and has length 30?

♦ Type 2, approximate values of line integrals

T4.^O **(a)** Express $\int_C x^7\,dy$ as a definite integral where $C : x = \sin t, y = \cos t, 0 \xrightarrow{t} \pi$. **(b)** Use either a calculator- or computer-algebra system to find the exact value of the integral in part (a) or a calculator or computer algorithm to give its approximate decimal value.

♦ Type 2, approximate value of a line integral

T5.^O What are the unit tangent vector **T** and the unit normal vector **n** at $(1, 1)$ to $y = x^2$, oriented from left to right?

♦ Type 1, **T** and **n**

♦The Tune-up Exercises and Problems are classified by type and content. The types are (1) basic, reactive; (2) basic reflective; (3) intermediate, reactive; (4) intermediate, reflective; (5) advanced, reactive; (6) advanced, reflective; and (7) advanced, theoretical.

13.1 Vector fields and line integrals

T6.^{O} Express **(a)** $\int_C [xy\mathbf{i} + (y-x)\mathbf{j}] \cdot \mathbf{T}\, ds$ and **(b)** $\int_C (\sin x\, \mathbf{i} + \cos y\, \mathbf{j}) \cdot \mathbf{n}\, ds$ in terms of integrals with respect to x and y, where $\mathbf{T}$ and $\mathbf{n}$ are the unit normal and tangent vectors to an oriented curve C in the xy-plane.

♦ Type 3, converting integrals involving $\mathbf{T}\,ds$ and $\mathbf{n}\,ds$ to integrals with respect to dx and dy

T7.^{O} What is the average value of $f(x,y,z) = -3$ on the curve $C: x = e^t, y = e^{2t}, z = e^{3t}, 0 \xrightarrow{t} 5$?

♦ Type 3, average value on a curve

Problems 13.1
^{A}Answer provided. ^{O}Outline of solution provided.

1. Sketch the vector fields **(a^O)** $\mathbf{A} = \langle 1, y \rangle$, **(b^A)** $\mathbf{B} = \langle y, y \rangle$, **(c)** $\mathbf{C} = \langle -\tfrac{1}{2}x, \tfrac{1}{2}y \rangle$, and **(d)** $\mathbf{D} = \langle y, 1 \rangle$.

♦ Type 1, drawing vector fields

2. Sketch the vector field $\tfrac{1}{2}(x+y)\mathbf{i} + \tfrac{1}{2}(y-x)\mathbf{j}$.

♦ Type 2, drawing a vector field

3. **(a)** Figures 34 through 36 show the velocity fields (i) $\mathbf{v} = \langle -1, \tfrac{1}{4}y \rangle$, (ii) $\mathbf{v} = \langle \tfrac{1}{2}y, \tfrac{1}{2}x \rangle$ and (iii) $\mathbf{v} = \langle \tfrac{1}{10}x - y, \tfrac{1}{10}y + x \rangle$. Match the drawings to the formulas. **(b)** Each of Figures 37 through 39 shows one or more streamlines for one of the vector fields in Figures 30 through 32. Match the streamlines to the vector fields. **(c)** Show that each of the curves (i) $x = -t, y = e^{t/4}$, (ii) $x = e^{t/2} + e^{-t/2}$, and (iii) $x = e^{t/10}\cos t, y = e^{t/10}\sin t$ is a streamline of one of the vector fields in Figures 34 through 36.

♦ Type 2, recognizing vector fields and streamlines

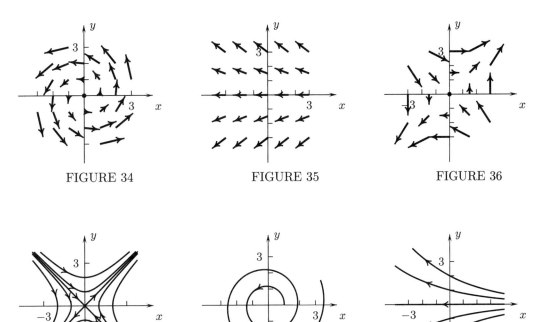

FIGURE 34   FIGURE 35   FIGURE 36

FIGURE 37   FIGURE 38   FIGURE 39

4. Show that the streamlines in Figure 8 for the velocity field $\mathbf{v} = \langle \tfrac{1}{2}x, -\tfrac{1}{2}y \rangle$ in Figure 7 are the positive and negative x- and y-axes and the half-hyperbolas $y = c/x$ with $c \neq 0$ and positive or negative x, all oriented as shown in Figure 8.

♦ Type 2, finding streamlines

Evaluate the line integrals in Problems 5 through 13.

5.º $\int_C (x+y)\,dx$ with $C: x = t^3, y = t^2, 0 \xrightarrow{t} 3$

♦ Type 1, evaluating line integrals

6.^A $\int_C e^{x+y-z}\,dy$ for C the line $x = 2+t, y = 3-2t, z = 4t, 0 \xrightarrow{t} 2$

♦ Type 1, evaluating line integrals

7. $\int_C xy\,dy$ for $C: x = t^3 + t, y = -t^5, 0 \xrightarrow{t} 1$

♦ Type 1, evaluating line integrals

8.º (a) $\int_C x\,dy$ and (b) $\int_C xy\,ds$ with C the line segment from $(1,1)$ to $(4,5)$

♦ Type 1, evaluating line integrals

9.^A (a) $\int_C e^{x+2y}\,ds$ and (b) $\int_C \sin x\,dy$ with C the line segment from $(0,0)$ to $(7,-6)$

♦ Type 1, evaluating line integrals

10. (a) $\int_C \sqrt{x+y}\,dz$ and (b) $\int_C \sqrt{x+z}\,ds$ with C the line segment from $(0,0,0)$ to $(4,5,6)$

♦ Type 1, evaluating line integrals

11.^A The integrals of $f = xy^2$ (a) with respect to x, (b) with respect to y, and (c) with respect to s for $C: x = t^3, y = t^2, 1 \xrightarrow{t} -1$

♦ Type 1, evaluating line integrals

12.^A The integrals of $f = xy/z$ (a) with respect to z and (b) with respect to s, where $C: x = t, y = t^2, z = \frac{2}{3}t^3, 2 \xrightarrow{t} 1$

♦ Type 1, evaluating line integrals

13. The integrals of $f = x$ (a) with respect to x, (b) with respect to y, and (c) with respect to s, where C is the semicircle $y = \sqrt{1-x^2}$, oriented from left to right

♦ Type 1, evaluating line integrals

14. Show that the integrals of $f(x,y) = y$ with respect to x, with respect to y, and with respect to s are equal, where C is the portion of the cubic $y = x^3$ for $-1 \leq x \leq 1$, oriented from left to right.

♦ Type 2, evaluating line integrals

15. Show that the value of $\int_C (xy\,dx - x^2\,dy)$ with $C: x = t^3, y = t^4 - 2t^2, -k \xrightarrow{t} k$ is independent of the value of the constant k.

♦ Type 2, evaluating line integrals

16.^A Give definite integrals that equal (a) $\int_C \sin(xy)\,dz$ and (b) $\int_C e^z\,ds$ with $C: x = \sin t, y = \sin(2t), z = \sin(3t), \pi \xrightarrow{t} 0$. Then use a calculator or computer algorithm to find their approximate decimal values.

♦ Type 3, approximate values of line integrals

17. Give definite integrals that equal (a) $\int_C \ln(x^2 + y^2 + z^2)\,ds$ and (b) $\int_C e^{xyz}\,dx$ with $C: x = t\cos t, y = t^2, z = t, 10 \xrightarrow{t} 1$. Then use a calculator or computer algorithm to find their approximate decimal values.

♦ Type 3, approximate values of line integrals

18.º Evaluate $\int_C x\,dy$, where C is the ellipse $x = 3\cos t, y = 5\sin t, 0 \xrightarrow{t} 2\pi$.

♦ Type 3, evaluating line integrals

19.^A What is the average value of $f = x^2 y$ on the line segment from the origin to the point $(2,3)$?

♦ Type 3, average values

13.1 Vector fields and line integrals

20.^A A space curve has length 20 and runs from $(2, 3, -4)$ to $(5, 0, 6)$. What are the approximate values of (a) $\int_C f\, dx$, (b) $\int_C f\, dz$, and (c) $\int_C f\, ds$ if $f(x, y, z) \approx -5$ on C?

♦ Type 2, estimating line integrals

21.^A What are the values of (a) $\int_{C_2} g\, dy$ and (b) $\int_{C_2} g\, ds$ if $\int_{C_1} g\, dy = 7$, $\int_{C_1} g\, ds = 14$, $\int_C g\, dy = 5$, $\int_C g\, ds = 24$, and $C = C_1 - C_2$ with C_1 and C_2 as in Figure 40?

♦ Type 3, evaluating line integrals with a drawing

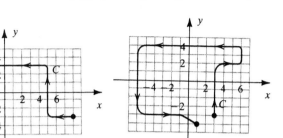

FIGURE 40   FIGURE 41   FIGURE 42

22.^A A function f satisfies $7.9 \leq f(x, y) \leq 8$ for (x, y) on the curve C in Figure 41. What are the approximate values of (a) $\int_C f\, dx$, (b) $\int_C f\, dy$, and (c) $\int_C f\, ds$?

♦ Type 3, estimating line integrals with a drawing

23. A function g satisfies $-11.01 \leq g(x, y) \leq -10.97$ for (x, y) on the curve C in Figure 42. Which of the numbers $0, \pm 10, \pm 20, \pm 80, \pm 250$, and ± 400 are closest to (a) $\int_C g\, dx$, (b) $\int_C g\, dy$, and (c) $\int_C g\, ds$?

♦ Type 3, estimating line integrals with a drawing

24. The function h satisfies $-4.03 \leq h(x, y) \leq -3.97$ for (x, y) on the curve C in Figure 43. Which of the numbers $0, \pm 15, \pm 30, \pm 80$, and ± 125 are closest to (a) $\int_C h\, dx$, (b) $\int_C h\, dy$, and (c) $\int_C h\, ds$?

♦ Type 3, estimating line integrals with a drawing

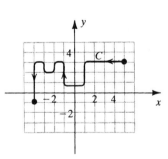 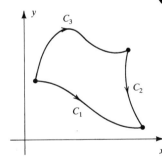 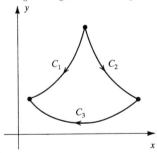

FIGURE 43   FIGURE 44   FIGURE 45

25.^A The curves C_1, C_2, and C_3 in Figure 44 form the boundary of a region and C is its entire boundary oriented counterclockwise. (a) Express C in terms of C_1, C_2, and C_3.
(b) If $\int_{C_1} f\, dy = 5$, $\int_{C_2} f\, dy = -10$, and $\int_C f\, dy = 12$, then what is the value of $\int_{C_3} f\, dy$?
(c) What is the value of $\int_{C_3} f\, ds$ if $\int_{C_1} f\, ds = -2$, $\int_{C_2} f\, ds = 6$, and $\int_C f\, ds = 14$?

♦ Type 2, line integrals over sums and differences of curves

26. The curves $C_1, C_2,$ and C_3 in Figure 45 form the boundary of a region and C is its entire boundary oriented clockwise. **(a)** Express C in terms of $C_1, C_2,$ and C_3.
(b) If $\int_{C_1} f\, dy = -1, \int_{C_2} f\, dy = 4,$ and $\int_C f\, dy = 3,$ then what is the value of $\int_{C_3} f\, dy$?
(c) What is the value of $\int_{C_3} f\, ds$ if $\int_{C_1} f\, ds = 10, \int_{C_2} f\, ds = -4,$ and $\int_C f\, ds = 0$?

◆ Type 2, line integrals over sums and differences of curves

27.[A] What are the approximate values of **(a)** $\int_C (p\, dx + q\, dy)$ and **(b)** $\int_C pq\, ds$ if C is a plane curve of length 25 that runs from $(2, -6)$ to $(4, 4)$ and if $p(x, y) \approx 6$ and $q(x, y) \approx -7$ on C?

◆ Type 3, estimating line integrals

28. Let C be the circle $x^2 + y^2 = 25$ oriented counterclockwise in an xy-plane with distances measured in centimeters. **(a)** Draw the vectors in the force field $\mathbf{F} = \langle \tfrac{1}{2}(x-y), \tfrac{1}{2}(x+y)\rangle$ at the eight points $(\pm 5, 0), (0, \pm 5),$ and $(\pm 4, \pm 3)$ on C. **(b)** Show that the tangential component of $\mathbf{F}$ is constant on C and use this value to find the work the force field does on an object that traverses it.

◆ Type 4, tangential components of a vector field and a line integral

29.[O] Evaluate $\int_C \mathbf{F} \cdot \mathbf{T}\, ds$ for $\mathbf{F} = \langle y^4, e^{x^2}, xy\rangle$ and $C: x = t^2, y = t^4, z = \ln t, 1 \xrightarrow[t]{} 2$. (Do not simplify your answer.)

◆ Type 3, evaluating a line integral in space

30.[A] What is the value of $\int_C \mathbf{F} \cdot \mathbf{T}\, ds$ for $\mathbf{F} = \langle x^3 y^4, x^4 y^3\rangle$ and $C: x = t^2, y = t^5, -1 \xrightarrow[t]{} 1$?

◆ Type 3, evaluating a line integral

31. Evaluate $\int_C \mathbf{F} \cdot \mathbf{T}\, ds$ for $\mathbf{F} = \langle e^x, \sin y + z, y\rangle$ and $C: x = t^3, y = e^t, z = 3, 1 \xrightarrow[t]{} 10$.

◆ Type 3, evaluating a line integral in space

32. What is the value of $\int_C \mathbf{F} \cdot \mathbf{T}\, ds$ for $\mathbf{F} = \langle xyz, 1, \tfrac{2}{3}y\rangle$ and $C: x = t^2, y = -3t, z = t, 1 \xrightarrow[t]{} 4$?

◆ Type 3, evaluating a line integral in space

33.[A] How much work is done by the force field $\mathbf{F} = xy^3\mathbf{i} - 5x^2 y\mathbf{j}$ (newtons) on an object that traverses $C: x = e^t, y = e^{2t}, 1 \xrightarrow[t]{} 0$ in an xy-plane with distances measured in meters?

◆ Type 3, work

34.[O] What is the flux of the velocity field $\mathbf{v} = xy\mathbf{i} + x^2 y^2\mathbf{j}$ across the quarter ellipse $C: x = 4\cos t, y = \sin t, 0 \xrightarrow[t]{} \tfrac{1}{2}\pi$?

◆ Type 3, flux

35.[A] Find the flux of the velocity field $\mathbf{v} = 2y\mathbf{i} - y^3\mathbf{j}$ across $C: y = e^x, 0 \xrightarrow[x]{} 2$.

◆ Type 3, flux

36.[A] A fluid has velocity $\mathbf{v} = \langle x + 2y, x - 3y\rangle$ (feet per second) at (x, y) in an xy-plane with distances measured in feet. What is the rate of flow of the fluid from right to left across the half ellipse $C: x = 3\cos t, y = 5\sin t, \tfrac{1}{2}\pi \xrightarrow[t]{} \tfrac{3}{2}\pi$?

◆ Type 3, flux

37. What is the flux of the velocity field $\mathbf{v} = e^{x+y}\mathbf{i} + e^{x-y}\mathbf{j}$ across the line segment PQ oriented from $P = (1, 2)$ to $Q = (-1, -40)$?

◆ Type 3, flux

38. A fluid has the velocity field $\mathbf{v} = x^3 y\mathbf{i} - xy^3\mathbf{j}$ (meters per minute) in an xy-plane with distances measured in meters. What is the rate of flow from left to right across the portion of $x = y^2$ for $-1 \leq y \leq 1$?

◆ Type 3, flux

13.1 Vector fields and line integrals

39.^A Figure 46 shows the velocity field **v** (yards per minute) of a fluid along a curve C where the scales on the axes are used to measure the lengths of the arrows and distances are given in yards What is the approximate rate of flow of the fluid across the curve in the direction of its unit normal vectors **n**?

♦ Type 4, estimating flux from a drawing

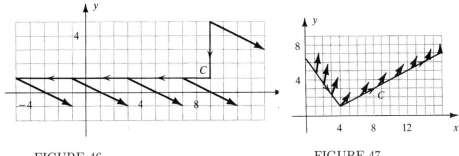

FIGURE 46   FIGURE 47

40. The velocity field **v** (meters per second) of a fluid is shown on a curve C in Figure 47. Distances are measured in meters and the scales on the axes are used to measure the lengths of the arrows. What is the approximate rate of flow of the fluid across the curve from above to below?

♦ Type 4, estimating flux from a drawing

41.^A What is the average value of $f(x,y) = xy^2$ on the line segment between $(1,0)$ and $(-3,-6)$?

♦ Type 3, average value

42. Find the average value of $g(x,y) = x^2$ on the circle $x^2 + y^2 = 1$.

♦ Type 3, average value

43. What is the average value of $h(x,y) = (1 + \frac{2}{3}y)^2$ on the curve $y = \frac{3}{2}x^{2/3}, 1 \leq x \leq 8$?

♦ Type 3, average value

Section 13.2

The fundamental theorems in the plane

OVERVIEW: *In this section we first use the Fundamental Theorem of Calculus to derive* GREEN'S THEOREM, *which relates double integrals of first-order partial derivatives of functions over bounded regions in the xy-plane to line integrals over the boundaries of the regions. It has two vector forms, which we refer to as the* DIVERGENCE THEOREM IN THE PLANE *and* STOKES' THEOREM IN THE PLANE. *They are used in studying force fields and fluid flow. We then use the Fundamental Theorem of Calculus and Green's Theorem to derive criteria for whether line integrals are* PATH INDEPENDENT *in the sense that their values depend only on the coordinates of the endpoints of the associated curves.*

Topics:

- *Green's Theorem*
- *Circulation and Stokes' Theorem in the plane*
- *The Divergence Theorem in the plane*
- *Gradient fields and path independence*
- *Conservative force fields*
- *Irrotational and incompressible fluid flow*

Green's Theorem

As we will see below, GREEN'S THEOREM is a consequence of (i) the Fundamental Theorem of Calculus, (ii) the procedures for evaluating double integrals as iterated definite integrals, and (iii) the definitions of line integrals from the last section. We begin by deriving the theorem for the rectangle $R: 1 \leq x \leq 5, 1 \leq y \leq 3$ shown in Figure 1, whose boundary, oriented counterclockwise, is the sum $C = C_1 + C_2 + C_3 + C_4$ of the line segments that form its sides. Notice that the sides can be given parametrizations with x and y as parameters by writing

$$C_1: y = 1, 1 \xrightarrow[x]{} 5 \qquad C_2: x = 5, 1 \xrightarrow[y]{} 3$$
$$C_3: y = 3, 5 \xrightarrow[x]{} 1 \qquad C_4: x = 1, 3 \xrightarrow[y]{} 1. \tag{1}$$

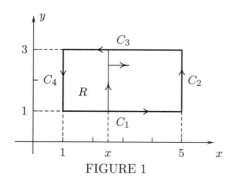
FIGURE 1

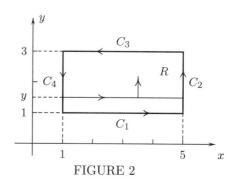
FIGURE 2

♦ SUGGESTIONS TO INSTRUCTORS: The main topics in this section can be divided into three parts to be covered in several lectures.

Part 1 (Green's Theorem and areas; Tune-up Exercises T1, T2; Problems 1–3, 6–13) Have students begin work on Questions 1 through 7 before class. Problems 1, 2, 6, 10, 13 are good for in-class discussion.

Part 2 (Circulation, scalar curl, Stokes' Theorem, divergence, and the Divergence Theorem in the plane; Tune-up Exercises T3–T7; Problems 4, 5, 14–34) Have students begin work on Question 8 and Tune-up Exercises T4, T5–T7 before class. Problems 14, 23, 25, 27, 32–34 are good for in-class discussion.

Part 3 (Gradient fields, simple connectivity, and path independence; Tune-up Exercises T8–T10; Problems 35–48) Have students begin work on Questions 9 through 12 before class. Problems 35, 36, 40, 42, 46, 47, 48 are good for in-class discussion.

13.2 The fundamental theorems in the plane

Example 1 Suppose that $f(x,y)$ has continuous first-order partial derivatives in the rectangle R of Figure 1. **(a)** Express $\iint_R \frac{\partial f}{\partial y}(x,y)\, dx\, dy$ as an iterated integral with the integration procedure shown in Figure 1 and then use the Fundamental Theorem of Calculus in the form

$$\int_a^b \frac{dF}{dy}(y)\, dy = F(b) - F(a) \tag{2}$$

to evaluate the inner integral and obtain a difference of definite integrals.

(b) Use the parametrizations **(1)** to express $\int_C f\, dx = \int_{C_1} f\, dx + \int_{C_2} f\, dx + \int_{C_3} f\, dx + \int_{C_4} f\, dx$ as a difference of definite integrals.

SOLUTION **(a)** By using the integration procedure in Figure 1 and the Fundamental Theorem **(2)** with $F(y) = f(x,y)$ and fixed x, we obtain

$$\iint_R \frac{\partial f}{\partial y}(x,y)\, dx\, dy = \int_{x=1}^{x=5} \int_{y=1}^{y=3} \frac{\partial f}{\partial y}(x,y)\, dy\, dx$$

$$= \int_{x=1}^{x=5} \Big[f(x,y)\Big]_{y=1}^{y=3} dx = \int_{x=1}^{x=5} [f(x,3) - f(x,1)]\, dx$$

$$= \int_{x=1}^{x=5} f(x,3)\, dx - \int_{x=1}^{x=5} f(x,1)\, dx. \tag{3}$$

(b) The line integral over the horizontal line $C_1 : y = 1, 1 \xrightarrow{x} 5$ is

$$\int_{C_1} f(x,y)\, dx = \int_{x=1}^{x=5} f(x,1)\, dx. \tag{4}$$

On the vertical line $C_2 : x = 5, 1 \xrightarrow{y} 3$, $dx = 0$ and hence

$$\int_{C_2} f(x,y)\, dx = 0. \tag{5}$$

The line integral over the horizontal line $C_3 : y = 3, 5 \xrightarrow{x} 1$ equals

$$\int_{C_3} f(x,y)\, dx = \int_{x=5}^{x=1} f(x,3)\, dx. \tag{6}$$

Finally, on the vertical line $C_4 : x = 1, 3 \xrightarrow{y} 1$, $dx = 0$ and therefore

$$\int_{C_4} f(x,y)\, dx = 0. \tag{7}$$

Adding **(4)** through **(7)** yields

$$\int_C f(x,y)\, dx = \int_{C_1} f\, dx + \int_{C_2} f\, dx + \int_{C_3} f\, dx + \int_{C_4} f\, dx$$

$$= \int_{x=1}^{x=5} f(x,1)\, dx + 0 + \int_{x=5}^{x=1} f(x,3)\, dx + 0$$

$$= \int_{x=1}^{x=5} f(x,1)\, dx - \int_{x=1}^{x=5} f(x,3)\, dx \tag{8}$$

where at the last step we used the fact that the definite integral from 5 to 1 is the negative of the integral from 1 to 5. $\square$

Since the number in **(8)** is the negative of the number on the right of **(3)**, we have shown that for the rectangle R in Figure 1,

$$\iint_R \frac{\partial f}{\partial y}(x,y)\, dx\, dy = -\int_C f(x,y)\, dx.$$

As we will see below, this is the first of two formulas in Green's Theorem in the special case of the rectangle R.

Question 1 Express $\iint_R \frac{\partial f}{\partial x}(x,y)\, dx\, dy$ as an iterated integral with the integration procedure shown in Figure 2. Then use the Fundamental Theorem of Calculus in the form

$$\int_{x=a}^{x=b} \frac{\partial F}{\partial x}(x)\, dx = F(b) - F(a)$$

to evaluate the inner integral and thereby express the double integral as a difference of definite integrals.

Question 2 Use the parametrizations **(1)** to express $\int_C f\, dy$ as a difference of definite integrals, where C is the boundary of the rectangle in Figure 2.

Because the final formulas from Questions 1 and 2 are the same, we have shown that for the rectangle in Figures 1 and 2,

$$\iint_R \frac{\partial f}{\partial x}(x,y)\, dx\, dy = \int_C f(x,y)\, dy.$$

This is the second statement in Green's Theorem for the rectangle R of Figures 1 and 2. Here is the theorem for general regions:

Theorem 1 (Green's Theorem) *Suppose that R is a closed, bounded region with a piecewise-smooth boundary in the xy-plane and that the first-order partial derivatives of $f(x,y)$ are continuous in R. Then*

$$\iint_R \frac{\partial f}{\partial y}(x,y)\, dx\, dy = -\int_C f(x,y)\, dx \tag{9}$$

$$\iint_R \frac{\partial f}{\partial x}(x,y)\, dx\, dy = \int_C f(x,y)\, dy \tag{10}$$

where C is the boundary of R oriented so that the region is on the left as the curve is traversed (Figure 3).

Equations **(9)** and **(10)** in Green's Theorem are often combined into the single equation

$$\int_C [p(x,y)\, dx + q(x,y)\, dy] = \iint_R \left[-\frac{\partial p}{\partial y}(x,y) + \frac{\partial q}{\partial x}(x,y) \right] dx\, dy. \tag{11}$$

The boundary of a region is said to be ORIENTED POSITIVELY if, as in the statement of Theorem 1, the region is on the left as the curve is traversed. Notice that this requirement causes the outer boundary of the region R in Figure 3 to be oriented counterclockwise and the boundary of the hole in R to be oriented clockwise.

The calculations in Example 1 can be modified to derive formula **(9)** for a region R as in Figure 4 that is bounded on the top and bottom by the graphs $y = h(x)$ and $y = g(x)$ of piecewise-continuous

13.2 The fundamental theorems in the plane

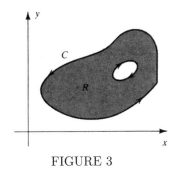

FIGURE 3

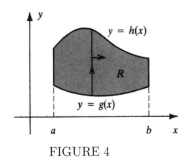

FIGURE 4

functions and extends from $x = a$ on the left to $x = b$ on the right. We first express the double integral as an iterated integral with the integration procedure in Figure 4 and use the Fundamental Theorem to evaluate the inner integral:

$$\iint_R \frac{\partial f}{\partial y}(x,y)\, dx\, dy = \int_{x=a}^{x=b} \int_{y=g(x)}^{y=h(x)} \frac{\partial f}{\partial y}(x,y)\, dy\, dx = \int_{x=a}^{x=b} \Big[f(x,y)\Big]_{y=g(x)}^{y=h(x)} dx$$

$$= \int_{x=a}^{x=b} [f(x,h(x)) - f(x,g(x))]\, dx \qquad (12)$$

$$= \int_{x=a}^{x=b} f(x,h(x))\, dx - \int_{x=a}^{x=b} f(x,g(x))\, dx.$$

On the other hand, the boundary C of the region R in Figure 4 consists of the four curves $C_1, C_2, C_3,$ and C_4 in Figure 5. With x as parameter, the top and bottom are given by $C_1 : y = h(x)$, $b \xrightarrow{x} a$ and $C_3 : y = g(x)$, $a \xrightarrow{x} b$. Also $dx = 0$ on the vertical line segments C_2 and C_4. Consequently,

$$\int_C f(x,y)\, dx = \int_{C_1} f\, dx + \int_{C_2} f\, dx + \int_{C_3} f\, dx + \int_{C_4} f\, dx$$

$$= \int_{x=b}^{x=a} f(x,h(x))\, dx + 0 + \int_{x=a}^{x=b} f(x,g(x))\, dx + 0 \qquad (13)$$

$$= \int_{x=a}^{x=b} f(x,g(x))\, dx - \int_{x=a}^{x=b} f(x,h(x))\, dx.$$

Since the expression on the right of **(13)** is the negative of the expression on the right of **(12)**, this establishes **(9)** for the type of region in Figures 4 and 5.

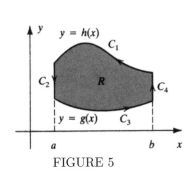

FIGURE 5

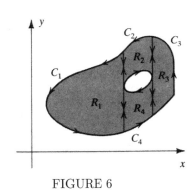

FIGURE 6

Formula **(9)** then follows for any region that can be partitioned into subregions of the type in Figures 4 and 5. Consider, for example, the region R of Figure 6, which consists of four subregions R_1, R_2, R_3, and R_4 of the type for which **(9)** has been established. Then

$$\iint_R f(x,y)\, dx\, dy = \iint_{R_1} f\, dx\, dy + \iint_{R_2} f\, dx\, dy + \iint_{R_3} f\, dx\, dy + \iint_{R_4} f\, dx\, dy$$
$$= -\left\{ \int_{C_1} f\, dy + \int_{C_2} f\, dy + \int_{C_3} f\, dy + \int_{C_4} f\, dy \right\} \tag{14}$$

where C_1, C_2, C_3, C_4 are the boundaries of R_1, R_2, R_3, R_4, respectively.

Question 3 Explain why the quantity on the right of **(14)** equals $-\int_C f(x,y)\, dy$ with C the boundary of R oriented positively, as in Figure 3.

Formula **(9)** can be derived for any region that can be partitioned into a finite number of subregions of the type in Figure 5 by using the reasoning in the response to Question 3 and for other regions by an approximation process.

The second statement **(10)** in Green's Theorem can be derived for a region of the type in Figure 7 by modifying the calculations in the responses to Questions 1 and 2.

Question 4 (a) Suppose $f(x,y)$ has continuous first-order derivatives in the region R in Figure 7. The integration procedure in the drawing gives

$$\iint_R \frac{\partial f}{\partial x}(x,y)\, dy\, dx = \int_{y=a}^{y=b} \int_{x=g(y)}^{x=h(y)} \frac{\partial f}{\partial x}(x,y)\, dy\, dx. \tag{15}$$

Use the Fundamental Theorem of Calculus to express **(15)** as a difference of definite integrals.

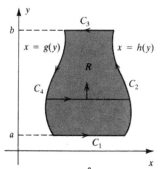

FIGURE 7

Question 5 Why are the line integrals $\int_{C_1} f\, dy$ and $\int_{C_3} f\, dy$ equal to zero for the curves C_1 and C_3 in Figure 7? (b) Use the parametrizations $C_2 : x = h(y), a \xrightarrow{y} b$ and $C_4 : x = g(y), b \xrightarrow{y} a$ to express $\int_C f\, dy$ as a difference of definite integrals, where $C = C_1 + C_2 + C_3 + C_4$ is the boundary of the region R in Figure 7, oriented positively.

Because the results of Questions 4 and 5b are equal, they establish **(10)** for regions of the type in Figure 7. Then this formula can be derived for general regions by partitioning and approximations.

13.2 The fundamental theorems in the plane

Example 2 Suppose that R is a bounded region in the xy-plane with piecewise-smooth boundary C, oriented positively. Express $\int_C (x \sin y \, dx + x^2 \cos^2 y \, dy)$ as a double integral over R.

SOLUTION By **(11)** with $p = x \sin y$ and $q = x^2 \cos^2 y$,

$$\int_C (x \sin y \, dx + x^2 \cos^2 y \, dy) = \iint_R \left[-\frac{\partial}{\partial y}(x \sin y) + \frac{\partial}{\partial x}(x^2 \cos^2 y) \right] dx \, dy$$

$$= \iint_R (-x \cos y + 2x \cos^2 y) \, dx \, dy. \quad \square$$

Question 6 Use Green's Theorem to express **(a)** $\int_C x \, dy$ and **(b)** $\int_C y \, dx$ as double integrals over R, where R is a bounded region in the xy-plane and C is its piecewise-smooth boundary oriented positively.

Since $\iint_R dx \, dy$ is equal to the area of the region R, the results of Question 6 show that

$$[\text{Area of } R] = \begin{cases} \int_C x \, dy \\ -\int_C y \, dx. \end{cases} \tag{16}$$

Question 7 Evaluate $\int_C x \, dy$ for C the circle $x^2 + y^2 = r^2$ of positive radius r oriented counterclockwise **(a)** by using formula **(16)** and **(b)** by using the parametrization $C: x = r \cos t, y = r \sin t, 0 \xrightarrow{t} 2\pi$ of C.

Circulation and Stokes' Theorem in the plane

Recall from the last section that for a force field $\mathbf{F}(x,y)$, the line integral $\int_C \mathbf{F} \cdot \mathbf{T} \, ds$ equals the work done by the force on an object that traverses the curve C. The same integral is used for the CIRCULATION of the velocity field of a fluid if the curve C is CLOSED, meaning that it is continuous and its beginning and endpoints coincide.

Definition 1 The CIRCULATION of a piecewise-continuous vector field $\mathbf{v}(x,y) = \langle p(x,y), q(x,y) \rangle$ around the closed, piecewise-smooth curve C in the xy-plane is

$$\int_C \mathbf{v} \cdot \mathbf{T} \, ds = \int_C [p(x,y) \, dx + q(x,y) \, dy]. \tag{17}$$

We used the formula $\mathbf{T} \, ds = \langle dx, dy \rangle$ from Rule 1 of Section 13.1 to obtain the second integral in **(17)**. If C is the boundary of a region R, then we can apply Green's Theorem in the form **(11)** to the second integral to show that the circulation is equal to $\iint_R [-p_y(x,y) + q_x(x,y)] \, dx \, dy$. This integrand is called the SCALAR CURL of the vector field v:[†]

Definition 2 The SCALAR CURL of $\mathbf{v}(x,y) = \langle p(x,y), q(x,y) \rangle$ is the real-valued function

$$\operatorname{curl}(x,y) = -\frac{\partial p}{\partial y}(x,y) + \frac{\partial q}{\partial x}(x,y). \tag{18}$$

[†] We refer to **(18)** as the "scalar curl" because for each (x,y) it is a number (scalar). This is to distinguish it from the vector-valued curl of a vector field in space that will be defined in the next section.

Theorem 2 (Stokes' Theorem in the plane) *If $\mathbf{v}(x,y)$ has continuous first-order derivatives in a closed, bounded region R in the xy-plane with piecewise-smooth boundary C, oriented positively, then*

$$\int_C \mathbf{v} \cdot \mathbf{T}\, ds = \iint_R \operatorname{curl} \mathbf{v}(x,y)\, dx\, dy. \tag{19}$$

Example 3 Use Stokes' theorem in the plane to calculate the circulation of $\mathbf{v} = (3x^2y^2 + 2y)\mathbf{i} + (9x + 2x^3y)\mathbf{j}$ around the circle $C : x^2 + y^2 = 1$, oriented counterclockwise.

SOLUTION By Definition 2,

$$\operatorname{curl} \mathbf{v} = -\frac{\partial}{\partial y}(3x^2y^2 + 2y) + \frac{\partial}{\partial y}(9x + 2x^3y)$$
$$= -(6x^2y + 2) + (9 + 6x^2y) = 7.$$

Since C is the positively oriented boundary of the disk $R = \{(x,y) : x^2 + y^2 \leq 1\}$, which has area π, Stokes' Theorem (19) yields

$$[\text{Circulation}] = \int_C \mathbf{v} \cdot \mathbf{T}\, ds = \iint_R \operatorname{curl} \mathbf{v}\, dx\, dy = \iint_R 7\, dx\, dy = 7\pi.\ \square$$

Example 4 Use Stokes' Theorem to find the work done by the force field $\mathbf{F} = \langle x \sin y, x^2 \cos y \rangle$ (pounds) on an object that traverses counterclockwise the boundary C of the square $R : 0 \leq x \leq 1, 0 \leq y \leq 1$ in an xy-plane with distances measured in feet.

SOLUTION We first calculate

$$\operatorname{curl} \mathbf{F} = -\frac{\partial}{\partial y}(x \sin y) + \frac{\partial}{\partial x}(x^2 \cos y) = -x \cos y + 2x \cos y = x \cos y.$$

The curve and square are shown in Figure 8. Since C is the boundary of R oriented positively, (19) with $\mathbf{F}$ in place of $\mathbf{v}$ gives

$$[\text{Work}] = \int_C \mathbf{F} \cdot \mathbf{T}\, ds = \iint_R \operatorname{curl} \mathbf{F}\, dx\, dy = \iint_R x \cos y\, dx\, dy$$
$$= \int_{x=0}^{x=1} \int_{y=0}^{y=1} x \cos y\, dy\, dx = \left[\int_{x=0}^{x=1} x\, dx\right]\left[\int_{y=0}^{y=1} \cos y\, dy\right]$$
$$= \left[\tfrac{1}{2}x^2\right]_{x=0}^{x=1}\left[\sin y\right]_{y=0}^{y=1} = \tfrac{1}{2}\sin(1) \text{ foot-pounds}.$$

Here we used the fact that R is a rectangle and $x \cos y$ is the product of a function of x and a function of y to simplify the evaluation of the double integral. $\square$

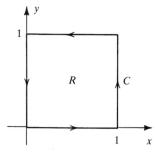

FIGURE 8

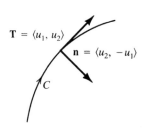

FIGURE 9

Divergence and the Divergence Theorem in the plane

Suppose that a fluid has velocity field $\mathbf{v}(x,y)$ defined in an open set containing an oriented curve C in the xy-plane and that $\mathbf{n}$ denotes the unit normal vectors to the curve that point to the right of the unit tangent vectors $\mathbf{T}$ in the direction of the orientation of the curve (Figure 9). As we saw in Section 13.1, the rate of flow of the fluid across C in the direction of the vectors $\mathbf{n}$ is equal to the flux of $\mathbf{v}$, defined by

$$[\text{Flux}] = \int_C \mathbf{v} \cdot \mathbf{n}\, ds. \tag{20}$$

We can apply Green's Theorem to this line integral if C is the boundary of a region R. First, we need another definition.

Definition 3 *The* DIVERGENCE *of a vector field* $\mathbf{v}(x,y) = \langle p(x,y), q(x,y)\rangle$ *in the xy-plane is the real-valued function*

$$\begin{aligned}\operatorname{div}\mathbf{v}(x,y) = \nabla \cdot \mathbf{v}(x,y) &= \left\langle \frac{\partial}{\partial x}, \frac{\partial}{\partial y}\right\rangle \cdot \langle p(x,y), q(x,y)\rangle \\ &= \frac{\partial p}{\partial x}(x,y) + \frac{\partial q}{\partial y}(x,y).\end{aligned} \tag{21}$$

With this definition can state the next result.

Theorem 3 (The Divergence Theorem in the plane) *If $\mathbf{v}$ has continuous first-order derivatives in a closed, bounded region R in the xy-plane with piecewise-smooth boundary C, oriented positively, then*

$$\int_C \mathbf{v}\cdot\mathbf{n}\, ds = \iint_R \operatorname{div}\mathbf{v}\, dx\, dy. \tag{22}$$

To derive this theorem, we use the relation $\mathbf{n}\, ds = \langle dy, -dx\rangle$ from Rule 2 of Section 13.1, followed by equations **(9)** and **(10)** in Green's Theorem with $f = p$ and $f = q$, respectively. We obtain

$$\int_C \mathbf{v}\cdot\mathbf{n}\, ds = \int_C (p\, dy - q\, dx) = \iint_R \left[\frac{\partial p}{\partial x} + \frac{\partial q}{\partial y}\right] dx\, dy.$$

This with Definition 3 gives **(22)**.

Question 8 (a) What is the divergence of $\mathbf{v} = \langle 8x - y^2, x - 5y\rangle$ (meters per second)? (b) What is the rate of flow of the fluid with the velocity field from part (a) out of a region with area 11 square meters?

Gradient vector fields

Recall that in Section 11.3 we used the vector differentiation operator $\nabla = \left\langle \dfrac{\partial}{\partial x}, \dfrac{\partial}{\partial y}\right\rangle$ to define the gradient

$$\nabla f(x,y) = \left\langle \frac{\partial f}{\partial x}(x,y), \frac{\partial f}{\partial x}(x,y)\right\rangle \tag{23}$$

of a function $f(x,y)$ with two variables. Thus, the gradient ∇f of f is a vector field whose components are the partial derivatives of f. Recall also from Theorem 5 of that section, that at each point where ∇f is not zero, it is perpendicular to the level curve of f through that point and its direction is that toward which f is increasing.

Example 5 Sketch the gradient field ∇f for $f(x,y) = xy$ at several points on the level curves $f = 0$, $f = 4$, and $f = -4$.

SOLUTION   The gradient field is $\nabla f(x,y) = \dfrac{\partial}{\partial x}(xy)\mathbf{i} + \dfrac{\partial}{\partial y}(xy)\mathbf{j} = \langle y, x\rangle$. The level curve $xy = 0$ consists of the x- and y-axes. The level curve $xy = 4$ is the hyperbola $y = 4/x$, and the level curve $xy = -4$ is the hyperbola $y = -4/x$. These are shown in Figure 10 with vectors in the gradient field shown as arrows whose lengths are measured by half the scales on the coordinate axes. Notice that the gradient vectors are perpendicular to the level curves and point in the direction in which $f = xy$ increases. $\square$

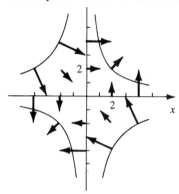

FIGURE 10

Path-independent line integrals

A region R in the xy-plane is said to be CONNECTED if any two points in it can be joined by a smooth curve that lies entirely in R. A line integral $\int_C [p(x,y)\,dx + q(x,y)\,dy]$ with C a continuous, piecewise-smooth curve in a connected region R is said to be PATH INDEPENDENT in R if its value depends only on the coordinates of the beginning and end of C and not on the particular path from the beginning to the end. We illustrate this concept first with an Example and a similar Question.

Example 6 The line $C_1 : y = x,\ 0 \xrightarrow[x]{} 1$ and the curve $C_2 : x = 5y^2 - 4y,\ 0 \xrightarrow[y]{} 1$ from $(0,0)$ to $(1,1)$ are shown in Figure 11. Evaluate **(a)** $\displaystyle\int_{C_1} \nabla(xy) \cdot \mathbf{T}\,ds$ and **(b)** $\displaystyle\int_{C_2} \nabla(xy) \cdot \mathbf{T}\,ds$.

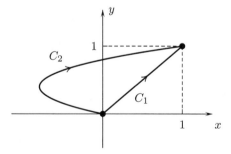

FIGURE 11

SOLUTION   **(a)** We saw in Example 5 that $\nabla(xy) = \langle y, x\rangle$. Since $y = x$, $dy = dx$ and $\mathbf{T}\,ds = \langle dx, dy\rangle$ on C_1, we obtain

$$\int_{C_1} \nabla(xy) \cdot \mathbf{T}\,ds = \int_{C_1} \langle y, x\rangle \cdot \langle dx, dy\rangle = \int_{C_1} (y\,dx + x\,dy)$$
$$= \int_{x=0}^{x=1} (x\,dx + x\,dx) = \int_{x=0}^{x=1} 2x\,dx = \Big[x^2\Big]_0^1 = 1. \tag{24}$$

13.2 The fundamental theorems in the plane

(b) On C_2, $x = 5y^2 - 4y$ and $dx = (10y - 4)\, dy$, so that

$$\int_{C_2} \nabla(xy) \cdot \mathbf{T}\, ds = \int_{C_2} (y\, dx + x\, dy) = \int_0^1 [y(10y-4)\, dy + (5y^2 - 4y)\, dy]$$
$$= \int_{y=0}^{y=1} (15y^2 - 8y)\, dy = \left[5y^3 - 4y^2\right]_0^1 = 1. \ \square \tag{25}$$

Notice that the integrals **(24)** and **(25)** have the same value.

Question 9 Evaluate (a) $\displaystyle\int_{C_1} \nabla(2x + 3y^2) \cdot \mathbf{T}\, ds$ and (b) $\displaystyle\int_{C_2} \nabla(2x + 3y^2) \cdot \mathbf{T}\, ds$, where C_1 and C_2 are the curves from Example 6.

The following theorem explains why the two line integrals in Example 6 are equal and the two line integrals in Question 9 are equal.

Theorem 4 *Suppose that $\mathbf{F}(x, y)$ is a continuous vector field in an open, connected region R in the xy-plane. Then the integral $\displaystyle\int_C \mathbf{F} \cdot \mathbf{n}\, ds$ is path independent in R if and only if there is a real-valued function $U(x, y)$ such that $\nabla U(x, y) = \mathbf{F}(x, y)$ in R. If this is the case, then for any continuous, piecewise-smooth curve C in R*

$$\int_C \mathbf{F} \cdot \mathbf{T}\, ds = U(Q) - U(P) \tag{26}$$

where P is the beginning and Q is the endpoint of the curve.

The function $U(x, y)$ in Theorem 4, if it exists, is called a POTENTIAL FUNCTION for the vector field $\mathbf{F}(x, y)$. A force field $\mathbf{F}$ that is the gradient ∇U of a potential function is called a CONSERVATIVE force field.

To prove the theorem, we first suppose that $\mathbf{F} = \nabla U$ for some function U so that

$$\mathbf{F}(x, y) = \langle U_x(x, y), U_y(x, y) \rangle. \tag{27}$$

We consider an arbitrary smooth curve $C : x = x(t), y = y(t), a \xrightarrow{t} b$ in the region R of the theorem. By the Chain-Rule from Section 11.2,

$$\frac{d}{dt}[U(x(t), y(t))] = U_x(x(t), y(t))\frac{dx}{dt}(t) + U_y(x(t), y(t))\frac{dy}{dt}(t). \tag{28}$$

Combining **(27)** and **(28)**, we obtain, with the Fundamental Theorem of Calculus from Section 4.3,

$$\int_C \mathbf{F} \cdot \mathbf{T}\, ds = \int_C [U_x(x, y)\, dx + U_y(x, y)\, dy]$$
$$= \int_{t=a}^{t=b} [U_x(x(t), y(t))\frac{dx}{dt}(t) + U_y(x(t), y(t))\frac{dy}{dt}(t)]\, dt \tag{29}$$
$$= \int_{t=a}^{t=b} \frac{d}{dt}[U(x(t), y(t))]\, dt = \Big[U(x(t), y(t))\Big]_{t=a}^{t=b}$$
$$= U(x(b), y(b)) - U(x(a), y(a)) = U(Q) - U(P)$$

where P is the beginning and Q the end of the curve C. This formula also holds for continuous, piecewise-smooth curves. Since the number on the right of **(29)** is the same for all such curves in R from P to Q, this calculation shows that the line integral is path independent and gives formula **(26)** if $\mathbf{F}$ is the gradient of $U(x, y)$.

To prove the converse statement in the theorem, we suppose that

$$\int_C \mathbf{F} \cdot \mathbf{T}\, ds = \int_C (p\, dx + q\, dy) \tag{30}$$

is path independent in the region R. We pick a fixed point (x_0, y_0) in R and define

$$U(x,y) = \int_{C_{(x,y)}} (p\, dx + q\, dy) \tag{31}$$

where $C_{(x,y)}$ is a continuous, piecewise-smooth curve in R from (x_0, y_0) to (x, y). The value of $U(x,y)$ does not depend on the choice of the curve from (x_0, y_0) to (x, y) because the line integral is path independent. We need to show that $U_x(x,y) = p(x,y)$ and $U_y(x,y) = q(x,y)$ for all (x,y) in R.

We first find the x-derivative of U. Because R is open, any point in R is on a horizontal line segment in R, so we consider points (x, y) on such a line segment and let C_2 be a horizontal path from a fixed point (x_1, y) on that line segment to (x, y). We let C_1 be a continuous, piecewise smooth curve in R from (x_0, y_0) to (x_1, y), as in Figure 12.

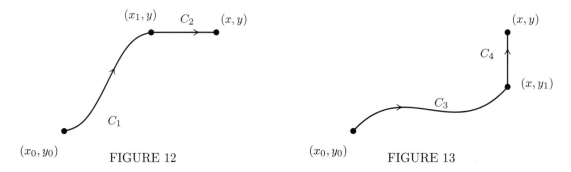

FIGURE 12   FIGURE 13

We can use $C_{(x,y)} = C_1 + C_2$ in (31) and write $C_2 : x = t, y = y$, $x_1 \xrightarrow{t} x$, for which $dy = 0$ and $dx = dt$ to obtain

$$U(x,y) = \int_{C_{(x,y)}} (p\, dx + q\, dy) = \int_{C_1} (p\, dx + q\, dy) + \int_{C_2} (p\, dx + q\, dy)$$

$$= \int_{C_1} (p\, dx + q\, dy) + \int_{t=x_1}^{t=x} p(t, y)\, dt.$$

The integral over C_1 is a constant and $p(t, y)$ is continuous, so by the Fundamental Theorem from Section 6.5, we have

$$\frac{\partial U}{\partial x}(x,y) = \frac{\partial}{\partial x} \int_{t=x_1}^{t=x} p(t,y)\, dt = p(x,y).$$

This shows that the x-component of ∇U equals the x-component of $\mathbf{F}$. A similar calculation with the y-components is given in the next Question.

Question 10 (a) Show that for $C_{(x,y)} = C_3 + C_4$ with $C_4 : x = x, y = t$, $y_1 \xrightarrow{t} y$ as in Figure 13,

$$U(x,y) = \int_{C_3} (p\, dx + q\, dy) + \int_{t=y_1}^{t=y} q(x,t)\, dt.$$ (b) Use the formula from part (a) to show that $U_y(x,y) = q(x,y)$.

Simply connected regions

A SIMPLE, CLOSED CURVE in the xy-plane is a continuous, piecewise-smooth curve whose endpoints coincide but which does not intersect itself at any other points. The curve in Figure 14 is simple and closed, whereas the curve in Figure 15 is simple but not closed and the curve in Figure 16 is closed but not simple. Every simple, closed curve in the xy-plane is the boundary of a bounded region.

13.2 The fundamental theorems in the plane

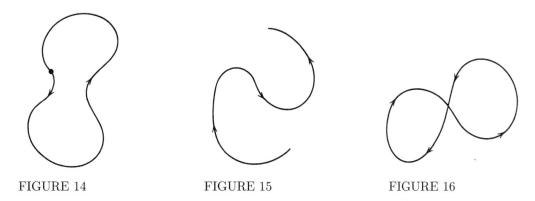

FIGURE 14   FIGURE 15   FIGURE 16

A connected open region R in the xy-plane is SIMPLY CONNECTED if every simple, closed curve in R is the boundary of a region that is entirely contained in R. This implies that R does not have any holes in it. The region R_1 in Figure 17 is simply connected because the region inside any simple, closed curve in R_1 is contained in R_1. The region R_2 in Figure 18, however, is not simply connected because the curve C around the hole in it is the boundary of a region that is not entirely contained in R_2.

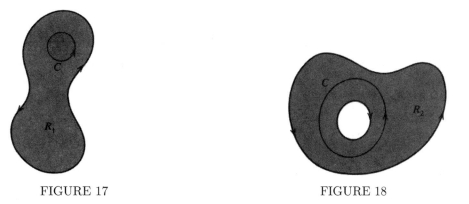

FIGURE 17   FIGURE 18

We need the concept of simple connectivity to state the next theorem.

Theorem 5 *Suppose that $p(x,y)$ and $q(x,y)$ have continuous first-order derivatives in an open region R in the xy-plane.* **(a)** *If the vector field $p\mathbf{i} + q\mathbf{j}$ is the gradient of a potential function $U(x,y)$ in R, then in R*

$$\frac{\partial p}{\partial y}(x,y) = \frac{\partial q}{\partial x}(x,y). \tag{32}$$

(b) *If* **(32)** *is satisfied and R is simply connected, then $p\mathbf{i} + q\mathbf{j}$ is the gradient of a potential function in R.*

To establish part (a) of Theorem 5, we suppose $\nabla U = \langle p, q \rangle$, then $U_x = p$ and $U_y = q$ and consequently

$$\frac{\partial p}{\partial y} = \frac{\partial}{\partial y}\left[\frac{\partial U}{\partial x}\right] = \frac{\partial}{\partial x}\left[\frac{\partial U}{\partial y}\right] = \frac{\partial q}{\partial x}$$

because the mixed, second-order derivatives of $U(x,y)$ are equal.

To prove part (b) we suppose that $p_y = q_x$ in a simply connected open region R and that C_1 and C_2 are two continuous, piecewise-smooth curves in R. If the curves do not intersect except at their endpoints, then $C_1 - C_2$ is a simple closed curve that forms the boundary of a bounded region R and by Green' Theorem,

$$\int_{C_1} (p\,dx + q\,dy) - \int_{C_2} (p\,dx + q\,dy) = \int_{C_1 - C_2} (p\,dx + q\,dy)$$
$$= \pm \iint_R \left[-\frac{\partial p}{\partial y} + \frac{\partial q}{\partial x} \right] dx\,dy = 0.$$

with the plus sign if C is oriented postively and the minus sign if C is oriented negatively.

The last equation shows that the line integrals over C_1 and C_2 are equal. If the curves intersect in more than one point, then they form the boundaries of more than one region. The line integrals over C_1 and C_2 are also equal in this case because the double integral of $p_y - q_x$ is zero in each of the regions. Because C_1 and C_2 are arbitrary continuous, piecewise-smooth paths in R with the same endpoints, the line integral is path independent. This completes the proof of the Theorem.

Example 7 Determine which the vector fields (a) $\langle xy^2, 2x \rangle$ and (b) $\langle y + \cos x, x - 1 \rangle$ is the gradient of a potential function $U(x, y)$ in the entire xy-plane and find such a function.

SOLUTION (a) In this case $\dfrac{\partial p}{\partial y} = \dfrac{\partial}{\partial y}(xy^2) = 2xy$ and $\dfrac{\partial q}{\partial x} = \dfrac{\partial}{\partial x}(2x) = 2$ are equal only along the curve $xy = 1$, so this vector field is not a gradient.

(b) Here $\dfrac{\partial p}{\partial y} = \dfrac{\partial}{\partial y}(y + \cos x) = 1$ and $\dfrac{\partial q}{\partial x} = \dfrac{\partial}{\partial x}(x - 1) = 1$ are equal in the entire xy-plane, which is simply connected, so this vector field is a gradient in the entire xy-plane.

A potential function $U(x, y)$ for this vector field must satisfy the two equations

$$U_x(x, y) = y + \cos x, \quad U_y(x, y) = x - 1. \tag{33}$$

We start with the first. (We could start with the second if we wanted.) We take the x-antiderivative of both sides of the equation $U_x(x, y) = y + \cos x$, to obtain

$$U(x, y) = \int U_x(x, y)\,dx = \int (y + \cos x)\,dx = xy + \sin x + \phi(y). \tag{34}$$

The function $\phi(y)$ is the constant of integration, which may be different for different values of y and hence is a function of y that we have to determine.

Differentiating **(34)** with respect to y yields

$$U_y(x, y) = \frac{\partial}{\partial y}[xy + \sin x + \phi(y)] = x + \phi'(y).$$

The second of equations **(33)** then shows that $\phi'(y) = -1$ and, hence $\phi(y) = -y + C$. We make this substitution in **(34)** to see that the suitable potential functions are $U(x, y) = xy + \sin x - y + C$. □

Theorem 5 can also be used to study path independence because, by Theorem 4, the line integral $\int_C (p\,dx + q\,dy)$ is path independent in a connected open set if and only if $\langle p, q \rangle$ is the gradient of a potential function.

Question 11 (a) Show that $\int_C [y \sin(xy)\,dx + x \sin(xy)\,dy]$ is path independent in the entire xy-plane. (b) Find a potential function for the vector field $y \sin(xy)\mathbf{i} + x \sin(xy)\mathbf{j}$.

13.2 The fundamental theorems in the plane

Example 8 Evaluate the integral from Question 11 for all curves from $(1,2)$ to $(3,4)$.

SOLUTION   You found the potential functions $U(x,y) = -\cos(xy) + C$ in answering Question 11. We need only one such function, so we set $C = 0$ and $U = -\cos(xy)$. By Theorem 4

$$\int_C [y\sin(xy)\,dy + x\sin(xy)\,dy] = U(3,4) - U(1,2) = -\cos(12) + \cos(2)$$

for all curves C from $(1,2)$ to $(3,4)$. $\square$

Irrotational and incompressible fluid flows

The flow of a fluid with velocity field $\mathbf{v} = \langle p(x,y), q(x,y) \rangle$ is said to be IRROTATIONAL in a simply connected region R in the xy-plane if its circulation $\int_C \mathbf{v} \cdot \mathbf{T}\,ds$ is zero for all simple, closed curves C in R.

The fluid is INCOMPRESSIBLE if its density cannot change. A SOURCE of a fluid is a process that creates fluid in the region and a SINK is a process that removes it. Sources and sinks can be caused, for example, by chemical reactions or condensation and evaporation. if a fluid is incompressible and has no sources or sinks in a region R, then the amount of fluid inside any simple, closed curve in R is constant. This means that the flux $\int_C \mathbf{v} \cdot \mathbf{n}\,ds$, which is the rate of flow out or into of the region bounded by C, is zero for all closed curves C in the region.

These definitions lead us to the next Theorem.

Theorem 6   Suppose that a fluid has the velocity field $\mathbf{v}(x,y)$ with continuous first-order derivatives in an open set R in the xy-plane. **(a)** If R is simply connected, then the fluid is irrotational in R if and only if curl $\mathbf{v}$ is zero in R. **(b)** If the fluid has no sources or sinks in R, then it is incompressble in R if and only if div $\mathbf{v}$ is zero in R.

The proof of this theorem is in four parts. Suppose first that curl $\mathbf{v} = \mathbf{0}$ in R and that C is a simple, closed curve in R and R_0 is the region inside C. By Stokes' Theorem in the plane (Theorem 2),

$$[\text{Circulation of } \mathbf{v} \text{ around } C] = \int_C \mathbf{v} \cdot \mathbf{T}\,ds = \pm\iint_{R_0} \text{curl}\,\mathbf{v}\,dx\,dy = 0$$

where the plus sign is used if C is oriented positively and the minus sign if C is oriented negatively. The circulation is zero for all such curves, so the fluid flow is irrotational.

Similarly, if div $\mathbf{v} = \mathbf{0}$ in R, then by the Divergence Theorem 3

$$[\text{Flux of } \mathbf{v} \text{ out of } C] = \int_C \mathbf{v} \cdot \mathbf{n}\,ds = \iint_{R_0} \text{div}\,\mathbf{v}\,dx\,dy = 0$$

for any simple, closed, piecewise-smooth curve C in R and, if there are no sources are sinks, then the fluid is incompresible.

To complete the derivation of Theorem 6, we need the following Lemma, whose proof requires the $\epsilon\delta$-definition of continuity of functions with two variables.[†]

[†]A possible proof: If $f(x_0,y_0) \neq 0$ and $f(x,y)$ is continuous in R, then there is, for each $\epsilon > 0$, a $\delta > 0$ such that the disk D of radius δ with its center at (x_0,y_0) is in R and $|f(x,y) - f(x_0,y_0)| < \epsilon$ for (x,y) in D. We use a δ for $\epsilon = |f(x_0,y_0)|$. Then $|f(x,y) - f(x_0,y_0)| < |f(x_0,y_0)|$ for (x,y) in D, and consequently, $f(x,y)$ is positive in D if $f(x_0,y_0)$ is positive and $f(x,y)$ is negative in D if $f(x_0,y_0)$ is negative.

Lemma 1 If $f(x,y)$ is continuous in an open set R and $f(x_0, y_0)$ is positive (or negative) at a point (x_0, y_0) in R, then there is an open disk D in R with its center at (x_0, y_0) such that $f(x,y)$ is positive (or negative, respectively) in all of D.

Now suppose that $\mathbf{v}$ is irrotational in R, and curl $\mathbf{v}$ is not zero at a point (x_0, y_0) in R, then with D as in Lemma 1, the integral $\iint_D \operatorname{curl} \mathbf{v} \, dx \, dy$ would not be zero. But this is impossible because the double integral equals, by Stokes' Theorem, the circulation around the boundary of D, which is zero. Therefore, curl $\mathbf{v}$ is zero throughout R.

The same reasoning with the Divergence Theorem, shows that if the fluid has no sources or sinks and is incompressible in R, then div $\mathbf{v} = 0$ in R. This completes the derivation of Theorem 6.

Question 12 Figure 19 shows the velocity field $\mathbf{v}(x,y) = \tfrac{1}{3}\langle x+y, x-y\rangle$ of a fluid with no sources or sinks that is defined in the entire xy-plane. Show that **(a)** the flow is irrotational and **(b)** the fluid is incompressible.

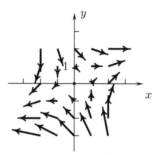

FIGURE 19

Principles and procedures

- To remember which of the formulas $\iint_R \dfrac{\partial f}{\partial y} \, dx \, dy = -\int_C f \, dx$ and $\iint_R \dfrac{\partial f}{\partial x} \, dx \, dy = \int_C f \, dy$ from Green's Theorem has the minus sign, notice that in the second formula, which does not have the minus sign, the "x" in the denominator of $\dfrac{\partial f}{\partial x}$ is adjacent to the "x" in the dx, whereas in the first formula, which has the minus sign, the "y" in the denominator of $\dfrac{\partial f}{\partial y}$ is separated from the "y" in dy by the dx.

- The formula [Area of R] $= -\int_C y \, dx$ in **(16)** was obtained from Green's Theorem. It also follows from the formula [Area of R] $= \int_{x=a}^{x=b} [h(x) - g(x)] \, dx$ from Section 4.6 for the area of a region of the type in Figure 20, because the line integrals $\int_{C_2} y \, dx$ and $\int_{C_4} y \, dx$ over the vertical sides of the region are zero and, with the parametrizations $C_1 : y = h(x)$, $a \xrightarrow{x} b$ and $C_3 : y = g(x)$, $b \xrightarrow{x} a$, we obtain $\int_{x=a}^{x=b} [h(x) - g(x)] \, dx = -\int_{C_1} y \, dx - \int_{C_3} y \, dx = -\int_C y \, dx.$

13.2 The fundamental theorems in the plane

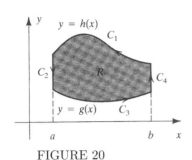

FIGURE 20

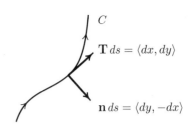

FIGURE 21

- To remember the symbolic equations $\mathbf{T}\,ds = \langle dx, dy\rangle$ and $\mathbf{n}\,ds = \langle dy, -dx\rangle$ from Rules 1 and 2 of the last section, imagine that these are tangential and normal vectors on a curve oriented up and to the right, as in Figure 21.

- To remember the condition $p_y = q_x$ in Theorem 5 for determining whether $\langle p, q\rangle$ is the gradient $\langle U_x, U_y\rangle$ of a potential function U in a simply connected region, note that if $p = U_x$ and $q = U_y$, then p_y and q_x are equal because they both equal the mixed partial derivative U_{xy} of U.

Tune-Up Exercises 13.2♦

A Answer provided. **O** Outline of solution provided.

T1.^O Verify the equation $\int_C f\,dy = \iint_R \dfrac{\partial f}{\partial x}\,dx\,dy$ from Green's Theorem for $f = x^3 y$ and R the triangle with vertices $(0,0), (1,0)$, and $(0,1)$ by evaluating the line and double integrals.
♦ Type 3, verifying Green's Theorem in a particular case

T2.^O Suppose that R is a bounded region in the xy-plane and C is its boundary, oriented positively. Express **(a)** $\int_C xe^y\,dx$ and **(b)** $\int_C y^3 \sin x\,dy$ as double integrals over R.
♦ Type 3, verifying Green's Theorem in a particular case

T3.^O Find the divergence and scalar curl of $\ln(xy)\mathbf{i} + x\tan y\,\mathbf{j}$.
♦ Type 1, finding a divergence and scalar curl

T4.^O Find the circulation $\int_C \langle y, -x\rangle \cdot \mathbf{T}\,ds$ of the vector field $\langle y, -x\rangle$ around the circle $C: x = 3\cos t$, $y = 3\sin t, 0 \xrightarrow{t} 2\pi$ by evaluating the line integral.
♦ Type 3, finding a circulation by evaluating the line integral

T5.^O Find the circulation from Tune-up Exercise 4 by using the Scalar Stokes' Theorem.
♦ Type 3, finding a circulation by using Stokes' Theorem

T6.^O Find the flux $\int_C \langle x, y\rangle \cdot \mathbf{n}\,ds$ of the vector field $\langle x, y\rangle$ out of the disk bounded by the circle $C: x = 3\cos t, y = 3\sin t, 0 \xrightarrow{t} 2\pi$ by evaluating the line integral.
♦ Type 3, finding a flux by evaluating the line integral

T7.^O Find the flux from Tune-up Exercise 6 by using the Divergence Theorem.
♦ Type 3, finding a flux by using the Divergence Theorem

T8.^O Sketch the gradient field of $f(x,y) = \tfrac{1}{4}(x^2 + y^2)$.
♦ Type 2, sketching a gradient field

T9.^O What is the value of $\int_C \nabla f(x,y)\cdot \mathbf{T}\,ds$ for all paths from $(0,0)$ to $(4,4)$ if $f(x,y) = \tfrac{1}{4}(x^2 + y^2)$ is the function of Tune-up Exercise 8?
♦ Type 3, evaluating an integral of $\nabla f \cdot \mathbf{T}\,ds$

T10.^O For what values of the constant k is $\mathbf{F} = 2xy^3\,\mathbf{i} + kx^2 y^2\,\mathbf{j}$ a conservative force field in the entire xy-plane?
♦ Type 3, determining whether a force field is conservative

♦The Tune-up Exercises and Problems are classified by type and content. The types are (1) basic, reactive; (2) basic reflective; (3) intermediate, reactive; (4) intermediate, reflective; (5) advanced, reactive; (6) advanced, reflective; and (7) advanced, theoretical.

Problems 13.2

A *Answer provided.* **O** *Outline of solution provided.*

1.^A Verify the equation $\iint_R f_y \, dx \, dy = -\int_C f \, dx$ in Green's Theorem for $f = x^2 y^5$ and R the rectangle $R : 0 \leq x \leq 1, 0 \leq y \leq 2$ by evaluating the line and double integrals.

♦ Type 3, verifying Green's Theorem in a particular case

2. Verify the equation $\iint_R f_x \, dx \, dy = \int_C f \, dy$ in Green's Theorem for f and R as in Problem 1 by evaluating the line and double integrals.

♦ Type 3, verifying Green's Theorem in a particular case

3.º Verify Green's Theorem for $f = xy^3$ and R the quarter-circle $0 \leq y \leq \sqrt{1-x^2}, 0 \leq x \leq 1$ by evaluating the line and double integrals. (Use the Pythagorean identity $\sin^2 t = 1 - \cos^2 t$ as required.)

♦ Type 4, verifying Green's Theorem in a particular case

4.^A Verify Stokes' Theorem in the plane (Theorem 2) for $\mathbf{v} = \langle 2xy, 3y^2 \rangle$ and R the region bounded by the parabola $y = 1 - x^2$ and the x-axis by evaluating the line and double integrals.

♦ Type 3, verifying Stokes' Theorem in a particular case

5. Verify the Divergence Theorem in the plane (Theorem 3) for $\mathbf{F} = \langle \cos y, \sin x \rangle$ and R the rectangle $0 \leq x \leq 3, 0 \leq y \leq 5$ by evaluating the line and double integrals.

♦ Type 3, verifying the Divergence Theorem in a particular case

In Problems 6 through 9 the curve C is the positively oriented, piecewise-smooth boundary of a bounded region R in the xy-plane. Use Green's theorem to express the line integrals as double integrals over R.

6.º $\int_C (xy^3 \, dx - 2x^3 y^2 \, dy)$

♦ Type 3, applying Green's Theorem

7.^A $\int_C [\sin(xy) \, dx + \cos(xy) \, dy]$

♦ Type 3, applying Green's Theorem

8. $\int_C (xe^{3y} \, dx - ye^{-4y} \, dy)$

♦ Type 3, applying Green's Theorem

9. $\int_C [2x \sin y \, dx + x^2 \cos y \, dy]$

♦ Type 3, applying Green's Theorem

In Problems 10 through 13 use one of formulas **(16)** to find the areas of the given regions.

10.º The region inside the loop of Tschirnhausen's cubic $x = t^2 - 3, y = \frac{1}{3}t^3 - t$ in Figure 22

♦ Type 4, using Green's Theorem to find an area

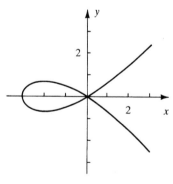

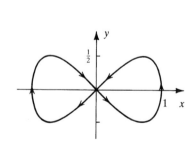

FIGURE 22   FIGURE 23

11.^A The region inside both loops of the eight curve $x = \cos t, y = \sin t \cos t, -\pi \xrightarrow{t} \pi$ in Figure 23

♦ Type 4, using Green's Theorem to find an area

13.2 The fundamental theorems in the plane

12. The region inside the piriform $x = 1 + \sin t, y = \cos t(1 + \sin t), -\frac{1}{2}\pi \xrightarrow{t} \frac{3}{2}\pi$

♦ Type 4, using Green's Theorem to find an area

13. The region inside the ellipse $x = a\cos t, y = b\sin t, 0 \xrightarrow{t} 2\pi$ with positive a and b

♦ Type 4, using Green's Theorem to find an area

Find **(a)** the divergence and **(b)** the scalar curl of the vector fields in Problems 14 through 21.

14.[A] $\sin(xy)\mathbf{i} + \cos(x-y)\mathbf{j}$

♦ Type 1, divergence and scalar curl

15. $\langle x^2y - 2xy, y^2 - xy^2 \rangle$

♦ Type 1, divergence and scalar curl

16. $7x^6y^6\mathbf{i} + 6y^7x^5\mathbf{j}$

♦ Type 1, divergence and scalar curl

17.[A] $x\sin y\,\mathbf{i} - y\cos x\,\mathbf{j}$

♦ Type 1, divergence and scalar curl

18.[A] $xy^3\mathbf{i} + (y^2 - x^3)\mathbf{j}$

♦ Type 1, divergence and scalar curl

19. $\langle x\ln y, y\ln x \rangle$

♦ Type 1, divergence and scalar curl

20. $\langle xe^{xy} + x, ye^{xy} - y^2 \rangle$

♦ Type 1, divergence and scalar curl

21.[A] $\nabla(x^2y^3 - x^3y^2)$

♦ Type 2, divergence and scalar curl of a gradient

22. Explain why $\int_{-C} \mathbf{F}\cdot\mathbf{T}\,ds = -\int_C \mathbf{F}\cdot\mathbf{T}\,ds$ and $\int_{-C} \mathbf{n}\cdot\mathbf{T}\,ds = -\int_C \mathbf{F}\cdot\mathbf{n}\,ds$ for all oriented curves C and vector fields $\mathbf{F}$ in the xy-plane.

♦ Type 3, line integrals over negatives of curves

23.[A] The scalar curl of $\mathbf{F}$ equals $\frac{1}{2}$ in the region R bounded by the curves C_1 and C_2 in Figure 24, and $\int_{C_1} \mathbf{F}\cdot\mathbf{T}\,ds = 10$. Which of the numbers $0, \pm 10, \pm 20, \pm 40, \pm 50$ is closest to the value of $\int_{C_2} \mathbf{F}\cdot\mathbf{T}\,ds$?

♦ Type 4, estimating a line integral with Stokes' Theorem and a drawing

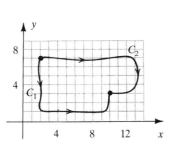

FIGURE 24

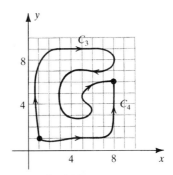

FIGURE 25

24. The scalar curl of $\mathbf{G}$ equals -2 in the region R bounded by the curves C_1 and C_2 in Figure 24, and $\int_{C_1} \mathbf{G}\cdot\mathbf{T}\,ds = -80$. Which of the numbers $0, \pm 40, \pm 80, \pm 120, \pm 160, \pm 200$ is closest to the value of $\int_{C_2} \mathbf{G}\cdot\mathbf{T}\,ds$?

♦ Type 4, estimating a line integral with Stokes' Theorem and a drawing

25.[A] The divergence of $\mathbf{A}$ equals -2 in the region R bounded by the curves C_3 and C_4 in Figure 25, and $\int_{C_3} \mathbf{A}\cdot\mathbf{n}\,ds = 35$. Which of the numbers $0, \pm 10, \pm 50, \pm 90, \pm 120$ is closest to the value of $\int_{C_4} \mathbf{A}\cdot\mathbf{n}\,ds$?

♦ Type 4, estimating a line integral with the Divergence Theorem and a drawing

26. The divergence of of **B** equals 3 in the region R bounded by the curves C_3 and C_4 in Figure 25, and $\int_{C_3} \mathbf{B} \cdot \mathbf{n} \, ds = 60$. Which of the numbers $0, \pm 15, \pm 70, \pm 105, \pm 150, \pm 1200$ is closest to the value of $\int_{C_4} \mathbf{B} \cdot \mathbf{n} \, ds$?

♦ Type 4, estimating a line integral with the Divergence Theorem and a drawing

27.[A] A fluid has the velocity field $(xy^3 - \sin y)\mathbf{i} + (yx^3 - \cos y)\mathbf{j}$ (feet per minute) at (x, y). Use the Divergence Theorem to find the rate of flow of the fluid out of the rectangle $R : 0 \leq x \leq 3, 0 \leq y \leq 2$.

♦ Type 4, using the Divergence Theorem

28. Use the Divergence Theorem to find the flux of the vector field $\mathbf{F} = \langle x \sin(x+y), x \cos(x-y) \rangle$ out of the square with corners $(0,0), (1,0), (1,1)$, and $(0,1)$.

♦ Type 4, using the Divergence Theorem

29.[A] Use Stokes' Theorem in the plane to find the work done by the force field $\mathbf{F} = \langle 2xy + x^4, x^2 - 3x \rangle$ (pounds) on an object that traverses the circle $x^2 + y^2 = 25$ once counterclockwise. Distances are measured in feet.

♦ Type 4, using Stokes' Theorem

30. Use Stokes' Theorem in the plane to find the circulation of the velocity field $\mathbf{v} = ye^x \mathbf{i} - xe^{-y} \mathbf{j}$ counterclockwise once around the boundary of the square $R : 0 \leq x \leq 4, 0 \leq y \leq 4$.

♦ Type 4, using Stokes' Theorem

31. An object is to be moved from $(0,0)$ to $(1,1)$ subject to the force field $\mathbf{F} = y^3 e^{x^2} \mathbf{i} - x^7 e^y \mathbf{j}$. Use Stokes' Theorem in the plane to determine whether it takes more work to move it along C_1 consisting of the line segments from $(0,0)$ to $(1,0)$ and from $(1,0)$ to $(1,1)$ or along C_2 consisting of the line segments from $(0,0)$ to $(0,1)$ and from $(0,1)$ to $(1,1)$.

♦ Type 4, work and Stokes' Theorem

32.[O] The velocity field of a fluid in Figure 26 has either positive or negative divergence. Which is it?

♦ Type 4, recognizing positive and negative divergence from a drawing

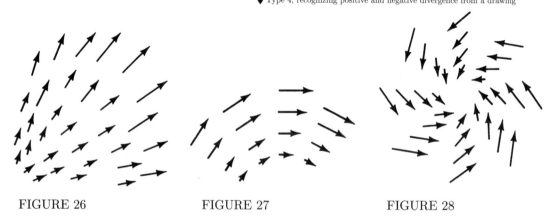

FIGURE 26   FIGURE 27   FIGURE 28

33.[A] The vector field in Figure 27 has positive or negative scalar curl. Which is it?

♦ Type 4, recognizing positive and negative scalar curl from a drawing

34. Are the scalar curl and divergence of the vector field in Figure 28 positive or negative?

♦ Type 4, recognizing positive and negative curl and divergence from a drawing

In Problems 35 through 40 determine whether the given vector fields are the gradients of potential functions. Then find potential functions that are gradients.

35.[O] $\langle \sin(3y) + x, 3x \cos(3y) - y \rangle$

♦ Type 1, gradient vector fields

36.[A] $\langle 3x \cos(3y) - 1, \sin(3x) + 1 \rangle$

♦ Type 1, gradient vector fields

37.[A] $\left\langle \dfrac{y}{1+xy}, \dfrac{x}{1+xy} \right\rangle$

♦ Type 1, gradient vector fields

38. $\left\langle \ln y + e^x, xy^{-1} - y^2 \right\rangle$

♦ Type 1, gradient vector fields

39.[A] $\langle ye^{xy} - 1, xe^{xy} \rangle$

♦ Type 1, gradient vector fields

40. $\langle 3x^2 y^4 + 2y, 4x^3 y^3 - x \rangle$

♦ Type 1, gradient vector fields

13.2 The fundamental theorems in the plane

In Problems 41 through 44 determine whether the line integrals are path independent in the entire xy-plane and give the values of those that are.

41.º $\int_C [(y^3 - 3x^2)\, dx + (3xy^2 + 2y)]\, dy$ with C running from $(0,0)$ to $(1,2)$

♦ Type 1, path-independent line integrals

42.^A $\int_C (x^2 y^3 \, dx + x^3 y^2 \, dy)$ with C running from $(2,3)$ to $(-3,-2)$

♦ Type 1, path-independent line integrals

43.^A $\int_C [(e^y - y^3 \cos x + 1)\, dx + (xe^y - 3y^2 \sin x)]\, dy$ with C running from $(20, 0)$ tp $(10, 0)$

♦ Type 1, path-independent line integrals

44. $\int_C [(2e^{2x+3y} - y)\, dx + (3e^{2x+3y} - x)]\, dy$ with C running from (z, z) to $(2, 2)$

♦ Type 1, path-independent line integrals

45. Is the force field $\mathbf{F} = \langle e^x + ye^{xy}, e^{-y} + e^{xy} \rangle$ conservative? If so, find a potential function for it and find the work it does on an object that moves from $(0, 1)$ to $(1, 0)$.

♦ Type 2, conservative force fields

46. Only one of the force fields sketched in Figures 29 through 31 is conservative. Show that the other two are not.

♦ Type 2, conservative force fields

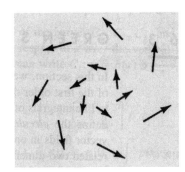
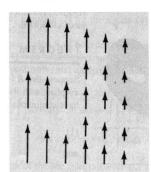

FIGURE 29   FIGURE 30   FIGURE 31

47. Does the nonconservative force field $\mathbf{F} = \langle x^2 y, xy^2 \rangle$ do more work on an object that moves from $(0, 0)$ to $(1, 1)$ if the object moves (i) on the parabola $y = x^2$, (ii) on the line $y = x$, or (iii) on the line segment from $(0, 0)$ to $(1, 0)$ followed by the line segment from $(1, 0)$ to $(1, 1)$?

♦ Type 2, conservative force fields

48. (a) Sketch the vector field $\mathbf{A} = \dfrac{\langle -y, x \rangle}{x^2 + y^2}$. (b) Show that the scalar curl of $\mathbf{A}$ is zero in the not-simply connected region R consisting of the xy-plane with the origin removed. (c) Show that $\int_C \mathbf{F} \cdot \mathbf{T} \, ds$ is not path independent in R by finding its value in the case where C a circle around the origin.

♦ Type 4, path independent line integrals

Section 13.3

Surface integrals and the fundamental theorems in space

OVERVIEW: *Green's Theorem, the Stokes' Theorem in the plane, and the Divergence Theorem in the plane from the last section relate double integrals of first-order partial derivatives of functions over bounded regions in the plane to line integrals over the boundaries of the regions. These are equivalent theorems that are given in three forms to fit different types of applications. There are, however, two distinct ways to extend these theorems to three dimensions We can replace the plane regions by solids and study how triple integrals of first derivatives of functions $f(x,y,z)$ relate to* SURFACE INTEGRALS *over the boundaries of the solids of integration. This leads to what we refer to as* GAUSS' THEOREM *and the equivalent* DIVERGENCE THEOREM IN SPACE. *Or, we can replace the plane region with a bounded surface with* STOKES' THEOREM IN SPACE, *which shows how certain surface integrals involving derivatives relate to line integrals around the boundaries of the surfaces. The Divergence Theorem in space, like the Divergence Theorem in the plane, is used to study rates of flow or* FLUX *of a fluid. Stokes' Theorem in space, like Stokes' Theorem in the plane, is used to study work done by force fields and circulation of velocity fields around closed paths. It is also needed to establish a criterion for whether line integrals in space are path independent. The Divergence Theorem uses the* DIVERGENCE *and Stokes' Theorem uses the* CURL *of vector fields in space.*

Topics:

- *Surface integrals*
- *Average values on surfaces*
- *Weights and centers of gravity of surfaces*
- *Gauss' Theorem*
- *Divergence and the Divergence Theorem in space*
- *The curl and Stokes' Theorem*
- *Gradient fields and path independence*

Surface integrals

Some surfaces can be described as graphs

$$\Sigma : z = g(x,y), \text{ for } (x,y) \text{ in } D$$

of functions of x and y, where D is a region in the xy-plane. Others can be described by equations of the form $y = g(x,z)$ with (x,z) in a region of the xz-plane or $x = g(y,z)$ with (y,z) in a region of the yz-plane.

Surfaces can also be given with parametric equations

$$\Sigma : x = x(u,v), y = y(u,v), z = z(u,v) \text{ for } (u,v) \text{ in } D \tag{1}$$

where D is a portion of the uv-plane. Cylindrical or spherical coordinates, in particular, can be used to obtain parametric equations of surfaces.

♦ SUGGESTIONS TO INSTRUCTORS: The main topics in this section can be divided into three parts to be covered in several class meetings.

Part 1 (Surface integrals and average values on surfaces; Tune-up Exercises T1–T4; Problems 1–13, 45–48, 61–63) Have students begin work on Questions 1 through 4 before class. Problems 1, 4, 11 are good for in-class discussion.

Part 2 (Gauss' Theorem, divergence, flux, the Divergence Theorem, centers of gravity; Tune-up Exercises T5–T6; Problems 14-18, 19–26 (parts (a)), 27–30, 51–55) Have students begin work on Questions 5 through 7 before class. Problems 14, 17, 19a, 54 are good for in-class discussion.

Part 3 (Oriented surfaces, curl, Stokes' Theorem, work, gradient fields, and path independence; Tune-up Exercises T7–T10; Problems 19–26 (parts (b)), 31–44, 49, 50, 56–60, 64–68) Have students begin work on Questions 8, 9, and 10 before class.Tune-up Exercise T7 and Problems 19b, 34, 40, 42, 43 are good for in-class discussion.

13.3 Surface integrals and the fundamental theorems in space

Example 1   Give the sphere $\Sigma = \{(x,y,z) : x^2 + y^2 + z^2 = 9\}$ in Figure 1 parametric equations by using spherical coordinates.

SOLUTION   We use the formulas $x = \rho \sin\phi \cos\theta, y = \rho \sin\phi \sin\theta, z = \rho \cos\phi$ for converting spherical coordinates $[\rho, \theta, \phi]$ to rectangular coordinates. Here ρ can be any nonnegative number, $0 \leq \phi \leq \pi$, and we can use $0 \leq \theta \leq 2\pi$. Because the sphere Σ consists of all points with $\rho = 3$, it can be given by the parametric equations

$$\Sigma : x = 3\sin\phi\cos\theta, y = 3\sin\phi\sin\theta, z = 3\cos\phi$$

where (θ, ϕ) varies over the rectangle $D = \{(\theta,\phi) : 0 \leq \theta \leq 2\pi, 0 \leq \phi \leq \pi\}$ in the $\theta\phi$-plane. $\square$.

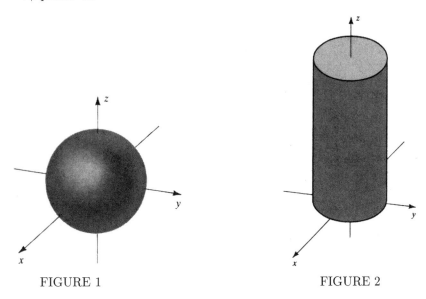

FIGURE 1   FIGURE 2

Question 1   Give the cylinder $\Sigma = \{(x,y,z) : x^2 + y^2 = 9, 0 \leq z \leq 6\}$ in Figure 2 parametric equations by using cylindrical coordinates.

Example 2   Describe the surface $\Sigma : x = r\cos\theta, y = r\sin\theta, z = \frac{1}{3}\theta$ for $0 \leq r \leq 1, 0 \leq \theta \leq 2\pi$.

SOLUTION   For each fixed θ, the set $\{(x,y,z) : x = r\cos\theta, r = r\sin\theta, z = \frac{1}{3}\theta, 0 \leq r \leq 1\}$ is a line segment of length 1 that extends in the direction of the angle θ horizontally from the point at $z = \frac{1}{3}\theta$ on the z-axis. The surface, therefore, has the shape of a helical ramp extending out from the z-axis (Figure 3). $\square$

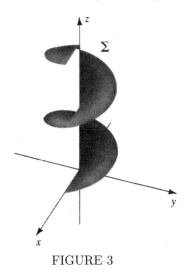
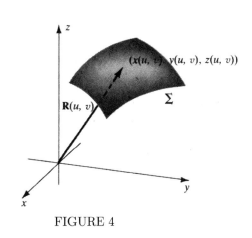

FIGURE 3   FIGURE 4

We will frequently give parametric equations **(1)** of a surface Σ in vector form by writing

$$\Sigma : \mathbf{R} = \mathbf{R}(u,v) \text{ for } (u,v) \text{ in } D \tag{2}$$

where $\mathbf{R}(u,v) = \langle x(u,v).y(u,v), z(u,v)\rangle$ is the position vector of the point $(x(u,v), y(u,v), z(u,v))$ on the surface and D is a region in the uv-plane (Figure 4).

We will say that a bounded surface Σ is SMOOTH if it is given by parametric equations **(2)**, where D is a closed, bounded region in the uv-plane with piecewise-smooth boundary and the vector derivatives $\dfrac{\partial \mathbf{R}}{\partial u}(u,v)$ and $\dfrac{\partial \mathbf{R}}{\partial v}(u,v)$ are continuous, nonzero, and nonparallel for (u,v) in D. A surface is PIECEWISE SMOOTH if it consists of a finite number of smooth surfaces.

Surface integrals

Suppose that $f(x,y,z)$ is piecewise continuous on a smooth surface $\Sigma : \mathbf{R} = \mathbf{R}(u,v)$ for (u,v) in D. The SURFACE INTEGRAL

$$\iint_\Sigma f(x,y,z)\,dS \tag{3}$$

is defined as follows. We choose constants $a, b, c,$ and d so that D is contained in the rectangle $Q : a \leq u \leq b, c \leq v \leq d$ in the uv-plane and partition Q into rectangles of width Δu and height Δv (Figure 5). We let Q_j for $j = 1, 2, \ldots, N$ be those rectangles in the partition that are entirely contained in D (the shaded rectangles for the region D and the partition of Q in Figure 5).

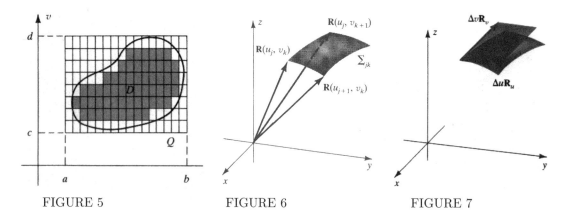

FIGURE 5   FIGURE 6   FIGURE 7

For each $j = 1, 2, \ldots, N$, we let (u_j, v_j) be the lower left corner of the rectangle Q_j. Then its lower right corner is $(u_j + \Delta u, v_j)$, its upper left corner is $(u_j, v_j + \Delta v)$, and the image of Q_j under the parametric equations is a subsurface Σ_j of Σ with three corners at the points with position vectors $\mathbf{R}(u_j, v_j), \mathbf{R}(u_j + \Delta u, v_j)$, and $\mathbf{R}(u_j, v_j + \Delta v)$ (Figure 6).

The displacement vectors $\mathbf{R}(u_j + \Delta u, v_j) - \mathbf{R}(u_j, v_j)$ and $\mathbf{R}(u_j, v_j) - \mathbf{R}(u_j, v_j + \Delta v)$ extend from one corner of the subsurface Σ_j to two others, and for small Δu and Δv

$$\begin{aligned}\mathbf{R}(u_j + \Delta u, v_j) - \mathbf{R}(u_j, v_j) &\approx \frac{\partial \mathbf{R}}{\partial u}(u_j, v_j)\,\Delta u \\ \mathbf{R}(u_j, v_j + \Delta v) - \mathbf{R}(u_j, v_j) &\approx \frac{\partial \mathbf{R}}{\partial v}(u_j, v_j)\,\Delta v.\end{aligned} \tag{4}$$

Consequently, for small Δu and Δv, the parallelogram in Figure 7 with sides formed by the vectors $\dfrac{\partial \mathbf{R}}{\partial u}(u_j, v_j)\,\Delta u$ and $\dfrac{\partial \mathbf{R}}{\partial v}(u_j, v_j)\,\Delta v$ approximates Σ_j.

13.3 Surface integrals and the fundamental theorems in space

We know from Section 10.2 that the area of the parallelogram with two adjacent sides formed by the vectors **A** and **B** is $|\mathbf{A} \times \mathbf{B}|$. Consequently the area of the parallelogram in Figure 7 is

$$\left| \left[\frac{\partial \mathbf{R}}{\partial u}(u_j, v_j) \Delta u \right] \times \left[\frac{\partial \mathbf{R}}{\partial v}(u_j, v_j) \Delta v \right] \right| = \left| \frac{\partial \mathbf{R}}{\partial u}(u_j, v_j) \times \frac{\partial \mathbf{R}}{\partial v}(u_j, v_j) \right| \Delta u \, \Delta v.$$

For small Δu and Δv, this parallelogram approximates the subsurface Σ_j, so we can expect that

$$[\text{Area of } \Sigma_j] \approx \left| \frac{\partial \mathbf{R}}{\partial u}(u_j, v_j) \times \frac{\partial \mathbf{R}}{\partial v}(u_j, v_j) \right| \Delta u \, \Delta v. \tag{5}$$

We pick, for each j, a point (u_j^*, v_j^*) in the rectangle Q_j such that $f(P_j)$ is defined, where P_j is the point on Σ_j with the position vector $\mathbf{R}(u_j^*, v_j^*)$. We would like to define the surface integral $\iint_\Sigma f \, dS$ to be the limit of the sum

$$\sum_{j=1}^{N} f(P_j)[\text{Area of } \Sigma_j]$$

of the values of f at the points in the subsurfaces Σ_j multiplied by the areas of the subsurfaces, but we have not yet defined the area of Σ_j. Because we anticipate that (5) will hold, we form instead the approximating sum

$$\sum_{j=1}^{N} f(P_j) \left| \frac{\partial \mathbf{R}}{\partial u}(u_j, v_j) \times \frac{\partial \mathbf{R}}{\partial v}(u_j, v_j) \right| \Delta u \, \Delta v. \tag{6}$$

The sum (6) is an approximating sum for the double integral

$$\iint_D f(\mathbf{R}(u, v)) \left| \frac{\partial \mathbf{R}}{\partial u}(u, v) \times \frac{\partial \mathbf{R}}{\partial v}(u, v) \right| du \, dv$$

over the set D in the uv-plane and approaches this integral as the number of rectangles in the partition of the rectangle Q in Figure 5 tend to infinity and their widths and heights Δu and Δv tend to zero. We define the surface integral of $f(x, y, z)$ over Σ to be this double integral so that it is also the limit of the approximating sums (6):

Definition 1 (a) Suppose that $\Sigma : \mathbf{R} = \mathbf{R}(u, v)$ for (u, v) in D is a bounded, smooth surface in xyz-space and that $f(x, y, z)$ is piecewise continuous on Σ. Then

$$\iint_\Sigma f(x, y, z) \, dS = \iint_D f(\mathbf{R}(u, v)) \left| \frac{\partial \mathbf{R}}{\partial u}(u, v) \times \frac{\partial \mathbf{R}}{\partial v}(u, v) \right| du \, dv. \tag{7}$$

(b) The surface integral over a piecewise-smooth surface is the sum of the surface integrals over its smooth subsurfaces.

We use the symbolic equation

$$dS = \left| \frac{\partial \mathbf{R}}{\partial u}(u, v) \times \frac{\partial \mathbf{R}}{\partial v}(u, v) \right| du \, dv \tag{8}$$

in applications of (7).

We note here for future reference that the cross product

$$\frac{\partial \mathbf{R}}{\partial u}(u,v) \times \frac{\partial \mathbf{R}}{\partial v}(u,v) \tag{9}$$

whose length appears in **(8)**, is a nonzero normal vector to the surface at the point P_0 with position vector $\mathbf{R}(u,v)$. To see why it is a normal vector, we consider the surface in Figure 8. The curve C_1 in the drawing is the curve on the surface through the point P_0 along which v is constant and, consequently, the velocity vector $\frac{\partial \mathbf{R}}{\partial u}$ at P_0 is tangent to this curve. Similarly, u is constant along the curve C_2, so that the velocity vector $\frac{\partial \mathbf{R}}{\partial v}$ at P_0 is tangent to this second curve. Because in our definition of a smooth surface we assume that these velocity vectors are not zero and nonparallel, the cross product **(9)** is a nonzero vector perpendicular to the two tangent vectors and is, in fact, a nonzero normal vector to the surface.

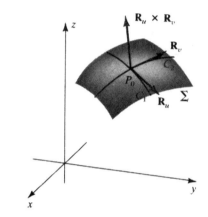

FIGURE 8

Formulas **(8)** and **(9)** take a simpler form if Σ is a portion

$$\Sigma : z = g(x,y), \text{ for } (x,y) \text{ in } D \tag{10}$$

of the graph of a function $g(x,y)$.

Question 2 Show that in case **(10)**, where $\mathbf{R}(x,y) = \langle x, y, g(x,y) \rangle$, the normal vector **(9)** to the surface is

$$\frac{\partial \mathbf{R}}{\partial x}(x,y) \times \frac{\partial \mathbf{R}}{\partial y}(x,y) = \langle -g_x(x,y), -g_y(x,y), 1 \rangle. \tag{11}$$

With **(11)**, formula **(8)** in the case of the graph $z = g(x,y)$ of a function becomes

$$dS = \sqrt{[g_x(x,y)]^2 + [g_y(x,y)]^2 + 1} \; dx \, dy \tag{12}$$

and for $f(x,y,.z)$ defined on $\Sigma : z = g(x,y)$ for (x,y) in D

$$\iint_\Sigma f(x,y,z) \, dS = \iint_D f(x,y,g(x,y)) \sqrt{[g_x(x,y)]^2 + [g_y(x,y)]^2 + 1} \; dx \, dy. \tag{13}$$

13.3 Surface integrals and the fundamental theorems in space

Example 3 Give an iterated integral that equals $\iint_\Sigma xyz^4 \, dS$ where Σ is the portion of the circular paraboloid $z = x^2 + y^2$ for $0 \leq x \leq 1, 0 \leq y \leq 1$.

SOLUTION We let D denote the square $D: 0 \leq x \leq 1, 0 \leq y \leq 1$ and set $g = x^2 + y^2$. Then $g_x = 2x$ and $g_y = 2y$, so that **(12)** gives

$$dS = \sqrt{(g_x)^2 + (g_y)^2 + 1} \ dx \, dy = \sqrt{(2x)^2 + (2y)^2 + 1} \ dx \, dy.$$

Then since $z = x^2 + y^2$ on the surface

$$\iint_\Sigma xyz^4 \, dS = \iint_D xy(x^2 + y^2)^4 \sqrt{(2x)^2 + (2y)^2 + 1} \ dx \, dy$$

$$= \int_{x=0}^{x=1} \int_{y=0}^{y=1} xy(x^2 + y^2)^4 \sqrt{(2x)^2 + (2y)^2 + 1} \ dy \, dx. \ \square$$

Surface area

Now that we have defined surface integrals, we define the area of a piecewise-smooth surface Σ to be the surface integral of the function with the constant value 1:

$$[\text{Area of } \Sigma] = \iint_\Sigma dS. \tag{14}$$

Question 3 Calculate the area of the portion Σ of the hyperbolic paraboloid $z = xy$ for $x^2 + y^2 \leq 9$. Use polar coordinates to evaluate the double integral.

Example 4 Use **(14)** and spherical coordinates to find the surface area of the sphere $\Sigma = \{(x, y, z) : x^2 + y^2 + z^2 = \rho^2\}$ with positive radius ρ.

SOLUTION The sphere is given by $\mathbf{R}(\theta, \phi) = \langle \rho \sin\phi \cos\theta, \rho \sin\phi \sin\theta, \rho \cos\phi \rangle$ for $0 \leq \theta \leq 2\pi$ and $0 \leq \phi \leq \pi$. We calculate

$$\mathbf{R}_\theta \times \mathbf{R}_\phi = \begin{vmatrix} \mathbf{i} & \mathbf{j} & \mathbf{k} \\ \dfrac{\partial}{\partial \theta}(\rho \sin\phi \cos\theta) & \dfrac{\partial}{\partial \theta}(\rho \sin\phi \sin\theta) & \dfrac{\partial}{\partial \theta}(\rho \cos\phi) \\ \dfrac{\partial}{\partial \phi}(\rho \sin\phi \cos\theta) & \dfrac{\partial}{\partial \phi}(\rho \sin\phi \sin\theta) & \dfrac{\partial}{\partial \phi}(\rho \cos\phi) \end{vmatrix}$$

$$= \begin{vmatrix} \mathbf{i} & \mathbf{j} & \mathbf{k} \\ -\rho \sin\phi \sin\theta & \rho \sin\phi \cos\theta & 0 \\ \rho \cos\phi \cos\theta & \rho \cos\phi \sin\theta & -\rho \sin\phi \end{vmatrix}$$

$$= \mathbf{i} \begin{vmatrix} \rho \sin\phi \cos\theta & 0 \\ \rho \cos\phi \sin\theta & -\rho \sin\phi \end{vmatrix} - \mathbf{j} \begin{vmatrix} -\rho \sin\phi \sin\theta & 0 \\ \rho \cos\phi \cos\theta & -\rho \sin\phi \end{vmatrix}$$

$$+ \mathbf{k} \begin{vmatrix} -\rho \sin\phi \sin\theta & \rho \sin\phi \cos\theta \\ \rho \cos\phi \cos\theta & \rho \cos\phi \sin\theta \end{vmatrix}$$

$$= -\rho^2 \sin^2\phi \cos\theta \, \mathbf{i} - \rho^2 \sin^2\phi \sin\theta \, \mathbf{j} - \rho^2 \sin\phi \cos\phi (\sin^2\theta + \cos^2\theta) \, \mathbf{k}$$

$$= -\rho^2 \sin\phi \langle \sin\phi \cos\theta, \sin\phi \sin\theta, \cos\phi \rangle.$$

Since $\langle \sin\phi \cos\theta, \sin\phi \sin\theta, \cos\phi \rangle$ is a unit vector and $0 \leq \phi \leq \pi$

$$dS = |\mathbf{R}_\theta \times \mathbf{R}_\phi| \ d\theta \, d\phi = \rho^2 \sin\phi \ d\theta \, d\phi. \tag{15}$$

Therefore

$$\iint_\Sigma dS = \int_{\theta=0}^{\theta=2\pi} \int_{\phi=0}^{\phi=\pi} \rho^2 \sin\phi \, d\phi \, d\theta = \rho^2 \int_{\theta=0}^{\theta=2\pi} d\theta \int_{\phi=0}^{\phi=\pi} \sin\phi \, d\phi$$

$$= \rho^2 \Big[\theta\Big]_{\theta=0}^{\theta=2\pi} \Big[-\cos\phi\Big]_{\phi=0}^{\phi=\pi} = 4\pi\rho^2. \quad \square$$

This next Question involves another application of spherical coordinates.

Question 4 Evaluate $\iint_\Sigma e^z \, dS$ for the sphere $\Sigma = \{(x,y,z) : x^2 + y^2 + z^2 = 9\}$.

Average values

Recall that the average value of a function $f(x,y)$ with two variables over a bounded region in the xy-plane equals the double integral of the function over the region, divided by the area of the region. Average values on surfaces have a similar definition, using surface integrals:

Definition 2 *The average value of the piecewise continuous $f(x,y,z)$ on the bounded, piecewise-smooth surface Σ is*

$$\frac{1}{[\text{Area of }\Sigma]} \iint_\Sigma f(x,y,z) \, dS. \tag{17}$$

Example 5 What is the average value of $f = (z - 3x)e^x$ on the parallelogram $\Sigma = \{(x,y,z) : z = 3x + 4y, 0 \le x \le 2, 0 \le y \le 2\}$?

SOLUTION Σ is a parallelogram because it is the portion of the plane $z = 3x + 4y$ above the square $D : 0 \le x \le 2, 0 \le y \le 2$ in the xy-plane. Also for $z = 3x + 4y$, $z_x = 3$, $z_y = 4$, and $dS = \sqrt{(z_x)^2 + (z_y)^2 + 1} \, dx \, dy = \sqrt{3^2 + 4^2 + 1} \, dx \, dy = \sqrt{26} \, dx \, dy$.

We first find the area of Σ. Because D is a square of width 2 and area 4,

$$[\text{Area of }\Sigma] = \iint_\Sigma dS = \iint_D \sqrt{26} \, dx \, dy = \sqrt{26} \, [\text{Area of }D] = 4\sqrt{26}. \tag{18}$$

To evaluate the surface integral of $(z - 3x)e^x$, we set $z = 3x + 4y$ to have $(z - 3x)e^x = 4ye^x$, and obtain

$$\iint_\Sigma (z-3x)e^x \, dS = \iint_R 4ye^x\sqrt{26} \, dx \, dy = 4\sqrt{26} \int_{x=0}^{x=2} \int_{y=0}^{y=2} ye^x \, dy \, dx$$

$$= 4\sqrt{26} \int_{x=0}^{x=2} e^x \, dx \int_{y=0}^{y=2} y \, dy = 4\sqrt{26} \Big[e^x\Big]_{x=0}^{x=2} \Big[\tfrac{1}{2}y^2\Big]_{y=0}^{y=2} \tag{19}$$

$$= 8\sqrt{26}(e^2 - 1).$$

The average value equals the surface integral (**19**), divided by the area (**18**) of the surface, so it is $\dfrac{8\sqrt{26}(e^2-1)}{4\sqrt{26}} = 2(e^2 - 1). \quad \square$

Weights and centers of gravity of surfaces

The reasoning that we used in Chapter 12 to obtain formulas for weights and centers of gravity of plane regions and solids leads us to the following definition for surfaces.

Definition 3 *If a bounded, piecewise-smooth surface Σ has piecewise continuous density $\rho(x, y, z)$ (weight per area) at (x, yz), then its weight is*

$$\iint_\Sigma \rho(x,y,z)\, dS \tag{20}$$

and its center of gravity is $(\overline{x}, \overline{y}, \overline{z})$, where

$$\overline{x} = \frac{1}{[\text{Weight of } \Sigma]} \iint_\Sigma x\, \rho(x,y,z)\, dS \tag{21}$$

$$\overline{y} = \frac{1}{[\text{Weight of } \Sigma]} \iint_\Sigma y\, \rho(x,y,z)\, dS \tag{22}$$

$$\overline{z} = \frac{1}{[\text{Weight of } \Sigma]} \iint_\Sigma z\, \rho(x,y,z)\, dS. \tag{23}$$

Gauss' Theorem

We now turn to GAUSS' THEOREM which relates triple integrals of first-order derivatives over bounded solids to surface integrals over the boundaries of the solids. We consider first the solid V in Figure 9, which is bounded on the top and bottom by the graphs $z = h(x,y)$ and $z = g(x,y)$ of piecewise-continuous functions and whose projection on the xy-plane is the bounded set D.

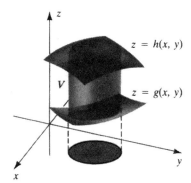

FIGURE 9

Question 5 Use the Fundamental Theorem in the form

$$\int_{z=a}^{z=b} \frac{dF}{dz}(z)\, dz = F(b) - F(a) \tag{24}$$

to show that for a function $f(x, y, z)$ with continuous first-order derivatives in V

$$\iiint_V \frac{\partial f}{\partial z}(x,y,z)\, dx\, dy\, dz = \iint_D f(x,y,h(x,y))\, dx\, dy - \iint_D f(x,y,g(x,y))\, dx\, dy. \tag{25}$$

We will show that the difference of integrals on the right of (25) is equal to a surface integral over the boundary Σ of the solid V that involves the UNIT EXTERIOR NORMAL VECTORS $\mathbf{n}$ to Σ. These are the unit normal vectors that point out of the solid.

The boundary Σ of V in Figure 9 consists of the top Σ_1, the bottom Σ_2, and the vertical sides Σ_3 of V. Because the top is formed by the graph $z = h(x,y)$, formula **(11)** with h in place of g shows that the vector

$$\langle -h_x(x,y), -h_y(x,y), 1\rangle \tag{26}$$

is normal to the top Σ_1 at $(x,y,h(x,y))$. The vector **(26)** is an exterior normal vector to the solid V because it has a positive z-component. It is not, however, a unit vector. To obtain the unit exterior normal vector, we divide the vector **(26)** by its length and obtain

$$\mathbf{n} = \frac{\langle -h_x, -h_y, 1\rangle}{\sqrt{(h_x)^2 + (h_y)^2 + 1}} \quad \text{on } \Sigma_1. \tag{27}$$

Similarly, because the bottom Σ_2 of V is formed by the graph $z = g(x,y)$ the vector

$$\langle -g_x(x,y), -g_y(x,y), 1\rangle \tag{28}$$

is normal to it. The vector **(28)** is an inward pointing normal vector to V because it has a positive z-component. Its negative $\langle g_x, g_y, -1\rangle$ is an exterior normal vector to Σ_2, so the unit exterior normal vector is

$$\mathbf{n} = \frac{\langle g_x, g_y, -1\rangle}{\sqrt{(g_x)^2 + (g_y)^2 + 1}} \quad \text{on } \Sigma_2. \tag{29}$$

The unit exterior normal vectors n on the sides Σ_3 of V point in different directions but are all horizontal.

Question 6 Write $\mathbf{n} = \langle n_1, n_2, n_3\rangle$, so that n_3 is the z-component of $\mathbf{n}$. Give formulas
(a) for n_3 on the top Σ_1 of V, (b) for n_3 on the bottom Σ_2 of V, and
(c) for n_3 on the sides Σ_3 of V.

Formula **(12)** gives $dS = \sqrt{(h_x)^2 + (h_y)^2 + 1}\, dx\, dy$ on Σ_1 and $dS = \sqrt{(g_x)^2 + (g_y)^2 + 1}\, dx\, dy$ on Σ_2. Therefore, the results of Question 6(a,b) show that

$$n_3\, ds = \begin{cases} \dfrac{1}{\sqrt{(h_x)^2 + (h_y)^2 + 1}}\sqrt{(h_x)^2 + (h_y)^2 + 1}\, dx\, dy = dx\, dy & \text{on } \Sigma_1 \\ \dfrac{-1}{\sqrt{(g_x)^2 + (g_y)^2 + 1}}\sqrt{(g_x)^2 + (g_y)^2 + 1}\, dx\, dy = -dx\, dy & \text{on } \Sigma_2. \end{cases}$$

Because, as you saw in part (c) of Question 6, $n_3 = 0$ on Σ_3, we obtain

$$\begin{aligned}
\iint_\Sigma &f(x,y,z)\, n_3\, dS \\
&= \iint_{\Sigma_1} f(x,y,z)\, n_3\, dS + \iint_{\Sigma_2} f(x,y,z)\, n_3\, dS + \iint_{\Sigma_3} f(x,y,z) n_3\, dS \\
&= \iint_D f(x,y,h(x,y))\, dx\, dy + \iint_D f(x,y,g(x,y))\, (-dx\, dy) + 0 \\
&= \iint_D f(x,y,h(x,y))\, dx\, dy - \iint_D f(x,y,g(x,y)) dx\, dy.
\end{aligned} \tag{30}$$

The right side of **(30)** is the same as the right side of **(25)**, so we have shown that for solids of the type in Figure 9

$$\iiint_V \frac{\partial f}{\partial z}(x,y,z)\, dx\, dy\, dz = \iint_\Sigma f(x,y,z)\, n_3\, dS.$$

This is the third statement in GAUSS' THEOREM for this type of solid:

13.3 Surface integrals and the fundamental theorems in space

Theorem 1 (Gauss' Theorem) *Suppose that V is a bounded solid in xyz-space with a piecewise-smooth boundary Σ and exterior unit normal vectors $\mathbf{n} = \langle n_1, n_2, n_3 \rangle$ (Figure 10). For any function $f(x, y, z)$ with continuous first-order partial derivatives in an open set containing V*

$$\iiint_V \frac{\partial f}{\partial x}(x,y,z)\,dx\,dy\,dz = \iint_\Sigma f(x,y,z)\,n_1\,dS \tag{31}$$

$$\iiint_V \frac{\partial f}{\partial y}(x,y,z)\,dx\,dy\,dz = \iint_\Sigma f(x,y,z)\,n_2\,dS \tag{32}$$

$$\iiint_V \frac{\partial f}{\partial z}(x,y,z)\,dx\,dy\,dz = \iint_\Sigma f(x,y,z)\,n_3\,dS. \tag{33}$$

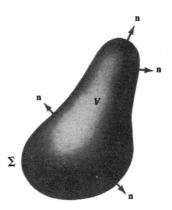

FIGURE 10

The reasoning above establishes equation **(33)** for solids of the type in Figure 9. Similar calculations would establish **(31)** for solids of the form $V = \{(x, y, z) : g(y, z) \leq x \leq h(y, z)$ for (y, z) in $D\}$ and **(32)** for solids of the form $V = \{(x, y, z) : g(x, z) \leq y \leq h(x, z)$ for (x, z) in $D\}$, and these formulas can be extended to general solids by partitioning and approximation to complete a proof of the Theorem.

Example 6   Use Gauss' Theorem to evaluate $\iint_\Sigma (x^2 - 3y + xz^3)\,n_2\,dS$, where Σ is the sphere $V = \{(x, y, z) : x^2 + y^2 + z^2 = 16\}$ and $\mathbf{n} = \langle n_1, n_2, n_3 \rangle$ are its outward pointing unit normal vectors.

SOLUTION   Let V be the ball $V = \{(x, y, z) : x^2 + y^2 + z^2 \leq 16\}$ of radius 4 whose boundary is Σ. By **(32)**

$$\iint_\Sigma (x^2 - 3y + xz^3)\,n_2\,dS = \iiint_V \frac{\partial}{\partial y}(x^2 - 3y + xz^3)\,dx\,dy\,dz$$

$$= \iiint_V (-3)\,dx\,dy\,dz = -3[\text{Volume of } V] = -3[\tfrac{4}{3}\pi(4^3)] = -256\pi. \;\square$$

Divergence and the Divergence Theorem

Recall that the divergence of a vector field $\mathbf{v} = \langle p(x,y), q(x,y)\rangle$ in the plane is defined as the dot product of $\nabla = \langle \partial/\partial x, \partial/\partial y\rangle$ with $\mathbf{v}$ by the equation $\operatorname{div}\langle p,q\rangle = \nabla \cdot \langle p,q\rangle = \dfrac{\partial p}{\partial x} + \dfrac{\partial q}{\partial y}$. The divergence of a vector field in space has an analogous definition using the differentiation operator $\nabla = \langle \partial/\partial x, \partial/\partial y, \partial/\partial z\rangle$ with three variables:

Definition 4 *The* DIVERGENCE *of* $\mathbf{v} = \langle p(x,y,z), q(x,y,z), r(x,y,z)\rangle$ *is the real-valued function*

$$\operatorname{div}\mathbf{v} = \nabla \cdot \mathbf{v} = \left\langle \frac{\partial}{\partial x}, \frac{\partial}{\partial y}, \frac{\partial}{\partial z}\right\rangle \cdot \langle p,q,r\rangle$$
$$= \frac{\partial p}{\partial x} + \frac{\partial q}{\partial y} + \frac{\partial r}{\partial z}. \tag{34}$$

Question 7 What is the divergence of $\mathbf{v}(x,y,z) = (x^3 y^2 - 3x^2)\mathbf{i} + (2yz - x^2 y^3)\mathbf{j} + (5 - z^2)\mathbf{k}$?

With Definition 4, Gauss' Theorem 1 yields the following result, which is used in the study of fluid flow and in topics of mathematical physics.

Theorem 2 (The Divergence Theorem in space) *Suppose that V is a bounded solid with piecewise-smooth boundary Σ and unit exterior normal vectors $\mathbf{n}$ and that $\mathbf{v}$ is a vector field with continuous first-order partial derivatives in an open set containing V. Then*

$$\iint_\Sigma \mathbf{v} \cdot \mathbf{n}\, dS = \iiint_V \operatorname{div}\mathbf{v}\, dx\, dy\, dz. \tag{35}$$

The Divergence Theorem follows from Gauss' Theorem because for $\mathbf{v} = \langle p,q,r\rangle$ and $\mathbf{n} = \langle n_1, n_2, n_3\rangle$,

$$\iint_\Sigma \mathbf{v}\cdot \mathbf{n}\, dS = \iint_\Sigma \langle p,q,r\rangle \cdot \langle n_1,n_2,n_3\rangle\, dS = \iint_\Sigma (p n_1 + q n_2 + r n_3)\, dS$$
$$= \iiint_V \left(\frac{\partial p}{\partial x} + \frac{\partial q}{\partial r} + \frac{\partial r}{\partial z}\right) dx\, dy\, dz$$

and $\dfrac{\partial p}{\partial x} + \dfrac{\partial q}{\partial r} + \dfrac{\partial r}{\partial z} = \operatorname{div}\mathbf{v}$.

In Section 16.2 we defined the flux of a vector field $\mathbf{v}$ across an oriented curve C in the xy-plane to be the rate of flow $\displaystyle\int_C \mathbf{v}\cdot\mathbf{n}\, ds$ of a fluid with this velocity field across C in the direction of the unit normal vectors $\mathbf{n}$. The flux of a vector field in space is analogous: For a solid V with unit exterior normal vectors $\mathbf{n}$ and a vector field $\mathbf{v}$ satisfying the conditions of Theorem 2, we define

$$[\text{The flux of } \mathbf{v} \text{ out of } V] = \iint_\Sigma \mathbf{v}\cdot\mathbf{n}\, dS. \tag{36}$$

If $\mathbf{v}$ is the velocity field of a fluid, then the flux (**36**) is the net rate of flow of the fluid out of V and the Divergence Theorem shows that the flux equals the triple integral of the divergence of $\mathbf{v}$ over V.

The sign of the divergence of a velocity field of a fluid in space has the same interpretation as we saw in Section 13.2 in the xy-plane. If $\operatorname{div}\mathbf{v}$ is positive at a point (x_0, y_0, z_0), then it is positive in small solids V containing the point, so that fluid is flowing out of such solids, and either the fluid is expanding or there is a source of fluid near the point. If $\operatorname{div}\mathbf{v}$ is negative at the point, then fluid is flowing into such solids and either the fluid is contracting or there is a sink of fluid near the point.

If the fluid has no sources or sinks and is incompressible in an open solid, there can be no net flow into or out of any parts of it and $\operatorname{div}\mathbf{v}$ is zero there.

13.3 Surface integrals and the fundamental theorems in space

Example 7 Find the rate of flow out of the cube $V = \{(x,y,z) : 0 \le x \le 1, 0 \le y \le 1, 0 \le z \le 1\}$ of a fluid that has the velocity field **v** of Question 7. The velocity field is measured in meters per minute and the scales in xyz-space are given in meters.

SOLUTION We let Σ denote the boundary of the cube and **n** its unit exterior normal vectors. You found in Question 7 that $\text{div}\,\mathbf{v} = -6x$. By the Divergence Theorem, the rate of flow is

$$\iint_\Sigma \mathbf{v} \cdot \mathbf{n}\, dS = \iiint_V \text{div}\,\mathbf{v}\, dx\, dy\, dz \iiint_V (-6x)\, dx\, dy\, dz$$

$$= \int_{x=0}^{x=1} \int_{y=0}^{y=1} \int_{z=0}^{z=1} (-6x)\, dz\, dy\, dx$$

$$= \left[\int_{x=0}^{x=1} (-6x)\, dx\right] \left[\int_{y=0}^{y=1} dy\right] \left[\int_{z=0}^{z=1} dz\right]$$

$$= -3 \text{ cubic meters per minute.} \square$$

Oriented surfaces

The BOUNDARY of a smooth surface

$$\Sigma : \mathbf{R} = \mathbf{R}(u,v) \text{ for } (u,v) \text{ in } D \tag{37}$$

is the curve C formed by the position vectors $\mathbf{R}(u,v)$ with (u,v) in the boundary of D. The INTERIOR of Σ consists of the points with position vectors $\mathbf{R}(u,v)$ with (u,v) in the interior of D.

The smooth surface (37) is said to be ORIENTED if unit nomal vectors **n** have been assigned to each point in the interior of Σ such that they vary continuously. In this case the boundary C of Σ is assumed to be oriented so that its positive direction is counterclockwise when viewed from the directions of nearby vectors **n** (Figure 11).

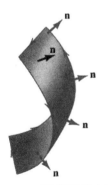

FIGURE 11

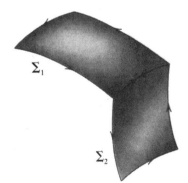

FIGURE 12

Two oriented smooth subsurfaces Σ_1 and Σ_2 of a piecewise-smooth surface that touch along parts of their boundaries have COMPATIBLE ORIENTATIONS if, as in Figure 12, the curves where they meet have one orientation as part of the boundary of Σ_1 and the opposite orientation as part of the boundary of Σ_2. This causes normal vectors on Σ_1 and Σ_2 near where the surfaces meet to be on the same side of Σ. A piecewise-smooth surface is oriented if all its smooth subsurfaces whose boundaries touch have compatible orientations.

We defined the flux of a vector field **v** across the boundary of a solid by using the exterior unit normals to the solid. If a surface Σ is not the boundary of a surface but is oriented, we use the normal vectors associated with the orientation in defining the flux of **v** by $\iint_\Sigma \mathbf{v} \cdot \mathbf{n}\, dS$.

The curl and Stokes Theorem in space

The (VECTOR) CURL of a vector field in space is found as its cross product with the three-dimensional operator $\nabla = \langle \frac{\partial}{\partial x}, \frac{\partial}{\partial y}, \frac{\partial}{\partial z} \rangle$:

Definition 5 The CURL of $\mathbf{v} = \langle p(x,y,z), q(x,y,z), r(x,y,z) \rangle$ is the vector field

$$\mathbf{curl\,v} = \nabla \times \langle p,q,r \rangle = \langle \frac{\partial}{\partial x}, \frac{\partial}{\partial y}, \frac{\partial}{\partial z} \rangle \times \langle p,q,r \rangle = \begin{vmatrix} \mathbf{i} & \mathbf{j} & \mathbf{k} \\ \frac{\partial}{\partial x} & \frac{\partial}{\partial y} & \frac{\partial}{\partial z} \\ p & q & r \end{vmatrix}$$

$$= \mathbf{i} \begin{vmatrix} \frac{\partial}{\partial y} & \frac{\partial}{\partial z} \\ q & r \end{vmatrix} - \mathbf{j} \begin{vmatrix} \frac{\partial}{\partial x} & \frac{\partial}{\partial z} \\ p & r \end{vmatrix} + \mathbf{k} \begin{vmatrix} \frac{\partial}{\partial x} & \frac{\partial}{\partial y} \\ p & q \end{vmatrix} \qquad (38)$$

$$= (r_y - q_z)\mathbf{i} - (r_x - p_z)\mathbf{j} + (q_x - p_y)\mathbf{k}.$$

How is the vector curl, defined by **(38)** related to the scalar curl of a vector field $\langle p(x,y), q(x,y) \rangle$ in the xy-plane, as defined in Section 13.2? To see the relationship, we convert $\langle p(x,y), q(x,y) \rangle$ into a vector $\mathbf{v} = \langle p(x,y), q(x,y), 0 \rangle$ in space by giving it a zero z-component. Then by **(38)**, we can see that its vector curl $[q_x(x,y) - p_y(x,y)]\mathbf{k}$, which equals the scalar curl

$$\mathrm{curl}\, \langle p(x,y), q(x,y) \rangle = -p_y(x,y) + q_x(x,y)$$

of the original vector field, multiplied by the unit vector $\mathbf{k}$.

Example 8 What is the curl of $\mathbf{v} = \mathbf{i} + xz\mathbf{j} + z^3\mathbf{k}$?

SOLUTION By definition **(38)**

$$\mathbf{curl\,v} = \nabla \times \langle 1, xz, z^3 \rangle = \begin{vmatrix} \mathbf{i} & \mathbf{j} & \mathbf{k} \\ \frac{\partial}{\partial x} & \frac{\partial}{\partial y} & \frac{\partial}{\partial z} \\ 1 & xz & z^3 \end{vmatrix}$$

$$= \mathbf{i} \begin{vmatrix} \frac{\partial}{\partial y} & \frac{\partial}{\partial z} \\ xz & z^3 \end{vmatrix} - \mathbf{j} \begin{vmatrix} \frac{\partial}{\partial x} & \frac{\partial}{\partial z} \\ 1 & z^3 \end{vmatrix} + \mathbf{k} \begin{vmatrix} \frac{\partial}{\partial x} & \frac{\partial}{\partial y} \\ 1 & xz \end{vmatrix}$$

$$= \left[\frac{\partial}{\partial y}(z^3) - \frac{\partial}{\partial z}(xz) \right] \mathbf{i} - \left[\frac{\partial}{\partial x}(z^3) - \frac{\partial}{\partial z}(1) \right] \mathbf{j} + \left[\frac{\partial}{\partial x}(xz) - \frac{\partial}{\partial y}(1) \right] \mathbf{k}$$

$$= -x\mathbf{i} + z\mathbf{k}. \square$$

Stokes' Theorem in space expresses the circulation $\int_C \mathbf{v} \cdot \mathbf{T}\, ds$ of a vector field $\mathbf{v}$ around the boundary of an oriented, piecewise-smooth bounded surface Σ as the surface integral of the normal component of $\mathbf{curl\,v}$:

Theorem 3 (Stokes' Theorem in space) Suppose that Σ is a bounded, piecewise-smooth, oriented surface with unit normal vectors $\mathbf{n}$ and oriented boundary C and that $\mathbf{v} = \langle p(x,y,z), q(x,y,z), r(x,y,z) \rangle$ has continuous first-order derivatives on Σ. Then

$$\int_C \mathbf{v} \cdot \mathbf{T}\, ds = \iint_\Sigma (\mathbf{curl\,v}) \cdot \mathbf{n}\, dS. \qquad (39)$$

We will discuss the derivation of **(39)** only for the case where Σ is the graph of a function, $\Sigma : z = h(x,y), (x,y)$ in D with D a closed set bounded by a simple, closed curve C in the xy-plane.

13.3 Surface integrals and the fundamental theorems in space

We assume that $h(x,y)$ has continuous second-order derivatives in D, and we orient Σ by assigning it unit normal vectors $\mathbf{n}$ with positive z-components and by orienting C counterclockwise as seen from above. We let $C_0 : x = x(t), y = y(t), a \xrightarrow[t]{} b$ be the boundary of D oriented counterclockwise so that

$$C : x = x(t), y = y(t), z = h(x(t), y(t)), a \xrightarrow[t]{} b.$$

We transform the line integral of $\mathbf{v} \cdot \mathbf{T}$ over C to a line integral over C_0, with the substitutions $z = h(x,y)$ and $dz = h_x(x,y)\,dx + h_y(x,y)\,dy$ and then apply Green's Theorem in the xy-plane from Section 13.2 to obtain,

$$\int_C \mathbf{v} \cdot \mathbf{T}\, ds = \int_C (p\, dx + q\, dy + r\, dz)$$
$$= \int_{C_0} [(p + rh_x)\, dx + (q + rh_y)\, dy] \qquad (40)$$
$$= \iint_D \left[-\frac{\partial}{\partial y}(p + rh_x) + \frac{\partial}{\partial x}(q + rh_y) \right] dx\, dy.$$

To find the derivatives in the last integral of **(40)**, we need to realize that p represents $p(x, y, h(x,y))$ and consequently[†]

$$\frac{\partial}{\partial y}(p) = \frac{\partial}{\partial y}[p(x, y, h(x,y)] = p_y(x, y, h(x,y)) + p_z(x, y, h(x,y))h_y(x,y) = p_y + p_z h_y. \qquad (41)$$

Similar formulas for the x-derivative of q and the x- and y-derivatives of r are also needed, along with the Product Rule. The calculations are left as Problem 68. The result is

$$-\frac{\partial}{\partial y}(p + rh_x) + \frac{\partial}{\partial x}(q + rh_y) = -(r_y - q_z)h_x - (p_z - r_x)h_y + (q_x - p_y). \qquad (42)$$

This with **(40)** yields

$$\int_C \mathbf{v} \cdot \mathbf{T}\, ds = \iint_D [-(r_y - q_z)h_x - (p_z - r_x)h_y + (q_x - p_y)]\, dx\, dy$$
$$= \iint_D \langle r_y - q_z, p_z - r_x, q_x - p_y \rangle \cdot \langle -h_x, -h_y, 1 \rangle\, dx\, dy$$
$$= \iint_\Sigma \operatorname{curl} \mathbf{v} \cdot \mathbf{n}\, dS$$

where at the last step we used $\mathbf{n} = \dfrac{\langle -h_x, -h_y, 1 \rangle}{\sqrt{(h_x)^2 + (h_y)^2 + 1}}$ from **(27)** and $dS = \sqrt{(h_x)^2 + (h_y)^2 + 1}\, dx\, dy$ from **(12)**. This establishes **(39)**.

[†]Since x is constant in **(41)**, we can obtain it from a formula for $\dfrac{d}{dy}[p(y, h(y))]$ with $p = p(y, z)$. For $\Delta y \neq 0$, the Mean-Value Theorem applied twice gives $\dfrac{p(y + \Delta y, h(y + \Delta y)) - p(y, h(y))}{\Delta y} = \dfrac{p(y + \Delta y, h(y + \Delta y)) - p(y, h(y + \Delta y))}{\Delta y}$
$-\dfrac{p(y, h(y + \Delta y)) - p(y, h(y))}{\Delta y} = p_y(y^*, h(y + \Delta y)) + p_z(y, z^*)\dfrac{h(y + \Delta y) - h(y)}{\Delta y}$ with y^* a point between y and $y + \Delta y$ and z^* a point between $h(y)$ and $h(y + \Delta y)$. Letting $\Delta y \to 0$ gives $\dfrac{\partial}{\partial y}[p(y(h(y))] = p_y(y, h(y)) + p_z(y, (h(y))h'(y)$, which becomes **(41)** when the variable x is put back in.

Example 9 Suppose that C is the boundary of $\Sigma: z = x^2y^3, 0 \leq x \leq 1, 0 \leq y \leq 1$, oriented counterclockwise when viewed from above. Use Stokes' Theorem to find the work done by the force field $\mathbf{F} = \langle 1, xz, z^3 \rangle$ (dynes) on an object that traverses C. Distances are measured in centimeters.

SOLUTION The orientation of C corresponds to orienting Σ with unit normal vectors having positive z-components, for which

$$\mathbf{n}\,dS = \left\langle -\frac{\partial}{\partial x}(x^2y^3), -\frac{\partial}{\partial y}(x^2y^3), 1 \right\rangle dx\,dy = \langle -2xy^3, -3x^2y^2, 1 \rangle\,dx\,dy.$$

We found $\mathbf{curl\,F} = \langle -x, 0, z \rangle$ in Example 8. On Σ, $z = x^2y^3$ and

$$(\mathbf{curl\,F}) \cdot \mathbf{n}\,dS = \langle -x, 0, x^2y^3 \rangle \cdot \langle -2xy^3, -3x^2y^2, 1 \rangle\,dx\,dy = 3x^2y^3\,dx\,dy.$$

Because the projection of Σ on the xy-plane is the square $D: 0 \leq x \leq 1$, $0 \leq y \leq 1$, the work equals

$$\int_C \mathbf{F} \cdot \mathbf{T}\,ds = \iint_\Sigma (\mathbf{curl\,v}) \cdot \mathbf{n}\,dS$$

$$= \iint_D 3x^2y^3\,dy\,dx = \int_{x=0}^{x=1} \int_{y=0}^{y=1} 3x^2y^3\,dy\,dx$$

$$= \left[\int_{x=0}^{x=1} 3x^2\,dx\right] \left[\int_{y=0}^{y=1} y^3\,dy\right] = \tfrac{1}{4} \text{ erg (dyne-centimeter)}. \square$$

Recall from Section 13.2 that if the scalar curl of a velocity field $\mathbf{v}$ in the xy-plane is positive in a region, then by Stokes' Theorem in the plane, the circulation $\int_C \mathbf{v} \cdot \mathbf{T}\,ds$ around every simple, closed curve C in the region that is oriented counterclockwise is positive. This means that the average value of the tangential component $\mathbf{v} \cdot \mathbf{T}$ of $\mathbf{v}$ is positive on all such curves C so that, in a sense, the fluid is rotating counterclockwise. (The fluid may not actually rotate because the normal component of $\mathbf{v}$ may be nonzero on the curves C and the fluid may also be flowing perpendicular to the curves.) Similarly, if the scalar curl is negative, then in the same sense, the fluid is rotating clockwise. The velocity field is called irrotational if the scalar curl is zero in the region.

In the case of a velocity field $\mathbf{v}$ of a fluid in xyz-space, the sense of circulation is determined by the direction of the vector curl. Suppose that $\mathbf{curl\,v}(x_0, y_0, z_0)$ is not zero at a point $P_0 = (x_0, y_0, z_0)$, and that C is a small, simple, closed curve in the plane through P_0 and perpendicular to $\mathbf{curl\,v}$ that goes around P_0 in the counterclockwise direction as viewed from the direction of $\mathbf{curl\,v}(P_0)$. Figure 13 shows how the the direction of the curl and the orientation of the curves where the circulation is positive are related by a right-hand rule. Let Σ be the bounded plane surface with boundary C. The unit normal vectors $\mathbf{n}$ to Σ have the direction of $\mathbf{v}(P_0)$, so that $\mathbf{v} \cdot \mathbf{n}$ is positive at P_0 and consequently throughout Σ if C is small enough. Then by Stokes' Theorem, the circulation of $\mathbf{v}$ around C is positive for such curves and, in this sense, the fluid rotates counterclockwise around the vector curl. A fluid is called IRROTATIONAL in a portion of xyz-space if its curl is zero there.

FIGURE 13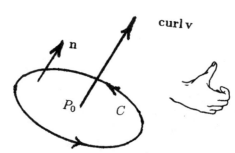

Path-independent line integrals in space

The derivation of Theorem 4 of Section 13.2 can be modified to establish the following similar result for line integrals in space.

Theorem 4 Suppose that $\mathbf{F}(x,y,z)$ is a continuous vector field in an open connected solid V in xyz-space. Then the integral $\int_C \mathbf{F} \cdot \mathbf{n}\, ds$ is path independent in V if and only if there is a real-valued function $U(x,y,z)$ such that $\nabla U(x,y,z) = \mathbf{F}(x,y,z)$ in V. If this is the case, then for any continuous, piecewise-smooth curve C in V

$$\int_C \mathbf{F} \cdot \mathbf{T}\, ds = \int_C \nabla U \cdot \mathbf{T}\, ds = U(Q) - U(P) \tag{43}$$

where P is the beginning and Q is the endpoint of the curve.

The second main result concerning path independence, like the corresponding result in the plane, uses the concept of simple connectivity.

We say that a solid V in xyz-space is CONNECTED if any two points in it can be joined by a smooth curve lying entirely in V. The solid is SIMPLY CONNECTED if every simple, closed curve C in V is the boundary of a piecewise-smooth, oriented surface that lies entirely in V. Recall that simply connected regions in the xy-plane cannot have any holes in them. A simply connected solid, in contrast, can have a hole inside it but not a hole through it. The solid V_1 in Figure 14 has a hole in the middle of it, but is simply connected, while the solid V_2 in Figure 15, which has a hole through it is not simply connected.

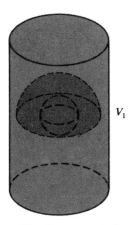

Simply connected

FIGURE 14

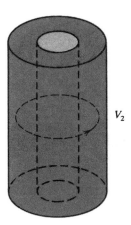

Not simply connected

FIGURE 15

Theorem 5 (a) Suppose that $p(x,y,z), q(x,y,z)$, and $r(x,y,z)$ have continuous first-order derivatives in an open solid V in xyz-space. If the vector field $\mathbf{F}(x,y,z) = \langle p, q, r \rangle$ is the gradient of a potential function $U(x,y,z)$ in V, then in V

$$\mathbf{curl\ F} = 0 \tag{44}$$

or equivalently

$$p_y = q_x,\ p_z = r_x,\ \text{and}\ q_z = r_y. \tag{45}$$

(b) If (44) (or equivalently (45)) holds and V is simply connected, then $\mathbf{F}$ is the gradient of a potential function in V.

To recall conditions **(45)**, notice that if $p = U_x$, $q = U_y$, and $r = U_z$ for a function $U(x, y, z)$, then equations **(45)** are the statements that the mixed partial derivatives $U_{xy} = U_{yx}, U_{xz} = U_{zx}$, and $U_{yz} = U_{zy}$ are equal.

Force fields in xyz-space that are the gradients of potential functions, like such force fields in the xy-plane, are called CONSERVATIVE.

The proof of Theorem 5 is short. To establish part (a), we suppose that $\mathbf{F} = \langle p, q, r \rangle = \nabla U(x, y, z) = \langle U_x(x, y, z), U_y(x, y, z), U_z(x, y, z) \rangle$ in V. Then $p = U_x, q = U_y, r = U_z$, and by definition **(38)**

$$\mathbf{curl\,F} = (r_y - q_z)\mathbf{i} - (r_x - p_z)\mathbf{j} + (q_x - p_y)\mathbf{k}$$
$$= \left[\frac{\partial}{\partial y}(U_z) - \frac{\partial U}{\partial z}(U_y)\right]\mathbf{i} - \left[\frac{\partial}{\partial x}(U_z) - \frac{\partial U}{\partial z}(U_x)\right]\mathbf{j} + \left[\frac{\partial}{\partial x}(U_y) - \frac{\partial U}{\partial y}(U_x)\right]\mathbf{k}.$$

This is the zero vector because the mixed second-order partial derivatives of $U(x, y, z)$ are equal.

To establish part (b) of Theorem 5, we suppose that V is simply connected and that C is any simple, closed curve in V. C is the boundary of an oriented piecewise-smooth surface Σ in V, and by Stokes' Theorem

$$\int_C \mathbf{F} \cdot \mathbf{T}\, ds = \iint_\Sigma (\mathbf{curl\,F}) \cdot \mathbf{n}\, dS = 0. \tag{46}$$

Consequently, if C_1 and C_2 are any piecewise-smooth curves in V with the same endpoints but no other intersections, then $C = C_1 - C_2$ is a simple, closed curve and by **(45)**

$$\int_{C_1} \mathbf{F} \cdot \mathbf{T}\, ds - \int_{C_2} \mathbf{F} \cdot \mathbf{T}\, ds = \int_C \mathbf{F} \cdot \mathbf{T}\, ds = 0$$

so the line integrals over C_1 and over C_2 are equal. We reach the same conclusion if C_1 and C_2 intersect at more points by appling Stokes' Theorem to all the regions they bound. Because the line integrals over any such curves C_1 and C_2 are equal, the line integral is path independent.

Example 10 Is $\mathbf{F} = \langle y^3, 3xy^2 + z, -y \rangle$ a conservative force field in xyz-space?

SOLUTION We set $p = y^3, q = 3xy^2 + z$, and $r = -y$. Then $p_y = \frac{\partial}{\partial y}(y^3) = 3y^2$ and $q_x = \frac{\partial}{\partial x}(3xy^2 + z) = 3y^2$ are equal for all (x, y); $p_z = \frac{\partial}{\partial z}(y^3) = 0$ and $r_x = \frac{\partial}{\partial x}(-y) = 0$ are equal for all (x, y) ; but $q_z = \frac{\partial}{\partial z}(3xy^2 + z) = 1$ and $r_y = \frac{\partial}{\partial y}(-y) = -1$ are not equal, so $\mathbf{F}$ is not the gradient of a potential function and is not conservative. □

Question 8 Show that $\mathbf{F} = (yz^2 - 3)\mathbf{i} + (xz^2 + 2y)\mathbf{j} + (2xyz + 3z^2)\mathbf{k}$ is a conservative force field in all of xyz-space.

Example 11 Find the potential functions $U(x, y, z)$ for the vector field $\mathbf{F}$ of Question 8.

SOLUTION In order that $\mathbf{F}$ equal ∇U, we must have $U_x = yz^2 - 3, U_y = xz^2 + 2y$, and $U_z = 2xyz + 2z^2$. To have $U_x = yz^2 - 3$ we set

$$U = \int (yz^2 - 3)\, dx = xyz^2 - 3x + \phi(y, z) \tag{47}$$

where $\phi(y, z)$ is the constant of integration that may have different values for different values of y and z and, hence, is a function of y and z.

Equation **(47)** gives all functions U with the required x-derivative. Differentiating that equation with respect to y yields a formula

$$U_y = \frac{\partial}{\partial y}[xyz^2 - 3x + \phi(y, z)] = xz^2 + \phi_y(y, z)$$

13.3 Surface integrals and the fundamental theorems in space

for the y-derivative of all functions U with the correct x-derivative. The condition $U_y = xz^2 + 2y$ then implies that $\phi_y(y,z) = 2y$ and hence

$$\phi(y,z) = \int 2y\, dy = y^2 + \psi(z) \tag{48}$$

where $\psi(x)$ is an unknown function of z. Substituting **(48)** in **(47)** gives a formula

$$U = xyz^2 - 3x + y^2 + \psi(z) \tag{49}$$

for all functions with the required x- and y-derivatives. Differentiating **(49)** with respect to z yields $U_z = 2xyz + \psi'(z)$, which with the requirement $U_z = 2xyz + 3z^2$ shows that $\psi'(z) = 3z^2$ and $\psi(z) = z^3 + C$ with a constant C. This formula with **(49)** shows that the potential functions are $U = xyz^2 - 3x + y^2 + z^3 + C$. □.

Question 9 Verify the formula for the potential functions from Example 11 by finding their x-, y-, and z-derivatives.

Question 10 Use the result of Example 11 to find the value of $\int_C \mathbf{F} \cdot \mathbf{T}\, ds$ for the vector field of Example 11 and for all curves C from $(0,0,0)$ to $(2,2,2)$.

Principles and procedures

- For surfaces that are parts of graphs $z = g(x, y)$ of functions

$$\mathbf{n} = \pm \frac{\langle -g_x(x,y), -g_y(x,y), 1\rangle}{\sqrt{[g_x(x,y)]^2 + [g_y(x,y)]^2 + 1}} \tag{50}$$

are the unit normal vectors to such a surface and

$$dS = \sqrt{[g_x(x,y)]^2 + [g_y(x,y)]^2 + 1}\ dx\, dy \tag{51}$$

is used for expressing surface integrals over such surfaces as double integrals with respect to x and y.

- Notice that the factor $\sqrt{[g_x(x,y)]^2 + [g_y(x,y)]^2 + 1}$ in **(51)** is greater than 1 unless $g_x(x,y)$ and $g_y(x,y)$ are zero. This reflects the fact that in most cases the area of a portion of a surface over a set D in the xy-plane is greater than the area of D because the surface is not horizontal.

- For parametrized surfaces $\Sigma : \mathbf{R} = \mathbf{R}(u,v)$,

$$\mathbf{n} = \pm \frac{\mathbf{R}_u(u,v) \times \mathbf{R}_v(u,v)}{|\mathbf{R}_u(u,v) \times \mathbf{R}_v(u,v)|} \tag{52}$$

and

$$dS = |\mathbf{R}_u(u,v) \times \mathbf{R}_v(u,v)|\ du\, dv \tag{53}$$

- If the divergence $\operatorname{div} \mathbf{v} = p_x + q_y + r_z$ of the velocity field $\mathbf{v} = \langle p, q, r\rangle$ of a fluid is positive in a portion V_0 of xyz-space, then fluid flows out of every solid V inside V_0, so that either the fluid is expanding or it has a source in V_0. If the divergence is negative, then fluid flows into every such solid V and it is either contracting or has a sink in V_0. If there are no sources or sinks, then a fluid is incompressible if and only if the divergence of its velocity field is zero.

- The evaluation of the surface integrals $\iint_\Sigma \mathbf{curl\, v} \cdot \mathbf{n}\, dS$ from Stokes' Theorem for Σ a portion of the graph $z = g(x,y)$ is relatively easy because when the surface integral is transformed to a double integral, the factor $\sqrt{(g_x)^2 + (g_y)^2 + 1}$ in formula **(50)** for $\mathbf{n}$ cancels with the same factor in the denominator of formula **(51)** for dS. A similar simplification occurs with formulas **(52)** and **(53)** for parametrized surfaces.

- If the vector curl of the velocity field $\mathbf{v}$ of a fluid is not zero at a point, then the circulation $\int_C \mathbf{v} \cdot \mathbf{T}\, ds$ is positive for simple, closed curves that go around the point counterclockwise as viewed from the direction of the vector curl. This means that the average value of the tangential component of $\mathbf{v}$ on such curves is positive, and in this sense the fluid "circulates" around the vector curl. The fluid does not necessarily actually circulate around the curl, however, because the normal component of the velocity field is not taken into account when the line integrals of $\mathbf{v} \cdot \mathbf{T}$ are calculated. A fluid whose velocity field has zero curl is called IRROTATIONAL.

- As in the corresponding theorem for curves in the xy-plane, the equation $\int_C \nabla U \cdot \mathbf{T}\, ds = U(Q) - U(P)$ for a curve C from the point P to the point Q comes directly from the Fundamental Theorem and the Chain Rule because if $C : x = x(t), y = y(t), z = z(t), a \xrightarrow[t]{} b$, then
$$\int_C \nabla U \cdot \mathbf{T}\, ds = \int_{t=a}^{t=b} \frac{d}{dt}[U(x(t), y(t), z(t))]\, dt = U(x(b), y(b), z(b)) - U(x(a), y(a), z(a)).$$

- To remember conditions **(45)**, $p_y = q_x$, $p_z = r_x$, $q_z = r_y$, in Theorem 5 for determining whether $v = \langle p, q, r \rangle$ is the gradient of a potential function $U(x, y, z)$, think of them as statements that the mixed second-order partial derivatives of U are equal if $p = U_x, q = U_y$, and $r = U_z$.

- When you have found a potential function U for a vector field $\langle p, q, r \rangle$, check your work by verifying that $U_x = p, U_y = q$, and $U_z = r$.

- Once you have determined that $\langle p, q, r \rangle$ is the gradient of a potential function by verifying conditions **(45)**, a separate (generally lengthy) calculation is required to find the potential function(s).

- A line integral $\int_C [p(x,y,z)\, dx + q(x,y,z)\, dy + r(x,y,z)\, dz]$ is path independent in a simply connected solid V if and only if conditions **(45)** in Theorem 5 are satisfied. If this is the case and you need to evaluate such an integral, find a potential function $U(x, y, z)$ for $\langle p, q, r \rangle$. You need only one such function, so set the constant of integration equal to zero in the formula for all potential functions. Then the line integral equals the value of the potential function at the end of the curve, minus its value at the beginning.

13.3 Surface integrals and the fundamental theorems in space

Tune-Up Exercises 13.3♦
A Answer provided. **O** Outline of solution provided.

T1.⁰ Figure 16 shows the rectangular surface $\Sigma : z = 2 - 2y, 0 \leq x \leq 2, 0 \leq y \leq 1$ and one of its unit normal vectors **n** with a positive z-component. **(a)** Find the unit normal vectors. **(b)** Express dS for the surface from part (a) in terms of $dx\,dy$. **(c)** Evaluate $\iint_\Sigma (x+z)\,dS$ for the surface from part (a).

♦ Type 3, normal vectors and dS for a rectangle

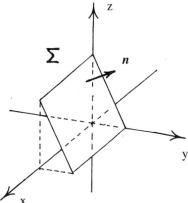

FIGURE 16

T2.⁰ **(a)** Find a formula for the unit normal vectors to the surface $\Sigma : z = x^2 + y^2$ that have negative z-components. **(b)** Express dS in terms of $dx\,dy$ for the surface in part (a).

♦ Type 3, normal vectors and dS

T3.⁰ **(a)** Give a formula for unit normal vectors to $\Sigma : \mathbf{R} = \langle r\cos\theta, r\sin\theta, \frac{1}{3}\theta \rangle$, $0 \leq r \leq 1, 0 \leq \theta \leq 2\pi$ from Example 2 and Figure 3. **(b)** Express dS in terms of $dr\,d\theta$ for the surface in part (a).

♦ Type 3, normal vectors and dS

T4.⁰ **(a)** What is the area of the surface Σ in Exercise T1? **(b)** Find the average value of $f(x,y,z) = x+z$ on that surface. (Use the result of Exercise T1c.)

♦ Type 3, area of a surface

T5.⁰ Verify the statement $\iiint_V f_z\,dx\,dy\,dz = \iint_\Sigma f\,n_3\,dS$ in Gauss' Theorem for $f(x,y,z) = \sin z$ and $V : 0 \leq x \leq 1, 0 \leq y \leq 1, 1 \leq z \leq 2$ by evaluating the triple and surface integrals.

♦ Type 3, verifying Green's Theorem in a particular case

T6.⁰ **(a)** Find the divergence of the velocity field $\mathbf{v} = \langle x^2+y, 1-2xy, xyz^2 \rangle$. **(b)** Use the result of part (a) to find the flux of **v** out of the cube $V : 0 \leq x \leq 1, 0 \leq y \leq 1, 0 \leq z \leq 1$.

♦ Type 3, using the Divergence Theorem

T7.⁰ **(a)** Find the curl of the force field $\mathbf{F} = \langle 4y, 3z, 2x \rangle$. **(b)** Use the result of part (a) to find the work done by **F** on an object that traverses the circle $C : x^2 + y^2 = 25$ in the xy-plane, oriented counterclockwise when viewed from above. Use the disk $\Sigma = \{(x,y,z) : x^2 + y^2 \leq 25, z = 0\}$.

♦ Type 3, using Stokes' Theorem

T8.⁰ Evaluate $\int_C \nabla U \cdot \mathbf{T}\,ds$, where $U(x,y,z) = xy^2z^3$ and C is a curve from $(0,0,0)$ to $(1,2,3)$.

♦ Type 3, evaluating an integral of $\nabla U \cdot \mathbf{T}$ with respect to ds

T9.⁰ Show that $\mathbf{A}(x,y,z) = yz\,\mathbf{i} + xz\,\mathbf{j} + xy\,\mathbf{k}$ is the gradient of a potential function $U(x,y,z)$ in all of xyz-space.

♦ Type 3, finding a potential function

T10.⁰ **(a)** Find a potential function for the vector field **A** of Exercise T9 and use it to evaluate $\int_C \mathbf{A} \cdot \mathbf{T}\,ds$, where C is a curve from $(1,1,1)$ to $(2,2,2)$.

♦ Type 3, evaluating an integral of $F \cdot \mathbf{T}$ with respect to ds by finding a potential function

---

♦The Tune-up Exercises and Problems are classified by type and content. The types are (1) basic, reactive; (2) basic reflective; (3) intermediate, reactive; (4) intermediate, reflective; (5) advanced, reactive; (6) advanced, reflective; and (7) advanced, theoretical.

Problems 13.3
[A]Answer provided. [O]Outline of solution provided.

1.[O] (a) Give a formula for the unit normal vectors to $\Sigma : z = xy^2, 0 \leq x \leq 2, 0 \leq y \leq 2$ that have positive z-components. (b) Express dS in terms of $dx\,dy$ for the surface in part (a). (c) Express $\iint_\Sigma xyz\,dS$ as an iterated integral, where Σ is the surface from part (a).

♦ Type 3, normal vectors, dS, and expressing a surface integral as a double integral

2.[A] Consider the surface $\Sigma : z = \sin y, 0 \leq y \leq x^2, 0 \leq x \leq 1$. (a) Give a formula for the unit normal vectors to Σ that have positive z-components. (b) Give a formula for dS in terms of $dx\,dy$ for the surface Σ. (c) Express $\iint_\Sigma xz^2\,dS$ as an iterated integral.

♦ Type 3, normal vectors, dS, and expressing a surface integral as a double integral

3. (a) Find the unit normal vectors to $\Sigma : z = e^x - e^y, 0 \leq x \leq 3, 1 \leq y \leq 4$ that have positive z-components. (b) Give a formula for dS in terms of $dx\,dy$ for the surface in part (a). (c) Write $\iint_\Sigma z^2\,dS$ as an iterated integral for the surface in part (a).

♦ Type 3, normal vectors, dS, and expressing a surface integral as a double integral

Evaluate the surface integrals in Problems 4 through 8.

4.[O] $\iint_\Sigma (4x + 2y + z)\,dS$ with the triangle $\Sigma : z = y, 0 \leq y \leq 1 - x, 0 \leq x \leq 1$

♦ Type 3, evaluating surface integrals

5.[A] $\iint_\Sigma x\,dS$ for the triangle $\Sigma : z = x^2 - 2y + 5, 0 \leq x \leq 1, 0 \leq y \leq 2$

♦ Type 3, evaluating surface integrals

6. $\iint_\Sigma x^3 yz\,dS$ for the parallelogram $\Sigma : z = 3x + 4y, 0 \leq x \leq 1, 0 \leq y \leq 3$

♦ Type 3, evaluating surface integrals

7.[A] $\iint_\Sigma z\,dS$ for the portion of a hemisphere $\Sigma : z = \sqrt{4 - x^2 - y^2}, x^2 + y^2 \leq 1$

♦ Type 3, evaluating surface integrals

8. $\iint_\Sigma \sqrt{x^2 + y^2 + 1}\,dS$ for the portion of a hyperbolic paraboloid $\Sigma : z = xy, 0, \leq x \leq 1, 0 \leq y \leq 1$

♦ Type 3, evaluating line integrals

9.[O] Find the flux of $\mathbf{v} = \langle xy, yz, xz \rangle$ across the parallelogram $\Sigma : z = x+y, 0 \leq x \leq 2, 0 \leq y \leq 3$, oriented so that its unit normal vectors have positive z-components.

♦ Type 4, finding a flux across a surface

10. What is the flux of $\mathbf{v} = \langle x \sin y, z \cos x, 16xy \rangle$ across the square $\Sigma : 0 \leq x \leq 1, 0 \leq y \leq 1, z = 0$ in the xy-plane with upward pointing unit normal vectors?

♦ Type 4, finding a flux across a surface

11.[O] What is the average value of $f = xyz$ on the portion of the plane $z = x - 2y$ for $-1 \leq x \leq 0, 0 \leq y \leq 1$?

♦ Type 3, average value on a surface

12.[A] Find the average value of $g = xz$ on the triangle $\Sigma : z = 2 - 2x - 2y, 0 \leq y \leq 1 - x, 0 \leq x \leq 1$.

♦ Type 3, average value on a surface

13. What is the average value of $h = yz$ on the portion of the plane $z = 2x + 3y$ for $0 \leq y \leq x^2, 0 \leq x \leq 1$?

♦ Type 3, average value on a surface

13.3 Surface integrals and the fundamental theorems in space

In Problems 14 through 16 verify the equation $\iiint_V f_z(x,y,z)\,dx\,dy\,dz = \iint_\Sigma f(x,y,z)\,n_3\,dS$ from Gauss' Theorem for the given functions and solids by evaluating the triple and surface integrals.

14.º $f(x,y,z) = xy^2z^3$ and the box $V: 0 \leq x \leq 3, 0 \leq y \leq 2, -1 \leq z \leq 0$.
♦ Type 3, verifying Gauss' Theorem in a partiucular case

15.^A $f(x,y,z) = \sin x \cos y \sin z$ and the box $V: 0 \leq x \leq 1, 0 \leq y \leq 2, 0 \leq z \leq 3$.
♦ Type 3, verifying Gauss' Theorem in a partiucular case

16. $f(x,y,z) = x^2 + y^3 + z^4$ and the cube $V: 0 \leq x \leq 5, 0 \leq y \leq 5, 0 \leq z \leq 5$.
♦ Type 3, verifying Gauss' Theorem in a partiucular case

17.^A Use Gauss' Theorem to evaluate $\iint_\Sigma (x^2yz^3\,n_1 - xy^2z^3\,n_2)\,dS$, where Σ is the boundary of the solid V between the xy-plane and $z = x^2 + y^4$ for $0 \leq x \leq 10, 0 \leq y \leq 10$ and $\mathbf{n} = \langle n_1, n_2, n_3 \rangle$ is its unit exterior normal vector.
♦ Type 3, applying Gauss' Theorem

18. Find the value of $\iint_\Sigma (x\,n_1 - 2y\,n_2 + 3z\,n_3)\,dS$ for the sphere $\Sigma : x^2 + y^2 + z^2 = 100$ and $\mathbf{n} = \langle n_1, n_2, n_3 \rangle$ is its unit exterior normal vector. Use Gauss' Theorem.
♦ Type 3, applying Gauss' Theorem

Find **(a)** the divergence and **(b)** the curl of the vector fields in Problems 19 through 26.

19.º $\mathbf{A} = \sin(xy)\,\mathbf{i} + \cos(yz)\,\mathbf{j} + \tan(z^2)\,\mathbf{k}$
♦ Type 1, finding a divergence and curl

20.^A $\mathbf{B} = \langle x^2y^3, y^3 - z^3, x + y - z \rangle$
♦ Type 1, finding a divergence and curl

21.^A $\mathbf{C} = xe^y\,\mathbf{i} - ye^z\,\mathbf{j} + ze^x\,\mathbf{k}$
♦ Type 1, finding a divergence and curl

22. $\mathbf{D} = x^2y^3z^4(\mathbf{i} + \mathbf{j} - 2\,\mathbf{k})$
♦ Type 1, finding a divergence and curl

23. $\mathbf{E} = \langle x^{-2}y, y^{-3}z, z^{-4}x \rangle$
♦ Type 1, finding a divergence and curl

24.^A $\mathbf{F} = \dfrac{\langle x, y, z \rangle}{x^2 + y^2 + z^2}$
♦ Type 1, finding a divergence and curl

25. $\mathbf{G} = \langle x^2y, y^3z^2, x^4z^3 \rangle$
♦ Type 1, finding a divergence and curl

26. $\mathbf{H} = \langle x^2 + y^2, y^2 + z^2, z^2 + x^2 \rangle$
♦ Type 1, finding a divergence and curl

27.º A fluid has the velocity field $\mathbf{v} = \langle 2 - x\cos z, y^2 z\sin x, \sin z \rangle$ (feet per minute). Use the Divergence Theorem to find its rate of flow out of the solid $V: 0 \leq z \leq y^2, 0 \leq x \leq 1, 0 \leq y \leq 1$ in xyz-space with distances measured in feet.
♦ Type 3, using the Divergence Theorem

28.^A What is the rate of flow of a fluid out of the box $V: 0 \leq x \leq 1, 1 \leq y \leq 2, 2 \leq z \leq 3$ in xyz-space with distances measured in meters if the fluid has the velocity field $\mathbf{v} = \langle x^2ye^z, xy^2e^z, x^2y^3 \rangle$ (meters per hour)? Use the Divergence Theorem.
♦ Type 3, using the Divergence Theorem

29. Use the Divergence Theorem to find the flux of $\mathbf{A} = x^2\sin y\,\mathbf{i} + 2x\cos y\,\mathbf{j} - 10z\,\mathbf{k}$ out of the ball $x^2 + y^2 + z^2 \leq 25$.
♦ Type 3, using the Divergence Theorem

30. What is the flux of $\mathbf{B} = (xy^2 + z^3)\,\mathbf{i} + (x^2y - x)\,\mathbf{j} + \tfrac{1}{3}z^3\,\mathbf{k}$ out of the ball $x^2 + y^2 + z^2 \leq 49$? Use the Divergence Theorem and spherical coordinates.
♦ Type 4, using the Divergence Theorem and spherical coordinates

31.º A fluid has the velocity field $\mathbf{v} = \langle xy^2, -z^2y, z^5 \rangle$ (centimeters per second) in xyz-space with distances measured in centimeters. The curve C is the boundary of the bounded surface $\Sigma : z = xy^2$, $0 \leq x \leq 1, 0 \leq y \leq 1$. Use Stokes' Theorem to find the circulation of the fluid around C, oriented counterclockwise when viewed from above.
♦ Type 3, using Stokes' Theorem

32.º Use Stokes' Theorem to find the work done by the force field $\mathbf{F} = \langle x, y^2z, -xy^2 \rangle$ (pounds) on an object that traverses the boundary C of the parallelogram $\Sigma : z = x + 2y, -1 \leq x \leq 0, 0 \leq y \leq 1$. The curve C is oriented counterclockwise when viewed from above, and distances are measured in feet.
♦ Type 3, using Stokes' Theorem

33. How much work is done by the force field $\mathbf{F} = \langle x+y, y-z, x-y \rangle$ (newtons) on an object that traverses the boundary C of $\Sigma : z = 4 - x^2 - y^2, 0 \leq x \leq 1, 0 \leq y \leq 1$. Use Stokes' Theorem. The curve C is oriented counterclockwise when viewed from above, and distances are measured in meters.
♦ Type 3, using Stokes' Theorem

In Problems 34 through 39 determine whether the given vector field is the gradient of a potential function, and if so find such a potential function.

34.º $\langle yz^2 - 3, xz^2 + 2y, 2xyz + 3z^2 \rangle$
♦ Type 3, recognizing gradient fields

35.º $\langle y^3, 3xy^2 + z, -y \rangle$
♦ Type 3, recognizing gradient fields

36.^A $\langle 2xy^3 + z^5, 3x^2y^2, 5xz^4 \rangle$
♦ Type 3, recognizing gradient fields

37.^A $\langle yz + 2xy^2, xz + 2x^2y, 2x^2y^2 \rangle$
♦ Type 3, recognizing gradient fields

38. $\langle \sin(y+z), x\cos(y+z), x\cos(y+z) + z \rangle$
♦ Type 3, recognizing gradient fields

39. $\langle yz, xz + e^y, xy \rangle$
♦ Type 3, recognizing gradient fields

40.º What is the value of $\int_C \nabla \sin(x+y+z) \cdot \mathbf{T} \, ds$ for curves C from $(1,1,1)$ to $(10, 10, 10)$?
♦ Type 3, evaluating an integral of $\nabla f \cdot \mathbf{T}$ with respect to ds

41. What is the value of $\int_C \nabla(x^3 y^2 z) \cdot \mathbf{T} \, ds$ for curves C from $(1,2,3)$ to $(1,10,100)$?
♦ Type 3, evaluating an integral of $\nabla f \cdot \mathbf{T}$ with respect to ds

42.^A Show that $\int_C (e^{yz} \, dx + xze^{yz} \, dy + xye^{yz} \, dz)$ is path independent in all of xyz-space and find its value for curves from $(0,0,0)$ to $(2,3,4)$.
♦ Type 3, proving path independence

43.^A Find the value of $\int_C (9 \, dx - 5 \, dy + 7 \, dz)$ for all curves from $(8, 10, -11)$ to $(4, -15, 2)$.
♦ Type 4, evaluating a path-independent line integral

44. Is $\int_C (yze^{xy} \, dx + xze^{xy} \, dy + xye^{xy} \, dz)$ path independent? If so, find its value for all curves from $(0,0,0)$ to $(10,10,10)$.
♦ Type 4, evaluating a path-independent line integral

45.º Evaluate $\iint_\Sigma 4z\sqrt{1+x^2+y^2} \, dS$ for Σ the portion of $z = xy$ for $0 \leq x \leq 1, 0 \leq y \leq 1$.
♦ Type 4, evaluating a surface integral

46.^A What is the average value of $f(x,y,z) = (1+4z)^3$ on the surface $\Sigma : z = x^2 + y^2, x^2 + y^2 \leq 1$?
♦ Type 3, average value on a surface

47. Find the average value of $g(x,y,z) = x^2yz$ on the portion of a hemisphere $\Sigma : z = \sqrt{9 - x^2 - y^2}$, $0 \leq y \leq \sqrt{1-x^2}$.
♦ Type 3, average value on a surface

48. What is the average value of $h(x,y,z) = \sin(x+y+z)$ on the parallelogram $\Sigma : z = 3x + 4y, -2 \leq x \leq 0, -2 \leq y \leq 0$?
♦ Type 3, average value on a surface

49.º Find the center of gravity of the surface $\Sigma : z = 3x + y^2, 0 \leq x \leq 1, 0 \leq y \leq 1$ with density $\rho(x,y,z) = \sqrt{10 + 4z - 12x}$ at (x,y,z).
♦ Type 3, center of gravity of a surface with variable density

50.^A The surface $\Sigma : z = x^2 + y^2, 0 \leq x \leq 1, 0 \leq y \leq 1$ has density $\rho(x,y,z) = \sqrt{1 + 2x^2 + 2y^2 + 2z}$ at (x,y,z). Find its center of gravity.
♦ Type 3, center of gravity of a surface with variable density

51. Verify the formula $\iiint_V f_z \, dx \, dy \, dz = \iint_\Sigma f n_3 \, dS$ in Gauss' Theorem for $f = \sin x + \cos y - z^3$ and V the solid bounded by $z = 0, z = 1 + x^2, y = 0, y = x$, and $x = 1$ by evaluating the triple and surface intergals.
♦ Type 4, verifying Gauss' Theorem in a particular case

13.3 Surface integrals and the fundamental theorems in space

52. Verify the formula $\iiint_V f_y \, dx \, dy \, dz = \iint_\Sigma f n_2 \, dS$ in Gauss' Theorem for $f = 8y^2 z$ and $V : 0 \leq x \leq 1, 0 \leq y \leq 1, 0 \leq z \leq xy$ by evaluating the triple and surface integrals.

♦ Type 4, verifying Gauss' Theorem in a particular case

53. Use Gauss' Theorem to find the value of $\iint_\Sigma (3x - y^2 + z^3) n_1 \, dS$, where Σ is the boundary of the half-ball $V : 0 \leq z \leq \sqrt{4 - x^2 - y^2}$.

♦ Type 3, applying Gauss' Theorem

54.$^\text{A}$ Use Gauss' Theorem to evaluate $\iint_\Sigma (x^2 y z^3 \, n_1 - x y^2 z^3 \, n_2) \, dS$, where Σ is the boundary of $V : 0 \leq z \leq x^2 + y^4, 0 \leq x \leq 10, 0 \leq y \leq 10$.

♦ Type 3, applying Gauss' Theorem

55. Use Gauss' Theorem to evaluate $\iint_\Sigma (x \, n_1 - 2y \, n_2 + 3z \, n_3) \, dS$, where Σ is the boundary of the ball $V : x^2 + y^2 + z^2 \leq 100$.

♦ Type 3, applying Gauss' Theorem

56. Find the work done by the force field $\mathbf{F} = \langle x^2 y, y^3 z^2, x^4 z^3 \rangle$ an object that traverses the boundary of the surface $\Sigma : z = xy, -1 \leq x \leq 1, -1 \leq y \leq 1$, where C is oriented counterclokwise when viewed from above, by using Stokes' Theorem.

♦ Type 3, applying Stokes' Theorem

57. Find the value of $\iint_\Sigma (\nabla \times \mathbf{v}) \cdot \mathbf{n} \, dS$ with Σ the portion of the paraboloid $z = x^2 + y^2 - 4$ for $x^2 + y^2 \leq 4$, where the unit normal vectors $\mathbf{n}$ have negative z-components, by evaluating a line integral around the boundary of the surface.

♦ Type 3, applying Stokes' Theorem

58.$^\text{A}$ Evaluate $\iint_\Sigma xy(4z+1) \, dS$ for Σ the portion of the paraboid $z = x^2 + y^2$ for (x,y) in the triangle with vertices $(0,0), (1,0)$, and $(1,1)$.

♦ Type 4, applying Stokes' Theorem

59.$^\text{A}$ Find the value of $\iint_\Sigma yz \cos x \, dS$ with Σ the portion of $z = \sin x$ for $0 \leq x \leq \frac{1}{2}\pi, 0 \leq y \leq 2$.

♦ Type 4, applying Stokes' Theorem

60. What is the value of $\iint_\Sigma xyz \, dS$, where Σ is the portion of $z = \sqrt{1 + x^2 + y^2}$ for $0 \leq x \leq 1$, $0 \leq y \leq 1$?

♦ Type 3, applying Stokes' Theorem

61. (a) Explain why the surface Σ with parametric equations $x = (3+\cos\alpha)\cos\theta, y = (3+\cos\alpha)\cos\theta, z = \sin\alpha, 0 \leq \theta \leq 2\pi, 0 \leq \alpha \leq 2\pi$ is a TORUS (a doughnut). (b) Evaluate $\iint_\Sigma z^2 \, dS$ for the surface of part (a).

♦ Type 5, describing a surface and evaluating a surface integral

62. A top-shaped surface Σ is described in cylindrical coordinates by the equation $r = z^2$ for $0 \leq z \leq 1, 0 \leq \theta \leq 2\pi$. (a) Sketch the surface. (b) Give parametric equations for the surface with z and θ as parameters. (c) Find a formula in terms of z and θ for the unit normal vectors $\mathbf{n}$ to Σ that have negative z-components. Then show that the projections of $\mathbf{n}$ on the xy-plane point away from the origin and draw one of these vectors with the surface. (c) Express the area of Σ as a definite integral and either use a calculator/computer algebra system to find the exact value of the integral or use a calculator/computer algorithm to obtain its approximate decimal value.

♦ Type 5, describing a surface and evaluating a surface integral

63. (a) Explain why $\Sigma : x = r\cos\theta, y = r\sin\theta, z = r + \theta, 0 \leq r \leq 10, 0 \leq \theta \leq 2\pi$ could be called a CORKSCREW SURFACE. (b) Express the area of Σ as a definite integral and either use a calculator/computer algebra system to find the exact value of the integral or use a calculator/computer algorithm to obtain its approximate decimal value.

♦ Type 5, describing a surface and evaluating a surface integral

64. If an object of mass m has the velocity vector $\mathbf{v}(t)$ at time t, then the quantity $\frac{1}{2}m|\mathbf{v}(b)|^2 - \frac{1}{2}m|\mathbf{v}(a)|^2$ is the CHANGE IN ITS KINETIC ENERGY from time $t = a$ to time $t = b$. **(a)** Show that the total work done by all forces on the object from time a to time b equals $\int_a^b m\dfrac{d\mathbf{v}}{dt} \cdot \mathbf{v}\, dt$. **(b)** Show that the total work equals the change in the kinetic energy.

♦ Type 6, work and kinetic energy

65. Some of the following expressions are defined and others are not. Give formulas in terms of $f = f(x,y,z)$, $\mathbf{F} = \mathbf{F}(x,y,z)$ and their derivatives for the expressions that are defined. **(a$^\mathbf{A}$)** div(∇f), **(b)** $\nabla(\text{div}\,\mathbf{F})$, **(c$^\mathbf{A}$)** curl$(\text{div}\,\mathbf{F})$, **(d)** curl(curl$\mathbf{F}$), **(e)** $\nabla(\nabla f)$, **(e)** curl(∇f) and **(f)** $\nabla(\mathbf{curl}\,\mathbf{F})$.

♦ Type 5, combining vector differention operators

66. Use Gauss' Theorem to derive the following GRADIENT THEOREM and CURL THEOREM: If $f(x,y,z)$ and $\mathbf{v}(x,y,z)$ have continuous first-order derivatives in a bounded solid V with piecewise-smooth boundary Σ and unit exterior normal vectors $\mathbf{n}$, then

$$\iiint_V \nabla f\, dx\, dy\, dz = \iint_\Sigma f\mathbf{n}\, dS \tag{52}$$

$$\iiint_V \text{curl}\,\mathbf{v}\, dx\, dy\, dz = \iint_\Sigma \mathbf{n} \times \mathbf{v}\, dS \tag{53}$$

♦ Type 5, deriving Gradient and Curl Theorems

67. **(a)** Explain why $\Sigma : x = \langle [1 + t\cos(\frac{1}{2}\theta)]\cos\theta, y = [1 + t\cos(\frac{1}{2}\theta)]\sin\theta, z = t\sin(\frac{1}{2}\theta)\rangle$ for $0 \leq \theta \leq 2\pi, -\frac{1}{2} \leq t \leq \frac{1}{2}$ is a MÖBIUS BAND, as in Figure 17, which could be formed by twisting one end of a rectangle $180°$ and gluing the sides together. **(b)** Show that the normal vector $\mathbf{R}_t \times \mathbf{R}_\theta$ switches direction when θ is increased by 2π to come back to the same point and, consequently, the surface is not orientable.

♦ Type 6, showing that a Möbius band is not orientable

A Möbius band

FIGURE 17

68. Derive equation **(42)** in the discussion of Stokes' Theorem.

♦ Type 6, supplying details for the proof of Stokes' Theorem

RESPONSES TO QUESTIONS IN CHAPTER 13

Responses 13.1

Response 1 Draw $\mathbf{B}(x,y) = \langle -1, -\frac{1}{2} \rangle$ at an array of points, as in Figure R1, where the scales on the axes are used to measure the arrows.

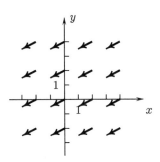

Figure R1

Response 2 The first of equations (2) gives $x - c_1 = \frac{3}{4}t$, so that $t = \frac{4}{3}(x - c_1)$. • The second of equations (2) then gives $y = \frac{4}{3}(x - c_1) + c_2$ or $y = \frac{4}{3}x - \frac{4}{3}c_1 + c_2$. (The streamlines are the lines of slope $\frac{4}{3}$.)

Response 3 (a) Solve $\dfrac{dx}{dt}(t) = -\frac{1}{3}x(t), x(0) = x_0$ and $\dfrac{dy}{dt}(t) = -\frac{1}{3}y(t), y(0) = y_0$. •
$x(t) = x_0 e^{-t/3}, y(t) = y_0 e^{-t/3}$
(b) If $x_0 = 0$ and $y_0 = 0$, then $x(t) = 0$ and $y(t) = 0$ for all t and the streamline is the origin. • If $x_0 = 0$ and $y_0 > 0$, then $x(t) = 0$ for all t, $y(t) > 0$ for all t, $y(t) \to 0$ as $t \to \infty$, $y(t) \to \infty$ as $t \to -\infty$, and the streamline is the positive y-axis oriented toward the origin. • Similarly, the streamline is the negative y-axis if $x_0 = 0, y_0 < 0$ • If $x_0 \neq 0$, then $e^{-t/3} = x/x_0$, $y = (y_0/x_0)x$ and the streamline is the ray of slope y_0/x_0 oriented toward the origin, as shown in Figure 6.

Response 4 Since $x = r\cos t$ and $y = r\sin t$: $\mathbf{v}(x,y) = \langle -r\sin t, r\cos t \rangle = \langle -y, x \rangle$

Response 5 (a) The width of the parallelogram is $\overline{PQ} = 20$ meters. •
Its height is $\left[50 \dfrac{\text{meters}}{\text{minute}} \right] \cos\left(\frac{1}{6}\pi \right) = 50(\frac{1}{2}\sqrt{3}) = 25\sqrt{3}\, \dfrac{\text{meters}}{\text{minute}}$.
Its area is $\left[25\sqrt{3}\, \dfrac{\text{meters}}{\text{minute}} \right] [20 \text{ meters}] = 500\sqrt{3}$ (square meters per minute) and equals the amount of water to cross PQ every minute.
(b) Because the rate of flow, $500\sqrt{3}$ square meters per minute, is constant it equals the instantaneous rate of flow.

Response 6 $\overrightarrow{PQ} = \langle 20 - 5, 30 - 10 \rangle = \langle 15, 20 \rangle$ centimeters •
$[\text{Work}] = \mathbf{F} \cdot \overrightarrow{PQ} = \langle 4, 7 \rangle \cdot \langle 15, 20 \rangle = 4(15) + 7(20) = 200$ dyne-centimeters (ergs)

Response 7 The curve can given by $C : y = x^3, 1 \xrightarrow{x} -1$. • Figure R7

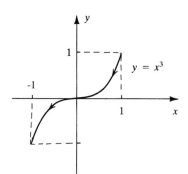

Figure R7

Response 8 For each partition, $\sum_{j=1}^{N} \Delta y_j$ equals the sum of the changes in y from P_{j-1} to P_j and this sum equals the value $y(b)$ at the end of the curve, minus the value $y(a)$ at the beginning of the curve. • $\int_C dy = \lim [y(b) - y(a)] = y(b) - y(a)$

For each partition, $\sum_{j=1}^{N} \Delta z_j$ equals the sum of the changes in z from P_{j-1} to P_j and this sum equals the value $z(b)$ at the end of the curve, minus the value $z(a)$ at the beginning of the curve. • $\int_C dz = \lim [x(b) - z(a)] = z(b) - z(a)$

Response 9 For $y = x^2$: $dy = \dfrac{d}{dx}(x^2)\, dx = 2x\, dx$ • $\int_C (x+y)\, dx = \int_1^0 (x + x^2)(2x\, dx)$
$= \int_1^0 (2x^2 + 2x^3)\, dx = \left[\tfrac{2}{3}x^3 + \tfrac{1}{2}x^4\right]_1^0 = [0] - [\tfrac{2}{3} + \tfrac{1}{2}] = -\tfrac{7}{6}$

Response 10 (a) $\int_C f\, dy = \int_C f\, u_2\, ds$ is positive because f and u_2 are both negative on C.

(b) $\int_C f\, ds$ is negative because f is negative on C.

Response 11 $\int_C x^2 y^2\, ds = \int_0^2 x^8 \sqrt{1 + 9x^4}\, dx \approx 561.58$ •

[Length] $= \int_C ds = \int_0^2 \sqrt{1 + 9x^4}\, dx \approx 8.63$ •

[Average value] $\approx \dfrac{561.58}{8.63} \doteq 65.07$

Responses 13.2

Response 1 $\iint_R \dfrac{\partial f}{\partial x}(x, y)\, dx\, dy = \int_{y=1}^{y=3} \int_{x=1}^{x=5} \dfrac{\partial f}{\partial x}(x, y)\, dx\, dy = \int_{y=1}^{y=3} \big[f(x, y)\big]_{x=1}^{x=5}\, dy$
$= \int_{y=1}^{y=3} [f(5, y) - f(1, y)]\, dy = \int_{y=1}^{y=3} f(5, y)\, dy - \int_{y=1}^{y=3} f(1, y)\, dy$

Response 2 With the notation in Figure 2: $dy = 0$ on C_1; $x = 5$ and $1 \xrightarrow{y} 3$ on C_2; $dy = 0$ on C_3; $x = 1$ and $3 \xrightarrow{y} 1$ on C_4 •

$\int_C f\, dy = \int_{C_1} f\, dy + \int_{C_2} f\, dy + \int_{C_3} f\, dy + \int_{C_4} f\, dy$
$= 0 + \int_{y=1}^{y=3} f(5, y)\, dy + 0 + \int_{y=3}^{y=1} f(1, y)\, dy = \int_{y=1}^{y=3} f(5, y)\, dy - \int_{y=1}^{y=3} f(1, y)\, dy$

Response 3 The sum of the integrals over C_1, C_2, C_3, and C_4 in Figure 6 equals the integral over the boundary of R in Figure 3, oriented positively, plus integrals in both directions over the vertical line segments that go through the interior of R. These last integrals cancel each other, so the integral over $C_1 + C_2 + C_3 + C_4$ equals the integral over the boundary C of the original region R.

Response 4
$$\iint_R \frac{\partial f}{\partial x}(x,y)\, dx\, dx = \int_{y=a}^{y=b} \int_{x=g(y)}^{x=h(y)} \frac{\partial f}{\partial x}(x,y)\, dx\, dy = \int_{y=a}^{y=b} \left[f(x,y) \right]_{x=g(y)}^{x=h(y)} dy$$
$$= \int_{y=a}^{y=b} [f(h(y),y) - f(g(y),y)]\, dy$$

Response 5 (a) $\int_{C_1} f\, dy$ and $\int_{C_3} f\, dy$ are zero because $dy = 0$ on the horizontal line segments C_1 and C_3.

(b) $\int_C f(x,y)\, dy = \int_{C_1} f\, dy + \int_{C_2} f\, dy + \int_{C_3} f\, dy + \int_{C_4} f\, dy$
$$= 0 + \int_{y=a}^{y=b} f(h(y), y)\, dy + 0 + \int_{y=b}^{y=a} f(g(y), y)\, dy$$
$$= \int_{y=a}^{y=b} [f(h(y),y) - f(g(y),y)]\, dy$$

Response 6 (a) $\int_C x\, dy = \iint_R \frac{\partial}{\partial x}(x)\, dx\, dy = \iint_R dx\, dy$

(b) $\int_C y\, dx = -\iint_R \frac{\partial}{\partial y}(y)\, dx\, dy = -\iint_R dx\, dy$

Response 7 (a) By (16): $\int_C x\, dy = [\text{Area of the circle of radius } r] = \pi r^2$

(b) On $C: x = r\cos t$ and $dy = \frac{d}{dt}(r\sin t)\, dt = r\cos t\, dt$ •
$$\int_C x\, dy = \int_{t=0}^{t=2\pi} (r\cos t)(r\cos t\, dt) = \int_{t=0}^{t=2\pi} r^2 \cos^2 t\, dt$$
$$= r^2 \int_{t=0}^{t=2\pi} \tfrac{1}{2}[1 + \cos(2t)]\, dt \;\; \text{(by the trigonometric identity } \cos^2 t = \tfrac{1}{2}[1+\cos(2t)]\text{)}$$
$$= r^2 \int_{t=0}^{t=2\pi} \tfrac{1}{2}\, dt + \tfrac{1}{4}r^2 \int_{t=0}^{t=2\pi} \cos u\, du \;\; (\text{with } u = 2t, du = 2\, dt)$$
$$= \pi r^2 + \tfrac{1}{4}r^2 \left[\sin(u)\right]_{t=0}^{t=2\pi} = \pi r^2 + \tfrac{1}{4}r^2 \left[\sin(2t)\right]_{t=0}^{t=2\pi} = \pi r^2$$

Response 8 (a) $\operatorname{div} \mathbf{v} = \operatorname{div} \langle 8x - y^2, x - 5y \rangle = \frac{\partial}{\partial x}(8x - y^2) + \frac{\partial}{\partial y}(x - 5y) = 8 - 5 = 3$

$[\text{Rate of flow out of } R] = \iint_R \operatorname{div} \mathbf{v}\, dx\, dy = \iint_R 3\, dx\, dy = 3[\text{Area of } R]$
$= 3(11) = 33$ square meters per second

Response 9 (a) $\nabla(2x + 3y^2) = \frac{\partial}{\partial x}(2x + 3y^2)\mathbf{i} + \frac{\partial}{\partial y}(2x + 3y^2)\mathbf{j} = \langle 2, 6y \rangle$ •
On $C_1: y = x, dy = dx$, and $0 \xrightarrow{x} 1$ •
$$\int_{C_1} \nabla(2x + 3y^2) \cdot \mathbf{T}\, ds = \int_{C_1} \langle 2, 6y \rangle \cdot \langle dx, dy \rangle = \int_{C_1} (2\, dx + 6y\, dy)$$
$$= \int_{x=0}^{x=1} (2\, dx + 6x\, dx) = \int_{x=0}^{x=1} (2 + 6x)\, dx = \left[2x + 3x^2\right]_0^1 = 5$$
(b) On $C_2: x = 5y^2 - 4y, dx = (10y - 4)\, dy$, and $0 \xrightarrow{y} 1$ •

$$\int_{C_2} \nabla(2x+3y^2)\cdot \mathbf{T}\,ds = \int_{C_2}(2\,dx+6y\,dy) = \int_{y=0}^{y=1}[2(10y-4)\,dy+6y\,dy]$$
$$= \int_{y=0}^{y=1}(26y-8)\,dy = \Big[13y^2-8y\Big]_0^1 = 5$$

Response 10 (a) On C_4 with parameter t and constant x: $x=x, y=t, dx=0$, and $dy=dt$ •
$$U(x,y) = \int_{C_{(x,y)}}(p\,dx+q\,dy) = \int_{C_3}(p\,dx+q\,dy)+\int_{C_4}(p\,dx+q\,dy)$$
$$= \int_{C_3}(p\,dx+q\,dy) + \int_{t=y_1}^{t=y}q(x,t)\,dt$$

(b) $\dfrac{\partial U}{\partial y}(x,y) = 0 + \dfrac{\partial}{\partial y}\int_{t=y_1}^{t=y}q(x,t)\,dy = q(x,y)$

Response 11 (a) For $p=y\sin(xy)$ and $q=x\sin(xy)$, the Product and Chain Rules give
$$\dfrac{\partial p}{\partial y} = \sin(xy)+y\cos(xy)\dfrac{\partial}{\partial y}(xy) = \sin(xy)+xy\cos(xy)$$
$$\dfrac{\partial q}{\partial x} = \sin(xy)+x\cos(xy)\dfrac{\partial}{\partial x}(xy) = \sin(xy)+xy\cos(xy) \; \bullet$$
$p_y=q_x$ for all (x,y) and the xy-plane is simply connected, so
$$\int_C [y\sin(xy)\,dx + x\sin(xy)\,dy]$$
is path independent in the entire xy-plane by Theorem 5.

(b) To have $U_x(x,y) = y\sin(xy)$, set $U(x,y)=\displaystyle\int y\sin(xy)\,dx$ with y constant.
• In the integral use the substitution $u=xy$, for which $du=y\,dx$:
$$U(x,y) = \int \sin(xy)(y\,dx) = \int \sin u\,du = -\cos u + \phi(y) = -\cos(xy)+\phi(y) \;\bullet$$
Take the y-derivative: $U_y(x,y) = \dfrac{\partial}{\partial y}[-\cos(xy)+\phi(y)] = \sin(xy)\dfrac{\partial}{\partial y}(xy) + \phi'(y)$
$= x\sin(xy)+\phi'(y)$ • To have $U_y(x,y)=x\sin(xy)$, set $\phi'(y)=0$ •
$\phi(y)=C$ and $U(x,y)=-\cos(xy)+C$

Response 12 (a) $\mathbf{v}=\langle p,q\rangle$ with $p=\tfrac{1}{3}(x+y)$ and $q=\tfrac{1}{3}(x-y)$ is irrotational because
$$\text{curl}\,\mathbf{v} = -\dfrac{\partial p}{\partial y}+\dfrac{\partial q}{\partial x} = -\dfrac{\partial}{\partial y}[\tfrac{1}{3}(x+y)] + \dfrac{\partial}{\partial x}[\tfrac{1}{3}(x-y)] = -\tfrac{1}{3}+\tfrac{1}{3} = 0 \text{ for all } (x,y).$$
(b) $\mathbf{v}$ is incompressible because $\text{div}\,\mathbf{v} = \dfrac{\partial p}{\partial x}+\dfrac{\partial q}{\partial y} = \dfrac{\partial}{\partial x}[\tfrac{1}{3}(x+y)] + \dfrac{\partial}{\partial y}[\tfrac{1}{3}(x-y)]$
$= \tfrac{1}{3}-\tfrac{1}{3} = 0$ for all (x,y).

Responses 13.3

Response 1 In cylindrical coordinates $[r,\theta,z]$, r can be any nonnegative number, z can be any number, and we can use $0\leq\theta\leq 2\pi$. • Because cylindrical coordinates are converted to rectangular coordinates by $x=r\cos\theta, y=r\sin\theta, z=z$, and $r=3$ on the cylinder, $\Sigma: x=3\cos\theta, y=3\sin\theta, z=z$ for (θ,z) in the rectangle $D=\{(\theta,z):0\leq\theta\leq 2\pi, 0\leq z\leq 6\}$ in the θz-plane.

Response 2 For $\mathbf{R} = \langle x, y, g(x,y) \rangle$: $\dfrac{\partial \mathbf{R}}{\partial x} \times \dfrac{\partial \mathbf{R}}{\partial y} = \begin{vmatrix} \mathbf{i} & \mathbf{j} & \mathbf{k} \\ \dfrac{\partial}{\partial x}(x) & \dfrac{\partial}{\partial x}(y) & \dfrac{\partial}{\partial x}[g(x,y)] \\ \dfrac{\partial}{\partial y}(x) & \dfrac{\partial}{\partial y}(y) & \dfrac{\partial}{\partial y}[g(x,y)] \end{vmatrix}$

$= \begin{vmatrix} \mathbf{i} & \mathbf{j} & \mathbf{k} \\ 1 & 0 & g_x(x,y) \\ 0 & 1 & g_y(x,y) \end{vmatrix} = \mathbf{i} \begin{vmatrix} 0 & g_x \\ 1 & g_y \end{vmatrix} - \mathbf{j} \begin{vmatrix} 1 & g_x \\ 0 & g_y \end{vmatrix} + \mathbf{k} \begin{vmatrix} 1 & 0 \\ 0 & 1 \end{vmatrix}$

$= \langle -g_x, -g_y, 1 \rangle$ • This is formula **(11)**.

Response 3 For $g = xy$: $g_x = y$ and $g_y = x$ • $dS = \sqrt{x^2 + y^2 + 1}\, dx\, dy$ •

[Surface area] $= \displaystyle\iint_D \sqrt{x^2 + y^2 + 1}\, dx\, dy$ with $D = \{(x,y) : x^2 + y^2 \le 9\}$ •

With polar coordinates: $x^2 + y^2 + 1 = r^2 + 1$ and $dx\, dy = r\, dr\, d\theta$ •

[Surface area] $= \displaystyle\int_{r=0}^{r=3} \int_{\theta=0}^{\theta=2\pi} r\sqrt{r^2+1}\, d\theta\, dr = \int_{\theta=0}^{\theta=2\pi} d\theta \int_{r=0}^{r=3} r\sqrt{r^2+1}\, dr$

$= \pi \displaystyle\int_{r=0}^{r=3} u^{1/2}\, du$ (with $u = r^2+1, du = 2r\, dr$)

$= \pi \left[\tfrac{2}{3} u^{3/2} \right]_{r=0}^{r=3} = \tfrac{2}{3}\pi \left[(r^2+1)^{3/2} \right]_{r=0}^{r=3} = \tfrac{2}{3}\pi (10^{3/2} - 1)$

Response 4 On the sphere Σ: $\rho = 3$, $z = 3\cos\phi$ and by **(15)**, $dS = 9\sin\phi\, d\theta\, d\phi$ •

$\displaystyle\iint_\Sigma e^z\, dS = \int_{\phi=0}^{\phi=\pi} \int_{\theta=0}^{\theta=2\pi} e^{3\cos\phi}\, 9\sin\phi\, d\theta\, d\phi = \int_{\phi=0}^{\phi=\pi} 18\pi e^{3\cos\phi} \sin\phi\, d\phi$

$= -6\pi \displaystyle\int_{\phi=0}^{\phi=\pi} e^{3\cos\phi}(-3\sin\phi\, d\phi)$

$= -6\pi \displaystyle\int_{\phi=0}^{\phi=\pi} e^u\, du$ (with $u = 3\cos\phi, du = -3\sin\phi\, d\phi$)

$= -6\pi \left[e^u \right]_{\phi=0}^{\phi=\pi} d\theta = -6\pi \left[e^{3\cos\phi} \right]_{\phi=0}^{\phi=\pi} = 6\pi(e^3 - e^{-3})$

Response 5 $\displaystyle\iiint_V \dfrac{\partial f}{\partial z}(x,y,z)\, dx\, dy\, dz = \iint_D \int_{z=g(x,y)}^{z=h(x,y)} \dfrac{\partial f}{\partial z}(x,y,z)\, dz\, dx\, dy$

$= \displaystyle\iint_D \left[f(x,y,z) \right]_{z=g(x,y)}^{z=h(x,y)} dx\, dy$

$= \displaystyle\iint_D f(x,y,h(x,y))\, dx\, dy - \iint_D f(x,y,g(x,y))\, dx\, dy$

Response 6 (a) $n_3 = \dfrac{1}{\sqrt{(h_x)^2 + (h_y)^2 + 1}}$ on Σ_1 by **(27)**

(b) $n_3 = \dfrac{-1}{\sqrt{(g_x)^2 + (g_y)^2 + 1}}$ on Σ_2 by **(29)**

(c) $n_3 = 0$ on Σ_3 because $\mathbf{n}$ is horizontal on the vertical surface Σ_3.

Response 7
$$\operatorname{div} \mathbf{v} = \operatorname{div} \langle x^3y^2 - 3x^2, 2yz - x^2y^3, 5 - z^2\rangle$$
$$= \frac{\partial}{\partial x}(x^3y^2 - 3x^2) + \frac{\partial}{\partial y}(2yz - x^2y^3) + \frac{\partial}{\partial z}(5 - z^2)$$
$$= (3x^2y^2 - 6x) + (2z - 3x^2y^2) + (-2z) = -6x$$

Response 8 Set $p = yz^2 - 3, q = xz^2 + 2y$, and $r = 2xyz + 3z^2$. •
$p_y = z^2$ and $q_x = z^2$ are equal for all (x, y, z)
$p_z = 2yz$ and $r_x = 2yz$ are equal for all (x, y, z)
$q_z = 2xz$ and $r_y = 2xz$ are equal for all (x, y, z) •
Because xyz-space is simply connected, $\mathbf{F} = \langle p, q, r\rangle$ is the gradient of a potential function and is conservative in all of xyz-space.

Response 9 For $U = xyz^2 - 3x + y^2 + z^3 + C$ and $\mathbf{F} = \langle p, q, r\rangle$:
$$U_x = \frac{\partial}{\partial x}(xyz^2 - 3x + y^2 + z^3 + C) = yz^2 - 3 = p$$
$$U_y = \frac{\partial}{\partial y}(xyz^2 - 3x + y^2 + z^3 + C) = xz^2 + 2y = q$$
$$U_z = \frac{\partial}{\partial z}(xyz^2 - 3x + y^2 + z^3 + C) = 2xyz + 3z^2 = r$$

Response 10 Set $C = 0$ in the formula $U = xyz^2 - 3x + y^2 + z^3 + C$ for the potential functions: $U = xyz^2 - 3x + y^2 + z^3$ • $\int_C \mathbf{F} \cdot \mathbf{T}\, ds = U(2,2,2) - U(0,0,0) = 2(2)(2^2) - 3(2) + 2^2 + 2^3 = 22$ for all curves C from $(0,0,0)$ to $(2,2,2)$, .

Answers 13.1

CHAPTER 13 ANSWERS

Tune-Up Exercises 13.1

T1. (a) $\langle -\frac{1}{2}, \frac{1}{2}|x|\rangle$ is in Figure 32 • One explanation: The y-component is zero for $x = 0$ and positive for $x \neq 0$, and the x-component is $-\frac{1}{2}$ for all (x,y).
(b) $\langle -\frac{1}{2}, 1\rangle$ is in Figure 31 • One explanation: It is a constant vector field that points up and to the left.
(c) $\langle \frac{1}{2}x, \frac{1}{2}x\rangle$ is in Figure 30 • One explanation: The vectors are zero for $x = 0$, point up to the right for $x > 0$, point down to the left for $x < 0$ and are longer for larger $|x|$.

T2. (a) $\int_C 5\,dx = 5([x \text{ at the end of } C] - [x \text{ at the beginning of } C]) = 5(5-1) = 20$

(b) $\int_C 10\,dy = 10([y \text{ at the end of } C] - [y \text{ at the beginning of } C]) = 10(1-3) = -20$

(c) $\int_C 15\,ds = 5([\text{Length of } C]) = 15(4+2) = 90$

T3. (a) $\int_C 5\,dx = 5([x \text{ at the end of } C] - [x \text{ at the beginning of } C]) = 5(6-1) = 25$

(b) $\int_C 10\,dy = 10([y \text{ at the end of } C] - [y \text{ at the beginning of } C]) = 10(-3-2) = -50$

(c) $\int_C 15\,dz = 15([z \text{ at the end of } C] - [z \text{ at the beginning of } C]) = 15(0-3) = -45$

(d) $\int_C 20\,ds = 20([\text{Length of } C]) = 20(30) = 600$

T4. (a) $dy = -\sin t\,dt$ • $\int_C x^7\,dy = \int_0^\pi \sin^7 t(-\sin t)\,dt = -\int_0^\pi \sin^8 t\,dt$

(b) $\int_C x^7\,dy = -\frac{35}{128}\pi \doteq 0.859029$

T5. $\dfrac{dy}{dx} = 2x$ • $\dfrac{dy}{dx}(1) = 2$ • $\langle 1, 2\rangle$ is a tangent vector pointing to the right. • $\mathbf{T} = \dfrac{\langle 1,2\rangle}{|\langle 1,2\rangle|} = \dfrac{\langle 1,2\rangle}{\sqrt{5}}$ •
$\mathbf{n} = \dfrac{\langle 2,-1\rangle}{\sqrt{5}}$

T6. (a) $\int_C [xy\,\mathbf{i} + (y-x)\,\mathbf{j}] \cdot \mathbf{T}\,ds = \int_C \langle xy, y-x\rangle \cdot \langle dx, dy\rangle = \int_C [xy\,dx + (y-x)\,dy]$

(b) $\int_C [\sin x\,\mathbf{i} + \cos y\,\mathbf{j}] \cdot \mathbf{n}\,ds = \int_C \langle \sin x, \cos y\rangle \cdot \langle dy, -dx\rangle = \int_C [\sin x\,dy - \cos y\,dx]$

T7. $[\text{Average value}] = \dfrac{\int_C -3\,ds}{\int_C ds} = \dfrac{-3\int_C ds}{\int_C ds} = -3$

Problems 13.1

1. (a) Figure A1a shows the vectors $\langle 1,-3\rangle$ at $(x,-3)$, the vectors $\langle 1,-1\rangle$ at $(x,-1)$, the vectors $\langle 1,0\rangle$ at $(x,0)$, the vectors $\langle 1,1\rangle$ at $(x,1)$, and the vectors $\langle 1,3\rangle$ at $(x,3)$, for $x=-6,-4,-2,0,2,4$.
   (b) Figure A1b

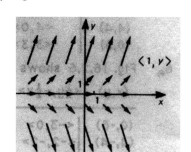

Figure A1a

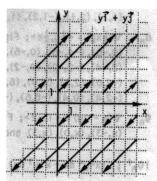

Figure A1b

5. $C: x=t^3, y=t^2, 0 \xrightarrow{t} 3 \bullet \int_C (x+y)\,dx = \int_0^3 (t^3+t^2)(3t^2)\,dt = \int_0^3 (3t^5+3t^4)\,dt$
   $= \frac{7}{10}(3)^6 = \frac{5103}{10}$

6. $\frac{2}{5}(e^{-5}-e^5)$

8. (a) $C: x=1+3t, y=1+4t, 0 \xrightarrow{t} 1 \bullet \int_C x\,dy = \int_0^1 (1+3t)(4\,dt) = \left[4t+6t^2\right]_0^1 = 10$
   (b) $ds = \sqrt{3^2+4^2}\,dt = 5\,dt \bullet \int_C xy\,ds = \int_0^1 (1+3t)(1+4t)(5\,dt) = 5\int_0^1 (1+7t+12t^2)\,dt$
   $= 5\left[t+\frac{7}{2}t^2+4t^3\right]_0^1 = \frac{85}{2}$

9. (a) $\frac{1}{5}\sqrt{85}(1-e^{-5})$ (b) $\frac{6}{7}[\cos(7)-1]$

11. (a) 0 (b) $-\frac{4}{9}$ (c) 0

12. (a) -7 (b) $\frac{17}{2}$

16. (a) $3\int_\pi^0 \sin[\sin t \sin(2t)]\cos(3t)\,dt \approx 2.164769$
    (b) $\int_0^\pi e^{\sin(3t)}\sqrt{\cos^2 t + 4\cos^2(2t) + 9\cos^2(3t)}\,dt \approx 11.062967$

18. $C: x=3\cos t, y=5\sin t, 0 \xrightarrow{t} 2\pi \bullet \int_C x\,dy = \int_0^{2\pi}(3\cos t)(5\cos t)\,dt = \int_0^{2\pi}(15\cos^2 t)\,dt$
    $= \frac{15}{2}\int_0^{2\pi}[1+\cos(2t)]\,dt = \frac{15}{2}\left[t+\frac{1}{2}\sin(2t)\right]_0^{2\pi} = 15\pi$

19. 3

20. (a) ≈ -15 (b) ≈ -50 (c) ≈ -100

21. (a) 2 (b) 10

22. Possible answers: (a) ≈ -112 (b) ≈ 8 (c) ≈ 200

25. (a) $C = C_1 - C_2 - C_3$ (b) 3 (c) 10

27. (a) ≈ -58 (b) ≈ -1050

Answers 13.1

29. $\int_C \mathbf{F}\cdot\mathbf{T}\,ds = \int_C (y^4\,dx + e^{x^2}\,dy + xy\,dz) = \int_C [(t^4)^4(2t) + e^{t^4}(4t^3) + (t^2)(t^4)(t^{-1})]\,dt$
$= \int_1^2 (2t^{17} + 4t^3 e^{t^4} + t^5)\,dt = \left[\tfrac{1}{9}t^{18} + e^{t^4} + \tfrac{1}{6}t^6\right]_1^2 = [\tfrac{1}{9}(2^{18}) + e^{16} + \tfrac{1}{6}(2^6)] - [\tfrac{1}{9} + e + \tfrac{1}{6}]$

30. 0

33. $\tfrac{5}{3}e^6 - \tfrac{1}{8}e^8 - \tfrac{37}{24}$

34. $[\text{Flux}] = \int_C \mathbf{v}\cdot\mathbf{n}\,ds = \int_C (p\,dy - q\,dx) = \int_C (xy\,dy - x^2y^2\,dx)$
$= \int_0^{\frac{1}{2}\pi} [(4\cos t)(\sin t)(\cos t) - (4\cos t)^2(\sin t)^2(-4\sin t)]\,dt$
$= \int_0^{\frac{1}{2}\pi} [68\cos^2 t - 64\cos^4 t]\sin t\,dt = \left[-\tfrac{68}{3}\cos^3 t + \tfrac{64}{5}\cos^5 t\right]_0^{\frac{1}{2}\pi} = \tfrac{148}{15}$

35. $e^4 + \tfrac{1}{3}e^6 - \tfrac{4}{3}$

36. -15π square feet per second

39. ≈ -44 square yards per minute

41. -24

Tune-Up Exercises 13.2

T1. The boundary C oriented positively is the sum $C_1 + C_2 + C_3$ of three line segments (Figure T1) •
$f = x^3 y = 0$ on C_1 and C_2 • $C_3 : y = 1 - x$, $1 \xrightarrow{x} 0$, on which $dy = -dx$ •

$\int_C f\,dy = \int_{C_3} x^3 y\,dy = \int_1^0 x^3(1-x)(-dx) = -\int_1^0 (x^3 - x^4)\,dx = -\left[\tfrac{1}{4}x^4 - \tfrac{1}{5}x^5\right]_1^0 = \tfrac{1}{20}$ •

$\iint_R \dfrac{\partial f}{\partial x}\,dx\,dy = \int_{x=0}^{x=1}\int_{y=0}^{y=1-x} 3x^2 y\,dy\,dx = \int_0^1 \left[\tfrac{3}{2}x^2 y^2\right]_{y=0}^{y=1-x}\,dx = \int_0^1 \tfrac{3}{2}x^2(1-x)^2\,dx$

$= \tfrac{3}{2}\int_0^1 (x^2 - 2x^3 + x^4)\,dx = \tfrac{3}{2}\left[\tfrac{1}{3}x^3 - \tfrac{1}{2}x^4 + \tfrac{1}{5}x^5\right]_0^1 = \tfrac{1}{20}$

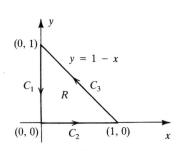

Figure T1

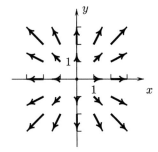

Figure T8

T2. (a) $\int_C xe^y\,dx = \iint_R -\dfrac{\partial}{\partial y}(xe^y)\,dx\,dy = -\iint_R xe^y\,dx\,dy$
(b) $\int_C y^3 \sin x\,dy = \iint_R \dfrac{\partial}{\partial x}(y^3 \sin x)\,dx\,dy = \iint_R y^3 \cos x\,dx\,dy$

T3. $\operatorname{div}\langle \ln(xy), x\tan y\rangle = \dfrac{\partial}{\partial x}[\ln(xy)] + \dfrac{\partial}{\partial y}(x\tan y) = \dfrac{1}{x} + x\sec^2 y$ •

$\operatorname{curl}\langle \ln(xy), x\tan y\rangle = -\dfrac{\partial}{\partial y}[\ln(xy)] + \dfrac{\partial}{\partial x}(x\tan y) = -\dfrac{1}{y} + \tan y$

T4. $x = 3\cos t$, $y = 3\sin t$, $dx = -3\sin t\, dt$, $dy = 3\cos t\, dt$ • $\int_C \langle y, -x\rangle \cdot \mathbf{T}\, ds = \int_C \langle y, -x\rangle \cdot \langle dx, dy\rangle$

$= \int_C (y\, dx - x\, dy) = \int_0^{2\pi} [(3\sin t)(-3\sin t) - (3\cos t)(3\cos t)]\, dt = -9\int_0^{2\pi} (\sin^2 t + \cos^2 t)\, dt$

$= -9\int_0^{2\pi} dx = -18\pi$

T5. The curve C is the boundary of $R: x^2 + y^2 \leq 9$ oriented positively. • $\int_C \langle y, -x\rangle \cdot \mathbf{T}\, ds$

$= \iint_R \operatorname{curl}\langle y, -x\rangle\, dx\, dy = \iint_R [\frac{\partial}{\partial x}(-x) - \frac{\partial}{\partial y}(y)]\, dx\, dy = \iint_R (-2)\, dx\, dy$

$= -2\,[\text{Area of } R] = -2\pi(3^2) = -18\pi$

T6. $x = 3\cos t$, $y = 3\sin t$, $dx = -3\sin t\, dt$, $dy = 3\cos t\, dt$ • $\int_C \langle x, y\rangle \cdot \mathbf{n}\, ds = \int_C \langle x, y\rangle \cdot \langle dy, -dx\rangle$

$= \int_C (x\, dy - y\, dx) = \int_0^{2\pi} [(3\cos t)(3\cos t) - (3\sin t)(-3\sin t)]\, dt = 9\int_0^{2\pi} (\sin^2 t + \cos^2 t)\, dt$

$= 9\int_0^{2\pi} dx = 18\pi$

T7. The curve C is the boundary of $R: x^2 + y^2 \leq 9$ oriented positively. •

$\int_C \langle x, y\rangle \cdot \mathbf{n}\, ds = \iint_R \operatorname{div}\langle x, y\rangle\, dx\, dy = \iint_R [\frac{\partial}{\partial x}(x) + \frac{\partial}{\partial y}(y)]\, dx\, dy = \iint_R 2\, dx\, dy$

$= 2\,[\text{Area of } R] = 2\pi(3^2) = 18\pi$

T8. $\nabla f = \langle \frac{\partial}{\partial x}[\frac{1}{4}(x^2 + y^2)], \frac{\partial}{\partial y}[\frac{1}{4}(x^2 + y^2)]\rangle = \langle \frac{1}{2}x, \frac{1}{2}y\rangle$ • Figure T8

T9. $\int_C \nabla f(x, y) \cdot \mathbf{T}\, ds = [f \text{ at the end of } C] - [f \text{ at the beginning of } C] = f(4, 4) - f(0, 0)$

$= \frac{1}{4}(4^2 + 4^2) - \frac{1}{4}(0^2 + 0^2) = 8$

T10. $p = 2xy^3$, $q = kx^2y^2$ • $p_y = 6xy^2$ and $q_x = 2kxy^2$ for all (x, y) • $\langle p, q\rangle$ is conservative
$\iff p_y = q_x \iff 6xy^2 = 2kxy^2 \iff k = 3$

Problems 13.2

1. $\iint_R f_y\, dx\, dy = \iint_R 5x^2y^4\, dx\, dy = \frac{32}{3}$ • $\int_C f\, dx = \int_C x^2y^5\, dx = -\frac{32}{3}$

3. For $f = xy^3$: $\iint_R f_y\, dx\, dy = \iint_R 3xy^2\, dx\, dy = \int_0^1 \int_0^{\sqrt{1-x^2}} 3xy^2\, dy\, dx = \int_0^1 x(1 - x^2)^{3/2}\, dx$

$= \left[-\frac{1}{5}(1 - x^2)^{5/2}\right]_0^1 = \frac{1}{5}$ • Since $f = 0$ on the coordinate axes, and $x = \cos t$, $y = \sin t$ on the

quarter-circle, $\int_C f\, dx = -\int_0^{\pi/2} \cos t \sin^4 t\, dt = \left[-\frac{1}{5}\sin^5 t\right]_0^{\pi/2} = -\frac{1}{5}$ •

$\iint_R f_x\, dx\, dy = \iint_R y^3\, dx\, dy = \int_0^1 \int_0^{\sqrt{1-x^2}} y^3\, dy\, dx = \frac{1}{4}\int_0^1 (1 - 2x^2 + x^4)\, dx$

$= \frac{1}{4}\left[x - \frac{2}{3}x^3 + \frac{1}{5}x^5\right]_0^1 = \frac{2}{15}$ • $\int_C f\, dy = \int_0^{\pi/2} \cos t \sin^3 t \cos t\, dt$

$= \int_0^{\pi/2} (\cos^2 t - \cos^4 t)\sin t\, dt = -\left[\frac{1}{3}\cos^3 t - \frac{1}{5}\cos^5 t\right]_0^{\pi/2} = \frac{1}{3} - \frac{1}{5} = \frac{2}{15}$

Answers 13.2

4. $\operatorname{curl}\mathbf{F} = -\dfrac{\partial}{\partial y}(2xy) + \dfrac{\partial}{\partial x}(3y^2) = -2x$ • $\iint_R \operatorname{curl}\mathbf{F}\, dx\, dy = \int_{-1}^{1}\int_{0}^{1-x^2}(-2x)\, dy\, dx = 0$ •

$\int_C \mathbf{F}\cdot\mathbf{T}\, ds = \int_C (2xy\, dx + 3y^2\, dy) = 0$

6. $\int_C (xy^3\, dx - 2x^3 y^2\, dy) = \iint_R [-\dfrac{\partial}{\partial y}(xy^3) + \dfrac{\partial}{\partial x}(-2x^3 y^2)]\, dx\, dy = \iint_R (-3xy^2 - 6x^2 y^2)\, dx\, dy$

7. $\iint_R [-x\cos(xy) - y\sin(xy)]\, dx\, dy$

10. The loop oriented counterclockwise is given by $C_1 : x = t^2 - 3,\ y = \tfrac{1}{3}t^3 - t,\ -\sqrt{3} \xrightarrow{t} \sqrt{3}$ •

[Area inside the loop] $= \int_{C_1} x\, dy = \int_{-\sqrt{3}}^{\sqrt{3}}(t^2 - 3)(t^2 - 1)\, dt = \int_{-\sqrt{3}}^{\sqrt{3}}(t^4 - 4t^2 + 3)\, dt$

$= 2\left[\tfrac{1}{5}t^5 - \tfrac{4}{3}t^3 + 3t\right]_0^{\sqrt{3}} = \tfrac{8}{5}\sqrt{3}$

11. $\tfrac{4}{3}$

14. (a) $y\cos(xy) + \sin(x - y)$ (b) $-x\cos(xy) - \sin(x - y)$

17. (a) $\sin y - \cos x$ (b) $-x\cos y + y\sin x$

18. (a) $y^3 + 2y$ (b) $-3xy^2 - 3x^2$

21. (a) $2y^3 - 6xy^2 + 6x^2 y - 2x^3$ (b) 0

23. -20

25. -50

27. $\tfrac{111}{2} - 3\cos(2)$ square feet per minute

29. -75π foot-pounds

32. Consider the region in Figure A32 with two sides parallel to the flow and a top and a bottom perpendicular to the flow. • No fluid crosses the sides and the rate of flow out, across the top, is greater than the rate of flow in, across the bottom, because $|\mathbf{v}|$ is greater on the top than on the bottom and the top is longer than the bottom. • The flux is positive, so if the divergence has costant sign, it must be positive.

Figure A32

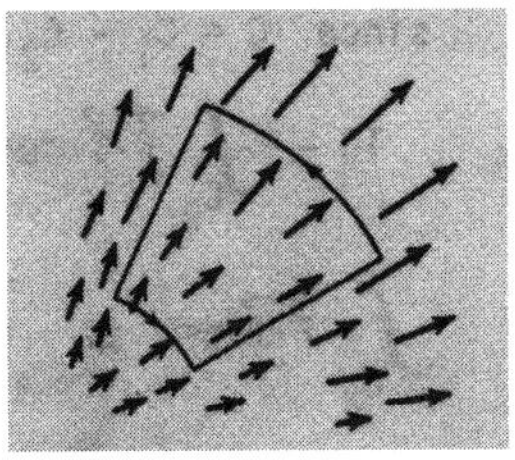

33. Negative

35. $p = \sin(3y) + x,\ q = 3x\cos(3y) - y$ • $p_y = 3\cos(3y) = q_x$ for all (x, y) • The vector field is the gradient of a potential function $U(x, y)$ in the entire xy-plane. •

$U(x, y) = \int [\sin(3y + x]\, dx = x\sin(3y) + \tfrac{1}{2}x^2 + \phi(y)$ • $U_y = 3x\cos(3y) + \phi'(y)$ •

$U_y = 3x\cos(3y) - y \implies \phi'(y) = -y$ • $\phi(y) = -\tfrac{1}{2}y^2 + C$ • $U(x, y) = x\sin(3y) + \tfrac{1}{2}x^2 - \tfrac{1}{2}y^2 + C$ for any C

36. Not the gradient of a potential function

37. $U(x, y) = \ln|1 + xy| + C$ for any C

39. $U(x, y) = e^{xy} - x + C$ for any C

41. $p = y^3 - 3x^2, q = 3x^2 + 2y$ • $p_y = 3y^2$ and $q_x = 3y^2$ are equal for all (x,y) • A gradient field •
$U(x,y) = \int (y^3 - 3x^2)\, dx = xy^3 - x^3 + \phi(y)$ • $U_y = 3xy^2 + \phi'(y)$ • $U_y = 3xy^2 + 2y \Longrightarrow \phi'(y) = 2y$
• $\phi(y) = y^2 + C$ • Take $C = 0$. • $U(x,y) = xy^3 - x^3 + y^2$ •
$\int_C [(y^3 - 3x^2)\, dx + (3xy^2 + 2y)]\, dy = U(1,2) - U(0,0) = 11$

42. The integral is not path independent.

43. -20

Tune-Up Exercises 13.3

T1. (a) $g(x,y) = 2 - 2y$ $g_x = 0$, $g_y = -2$ • $\mathbf{n} = \dfrac{\langle -g_x, -g_y, 1\rangle}{\sqrt{(g_x)^2 + (g_y)^2 + 1}} = \dfrac{\langle 0, 2, 1\rangle}{\sqrt{5}}$

(b) $dS = \sqrt{(g_x)^2 + (g_y)^2 + 1}\, dx\, dy = \sqrt{5}\, dx\, dy$

(c) $\displaystyle\iint_\Sigma (x + z)\, dS = \int_{x=0}^{x=2} \int_{y=0}^{y=1} (x + 2 - 2y)\sqrt{5}\, dy\, dx = \sqrt{5}\int_{x=0}^{x=2}\Big[xy + 2y - y^2\Big]_{y=0}^{y=1} dx$

$= \sqrt{5}\displaystyle\int_{x=0}^{x=2} (x + 1)\, dx = \sqrt{5}\Big[\tfrac{1}{2}x^2 + x\Big]_{x=0}^{x=2} = 4\sqrt{5}$

T2. (a) $\Sigma: z = x^2 + y^2$ • $z_x = 2x$, $z_y = 2y$ • $\mathbf{n} = \dfrac{\langle z_x, z_y, -1\rangle}{\sqrt{(z_x)^2 + (z_y)^2 + 1}} = \dfrac{\langle 2x, 2y, -1\rangle}{\sqrt{4x^2 + 4y^2 + 1}}$

(b) $dS = \sqrt{(z_x)^2 + (z_y)^2 + 1}\, dx\, dy = \sqrt{4x^2 + 4y^2 + 1}\, dx\, dy$

T3. (a) $\mathbf{R} = \langle r\cos\theta, r\sin\theta, \tfrac{1}{3}\theta\rangle$ • $\mathbf{R}_r = \langle\cos\theta, \sin\theta, 0\rangle$, $\mathbf{R}_\theta = \langle -r\sin\theta, r\cos\theta, \tfrac{1}{3}\rangle$ •

$\mathbf{R}_r \times \mathbf{R}_\theta = \begin{vmatrix} \mathbf{i} & \mathbf{j} & \mathbf{k} \\ \cos\theta & \sin\theta & 0 \\ -r\sin\theta & r\cos\theta & \tfrac{1}{3} \end{vmatrix} = \mathbf{i}\begin{vmatrix}\sin\theta & 0 \\ r\cos\theta & \tfrac{1}{3}\end{vmatrix} - \mathbf{j}\begin{vmatrix}\cos\theta & 0 \\ -r\sin\theta & \tfrac{1}{3}\end{vmatrix} + \mathbf{k}\begin{vmatrix}\cos\theta & \sin\theta \\ -r\sin\theta & r\cos\theta\end{vmatrix}$

$= \langle\tfrac{1}{3}\sin\theta, -\tfrac{1}{3}\cos\theta, r\rangle$ • $|\mathbf{R}_r \times \mathbf{R}_\theta|^2 = (\tfrac{1}{3}\sin\theta)^2 + (-\tfrac{1}{3}\cos\theta)^2 + r^2 = \tfrac{1}{9} + r^2$ •

$\mathbf{n} = \dfrac{\langle\tfrac{1}{3}\sin\theta, -\tfrac{1}{3}\cos\theta, r\rangle}{\sqrt{\tfrac{1}{9} + r^2}}$ (b) $dS = \sqrt{\tfrac{1}{9} + r^2}\, dr\, d\theta$

T4. (a) From the solution of T1: [Area of Σ] $= \displaystyle\iint_\Sigma dS = \int_{x=0}^{x=2}\int_{y=0}^{y=1} \sqrt{5}\, dy\, dx = 2\sqrt{5}$

(b) [Average value] $= \dfrac{1}{\text{Area of }\Sigma}\displaystyle\iint_\Sigma (x + z)\, dS = \dfrac{4\sqrt{5}}{2\sqrt{5}} = 2$

T5. $f = \sin z$ • $\displaystyle\iiint_V f_z\, dx\, dy\, dz = \iiint \cos z\, dx\, dy\, dz = \int_{x=0}^{x=1} dx \int_{y=0}^{y=1} dy \int_{z=1}^{z=2} \cos z\, dx$

$= \Big[\sin z\Big]_1^2 = \sin(2) - \sin(1)$ • $n_3 = 1$ and $dS = dx\, dy$ on the top $\Sigma_1: z = 2$, $0 \le x \le 1$, $0 \le y \le 1$ of V • $n_3 = -1$ and $dS = dx\, dy$ on the bottom $\Sigma_2: z = 1$, $0 \le x \le 1$, $0 \le y \le 1$ of V • $n_3 = 0$ on the sides Σ_3 of V • The square $D: 0 \le x \le 1$, $0 \le y \le 1$ is the projection of Σ_1 and Σ_2 on the xy-plane. •

$\displaystyle\iint_\Sigma \sin z\, n_3\, dS = \left\{\iint_{\Sigma_1} + \iint_{\Sigma_2} + \iint_{\Sigma_3}\right\}\sin z\, n_3\, dS = \iint_D \sin(2)\, dx\, dy - \iint_D \sin(1)\, dx\, dy$

$= \sin(2)[\text{Area of }D] - \sin(1)[\text{Area of }D] = \sin(2) - \sin(1)$

Answers 13.3

T6. (a) $\text{div}\langle x^2 + y, 1 - 2xy, xyz^2 \rangle = \frac{\partial}{\partial x}(x^2 + y) + \frac{\partial}{\partial y}(1 - 2xy) + \frac{\partial}{\partial z}(xyz^2) = 2x - 2x + 2xyz = 2xyz$

(b) The flux out of $V: 0 \leq x \leq 1,\ 0 \leq y \leq 1,\ 0 \leq z \leq 1$ is $\iint_\Sigma \mathbf{v} \cdot \mathbf{n}\, dS =$

$\iiint_V \text{div}\,\mathbf{v}\, dx\, dy\, dx = \int_{x=0}^{x=1} \int_{y=0}^{y=1} \int_{z=0}^{z=1} 2xyz\, dz\, dy\, dx = \int_{x=0}^{x=1} 2x\, dx \int_{y=0}^{y=1} y\, dy \int_{z=0}^{z=1} z\, dz = \frac{1}{4}.$

T7. (a) $\text{curl}\,\langle 4y, 3z, 2x \rangle = \begin{vmatrix} \mathbf{i} & \mathbf{j} & \mathbf{k} \\ \frac{\partial}{\partial x} & \frac{\partial}{\partial y} & \frac{\partial}{\partial z} \\ 4y & 3z & 2x \end{vmatrix} = \mathbf{i}\begin{vmatrix} \frac{\partial}{\partial y} & \frac{\partial}{\partial z} \\ 3z & 2x \end{vmatrix} - \mathbf{j}\begin{vmatrix} \frac{\partial}{\partial x} & \frac{\partial}{\partial z} \\ 4y & 2x \end{vmatrix} + \mathbf{k}\begin{vmatrix} \frac{\partial}{\partial x} & \frac{\partial}{\partial y} \\ 4y & 3z \end{vmatrix} = \langle -3, -2, -4 \rangle$

(b) $\mathbf{n} = \langle 0, 0, 1 \rangle$ on Σ • $[\text{Work}] = \int_C \mathbf{F} \cdot \mathbf{T}\, dS = \iint_\Sigma (\text{curl}\,\mathbf{F}) \cdot \mathbf{n}\, dS$

$= \iint_\Sigma \langle -3, -2, -4 \rangle \cdot \langle 0, 0, 1 \rangle\, dS = \iint_\Sigma -4\, dS = -4\,[\text{Area of } \Sigma] = -4(25\pi) = -100\pi$

T8. $U(x, y, z) = xy^2z^3$ • C runs from $(0, 0, 0)$ to $(1, 2, 3)$ •

$\int_C \nabla U \cdot \mathbf{T}\, ds = U(1, 2, 3) - U(0, 0, 0) = 1(2)^2(3)^3 - 0 = 108$

T9. $\mathbf{A} = \langle p, q, r \rangle = \langle yz, xz, xy \rangle$ • $p_y = \frac{\partial}{\partial y}(yz) = z$ and $q_x = \frac{\partial}{\partial x}(xz) = z$ are equal for all (x, y, z). •

$p_z = \frac{\partial}{\partial z}(yz) = y$ and $r_x = \frac{\partial}{\partial x}(xy) = y$ are equal for all (x, y, z). • $q_z = \frac{\partial}{\partial z}(xz) = x$ and

$r_y = \frac{\partial}{\partial y}(xy) = x$ are equal for all (x, y, z). • xyz-space is simply connected. •

$\mathbf{A}$ is the gradient of a potential function in all of xyz-space.

T10. (a) $U_x = yz \implies U(x, y, z) = \int yz\, dx = xyz + \phi(y, z)$ • $U_y = xz + \phi_y(y, z)$ equals

$xz \iff \phi_y(y, z) = 0$ • $\phi(y, z) = \psi(z)$ and $U(x, y, z) = xyz + \psi(z)$ • $U_z = xy + \psi'(z)$ equals

$xy \iff \psi'(z) = 0$ • $\psi(z) = C$ • Take $C = 0$. • $U(x, y, z) = xyz$

(b) $\int_C (yz\, dx + xz\, dy + xy\, dz) = U(2, 2, 2) - U(1, 1, 1) = 8 - 1 = 7$

Problems 13.3

1. (a) $\Sigma: z = xy^2,\ 0 \leq x \leq 2,\ 0 \leq y \leq 2$ • $z_x = y^2$, $z_y = 2xy$ • $\mathbf{n} = \dfrac{\langle -y^2, -2xy, 1 \rangle}{\sqrt{y^4 + 4x^2y^2 + 1}}$

(b) $dS = \sqrt{y^4 + 4x^2y^2 + 1}\, dx\, dy$ (c) $\iint_\Sigma xyz\, dS = \int_{x=0}^{x=2} \int_{y=0}^{y=2} xy(xy^2)\sqrt{y^4 + 4x^2y^2 + 1}\, dy\, dx$

2. (a) $\mathbf{n} = \dfrac{\langle 0, -\cos y, 1 \rangle}{\sqrt{\cos^2 y + 1}}$ (b) $dS = \sqrt{\cos^2 y + 1}\, dx\, dy$

(c) $\iint_\Sigma xz^2\, dS = \int_{x=0}^{x=1} \int_{y=0}^{y=x^2} x\sin^2 y \sqrt{\cos^2 y + 1}\, dy\, dx$

4. $\Sigma: z = y$, $0 \le y \le 1-x$, $0 \le x \le 1$ • $z_x = 0$, $z_y = 1$ • $dS = \sqrt{2}\, dx\, dy$ • $\iint_\Sigma (4x + 2y + z)\, dS$

$= \sqrt{2} \iint_D (4x+2y+y)\, dy\, dx = \sqrt{2} \int_{x=0}^{x=1} \int_{y=0}^{y=1-x} (4x+3y)\, dy\, dx = \sqrt{2} \int_{x=0}^{x=1} \left[4xy + \tfrac{3}{2}y^2\right]_{y=0}^{y=1-x} dx$

$= \sqrt{2} \int_{x=0}^{x=1} [4x(1-x) + \tfrac{3}{2}(1-x)^2]\, dx = \sqrt{2} \int_{x=0}^{x=1} (\tfrac{3}{2} + x - \tfrac{5}{2}x^2)\, dx = \sqrt{2}(\tfrac{3}{2} + \tfrac{1}{2} - \tfrac{5}{6}) = \tfrac{7}{6}\sqrt{2}$

5. $\tfrac{1}{6}(27 - 5^{3/2})$

7. 2π

9. $\Sigma: z = x+y$, $0 \le x \le 2$, $0 \le y \le 3$ • $z_x = 1$, $z_y = 1$ • $\mathbf{n} = \dfrac{\langle -1, -1, 1\rangle}{\sqrt{3}}$ and $dS = \sqrt{3}\, dx\, dy$ •

$[\text{Flux}] = \iint_\Sigma \mathbf{v} \cdot \mathbf{n}\, dS = \iint_\Sigma \dfrac{-xy - yz + xz}{\sqrt{3}}\, dS = \iint_R [-xy - y(x+y) + x(x+y)]\, dx\, dy$

$= \int_{x=0}^{x=2} \int_{y=0}^{y=3} (-xy - y^2 + x^2)\, dy\, dx = \int_{x=0}^{x=2} \left[-\tfrac{1}{2}xy^2 - \tfrac{1}{3}y^3 + x^2 y\right]_0^3 dx$

$= \int_{x=0}^{x=2} (-\tfrac{9}{2}x - 9 + 3x^2)\, dx = -19$

11. $\Sigma: z = x - 2y$, $-1 \le x \le 0$, $0 \le y \le 1$ • $z_x = 1, z_y = -2$ • $dS = \sqrt{1^2 + 2^2 + 1}\, dx\, dy$
$= \sqrt{6}\, dx\, dy$ •

$[\text{Area of } \Sigma] = \iint_\Sigma dS = \int_{x=-1}^{x=0} \int_{y=0}^{y=1} \sqrt{6}\, dy\, dx = \sqrt{6}$ • $[\text{Average value}] = \dfrac{1}{\sqrt{6}} \iint_\Sigma xyz\, dS$

$= \int_{x=-1}^{x=0} \int_{y=0}^{y=1} xy(x - 2y)\, dy\, dx = \int_{x=-1}^{x=0} \left[\tfrac{1}{2}x^2 y^2 - \tfrac{2}{3}xy^3\right]_{y=0}^{y=1} dx = \int_{x=-1}^{x=0} (\tfrac{1}{2}x^2 - \tfrac{2}{3}x)\, dx$

$= \left[\tfrac{1}{6}x^3 - \tfrac{1}{3}x^2\right]_{-1}^{0} = \tfrac{1}{2}$

12. $[\text{Average value}] = \tfrac{1}{6}$

14. $n_3\, dS = dx\, dy$ on the top of the box, $n_3\, dS = -dx\, dy$ on the bottom, and $n_3 = 0$ on the sides.
• $f = xy^2 z^3$ equals 0 on the top where $z = 0$ and equals $-xy^2$ on the bottom where $z = -1$ •

$\iint_\Sigma f n_3\, dS = \iint_{\text{Bottom}} f n_3\, dS = \int_{x=0}^{x=3} x\, dx \int_{y=0}^{y=2} y^2\, dy = (\tfrac{9}{2})(\tfrac{8}{3}) = 12$ •

$\iiint_V \dfrac{\partial f}{\partial z}\, dx\, dy\, dz = \iiint_V 3xy^2 z^2\, dx\, dy\, dz = 3 \int_{x=0}^{x=3} x\, dx \int_{y=0}^{y=2} y^2\, dy \int_{z=-1}^{z=0} z^2\, dz = 12$

15. Both integrals equal $[1 - \cos(1)]\sin(2)\sin(3)$.

17. 0

19. (a) $\text{div}\,\mathbf{A} = \dfrac{\partial}{\partial x}[\sin(xy)] + \dfrac{\partial}{\partial y}[\cos(yz)] + \dfrac{\partial}{\partial z}[\tan(z^2)] = y\cos(xy) - z\sin(yz) + 2z\sec^2(z^2)$

(b) $\text{curl}\,\mathbf{A} = \begin{vmatrix} \mathbf{i} & \mathbf{j} & \mathbf{k} \\ \dfrac{\partial}{\partial x} & \dfrac{\partial}{\partial y} & \dfrac{\partial}{\partial z} \\ \sin(xy) & \cos(yz) & \tan(z^2) \end{vmatrix}$

$= \mathbf{i}\{\dfrac{\partial}{\partial y}[\tan(z^2)] - \dfrac{\partial}{\partial z}[\cos(yz)]\} - \mathbf{j}\{\dfrac{\partial}{\partial x}[\tan(z^2)] - \dfrac{\partial}{\partial z}[\sin(xy))]\} + \mathbf{k}\{\dfrac{\partial}{\partial x}[\cos(yz)] - \dfrac{\partial}{\partial y}[\sin(xy)]\}$
$= \langle y\sin(yz), 0, -x\cos(xy)\rangle$

20. (a) $\text{div}\,\mathbf{B} = 2xy^3 + 3y^2 - 1$ (b) $\text{curl}\,\mathbf{B} = \langle 1 + 3z^2, -1, -3x^2 y^2\rangle$

21. (a) $\text{div}\,\mathbf{C} = e^y - e^z + e^x$ (b) $\text{curl}\,\mathbf{C} = \langle ye^z, -ze^x, -xe^y\rangle$

24. (a) $\text{div}\,\mathbf{F} = \dfrac{1}{x^2 + y^2 + z^2}$ (b) $\text{curl}\,\mathbf{F} = \langle 0, 0, 0\rangle$

Answers 13.3

27. $\text{div}\,\mathbf{v} = \dfrac{\partial}{\partial x}(2 - x\cos z) + \dfrac{\partial}{\partial y}(y^2 z \sin x) + \dfrac{\partial}{\partial z}(\sin z) = -\cos z + 2yz\sin x + \cos z = 2yz\sin x$ •

$[\text{Flux}] = \displaystyle\iint_\Sigma \mathbf{v}\cdot\mathbf{n}\,dS = \iiint_V \text{div}\,\mathbf{v}\,dx\,dy\,dz = \int_{x=0}^{x=1}\int_{y=0}^{y=1}\int_{z=0}^{z=y^2} 2yz\sin x\,dz\,dy\,dx$

$= \displaystyle\int_{x=0}^{x=1}\sin x\,dx \int_{y=0}^{y=1} y\Big[z^2\Big]_{z=0}^{z=y^2}dy = \Big[-\cos x\Big]_{x=0}^{x=1}\int_{y=0}^{y=1} y^5\,dy = \tfrac{1}{6}[1-\cos(1)]$

28. $3(e^3 - e^2)$

31. $\Sigma: z = xy^2,\ 0 \le x \le 1,\ 0 \le y \le 1$ • $z_x = y^2,\ z_y = 2xy$ •

$\mathbf{n} = \dfrac{\langle -y^2, -2xy, 1\rangle}{\sqrt{y^4 + 4x^2y^2 + 1}}$ • $dS = \sqrt{y^4 + 4x^2y^2 + 1}\,dx\,dy$ • $\mathbf{v} = \langle xy^2, -z^2 y, z^5\rangle$ •

$\text{curl}\,\mathbf{v} = [\dfrac{\partial}{\partial y}(z^5) - \dfrac{\partial}{\partial z}(-z^2 y)]\mathbf{i} - [\dfrac{\partial}{\partial x}(z^5) - \dfrac{\partial}{\partial z}(xy^2)]\mathbf{j} + [\dfrac{\partial}{\partial x}(-z^2 y) - \dfrac{\partial}{\partial y}(xy^2)]\mathbf{k} = \langle 2yz, 0, -2xy\rangle$

• $(\text{curl}\,\mathbf{v})\cdot\mathbf{n}\,dS = (-2xy^5 - 2xy)\,dx\,dy$ • $[\text{Circulation}] = \displaystyle\int_C \mathbf{v}\cdot\mathbf{T}\,ds = \iint_\Sigma (\text{curl}\,\mathbf{v})\cdot\mathbf{n}\,dS$

$= \displaystyle\int_{x=0}^{x=1}\int_{y=0}^{y=1}(-2xy^5 - 2xy)\,dy\,dx = -\tfrac{4}{3}\int_{x=0}^{x=1} x\,dx = -\tfrac{2}{3}$

32. $\mathbf{F} = \langle x, y^2 z, -xy^2\rangle$ • $\text{curl}\,\mathbf{F} = \begin{vmatrix} \mathbf{i} & \mathbf{j} & \mathbf{k} \\ \dfrac{\partial}{\partial x} & \dfrac{\partial}{\partial y} & \dfrac{\partial}{\partial z} \\ x & y^2 z & -xy^2 \end{vmatrix}$

$= [\dfrac{\partial}{\partial y}(-xy^2) - \dfrac{\partial}{\partial z}(y^2 z)]\mathbf{i} - [\dfrac{\partial}{\partial x}(-xy^2) - \dfrac{\partial}{\partial z}(x)]\mathbf{j} + [\dfrac{\partial}{\partial x}(y^2 z) - \dfrac{\partial}{\partial y}(x)]\mathbf{k} = \langle -2xy - y^2, y^2, 0\rangle$ •

$\Sigma: z = x + 2y,\ -1 \le x \le 0,\ 0 \le y \le 1$ • $z_x = 1,\ z_y = 2$ • $\mathbf{n} = \dfrac{\langle -1, -2, 1\rangle}{\sqrt{6}}$ • $dS = \sqrt{6}\,dx\,dy$ •

$(\text{curl}\,\mathbf{v})\cdot\mathbf{n}\,dS = (2xy - y^2)\,dx\,dy$ • $[\text{Work}] = \displaystyle\int_C \mathbf{F}\cdot\mathbf{n}\,ds$

$= \displaystyle\iint_\Sigma (\text{curl}\,\mathbf{v})\cdot\mathbf{n}\,dS = \int_{-1}^0 \int_0^1 (2xy - y^2)\,dy\,dx = \int_{-1}^0 (x - \tfrac{1}{3})\,dx = -\tfrac{5}{6}$ newton-meters (joules)

34. $\langle p, q, r\rangle = \langle yz^2 - 3, xz^2 + 2y, 2xyz + 3z^2\rangle$ • $p_y = z^2$ and $q_x = z^2$ are equal for all (x, y, z). • $p_z = 2yz$ and $r_x = 2yz$ are equal for all (x, y, z). • $q_z = 2xz$ and $r_y = 2xz$ are equal for all (x, y, z). • xyz-space is simply connected. • The field is a gradient field. • $U_x = yz^2 - 3 \implies U = xyz^2 - 3x + \phi(y, z)$ • $U_y = xz^2 + \phi_y(y, z)$ • $U_y = xz^2 + 2y \implies \phi_y = 2y$ • $\phi = y^2 + \psi(z)$ and $U = xyz^2 - 3x + y^2 + \psi(z)$ • $U_z = 2xyz + \psi'(z)$ • $U_z = 2xyz + 3z^2 \implies \psi'(z) = 3z^2$ • $\psi(z) = z^3 + C$ • $U = xyz^2 - 3x + y^2 + z^3 + C$

35. $\langle p, q, r\rangle = \langle y^3, 3xy^2 + z, -y\rangle$ • $p_y = 3y^2$ and $q_x = 3y^2$ are equal for all (x, y, z). • $p_z = 0$ and $r_x = 0$ are equal for all (x, y, z). • $q_z = 1$ and $r_y = -1$ are not equal. • The field is not a gradient field.

36. $U(x, y, z) = x^2 y^3 + xz^5 + C$.

37. Not a gradient field

40. For curves C from $(1, 1, 1)$ to $(10, 10, 10)$, $\displaystyle\int_C (\nabla \sin(x + y + z))\cdot \mathbf{T}\,ds = \Big[\sin(x + y + z)\Big]_{(1,1,1)}^{(10,10,10)}$
$= \sin(30) - \sin(3)$

42. $\dfrac{\partial}{\partial y}(e^{yz}) = ze^{yz}$ equals $\dfrac{\partial}{\partial x}(xze^{yz}) = ze^{yz}$, $\dfrac{\partial}{\partial z}(e^{yz}) = ye^{yz}$ equals $\dfrac{\partial}{\partial x}(xye^{yz}) = ye^{yz}$, $\dfrac{\partial}{\partial z}(xze^{yz}) = (x + xyz)e^{yz}$ equals $\dfrac{\partial}{\partial y}(xye^{yz}) = (x + xyz)e^{yz}$ for all (x, y, z), and xyz-space is simply connected. • The integrals equal $2e^{12}$.

43. 180

45. $\Sigma: z = xy$ for $0 \leq x \leq 1$, $0 \leq y \leq 1$ • $z_x = y, z_y = x$ • $dS = \sqrt{1 + x^2 + y^2}\, dx\, dy$ •
$$\iint 4z\sqrt{1 + x^2 + y^2}\, dS = \int_{x=0}^{x=1} \int_{y=0}^{y=1} 4xy(1 + x^2 + y^2)\, dy\, dx$$
$$= \int_{x=0}^{x=1} \left[2xy^2 + 2x^3y^2 + xy^4\right]_{y=0}^{y=1} dx = \int_{x=0}^{x=1} (3x + 2x^3)\, dx = 2$$

46. $\dfrac{5^{9/2} - 1}{3(5^{3/2} - 1)}$

49. $\Sigma: z = 3x + y^2, 0 \leq x \leq 1, 0 \leq y \leq 1$ • $z_x = 3, z_y = 2y$ • $dS = \sqrt{10 + 4y^2}\, dx\, dy$ •
On Σ, $\rho = \sqrt{10 + 4y^2}$ and $\rho\, dS = (10 + 4y^2)\, dx\, dy$ •
$$[\text{Weight}] = \iint_\Sigma \rho\, dS = \int_{x=0}^{x=1} \int_{y=0}^{y=1} (10 + 4y^2)\, dy\, dx = \tfrac{34}{3}$$
$$\bar{x} = \frac{1}{[\text{Weight}]} \iint_\Sigma x\rho\, dS = \tfrac{3}{34} \int_{x=0}^{x=1} \int_{y=0}^{y=1} x(10 + 4y^2)\, dy\, dx = \tfrac{1}{2}$$
$$\bar{y} = \frac{1}{[\text{Weight}]} \iint_\Sigma y\rho\, dS = \tfrac{3}{34} \int_{x=0}^{x=1} \int_{y=0}^{y=1} y(10 + 4y^2)\, dy\, dx = \tfrac{9}{17}$$
$$\bar{z} = \frac{1}{[\text{Weight}]} \iint_\Sigma z\rho\, dS = \tfrac{3}{34} \int_{x=0}^{x=1} \int_{y=0}^{y=1} (3x + y^2)(10 + 4y^2)\, dy\, dx$$
$$= \tfrac{3}{34} \int_{x=0}^{x=1} \int_{y=0}^{y=1} (30x + 10y^2 + 12xy^2 + 4y^4)\, dy\, dx = \tfrac{317}{170}$$

50. $\bar{x} = \tfrac{13}{22}$, $\bar{y} = \tfrac{13}{22}$ $\bar{z} = \tfrac{142}{165}$

54. 0

58. $\tfrac{1}{560}(1094 - 5^{7/2})$

59. $\tfrac{2}{3}(2^{3/2} - 1)$

65. (a) $f_{xx} + f_{yy} + f_{zz}$
(c) **curl**(div **F**) is not defined because div **F** is not a vector.

CHAPTER 14
FURTHER TOPICS IN DIFFERENTIAL EQUATIONS

In Section 7.1 we studied separable, first-order, linear differential equations. In the first section of this chapter we discuss GENERAL FIRST-ORDER LINEAR and FIRST-ORDER EXACT differential equations. Section 14.2 deals with SECOND-ORDER LINEAR differential equations with constant coefficients and their applications to the study of vibrating springs and simple electric circuits. Section 14.3 contains an introduction to the techniques of finding power-series solutions of linear differential equations.

Section 14.1

First-order linear and exact differential equations

OVERVIEW: *In this section we will see how the Product Rule for derivatives can be employed to solve first-order linear differential equations and how the results from Section 13.2 concerning gradient vector fields can be applied to study first-order exact differential equations.*

Topics:

- **First-order linear differential equations**
- **A mixing problem**
- **First-order exact differential equations**

First-order linear differential equations

A first-order differential equation is LINEAR if it is of the form

$$a(x)\frac{dy}{dx} + b(x)y = f(x) \tag{1}$$

where $a(x), b(x)$, and $f(x)$ are given functions and $y = y(x)$ is the unknown function. Before we discuss the general procedure for solving such equations, we will see in Example 1 and Question 2 below, how in certain cases they can be solved by direct applications of the Product Rule.

Example 1   Use the Product Rule to find the solution of the initial-value problem

$$x^2\frac{dy}{dx} + 2xy = 3x^2, \ y(1) = 2. \tag{2}$$

SOLUTION   We can use the Product Rule here because the coefficent $2x$ of y in (2) is the derivative of the coefficient x^2 of dy/dx. Consequently

$$\frac{d}{dx}(x^2 y) = x^2\frac{dy}{dx} + y\frac{d}{dx}(x^2) = x^2\frac{dy}{dx} + 2xy$$

and the left side of (2) can be written in the form

$$\frac{d}{dx}(x^2 y) = 3x^2.$$

Taking the x-antiderivative of both sides of this equation yields $x^2 y = \int 3x^2 \ dx$ and then

$$x^2 y = x^3 + C. \tag{3}$$

♦ SUGGESTIONS TO INSTRUCTORS: This section contains two independent topics, which can be covered in two or more lectures.
Part 1 (First-order linear differential equations; Tune-up Exercises T1–T4, Problems 1–12, 27–29 32–43) Have students begin work on Questions 1–6 before class. Problems 2, 9, are good for in-class discussion.
Part 2 (First-order exact differential equations; Tune-up Exercises T5–T6, Problems 13–26, 39, 31) Have students begin work on Question 7 before class. Problems 13, 22 are good for in-class discussion.

The constant of integration C in **(3)** is determined by the initial value $y(1) = 2$ in **(2)**. Setting $x = 1$ and $y = 2$ in **(3)** gives $(1^2)(2) = 1^3 + C$, so that $C = 1$ and $x^2 y = x^3 + 1$. Finally, dividing by x^2 gives the solution

$$y = x + x^{-2}. \quad \Box \tag{4}$$

Question 1 Verify that $y = x + x^{-2}$, given in **(4)**, is a solution of the initial-value problem **(2)**
(a) by finding $x^2 \dfrac{dy}{dx} + 2xy$ and (b) by determining the value $y(1)$.

Question 2 Use the Product Rule to find all solutions of the differential equation

$$e^{x^2} \frac{dy}{dx} + 2x e^{x^2} y = \cos x. \tag{5}$$

We could solve **(2)** and **(5)** by direct applications of the Product Rule because of their particular forms. We might, however, be given **(2)** in the form

$$x \frac{dy}{dx} + 2y = 3x \tag{6}$$

which is obtained by dividing both sides of **(2)** by x. Or, we might be given **(5)** as

$$\frac{dy}{dx} + 2x\, y = e^{-x^2} \cos x \tag{7}$$

which comes from dividing both sides of **(5)** by e^{x^2}. In these cases we would have to multiply both sides of the differential equation by the appropriate INTEGRATING FACTOR to put it in a form to which the Product Rule can be applied. The essential properties of integrating factors are explored in the next Question.

Question 3 Suppose that $P(x)$ is continuous in an open interval. Define

$$I(x) = e^{\int P(x)\, dx} \tag{8}$$

for x in the interval, where $\displaystyle\int P(x)\, dx$ denotes one anti-derivative of $P(x)$.
(a) Show that for x in the interval,

$$\frac{dI}{dx}(x) = I(x) P(x). \tag{9}$$

(b) Use **(9)** to show that for any differentiable function $y = y(x)$ in the interval,

$$\frac{d}{dx}[I(x) y(x)] = I(x) \frac{dy}{dx}(x) + I(x) P(x) y(x). \tag{10}$$

As is explained in the next Rule, formula **(8)** is used to construct integrating factors for general linear, first-order equations. It is property **(10)** that makes the integrating factors work.

14.1 First-order linear and exact differential equations

Rule 1 To solve a first-order linear differential equation **(1)**, first divide by the coefficient $a(x)$ of $\dfrac{dy}{dx}$, if necessary, to put the differential equation in the form

$$\frac{dy}{dx}(x) + P(x)y(x) = Q(x). \tag{11}$$

Next, find an antiderivative $\int P(x)\,dx$ of the function $P(x)$, form the integrating factor **(8)**, and multiply both sides of **(11)** by it to obtain.

$$I(x)\frac{dy}{dx}(x) + I(x)P(x)y(x) = I(x)Q(x).$$

Then, because of **(10)**, the differential equation can be written in the form

$$\frac{d}{dx}[I(x)y(x)] = I(x)Q(x) \tag{12}$$

which can be solved by taking antiderivatives of both sides.

Example 2 Apply the procedure in Rule 1 to the initial-value problem

$$x\frac{dy}{dx} + 2y = 3x,\ y(1) = 2. \tag{13}$$

SOLUTION   Dividing **(13)** by the coefficient x of $\dfrac{dy}{dx}$ gives

$$\frac{dy}{dx} + \frac{2}{x}y = 3 \tag{14}$$

which is of form **(11)** with $P(x) = 2/x$. We consider only $x > 0$, for which

$$\int P(x)\,dx = \int \frac{2}{x}\,dx = 2\ln x + C. \tag{15}$$

We need only one integrating factor, so we set $C = 0$ in **(15)** and define

$$I(x) = e^{\int P(x)\,dx} = e^{2\ln x} = (e^{\ln x})^2 = x^2.$$

Multiplying both sides of **(14)** by this integrating factor yields

$$x^2\frac{dy}{dx} + 2xy = 3x^2$$

which is the form **(2)** that we solved in Example 1 by using the Product Rule, so the solution of **(13)** is the solution $y = x + x^{-2}$ of **(2)** that we found in Example 1. □

Question 4 Find the integrating factor for

$$\frac{dy}{dx} + 2xy = e^{-x^2}\cos x. \tag{16}$$

Example 3 Find all solutions of **(16)**.

SOLUTION Multiplying both sides of **(16)** by the integrating factor e^{x^2} from Question 4 yields

$$e^{x^2}\frac{dy}{dx} + 2xe^{x^2}y = \cos x. \tag{17}$$

By the construction of the integrating factor, the left side of **(17)** should be the derivative $\frac{d}{dx}(e^{x^2}y)$ of the product of the integrating factor and the unknown function. This is the case because, by the Product and Chain Rules,

$$\frac{d}{dx}(e^{x^2}y) = e^{x^2}\frac{dy}{dx} + y\frac{d}{dx}(e^{x^2})$$

$$= e^{x^2}\frac{dy}{dx} + ye^{x^2}\frac{d}{dx}(x^2)$$

$$= e^{x^2}\frac{dy}{dx} + 2xe^{x^2}y.$$

Consequently, **(17)** can be written

$$\frac{d}{dx}(e^{x^2}y) = \cos x.$$

As you saw in answering Question 2, taking antiderivatives of both sides of this equation gives $e^{x^2}y = \sin x + C$ and then $y = (\sin x + C)e^{-x^2}$. □

A mixing problem

Imagine that a large tank contains 10 gallons of a sugar solution with 5 pounds of sugar dissolved in it at time $t = 0$ (minutes). A solution containing 4 pounds of sugar per gallon is then added at the rate of 3 gallons per minute, the solutions mix instantly, and the mixture is drained off at the rate of 2 gallons per minute. Let $y(t)$ (pounds) be the amount of sugar in the tank at time $t \geq 0$.

Question 5 (a) What is the volume of solution at time $t \geq 0$? (b) What is $y(0)$?

Example 4 Give an initial-value problem for the amount $y(t)$ of sugar in the tank.

SOLUTION The rate (pounds per minute) at which sugar is added is equal to the concentration (pounds per gallon) of the solution that is being added, multiplied by the rate (gallons per minute) at which the solution is added. Similarly, the rate at which sugar is removed is equal to the concentration of the mixture in the tank, multiplied by the rate at which it is removed.

The concentration of the solution being added is given as 4 pounds per gallon and the concentration of the mixture at time t is equal to the weight $y(t)$ of sugar in the tank, divided by the volume $10+t$ of the solution that you found in Question 5a. The solution is added at the rate of 3 gallons per minute and the mixture drained off at the rate of 2 gallons per minute. Therefore,

$$\frac{dy}{dt}(t) = [\text{Rate in}] - [\text{Rate out}]$$

$$= \left[4\,\frac{\text{pounds}}{\text{gallon}}\right]\left[3\,\frac{\text{gallons}}{\text{minute}}\right] - \left[\frac{y(t)}{10+t}\,\frac{\text{pounds}}{\text{gallon}}\right]\left[2\,\frac{\text{gallons}}{\text{minute}}\right]$$

$$= 12 - \frac{2y(t)}{10+t}\,\frac{\text{pounds}}{\text{minute}}.$$

We put this equation in the standard form of a linear differential equation and add the initial condition from Question 5b to obtain the initial-value problem

$$\frac{dy}{dt} + \left(\frac{2}{10+t}\right)y = 12,\ y(0) = 5.\ \square \tag{18}$$

14.1 First-order linear and exact differential equations

Example 5 Solve the initial-value problem (18).

SOLUTION In equation (18), $P(t) = \dfrac{2}{10+t}$ and

$$\int P(t)\,dt = \int \frac{2}{10+t}\,dt = \int \frac{2}{u}\,du = 2\ln|u| + C = 2\ln|10+t| + C.$$

We used the substitution $u = 10 + t, du = dt$ to carry out this integration. We can take $C = 0$ and can replace the absolute value signs by parentheses because $t \geq 0$. This gives the integrating factor

$$I(t) = e^{\int P(t)\,dt} = e^{2\ln(10+t)} = \left[e^{\ln(10+t)}\right]^2 = (10+t)^2.$$

Multiplying the differential equation in (18) by this integrating factor gives

$$(10+t)^2 \frac{dy}{dt} + 2(10+t)y = 12(10+t)^2. \tag{19}$$

The left side of this equation should be the derivative $\dfrac{d}{dt}[(10+t)^2 y]$ of the product of the integrating factor and the unknown function. This is the case, as can be seen with the Product Rule. Consequently, (19) can be written

$$\frac{d}{dt}[(10+t)^2 y] = 12(10+t)^2. \tag{20}$$

We take the t-antiderivatives of both sides of (20) to obtain

$$(10+t)^2 y = 12 \int (10+t)^2\,dt. \tag{21}$$

With the substitution $u = 10+t, du = dt$ in the integral, we have

$$12 \int (10+t)^2\,dt = 12 \int u^2\,du = 4u^3 + C = 4(10+t)^3 + C$$

and (21) becomes

$$(10+t)^2 y = 4(10+t)^3 + C.$$

To use the initial condition $y(0) = 5$, we set $t = 0$ and $y = 5$ in the last equation and obtain $10^2(5) = 4(10^3) + C$, so that $C = 500 - 4000 = -3500$ and

$$(10+t)^2 y = 4(10+t)^3 - 3500.$$

Dividing by $(10+t)^2$ gives our final formula $y(t) = 4(10+t) - 3500(10+t)^{-2}$ (pounds) for the amount of sugar in the tank. □

Question 6 (a) What is the sugar concentration $C(t)$ in the tank at time $t \geq 0$? (b) What is the limit of the sugar concentration in the tank as $t \to \infty$ and why is this plausible?

Exact, first-order differential equations

Recall from Section 7.1 that differentials dx and dy to a curve C at a point where it has a tangent line can be defined as corresponding changes in x and y along the tangent line (Figure 1). Recall also that a curve C is a solution of a differential equation in the form

$$p(x,y)\,dx + q(x,y)\,dy = 0 \tag{22}$$

if the equation is satisfied by the differentials at each point (x,y) on it, and that a function $y = y(x)$ is a solution if the equation is satisfied by the differentials at each point on its graph.

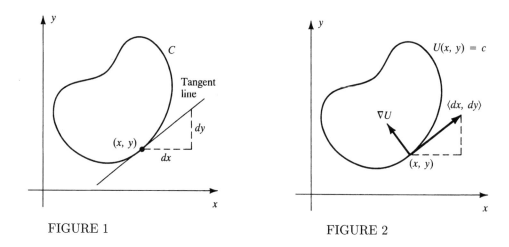

FIGURE 1   FIGURE 2

We are interested here in the case where the curve C is a level curve $U(x,y) = c$ of a function $U(x,y)$ with continuous first-order derivatives and nonzero gradient vector. Recall from Section 11.3 (Theorem 5) that the gradient vector $\nabla U = \langle U_x, U_y \rangle$ is perpendicular to the level curve (Figure 2), and, since the vector $\langle dx, dy \rangle$ is tangent to the level curve, the dot product $\langle U_x, U_y \rangle \cdot \langle dx, dy \rangle = U_x\,dx + U_y\,dy$ is zero. Therefore, the curve is a solution of the differential equation

$$U_x(x,y)\,dx + U_y(x,y)\,dy = 0.$$

This establishes the following theorem:

Theorem 1 *The level curves $U(x,y) = c$ of a function $U(x,y)$ with continuous first-order derivatives and functions $y = y(x)$ defined implicitly by the equations $U(x,y) = c$ are solutions of the differential equation*

$$U_x(x,y)\,dx + U_y(x,y)\,dy = 0 \tag{23}$$

at any point where the gradient $\nabla U = \langle U_x, U_y \rangle$ of U is not zero.

To apply this Theorem to solve a differential equation $p(x,y)\,dx + q(x,y)\,dy = 0$ we need to find a function $U(x,y)$ such that $U_x = p$ and $U_y = q$. If this is possible, the differential equation is called EXACT:

Definition 1 *The differential equation $p(x,y)\,dx + q(x,y)\,dy = 0$ is EXACT in an open region if there is a function $U(x,y)$ such that $p(x,y) = U_x(x,y)$ and $q(x,y) = U_y(x,y)$ for (x,y) in the region.*

14.1 First-order linear and exact differential equations

If the differential equation (22) is exact so that $p = U_x$ and $q = U_y$ for some function U, and if p and q have continuous first-order derivatives, then $\dfrac{\partial p}{\partial y}$ and $\dfrac{\partial q}{\partial x}$ are both equal to the mixed second derivative $\dfrac{\partial^2 U}{\partial x \partial y}$ of U and consequently are equal. This gives part (a) of the next Theorem.

Theorem 2 (a) *Suppose that $p(x,y)$ and $q(x,y)$ have continuous first-order derivatives in an open region R in the xy-plane and that the differential equation $p(x,y)\,dx + q(x,y)\,dy = 0$ is exact. Then*

$$\frac{\partial p}{\partial y}(x,y) = \frac{\partial q}{\partial x}(x,y) \text{ in } R. \tag{24}$$

(b) *If condition (24) holds and R is simply connected†, then the differential equation is exact.*

Part (b) of the Theorem was established in Section 13.3 (Theorem 5), where we studied path-independent line integrals. In that section we referred to the function $U(x,y)$ in Definition 1 as a potential function for the vector field $\langle p, q \rangle$. The technique for finding functions $U(x,y)$ to solve exact differential equations is the same as we used in Section 13.2 to find potential functions of vector fields. It is illustrated in the next Example.

Example 6 Is the differential equation

$$(2xy^2 - 1)\,dx + (2x^2 y + 2y)\,dy = 0 \tag{25}$$

exact? If so, find the solution curve that passes through the point $(1, 2)$.

SOLUTION Set $p(x,y) = 2xy^2 - 1$ and $q(x,y) = 2x^2 y + 2y$. Then $p_y = 4xy$ and $q_x = 4xy$ are equal for all (x,y) in the simply connected xy-plane, so the differential equation is exact for all (x,y).

The function $U(x,y)$ whose level curves give the solutions of the differential equation must satisfy

$$U_x = 2xy^2 - 1, \text{ and } U_y = 2x^2 y + 2y. \tag{26}$$

The first of equations (26) implies that

$$U(x,y) = \int (2xy^2 - 1)\,dx = x^2 y^2 - x + \phi(y) \tag{27}$$

where $\phi(y)$ is the constant of integration, which can have different values for different values of y. Taking the y-derivative of both sides of (27) gives $U_y = 2x^2 y + \phi'(y)$ and then to satisfy the second of equations (26), we take $\phi'(y) = 2y$, for which $\phi(y) = y^2 + C$. We set $C = 0$ (because we need only one function U) and substitute the result in (27) to obtain $U(x,y) = x^2 y^2 - x + y^2$. By Theorem 1, the solution curves of (25) are the level curves

$$x^2 y^2 - x + y^2 = c \tag{28}$$

of this function U. To satisfy the initial condition that the curve pass through the point $(1, 2)$, we set $x = 1$ and $y = 2$ in (28). We conclude that $c = 7$ and the curve is $x^2 y^2 - x + y^2 = 7$. □

Question 7 Find the function $y = y(x)$ that satisfies the differential equation (25) of Example 6 with the initial condition $y(1) = 2$.

---

†Recall that a region in the xy-plane is simply conncected if there are no holes in it (see Section 13.2).

Principles and procedures

- As we saw in Section 7.1, a first-order linear differential equation is separable if it can be written in the form $\frac{dy}{dx}(x) = r(x)y(x)$. Writing this equation as $\frac{dy}{dx}(x) - r(x)y(x) = 0$ shows that it has the integrating factor $I(x) = e^{-R(x)}$, where $R(x) = \int r(x)\,dx$ is an antiderivative of $r(x)$. Multiplying both sides of the equation by the integrating factor then gives $\frac{d}{dx}[y(x)e^{-R(x)}] = 0$ so that $y(x)e^{-R(x)} = C$ with a constant C, and solving for $y(x)$ gives the formula $y(x) = Ce^{R(x)}$ for solutions from Rule 1 of Section 7.1.

- If you find a suitable integrating factor $I(x)$ for $\frac{dy}{dx}(x) + P(x)y(x) = Q(x)$, then the left side of the equation $I(x)\frac{dy}{dx}(x) + I(x)P(x)y(x) = I(x)Q(x)$ should be the derivative $\frac{d}{dx}[I(x)y(x)]$ of the product of the integrating factor and the unknown function. In each case, use the Product Rule to find this derivative to see whether your formula for $I(x)$ is correct.

- If $P(x)$ in $\frac{dy}{dx}(x) + P(x)y(x) = Q(x)$ is of the form $P(x) = \frac{kg'(x)}{g(x)}$ with a positive function $g(x)$ and a nonzero constant k, then $\int P(x)\,dx = k\ln[g(x)] + C$ and the exponential function and logarithm can be eliminated in the integrating factor $I(x) = e^{k\ln[g(x)]} = \{e^{\ln[g(x)]}\}^k = [g(x)]^k$. If $g(x)$ is negative, this calculation gives the integrating factor $|g(x)|^k$ with an absolute value, but if k is an integer or a fraction with an odd denominator, you can use $[g(x)]^k$ instead.

- To remember the condition $p_y = q_x$ for an exact differential equation, recall that it must hold if $p = U_x$ and $q = U_y$ because then p_y and q_x both equal the mixed second derivative U_{xy} of U.

- You need to review Section 13.3 to understand the proof of Theorem 1 on exact differential equations, but you should be able to learn how to apply it from Example 6 and the outlines of solutions of Tune-up Exercises and Problems in this section.

- If you find that the level curves $U(x, y) = c$ are solution curves of an exact differential equation, $p(x, y)\,dx + q(x, y)\,dy = 0$, you can check your work by verifying that $U_x(x, y) = p(x, y)$ and $U_y(x, y) = q(x, y)$.

Tune-Up Exercises 14.1♦

A Answer provided. **O** Outline of solution provided.

T1.^O Use the techniques of this section to find all solutions of $\frac{dy}{dx} - y = 2$.

♦ Type 1, solving a separable linear equation with an integrating factor

T2.^O Find all solutions of the differential equation of Exercise T1 by rewriting it in the form $\frac{d}{dx}(y+2) = y+2$ and using Rule 1 of Section 7.1.

♦ Type 1, solving a separable linear equation directly

T3.^O Find all solutions of $\frac{dy}{dx} - (\sin x)\,y = 5\sin x$.

♦ Type 1, a linear equation

T4.^O Solve the initial-value problem $x\frac{dy}{dx} + 5y = 9x^4$, $y(1) = 3$.

♦ Type 2, a linear equation

♦The Tune-up Exercises and Problems are classified by type and content. The types are (1) basic, reactive; (2) basic reflective; (3) intermediate, reactive; (4) intermediate, reflective; (5) advanced, reactive; (6) advanced, reflective; and (7) advanced, theoretical.

14.1 First-order linear and exact differential equations

T5.^{O} Is $(y+1)\,dx + (x-1)\,dy = 0$ exact? If so, find its solution curves.
◆ Type 1, exact equations

T6.^{O} Is $(xy+x)\,dx + (y^2-xy)\,dy = 0$ exact? If so, find its solution curves.
◆ Type 1, exact equations

Problems 14.1
^{A}Answer provided. ^{O}Outline of solution provided.

Solve the differential equations and initial-value problems in Problems 1 through 8.

1.^{O} $\dfrac{dy}{dx} - x^{-2}y = 6x^{-2}$
◆ Type 1, a linear equation

2.^{O} $\dfrac{dy}{dx} - \dfrac{xy}{x^2+1} = (x^2+1)^{3/2}$
◆ Type 1, a linear equation

3.^{A} $\dfrac{dy}{dx} + \dfrac{y\sin x}{\cos x} = \cos^2 x$
◆ Type 1, a linear equation

4. $\dfrac{dy}{dx} - \dfrac{2y}{x+1} = (x+1)^2$
◆ Type 1, a linear equation

5.^{A} $\dfrac{dy}{dx} - 3x^2 y = 6x^2,\, y(0) = 0$
◆ Type 1, a linear equation

6. $\dfrac{dy}{dx} - \dfrac{y}{2x} = x^{3/2},\, y(1) = 2$
◆ Type 1, a linear equation

7.^{A} $\dfrac{dy}{dx} + \dfrac{10y}{2x+1} = 1,\, y(0) = 1$
◆ Type 1, a linear equation

8.^{A} $\dfrac{dy}{dx} - \dfrac{5y}{x} = x^3,\, y(1) = 0$
◆ Type 1, a linear equation

9.º At $t=0$ (minutes) a tank contains 8 gallons of a salt solution in which 5 pounds of salt are dissolved. A solution with one-half pound of salt per gallon is then added at the rate of 4 gallons per minute. The solutions mix instantly and the mixture is drained off at the rate of one gallon per minute. **(a)** How much salt is in the tank at time t? **(b)** What is the salt concentration in the tank at time t?
◆ Type 3, a mixing problem with a linear equation

10.^{A} A fifty-gallon tank contains a gallon of a sugar solution with a concentration of two pounds of sugar per gallon. Then a solution with three pounds of sugar per gallon is added at the rate of ten gallons per minute. The solutions mix instantly and the mixture is drained off at the rate of three gallons per minute. **(a)** When is the tank full? **(b)** Give a formula for the amount of sugar in the tank up to the time when it is full.
◆ Type 3, a mixing problem with a linear equation

11. A large tank contains 25 gallons of a brine solution with 48 pounds of salt dissolved in it. Brine containing 2 pounds of salt per gallon is then added at the rate of 5 gallons per hour, the solutions mix, and the mixture is drained off at the rate of 3 gallons per hour. How much salt is in the tank after 4 hours?
◆ Type 3, a mixing problem with a linear equation

12. A hundred gallons of a 20% solution of alcohol in water is in a tank. A 30% solution is added at the rate of five gallons per minute and the mixture is drained of at the rate of five gallons per minute. How long does it take until the mixture in the tank is 29% alcohol? Give the exact answer and its approximate decimal value.
◆ Type 3, a mixing problem with a linear equation

In problems 13 through 20, determine whether the differential equations are exact and find the solutions curves of those that are exact.

13.º $(2x-6y)\,dx + (4y^3 - 6x)\,dy = 0$
◆ Type 1, exact equations

14.^{A} $(2xe^y - e^{-x})\,dx + (x^2 e^y - 2y)\,dy = 0$
◆ Type 1, exact equations

15. $(5x^3 y^4 - 2y)\,dx + (3x^2 y^5 + x)\,dy = 0$
◆ Type 1, exact equations

16. $\cos x \sin y\,dx + \sin x \cos y\,dy = 0$
◆ Type 1, exact equations

17.^{A} $[y\cos(xy) + 1]\,dx + [x\cos(xy) + \sin y]\,dy = 0$
◆ Type 1, exact equations

18. $(x^2 - 3y^2)\,dx + (2xy + 4x^3)\,dy = 0$
◆ Type 1, exact equations

19.^{A} $(3x^2 y^2 + 2x)\,dx + (2x^3 y - 3y^2)\,dy = 0$
◆ Type 1, exact equations

20. $(18x - 2xy^2)\,dx + (8y - 2x^2 y)\,dy = 0$
◆ Type 1, exact equations

In Problems 21 through 26 determine whether the differential equations are exact and in the exact cases find explicit formulas for the functions $y(x)$ that satisfy the initial-value problems.

21.[O] $(2x^3y^2 - \sin x \cos x) \, dx + (x^4y + 2y) \, dy = 0$, $y(0) = 3$
♦ Type 1, exact equations

22.[A] $(2xy - 2x) \, dx + (x^2 + 1) \, dy = 0$, $y(0) = 2$
♦ Type 1, exact equations

23. $(y^{1/3} - \cos x) \, dx + \frac{1}{3}xy^{-2/3} \, dy = 0$, $y(\pi) = 0$
♦ Type 1, exact equations

24. $(e^x - y \sin x) \, dx + (3 + \cos x) \, dy = 0$, $y(0) = 1$
♦ Type 1, exact equations

25.[A] $(x^2y - xy^2) \, dx + (xy^3 - x^3y) \, dy = 0$, $y(1) = 2$
♦ Type 1, exact equations

26. $y \sin(xy) \, dx - x \cos(xy) \, dy = 0$, $y(\pi) = 4$
♦ Type 1, exact equations

27. Find a formula for the function $y(x)$ whose graph passes through the origin and whose tangent line at (x, y) has slope $x - y$.
♦ Type 2, a linear equation

28. Find the general solution of $\dfrac{dy}{dx} + \dfrac{ny \cos x}{\sin x} = \cos x$, where n is a positive integer.
♦ Type 2, a linear equation

29.[A] Give a formula in terms of x and the parameter k for the solution of the initial-value problem $(x^3 + 1)\dfrac{dy}{dx} + x^2 y = (x^3 + 1)^{5/3}$, $y(0) = k$.
♦ Type 2, a linear equation

30. Show that the function $y = y(x)$ that satisfies $(2x - y) \, dx - (x + 1) \, dy = 0$, $y(3) = 2$ is linear.
♦ Type 2, exact equations

31. Why does the initial-value problem $\left(5x^4 + \dfrac{1}{x-y}\right) dx - \left(6y^5 + \dfrac{1}{x-y}\right) dy = 0$, $y(3) = 3$ not have any solutions?
♦ Type 2, exact equations

32. Find the functions $G(z)$ such that $G'(z) = G(z) \sin z + 3 \sin z$.
♦ Type 2, a linear equation

Use a calculator/computer algebra system or a table of integrals to solve the initial-value problems in Problems 33 through 36.

33.[A] $\dfrac{dy}{dx} + 2y = \cos x$, $y(\pi) = 0$.
♦ Type 3, a linear equation

34. $\dfrac{dy}{dx} - y = x$, $y(0) = 10$.
♦ Type 3, a linear equation

35. $\dfrac{dy}{dx} + \dfrac{2y \cos x}{\sin x} = \sin x$, $y(\frac{1}{2}\pi) = 4$
♦ Type 3, a linear equation

36. $\dfrac{dy}{dx} - y \tan x = x^4$, $y(0) = 2$
♦ Type 3, a linear equation

Use definite integrals to give the solutions of the initial-value problems in Problems 37 through 40.

37.[O] $\dfrac{dy}{dx} + \frac{1}{2}x^{-1/2}y = \sin x$, $y(1) = 4$
♦ Type 5, expressing solutions with definite integrals

38.[A] $\dfrac{dy}{dx} - y = \tan x$, $y(0) = 0$
♦ Type 5, expressing solutions with definite integrals

39. $\dfrac{dy}{dx} - y \tan x = x^{4/3}$, $y(0) = 1$
♦ Type 5, expressing solutions with definite integrals

40. $\dfrac{dy}{dx} - \dfrac{y}{1 + x^2} = 1$, $y(1) = 0$
♦ Type 5, expressing solutions with definite integrals

41.[O] Find the orthogonal trajectories of the family of curves $y = x + Ce^y - 1$.
♦ Type 3, orthogonal trajectories with a linear equation

42.[A] Find the orthogonal trajectories of the curves $x^2 - y^2 = Cy$. (Use the substitution $u = y^2$.)
♦ Type 3, orthogonal trajectories with a linear equation

43. What are the orthogonal trajectories of the curves $y - x + Ce^{-y} = 1$?
♦ Type 3, orthogonal trajectories with a linear equation

Section 14.2

Second-order linear equations with constant coefficients

OVERVIEW: *In this section we study second-order linear differential equations with constant coefficients and their applications to the study of vibrating springs and simple electric circuits. We first derive the relevant differential equation for a vibrating spring. Then we see how solutions of the differential equations can be found and use these techniques in the applications.*

Topics:
- **Vibrating springs**
- **Homogeneous equations**
- **Inhomogeneous equations**
- **The Method of Undetermined Coefficients**
- **Simple electrical circuits**

Vibrating springs

A SECOND-ORDER, LINEAR DIFFERENTIAL EQUATION WITH CONSTANT COEFFICIENTS is of the form

$$ay''(t) + by'(t) + cy(t) = f(t) \tag{1}$$

where $a, b,$ and c are constants with $a \neq 0$, $f(t)$ is a given function, and $y(t)$ is the unknown solution. INITIAL CONDITIONS for this differential equation are obtained by prescribing the values $y(t_0)$ and $y'(t_0)$ of the solution and its first derivative at one value of the variable t. Our main application of such differential equations will be to the study of vibrating springs, as in Figures 1 or 2.

In Figure 1 the spring is fixed at its left end and is fastened at the right end to an object with mass M that moves horizontally along a y-axis. We disregard the effect of gravity, so that the only forces on the object are (i) the force of the spring, (ii) a possible force of air resistance, and (iii) a possible external force, all of which act parallel to the y-axis.

We put the origin of the y-axis at the object's REST POSITION—its location when the spring exerts no force to the left or right—and let $y(t)$ be the object's y-coordinate at time t, so that $y'(t)$ is the object's velocity toward the right and $y''(t)$ is its acceleration toward the right at time t.

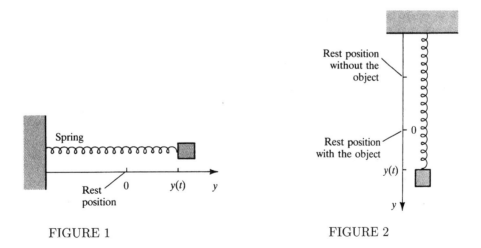

FIGURE 1   FIGURE 2

♦ SUGGESTIONS TO INSTRUCTORS: The main topics in this section can be divided into three parts, which could be covered in four or more lectures.
Part 1 (The differential equation for a vibrating spring and homogeneous equations; Tune-up Exercises T1, T2, T4; Problems 1–8, 16–26, 32, 34, 35, 39) Have students begin work on Questions 1–6 before class.
Part 2 (The method of undetermined coefficients and forced springs; Tune-up Exercises T3, T5, T6; Problems 9–15, 27–31, 33) Have students begin work on Questions 7 and 8 before class.
Part 3 (Simple electric circuits (Problems 36–38)

We measure the time in seconds. Distance, mass, and force are measured either in feet, slugs, and pounds (the British system); in meters, kilograms, and newtons (the mks system); or in centimeters, grams and dynes (the cgs system) so that Newton's law, [Force] = [Mass][Acceleration] can be applied.

We assume that the spring obeys HOOKE'S LAW, which states that the force exerted by the spring is proportional to the displacement of the end from its rest position. Consequently, at time t, when the object is at $y(t)$, the force exerted on it toward the right by the spring is

$$[\text{Force exerted by the spring}] = -ky(t) \tag{2}$$

where k is a positive constant. We also assume that either there is no air resistance or that the force of air resistance is proportional to the object's velocity. Then the force exerted toward the right by air resistance is

$$[\text{Force exerted by air resistance}] = -Ry'(t) \tag{3}$$

where R is zero if there is no air resistance and positive otherwise.

Finally, we suppose that the external force on the spring toward the right at time t is given by a function $f(t)$, which is the zero function if there is no external force:

$$[\text{External force}] = f(t). \tag{4}$$

Question 1 (a) Why is there a minus sign in **(2)**? (b) Why is there a minus sign in **(3)**?

Adding equations **(2)** through **(4)** shows that the total force toward the right on the object in Figure 1 is $-ky(t) - Ry'(t) + f(t)$. Since, by Newton's law, this equals the object's mass M, multiplied by its acceleration $y''(t)$ toward the right, we have $-ky(t) - Ry'(t) + f(t) = My''(t)$ or

$$My''(t) + Ry'(t) + ky(t) = f(t). \tag{5}$$

This equation is of the form **(1)** with the constants labeled M, R, and k instead of a, b, and c.

If the spring is vertical, we use a downward-pointing y-axis as in Figure 2 with its origin at the rest position of the end of the spring when the object is fastened to it. This point is Mg/k units below the rest position of the end of the spring if the object is not fastened to it. Consequently, when the end of the spring is at y, the combined force of gravity and the spring is $Mg - k(y + Mg/k) = -ky$ and we again obtain **(5)**.

Homogeneous equations

If the function on the right of **(1)** is zero, then the differential equation takes the form

$$ay''(t) + by'(t) + cy(t) = 0 \tag{6}$$

and is called HOMOGENEOUS.

We say that two functions $y_2(t)$ and $y_2(t)$ are INDEPENDENT or LINEARLY INDEPENDENT if neither equals a constant multiplied by the other. The functions $y_1(t) = t$ and $y_2(t) = t^2$, for example, are independent, as are $y_1(t) = \cos t$ and $y_2(t) = \sin t$, whereas $y_1(t) = t$ and $y_2(x) = 2t$ are not independent because in this case $y_2(t)$ equals 2 times $y_1(t)$ for all t.

With this terminology, we can state the following result:

Theorem 1 *Suppose that $y_1(t)$ and $y_2(t)$ are independent solutions of the homogeneous equation* **(6)**. *Then all solutions are given by linear combinations*

$$y(t) = Ay_1(t) + By_2(t) \tag{7}$$

of $y_1(t)$ and $y_2(t)$ with constants A and B. Furthermore, a unique solution of **(6)** *is determined by specifying the values $y(t_0)$ and $y'(t_0)$ of the solution and its first derivative at any point $t = t_0$.*

A short calculation shows that $y(t)$, given by **(7)** is a solution of **(6)** if $y_1(t)$ and $y_2(t)$ are solutions. The rest of Theorem 1 is established in courses on the theory of differential equations.

14.2 Second-order linear equations with constant coefficients

Free, undamped motion of a spring

The differential equation **(5)** for the motion of an object on a spring is homogeneous if there is no external force and $f(t) = 0$. In this case we say that the motion of the spring is FREE. If there is also no resistance, so that $R = 0$, we say that the motion is UNDAMPED, and the differential equation **(5)** takes the form

$$My''(t) + ky(t) = 0. \tag{8}$$

Example 1 Give an initial-value problem for the displacement $y(t)$ (centimeters) from rest position of a a five-gram ball on a spring as in Figure 1 that exerts a force of $-20y$ dynes on the ball when it is at y, under the conditions that there is no external force or resistance and that at time $t = 0$ (seconds) the spring is stretched one centimeter and the ball is moving at the rate of two centimeters per second away from its rest position.

SOLUTION We use **(8)** with $M = 5$ and $k = 20$ to obtain $5y''(t) + 20y(t) = 0$. We also have $y(0) = 1$ because the spring is stretched one centimeter at $t = 0$ and $y'(0) = 2$ because the ball is moving at the rate of two centimeters per second toward the right at $t = 0$. We then divide both sides of the differential equation by 5 to obtain the initial-value problem,

$$\begin{cases} y''(t) + 4y(t) = 0 \\ y(0) = 1, y'(0) = 2. \end{cases} \square \tag{9}$$

As is shown in the next Question, the functions $y = \cos(\omega t)$ and $y = \sin(\omega t)$ are solutions of the differential equation in **(9)** with a suitable choice of the constant ω.

Question 2 Find a positive constant ω such that $y = \cos(\omega t)$ and $y = \sin(\omega t)$ are solutions of the differential equation in **(9)**.

Example 2 Find the solution of the initial-value problem **(9)**.

SOLUTION The results of Question 2 show that $y_1(t) = \cos(2t)$ and $y_2(t) = \sin(2t)$ are independent solutions of the differential equation, so that by Theorem 1, all of the solutions are in the form

$$y(t) = A\cos(2t) + B\sin(2t) \tag{10}$$

with constants A and B. Formula **(10)** gives $y(0) = A\cos(0) + B\sin(0) = A$, so the initial condition $y(0) = 1$ implies that $A = 1$. Also with **(10)**

$$y'(t) = A\frac{d}{dt}[\cos(2t)] + B\frac{d}{dt}[\sin(2t)]$$

$$= -A\sin(2t)\frac{d}{dt}(2t) + B\cos(2t)\frac{d}{dt}(2t) = -2A\sin(2t) + 2B\cos(2t).$$

Therefore, $y'(0) = -2A\sin(0) + 2B\cos(0) = 2B$ and the condition $y'(0) = 2$ requires that $B = 1$. The solution of **(9)** is the function

$$y(t) = \cos(2t) + \sin(2t). \square \tag{11}$$

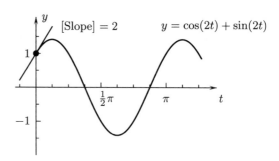

FIGURE 3

The graph of the solution **(11)** of the initial-value problem **(9)** is shown in Figure 3. The initial conditions $y(0) = 1$ and $y'(0) = 2$ cause the graph to pass through the point $(0,1)$ and be tangent to the line of slope 2 through that point. We can use the following Rule to put the formula for the solution in a form that makes it easy to understand the rest of the shape of the graph.

Rule 1 *If at least one of the constants A and B is not zero, then*

$$A\cos(\omega t) + B\sin(\omega t) = C\sin(\omega t + \psi) \tag{12}$$

where $C = \sqrt{A^2 + B^2}$ and ψ is an angle such that $\sin\psi = A/C$ and $\cos\psi = B/C$.

We can find an angle ψ that satisfies the conditions in this Rule because the point $(A/C, B/C)$ is on the unit circle. The Rule is then a consequence of the trigonometric identity

$$\sin\alpha\cos\beta + \cos\alpha\sin\beta = \sin(\alpha + \beta)$$

with C and ψ defined as stated, because multiplying and dividing the expression on the left of **(12)** by C gives

$$A\cos(\omega t) + B\sin(\omega t) = C\left[\frac{A}{C}\cos(\omega t) + \frac{B}{C}\sin(\omega t)\right]$$
$$= C[\sin\psi\cos(\omega t) + \cos\psi\sin(\omega t)] = C\sin(\omega t + \psi).$$

Functions of the form **(12)** are called SINUSOIDAL FUNCTIONS. The constant C is called the AMPLITUDE of the function if $\omega > 0$ because its values oscillate between C and $-C$ as t ranges over all values. Because its values repeat whenever ωt increases by 2π, this function has PERIOD $T = 2\pi/\omega$. The constant ψ/ω is called a PHASE SHIFT because $\sin(\omega t)$ can be converted into $\sin(\omega t + \psi)$ by replacing t with $t + \psi/\omega$.

Example 3   Apply Rule 1 to formula **(11)** and use the result to explain the shape of the graph in Figure 3.

SOLUTION   The constants A and B in **(11)** are both 1, so $C = \sqrt{1^2 + 1^2} = \sqrt{2}$ and we need an angle ψ with $\sin\psi = 1/\sqrt{2}$ and $\cos\psi = 1/\sqrt{2}$. We can use $\psi = \frac{1}{4}\pi$ and then have

$$y = \sqrt{2}\sin(2t + \tfrac{1}{4}\pi).$$

This formula shows that the graph is a sine curve that has amplitude $\sqrt{2} \doteq 1.414$ and that crosses the t-axis where $2t + \frac{1}{4}\pi$ equals $0, \pi, 2\pi, \ldots$, which is at $t = -\frac{1}{8}\pi, \frac{3}{8}\pi, \frac{7}{8}\pi, \ldots$. □

14.2 Second-order linear equations with constant coefficients

Question 3 The graph of the position of the ball in Example 1 with the initial conditions changed to $y(0) = 0, y'(0) = -2$ is shown in Figure 4. **(a)** How do the initial conditions affect the graph? **(b)** Find a formula for the solution. **(c)** What are the amplitude and period of the solution?

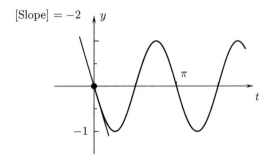

FIGURE 4

Free, damped motion of a spring

Suppose that an object is at the end of a spring that satisfies Hooke's Law and that is oriented as in Figure 1. If there is no external force, but there is air resistance, then the differential equation **(5)** is

$$My''(t) + Ry'(t) + ky(t) = 0 \tag{13}$$

with positive constants $M, R,$ and k. In this case we say that the spring is free and DAMPED.

Example 4 Solve the initial-value problem

$$\begin{cases} y''(t) + 5y'(t) + 4y(t) = 0 \\ y(0) = 1, y'(0) = 2 \end{cases} \tag{14}$$

for the free, damped motion of an object on a spring.

SOLUTION In this case we first look for solutions of the differential equation of the form

$$y(t) = e^{rt} \tag{15}$$

with constants r. The Chain Rule gives

$$y'(t) = \frac{d}{dt}(e^{rt}) = e^{rt}\frac{d}{dt}(rt) = re^{rt}$$

$$y''(t) = \frac{d}{dt}(y'(t)) = \frac{d}{dt}(re^{rt}) = re^{rt}\frac{d}{dt}(rt) = r^2 e^{rt}.$$

Consequently, the left side of the differential equation in **(14)** is

$$y''(t) + 5y'(t) + 4y(t) = r^2 e^{rt} + 5re^{rt} + 4e^{rt} = (r^2 + 5r + 4)e^{rt}$$

and the differential equation is satisfied if

$$r^2 + 5r + 4 = 0. \tag{16}$$

Equation **(16)** is called the CHARACTERISTIC EQUATION of the differential equation in **(14)**. Notice that it is obtained from the differential equation by replacing $y''(t)$ with r^2, $y'(t)$ with r and $y(t)$ with 1. We can factor its left side to obtain $(r+1)(r+4) = 0$, which shows that its solutions are $r = -1$ and $r = -4$. We could also obtain the same result with the Quadratic Formula by writing

$$r = \frac{-(-5) \pm \sqrt{(-5)^2 - 4(1)(4)}}{2(1)} = \frac{-5 \pm 3}{2} = -1, -4.$$

With either approach, we find that $y_1(t) = e^{-t}$ and $y_2(t) = e^{-4t}$ are solutions of the differential equation in **(14)**.

According to Theorem 1, all solutions of the differential equation are of the form $y = Ay_1(t) + By_2(t) = Ae^{-t} + Be^{-4t}$, for which $y'(t) = -Ae^{-t} - 4Be^{-4t}$. Therefore, $y(0) = A + B, y'(0) = -A - 4B$ and the initial conditions $y(0) = 1, y'(0) = 2$ will be satisfied if

$$\begin{cases} A + B &= 1 \\ -A - 4B &= 2 \end{cases}$$

Adding these equations gives $-3B = 3$, which implies that $B = -1$, and then either equation gives $A = 2$. The solution of **(14)** is $y(t) = 2e^{-t} - e^{-4t}$. $\square$

The graph of the solution of Example 4 is in Figure 5. The initial conditions $y(0) = 1$ and $y'(0) = 2$ cause the graph to pass through $(0, 1)$ and be tangent to the line of slope 2 through that point. The function tends to 0 as $t \to \infty$ because both e^{-t} and e^{-4t} tend to zero. The function is positive for all positive t because it is zero where $2e^{-t} = e^{-4t}$, which is where $2 = e^{-3t}$ and is at a negative value of t.

Because $y_1(t) = e^{-t}$ and $y_2(t) = e^{-4t}$ are both exponential functions, the motion of the spring is said to be OVERDAMPED.

FIGURE 5

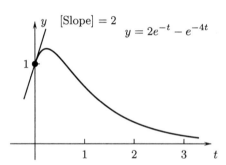

Question 4 Describe the motion for $t \geq 0$ of the object at the end of the spring in Example 4.

The damping force of air resistance in Example 4 was so large compared to the force of the spring that the spring did not oscillate. The differential equation

$$y''(t) + 2y'(t) + 37y(t) = 0 \tag{17}$$

in contrast, is a mathematical model of a free spring with damping that is so small, relative to the force of the spring, that the solution oscillates.

We again look for solutions in the form $y(t) = e^{rt}$, for which $y''(t) + 2y'(t) + 37y(t) = (r^2 + 2r + 37)e^{rt}$. For $y(t)$ to be a solution of **(17)**, the constant r must satisfy the corresponding characteristic equation, which is

$$r^2 + 2r + 37 = 0. \tag{18}$$

The Quadratic Formula gives the solutions of this equation:

$$r = \frac{-2 \pm \sqrt{(2)^2 - 4(37)}}{2(1)} = \frac{-2 \pm \sqrt{-144}}{2} = \frac{-2 \pm 12i}{2} = -1 \pm 6i. \tag{19}$$

14.2 Second-order linear equations with constant coefficients

Because these involve the square root of a negative number, they are nonreal, COMPLEX numbers which we have expressed here by using the symbol i for the IMAGINARY square root of -1. Because the numbers **(19)** are complex, we need the following definition of an exponential function with a complex variable in order to construct the solutions of the differential equation,

$$y = e^{(-1+6i)t} \text{ and } y = e^{(-1-6i)t}. \tag{20}$$

Definition 1 For any real number $\theta^\dagger$

$$e^{i\theta} = \cos\theta + i\sin\theta \tag{21}$$

and for any real numbers α and θ

$$e^{\alpha+i\theta} = e^\alpha e^{i\theta} = e^\alpha(\cos\theta + i\sin\theta). \tag{22}$$

With definition **(22)** applied to the functions **(20)**, we obtain

$$\begin{cases} e^{(-1+6i)t} = e^{-t}[\cos(6t) + i\sin(6t)] \\ e^{(-1-6i)t} = e^{-t}[\cos(6t) - i\sin(6t)]. \end{cases}$$

Straightforward calculations show that these are, in fact, solutions of the differential equation **(17)**. Then, because their sum divided by 2 is $e^{-t}\cos(6t)$ and their difference divided by $2i$ equals $e^{-t}\sin(6t)$, the functions $y_1(t) = e^{-t}\cos(6t)$ and $y_2(t) = e^{-t}\sin(6t)$ are also solutions, so that by Theorem 1 all solutions of **(17)** have the form

$$y(t) = e^{-t}[A\cos(6t) + B\sin(6t)]. \tag{23}$$

Example 5 Solve the following initial-value problem for the differential equation **(17)**:

$$\begin{cases} y''(t) + 2y'(t) + 37y(t) = 0 \\ y(0) = 1, y'(0) = -1. \end{cases} \tag{24}$$

SOLUTION Because all solutions of the differential equation have the form **(23)**, we need only to find the constants A and B so the initial conditions are satisfied. The Product Rule gives

$$y'(t) = \frac{d}{dt}\{e^{-t}[A\cos(6t) + B\sin(6t)]\}$$

$$= e^{-t}\frac{d}{dt}[A\cos(6t) + B\sin(6t)] + [A\cos(6t) + B\sin(6t)]\frac{d}{dt}(e^{-t})$$

$$= e^{-t}[-A\sin(6t)\frac{d}{dt}(6t) + B\cos(6t)\frac{d}{dt}(6t)] + [A\cos(6t) + B\sin(6t)]e^{-t}\frac{d}{dt}(-t)$$

$$= e^{-t}[-6A\sin(6t) + 6B\cos(6t) - A\cos(6t) - B\sin(6t)].$$

---

†We can obtain **(21)** by replacing x by $i\theta$ in the MacLaurin series $e^x = 1 + x + \frac{1}{2!}x^2 + \frac{1}{3!}x^3 + \frac{1}{4!}x^4 + \frac{1}{5!}x^5 + \cdots$
$= \sum_{j=0}^\infty \frac{1}{j!}x^j$ for e^x. Then with the facts that $i^2 = -1, i^3 = -i, i^4 = 1$, etc., we obtain $e^{i\theta} = 1 + i\theta + \frac{1}{2!}(i\theta)^2 + \frac{1}{3!}(i\theta)^3 + \frac{1}{4!}(i\theta)^4 + \frac{1}{5!}(i\theta)^5 + \cdots = 1 + i\theta - \frac{1}{2!}\theta^2 - i\frac{1}{3!}\theta^3 + \frac{1}{4!}\theta^4 + i\frac{1}{5!}\theta^5 + \cdots = [1 - \frac{1}{2!}\theta^2 + \frac{1}{4!}\theta^4 + \cdots] + i[\theta - \frac{1}{3!}\theta^3 + \frac{1}{5!}\theta^5 + \cdots] = \cos\theta + i\sin\theta$.

Setting $t = 0$ in the formulas for $y(t)$ and $y'(t)$ then gives $y(0) = A$ and $y'(0) = 6B - A$, so the initial condition $y(0) = 1$ implies that $A = 1$. Then the initial condition $y'(0) = -1$ implies that $B = 0$, so the solution is $y(t) = e^{-t}\cos(6t)$. □

Figure 6 shows the graph of the solution $y(x) = e^{-t}\cos(6t)$ from Example 5, drawn with a heavy line, and the curves $y =, \pm e^{-t}$, drawn with finer lines. Because $y(x) = e^{-t}\cos(6t)$ equals e^{-t} wherever $\cos(6t) = 1$ and equals $-e^{-t}$ wherever $\cos(6t) = -1$, the graph of $y(x)$ oscillates between the curves $y = e^{-t}$ and $y = -e^{-t}$, so that $y(t)$ oscillates between positive and negative values and tends to 0 as $t \to \infty$. Because the spring is damped but oscillates, it is said to be UNDERDAMPED.

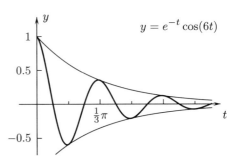

FIGURE 6

Examples 2, 4, and 5 illustrate parts (a) and (b) of the following Rule for solving the homogeneous differential equation

$$ay''(t) + by'(t) + cy(t) = 0 \tag{25}$$

with constants $a \neq 0, b,$ and c. To use the Rule, we first calculate the solutions

$$r = \frac{-b \pm \sqrt{b^2 - 4ac}}{2a} \tag{26}$$

of the corresponding characteristic equation

$$ar^2 + br + c = 0. \tag{27}$$

There are two, distinct, real solutions if the DISCRIMINANT $b^2 - 4ac$ is positive, two complex solutions if $b^2 - 4ac$ is negative, and one real solution if $b^2 - 4ac$ is zero. The Rule is given in three cases that correspond to these three types of solutions of the characteristic equation.

Rule 2 To find the solutions of the homogeneous differential equation (**25**), calculate, by factoring or the Quadratic Formula, the solutions r of the corresponding characteristic equation (**27**).

(a) If $b^2 - 4ac > 0$, then (**27**) has two, distinct, real solutions r_1 and r_2, and all solutions of (**25**) are in the form

$$y(t) = Ae^{r_1 t} + Be^{r_2 t}. \tag{28}$$

(b) If $b^2 - 4ac < 0$, then the solutions of the characteristic equation (**27**) are complex numbers $r = \tau \pm i\omega$ with real numbers τ and $\omega \neq 0$ and all solutions of (**25**) are in the form

$$y(t) = \begin{cases} e^{\tau t}[A\cos(\omega t) + B\sin(\omega t)] \\ \quad\text{or} \\ Ce^{\tau t}\sin(\omega t + \psi). \end{cases} \tag{29}$$

14.2 Second-order linear equations with constant coefficients

(c) If $b^2 - 4ac = 0$, then **(27)** has only one real solution r_0 and the solutions of **(25)** are

$$y(t) = e^{r_0 t}(A + Bt). \tag{30}$$

The letters $A, B, C,$ and ψ in this Rule represent arbitrary constants. The second of formulas **(29)** follows from the first because of Rule 1. Also, if $b = 0$ in **(25)**, as for the free, undamped spring of Example 2, then $r = \pm i\omega$ and the solution is covered by part (b) of the Rule with $\tau = 0$.

Parts (a) and (b) of Rule 2 can be established by modifying the reasoning in Examples 4 and 5. In case (c), where the charcteristic equation **(27)** has only one solution r_0, the characteristic equation is $a(r - r_0)^2 = 0$ or $a(r^2 - 2r_0 r + r_0^2) = 0$, and the differential equation **(25)** can be written

$$a[y''(t) - 2r_0 y'(t) + r_0^2 y(t)] = 0. \tag{31}$$

The function $y(t) = e^{r_0 t}$ satisfies this equation because r_0 is a solution of the characteristic equation. A calculation, to be carried out in the next Question for $a = 1$ and $r_0 = -1$, shows that $y(t) = te^{r_0 t}$ is also a solution of **(31)**. This with Theorem 1 establishes part (c) of the Rule.

Question 5 Show that $y(t) = te^{-t}$ is a solution of the differential equation $y''(t) + 2y'(t) + y(t) = 0$.

Case (c) of Rule 2 for a vibrating spring is illustrated in the next Example.

Example 6 An object of mass 1 slug is at the end of a spring as in Figure 1 with distances measured in feet. The spring exerts the force of $-y$ pounds to the right when the object is at y and air resistance exerts a force of $-2y'$ toward the right when the object's velocity toward the right is y' feet per second. Also, at $t = 0$ (seconds) the object is 1 foot to the right of its rest position and is moving toward the left at the rate of 2 feet per second. Find a formula for the object's position as a function of $t \geq 0$.

SOLUTION We obtain the following initial-value problem from **(5)** with $M = 1, R = 2$, and $k = 1$ and from the conditions on the spring at $t = 0$:

$$\begin{cases} y''(t) + 2y'(t) + y(t) = 0 \\ y(0) = 1, y'(0) = -2. \end{cases} \tag{32}$$

The characteristic equation is $r^2 + 2r + 1 = 0$ and can be factored as $(r+1)^2 = 0$ to show that it has one real solution $r = -1$. By part (c) of Rule 2, the solution of the differential equation is in the form

$$y(t) = e^{-t}(A + Bt) \tag{33}$$

with some constants A and B. Then by the Product Rule, $y'(t) = e^{-t}(-A - Bt + B)$, so that $y(0) = A, y'(0) = B - A$, and the initial conditions $y(0) = 1, y'(0) = -2$ imply that $A = 1$ and $B = -1$. The solution is $y(t) = e^{-t}(1 - t)$. □

Question 6 Figure 7 shows the graph of the solution $y(t)$ from Example 6. **(a)** What is its t-intercept? **(b)** Describe the motion of the object on the spring.

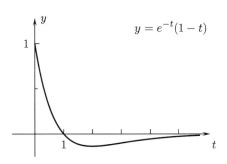

FIGURE 7

If, as Example 6, the discriminant of the characteristic equation for a free, damped motion of a spring is zero and the characteristic equation has only one real solution, then the spring is said to be CRITICALLY DAMPED. This terminology can be explained in the context of car shock absorbers. The differential equation $My''(t) + Ry'(t) + ky(t) = 0$ can be used as a mathematical model of a shock absorber that contains a spring that compresses and stretches as the car bounces up and down, and fluid that resists the compression or stretching of the spring. A good shock absorber is somewhat overdamped, so that if a car's fender is moved up or down, it will approach its rest position without oscillating. When the shock absorber wears out, however, the resistance drops so that the motion becomes underdamped and the fender oscillates. The transition occurs at the point when the shock absorber is critically damped, and this is called "critical" because any further loss of resistance would allow oscillations.

The inhomogeneous equation

If the funtion $f(t)$ in the differential equation

$$ay''(t) + by'(t) + cy(t) = f(t) \tag{34}$$

is not the zero function, we say that the differential equation is INHOMOGENEOUS. If we can find one solution $y_p(t)$ of this equation, then we can find all of them by using the next result.

Theorem 3 *If $y_p(t)$ is one solution of the inhomogeneous differential equation* **(34)**, *then all solutions are of the form*

$$y(t) = y_p(t) + Ay_1(t) + By_2(t) \tag{35}$$

where $y_1(t)$ and $y_2(t)$ are independent solutions of the corresponding homogeneous equation $ay''(t) + by'(t) + cy(t) = 0$.

We refer to the function $y_p(t)$ in this Theorem as a PARTICULAR solution of the inhomogeneous equation.

The proof of Theorem 3 can be given in two parts. We let $y_h(t) = Ay_1(t) + By_2(t)$ denote the general solution of the homogeneous equation. Then the sum $y(t) = y_p(t) + y_h(t)$ in **(35)** is a solution of the inhomogeneous equation because

$$ay'' + by' + cy = a(y_p + y_h)'' + b(y_p + y_h)' + c(y_p + y_h)$$
$$= (ay_p'' + by_p' + cy_p) + (ay_h'' + by_h' + cy_h) = f(x) + 0 = f(x).$$

On the other hand, if $y(t)$ is any solution of the inhomogeneous equation **(34)**, then

$$a(y - y_p)'' + b(y - y_p)' + c(y - y_p)$$
$$= (ay'' + by' + cy) - (ay_p'' + by_p' + cy_p) = f(x) - f(x) = 0$$

so that $y - y_p$ is a solution of the homogeneous equation and, by Theorem 1, is of the form $Ay_1 + By_2$, so that $y = y_p + Ay_1 + By_2$, as stated in **(35)**.

Because of Theorem 3, we can solve the inhomogeneous equation **(34)** by finding a particular solution, by using Rule 2 to find the general solution of the homogeneous equation, and by adding the results.

The Method of Undetermined Coefficients

The METHOD OF UNDETERMINED COEFFIIENTS is one technique for finding solutions of certain inhomogeneous equations. We illustrate it with an Example and a Question before we describe the general procedure.

Example 7 (a) Find a particular solution in the form $y_p(x) = c_1 \sin t + c_2 \cos t$ of the differential equation

$$y''(t) + y'(t) = 2\sin t. \tag{36}$$

(b) Give the general solution of (36).

SOLUTION (a) The first and second derivatives of $y_p(t) = c_1 \sin t + c_2 \cos t$ are $y_p'(t) = c_1 \cos t - c_2 \sin t$ and $y_p''(t) = -c_1 \sin t - c_2 \cos t$, so that

$$y_p''(t) + y_p'(t) = -c_1 \sin t - c_2 \cos t + c_1 \cos t - c_2 \sin t$$
$$= (-c_1 - c_2)\sin t + (c_1 - c_2)\cos t.$$

For y_p to satisfy (36), we must have

$$\begin{cases} -c_1 - c_2 = 2 \\ c_1 - c_2 = 0 \end{cases}$$

The second of these equations shows that $c_1 = c_2$ and then the first implies that c_1 and c_2 are both -1. The particular solution is $y_p(t) = -\sin t - \cos t$.

(b) The characteristic equation of the homogeneous equation $y''(t) + y'(t) = 0$ is $r^2 + r = 0$, which can be written $r(r+1) = 0$ and has the distinct real solutions $r = 0$ and $r = -1$. By Part (a) of Rule 2, $y_1(t) = e^0 = 1$ and $y_2(t) = e^{-t}$ are independent solutions of the inhomogenous equation and the general solution of the inhomogeneous equation (36) is

$$y(t) = y_p(t) + Ay_1(t) + By_2(t) = -\sin t - \cos t + A + Be^{-t}. \;\square \tag{37}$$

The constants A and B in (37) can be determined by prescribing initial values of the solution and its first derivative at one value of x.

Question 7 (a) Find a solution in the form $y_p(t) = c_1 + c_2 t$ of the differential equation

$$y''(t) + 4y(t) = t. \tag{38}$$

(b) Give the general solution of (38).

The Method of Undetermined Coefficients can be used to find a particular solution of the inhomogeneous differential equation (34) for any function $f(t)$ that is a product of functions of the forms

$$t^n, e^{\alpha t}, \sin(\alpha t + \beta), \cos(\alpha t + \beta) \tag{39}$$

with nonnegative integers n and any constants α and β. The technique can then be applied to linear combinations of such functions $f(t)$ by treating the functions separately and combining the results. In each case the particular solution is found as a linear combination

$$y_p(t) = c_1 z_1(t) + c_2 z_2(t) + \cdots + c_k z_k(t) \tag{40}$$

of functions $z_1(t), z_2(t), \ldots, z_k(t)$ in what we call a UC SET for the function $f(x)$. (The "UC" here stands for "undetermined coefficients.") The next Rule describes how UC sets are found.

Rule 3 (The Method of Undetermined Coefficients) To find a particular solution of

$$ay''(t) + by'(t) + cy(t) = f(t) \tag{41}$$

with constants $a \neq 0, b$, and c and $f(t)$ a product of functions of the types in (40), choose functions

$$w_1(t), w_2(t), \ldots, w_k(t) \tag{42}$$

such that none is a linear combination of the others and such that $f(t)$ and each of its derivatives can be expressed as a linear combination of these functions, If none of the functions (42) is a solution of the homogeneous equation

$$ay''(t) + by'(t) + cy(t) = 0 \tag{43}$$

then set $z_j(t) = w_j(t)$ for each j. If any of the w_j's are solutions of (43), then set $z_1(t) = t^n w_1(t), z_2(t) = t^n w_2(t), \ldots, z_k(t) = t^n w_k(t)$, where n is the smallest positive integer such that none of the resulting functions $z_j(t)$ is a solution of the homogeneous equation. The functions $z_1(t), z_2(t), \ldots, z_k(t)$ form a UC set for $f(t)$.

Once a UC set for $f(t)$ has been selected, the particular solution is found by substituting (40) into the differential equation (41) and solving for the numbers $c_1, c_2, \ldots, c_k$ by matching the coefficients of z_j on the two sides of the equation.

This procedure was used in Example 7, where $f(t) = 2\sin t$. This function and each of its derivatives is equal either to a constant times $w_1(t) = \sin t$ or a constant times $w_2(t) = \cos t$. Moreover, neither $\sin t$ nor $\cos t$ is a solution of the homogeneous equation $y''(t) + y'(t) = 0$, so that we could use the UC set consisting of $z_1(t) = \sin t$ and $z_2(t) = \cos t$ to find a particular solution of the form $y_p(t) = c_1 z_1(t) + c_2 z_2(t) = c_1 \cos t + c_2 \sin t$.

In the case of Question 7, $f(t) = t$, for which $f'(t) = 1$ and all higher-order derivatives are zero. Consequently, $f(t)$ and all of its derivatives are linear combinations of $w_1(t) = 1$ and $w_2(t) = t$. Neither of these is a solution of the homogeneous equation $y'' + 4y = 0$ (whose solutions are $A\cos(2t) + B\sin(2t)$), so $z_1(t) = 1$ and $z_2(t) = t$ form a UC set and there is a particular solution in the form $y_p = c_1 + c_2 t$.

The next Example is a case where the functions w_j in Rule 3 need to be modified to form the UC set.

Example 8 Find all solutions of

$$\frac{d^2 y}{dt^2} - 2\frac{dy}{dt} = -4t. \tag{44}$$

SOLUTION   The first derivatives of $f(t) = -4t$ is -4 and its higher-order derivatives are zero so $f(t)$ and its derivatives are linear combinations of $w_1(t) = 1$ and $w_2(t) = t$.

The homogeneous equation $y'' - 2y' = 0$ has the characteristic equation $r^2 - 2r = 0$ or $r(r-2) = 0$ with solutions $r = 0$ and $r = 2$, so the solutions of the homogeneous equation are

$$y_h(t) = A + Be^{2t}. \tag{45}$$

Because $w_1(t) = 1$ is a solution of the homogeneous equation but $z_1(t) = tw_1(t) = t$, and $z_2(t) = tw_2(t) = t^2$ are not, we use the latter two as the UC set and look for a particular solution in the form

$$y_p(t) = c_1 t + c_2 t^2. \tag{46}$$

14.2 Second-order linear equations with constant coefficients

Then $y'_p(t) = c_1 + 2c_2 t$ and $y''_p(x) = 2c_2$, so that

$$y''_p(t) - 2y'_p(t) = (2c_2) - 2(c_1 + 2c_2 t) = -4c_2 t + (2c_2 - 2c_1).$$

For this to equal $-4t$, we must have $c_2 = 1$, and $2c_2 - 2c_1 = 0$. so that $c_1 = 1$ and $c_2 = 1$, and, by **(46)**, $y_p(x) = t + t^2$. Adding the general solution **(45)** of the homogeneous equation gives the general solution $y(x) = t + t^2 + A + Be^{2t}$ of **(43)**. □

Forced springs

If there is an external force on the object at the end of the spring in Figure 1 or 2, then the function $f(t)$ is not zero and the differential equation

$$My''(t) + Ry'(t) + ky(t) = f(t) \tag{47}$$

for the position $y(t)$ of the object is inhomogeneous and we say that the spring is FORCED. We will consider one such case in the following Example and others in the Tune-Up Exercises and Problems below.

Example 9   An object with a mass of 10 grams is at the end of a spring as in Figure 1 that exerts a force of $-40y$ dynes toward the right when the object is y centimeters to the right its rest position. There is no air resistance and the external force on the spring at time t (seconds) toward the right is $10t$ dynes. At time $t = 0$ the weight is at its rest position and its velocity is 2 centimeters per second toward the right. Find a formula for the position of the object as a function of t.

SOLUTION   The differential equation **(47)** with $M = 10, R = 0, k = 40$, and $f(t) = 10t$ is $10y'' + 40y = 10t$, which when divided by 10 becomes $y'' + 4y = t$. You found the general solution

$$y(t) = \tfrac{1}{4} t + A\cos(2t) + B\sin(2t)$$

in Question 7. Since $y'(t) = \tfrac{1}{4} - 2A\sin(2t) + 2B\cos(2t)$, we have $y(0) = A, y'(0) = \tfrac{1}{4} + 2B$ and the initial conditions $y(0) = 0$ and $y'(0) = 2$ require that $A = 0$ and $B = \tfrac{7}{8}$. The solution is $y(t) = \tfrac{1}{4} t + \tfrac{7}{8} \sin(2t)$. □

Question 8   The graph of the solution from Example 9 is shown in Figure 8. **(a)** Descibe the motion of the object for $t \geq 0$ and explain why its motion is a plausible consequence of having the external force $10t$ on the spring. **(b)** Why is this not a reasonable mathematical model of an object on a apring as $t \to \infty$?

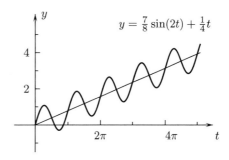

FIGURE 8

Simple electrical circuits

A simple electrical circuit, as in Figure 9, consisting of a generator or other source of current, a resistor, a capacitor, and an inductor, is studied with the same type of differential equation as for a vibrating spring. We can think of the inductor as a coil of wire and consider the capacitor as parallel plates across which no current can flow. We assume that the wire connecting the elements of the circuit has no resistance and has no inductive effect, so that all of the resistance is in the resistor and all of the inductance is in the inductor.

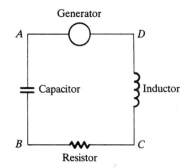

FIGURE 9

We let $q(t)$ denote the excess charge on the lower side of the capacitor at time t (seconds). Then the current $i(t)$ in the circuit at time t is the derivative $q'(t)$. We let $V_A, V_B, V_C,$ and V_D denote the voltage at points $A, B, C,$ and D in Figure 9. We assume that the voltage drop across the capacitor is proportional to the charge $q(t)$; that the voltage drop across the resistor is proportional to the current $i(t) = q'(t)$; and that the voltage drop across the inductor is proportional to the rate of change $i'(t) = q''(t)$ of the current. This gives the equations

$$\begin{aligned} V_A - V_B &= \frac{1}{C} q(t) \\ V_B - V_C &= R q'(t) \\ V_C - V_D &= L q''(t) \end{aligned} \qquad (48)$$

with positive constants C and L and a nonnegative constant R. C is called the CAPACITANCE, R is called the RESISTANCE, and L is called the INDUCTANCE of the circuit.

The voltage difference $V_A - V_D$ created by the generator is called the ELECTROMOTIVE FORCE, which we suppose is given by function $f(t)$, which is zero if there is no generator. Then, since $V_A - V_D$ is the sum of the quantities in (48), we obtain the differential equation

$$L q''(t) + R q'(t) + \frac{1}{C} q(t) = f(t) \qquad (49)$$

for the charge $q(t)$. Also, differentiating this equation with respect to t and noting that $q'(t)$ is the current $i(t)$ in the circuit, we obtain a differential equation

$$L i''(t) + R i'(t) + \frac{1}{C} i(t) = g(t) \qquad (50)$$

for the current, where $g(t)$ is the derivative $f'(t)$ of the electromotive force.

In these equations, time is measured in seconds, charge in COULOMBS, voltage in VOLTS, current in AMPERES, inductance in HENRIES, resistance in OHMS, and capacitance in FARADS.

Because (49) and (50) are second-order, linear differential equations with constant, positive L and C and nonnegative R, the techniques of this section can be applied to them, and the solutions exhibit the same sort of behavior as in problems dealing with vibrating springs.

14.2 Second-order linear equations with constant coefficients

Principles and procedures

- With the mathematical model $My'' + Ry' + ky = 0$ for the motion of an object on a free spring, the solutions either oscillate without dampling if $R = 0$ (the undamped case), oscillate and tend to 0 as $t \to \infty$ if $R > 0$ and $R^2 - 4Mk < 0$ (the underdamped case), or tend to 0 without oscillation if $R^2 - 4Mk \geq 0$ (the critically damped and overdamped cases). There are no other possibilities because M and k are positive and $R \geq 0$. Solutions of a differential equation $ay'' + by' + cy = 0$ that does not arise as a model of a vibrating spring can, however, exhibit different behavior if any of the constants $a, b,$ and c are negative.

- Theorem 3 shows that you can find all solutions of an inhomogeneous equation $ay'' + by' + cy = f$ by finding one solution y_p (which we call a particular solution) and adding it to the general solution $y_h = Ay_1 + By_2$ of the homogeneous equation $ay'' + by' + cy = 0$ to obtain the general solution $y = y_p + Ay_1 + By_2$ of the inhomogeneous equation. For the Problems in this section, you generally find y_1 and y_2 with Rule 2 first because you need to know the solutions of the homogeneous equation before you can use the Method of Undetermined Coefficients (Rule 3) to find a particular solution. Then you find the constants A and B from initial conditions, if required.

Tune-Up Exercises 14.2♦

A Answer provided. **O** Outline of solution provided.

T1.⁰ Solve the initial-value problem $y''(t) + 3y'(t) + 2y(t) = 0, y(0) = 2, y'(0) = -3$
♦ Type 1, a homogeneous initial-value problem

T2.⁰ Give the general solution of $\dfrac{d^2y}{dt^2} - \dfrac{dy}{dt} = 0$.
♦ Type 1, a homogeneous equation

T3.⁰ Give the general solution of $y''(t) - y(t) = t$.
♦ Type 1, variation of parameters

T4.⁰ Determine whether the differential equations for a free spring (a) $y''(t) + 3y'(t) + 2y(t) = 0$, (b) $y''(t) + 9y(t) = 0$, and (c) $y''(t) + 2y'(t) + 10y(t) = 0$ are undamped, overdamped, or underdamped. Then match the differential equations with graphs of their solutions in Figures 10 through 12.
♦ Type 1, recognizing equations for damped, free motion of a spring

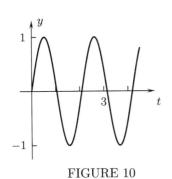
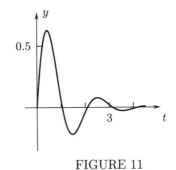
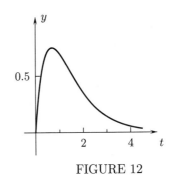

FIGURE 10 FIGURE 11 FIGURE 12

T5.⁰ (a) Show that $y(t) = \sin t - \frac{1}{6}\sin(6t)$ is the solution of $y'' + 36y = 35\sin t, y(0) = 0, y'(0) = 0$.
(b) Use the formula for $y(t)$ to explain the shape of its graph in Figure 13.
♦ Type 1, verifying a solution and explaining its graph

♦The Tune-up Exercises and Problems are classified by type and content. The types are (1) basic, reactive; (2) basic reflective; (3) intermediate, reactive; (4) intermediate, reflective; (5) advanced, reactive; (6) advanced, reflective; and (7) advanced, theoretical.

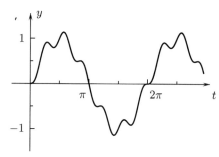

FIGURE 13

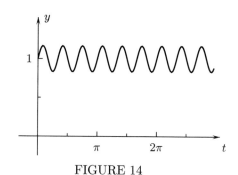

FIGURE 14

T6.^{O} (a) Show that $y(t) = 1 + \frac{1}{6}\sin(6t)$ is the solution of solution of $y''(t) + 36y(t) = 36$, $y(0) = 1$, $y'(0) = 1$.
(b) Use the formula for $y(t)$ to explain the shape of its graph in Figure 14.

♦ Type 1, verifying a solution and explaining its graph

Problems 14.2

^{A}*Answer provided.* ^{O}*Outline of solution provided.*

Solve the differential equations and initial-value problems in Problems 1 through 8.

1.^{O} $2y''(t) - 5y'(t) - 3y(t) = 0$,
$y(0) = 1, y'(0) = -4$

♦ Type 1, a homogeneous equation

2.^{O} $9\frac{d^2y}{dt^2} - 12\frac{dy}{dt} + 4y = 0$.

♦ Type 1, a homogeneous equation

3.^{A} $y''(t) - 3y'(t) + 2y(t) = 0$,
$y(0) = 1, y'(0) = 0$

♦ Type 1, a homogeneous equation

4. $y''(t) - y'(t) - 6y(t) = 0$,
$y(0) = 1, y'(0) = 8$

♦ Type 1, a homogeneous equation

5.^{A} $y''(t) + 2y'(t) - 3y(t) = 0$

6.º $y''(t) - 2y'(t) + 2y(t) = 0$,
$y(0) = 0, y'(0) = 1$

♦ Type 1, a homogeneous equation

7.^{A} $y''(t) + 4y'(t) + 5y(t) = 0$,
$y(0) = -1, y'(0) = 0$

♦ Type 1, a homogeneous equation

8. $y''(t) - 2y'(t) + 5y(t) = 0$,
$y(0) = 1, y'(0) = -1$

♦ Type 1, a homogeneous equation

9.º $y''(t) - y(t) = t$,
$y(0) = 0, y'(0) = 1$

♦ Type 1, an inhomogeneous equation

10.^{A} $y''(t) + 4y(t) = 5e^t$

♦ Type 1, an inhomogeneous equation

11. $y''(t) - y(t) = 2\sin t$,
$y(0) = 5, y'(0) = 1$

♦ Type 1, an inhomogeneous equation

12.º $y''(t) + y(t) = 2\cos t$,
$y(0) = 0, y'(0) = 1$

♦ Type 1, an inhomogeneous equation

13.^{A} $\frac{d^2y}{dt^2}(t) - y(t) = 2e^t$

14. $y'' + 2y' = 6$,
$y(0) = -4, y'(0) = 3$

♦ Type 1, an inhomogeneous equation

15. $y'' + 4y' + 4y = 2e^{-2t}$,
$y(0) = 0, y'(0) = 1$

♦ Type 1, an inhomogeneous equation

16.^{A} $y'' - 2y' + y = 2e^t$,
$y(0) = 0, y'(0) = 0$

♦ Type 1, an inhomogeneous equation

17.^{A} $y'' - 4y' + 4y = 0$,
$y(1) = 3, y'(1) = -5$

♦ Type 2, a homogeneous equation

18. $y'' - 2y' + y = 0$,
$y(2) = 3, y'(2) = 4$

♦ Type 2, a homogeneous equation

19.^{A} $9y'' - 6y' + 2y = 0$

♦ Type 2, a homogeneous equation

20. $4y'' + 5y = 0$,
$y(0) = A, y'(0) = 0$

♦ Type 2, a homogeneous equation

21.^{A} An object with a mass of 2 slugs is fastened to the end of a spring as in Figure 1 that exerts a force of $-5y$ pounds toward the right when the object is at y (feet). The force of air resistance is $-2y'$ toward the right when the object's velocity is y' feet per second. There is no external force. At $t = 0$ (seconds) the object is 2 feet to the right of its rest position and is moving toward the left at the rate of 1 foot per second. **(a)** Give an initial-value problem that is satisfied by the object's position $y(t)$. **(b)** Is the motion free or forced? Is it undamped, overdamped, underdamped, or critically damped? **(c)** Solve the initial-value problem. Then generate the graph of its solution and copy it on your paper.

♦ Type 2, a spring

14.2 Second-order linear equations with constant coefficients

In each of Problems 22 through 31 an initial-value problem is given for the motion of an object at the end of spring as in Figure 1. (a) Determine whether the motion is free or forced and whether is it is undamped, ovedrdamped, underdamped, or critcially damped. (b) Find the solution of the initial-value problem. Then generate its graph and copy it on your paper.

22.[A] $y'' + 9y = 0$,
$y(0) = 1, y'(0) = 3$
♦ Type 2, a spring with a graph

23.[A] $y'' + 3y' + 2y = 0$,
$y(0) = -1, y'(0) = 4$
♦ Type 2, a spring with a graph

24. $y'' + 2y' + 10y = 0$,
$y(0) = 0, y'(0) = 3$
♦ Type 2, a spring with a graph

25.[A] $4y'' + 4y' + y = 0$,
$y(0) = 2, y'(0) = -2$
♦ Type 2, a spring with a graph

26. $9y'' + y = 0$,
$y(0) = 4, y'(0) = -1$
♦ Type 2, a spring with a graph

27.[O] $y'' + y = \frac{2}{5}\sin t$,
$y(0) = 0, y'(0) = \frac{2}{5}$
♦ Type 2, a spring with a graph

28.[A] $y'' + 100y = 99\sin t$,
$y(0) = \frac{1}{6}, y'(0) = 1$
♦ Type 2, a spring with a graph

29.[A] $y'' + 2y' + 10y = 9\cos t - 2\sin t$,
$y(0) = 2, y'(0) = -1$
♦ Type 2, a spring with a graph

30. $y'' + 3y' + 2y = \sin t + 3\cos t$,
$y(0) = 5, y'(0) = -4$
♦ Type 2, a spring with a graph

31. $4y'' + 36y = 9t$,
$y(0) = 0, y'(0) = 1$
♦ Type 2, a fored spring with a graph

32. Match the differential equations (a) $y''(t) - 3y'(t) + 2y(t) = 0$ and (b) $y''(t) - 2y' + 65y(t) = 0$ to the graphs of their solutions in Figures 15 and 16.
♦ Type 2, homogeneous equations (not springs) with graphs

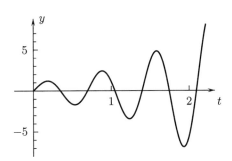

FIGURE 15

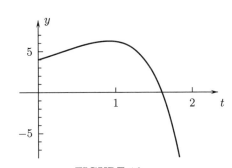

FIGURE 16

33. (a) Show that in the case of a differential equation $My'' + Ry' + ky = \sin(at)$ for a damped spring with a sinusoidal external force all solutions approach the particular solution given by Rule 3 as $t \to \infty$. For this reason, the particular solution is called the STEADY-STATE SOLUTION and the solutions of the homogeneous equation are called TRANSIENT SOLUTIONS. What are the steady-state and transient parts of the solutions (b) in Problem 29 and (c) in Problem 30?
♦ Type 4, steady-state and transient solutions

34. Show that (a) for a free, overdamped spring and (b) for a free, critically damped spring, each solution of the differential equation is either nonzero for $t \geq 0$ or is zero at only one positive value of t.
♦ Type 4, zeros of overdamped and critically damped solutions

35. (a[A]) Find the solution of the initial value problem $y'' + (1+\lambda)y' + \lambda y = 0, y(0) = 0, y'(0) = 1$, where λ is a positive constant $\neq 1$. (b) Use l'Hopital's Rule with λ as variable to find the limit as $\lambda \to 1$ of the solution from part (a). (c) Show that the function in part (b) is the solution of the initial-value problem $y'' + 2y' + y = 0, y(0) = 0, y'(0) = 1$ that is given by part (c) of Rule 3.
♦ Type 5, using l'Hopital's rule to treat the case of zero discriminant

36.[A] An electrical circuit as in Figure 9 but with no generator has inductance of 2 henries, resistance of 4 ohms, and capacitance of $\frac{1}{10}$ farad. At time $t = 0$ (seconds) there is a charge of 5 coulombs on the lower side of the capacitor and no current flow. Find a formula for the charge $q(t)$ on the capacitor as a function of t
♦ Type 3, an electrical circuit

37. An electrical circuit as in Figure 9 has inductance of 2 henries, resistance of 6 ohms, and capacitance of $\frac{1}{3}$ farad. The generator produces a voltage difference $f(t) = \sin t$ volts at time t. At time $t = 0$ (seconds) there is no charge on the capacitor and the current is 1 ampere. Find formulas for the charge $q(t)$ on the capacitor and the current as function of t

♦ Type 3, an electrical circuit

38. Give differential equations for the current in the circuit of Figure 9 but **(a)** with no capacitor and **(b)** with no inductor.

♦ Type 4, an electrical circuit

39. Use the fact that the initial value problems $y'' + y = 0, y(0) = \sin \alpha, y'(0) = \cos \alpha$ and $y'' + y = 0, y(0) = \cos \alpha, y'(0) = -\sin \alpha$ have unique solutions to derive the trigonometric identities **(a)** $\sin(t + \alpha) = \sin \alpha \cos t + \cos \alpha \sin t$ and **(b)** $\cos(t + \alpha) = \cos \alpha \cos t - \sin \alpha \sin t$.

♦ Type 3, using a differential equation to derive trigonometric identities

Section 14.3

Power-series solutions

OVERVIEW: *In this section we show how power series can be used to solve linear differential equations. This technique is used to construct solutions of a variety of differential equations that are used in mathematical physics and engineering. Because the differential equations in the applications generally require fairly complicated calculations, we will illustrate the technique primarily in less complex cases where the solutions could also be found by other means. As a further simplification of the calculations, we will only use power series centered at $x = 0$.*

Topics:

- **Using MacLaurin Series by finding derivatives**
- **Finding coefficients directly from the power series**

Using MacLaurin series by finding derivatives

We will use two methods to construct power-series solutions of initial-value problems for linear differential equations. With the first method, which is illustrated in this first subsection, we use the differential equation and initial conditions to find formulas in terms of j for the values $\{y^{(j)}(0)\}_{j=0}^{\infty}$ of the unknown function and all of its derivatives at $x = 0$.† Then we use these formulas in the MacLaurin series

$$y(x) = \sum_{j=0}^{\infty} \frac{1}{j!} y^{(j)}(0) x^j \tag{1}$$

for the solution $y(x)$.

With the second method, which is illustrated in the next subsection, the coefficients a_j in the power series

$$y(x) = \sum_{j=0}^{\infty} a_j x^j \tag{2}$$

for the solution are determined by applying the differential equation and initial conditions directly to the infinite series (2).

The first method works well if the coefficients in the differential equation are constant, but the second method is preferable in other cases. We will use the second method in most of the Problems in this section.

Both methods generally require the use of MATHEMATICAL INDUCTION to derive the necessary formulas. We will use this principle in the following two forms:

Rule 1 (Mathematical Induction) (a) *Suppose that $\mathcal{P}_j$ is a statement involving the parameter j for every $j \geq j_0$, where j_0 is a fixed integer. Then if $\mathcal{P}_{j_0}$ is true and if statement $\mathcal{P}_j$ implies $\mathcal{P}_{j+1}$ for every $j \geq j_0$, then $\mathcal{P}_j$ is true for all $j \geq j_0$.*
(b) *If $\mathcal{P}_{j_0}$ is true and if statements $\mathcal{P}_k$ for $j_0 \leq k \leq j$ imply $\mathcal{P}_{j+1}$ for every $j \geq j_0$, then $\mathcal{P}_j$ is true for all $j \geq j_0$.*

---

†Recall that $y^{(j)}(x)$ denotes the jth derivative of the function $y(x)$ so that $y^{(0)} = y(x), y^{(1)}(x) = y'(x)$, and $y^{(2)}(x) = y''(x)$. Also recall that the MacLaurin series (1) is the Taylor series $\sum_{j=0}^{\infty} \frac{1}{j!} y^{(j)}(x_0)(x - x_0)^j$ centered at x_0 with $x_0 = 0$.

♦ SUGGESTIONS TO INSTRUCTORS: The main topics in this section can be divided in three parts. The first and third parts scould be covered in one class meeting and Part 2 in one or more meetings.
Part 1 (Finding MacLaurin series by finding all derivatives at $x = 0$; Tune-up Exercises T1, T3) Have students begin Question 1 before class.
Part 2 (Finding MacLaurin series by operating with the series; Tune-up Exercises T2, T4; Problems 1–10, 15–30) Have students begin Questions 2–4 before class.
Part 3 (Finding a few terms in a MacLaurin series by finding derivatives at $x = 0$; Problems 11–14)

Example 1 (a) Use the initial conditions and differential equation in the initial-value problem

$$y'(x) + y(x) = 0, y(0) = 1 \tag{3}$$

to find a formula for the values $y(0), y'(0), y''(0), y^{(3)}(0), \ldots$ of the solution and its derivatives at $x = 0$. (b) Then use the MacLaurin series (1) to give a power-series representation of the solution.

SOLUTION (a) We rewrite the differential equation in (3) as $y'(x) = -y(x)$. When we differentiate both sides of this equation j times, where j is any nonnegative integer, we obtain $y^{(j+1)}(x) = -y^{(j)}(x)$, so that

$$y^{(j+1)}(0) = -y^{(j)}(0) \text{ for } j \geq 0. \tag{4}$$

Because $y^{(0)}(0) = y(0) = 1$ by the initial condition, equation (4) gives $y^{(1)}(0) = -y^{(0)}(0) = -1, y^{(2)}(0) = -y^{(1)}(0) = -(-1) = 1$, and $y^{(3)}(0) = -y^{(2)}(0) = -1$. This suggests that $y^{(j)}(0)$ equals 1 for all nonnegative, even integers j and equals -1 for all positive, odd integers j. Since $(-1)^j$ equals 1 for even integers j and -1 for odd integers j, we can phrase this prediction as statements $\mathcal{P}_j$ by writing for $j = 0, 1, 2, 3, \cdots$,

$$\mathcal{P}_j : y^{(j)}(0) = (-1)^j. \tag{5}$$

Statement $\mathcal{P}_0 : y^{(0)} = 1$ is given as the initial condition in (3), so it is true. Moreover, if $\mathcal{P}_j$ is true for any $j \geq 0$, then $y^{(j)}(0) = (-1)^j$ and by (4), $y^{(j+1)}(0) = -y^{(j)}(0) = -(-1)^j = (-1)^{j+1}$, which is $\mathcal{P}_{j+1}$. Thus $\mathcal{P}_j$ implies $\mathcal{P}_{j+1}$ for each integer $j \geq 0$ and, the principle of Mathematical Induction as stated in part (a) of Rule 1, $\mathcal{P}_j$ in (5) is true for all $j \geq 0$.

(b) Substituting formula (5) in the MacLaurin series (1) yields

$$y(x) = \sum_{j=0}^{\infty} \frac{1}{j!}(-1)^j x^j. \quad \Box \tag{6}$$

We can rewrite the sum in (6) as $\sum_{j=0}^{\infty} \frac{1}{j!}(-x)^j$ to see that it is the MacLaurin series of $y = e^{-x}$. We could also have found this solution by applying Rule 1 in Section 7.1 since the differential equation in (3) has the solutions $y = Ce^{-x}$ by that Rule, and the initial condition implies that $C = 1$.

As a second illustration of the use of MacLaurin series, we consider the initial-value problem

$$y''(x) + y(x) = 0, y(0) = 0, y'(0) = 1. \tag{7}$$

We rewrite this differential equation as $y''(x) = -y(x)$ and differentiate both sides j times, where j is any nonnegative integer, to obtain $y^{(j+2)}(x) = -y^{(j)}(x)$. Then we set $x = 0$ to have

$$y^{(j+2)}(0) = -y^{(j)}(0) \text{ for } j \geq 0. \tag{8}$$

Question 1 Use (8) and the initial conditions in (7) to find the values of $y^{(j)}(0)$ for $j = 0, 1, 2, 3, 4, 5, 6, 7$.

14.3 Power-series solutions

Example 2 Find formulas for all of the derivatives at $x = 0$ of the solution $y(x)$ of the initial-value problem **(7)** and use them to give the MacLaurin series of the solution.

SOLUTION The calculations in the Response to Question 1 suggest that $y^{(j)}(0)$ is zero for all even, nonnegative integers j and that for odd j's it alternates between 1 and -1 with the value 1 for $j = 1$. We express the even integers as $2k$ and the odd integers as $2k+1$ with integers k and write these predictions as the statements for $k = 0, 1, 2, \ldots$,

$$\mathcal{P}_k : \begin{cases} y^{(2k)}(0) = 0 \\ y^{(2k+1)}(0) = (-1)^k. \end{cases} \tag{9}$$

The initial conditions show that $\mathcal{P}_0$ is true. (In fact, the Response to Question 1 shows that $\mathcal{P}_k$ is true for $k = 0, 1, 2, 3$.) To use Mathematical Induction, we need to show that $\mathcal{P}_k$ implies $\mathcal{P}_{k+1}$ for every nonnegative integer k.

This is the case because, if $\mathcal{P}_k$ is true for any $k \geq 0$, then $y^{(2k)}(0) = 0$ and $y^{(2k+1)}(0) = (-1)^k$, and by **(8)**,

$$y^{(2(k+1))}(0) = y^{(2k+2)}(0) = -y^{(2k)}(0) = -0 = 0$$
$$y^{(2(k+1)+1)}(0) = y^{(2k+3)}(0) = -y^{(2k+1)}(0) = -(-1)^k = (-1)^{k+1}$$

and these equations are $\mathcal{P}_{k+1}$. Therefore, $\mathcal{P}_k$ is true for all $k \geq 0$.

With the values **(9)** for the derivatives, **(1)** gives

$$y(x) = \sum_{k=0}^{\infty} \frac{1}{(2k+1)!} (-1)^k x^{2k+1}. \; \square \tag{10}$$

The sum in **(10)** is the MacLaurin series for $y(x) = \sin x$. We could also find this solution by the techniques of Section 14.2: the solutions of the differential equation in **(7)** are $y = A \cos x + B \sin x$, and the initial conditions require that $A = 0$ and $B = 1$.

Finding coefficients directly from the power series

We could find power-series solutions of the initial-value problems in Examples 1 and 2 fairly easily by finding all of the derivatives of the solutions at $x = 0$ and using a MacLaurin series **(1)** because the differential equations had constant coefficients. In other cases it is easier to use the differential equation and initial conditions to study directly the power series

$$y(x) = a_0 + a_1 x + a_2 x^2 + a_3 x^3 + \cdots = \sum_{j=0}^{\infty} a_j x^j. \tag{11}$$

We will illustrate this approach first with the initial-value problem of Example 1.

Example 3 Suppose that $y(x)$, given by **(11)**, is the solution of the initial-value problem

$$y'(x) + y(x) = 0, \; y(0) = 1. \tag{12}$$

Find a formula for a_{j+1} in terms of a_j for every $j \geq 0$.

SOLUTION   We assume that the power series **(11)** converges absolutely, at least for x near 0, so that we can find its derivatives by differentiating term by term. Then the first derivative of $y(x)$ is

$$y'(x) = \frac{d}{dx}(a_0 + a_1 x + a_2 x^2 + a_3 x^3 + \cdots) = \sum_{j=0}^{\infty} \frac{d}{dx}(a_j x^j)$$

$$= 0 + a_1 + 2a_2 x + 3a_3 x^2 + \cdots = \sum_{j=0}^{\infty} j a_j x^{j-1} \tag{13}$$

$$= a_1 + 2a_2 x + 3a_3 x^2 + \cdots = \sum_{j=1}^{\infty} j a_j x^{j-1}.$$

We can begin the sum on the right of **(13)** at $j = 1$ because the term for $j = 0$ is zero.

We need to rewrite the sum in **(13)** so it involves x^j instead of x^{j-1}, so we can use it with the sum in **(11)**. We do this by setting $k = j - 1$, so that $j = k + 1$ and the last sum in **(13)** runs from $k = 0$ to ∞. This gives

$$y'(x) = \sum_{k=0}^{\infty} (k+1) a_{k+1} x^k.$$

Then we replace the letter k by the letter j to have

$$y'(x) = \sum_{j=0}^{\infty} (j+1) a_{j+1} x^j. \tag{14}$$

With formulas **(11)** and **(14)**, the differential equation $y'(x) + y(x) = 0$ in **(12)** reads

$$\sum_{j=0}^{\infty} (j+1) a_{j+1} x^j + \sum_{j=0}^{\infty} a_j x^j = 0$$

or

$$\sum_{j=0}^{\infty} [(j+1) a_{j+1} + a_j] x^j = 0. \tag{15}$$

In order for **(15)** to be satisfied for all x in an open interval, the coefficients of all powers of x must be zero and we must have $(j+1) a_{j+1} + a_j = 0$ for $j \geq 0$. Solving this equation for a_{j+1} yields

$$a_{j+1} = \frac{-a_j}{j+1} \quad \text{for } j \geq 0. \ \square \tag{16}$$

Formula **(11)** shows that $a_0 = y(0)$ and this is equal to 1 by the initial condition in **(12)**. Therefore, we can find the power series **(11)** for the solution by using equations **(16)** to find a_j for all $j \geq 1$. We first calculate some of these numbers to discern the patterns in their formulas.

Question 2   Give expressions for a_1, a_2, a_3, a_4, and a_5, where $a_0 = 1$ and the other numbers are determined by equations **(16)**. Give the results in terms of factorials instead of simplifying the expressions by multiplication.

The results of Question 2 suggest that a_j equals $1/j!$ for even, nonnegative integers j and equals $-1/j!$ for odd, positive integers j. We give these predictions as statements $\mathcal{P}_j$ by writing for $j \geq 0$,

$$\mathcal{P}_j : a_j = \frac{(-1)^j}{j!}. \tag{17}$$

14.3 Power-series solutions

Question 3 Use Mathematical Induction with the value $a_0 = 1$ and equations **(16)** to show that **(17)** is correct.

With expressions **(17)** for the coefficients in the power series **(11)** we obtain

$$y(x) = \sum_{j=0}^{\infty} \frac{1}{j!}(-1)^j x^j$$

which is the power series for e^{-x} that we found in Example 1.

Next we will construct the solution of the initial-value problem from Example 2,

$$y''(x) + y(x) = 0,\ y(0) = 0,\ y'(0) = 1 \tag{18}$$

by calculations with the power series **(11)**. To find the power series of $y''(x)$, we differentiate the power series **(13)** of $y'(x)$. We obtain

$$y''(x) = \frac{d}{dx}[y'(x)] = \frac{d}{dx}\sum_{j=1}^{\infty} j a_j x^{j-1} = \sum_{j=1}^{\infty} \frac{d}{dx}(j a_j x^{j-1}) = \sum_{j=2}^{\infty} j(j-1) a_j x^{j-2}.$$

The last sum can begin with $j = 2$ because the term for $j = 1$ in it is zero.

To rewrite this sum so it contains the symbols x^j instead of x^{j-2}, we first set $k = j - 2$, so that $j = k + 2$ and $j - 1 = k + 1$. Then the sum begins at $k = 0$, and we obtain

$$y''(x) = \sum_{k=0}^{\infty} (k+2)(k+1) a_{k+2} x^k.$$

Finally we replace the letter k by j to have

$$y''(x) = \sum_{j=0}^{\infty} (j+2)(j+1) a_{j+2} x^j. \tag{19}$$

With **(11)** and **(19)**, the equation $y''(x) + y(x) = 0$ reads

$$\sum_{j=0}^{\infty} [(j+2)(j+1)a_{j+2} + a_j] x^j = 0$$

and implies that $(j+2)(j+1)a_{j+2} + a_j = 0$ for $j \geq 0$. Solving for a_{j+2} gives

$$a_{j+2} = \frac{-a_j}{(j+2)(j+1)} \quad \text{for } j \geq 0. \tag{20}$$

Formula **(11)** implies that $y(0) = a_0$ and **(13)** gives $y'(0) = a_1$, so the initial conditions in **(18)** imply that $a_0 = 0$ and $a_1 = 1$. Then with **(20)** for $j = 0, 1, 2$, and 3 we obtain

$$a_0 = 0$$
$$a_1 = 1$$
$$a_2 = \frac{-a_0}{2(1)} = \frac{0}{2(1)} = 0$$
$$a_3 = \frac{-a_1}{3(2)} = \frac{-1}{3(2)} = -\frac{1}{3!} \tag{21}$$
$$a_4 = \frac{-a_2}{4(3)} = \frac{0}{4(3)} = 0$$
$$a_5 = \frac{-a_3}{5(4)} = \frac{1/3!}{5(4)} = \frac{1}{5!}.$$

Question 4 (a) Use the values **(21)** to predict formulas for a_j for all $j \geq 0$ and verify it by using Mathematical Induction with **(20)** and the initial conditions in **(18)**. (b) Use the results of part (a) to give the MacLaurin series of the solution.

In the next Example we use power series to solve an initial-value problem we could not solve by other means.

Example 4 Find a power-series solution of the initial-value problem

$$y''(x) + 2xy'(x) + 2y(x) = 0, \; y(0) = 1, y'(0) = 0. \tag{22}$$

SOLUTION We set $y(x) = \sum_{j=0}^{\infty} a_j x^j$ for which

$$y''(x) = \sum_{j=0}^{\infty} j(j-1)a_j x^{j-2} = \sum_{j=0}^{\infty} (j+2)(j+1)a_{j+2} x^j$$

$$2xy'(x) = 2x \sum_{j=0}^{\infty} j a_j x^{j-1} = \sum_{j=0}^{\infty} 2j a_j x^j$$

$$2y(x) = 2 \sum_{j=0}^{\infty} a_j x^j = \sum_{j=0}^{\infty} 2a_j x^j.$$

We obtained the second sum in the first of these equations by the reasoning that led to **(19)**. Adding these equations gives

$$y''(x) + 2xy'(x) + 2y(x) = \sum_{j=0}^{\infty} [(j+2)(j+1)a_{j+2} + 2(j+1)a_j]x^j = 0$$

and in order for this to hold for x in an open interval we must have $(j+2)(j+1)a_{j+2} + 2(j+1)a_j = 0$ for $j \geq 0$. This yields

$$a_{j+2} = -\frac{2(j+1)a_j}{(j+2)(j+1)} = \frac{-2a_j}{j+2} \quad \text{for } j \geq 0. \tag{23}$$

Since $a_0 = y(0) = 1$ and $a_1 = y'(0) = 0$, formula **(23)** gives

$$a_2 = \frac{-2a_0}{0+2} = \frac{-2(1)}{2} = -1$$

$$a_3 = \frac{-2a_1}{1+2} = \frac{-2(0)}{3} = 0$$

$$a_4 = \frac{-2a_2}{2+2} = \frac{-2(-1)}{4} = \frac{1}{2}$$

$$a_5 = \frac{-2a_3}{3+2} = \frac{-2(0)}{5} = 0$$

$$a_6 = \frac{-2a_4}{4+2} = \frac{-2(1/2)}{6} = \frac{-1}{6} = -\frac{1}{3!}.$$

These values lead us to predict that for $k \geq 0$

$$\mathcal{P}_k : \begin{cases} a_{2k} = \dfrac{(-1)^k}{k!} \\ a_{2k+1} = 0 \end{cases}. \tag{24}$$

14.3 Power-series solutions

$\mathcal{P}_0$ is true because the initial conditions give $a_0 = y(0) = 1$ and $a_1 = y'(0) = 0$. Also, if $\mathcal{P}_k$ is true for any $k \geq 0$, then $a_{2k} = (-1)^k/k!$ and $a_{2k+1} = 0$, so that by **(23)** with $j = 2k$ and $j = 2k+1$,

$$a_{2k+2} = \frac{-2a_{2k}}{2k+2} = \frac{-a_{2k}}{k+1} = \frac{-1}{k+1}\frac{(-1)^k}{k!} = \frac{(-1)^{k+1}}{(k+1)!}$$

$$a_{2k+3} = \frac{-2a_{2k+1}}{2k+2} = \frac{0}{2k+2} = 0.$$

These equations are $\mathcal{P}_{k+1}$. Therefore, $\mathcal{P}_k$ is true for all $k \geq 0$, and

$$y(x) = \sum_{k=0}^{\infty} \frac{(-1)^k}{k!} x^{2x}. \square$$

Question 5 (a) Explain why the solution in Example 4 is $y(x) = e^{-x^2}$. (b) Show directly that $y(x) = e^{-x^2}$ satisfies the initial-value problem **(22)**.

Principles and procedures

- The technique of finding MacLaurin series of solutions of linear differential equations that is used in Examples 1 and 2 and Tune-up Exercises T1 and T3 is included to make a transition from the concept of MacLaurin series to the techniques of operating directly with the power series that is usually used.

- Be sure you understand how the expression $\sum_{j=1}^{\infty} j a_j x^{j-1}$ for $y'(x)$ is converted to $\sum_{j=0}^{\infty} (j+1)a_{j+1} x^j$ in the solution of Example 3 and how the expression $\sum_{j=1}^{\infty} (j-1)j a_j x^{j-2}$ for $y''(x)$ is converted to $\sum_{j=0}^{\infty} (j+2)(j+1)a_{j+2} x^j$ before Question 4. One or both of these steps are needed in most of the Problems that call for solving initial-value problems by manipulating power series.

- In some cases involving second-order differential equations, you need to have the induction step deal with two coefficients, generally by having $\mathcal{P}_k$ for $k = 0, 1, 2, 3, \ldots$ give formulas for a_{2k} and a_{2k+1}, as in the solution of Example 4. In other cases you can have $\mathcal{P}_j$ deal only with a_j, verify $\mathcal{P}_0$ and $\mathcal{P}_1$, and use the version of Mathematical Induction in part (b) of Rule 1 (see the outline of solution of Tune-up Exercise T4).

- It is usually fairly difficult to use the differential equation to find all of the derivatives of the solution at $x = 0$ if the equation has nonconstant coefficients. This technique can be used, however, in some cases with nonconstant coefficients to find a finite number of derivatives, which then can be used to give a finite number of terms in the MacLaurin series (see Problems 11 through 14).

Tune-Up Exercises 14.3♦

A Answer provided. **O** Outline of solution provided.

T1.^O (a) Give the MacLaurin series for the solution of the initial-value problem $y'(x) - y(x) = 0, y(0) = 1$ by using the differential equation and initial condition to find all derivatives of the solution at $x = 0$. (Mathematical Induction is not required in this case.) (b) What is $y(x)$?
♦ Type 1, using derivatives to form a MacLaurin series

T2.^O Find the power series $y(x) = \sum_{j=0}^{\infty} a_j x^j$ centered at $x = 0$ for the solution of the initial-value problem in Exercise T1 by applying the differential equation and initial condition to the series.
♦ Type 1, finding coefficients in a power series

T3.^O (a) Give the MacLaurin series for the solution of the initial-value problem $y''(x) - y(x) = 0$, $y(0) = 1, y'(0) = -1$ by using the differential equation and initial condition to find all derivatives of the solution at $x = 0$. (b) What is $y(x)$?
♦ Type 1, using derivatives to form a MacLaurin series

T4.^O Find the power series $y(x) = \sum_{j=0}^{\infty} a_j x^j$ centered at $x = 0$ for the solution of the initial-value problem in Exercise T3 by applying the differential equation and initial condition to the series.
♦ Type 1, finding coefficients in a power series

Problems 14.3

A Answer provided. **O** Outline of solution provided.

Use power series centered at $x = 0$ to find solutions of the initial-value problems in Problems 1 through 10. Then identify the solutions.

1.^O $y''(x) + 2y'(x) + y(x) = 0$, $y(0) = 0, y'(0) = 1$
♦ Type 1, finding coefficients in a power series

2.^O $y''(x) - y'(x) - 2y(x) = 0$, $y(0) = 1, y'(0) = 2$
♦ Type 1, finding coefficients in a power series

3. $y''(x) - 3y'(x) = 0, y(0) = 3, y'(0) = 3$
♦ Type 1, finding coefficients in a power series

4.^A $y'(x) - 2xy(x) = 0, y(0) = 1$
♦ Type 1, finding coefficients in a power series

5. $y'(x) + 4y(x) = 0, y(0) = 3$
♦ Type 1, finding coefficients in a power series

6. $y''(x) - xy'(x) - y(x) = 0$, $y(0) = 1, y'(0) = 0$
♦ Type 2, finding coefficients in a power series

7.^A $(x-1)y'(x) + y(x) = 0, y(0) = 1$
♦ Type 2, finding coefficients in a power series

8.º $y'(x) - y(x) = -x^2, y(0) = 2$
♦ Type 2, finding coefficients in a power series

9 $y'(x) - y(x) = x - 1, y(0) = 1$
♦ Type 2, finding coefficients in a power series

10. $y'(x) - 2xy(x) = 2x, y(0) = 0$
♦ Type 2, finding coefficients in a power series

In Problems 11 through 14, find the values at $x = 0$ of the derivatives up to order 4 of the solutions of the initial-value problems and then give the terms up to degree 4 in the MacLaurin series of the solutions.

11.º $y' = x^2 + y^2, y(0) = 1$
♦ Type 3, finding terms in a MacLaurin series

12. $y' = \cos y, y(0) = 0$
♦ Type 3, finding terms in a MacLaurin series

13.^A $y'' = y^2, y(0) = 1, y'(0) = 2$
♦ Type 4, finding terms in a MacLaurin series

14. $y'' = \sqrt{y+1}, y(0) = 0, y'(0) = 2$
♦ Type 4, finding terms in a MacLaurin series

♦The Tune-up Exercises and Problems are classified by type and content. The types are (1) basic, reactive; (2) basic reflective; (3) intermediate, reactive; (4) intermediate, reflective; (5) advanced, reactive; (6) advanced, reflective; and (7) advanced, theoretical.

14.3 Power-series solutions

In Problems 15 through 24, use power series centered at $x = 0$ to find solutions of the initial-value problems.

15.[A] $y'' - 2xy' - 2y = 0$,
$y(0) = 1, y'(0) = 0$

16.[A] $(x-1)y'' + (2-x)y' + 2y = 0$,
$y(0) = 7, y'(0) = -6$

17.[A] $y'' - 2xy' + 8y = 0, y(0) = 3, y'(0) = 0$
(A HERMITE POLYOMIAL)

18. $(1-x^2)y'' - 4xy' + 18y = 0$,
$y(0) = 0, y'(0) = 3$

19. $(x+1)y'' + y' - xy = 0$,
$y(0) = 1, y'(0) = -1$

20.[A] $xy'' + y' + xy = 0, y(0) = 1, y'(0) = 0$
(A BESSEL FUNCTION)

21. $xy'' + (1-x)y' + 3y = 0$,
$y(0) = 6, y'(0) = -18$ (A LAGUERRE POLYNOMIAL).

22.[A] $xy'' - (x^2 + 2)y' + xy = 0$,
$y(0) = 1, y'(0) = 0$

23. $xy'' + y' + 2y = 0$,
$y(0) = 1, y'(0) = -2$

24. $xy'' - (1+x)y' + y = 0$,
$y(0) = 1, y'(0) = 1$

25. Only the value of $y(0)$ and not the value of $y'(0)$ has to be prescribed for the initial-value problems in Problems 20 through 24. Explain.

In Problems 26 through 29 give equations expressing a_j in terms of two of a_k for $k < j$ that would be used to find all of the coefficients in a power-series solution $\sum_{j=0}^{\infty} a_j x^j$ of the initial-value problem. Then find the first five nonzero terms.

26.[A] $y'' + y' + xy = 0, y(0) = 0, y'(0) = 1$
♦ Type 4, finding coefficients in a power series.

27 $y'' + xy' - xy = 0, y(0) = 1, y'(0) = 0$
♦ Type 4, finding coefficients in a power series.

28. $2y'' + xy' - 4y = 0, y(0) = 0, y'(0) = 6$
♦ Type 4, finding coefficients in a power series.

29. $y'' - xy' + x^2 y = 0, y(0) = 1, y'(0) = 0$
♦ Type 4, finding coefficients in a power series.

30. The solutions of $y''(x) - xy(x) = 0$ arise in mathematical physics and are called AIRY FUNCTIONS.
(a) Show that $y = x + \sum_{k=1}^{\infty} \frac{((2)(5)(8) \cdots (3k-1))}{(3k+1)!} x^{3k+1}$ is the solution with $y(0) =, 0, y'(0) = 1$.
(b) Use the Ratio Test to show that the series in part (a) converges for all x, (b) Find the solution of the differential equation in part (a) with the initial conditions $y(0) = 1, y'(0) = 0$.
♦ Type 6, finding power-series solutions

RESPONSES TO QUESTIONS IN CHAPTER 14

Responses 14.1

Response 1 (a) $y = x + x^{-2}$ • $\dfrac{dy}{dx} = 1 - 2x^{-3}$ •

$x^2 \dfrac{dy}{dx} + 2xy = x^2(1 - 2x^{-3}) + 2x(x + x^{-2}) = x^2 - 2x^{-1} + 2x^2 + 2x^{-1} = 3x^2$

(b) $y(1) = \left[x + x^{-2}\right]_{x=1} = 2$

Response 2 Since by the Chain Rule, $\dfrac{d}{dx}(e^{x^2}) = e^{x^2}\dfrac{d}{dx}(x^2) = 2xe^{x^2}$, equation **(5)** can be written $e^{x^2}\dfrac{dy}{dx} + y\dfrac{d}{dx}(e^{x^2}) = \cos x$ and then, by the Product Rule, as $\dfrac{d}{dx}(e^{x^2} y) = \cos x$. • Take antiderivatives of both sides: $e^{x^2} y = \sin x + C$ •
Divide by e^{x^2}: $y = (\sin x + C)e^{-x^2}$

Response 3 (a) Because the derivative of the antiderivative $\int P(x)\, dx$ of $P(x)$ is $P(x)$, the Chain Rule gives $\dfrac{dI}{dx}(x) = \dfrac{d}{dx}\left(e^{\int P(x)\, dx}\right) = \left(e^{\int P(x)\, dx}\right)\dfrac{d}{dx}\int P(x)\, dx$
$= \left(e^{\int P(x)\, dx}\right) P(x) = I(x)P(x)$.

(b) The Product Rule and **(9)** yield
$\dfrac{d}{dx}[I(x)y(x)] = I(x)\dfrac{dy}{dx}(x) + y(x)\dfrac{dI}{dx}(x) = I(x)\dfrac{dy}{dx}(x) + I(x)P(x)y(x)$,
which is **(9)**.

Response 4 Because the coefficient of $\dfrac{dy}{dx}$ in **(16)** is 1, it is of the form **(11)** with $P(x) = 2x$. •
$\int P(x)\, dx = \int 2x\, dx = x^2 + C$ •
Take $C = 0$ and $I(x) = e^{\int P(x)\, dx} = e^{x^2}$.

Response 5 (a) Solution is added at the rate of 3 gallons per minute and drained off at the rate of 2 gallons per minute so the volume V increases at the constant rate of 1 gallon per minute. • $V(0) = 10$ gallons at $t = 0$ • $V(t) = 10 + t$ gallons from $t = 0$ to when the tank is full.
(b) $y(0) = 5$ pounds

Response 6 (a) The tank contains $10 + t$ gallons of solution with
$y(t) = 4(10 + t) - 3500(10 + t)^{-2}$ pounds of sugar in it, so the concentration is
$C(t) = \dfrac{y(t)}{10 + t} = 4 - 3500(10 + t)^{-3}$ pounds per gallon at time t.
(b) $C(t) \to 4$ as $t \to \infty$ • This is plausible because 4 is the concentration of the incoming solution and if the tank had infinite volume, the volume of incoming solution would tend to ∞ and the effect of the original solution on the concentration would disappear.

Response 7   For the function $y(x)$ to satisfy the differential equation its graph must be a part of one of the solution curves $x^2y^2 - x + y^2 = c$ from Example 6. • Set $x = 1, y = 2$: $1^2(2^2) - 1 + 2^2 = c$ • $c = 7$ • $y(x)$ is defined implicitly by the solution curve $x^2y^2 - x + y^2 = 7$ from Example 6. • $y^2(x^2 + 1) = x + 7$ • $y = \pm\sqrt{\dfrac{x+7}{x^2+1}}$ •

Take the plus sign to have $y(1) = 2$. • $y(x) = \sqrt{\dfrac{x+7}{x^2+1}}$

Responses 14.2

Response 1   (a) There is a minus sign in **(2)** because k is positive and the force is to the left (in the negative s-direction) in Figure 1 when $y(t)$ is positive and to the right (in the positive s-direction) when $y(t)$ is negative.
(b) There is a minus sign in **(3)** because if there is resistance, then R is positive and the force is to the left when $y'(t)$ is positive and the object is moving toward the right and the force is to the right when $y'(t)$ is negative and the object is moving toward the left.

Response 2   For $y = \cos(\omega t)$: $y' = \dfrac{d}{dt}[\cos(\omega t)] = -\sin(\omega t)\dfrac{d}{dt}(\omega t) = -\omega\sin(\omega t)$;

$y'' = \dfrac{d}{dt}[-\omega\sin(\omega t)] = -\omega\cos(\omega t)\dfrac{d}{dt}(\omega t) = -\omega^2\cos(\omega t)$; and

$y'' + 4y = (-\omega^2 + 4)\cos(\omega t)$ • $y'' + 4y = 0 \iff -\omega^2 + 4 = 0$

For $y = \sin(\omega t)$: $y' = \dfrac{d}{dt}[\sin(\omega t)] = \omega\cos(\omega t)\dfrac{d}{dt}(\omega t) = \omega\cos(\omega t)$;

$y'' = \dfrac{d}{dt}[\omega\cos(\omega t)] = -\omega\sin(\omega t)\dfrac{d}{dt}(\omega t) = -\omega^2\sin(\omega t)$; and

$y'' + 4y = (-\omega^2 + 4)\sin(\omega t)$ • $y'' + 4y = 0 \iff -\omega^2 + 4 = 0$ •
In both cases $\omega = 2$ since $\omega > 0$.

Response 3   (a) The graph passes through the origin and is tangent to the line of slope -2 through the origin because of the initial conditions.
(b) $y = A\cos(2t) + B\sin(2t)$ by **(10)** • $y(0) = A$ and $y(0) = 0 \implies A = 0$ •

$y' = -A\sin(2t)\dfrac{d}{dt}(2t) + B\cos(2t)\dfrac{d}{dt}(2t) = -2A\sin(2t) + 2B\cos(2t)$ •

$y'(0) = -2B$ and $y'(0) = -2 \implies B = -1$ • $y = -\sin(2t)$
(c) The solution has amplitude 1 because its values range between -1 and 1. It has period π because $2t$ increases by 2π when t increases by π.

Response 4   One answer: Suppose the spring is oriented as in Figure 1 and that distances are measured in feet and time in seconds. • At $t = 0$ the object is 1 unit to the right of its rest position and is moving away from it with the velocity of 2 feet per second. Then, after about one-fourth second, the object starts moving toward its rest position, which it approaches as $t \to \infty$.

Response 5   For $y(t) = te^{-t}$:

$y'(t) = \dfrac{d}{dt}(te^{-t}) = t\dfrac{d}{dt}(e^{-t}) + e^{-t}\dfrac{d}{dt}(t) = te^{-t}\dfrac{d}{dt}(-t) + e^{-t} = (-t+1)e^{-t}$ •

$y''(t) = \dfrac{d}{dt}[(-t+1)e^{-t}] = (-t+1)\dfrac{d}{dt}(e^{-t}) + e^{-t}\dfrac{d}{dt}(-t+1)$

$= (-t+1)e^{-t}\dfrac{d}{dt}(-t) + -e^{-t} = (t-2)e^{-t}$ •

$y''(t) + 2y'(t) + y(t) = [(t-2) + 2(-t+1) + t]e^{-t} = 0$

Response 6   (a) $e^{-t}(1-t) = 0 \iff t = 1$. • The t-intercept is 1
(b) One answer: At $t = 0$ the object is one foot to the right of its rest position and is moving toward it with velocity two feet per second. At $t = 1$ it passes through its rest position. It moves to the left for approximately one second and then moves to the right and approaches its rest position as $t \to \infty$.

Response 7   (a) For $y_p(t) = c_1 + c_2 t$: $y_p'(t) = c_2$ and $y_p''(t) = 0$ • $y_p''(t) + 4y_p(t) = 4c_1 + 4c_2 t$ • $y_p''(t) + 4y_p(t) = t \iff c_1 = 0$ and $c_2 = \frac{1}{4}$ • $y_p(t) = \frac{1}{4}t$
(b) The chacteristic equation of $y'' + 4y = 0$ is $r^2 + 4 = 0$. • $r = \pm 2i$ • $y_h(t) = A\cos(2t) + B\sin(2t)$ • $y = y_p + y_h = \frac{1}{4}t + A\cos(2t) + B\sin(2t)$

Response 8   (a) One answer: The object starts at its rest position, moves to the right, and then moves repeatedly to the right and left, where at each stage it moves more to the right than to the left. This is plausible because the force on the spring is constantly increasing, but the spring resists being stretched.
(b) One answer: The length of a real spring could not tend to ∞.

Responses 14.3

Response 1   The initial conditions give
$y^{(0)}(0) = y(0) = 0$ and $y^{(1)}(0) = y'(0) = 1$. •
Use **(8)** with $j = 0, 1, 2, 3, 4, 5$:
$y^{(2)}(0) = -y^{(0)}(0) = 0$ • $y^{(3)}(0) = -y^{(1)}(0) = -1$ •
$y^{(4)}(0) = -y^{(2)}(0) = 0$ • $y^{(5)}(0) = -y^{(3)}(0) = -(-1) = 1$ •
$y^{(6)}(0) = -y^{(4)}(0) = 0$ • $y^{(7)}(0) = -y^{(5)}(0) = -1$

Response 2   By **(16)** with $j = 0, 1, 2, 3, 4$ and $a_0 = 1$:

$a_1 = a_{0+1} = \dfrac{-a_0}{0+1} = \dfrac{-1}{1} = -1$ •

$a_2 = a_{1+1} = \dfrac{-a_1}{1+1} = \dfrac{-(-1)}{2} = \dfrac{1}{2}$ •

$a_3 = a_{2+1} = \dfrac{-a_2}{2+1} = \dfrac{-1/2}{3} = -\dfrac{1}{3!}$ •

$a_4 = a_{3+1} = \dfrac{-a_3}{3+1} = \dfrac{-1/3!}{4} = \dfrac{1}{4!}$ •

$a_5 = a_{4+1} = \dfrac{-a_4}{4+1} = \dfrac{-1/4!}{5} = -\dfrac{1}{5!}$

Response 3 $\mathcal{P}_0$ reads $a_0 = \dfrac{(-1)^0}{0!}$ and is correct because $a_0 = 1$, $(-1)^0 = 1$, and $0! = 1$. •
If for any $j \geq 0$, $\mathcal{P}_j$ is true, then $a_j = \dfrac{(-1)^j}{j!}$ and by **(16)** $a_{j+1} = -\dfrac{1}{j+1}\left(\dfrac{(-1)^j}{j!}\right) = \dfrac{(-1)^{j+1}}{(j+1)!}$ and this is $\mathcal{P}_{j+1}$. • $\mathcal{P}_0$ is true and $\mathcal{P}_j$ implies $\mathcal{P}_{j+1}$ for any $j \geq 0$, so $\mathcal{P}_j$ is true for all $j \geq 0$.

Response 4   **(a)** Calculations **(21)** suggest that a_j is 0 for even, positive integers j and equals $\pm 1/j!$ for odd, positive integers j with alternating plus and minus signs starting with a plus sign for a_1. • Prediction $\mathcal{P}_k :$ $\left\{ \begin{array}{l} a_{2k} = 0 \\ a_{2k+1} = \dfrac{(-1)^k}{(2k+1)!} \end{array} \right\}$ for $k \geq 0$ • $\mathcal{P}_0$ states $a_0 = 0$ and $a_1 = \dfrac{(-1)^0}{(2(0)+1)!} = 1$ and is true. • If $\mathcal{P}_k$ is true for any $k \geq 0$, then $a_{2k} = 0$ and $a_{2k+1} = \dfrac{(-1)^k}{(2k+1)!}$ so that by **(20)**,

$$a_{2(k+1)} = a_{2k+2} = -\frac{a_{2k}}{(2k+2)(2k+1)} = 0 \text{ and } a_{2(k+1)+1} = a_{2k+3} =$$

$$-\frac{a_{2k+1}}{(2k+3)(2k+2)} = -\frac{1}{(2k+3)(2k+1)}\left(\frac{(-1)^k}{(2k+1)!}\right) = \frac{(-1)^{k+1}}{(2k+3)!}$$

and this is $\mathcal{P}_{k+1}$. • $\mathcal{P}_k$ is true for $k \geq 0$ by Mathematical Induction.

(b) $y(x) = \displaystyle\sum_{k=0}^{\infty} \dfrac{1}{(2k+1)!} (-1)^k x^k = \sin x$

Response 5   **(a)** Replacing x by $-x^2$ in the MacLaurin series $e^x = \displaystyle\sum_{k=0}^{\infty} \dfrac{1}{k!} x^k$ gives

$$e^{-x^2} = \sum_{k=0}^{\infty} \frac{1}{k!}(-x^2)^k = \sum_{k=0}^{\infty} \frac{(-1)^k}{k!} x^{2k}.$$

(b) $y(0) = e^0 = 1$ • $y'(x) = -2xe^{-x^2}$ • $y'(0) = 0$ • $y''(x) = (-2 + 4x^2)e^{-x^2}$ •

$y''(x) + 2xy'(x) + 2y(x) = [(-2+4x^2) + 2x(-2x) + 2]e^{-x^2} = 0$

CHAPTER 14 ANSWERS

Tune-up exercises 14.1

T1. $\dfrac{dy}{dx} - y = 2$ • $P = -1$ • $\displaystyle\int P\,dx = -x + C$ • Take $C = 0$. •
$I = e^{\int P\,dx} = e^{-x}$ • $e^{-x}\dfrac{dy}{dx} - e^{-x}y = 2e^{-x}$ • $\dfrac{d}{dx}(e^{-x}y) = 2e^{-x}$ • $e^{-x}y = 2\displaystyle\int e^{-x}\,dx$ •
$e^{-x}y = -2e^{-x} + C$ • $y = -2 + Ce^{x}$

T2. $\dfrac{d}{dx}(y+2) = y + 2$ • $y + 2 = Ce^{x}$ • $y = -2 + Ce^{x}$

T3. $P(x) = -\sin x$ • $\displaystyle\int P(x)\,dx = \cos x + C$ • Take $C = 0$. • $I(x) = e^{\cos x}$ •
$e^{\cos x}\dfrac{dy}{dx} - (\sin x)e^{\cos x}y = 5(\sin x)e^{\cos x}$ • $\dfrac{d}{dx}[e^{\cos x}y] = 5(\sin x)e^{\cos x}$ •
$e^{\cos x}y = 5\displaystyle\int(\sin x)e^{\cos x}\,dx$ • Use $u = \cos x,\, du = -\sin x\,dx$. •
$5\displaystyle\int(\sin x)e^{\cos x}\,dx = -5\displaystyle\int e^{u}\,du = -5e^{u} + C = -5e^{\cos x} + C$ •
$e^{\cos x}y = -5e^{\cos x} + C$ • $y = -5 + Ce^{-\cos x}$

T4. Divide by x: $\dfrac{dy}{dx} + \dfrac{5}{x}y = 9x^{3}$ • $P = \dfrac{5}{x}$ • $\displaystyle\int P\,dx = 5\ln|x| + C$ • Take $C = 0$ and consider only $x > 0$. • $I = e^{5\ln x} = (e^{\ln x})^{5} = x^{5}$ • $x^{5}\dfrac{dy}{dx} + 5x^{4}y = 9x^{8}$ • $\dfrac{d}{dx}[x^{5}y] = 9x^{8}$ • $x^{5}y = \displaystyle\int 9x^{8}\,dx$ •
$x^{5}y = x^{9} + C$ • Set $x = 1, y = 3$: $3 = 1 + C$ • $C = 2$ • $x^{5}y = x^{9} + 2$ • $y = x^{4} + 2x^{-5}$

T5. $p = y + 1$, $q = x - 1$ • $p_{y} = 1$ and $q_{x} = 1$ are equal. • The differential equation is exact. •
$U_{x} = y + 1 \implies U = xy + x + \phi(y) \implies U_{y} = x + \phi'(y)$ • $U_{y} = x - 1 \implies \phi'(y) = -1$ •
$\phi(y) = -y + C$ • Use $C = 0$. • $U = xy + x - y$ • Solution curves: $xy + x - y = c$

T6. $p = xy + x$, $q = y^{2} - xy$ • $p_{y} = x$ and $q_{x} = -y$ are not equal in any open set. • The differential equation is not exact.

Problems 14.1

1. $\displaystyle\int P\,dx = -\displaystyle\int x^{-2}\,dx = x^{-1} + C$ • Use $I = e^{1/x}$. • $e^{1/x}\dfrac{dy}{dx} - x^{-2}e^{1/x}y = 6x^{-2}e^{1/x}$ •
$\dfrac{d}{dx}(ye^{1/x}) = 6x^{-2}e^{1/x}$ • $ye^{1/x} = -6e^{1/x} + C$ • $y = -6 + Ce^{-1/x}$

2. $\displaystyle\int P\,dx = -\displaystyle\int[x/(x^{2}+1)]\,dx = -\tfrac{1}{2}\ln(x^{2}+1) + C$ • Use $I = (x^{2}+1)^{-1/2}$. •
$(x^{2}+1)^{-1/2}\dfrac{dy}{dx} - x(x^{2}+1)^{-3/2}y = x^{2}+1$ • $\dfrac{d}{dx}[y(x^{2}+1)^{-1/2}] = x^{2}+1$ • $y(x^{2}+1)^{-1/2} = \tfrac{1}{3}x^{3} + x + C$
• $y = \sqrt{x^{2}+1}\left(\tfrac{1}{3}x^{3} + x + C\right)$

3. $y = \sin x \cos x + C\cos x$

5. $y = -2 + 2e^{x^{3}}$

7. $y = \tfrac{1}{12}(2x+1) + \tfrac{11}{12}(2x+1)^{-5}$

8. $y = x^{5} - x^{4}$

9. **(a)** [Volume of solution at time t] $= 8 + 3t$. • Set $y =$ [Weight (pounds) of salt]. •
$\dfrac{dy}{dt} = 2 - \dfrac{y}{8+3t}$ or $\dfrac{dy}{dt} + \dfrac{y}{8+3t} = 2$ • $\displaystyle\int P\,dt = \displaystyle\int\dfrac{1}{8+3t}\,dt = \tfrac{1}{3}\ln|8+3t| + C$ •
Use $I = (8+3t)^{1/3}$. • $\dfrac{d}{dt}[y(8+3t)^{1/3}] = 2(8+3t)^{1/3}$ • $y(8+3t)^{1/3} = \tfrac{2}{3}(\tfrac{3}{4})(8+3t)^{4/3} + C$ •
Set $t = 0$ and $y = 5$: $C = 2$. • $y = \tfrac{1}{2}(8+3t) + 2(8+3t)^{-1/3}$ pounds.

(b) [Concentration] $= \dfrac{y}{8+3t} = \dfrac{\tfrac{1}{2}(8+3t) + 2(8+3t)^{-1/3}}{8+3t} = \tfrac{1}{2} + 2(8+3t)^{-4/3}$ pounds per gallon.

10. (a) The tank is full at $t = 7$ minutes. (b) $y = 3(1 + 7t) - (1 + 7t)^{-3/7}$ pounds

13. $\frac{\partial}{\partial y}(2x - 6y) = -6$ and $\frac{\partial}{\partial x}(4x^3 - 6x) = -6$ for all (x, y) • Exact •
$U = \int (2x - 6y)\, dx = x^2 - 6xy + \phi(y)$ • $U_y = -6x + \phi'(y) = 4y^3 - 6x$ • $\phi'(y) = 4y^3$ •
$\phi(y) = y^4 + C$ • $U = x^2 - 6xy + y^4 + C$ • $x^2 - 6xy + y^4 = c$

14. $x^2 e^y + e^{-x} - y^2 = c$

17. $\sin(xy) + x - \cos y = c$

19. $x^3 y^2 + x^2 - y^3 = c$

21. $\frac{\partial}{\partial y}(2x^3 y^2 - \sin x \cos x) = 4x^3 y$ and $\frac{\partial}{\partial x}(x^4 y + 2y) = 4x^3 y$ for all (x, y) • Exact •
$U = \int (2x^3 y^2 - \sin x \cos x)\, dx = \frac{1}{2}x^4 y^2 - \frac{1}{2}\sin^2 x + \phi(y)$ • $U_y = x^4 y + \phi'(y) = x^4 y + 2y$ •
$\phi'(y) = 2y$ • $\phi(y) = y^2 + C$ • $U = \frac{1}{2}x^4 y^2 - \frac{1}{2}\sin^2 x + y^2 + C$ • $\frac{1}{2}x^4 y^2 - \frac{1}{2}\sin^2 x + y^2 = c$ • Set $x = 0$ and $y = 3$: $c = 9$ • $\frac{1}{2}x^4 y^2 - \frac{1}{2}\sin^2 x + y^2 = 9$ • $y = \sqrt{(18 + \sin^2 x)/(x^4 + 2)}$

22. $y = \dfrac{x^2 + 2}{x^2 + 1}$

25. Not exact.

29. $y = (\frac{1}{4}x^4 + x + k)(x^3 + 1)^{-1/3}$

33. $y = \frac{2}{5}\cos x + \frac{1}{5}\sin x + \frac{2}{5}e^{2\pi - 2x}$

37. $\int P\, dx = x^{1/2} + C$ • $I = e^{\sqrt{x}}$ • $\frac{d}{dx}[ye^{\sqrt{x}}] = e^{\sqrt{x}}\sin x$ • $ye^{\sqrt{x}} = C + \int_1^x e^{\sqrt{t}}\sin t\, dt$ •
Set $x = 1$ and $y = 4$: $4e = C$ • $y = e^{-\sqrt{x}}\left(4e + \int_1^x e^{\sqrt{t}}\sin t\, dt\right)$

38. $y = e^x \int_0^x e^{-t}\tan t\, dt$

41. $y = x + Ce^y - 1 \Rightarrow \frac{dy}{dx} = 1 + Ce^y \frac{dy}{dx}$ • $Ce^y = y - x + 1$ so $\frac{dy}{dx} = \dfrac{1}{x - y}$ •
Orthogonal trajectories: $\frac{dy}{dx} = y - x$ • $\int P\, dx = \int -1\, dx = -x + c$ and $I = e^{-x}$ •
$e^{-x} y = (x + 1)e^{-x} + c$ • $y = x + 1 + ce^x$

42. $xy^2 + \frac{1}{3}x^3 = c$

Tune-up exercises 14.2

T1. $y'' + 3y' + 2y = 0$ • Characteristic equation: $r^2 + 3r + 2 = 0$ • $(r + 2)(r + 1) = 0$ • $r = -1, -2$ •
General solution: $y = Ae^{-t} + Be^{-2t}$ • $y' = -Ae^{-t} - 2Be^{-2t}$ • $y(0) = A + B, y'(0) = -A - 2B$ •
$y(0) = 2, y'(0) = -3$ • $\begin{cases} A + B = 2 \\ -A - 2B = -3 \end{cases}$ • Add: $B = 1, A = 1$ • $y = e^{-t} + e^{-2t}$

T2. $y'' - y' = 0$ • Characteristic equation: $r^2 - r = 0$ • $r(r - 1) = 0$ • $r = 0, 1$ • General solution: $y = Ae^0 + Be^t$ • $y = A + Be^t$

T3. $y'' - y = t$ • Homogeneous equation: $y'' - y = 0$ • Characteristic equation: $r^2 - 1 = 0$ •
$r = \pm 1$ • $y_h = Ae^t + Be^{-t}$ • UC set: $z_1 = t, z_2 = 1$ • $y_p = c_1 t + c_2$ • $y_p' = c_1$ • $y_p'' = 0$ •
$y_p'' - y_p = -c_1 t - c_2$ • $y_p'' - y_p = t \Rightarrow c_1 = -1, c_2 = 0$ • $y_p = -t$ •
General solution: $y = y_p + y_h = -t + Ae^t + Be^{-t}$

T4. (a) For $y'' + 3y' + 2y = 0$: $a = 1, b = 3, c = 2$ • $b^2 - 4ac = 1 > 0$ • Overdamped • Figure 12
(b) For $y'' + 9y = 0$: $a = 1, b = 0, c = 9$ • Undamped • Figure 10
(c) For $y'' + 2y' + 10y = 0$: $a = 1, b = 3, c = 10$ • $b^2 - 4ac = -31 < 0$ • Underdamped • Figure 11

Answers 14.2

T5. (a) $y = \sin t - \frac{1}{6}\sin(6t)$ • $y' = \cos t - \cos(6t)$ • $y'' = -\sin t + 6\sin(6t)$ •
$y'' + 36y = [-\sin t + 6\sin(6t)] + [36\sin t - 6\sin(6t)] = 35\sin t$ •
$y(0) = \sin(0) - \frac{1}{6}\sin(0) = 0$ • $y'(0) = \cos(0) - \cos(0) = 0$
(b) One answer: $y = \sin t - \frac{1}{6}\sin(6t)$ is a sinusoidal function of period $\frac{1}{3}\pi$ and amplitude $\frac{1}{6}$ added to a sinusoidal function of period 2π and amplitude 1.

T6. (a) $y = 1 + \frac{1}{6}\sin(6t)$ • $y' = \cos(6t)$ • $y'' = -6\sin(6t)$ •
$y'' + 36y = [-6\sin(6t)] + [36 + 6\sin(6t)] = 36$ • $y(0) = 1 + \frac{1}{6}\sin(0) = 1$ • $y'(0) = \cos(0) = 1$
(b) One answer: $y = 1 + \frac{1}{6}\sin(6t)$ is a sinusoidal function of period $\frac{1}{3}\pi$ and amplitude $\frac{1}{6}$ added to the constant function 1.

Problems 14.2

1. $2y'' - 5y' - 3y = 0$ • $2r^2 - 5r - 3 = 0$ • $r = \dfrac{-(-5) \pm \sqrt{(-5)^2 - 4(2)(-3)}}{2(2)} = \frac{1}{4}(5 \pm 7) = 3, -\frac{1}{2}$ •
$y = Ae^{3t} + Be^{-t/2}$ • $y' = 3Ae^{3t} - \frac{1}{2}Be^{-t/2}$ • $y(0) = A + B$ • $y'(0) = 3A - \frac{1}{2}B$ •
$y(0) = 1, y'(0) = -4 \implies A = -1, B = 2$ • $y = -e^{3t} + 2e^{-t/2}$

2. $9y'' - 12y' + 4y = 0$ • $9r^2 - 12r + 4 = 0$ • $r = \dfrac{-(-12) \pm \sqrt{(-12)^2 - 4(9)(4)}}{2(9)} = \frac{2}{3}$ •
$y = (A + Bt)e^{2t/3}$

3. $y = 2e^t - e^{2t}$

5. $y = Ae^t + Be^{-3t}$

6. $y'' - 2y' + 2y = 0$ • $r^2 - 2r + 2 = 0$ • $r = \dfrac{-(-2) \pm \sqrt{(-2)^2 - 4(2)(1)}}{2(1)} = 1 \pm i$ •
$y = e^t(A\cos t + B\sin t)$ • $y' = e^t[(A+B)\cos t + (B-A)\sin t]$ • $y(0) = A$ • $y'(0) = A + B$ •
$y(0) = 0, y'(0) = 1 \implies A = 0, B = 1$ • $y = e^t \sin t$

7. $y = e^{-2t}(-\cos t - 2\sin t)$

9. $y'' - y = t$ • Homogeneous equation: $y'' - y = 0$ • $r^2 - 1 = 0$ • $r = \pm 1$ •
$y_h = Ae^t + Be^{-t}$ • $z_1 = 1$ and $z_2 = t$ for a UC set. • $y_p = c_1 + c_2 t$ • $y_p' = c_2$ • $y_p'' = 0$ •
$y_p'' - y_p = -c_1 - c_2 t$ • $y_p'' - y_p = t \implies c_1 = 0, c_2 = -1$ • $y_p = -t$
$y = -t + Ae^t + Be^{-t}$ • $y' = -1 + Ae^t - Be^{-t}$ • $y(0) = A + B$ • $y'(0) = -1 + A - B$ •
$y(0) = 0, y'(0) = 1 \implies A = 1, B = -1$ • $y = -t + e^t - e^{-t}$

10. $y = e^t + A\cos(2t) + B\sin(2t)$

12. $y'' + y = 2\cos t$ • Homogeneous equation: $y'' + y = 0$ • $r^2 + 1 = 0$ • $r = \pm i$ •
$y_h = A\cos t + B\sin t$ • $z_1 = t\cos t$ and $z_2 = t\sin t$ form a UC set. • $y_p = c_1 t\cos t + c_2 t\sin t$ •
$y_p' = -c_1 t\sin t + c_2 t\cos t + c_1 \cos t + c_2 \sin t$ • $y_p'' = (-c_1 t\cos t - c_2 t\sin t) + 2(-c_1 \sin t + c_2 \cos t)$ •
$y_p'' + y_p = -2c_1 \sin t + 2c_2 \cos t$ • $y_p'' + y_p = 2\cos t \implies c_1 = 0, c_2 = 1$ • $y_p = t\sin t$ •
$y = t\sin t + A\cos t + B\sin t$ • $y' = t\cos t + \sin t - A\sin t + B\cos t$ • $y(0) = A$ • $y'(0) = B$ •
$y(0) = 0, y'(0) = 1 \implies A = 0, B = 1$ • $y = t\sin t + \sin t$

13. $y = te^t + Ae^t + Be^{-t}$

16. $y = t^2 e^t$

17. $y = (14 - 11t)e^{2t-2}$

19. $y = e^{t/3}[A\cos(\frac{1}{3}t) + B\sin(\frac{1}{3}t)]$

21. (a) $My'' + Ry' + ky = 0$ with $M = 2, R = 2$, and $k = 5$ • $2y'' + 2y' + 5y = 0, y(0) = 2, y'(0) = -1$
(b) Free because $f = 0$ • Damped because $R > 0$ • $R^2 - 4Mk = 2^2 - 4(2)(5) = -36 < 0$ • Underdamped
(c) $2r^2 + 2r + 5 = 0$ • $r = \dfrac{-2 \pm \sqrt{2^2 - 4(2)(5)}}{2(2)} = -\frac{1}{2} \pm \frac{3}{2}i$ • $y = e^{-t/2}[A\cos(\frac{3}{2}t) + B\sin(\frac{3}{2}t)]$ •
$y' = e^{-t/2}[(-\frac{1}{2}A + \frac{3}{2}B)\cos(\frac{3}{2}t) + (-\frac{1}{2}B - \frac{3}{2}A)\sin(\frac{3}{2}t)]$ • $y(0) = A$ • $y'(0) = -\frac{1}{2}A + \frac{3}{2}B$ •
$y(0) = 2, y'(0) = -1 \implies A = 2, B = 0$ • $y = 2e^{-t/2}\cos(\frac{3}{2}t)$ • Figure A21

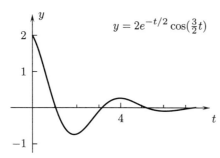

Figure A21

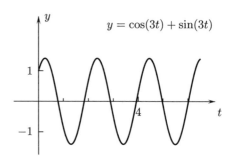

Figure A22

22. (a) Free because $f = 0$ • Undamped because $R = 0$
    (b) $y = \cos(3t) + \sin(3t)$ • Figure A22

23. (a) Free because $f = 0$ • Overdamped because $R^2 - 4Mk = 3^2 - 4(1)(2) = 1 > 0$
    (b) $y = 2e^{-t} - 3e^{-2t}$ • Figure A23

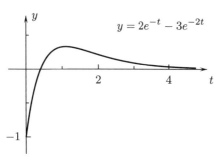

Figure A23

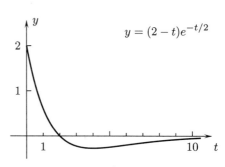

Figure A25

25. (a) Free because $f = 0$ • Critically damped because $R^2 - 4Mk = 4^2 - 4(4)(1) = 0$
    (b) $y = (2-t)e^{-t/2}$ • Figure A25

27. (a) Forced because $f \neq 0$ • Undamped because $R = 0$
    (b) Homogeneous equation: $y'' + y = 0$ • $r^2 + 1 = 0$ • $r = \pm i$ • $y_h = A\cos t + B\sin t$ •
    For $f(t) = \frac{2}{5}\sin t$: $w_1 = \sin t, w_2 = \cos t$ are solutions of the homogeneous equation. •
    $z_1 = t\cos t, z_2 = t\sin t$ form a UC set. • $y_p = c_1 t\cos t + c_2 t\sin t$ •
    $y_p' = -c_1 t\sin t + c_2 t\cos t + c_1 \cos t + c_2 \sin t$ • $y_p'' = -c_1 t\cos t - c_2 t\sin t - 2c_1 \sin t + 2c_2 \cos t$ •
    $y_p'' + y_p = -2c_1 \sin t + 2c_2 \cos t$ • $y_p'' + y_p = \frac{2}{5}\sin t \implies c_1 = -\frac{1}{5}, c_2 = 0$ •
    $y_p = -\frac{1}{5}t\cos t$ • $y = -\frac{1}{5}t\cos t + A\cos t + B\sin t$ • $y' = \frac{1}{5}t\sin t - \frac{1}{5}\cos t - A\sin t + B\cos t$ •
    $y(0) = A, y'(0) = B - \frac{1}{5}$ • $y(0) = 0, y'(0) = \frac{2}{5} \implies A = 0, B = \frac{3}{5}$ • $y = \frac{3}{5}\sin t - \frac{1}{5}t\cos t$ •
    Figure A27

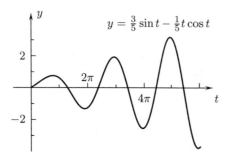

Figure A27

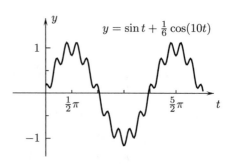

Figure A28

Answers 14.3

28. (a) Forced because $f \neq 0$ • Undamped because $R = 0$
(b) $y = \sin t + \frac{1}{6}\cos(10t)$ • Figure A28

29. (a) Forced because $f \neq 0$ • Underdamped because $R = 2 > 0$ and $R^2 - 4Mk = 2^2 - 4(1)(10) = -36 < 0$
(b) $y = e^{-t}\cos(3t) + \cos t$ • Figure A29

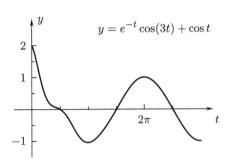

Figure A29

35a. $y = \dfrac{e^{-t} - e^{-\lambda t}}{\lambda - 1}$

36. $q(t) = e^{-t}[5\cos(2t) + \frac{5}{2}\sin(2t)]$

Tune-up exercises 14.3

T1. (a) $y'(x) = y(x) \implies y^{(j)}(x) = y(x)$ for $j = 1, 2, 3, \ldots$ •
$y(0) = 1 \implies y^{(j)}(0) = 1$ for all $j \geq 0$ • $y(x) = \sum_{j=0}^{\infty} \frac{1}{j!} y^{(j)}(0) x^j = \sum_{j=0}^{\infty} \frac{1}{j!} x^j$
(b) $y(x) = e^x$

T2. $y(x) = \sum_{j=0}^{\infty} a_j x^j$ • $y'(x) = \sum_{j=0}^{\infty} j a_j x^{j-1} = \sum_{j=1}^{\infty} j a_j x^{j-1}$ •
Set $j = k+1$: $y'(x) = \sum_{k=0}^{\infty} (k+1)a_{k+1} x^k$ • Replace k by j: $y'(x) = \sum_{j=0}^{\infty}(j+1)a_{j+1}x^j$ •
$y'(x) - y(x) = \sum_{j=0}^{\infty}[(j+1)a_{j+1} - a_j] x^j$ •
$y'(x) - y(x) = 0$ for all $x \implies (j+1)a_{j+1} - a_j = 0$ for $j \geq 0$ •
$(*)\; a_{j+1} = \dfrac{a_j}{j+1}$ for $j \geq 0$ •
$y(0) = 1 \implies a_0 = 1$ • $a_1 = \dfrac{a_0}{1} = 1$ • $a_2 = \dfrac{a_1}{2} = \dfrac{1}{2}$ • $a_3 = \dfrac{a_2}{3} = \dfrac{1}{3(2)} = \dfrac{1}{3!}$ • $a_4 = \dfrac{a_3}{4} = \dfrac{1}{4!}$ •
Prediction $\mathcal{P}_j : a_j = \dfrac{1}{j!}$ for $j \geq 0$ • $\mathcal{P}_0$ is true. •
If $\mathcal{P}_j$ is true for any $j \geq 0$, then $a_j = \dfrac{1}{j!}$ and by $(*)$, $a_{j+1} = \dfrac{a_j}{j+1} = \dfrac{1}{(j+1)j!} = \dfrac{1}{(j+1)!}$ so that
$\mathcal{P}_{j+1}$ is true. • $\mathcal{P}_j$ is true for $j \geq 0$ by Mathematical Induction. • $y(x) = \sum_{j=0}^{\infty} \dfrac{1}{j!} x^j = e^x$

T3. (a) $y''(x) = y(x) \implies y^{(j+2)}(x) = y^{(j)}(x)$ for $j \geq 0$ •
$(*)\; y^{(j+2)}(0) = y^{(j)}(0)$ for $j \geq 0$ •
Initial conditions: $y(0) = 1, y'(0) = -1$ • $y''(0) = y(0) = 1$ • $y^{(3)}(0) = y'(0) = -1$ •
$y^{(4)}(0) = y''(0) = 1$ • $y^{(5)}(0) = y^{(3)}(0) = -1$
Prediction $\mathcal{P}_j : y^{(j)}(0) = (-1)^j$ for $j \geq 0$ • $\mathcal{P}_0$ and $\mathcal{P}_1$ are true. •
If for any $j \geq 1$ $\mathcal{P}_k$ is true for $0 \leq k \leq j$, then $y^{(j-1)(0)} = (-1)^{j-1}$ and by $(*)$,
$y^{(j+1)}(0) = (-1)^{(j-1)} = (-1)^{(j+1)}$ and this is $\mathcal{P}_{j+1}$. •

$\mathcal{P}_j$ is true for $j \geq 0$ by Mathematical Induction, as stated in part (b) of Rule 1. •

$$y(x) = \sum_{j=0}^{\infty} \frac{1}{j!} y^{(j)}(0) x^j = \sum_{j=0}^{\infty} \frac{1}{j!} (-1)^j x^j = \sum_{j=0}^{\infty} \frac{1}{j!} (-x)^j$$

(b) $y(x) = e^{-x}$

T4. $y(x) = \sum_{j=0}^{\infty} a_j x^j$ • $y'(x) = \sum_{j=0}^{\infty} (j+1) a_{j+1} x^j$ (See the solution of T2.) •

$y''(x) = \sum_{j=2}^{\infty} j(j-1) a_j x^{j-2} = \sum_{j=0}^{\infty} (j+2)(j+1) a_{j+2} x^j$ (Set $k = j - 2$ and then replace k by j.) •

$y''(x) - y(x) = \sum_{j=0}^{\infty} [(j+2)(j+1) a_{j+2} - a_j] x^j$ •

$y''(x) - y(x) = 0$ for all $x \implies (j+2)(j+1) a_{j+2} - a_j = 0$ for $j \geq 0$ •

(∗) $a_{j+2} = \dfrac{a_j}{(j+2)(j+1)}$ for $j \geq 0$ •

$y(0) = 1 \implies a_0 = 1$ • $y'(0) = -1 \implies a_1 = -1$ • $a_2 = \dfrac{a_0}{1(2)} = \dfrac{1}{2!}$ • $a_3 = \dfrac{a_1}{3(2)} = \dfrac{-1}{3(2)} = \dfrac{-1}{3!}$ •

$a_4 = \dfrac{a_2}{4(3)} = \dfrac{1/2}{4(3)} = \dfrac{1}{4!}$ • $a_5 = \dfrac{a_3}{5(4)} = \dfrac{-1/3!}{5(4)} = \dfrac{-1}{5!}$ •

Prediction $\mathcal{P}_j : a_j = \dfrac{(-1)^j}{j!}$ for $j \geq 0$ • $\mathcal{P}_0$ and $\mathcal{P}_1$ are true. •

Use part (b) of Rule 1. • If for any $j \geq 1$, $\mathcal{P}_k$ is true for $0 \leq k \leq j$, then $a_k = \dfrac{(-1)^k}{k!}$ for $0 \leq k \leq j$

and by (∗), $a_{j+1} = \dfrac{a_{j-1}}{(j+1)j} = \dfrac{(-1)^{j-1}/(j-1)!}{(j+1)j} = \dfrac{(-1)^{j+1}}{(j+1)!}$ so that $\mathcal{P}_{j+1}$ is true. •

$\mathcal{P}_j$ is true for $j \geq 0$ by Mathematical Induction. • $y(x) = \sum_{j=0}^{\infty} \dfrac{(-1)^j}{j!} x^j = e^{-x}$

Problems 14.3

1. $y = \sum_{j=0}^{\infty} a_j x^j$ • $y'' + 2y' + y = \sum_{j=0}^{\infty} [(j+2)(j+1) a_{j+2} + 2(j+1) a_{j+1} + a_j] x^j$ •

(∗) $a_{j+2} = -\dfrac{2}{j+2} a_{j+1} - \dfrac{1}{(j+2)(j+1)} a_j$ for $j \geq 0$ • $y(0) = 0 \implies a_0 = 0$ •

$y'(0) = 1 \implies a_1 = 1$ • $a_2 = -a_1 - \frac{1}{2} a_0 = -1$ • $a_3 = \dfrac{1}{2!}$ • $a_4 = -\dfrac{1}{3!}$ •

Prediction $\mathcal{P}_j : a_j = \dfrac{(-1)^{j-1}}{(j-1)!}$ for $j \geq 1$ • $\mathcal{P}_1$ and $\mathcal{P}_2$ are true. • If for any $j \geq 1$, $\mathcal{P}_k$ is true for $1 \leq k \leq j$, then by (∗),

$a_{j+1} = -\dfrac{2}{j+1} a_j - \dfrac{1}{(j+1)j} a_{j-1} = \dfrac{-2}{j+1} \left(\dfrac{(-1)^{j-1}}{(j-1)!} \right) - \dfrac{1}{(j+1)j} \left(\dfrac{(-1)^{j-2}}{(j-2)!} \right)$

$= (-1)^j \left[\dfrac{2j}{(j+1)!} - \dfrac{j-1}{(j+1)!} \right] = (-1)^j \left[\dfrac{j+1}{(j+1)!} \right] = \dfrac{(-1)^j}{j!}$ and this is $\mathcal{P}_{j+1}$. •

$\mathcal{P}_j$ is true for $j \geq 0$ by Mathematical Induction. • $y = \sum_{j=1}^{\infty} \dfrac{(-1)^{j-1}}{(j-1)!} x^j = \sum_{j=0}^{\infty} \dfrac{(-1)^j}{j!} x^{j+1} = x e^{-x}$

2. $y = \sum_{j=0}^{\infty} a_j x^j$ • $y'' - y' - 2y = \sum_{j=0}^{\infty} [(j+2)(j+1) a_{j+2} - (j+1) a_{j+1} - 2 a_j] x^j$ •

(∗) $a_{j+2} = \dfrac{a_{j+1}}{j+2} + \dfrac{2 a_j}{(j+2)(j+1)}$ for $j \geq 0$ • $y(0) = 1 \implies a_0 = 1$ and $y'(0) = 2 \implies a_1 = 2$ •

$a_2 = \dfrac{4}{2!}$ • $a_3 = \dfrac{8}{3!}$ • Prediction $P_j : a_j = \dfrac{2^j}{j!}$ for $j \geq 0$ • P_0 and P_1 are true. • If for any $j \geq 0$,

P_j holds for $0 \leq k \leq j$, then by (∗), $a_{j+1} = \dfrac{1}{j+1} a_j + \dfrac{2}{(j+1)j} a_{j-1} = \dfrac{2^j}{(j+1)j!} + \dfrac{2(2)^{j-1}}{(j+1)j(j-1)!}$

Answers 14.3

$$= \frac{2^{j+1}}{(j+1)!} \text{ and } P_{j+1} \text{ holds.} \bullet P_j \text{ holds for all } j \geq 0 \text{ by Mathematical Induction.} \bullet$$
$$y = \sum_{j=0}^{\infty} \frac{2^j}{j!} x^j = \sum_{j=0}^{\infty} \frac{1}{j!} (2x)^j = e^{2x}$$

4. $y = \sum_{k=0}^{\infty} \frac{1}{k!} x^{2k} = e^{x^2}$

7. $y = \sum_{j=0}^{\infty} x^j = \frac{1}{1-x}$ for $|x| < 1$

8. $y = \sum_{j=0}^{\infty} a_j x^j \bullet y' - y = \sum_{j=0}^{\infty} [(j+1)a_{j+1} - a_j] x^j = -x^2 \bullet$
$(j+1)a_{j+1} - a_j$ equals -1 for $j=2$ and equals 0 for $j=0,1,$ and $j \geq 3.$ $\bullet$
$y(0) = 2 \implies a_0 = 2 \bullet a_1 - a_0 = 0 \implies a_1 = 2 \bullet 2a_2 - a_1 = 0 \implies a_2 = 1 \bullet$
$3a_3 - a_2 = -1 \implies a_3 = 0 \bullet (j+1)a_{j+1} = a_j$ for $j \geq 3 \implies a_j = 0$ for $j \geq 4 \bullet y = 2 + 2x + x^2$

11. $y' = x^2 + y^2 \bullet y(0) = 1 \implies y'(0) = 1 \bullet$ Differentiate both sides of the differential equation:
$y'' = 2x + 2yy' \bullet y''(0) = 0 + 2y(0)y'(0) = 2 \bullet$ Differentiate again: $y^{(3)} = 2 + 2(y')^2 + 2yy'' \bullet$
$y^{(3)}(0) = 8 \bullet$ Differentiate once more: $y^{(4)} = 6y'y'' + 2yy^{(3)} \bullet y^{(4)}(0) = 12 + 16 = 28. \bullet$
By Taylor's formula: $y = 1 + x + \frac{1}{2}(2)x^2 + \frac{1}{3!}(8)x^3 + \frac{1}{4!}(28)x^4 + \cdots = 1 + x + x^2 + \frac{4}{3}x^3 + \frac{7}{6}x^4 \cdots$

13. $y = 1 + 2x + \frac{1}{2}x^2 + \frac{2}{3}x^3 + \frac{5}{12}x^4 + \cdots$

15. $y = \sum_{k=0}^{\infty} \frac{1}{k!} x^{2k} = e^{x^2}$

16. $y = 7 - 6x + x^2$

17. $y = 3 - 12x^2 + 4x^4$

20. $y = \sum_{j=0}^{\infty} \frac{(-1)^j}{(j!)^2} (\frac{1}{2}x)^{2j}$

22. $y = \sum_{j=0}^{\infty} \frac{1}{j!} (\frac{1}{2}x^2)^j = e^{x^2/2}$

26. $y = x - \frac{1}{2}x^2 + \frac{1}{6}x^3 - \frac{1}{8}x^4 + \frac{1}{20}x^5 + \cdots$

Index

$b < a$ in parametrized curves, **375**
$e^{i\theta}$, **471**
$e^{\alpha+i\theta}$, **471**
nabla
    with three variables, **424**
$\langle A_1, A_2, A_3 \rangle$, **25**
$\langle A_1, A_2 \rangle$, **3**
A, **3**
i, **7**, **29**
j, **7**, **29**
k, **29**

Acceleration
    normal and tangential components of, **92**
Acceleration vector, **86**
Adding three vectors in space, **48**
Adding vectors, **4**
Affine transformation
    and change in area, **334**
    in double integrals, 336, **337**
    in the plane, **332**
    inverse, **335**
Angle of inclination
    of a vector, **3**
Arclength, **67**
Area
    of a parametrized surface, **419**
Area of a parallelogram
    with the cross product, **46**
Area of a triangle
    with the cross product, **46**
Average value
    of $f(x,y)$, **290**
    of a function on a curve, **386**
    on a surface, **420**
    with three variables, 305

Boundary
    of a surface, **425**
Boundary of a set
    in xyz-space, **214**
    in an xy-plane, **147**
Center of gravity
    of a plate, **288**
    of a plate, 287
    of a surface, **421**
    with three variables, 304
Chain Rule
    with three variables, **215**
    with two variables, 171, 172, **173**, 174
Circle of curvature
    of a plane curve, **91**
Circulation
    of a vector field, **399**
Closed set
    in xyz-space, **214**
    in an xy-plane, **147**
Closure of a set
    in xyz-space, **214**
    in an xy-plane, **147**
Compatible orientation
    of surfaces, **425**
Component
    of a vector, 12, **14**
Component of a vector, **3**
Components
    of a vector, in the direction of another, **29**
Conservative force field
    in the plane, **403**
Constraint curve, 234
Continuity
    of a function of three variables, **214**
    of a function of two variables, **149**, 150

Contour curve
    of a function of two variables, **145**
Coordinate surface
    in cylindrical coordinates, **316**
    in spherical coordinates, **321**
Cramer's rule, 54
Critical point
    with two or three variables, **226**
Cross product
    of two vectors, **39**, 42
Cross section
    of $f(x,y)$, **187**
Curl
    in xyzspace, **426**
Curl of a two-dimensional vectorfield, **399**
Curvature
    of a plane curve, **89**, 90
Curve
    length of, **67**
    oriented, 56
    parametric equations of, 56
    sum of two, **377**
    vector representation of, **59**
Cylindrical coordinates, **316**

Damped motion
    of a spring, **467**
Density
    two dimensional and weight, **286**
Density and weight
    with three variables, **303**
Derivative
    of a vector, **62**
    partial, of $f(x,y)$, **165**
Determinant, 41
Diameter
    of a solid, **299**
Differential
    total, with three variables, **218**
    with three variables, **218**
Differential equation
    homogeneous, **472**
    first-order, exact, **460**, 461
    first-order, linear, **455**, 457
    homogeneous linear, **466**
    power series solution of, **483**
    second-order, linear, **465**
Differentials
    with two variables, **201**
Direction angles of vectors, **30**
Direction cosines of vectors, **30**
Directional derivative
    maximum and minimum, **199**
    with three variables, **216**
    with two variables, **190**
Displacement vector, **5**
Distance
    from a point to a line in the xy-plane, **45**
    from a point to a plane, **43**
Divergence
    inspace, **424**
    of a vector field in the plane, **401**
Divergence Theorem
    in space, **424**
    in the plane, **401**
Dot product
    in xyz-space, **28**
Dot product of vectors, 13, **14**
Double integral, **276**
    in polar coordinates, **311**, **314**
    in terms of volumes, **278**
    Riemann sum for, **276**
Double integrals
    changes of variables in, **340**

Electrical circuit
    simple, **478**
Equation
    of plane, **38**
Estimating directional derivatives
    from level curves, **191**
Estimating gradient vectors
    from level curves, **201**
Exact, first-order differential equation, **460**, 461

First-Derivative Test
    with two or three variables, **225**
First-order, exact differential equation, **460**, 461
First-order, linear differential equation, **455**, 457
Flux
    of a constant velocity field, **374**
    of a velvoity field across a plane curve, **383**

Index

Force field
    conservative, in the plane, **403**
Force vector, **7**
Forced motion
    of a spring, **467**
Forced spring, **477**
Free motion
    of a spring, **467**
Free, damped motion
    of a spring, 469
Function
    of three variables, **210**
    of two variables, **137**
    of two variables, domain ofs, **137**
    of two variables, graph of, **138**
    of two variables, level curve of, **145**
    of two variables, range of, **137**
    of two variables, value of, **137**
    piecewise continuous of two variables, **276**
    with more than three variables, **219**

Gauss' Theorem, **423**
Gradient
    with three variables, **217**
Gradient field
    in space, **429**
    in the xy-plane, 405
Gradient vector
    of $f(x,y)$, **197**
Gradient vector field
    with two variables, **401**
Gravitational force field, **373**
Great-circle distance
    and spherical coordinates, 326
Green's theorem, **396**

Homogeneous differential equation, **472**
Homogeneous linear differential equation, **466**
Hyperboloid
    of one sheet, **213**
    of two sheets, **213**

Implicit Function Theorem
    with three variables, **217**
    with two variables, **200**
Incompressible fluid, **407**

Induction
    Mathematical, **483**
Inhomogeneous differential equation, **474**
Integrals
    in spherical coordinates, **322**, **323**
Integrating vector-valued functions, **84**
Intercept equation
    of a plane, **44**
Interior of a set
    in xyz-space, **214**
    in an xy-plane, **147**
Irrotational fluid flow, **407**
Iterated double integral, 278–280
Iterated triple integrals, 300, **301**, 302

Jacobian
    in double integrals, **340**
    in triple integrals, **342**

Kepler's Laws, **94**, 107

Lagrange multipliers
    with three variables, **237**
    with two variables, **235**
Length
    of a curve, **67**
Length of a vector, **2**, 26
Level curve
    of a function of two variables, **145**
Level surface
    of $f(x,y,z)$, **210**
Limit
    of a function of three variables, **214**
    of a function of two variables, **148**
    of a function of two variables , 151
    of a vector, **61**
Line
    parametric equations of, **36**
Line integral
    evaluating, **379**
    path independent, in space, **429**
    with respect to ds, in terms of areas, **381**
Line integrals, **376**
    path independent, in the plane, **403**
Line of force, **9**
Linear function

with two variables, **192**
Linear, first-order differential equation, **455**, 457
Linear, second-order differential equation, **465**
Lines
    skew, **43**
Lines of projection
    in xyz-space, **23**
Local maximum
    with two or three varaibles, **225**
Local minimum
    with two or three varaibles, **225**

Magnitude of a vector, **2**
Mathematical Induction, **483**
Maxima and minima
    local, on a curve, **234**
    with two or three variables, 224
Maximum
    local, with two or three varaibles, **225**
Method of Undetermined Coefficients, **476**
Minimum
    local, with two or three varaibles, **225**
Moment
    of a plane, **288**
    with three variables, 304
Multiplying vectors by numbers, **4**

Newton's Law of gravity, **94**, 107
Newton's Law of motion, **86**
Normal vector
    to a plane curve, **87**
    of a line in the xy-plane, **45**
    to a plane, **37**
    to an oriented curve, **383**

Oblique projection
    in xyz-space, **24**
Octant, **23**
One-to-one mapping, **339**
Onto mapping, **339**
Open set
    in xyz-space, **214**
    in an xy-plane, **147**
Oriented curve, 56
    normal vector to, 383
Oriented surface, **425**

Orthogonal projection
    in xyz-space, **24**
    of a vector on a plane, **49**
    of a vectortextbf, 11

Paraboloid
    circular, **145**
    hyperbolic, **146**
Parallel projection
    in xyz-space, **23**
Parametric equations
    of a line, **36**
    of a curve, **56**
Parametrized surface, **414**
Partial derivative
    estimating from a table, **168**
    estimating from level curves, **170**
    of $f(x,y)$, **165**
    of higher orders, **175**
    with three variables, **214**
Particular solution
    of an inhomogeneous differential equation, **474**
Partition
    of a region in the xy-plane, **275**
    of a solid, **299**
Path-independent line integral
    in space, **429**
Path-independent line integrals
    in the plane, **403**
Perpendicular vectors, **15**
Piecewise continuous function
    of three variables, **299**
    of two variables, **276**
    on a curve, **376**
Piecewise smooth function
    of two variables, **275**
Plane
    equation of, **38**
    intercept equation of, **44**
Polar coordinates
    in double integrals, 311, **314**
Polar forms
    of velocity and acceleration vectors, **97**
Position vector, **5**
Potential function
    of a vector field in the plane, **403**

Index

Power-series solution
    of differential equations, **483**
Projection
    lines of
        in xyz-space, **23**
    oblique, in xyz-space, **24**
    of a vector on a line, **14**
    orthogonal
        , in xyz-space, **24**
    orthogonal, of a vector, **11**
    orthogonal, of a vector on a plane, **49**
    parallel
        in xyz-space, **23**
Projection of a vector
    on a line, **29**
Pythagorean Theorem, **24**

Quadric surface, **210**

Radius of curvature
    of a plane curve, **91**
Rectangular coordinates in space, **23**
Relative velocity, **68**
Resultant
    of force force vectors, **9**
Riemann sum
    for a double integral, **283**
    for a double integral, 276
    for a triple integral, **299**

Saddle point, **232**
Scalar curl of a two-dimensional vectorfield, **399**
Scalar triple product of vectors, **46**
Second-Derivative Test
    with two variables, 230, **232**
Second-order, linear differential equation, **465**
Simple electrical circuit
    textbf, 478
Simple, closed curve, **404**
Simply connected region
    in the xy-plane, **405**
Sink
    of a fluid, **407**
Skew lines, **43**
Source
    of a fluid, **407**

Speed, **64**
Spherical coordinates, **319**
    and geography, **325**
    and great-circle distances, 326
    integrals in, 322, **323**
Spring
    damped motion of, **467**
    forced, **477**
    forced motion of, **467**
    free motion of, **467**
    free, damped motion of, 469
    undamped motion of, **467**
    vibrating, **465**
Step function
    of three variables, **299**
Stokes' Theorem
    in space, **426**
    in the plane, **400**
Streamline
    of a velocity field, **370**
Sum of vectors, **4**
Surface
    boundary of, **425**
    compatible orientation of two, **425**
    oriented, **425**
    parametrized, **414**
Surface area
    of a parametrized surface, **419**
Surface integral, **417**

Tangent plane
    to the graph of $f(x,y)$, **196**
Tangent vector
    to a plane curve, **87**
The difference of two oriented curves, **377**
The negative of an oriented curve, **377**
The sum of two oriented curves, **377**
Total differential
    with two variables, **201**
Triple integral, **299**
Triple integrals
    changes of variables in, **342**

Undamped motion
    of a spring, **467**
Undetermined coefficients

Method of, **476**
Unit vector
  i and **j**, **7**
  i,j and **j**, **29**

Vector
  xy-components of, **3**
  acceleration, **86**
  adding three in space, **48**
  angle of inclination of, **3**
  component
    in the direction of another vector, **29**
  component of, 12, **14**
  derivative of, **62**
  direction angles of, **30**
  direction cosines of, **30**
  displacement, **5**
  dot product in xyz-space, **28**
  dot product of, 13, **14**
  force, **7**
  in xyz-space, **25**
  in the xy-plane, **2**
  length of, **2**, 26
  limit of, **61**
  multiplying by a number, **26**
  multiplying by numbers, **4**
  normal of a line in the xy-plane, **45**
  normal to a plane, **37**
  perpendicular, **15**
  position, **5**
  projection of, on a line, **14**
  projection on a line, **29**
  scalar triple product of three, **46**
  sums of, **4**
Vector field, **369**
Vector representation
  of a curve, **59**
Vector, magnitude of, **2**
Vectors
  addition of, **26**
  cross product of, 42
  cross product of of, **39**
Velocity
  relative, **68**
Velocity field
  of a fluid, **370**
  streamline of, **370**
Velocity vector, **64**
Vibrating spring, **465**
Volume
  as a double integral, **277**
  from double integrals, 283
Volume of a parallelepiped
  with the scalar triple product, **47**
Volume of a tetrahedron
  with the scalar triple product, **47**

Weight
  of a surface, **421**
Work
  by a constant force field, 374
  on an object that traverses a curve, **384**